## Classical and Quantum Information Theory

### An Introduction for the Telecom Scientist

Information theory lies at the heart of modern technology, underpinning all communications, networking, and data storage systems. This book sets out, for the first time, a complete overview of both classical and quantum information theory. Throughout, the reader is introduced to key results without becoming lost in mathematical details.

The opening chapters deal with the basic concepts and various applications of Shannon's entropy. The core features of quantum information and quantum computing are then presented. Topics such as coding, compression, error correction, cryptography, and channel capacity are covered from both classical and quantum viewpoints. Employing an informal yet scientifically accurate approach, Desurvire provides the reader with the knowledge to understand quantum gates and circuits.

Highly illustrated, with numerous practical examples and end-of-chapter exercises, this text is ideal for graduate students and researchers in electrical engineering and computer science, and also for scientists and practitioners in the telecommunications industry.

Further resources and instructor-only solutions are available at www.cambridge.org/desurvire.

**Emmanuel Desurvire** is Director of the Physics Research Group at Thales Research and Technology, and has held previous positions at Stanford University, AT&T Bell Laboratories, Columbia University, and Alcatel. With over 25 years' experience in the field of optical communications, he has received numerous recognitions for his scientific contributions, including the 1994 Prize from the International Commission for Optics, the 1998 Benjamin Franklin Medal in Engineering, the 2005 William Streifer Scientific Achievement Award, and, in 2007, the IEEE/LEOS John Tyndall Award, Engineer of the Year Award, and the France-Telecom Prize of the Académie des Sciences. He is also Laureate of the 2008 Millennium Technology Prize.

# Classical and Quantum Information Theory

## An Introduction for the Telecom Scientist

EMMANUEL DESURVIRE
*Thales Research & Technology, France*

CAMBRIDGE UNIVERSITY PRESS
Cambridge, New York, Melbourne, Madrid, Cape Town, Singapore, São Paulo, Delhi

Cambridge University Press
The Edinburgh Building, Cambridge CB2 8RU, UK

Published in the United States of America by Cambridge University Press, New York

www.cambridge.org
Information on this title: www.cambridge.org/9780521881715



First published 2009

Printed in the United Kingdom at the University Press, Cambridge

*A catalog record for this publication is available from the British Library*

*Library of Congress Cataloging in Publication data*
Desurvire, Emmanuel, 1955–
Classical and quantum theory : an introduction for the telecom scientist / Emmanuel Desurvire.
p. cm.
Includes index.
ISBN 978-0-521-88171-5
1. Quantum theory. 2. Information measurement. I. Title.
QC174.12.D455 2009
530.12–dc22 2008038909

ISBN 978-0-521-88171-5 hardback

# Contents

# Foreword

It is always a great opportunity and pleasure for a professor to introduce a new textbook. This one is especially unusual, in a sense that, first of all, it concerns two fields, namely, classical and quantum information theories, which are rarely taught altogether with the same reach and depth. Second, as its subtitle indicates, this textbook primarily addresses the telecom scientist. Being myself a quantum-mechanics teacher but not being conversant with the current Telecoms paradigm and its community expectations, the task of introducing such a textbook is quite a challenge. Furthermore, both subjects in information theory can be regarded by physicists and engineers from all horizons, including in telecoms, as essentially academic in scope and rather difficult to reconcile in their applications. How then do we proceed from there?

I shall state, firsthand, that there is no need to convince the reader (telecom or physicist or both) about the benefits of Shannon's classical theory. Generally unbeknown to millions of telecom and computer users, Shannon's principles pervade all applications concerning data storage and computer files, digital music and video, wireline and wireless broadband communications altogether. The point here is that classical information theory is not only a must to know from any academic standpoint; it is also a key to understanding the mathematical principles underlying our information society.

Shannon's theory being reputed for its completeness and societal impact, the telecom engineer (and physicist within!) may, therefore, wonder about the benefits of quantum mechanics (QM), when it comes to *information*. Do we really need a quantum information theory (QIT), considering? What novel concepts may be hiding in there, really, that we should be aware of? Is quantum information theory a real field with any engineering worth and perspectives to shape the future, or some kind of fashionable, academic fantasy?

The answer to the above questions first comes from realizing the no-less phenomenal impact of *quantum physics* in modern life. As of today, indeed, there is an amazing catalog of paradigms, inventions, applications, that have been derived from the quantum physics of the early twentieth century. Suffice it to mention the *laser*, whose extraordinary diversity of applications (global communications, data storage, reprography, imaging, machining, robotics, surgery, energy, security, aerospace, defense . . . ) has truly revolutionized our society and – already – information society. As basic or innocuous as it may now seem to anyone, the laser invention yet remains a quantum physics jewel, a man-made wonder, which finds no explanation outside quantum mechanics principles. How did all this happen?

Following some 20 years of experimental facts, intuitions and hypotheses, and first foundations, by mind giants, such as Planck, Einstein, or Bohr, the structure of quantum mechanics was finally laid down within a pretty short period of time (1925–1927). At this time, the actual fathers of this revolutionary "worldview" formalism, e.g., de Broglie, Heisenberg, Schrödinger, or Dirac, could certainly not foresee that future armies of physicists and engineers would use quantum mechanics as an "Everest base camp" to conquer many higher summits of knowledge and breakthroughs.

There is practically no field of physics and advanced engineering that has not been revolutionized from top to bottom by quantum mechanics. Nuclear and particle physicists used quantum mechanics principles to foresee (and then discover experimentally) the existence of new elementary particles, thus lifting some of the microscopic world mysteries. Astrophysics and cosmology were also completely rejuvenated as quantum mechanics formalism proposed explanations for new macroscopic objects, such as white dwarfs or supernovae. Black body emission, one of the earliest experimental evidences of the very origin of quantum physics, was also found to explain the electromagnetic signature of the background of our Universe, telling us about the history of the Big Bang. The discipline where quantum mechanics had more impact on today's life was, however – by and large – solid-state physics. Quantum theory led to the understanding of how electrons and nucleons are organized in solids, how this microscopic world can evolve, interact with light or X-rays, transport heat, respond to magnetic fields, or self-organize at atomic scales. Nowadays, quantum chemistry explores the energy levels of electrons in complex molecules, and explains its spectroscopic properties in full intimacy. Mechanical, thermal, electric, magnetic, and optical properties of matter were first understood and then engineered. In the second half of the last century, transistors, storage disks (magnetic and optical), laser diodes, integrated semiconductor circuits and processors were developed according to an exponential growth pattern. Computers, telecommunication networks, and cellular phones changed everyone's life. All sectors of human activity were deeply influenced by the above technologies. Globalization and a booming of economy were observed during these decades. Neither a physicist, nor an economist, nor the last mad sci-fi novel writer, could have foreseen, one century ago, such a renewal of knowledge, of production means, and of global information sharing. This consideration illustrates how difficult it is to anticipate the future of mankind, since major changes can originate from the most basic or innocuous academic discoveries.

In spite of the difficulty of safe predictions, it is the unwritten duty of a physicist to try to probe this dark matter: the future. While quantum mechanics were revealed to be phenomenally beneficial to humankind, some physicists believe today that all this history is nothing but a first, inaugural, chapter. The first chapter would have "only" consisted of rethinking our world and engineering by introducing a first class of quantum ingredients: quantification of energy or momentum, wave functions, measurement probabilities, spin, quarks . . . Alain Aspect from France's Institut d'Optique, for example, envisions a "second revolution" of quantum mechanics. This second revolution paradigm will move the perspective one step further thanks to the ambitious introduction of a new stage of *complexity*. A way to approach such a complexity is *entanglement*, as I shall further explain.

Entanglement, which is the key to understanding the second quantum mechanics paradigm, is, in fact, an old concept that resurfaced only recently. Although entanglement was questioned by the famous 1935 joint paper by Einstein, Podolski, and Rosen,[1] it became clear over recent years that the matter represented far more than an academic discussion, and, furthermore, that it offered new perspectives. What is entanglement? This property concerns a group of particles that cannot be described separately, despite their physical separation or difference. A classical view of entanglement is provided by the picture of two magic dice, which always show up the same face. You may roll the pair of dice at random as many times as wished, but you will get the same result as with any single die, namely, a probability of 1/6 each to show up any spot patterns between 1 and 6, but with a strange property: the same random result is obtained by the two dice altogether. If one die comes out with six spots, so does the other one. Although this phenomenon of entanglement has no equivalent in the classical world as we normally experience it, it becomes real and tangible at the atomic scale. And unbeknown to large and even scientifically cultivated audiences, physicists have been playing with entanglement for about 20 years.

By means of increasingly sophisticated tools making it possible to manipulate single atoms, electrons, or photons, physicists are now beginning, literally, to "engineer" entangled states of matter. The Holy Grail they are after is building a practical toolbox for quantum entanglement. It is not clear at present which approach may show efficient, resilient, and environment-insensitive entanglement, while at the same time remaining "observable" and, furthermore, lending itself to external manipulation. To darken the picture, it is not at all clear either what could be the maximum size of an entangled system. Such questions come close to the actual definition of the boundary between the quantum and classical worlds, as emphasized by the famous *Schrödinger's cat paradox*. We may find ourselves in a situation similar to that of solid-state physics after World War II: quantum physics and many solid-state physics concepts were duly established, but the transistor remained yet to be invented. To the same extent that the revolutionary concepts of electronic wave functions, band theory, and conductivity led to the development of modern electronics and computers, the concept of entanglement, which stands at the core of quantum information, is now waiting for a revolutionary outcome. The parallel evolution between the different constitutive elements of entangled systems indeed offers huge opportunities to build radically new computing machines, with unprecedented characteristics and performance.

What does entanglement have to do with *complexity*? Whereas basic mechanics laws can predict the trajectory of a ball, the oscillation period of a pendulum, the lift of a plane, complexity characterizes systems where the overall properties cannot be derived from that of the constituent subsystems. For the philosopher Edgar Morin, it is not the number of the components that defines the complexity of any system. More components certainly call for more computing power to calculate the system's behavior, but the problem remains tractable in polynomial time (e.g., quadratic in the number of components): it

[1] A. Einstein, B. Podolsky, and N. Rosen, Can quantum-mechanical description of physical reality be considered complete? *Phys. Rev.*, **47** (1935), 777–80.

is referred to as a "P" problem. Complexity is another story: "complex" has a different meaning here from "complicated." It is, rather, the intimate nature of the interaction between the different components (including, for instance, recursion) that governs the emergence of novel types and classes of macroscopic behavior. Complex systems show properties that are not predictable from the single analysis of their constitutive elements, just as the properties of entangled particles cannot be understood from the simple inference of single particule behaviors.

In the last decades, complex systems have caused many developments in fields as varied as physics, astrophysics, chemistry, and biology. New mathematical tools, chaos, nonlinear physics, have been introduced. From the dynamics of sand dunes to schools of fishes, from ferrofluids to traffic jams, complexity never results from a simple extrapolation of classical individual behaviors. Hence, the challenge of understanding and harnessing entanglement is the possibility of extending the perspectives of quantum mechanics in the same way that macroscopic physics was renewed by the introduction of complexity. *Entanglement is the complexity of quantum mechanics.*

Considering this, it is not surprising that Shannon's *classical* theory of information (CIT) and quantum physics, with the emerging field of *quantum* information theory (QIT), have many background concepts in common. The former classical theory of information was a revolution in its own times, just as quantum mechanics, but with neither conceptual links, nor the least parallelism whatsoever with the latter. The great news is that the two fields have finally reached each other, in most unexpected and elegant ways. It is at the very interface of these two fields and cultures, classical and quantum information theory, that this textbook takes a crucial place and also innovates in the descriptive approach. I have spent so much time as Head of the Physics Department in my University convincing students and researchers to kick against the partitioning between physics and science in general, that I am very pleased to welcome this work of Emmanuel Desurvire, which is a model of scientific "hybridization."

Combining the cultures of a physicist (as a researcher), an academic (as a former professor and author of several books), and an engineer (as a developer and project manager) from the telecom industry, Emmanuel Desurvire attempts here to bridge the gap between the CIT and QIT cultures. On the CIT side, fundamentals, such as information, entropy, mutual entropy, and Shannon capacity theorems, are reviewed in detail, using a wealth of practical and original application examples. Worth mentioning are the reputedly difficult notions of *Kolmogorov complexity* and *Turing machines*, which were developed independently from Shannon during the same historical times, described herewith with thrust and clarity, again with original examples and illustrations. The mind-boggling (and little-known) conclusion to be retained is that Kolmogorov complexity and Shannon entropy asymptotically converge towards each other, despite fundamentally different ground assumptions. Then, under any expectation for a textbook in this subject, comes a detailed (and here quite vivid) description of various principles of *data compression* (coding optimality, integer, arithmetic, and adaptive compression) and *error-correction coding* (block and cyclic codes). Shannon's classical theory of information then moves on and concludes with the *channel-capacity theorems*, including the most elegant *Shannon–Hartley theorem* of incredibly simple and universal formulation, $C = \log(1 + SNR)$, which relates the channel capacity ($C$) to the signal-to-noise ratio ($SNR$) available at the channel's end.

The second part of Emmanuel Desurvire's book is about quantum information theory. This is where the telecom scientist, together with the author, is taken out to a work tour that she or he may not forget, hopefully a most stimulating and pleasurable one. With the notion of *reversible computation* and the *Landauer Principle*, the reader gets a first hint that "information is physical." It takes a quantum of heat $kT$ to tamper with a single classical bit. From this point on, we begin to feel that quantum mechanics realities are standing close behind. Then come the notions of *quantum bits* or *qubits* and their logic gates to form elementary quantum circuits. Such an innocuous introduction, in fact, represents the launching pad of a rocket destined to send the reader into QIT orbit. In this adventurous journey, no spot of interest is neglected, from *superdense coding*, *teleportation*, the *Deutsch–Jozsa algorithm*, *quantum Fourier transform* and *Grover's Quantum Database Search*, to the mythical *Shor factorization* algorithm. Here, the demonstration of Shor's algorithm turns out to be very interesting and useful. Most physicists have heard about this incredible possibility, offered by quantum computing, of factorizing huge numbers within a short time, but have rarely gone into the explanatory detail. Shor's algorithm resembles the green flash: heard of by many, seen by some, but understood by few. The interest continues with a discussion of the computing times required for factorization with classical means, and to meet the various RSA challenges offered on the Internet.

The conclusive chapter on cryptography is also quite original in its approach and conclusions. First, it includes *both* classical and quantum cryptography concepts, according to the author's view that there is no point in addressing the second if one has not mastered the first. Cryptography, a serious matter for network security and privacy, is treated here with the very instructive and specific view of a telecom scientist. Forcefully and crudely stated, "The world is ugly out there," in spite of Alice and Bob's "provably secure" key exchanges (quantum key distribution, QKD). Let one not be mistaken as to the author's intent. Quantum key distribution is most precious as an element in the network security chain; Emmanuel Desurvire is only reminding the community, now with the authority of a telecom professor, that Alice and Bob are exposed, in turn, to higher-level network attacks, and that unless the Internet becomes quantum all the way through, there is no such a thing as "absolute" network security. It is only with this type of cross-disciplined book that elementary truths of the like may be spelled out.

A pervasive value and flavor of this book is that the many practical examples and illustrations provided help the reader to *think concrete*. Both the classical and quantum sides of information theory may seem difficult, rusty, oblivious, if not forthright mysterious to many engineers and scientists since long-past school graduation. More so with the quantum side, which is actually a recent expansion of knowledge (as dated after the Shor algorithm "milestone"), and that only a few engineers and scientists had the privilege to be exposed to so far, prior to beginning their professional careers. Hence, this book represents a first attempt at reconciling old with new knowledge, as destined primarily to mature engineers and scientists, particularly from, but not limited to, the telecom circle. Decision makers from government and industry, investors, and entrepreneurs may also reap some benefit by being better acquainted with the reality of quantum mechanics and the huge application potentials of QIT, apart from any timeliness consideration. Progress in quantum information theory may be a (very) long-term view indeed, but its future is confined to today's humble steps; called awareness, discipline, imagination, creativity and patience.

Thanks to Emmanuel Desurvire's book, many concepts such as quantum information theory, and the reconciliation and familiarity thereof, will be shared by both engineers and physicists, within the telecom community and hopefully far beyond. It is our deep conviction that such cross-border knowledge sharing is necessary to engage in this second revolution of quantum physics.

Professor Vincent Berger
Université Paris-Diderot, Paris 7
February 29, 2008

# Introduction

In the world of telecoms, the term *information* conveys several levels of meaning. It may concern individual bits, bit sequences, blocks, frames, or packets. It may represent a message payload, or its overhead; the necessary extra information for the network nodes to transmit the message payload practically and safely from one end to another. In many successive stages, this information is encapsulated altogether to form larger blocks corresponding to higher-level network protocols, and the reverse all the way down to destination. From any telecom-scientist viewpoint, information represents this uninterrupted *flow of bits*, with network intelligence to process it. Once converted into characters or pixels, the remaining message bits become meaningful or valuable in terms of acquisition, learning, decision, motion, or entertainment. In such a larger network perspective, where information is well under control and delivered with the quality of service, what could be today's need for any *information theory* (IT)?

In the telecom research community indeed, there seems to be little interest for information theory, as based on the valid perception that there is nothing new to worry about. While the occasional evocation of *Shannon* invariably raises passionate group discussions, the professional focus is about the exploitation of bandwidth and network deployment issues. The telecom scientist may, however, wonder about the potentials of *quantum information* and *computing*, and their impact. But not only does the field seem intractable to the nonspecialist, its applications are widely believed to belong to the far-distant future. Then what could be this community's need for any *quantum information theory* (QIT)? While some genuine interest has been raised by the outcome of *quantum cryptography*, or more accurately, *quantum key distribution* (QKD), there is at present not enough matter of concern or driving market factor to bring QIT into the core of telecoms.

The situation is made even more confused through the fact that information theory and quantum information theory appear to have little in common, or that the parallels between the two can be established only at the expense of advanced specialization. The telecom scientist is thus left with unsolved questions. For instance, what is quantum information, and how is it different from Shannon's theorem? How is information carried by *qubits*, as opposed to classical bits? How do IT theorems translate into QIT? What are the ultimate algorithms for quantum information compression, error correction, and encryption, and what benefit do they provide, compared with classical approaches? What are the main conceptual realizations of quantum information processing? The curious might peruse reference books, key papers, or Internet cross-references and tutorials, but this endeavor leaves little chance of reaching satisfying conclusions, let alone acquiring solid grounds for pointing to future research directions.

To summarize, on one hand, we find the old-and-forgotten IT field, with its wealth of very mature applications in all possible areas of information processing. On the other hand, we find the more recent and poorly known QIT field, showing high promise, but little potential of application within reasonable sight. In between, the difficulty for nonspecialists to make sense of any parallels between the two, and the lack of motivation to dig into what appears an austere or intractable bunch of mathematical formalism.

The above description suggests the reason why this book was written, and its key purpose. Primarily, it is my belief that IT is incomplete without QIT, and that the second should not be approached without a fair assimilation of the first. Secondly, the mathematical difficulties of IT and QIT can, largely, be alleviated by making the presentation less formal than in the usual academic reference format. This does not mean oversimplification, but rather skipping many academic caveats, which flourish in most reference textbooks, and which make progression a tedious and risky adventure. Our portrayed telecom scientist only needs the fundamental concepts, along with supporting proof at a satisfactory level. Also, IT and QIT can be made far more interesting and entertaining by use of many illustrations and application examples.

With these goals in mind, this book has been organized as a sequence of *chapters*, each of which can be presented in two or three hour courses or seminars, and which the reader should be able to teach in turn! Except at the beginning, the sequence of chapters presents a near-uniform level of difficulty, which rapidly assures the reader that she or he will be able to make it to the very end. For the demanding, or later reference, the most advanced demonstrations have been relegated into as many Appendices. To lighten the text, an extensive use of footnotes is made. These footnotes also contain useful Internet links, and sometimes bibliographical references. Finally, lots of original exercises with difficulty levels graded as basic (B), medium (M), or tricky (T) are proposed, the set of solutions being available to class teachers from Cambridge University Press. As to the Internet links, one is aware that they do not have the value of permanent references, owing to the finite lifetime of most websites or their locators or addresses (URL). To alleviate this problem, the Publisher has agreed with the author to keep up an updated list of URLs on the associated website: www.cambridge.org/9780521881715, along with errata information.

What about the book contents?

The first two chapters (1 and 2) concern basic recalls of *probability theory*. These are purposefully entertaining to read, while the advanced reader might find useful teaching ideas for undergraduate courses.

Chapter 3 addresses the tricky concept of *information measure*. We learn something that everyone intuitively knows, namely, that there is no or little information in events that are certain or likely to happen. Uncertainty, on the other hand, is associated with high information contents.

When several possible events are being considered, the correct information measure becomes *entropy* (Chapters 4–6). As shown, Shannon's entropy concept in IT is not without strong but subtle connections with the world of Boltzmann's thermodynamics. But

IT goes a step further with the key notion of *mutual information*, and other useful entropy definitions (joint, conditional, relative), including those related to continuous random variables (differential). Chapter 7, on *algorithmic entropy* (or equivalently, *Kolmogorov complexity*), is meant to be a real treat. This subject, which comes with its strange *Turing machines*, is, however, reputedly difficult. Yet the reader should not find the presentation level different from preceding material, thanks to many supporting examples. The conceptual beauty and reward of the chapter is the asymptotic convergence between Shannon's entropy and Kolmogorov's complexity, which were derived on completely independent assumptions!

Chapters 8–10 take on a tour of *information coding*, which is primarily the art of compressing bits into shorter sequences. This is where IT finds its first and everlasting success, namely, *Shannon's source coding theorem*, leading to the notion of *coding optimality*. Several coding algorithms (Huffmann, integer, arithmetic, adaptive) are reviewed, along with a daring appendix (Appendix G), attempting to convey a comprehensive flavor in both *audio* and *video standards*.

With Chapter 11, we enter the magical world of *error correction*. For the scientist, unlike the telecom engineer, it is phenomenal that bit errors coming from random physical events can be corrected with 100% accuracy. Here, we reach the concept of a *communication channel*, with its own imperfections and intrinsic *noise*. The chapter reviews the principles and various families of *block codes* and *cyclic codes*, showing various capabilities of error-correction performance.

The communication channel concept is fully disclosed in the description going through Chapters 12–14. After reviewing *channel entropy* (or mutual information in the channel), we reach Shannon's most famous *channel-coding theorem*, which sets the ultimate limits of *channel capacity* and error-correction potentials. The case of the *Gaussian channel*, as defined by continuous random variables for signal and noise, leads to the elegant *Shannon–Hartley theorem*, of universal implications in the field of telecoms. This closes the first half of the book.

Next we approach QIT by addressing the issue of *computation reversibility* (Chapter 15). This is where we learn that information is "physical," according to *Landauer's principle* and based on the fascinating "Maxwell's demon" (thought) experiment. We also learn how *quantum gates* must differ from classical *Boolean logic gates*, and introduce the notion of *quantum bit*, or *qubit*, which can be manipulated by a "zoo" of elementary quantum gates and circuits based on *Pauli matrices*.

Chapters 17 and 18 are about *quantum measurements* and *quantum entanglement*, and some illustrative applications in *superdense coding* and *quantum teleportation*. In the last case, an appendix (Appendix P) describes the algorithm and quantum circuit required to achieve the *teleportation of two qubits* simultaneously, which conveys a flavor of the teleportation of more complex systems.

The two former chapters make it possible in Chapters 19 and 20 to venture further into the field of *quantum computing (QC)*, with the *Deutsch–Jozsa* algorithm, the *quantum Fourier transform,* and, overall, two famous QC algorithms referred to as the *Grover Quantum Database Search* and *Shor's factorization*. If, some day it could be implemented in a physical quantum computer, Grover's search would make it possible to explore databases with a quadratic increase in speed, as compared with any classical computer. As to Shor's factorization, it would represent the end of classical cryptography in global use today. It is, therefore, important to gain a basic understanding of both Grover and Shor QC algorithms, which is not a trivial task altogether! Such an understanding not only conveys a flavor of QC power and potentials (as due to the property of quantum parallelism), but it also brings an awareness of the high complexity of quantum-computing circuits, and thus raises true questions about practical hardware, or massive or parallel quantum-gates implementation.

Quantum information theory really begins with Chapter 21, along with the introduction of *von Neumann entropy*, and related variants echoing the classical ones. With Chapters 22 and 23, the elegant analog of Shannon's channel source-coding and channel-capacity theorems, this time for quantum channels, is reached with the *Holevo bound* concept and the so-called *HSW theorem*.

Chapter 24 is about quantum error correction, in which we learn that various types of single-qubit errors can be effectively and elegantly corrected with the *nine-qubit Shor code* or more powerfully with the equally elegant, but more universal *seven-qubit CSS code*.

The book concludes with a hefty chapter dedicated to *classical* and *quantum cryptography* together. It is the author's observation and conviction that quantum cryptography cannot be safely approached (academically speaking) without a fair education and awareness of what cryptography, and overall, network security are all about. Indeed, there is a fallacy in believing in "absolute security" of one given ring in the security chain. Quantum cryptography, or more specifically as we have seen earlier, *quantum key distribution* (QKD), is only one constituent of the security issue, and contrary to common belief, it is itself exposed to several forms of potential attacks. Only with such a state of mind can cryptography be approached, and QKD be appreciated as to its relative merits.

Concerning the QIT and QC side, it is important to note that this book purposefully avoids touching on two key issues: the effects of *quantum decoherence*, and the *physical implementation of quantum-gate circuits*. These two issues, which are intimately related, are of central importance in the industrial realization of practical, massively parallel *quantum computers*. In this respect, the experimental domain is still at a stage of infancy, and books describing the current or future technology avenues in QC already fill entire shelves.

Notwithstanding long-term expectations and coverage limitations, it is my conviction that this present book may largely enable telecom scientists to gain a first and fairly complete appraisal of both IT and QIT. Furthermore, the reading experience should substantially help one to acquire a solid background for understanding QC applications and experimental realizations, and orienting one's research programs and proposals accordingly. In large companies, such a background should also turn out to be helpful to propose related positioning and academic partnership strategy to the top management, with confident knowledge and conviction.

# Acknowledgments

The author is indebted to Dr. Ivan Favero and Dr. Xavier Caillet of the Université Paris-Diderot and Centre National de la Recherche Scientifique (CNRS, www.cnrs.fr/index.html) for their critical review of the manuscript and very helpful suggestions for improvement, and to Professor Vincent Berger of the Université Paris-Diderot and Centre National de la Recherche Scientifique (CNRS, www.cnrs.fr/index.html) for his Foreword to this book.

# 1 Probability basics

Because of the reader's interest in *information theory*, it is assumed that, to some extent, he or she is relatively familiar with *probability theory*, its main concepts, theorems, and practical tools. Whether a graduate student or a confirmed professional, it is possible, however, that a good fraction, if not all of this background knowledge has been somewhat forgotten over time, or has become a bit rusty, or even worse, completely obliterated by one's academic or professional specialization!

This is why this book includes a couple of chapters on *probability basics*. Should such basics be crystal clear in the reader's mind, however, then these two chapters could be skipped at once. They can always be revisited later for backup, should some of the associated concepts and tools present any hurdles in the following chapters. This being stated, some expert readers may yet dare testing their knowledge by considering some of this chapter's (easy) problems, for starters. Finally, any parent or teacher might find the first chapter useful to introduce children and teens to probability.

I have sought to make this review of probabilities basics as simple, informal, and practical as it could be. Just like the rest of this book, it is definitely not intended to be a math course, according to the canonic theorem–proof–lemma–example suite. There exist scores of rigorous books on probability theory at all levels, as well as many Internet sites providing elementary tutorials on the subject. But one will find there either too much or too little material to approach Information Theory, leading to potential discouragement. Here, I shall be content with only those elements and tools that are needed or are used in this book. I present them in an original and straightforward way, using fun examples. I have no concern to be rigorous and complete in the academic sense, but only to remain accurate and clear in all possible simplifications. With this approach, even a reader who has had little or no exposure to probability theory should also be able to enjoy the rest of this book.

## 1.1 Events, event space, and probabilities

As we experience it, reality can be viewed as made of different environments or situations in time and space, where a variety of possible *events* may take place. Consider dull and boring life events. Excluding future possibilities, basic events can be anything like:

- It is raining,
- I miss the train,
- Mom calls,
- The check is in the mail,
- The flight has been delayed,
- The light bulb is burnt out,
- The client signed the contract,
- The team won the game.

Here, the events are defined in the present or past tense, meaning that they are known facts. These known facts represent something that is either true or false, experienced or not, verified or not. If I say, "Tomorrow *will* be raining," this is only an assumption concerning the future, which may or may not turn out to be true (for that matter, weather forecasts do not enjoy universal trust). Then tomorrow will tell, with rain being a more likely possibility among other ones. Thus, future events, as we may expect them to come out, are well defined facts associated with some degree of likelihood. If we are amidst the Sahara desert or in Paris on a day in November, then rain as an event is associated with a very low or a very high likelihood, respectively. Yet, that day precisely it may rain in the desert or it may shine in Paris, against all preconceived certainties. To make things even more complex (and for that matter, to make life exciting), a few other events may occur, which weren't included in any of our predictions.

Within a given environment of causes and effects, one can make a list of all possible events. The set of events is referred to as an *event space* (also called *sample space*). The event space includes anything that can possibly happen.[1] In the case of a sports match between opposing two teams, A and B, for instance, the basic event space is the four-element set:

$$S = \left\{ \begin{array}{c} \text{team A wins} \\ \text{team A loses} \\ \text{a draw} \\ \text{game canceled} \end{array} \right\}, \tag{1.1}$$

with it being implicit that if team A wins, then team B loses, and the reverse. We can then say that the events "team A wins" and "team B loses" are strictly equivalent, and need not be listed twice in the event space. People may take bets as to which team is likely to win (not without some local or affective bias). There may be a draw, or the game may be canceled because of a storm or an earthquake, in that order of likelihood. This pretty much closes the event space.

When considering a trial or an experiment, events are referred to as *outcomes*. An experiment may consist of picking up a card from a 32-card deck. One out of the 32 possible outcomes is the card being the Queen of Hearts. The event space associated

[1] In any environment, the list of possible events is generally infinite. One may then conceive of the event space as a limited set of well defined events which encompass all known possibilities at the time of the inventory. If other unknown possibilities exist, then an event category called "other" can be introduced to close the event space.

with this experiment is the list of all 32 cards. Another experiment may consist in picking up two cards successively, which defines a different event space, as illustrated in Section 1.3, which concerns *combined* and *joint* events.

The *probability* is the mathematical measure of the likelihood associated with a given *event*. This measure is called $p$(event). By definition, the measure ranges in a zero-to-one scale. Consistently with this definition, $p(\text{event}) = 0$ means that the event is *absolutely unlikely* or "impossible," and $p(\text{event}) = 1$ is *absolutely certain*.

Let us not discuss here what "absolutely" or "impossible" might really mean in our physical world. As we know, such extreme notions are only relative ones! Simply defined, without purchasing a ticket, it is impossible to win the lottery! And driving 50 mph above the speed limit while passing in front of a police patrol leads to absolute certainty of getting a ticket. Let's leave alone the weak possibilities of finding by chance the winning lottery ticket on the curb, or that the police officer turns out to be an old schoolmate. That's part of the event space, too, but let's not stretch reality too far. Let us then be satisfied here with the intuitive notions that impossibility and absolute certainty do actually exist.

Next, formalize what has just been described. A set of different events in a family called $x$ may be labeled according to a series $x_1, x_2, \ldots, x_N$, where $N$ is the number of events in the event space $S = \{x_1, x_2, \ldots, x_N\}$. The probability $p(\text{event} = x_i)$, namely, the probability that the outcome turns out to be the event $x_i$, will be noted $p(x_i)$ for short.

In the general case, and as we well know, events are neither "absolutely certain" nor "impossible." Therefore, their associated probabilities can be any real number between 0 and 1. Formally, for all events $x_i$ belonging to the space $S = \{x_1, x_2, \ldots, x_N\}$, we have:

$$0 \leq p(x_i) \leq 1. \tag{1.2}$$

Probabilities are also commonly defined as percentages. The event is said to have anything between a 0% chance (impossible) and a 100% chance (absolutely certain) of happening, which means strictly the same as using a 0–1 scale. For instance, an election poll will give a 55% chance of a candidate winning. It is equivalent to saying that the odds for this candidate are 55:45, or that $p(\text{candidate wins}) = 0.55$.

As a fundamental rule, the sum of all probabilities associated with an event space $S$ is equal to unity. Formally,

$$p(x_1) + p(x_2) + \cdots p(x_N) = \sum_{i=1}^{i=N} p(x_i) = 1. \tag{1.3}$$

In the above, the symbol $\Sigma$ (in Greek, capital S or *sigma*) implies the summation of the argument $p(x_i)$ with index $i$ being varied from $i = 1$ to $i = N$, as specified under and above the sigma sign. This concise math notation is to be well assimilated, as it will be used extensively throughout this book. We can interpret the above summation rule according to:

It is absolutely certain that one event in the space will occur.

This is another way of stating that the space includes *all* possibilities, as for the game space defined in Eq. (1.1). I will come back to this notion in Section 1.3, when considering combined probabilities.

But how are the probabilities calculated or estimated? The answer depends on whether or not the event space is well or completely defined. Assume first for simplicity the first case: we know for sure all the possible events and the space is complete. Consider then two familiar games: coin tossing and playing dice, which I am going to use as examples.

## Coin tossing

The coin has two sides, heads and tails. The experiment of tossing the coin has two possible outcomes (heads or tails), if we discard any possibility that the coin rolls on the floor and stops on its edge, as a third physical outcome! To be sure, the coin's mass is also assumed to be uniformly distributed into both sides, and the coin randomly flipped, in such a way that no side is more likely to show up than the other. The two outcomes are said to be *equiprobable*. The event space is $S = \{\text{heads, tails}\}$, and, according to the previous assumptions, $p(\text{heads}) = p(\text{tails})$. Since the space includes all possibilities, we apply the rule in Eq. (1.3) to get $p(\text{heads}) = p(\text{tails}) = 1/2 = 0.5$. The odds of getting heads or tails are 50%. In contrast, a realistic coin mass distribution and coin flip may not be so perfect, so that, for instance, $p(\text{heads}) = 0.55$ and $p(\text{tails}) = 0.45$.

## Rolling dice (game 1)

Play first with a single die. The die has six faces numbered one to six (after their number of spots). As for the coin, the die is supposed to land on one face, excluding the possibility (however well observed in real life!) that it may stop on one of its eight corners after stopping against an obstacle. Thus the event space is $S = \{1, 2, 3, 4, 5, 6\}$, and with the equiprobability assumption, we have $p(1) = p(2) = \cdots = p(6) = 1/6 \approx 0.166\,666\,6$.

## Rolling dice (game 2)

Now play with two dice. The game consists in adding the spots showing in the faces. Taking successive turns between different players, the winner is the one who gets the highest count. The sum of points varies between $1 + 1 = 2$ to $6 + 6 = 12$, as illustrated in Fig. 1.1. The event space is thus $S = \{2, 3, 4, 5, 6, 7, 8, 9, 10, 11, 12\}$, corresponding to 36 possible outcomes. Here, the key difference from the two previous examples is that the events (sum of spots) are not equiprobable. It is, indeed, seen from the figure that there exist six possibilities of obtaining the number $x = 7$, while there is only one possibility of obtaining either the number $x = 2$ or the number $x = 12$. The count of possibilities is shown in the graph in Fig. 1.2(a).

Such a graph is referred to as a *histogram*. If one divides the number of counts by the total number of possibilities (here 36), one obtains the corresponding probabilities. For instance, $p(x = 2) = p(x = 22) = 1/36 = 0.028$, and $p(x = 7) = 6/36 = 0.167$. The different probabilities are plotted in Fig. 1.2(b). To complete the plot, we have

| | ⚀ | ⚁ | ⚂ | ⚃ | ⚄ | ⚅ |
|---|---|---|---|---|---|---|
| ⚀ | 2 | 3 | 4 | 5 | 6 | 7 |
| ⚁ | 3 | 4 | 5 | 6 | 7 | 8 |
| ⚂ | 4 | 5 | 6 | 7 | 8 | 9 |
| ⚃ | 5 | 6 | 7 | 8 | 9 | 10 |
| ⚄ | 6 | 7 | 8 | 9 | 10 | 11 |
| ⚅ | 7 | 8 | 9 | 10 | 11 | 12 |

**Figure 1.1** The 36 possible outcomes of counting points from casting two dice.

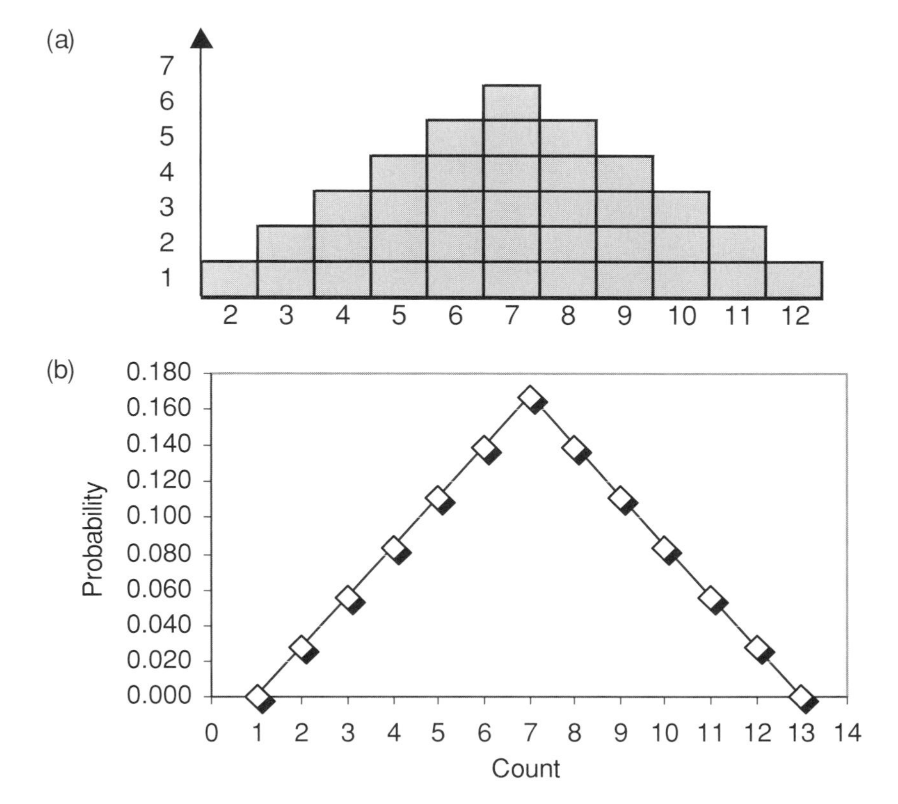


**Figure 1.2** (a) Number of possibilities associated with each possible outcome of casting two dice, (b) corresponding probability distribution.

included the two count events $x = 1$ and $x = 13$, which both have zero probability. Such a plot is referred to as the *probability distribution*; it is also called the *probability distribution function* (PDF). See more in Chapter 2 on PDFs and examples. Consistently with the rule in Eq. (1.3), the sum of all probabilities is equal to unity. It is equivalent to say that the *surface* between the PDF curve linking the different points $(x, p(x))$ and the horizontal axis is unity. Indeed, this surface is given by $s = (13 - 1)^* p(x = 7)/2 = 12^*(6/36)/2 \equiv 1$.

The last example allows us to introduce a fundamental definition of the probability $p(x_i)$ in the general case where the events $x_i$ in the space $S = \{x_1, x_2, \ldots, x_N\}$ do not have equal likelihood:

$$p(x_i) = \frac{\text{number of possibilities for event } i}{\text{number of possibilities for all events}}. \tag{1.4}$$

This general definition has been used in the three previous examples. The single coin tossing or single die casting are characterized by equiprobable events, in which case the PDF is said to be *uniform*. In the case of the two-dice roll, the PDF is nonuniform with a triangular shape, and peaks about the event $x = 7$, as we have just seen.

Here we are reaching a subtle point in the notion of probability, which is often mistaken or misunderstood. The known fact that, in principle, a flipped coin has equal chances to fall on heads or tails *provides no clue as to what the outcome will be*. We may just observe the coin falling on tails several times in a row, before it finally chooses to fall on heads, as the reader can easily check (try doing the experiment!). Therefore, the meaning of a probability is not the prediction of the outcome (event $x$ being verified) but the measure of how likely such an event is. Therefore, it actually takes quite a number of trials to measure such likelihood: one trial is surely not enough, and worse, several trials could lead to the wrong measure. To sense the difference between probability and outcome better, and to get a notion of how many trials could be required to approach a good measure, let's go through a realistic coin-tossing experiment.

First, it is important to practice a little bit in order to know how to flip the coin with a good feeling of randomness (the reader will find that such a feeling is far from obvious!). The experiment may proceed as follows: flip the coin then record the result on a piece of paper (heads = H, tails = T), and make a pause once in a while to enter the data in a computer spreadsheet (it being important for concentration and expediency not to try performing the two tasks altogether). The interest of the computer spreadsheet is the possibility of seeing the statistics plotted as the experiment unfolds. This creates a real sense of fun. Actually, the computer should plot the cumulative count of heads and tails, as well as the experimental PDF calculated at each step from Eq. (1.4), which for clarity I reformulate as follows:

$$\begin{cases} p(\text{heads}) = \dfrac{\text{number of heads counts}}{\text{number of trials}} \\ p(\text{tails}) = \dfrac{\text{number of tails counts}}{\text{number of trials}}. \end{cases} \tag{1.5}$$

The plots of the author's own experiment, by means of 700 successive trials, are shown in Fig. 1.3. The first figure shows the cumulative counts for heads and tails, while the second figure shows plots of the corresponding experimental probabilities $p$(heads), $p$(tails) as the number of trials increases. As expected, the counts for heads and tails are seemingly equal, at least when considering large numbers. However, the detail shows that time and again, the counts significantly depart from each other, meaning that there are more heads than tails or the reverse. But eventually these discrepancies seem to correct

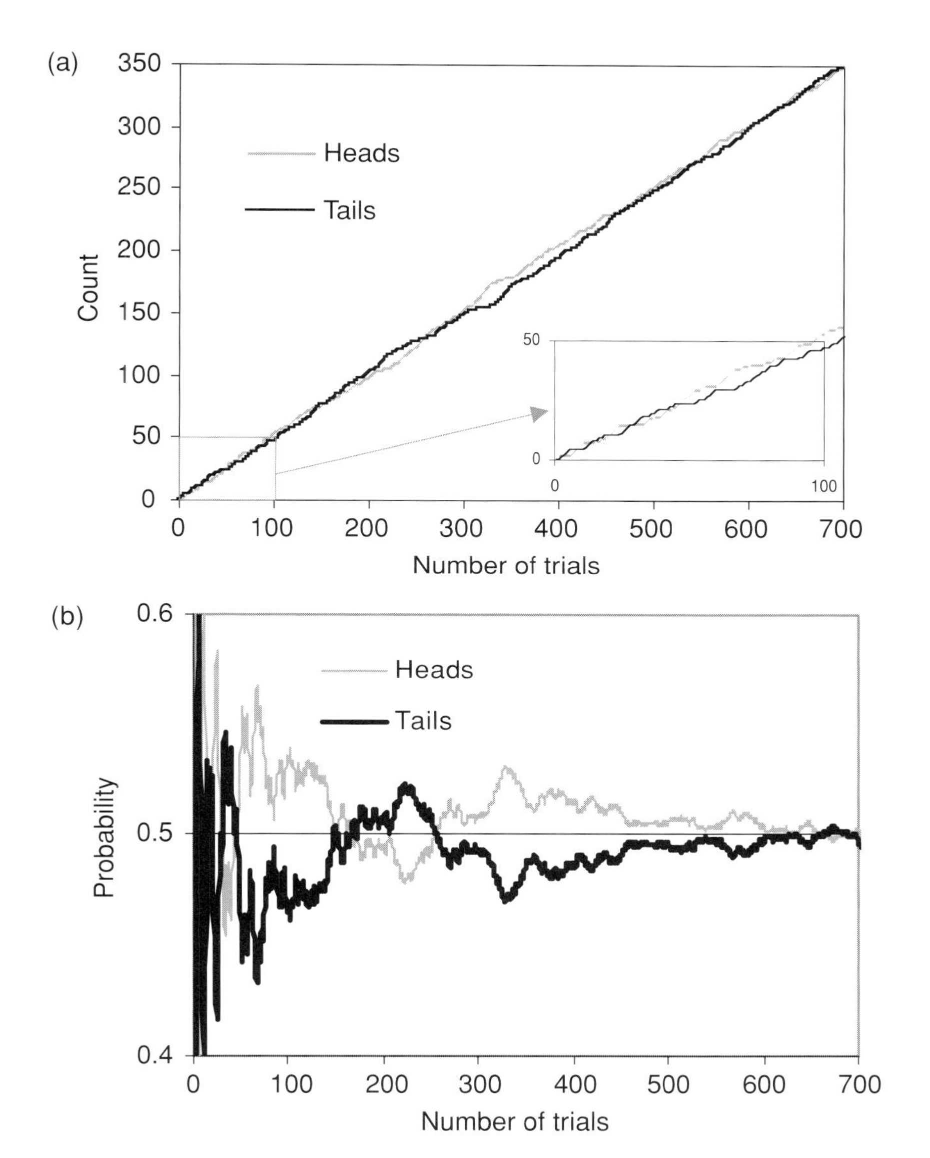


**Figure 1.3** Experimental determination of the probability distribution of coin flipping, by means of 700 successive trials: (a) cumulative count of head and tails outcomes with inset showing detail for the first 100 trials, (b) corresponding probabilities.

themselves as the game progresses, as if the coin would "know" how to come back to the 50:50 odds rule. Strange isn't it? The discrepancies between counts are reflected by the wide oscillations of the PDF (Fig. 1.3(b)). But as the experiment progresses, the oscillations are damped to the point where $p(\text{heads}) \approx p(\text{tails}) \approx 0.5$, following an asymptotic behavior.[2]

[2] Performing this experiment and obtaining such results is not straightforward. Different types of coins must be tested first, some being easier to flip than others, because of size or mass. Some coins seem not to lead

The above experiment illustrates the difference between event probabilities and their actual outcomes in the physical world. The nice thing about probability theory is that the PDF gives one a sense of the unknown when it comes to a relatively large number of outcomes, as if the unknown, or "chance," were domesticated by underlying mythical principles. On the other hand, a known probability gives no clue about a single event, just a sense of what it is most likely to be. A fair way to conceive of a 10% odds is that the corresponding event "should be observed" at least once after ten outcomes, and at least ten times after 100 outcomes, and very close to $Q/10$ times after $Q$ outcomes, the closeness being increasingly accurate as $Q$ becomes larger. The expression "should be observed" progresses towards "should be absolutely certain." To top off this statement, we can say that *an event with a finite, nonzero probability is absolutely certain to occur at least once in the unbounded future*. Such a statement is true provided the physics governing the event (and its associated PDF) remain indefinitely the same, or is "invariant by time translation" in physicists' jargon.

## 1.2 Combinatorics

The examples in the previous section concerned events whose numbers of possible occurrences in a given trial are easily defined. For instance, we know for certain that a single die has exactly six faces, with their corresponding numbers of spots. But when adding spots from two dice (rolling dice, game 2), we had to go through a kind of inventory in order to figure out all the different possibilities. While this inventory was straightforward in this example, in the general case it can be much more tedious if not very complex. Studying the number of different ways of counting and arranging events is called *combinatorial analysis* or *combinatorics*. Instead of formalizing combinatorics at once, a few practical examples are going to be used to introduce its underlying concepts progressively.

### Arranging books on a shelf

This is a recurrent problem when moving into a new home or office. A lazy solution is to unpack the books and arrange them on the shelf from left to right, as we randomly pick them up from the box. How many different ways can we do this?

*Answer:* Say the shelf can hold ten books, and number the book position from the left. Assume that the box fits in ten different (or distinguishable) books. The number of ways to pick up the first book to place in position 1 is, therefore, $Q = 10$. For the second book to put in position 2, there remain nine possibilities. So far, there have been $Q = 10 \times 9$ possibilities. Positioning next the third book, the count of possibilities becomes $Q = 10 \times 9 \times 8$. And so on, until the last book, for which there is only a single

to equiprobable outcomes, even over a number of trials as high as 1000 and after trying different tossing methods. If the coin is not 100% balanced between the two sides in terms of mass, nonuniform PDFs can be obtained.

choice left, which gives for the total number of possibilities: $Q = 10 \times 9 \times 8 \times 7 \times 6 \times 5 \times 4 \times 3 \times 2 \times 1$. By definition, this product is called the *factorial* of 10 and is noted 10!

More generally, for any integer number $n$ we have the factorial definition:

$$n! = n \times (n-1) \times (n-2) \times \cdots \times 3 \times 2 \times 1. \tag{1.6}$$

Thus

$$\begin{aligned} 1! &= 1, \\ 2! &= 2 \times 1 = 2, \\ 3! &= 3 \times 2 \times 1 = 6, \end{aligned}$$

and so on. One can easily compute 10! (using, for instance, the factorial function of a scientific pocket calculator) to find

$$10! = 3\,628\,800 \approx 3.6\,\text{million}.$$

There is an impressive number of ways indeed to personalize one's office shelves!

When it comes to somewhat greater arguments (e.g., $n = 100$), the factorials cannot be computed by hand or through a pocket calculator, owing to their huge size. To provide an idea (as computed through a spreadsheet math function up to a maximum of 170!):

$$\begin{aligned} 50! &\approx 3.0 \times 10^{64}, \\ 100! &\approx 9.3 \times 10^{157}, \\ 170! &\approx 7.2 \times 10^{306}. \end{aligned}$$

In this case it is possible to use *Stirling's approximation theorem*, which is written as

$$n! \approx n^n \mathrm{e}^{-n} \sqrt{2\pi n}. \tag{1.7}$$

Remarkably, the Stirling theorem is accurate within 1.0% for arguments $n \geq 10$ and within 0.20% for arguments $n \geq 40$.

To summarize, *the factorial of n is the number of ways to arrange n distinguishable elements into a given orderly fashion*. Such an arrangement is also called a *permutation*. What about the factorial of the number *zero*? By convention, mathematicians have set the odd property

$$0! = 1.$$

We shall accept here that $0! = 1$ without advanced justification. Simply put, if not a satisfactory explanation, there is only one way to arrange/permute a set containing zero element.[3]

Consider next a second example.

[3] For the math-oriented reader, it is interesting to mention that the factorial function can be generalized to *any real or even complex arguments* $x$, i.e., $x! = \Gamma(x+1)$, where $\Gamma$ is the gamma function, which can be computed numerically.

## Arranging books on a shelf, with duplicates

Assume that some of the books have one or even several duplicate copies, which cannot be distinguished from each other, as in a bookstore. For instance, the series of 10 books includes two brand-new English–Russian dictionaries. Having these two in shelf positions $(a, b)$ or $(b, a)$ represents the same arrangement. So as not to count this arrangement twice, we should divide the previous result (10!) by two (2!), which is the number of possible permutations for the duplicated dictionaries. If we had three identical books in the series, we should divide the result by six (3!), and so on. It is clear, then, that *the number of ways of arranging n elements containing p indistinguishable elements and n − p distinguishable ones is given by the ratio:*

$$A_n^p = \frac{n!}{p!}. \tag{1.8}$$

The above theorem defines the number of possible arrangements "without repetition" (of the indistinguishable copies). Consistently, if the series contains $n$ indistinguishable duplicates, the number of arrangements is simply $A_n^n = n!/n! = 1$, namely, leaving a unique possibility.

The next example will make us progress one step more. Assume that we must make a selection from a set of objects. The objects can be all different, all identical, or partly duplicated, which does not matter. We would only like to know the number of possibilities there are to make any random selection of these objects.

## Fruit-market shopping

In a fruit market, the stall displays 100 fruits of various species and origins. We have in mind to pick at random up to five fruits, without preference. There are 100 different possibilities to pick up the first fruit, 99 possibilities to pick up the second, and so on until the last fifth, which has 96 possibilities left. The total number of possibilities to select five specific fruits out of 100 is therefore $Q = 100 \times 99 \times 98 \times 97 \times 96$. Based on the definition of factorials in Eq. (1.6), we can write this number in the form $Q = 100!/95! = 100!/(100 - 5)!$ But in each selection of five fruits we put into the bag, it does not matter in which order they have been selected. All permutations of these five specific fruits, (which are 5!), represent the same final bag selection. Therefore, the above count should be divided by 5! because of the 5! possible redundancies. The end result is, therefore, $Q = 100!/[5!(100 - 5)!]$. Most generally, *the number of ways to pick up p unordered samples out of a set of n items* is

$$C_n^p = \frac{n!}{p!(n-p)!}. \tag{1.9}$$

The number $C_n^p$, which is also noted $\binom{n}{p}$ or ${}_nC_p$ or $C(n, p)$ is called the *binomial coefficient.*[4]

[4] Since the factorial is expandable to a continuous function (see previous note), the binomial coefficient is most generally defined for any real/complex $x$, $y$ numbers as $C_x^y = \Gamma(x+1)/\Gamma(y)[\Gamma(x-y+1)]$. Beautiful plots of $C_x^y$ in the real $x$, $y$ plane, and more on the very rich binomial

A final example is going to show how we can use the binomial coefficient in probability analysis.

### Scooping jellybeans

A candy jar contains 20 jellybeans, of which three quarters are green and one quarter is red. If one picks up ten jellybeans at random, *what is the probability that all beans in the selection will be green*?

*Answer:* By definition, the number of ways to pick up 10 green beans out of 15 is $C_{15}^{10}$. The total number of possible picks, i.e., a selection of 10 out of 20, is $C_{20}^{10}$. Applying the probability definition in Eq. (1.4), with the event being "all picked beans are green," we obtain:

$$\begin{aligned} p(\text{all green}) &= \frac{C_{15}^{10}}{C_{20}^{10}} = \frac{15!}{10!5!} \times \frac{10!10!}{20!} = \frac{15!10!}{20!5!} \\ &= \frac{10 \times 9 \times 8 \times 7 \times 6}{20 \times 19 \times 18 \times 17 \times 16} = 0.016. \end{aligned} \tag{1.10}$$

The result shows that the likelihood of getting only green jellybeans is relatively low (near 1.5%), even if ¾ of the jar are of this type. Interestingly, if the jar had only one red jelly bean and 19 green ones, the event probability would be $p(\text{all green}) = 0.5$, as the reader should easily verify. It is also easy to show as an exercise that for a jar of one red jelly bean and $N - 1$ green jelly beans, the probability $p(\text{all green})$ becomes asymptotically close to unity as $N$ reaches infinity, which is the expected result.

## 1.3 Combined, joint, and conditional probabilities

We have learnt that probabilities are associated with certain events $x_i$ belonging to an event space $S = \{x_1, x_2, \ldots, x_N\}$. Here, we further develop the analysis by associating events from different subspaces of $S$, and establish a new set of rules for the corresponding probabilities.

Assume first two subspaces containing a single event, which we note $A = \{a\}$ and $B = \{b\}$, with $a$, $b$ being included in the space $S = \{a, b, \ldots\}$. Let me now introduce two definitions:[5]

- The *combined event* is the event corresponding to the occurrence of *either $a$ or $b$ or both*; it is also called the *union* of $a$ and $b$ and is equivalently noted $a \cup b$ or $a \vee b$;
- The *joint event* is the event corresponding to the occurrence of *both $a$ and $b$*; it is also called the *intersection* of $a$ and $b$, and is equivalently noted $a \cap b$ or $a \wedge b$.

coefficient properties can be found in http://mathworld.wolfram.com/BinomialCoefficient.html or www.math.sdu.edu.cn/mathency/math/b/b219.htm.

[5] The symbols ∪ and ∩ are generally used to describe union and intersection of *sets*, while the symbols ∨ and ∧ apply to *logical or Boolean variables*. If $a$ is an element of the event set, then the Boolean correspondence is "event $a$ = verified or true" For simplicity, I shall use the two notations indifferently.

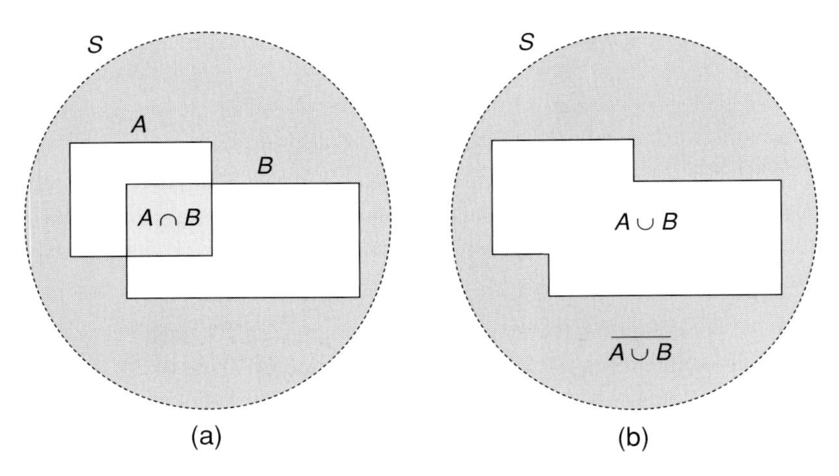


**Figure 1.4** Venn diagrams showing event space $S$ and its subspaces $A$ and $B$, with: (a) the intersection $A \cap B$ and (b) the union $A \cup B$ and its complement $\overline{A \cup B}$.

A visual representation of these two definitions is provided by *Venn diagrams*, as illustrated in Fig. 1.4.

In the figure, the circle $S$ represents the set of all possible occurrences. Its surface is equal to unity. The rectangles A and B represent the set of occurrences for the two events $a$, $b$, respectively, and their surfaces correspond to the associated probabilities. The intersection of the two rectangles, $A \cap B$, defines the number of occurrences (the probability) for the *joint event*, $a \cap b$. The union of the two rectangles, $A \cup B$, defines the number of occurrences (the probability) of the *combined event*, $a \cup b$. If we note $n(x)$, the number of occurrences for the event $x$, then we have

$$n(a \cup b) = n(a) + n(b) - n(a \cap b). \tag{1.11}$$

The term that is subtracted in the right-hand side in Eq. (1.11) corresponds to events that would be otherwise counted twice.

Two events are said to be *mutually exclusive* if $n(a \cap b) = 0$, meaning that the second cannot be observed if the first occurs, and the reverse. In this case, we have the property $n(a \cup b) = n(a) + n(b)$.

As a third definition, the *complementary event* of $x$ is the event that occurs if $x$ does not occur. It is noted $\bar{x}$ or sometimes $\neg x$. In Fig. 1.4, the complementary event of $a \cup b$, which is noted $\overline{a \cup b}$, is represented by the shaded surface spreading inside the space $S$ and outside the surface of $a \cup b$.

Since any event probability $p(x)$ is proportional to its number of occurrences, $x$, the previous relation can also be written under the form:

$$p(a \cup b) = p(a) + p(b) - p(a \cap b), \tag{1.12}$$

which is also usually written as

$$p(a + b) = p(a) + p(b) - p(a, b), \tag{1.13}$$

where $p(a \cup b) \equiv p(a + b)$ is the *combined probability* of the two events $a$, $b$ and $p(a \cap b) \equiv p(a, b)$ is the corresponding *joint probability* (one must be careful to read the + sign in $p(a + b)$ as meaning "or" and not "and").

Finally, the probability $p(\bar{x})$ of the complementary event of $x$ ($x = a, b, a \cup b, a \cap b..$) is given by

$$p(\bar{x}) = 1 - p(x). \tag{1.14}$$

Let us apply now the above properties with two illustrative examples.

## Taking exams

Two students, John and Peter, must take an exam. The probability that John passes is $p(\text{John pass}) = 0.7$ and for Peter it is $p(\text{Peter pass}) = 0.4$. The probability that they both pass is $p(\text{John pass} \cap \text{Peter pass}) = 0.3$. What are the probabilities that:

(a) *At least one* of the two students passes the exam?
(b) *The two* students fail the exam?
(c) *At least one* of the two students fails the exam?

*Answer:* In the event (a), we have

$$\begin{aligned} p(\text{John pass} \cup \text{Peter pass}) &= p(\text{John pass}) + p(\text{Peter pass}) - p(\text{John pass} \cap \text{Peter pass}) \\ &= 0.7 + 0.4 - 0.3 = 0.8. \end{aligned}$$

The event (b) is the complement of event (a), since the two are mutually exclusive (at least one passing excludes both failures). Therefore

$$\begin{aligned} p(\text{John fail} \cap \text{Peter fail}) &= 1 - p(\text{John pass} \cup \text{Peter pass}) \\ &= 1 - 0.8 = 0.2. \end{aligned}$$

The probability of the event (c) can be calculated as follows:

$$\begin{aligned} p(\text{John fail} \cup \text{Peter fail}) &= p(\text{John fail}) + p(\text{Peter fail}) - p(\text{John fail} \cap \text{Peter fail}) \\ &= (1 - 0.7) + (1 - 0.4) - 0.2 = 0.7. \end{aligned}$$

To conclude this example, we should take note of an interesting feature. Indeed, we have

$$\begin{aligned} & p(\text{John pass}) \cup \text{Peter pass} = 0.3 \\ & > p(\text{John pass}) \times p(\text{Peter pass}) = 0.7 \times 0.3 = 0.21, \end{aligned}$$

which means that the probability of both students passing the exam is greater than the product of their respective probabilities of succeeding. This means that the events of their individual successes are *correlated*. This notion of *event correlation* will be clarified further on.

### Sharing birthdays

Given a group of people, the issue is to find how many persons may share your birthday (assuming that birth events are independent):

(a) In a group of $n$ people, what is the probability of finding at least one person who shares your birthday?
(b) How many people $n$ should be in the group so that this probability is 50%?
(c) How many people $n$ should be in the group so that the probability to find at least two persons sharing the same birthday is 50%?

*Answer:* The probability that any person met at random shares your birthday is $1/365$ (since your birthday is one date out of 365 possibilities[6]), which is relatively small ($1/365 = 0.27\%$). The probability that this person *does not* share your birthday is thus $364/365 = 99.7\%$, which is relatively high, as expected.

To answer question (a), we consider the complementary event $\bar{x}$ = nobody in this group shares your birthday. The probability that two persons selected at random *do not* share your birthday is $(364/365)^2$, thus for a group of $n$ people we have $p(\bar{x}) = (364/365)^n$. The probability that *at least one* person shares your birthday is, therefore, $p(x) = 1 - p(\bar{x}) = 1 - (364/365)^n$.

Question (b) is answered by finding the value of $n$ for which $p(x) \approx 0.5$. With a pocket calculator, one easily finds after a few trials that $n = 253$ (or alternatively by computing $n = \log(0.5)/\log(364/365)$). As expected, this is a relatively large group.

Question (c) leads to a nonintuitive result, as we shall see. The number of ways of selecting a pair of persons in a group of $n$ people is $C_n^2 = n(n-1)/2$. The probability that two people *do not* share birthdays is $364/365$. Thus, the probability of finding no matching pair in the whole group is $p(\bar{x}) = (364/365)^{n(n-1)/2}$, and the probability of finding at least one matching pair is $p(x) = 1 - p(\bar{x}) = 1 - (364/365)^{n(n-1)/2}$. Solving for $p(x) \approx 0.5$ (or $n(n-1)/2 = 253$) yields $n = 23$. Thus, a group as small as 23 people has a 50% chance of including *at least two persons* sharing the same birthday. Such a result is indeed far from intuitive!

I address next the concept of *conditional probabilities*. By definition, the conditional probability $p(b\,|a)$ is *the probability that event b occurs given the fact that event a has occurred*. The factual knowledge regarding the first event thus provides a clue regarding the likelihood of the second.

The conditional probability is calculated according to what is called *Bayes's theorem*:

$$p(b\,|a) = \frac{p(a,b)}{p(a)}. \tag{1.15}$$

[6] For simplicity, we excluded here people born on February 29th (which only happens every four years). It is also assumed that birthday events are independent, which is not true in real life, for instance, due to seasonal peaks (i.e., observed peaks in the summer months) and effects of induced labor (i.e., a majority of births happen on weekdays). A full analysis of the different ways to analyze and solve the Birthday paradox can be found in http://en.wikipedia.org/wiki/Birthday_paradox.

Alternatively, we can write

$$p(a, b) = p(b\,|a\,)p(a) = p(a\,|b\,)p(b), \tag{1.16}$$

which defines the fundamental relation between *joint* and *conditional probabilities*.

If there is no correlation whatsoever between two events $a$ and $b$, the events are said to be *uncorrelated* or *independent*. This means that the probability of $b$ knowing $a$ (or any other event different from $b$) is unchanged, which implies that $p(b\,|a) \equiv p(b)$. Likewise, the probability of $a$ knowing $b$ must also be unchanged, or $p(a\,|b) \equiv p(a)$. Replacing these two results in Eq. (1.16) yields a fundamental property for *the joint probability of independent/uncorrelated events*:

$$p(a, b) = p(a)p(b). \tag{1.17}$$

In the opposite case to correlated or dependent events, we have $p(a, b) \neq p(a)p(b)$, which means that either $p(a, b) > p(a)p(b)$ or $p(a, b) < p(a)p(b)$, representing two possibilities (the second case is being referred to as *anti-correlation*).

Let us play with conditional probabilities through an illustrative example.

### Party meetings

Assume that the probabilities of meeting two old friends, Alice and Bob, in a party are $p(\text{Alice}) = 0.6$ and $p(\text{Bob}) = 0.4$, respectively. The probability of meeting Bob if we know that Alice is there is $p(\text{Bob}\,|\text{Alice}\,) = 0.5$. Then what are the probabilities of:

(a) Meeting at least one of them?
(b) Meeting both Alice and Bob?
(c) Meeting Alice if we know that Bob is there?

*Answer:* For question (a), let's apply the theorem in Eq. (1.13) combined with Bayes's theorem in Eq. (1.16) to get:

$$\begin{aligned} p(\text{Alice} + \text{Bob}) &= p(\text{Alice}) + p(\text{Bob}) - p(\text{Alice, Bob}) \\ &= p(\text{Alice}) + p(\text{Bob}) - p(\text{Bob}\,|\text{Alice}\,)p(\text{Alice}). \\ &= 0.6 + 0.4 - 0.5^{*}0.6 = 0.7 \end{aligned}$$

For question (b), we directly calculate the joint probabilities as follows:

$$p(\text{Alice, Bob}) = p(\text{Bob}\,|\text{Alice}\,)p(\text{Alice}) = 0.5^{*}0.6 = 0.3.$$

For question (c), we apply Bayes's theorem to get:

$$p(\text{Alice}\,|\text{Bob}\,) = \frac{p(\text{Alice, Bob})}{p(\text{Bob})} = \frac{0.3}{0.4} = 0.75.$$

We see from the results that the most likely outcome is to meet at least one of them, ($p(\text{Alice} + \text{Bob}) = 0.7$), meaning either Alice or Bob or both. The less likely outcome is to meet both ($p(\text{Alice, Bob}) = 0.3$). But if Alice is first seen there, it is much more likely to see Bob ($p(\text{Bob}\,|\text{Alice}\,) = 0.7 > p(\text{Bob}) = 0.4$) and if Bob is first seen there, it is a bit more likely to see Alice ($p(\text{Alice}\,|\text{Bob}\,) = 0.75 > p(\text{Alice}) = 0.6$). These last two facts illustrate the correlation between the respective presences of Alice and Bob. If their presence were not correlated, we would have $p(\text{Alice, Bob}) = p(\text{Alice}) \times p(\text{Bob}) = 0.6^*0.4 = 0.24$, which is a lower probability than found here ($p(\text{Alice, Bob}) = 0.3$). This shows that the known presence of either one increases the chances of seeing the other, a fact that is usually verified in parties or social life! Incidentally, the probability that neither Alice nor Bob be seen at the party is $p(\overline{\text{Alice} + \text{Bob}}) = 1 - p(\text{Alice} + \text{Bob}) = 1 - 0.7 = 0.3$. Thus, there exist equal chances of seeing them both or of seeing neither of them, but this is only coincidental to this example.

I shall conclude this section and chapter by considering the probabilities associated with a *succession of independent or uncorrelated events*. As we have seen, if $x_1, x_2$ are independent events with probabilities $p(x_1)$, $p(x_2)$, their joint probability is simply given by the product $p(x_1, x_2) = p(x_1) \times p(x_2)$. Thus for three independent events, we have $p(x_1, x_2, x_3) = p(x_1, x_2) \times p(x_3) = p(x_1)p(x_2)p(x_3)$, and so on.

Consider a few examples to illustrate the point.

## Tossing the coin

What is the probability that a flipped coin lands three times on the same side?

*Answer:* The probability of getting either three tails or three heads is $q = (1/2)^3 = 0.125$. But there is a trick: since either succession of events is possible (i.e., getting three heads or three tails), the total probability is actually $0.125 + 0.125 = 0.250 = 1/4$.

## Double six

Rolling two dice, what is the number of trials required to obtain a double six with a chance of at least 50%?

*Answer:* The probability of getting a six from a single die is $p(6) = 1/6$. Since the two dice are independent (their outcome is uncorrelated), the joint probability for a double six is $p(6, 6) = p(6) \times p(6)(1/6)^2 = 1/36$. The probability of not getting a double six is therefore $q = 1 - p(6, 6) = 1 - 1/36 = 35/36 = 0.97222\ldots$ The probability of *not* getting a double six after $N$ trials is $q^N$. We must then find the number $N$, such that $q^N \leq 0.5$, meaning that the complementary event (not getting any double six) has the probability $1 - q^N \geq 0.5$, or greater than 50%. Solving this equation for $N$ is straightforward using the successive steps: $N \log q = N \log(35/36) \leq 0.5$, then $-N \log(36/35) \leq 0.5$, then $N \geq 0.5/\log(36/35)$. The equation can also be solved using a computer spreadsheet. Either method yields the solution $N = 25$, for which $1 - q^N = 0.5055$.

## Drawing cards

Five cards are successively drawn at random from a 32-card deck. What are the probabilities for the resulting hand to:

(a) Have only red cards?
(b) Include at least one ace?

*Answer*: Recall first that a 32-card deck has four series (red hearts, red diamonds, black clubs, and black spades) of eight cards each, including an ace, a king, a queen, a jack and four cards having the values 10, 9, 8, and 7.

Considering question (a), we must calculate the probability of drawing five red cards successively. We first observe that half of the cards (16) have the same (red or black) color. The probability of drawing a red card is thus ½, or, more formally, $p(x_1 = \text{red}) = C_{16}^1/C_{32}^1 = 16/32 = 1/2$. For the second draw, we have 31 cards left, which by definition include only 15 red cards. Thus, $p(x_2 = \text{red}) = C_{15}^1/C_{31}^1 = 15/31$, and so on for the other three selections. The final answer for the joint probability is, therefore:

$$p(\text{red, red, red, red, red}) = \frac{16}{32}\frac{15}{31}\frac{14}{30}\frac{13}{29}\frac{12}{28} = 0.021.$$

Question (b) concerns a hand having *at least* one ace. We could go through all the possibilities of successively drawing five cards with one, two, three, or four aces, but we can proceed much more quickly. Indeed, we can instead calculate the probability of the complementary event "no ace in hand," then use the property $p(\text{at least one ace}) = 1 - p(\text{no ace})$. The task is to use combinatorics to evaluate the probabilities of having no ace at each draw. The number of ways of drawing a first card that is not an ace is $C_{32-4}^1 = C_{28}^1 = 28$. The probability of having no ace in the first draw is, therefore, 28/32. With similar reasoning for the next four draws, it is straightforward to obtain finally:

$$p(\text{no ace}) = \frac{28}{32}\frac{27}{31}\frac{26}{30}\frac{25}{29}\frac{24}{28} = 0.488,$$

and $p(\text{at least one ace}) = 1 - 0.488 = 0.511$. We see that the probability of eventually having *at least* one ace in the five-card hand is a little over 50%, which was not an intuitive result. The lesson learnt is that successive events may be independent, but their respective probabilities keep evolving according to the outcome of the preceding events. Put otherwise, each occurring event affects the space of the next series of events, even if all events are 100% independent.

To complete this first chapter, I must introduce the summing rule of conditional probabilities, also called the law of total probability. Assume two complete and disjoint event spaces $S = \{a, b\}$ and $T = \{x, y, z\}$, with $S \cap T = \emptyset$ (the symbol $\emptyset$ meaning an empty set or a set having zero elements) and with, by definition of probabilities $p(a) + p(b) = 1$, and $p(x) + p(y) + p(z) = 1$. Then the following summing relations hold:

$$\begin{cases} p(x) = p(x\,|a\,)p(a) + p(x\,|b\,)p(b) \\ p(y) = p(y\,|a\,)p(a) + p(y\,|b\,)p(b) \end{cases} \tag{1.18}$$

and

$$\begin{cases} p(a) = p(a\,|x\,)p(x) + p(a\,|y\,)p(y) + p(a\,|z\,)p(z) \\ p(b) = p(b\,|x\,)p(x) + p(b\,|y\,)p(y) + p(b\,|z\,)p(z). \end{cases} \tag{1.19}$$

Thus with the knowledge of the conditional probabilities $p(\cdot\,|\cdot)$ relating the two event space causalities, one is able to compute the probabilities from one space to the other.[7] It is important to note that while the conditional $p(\cdot\,|\cdot)$ are true *probabilities*, they do not sum up to unity, a mistake to avoid by firmly memorizing any of the above summing relations.

## 1.4 Exercises

**1.1** (B): Flipping two coins simultaneously, what are the probabilities associated with the following events:
(a) Getting two heads?
(b) Getting one heads and one tails?
(c) Getting either two heads or two tails?

**1.2** (B): Rolling three dice, one wins if the outcome is 4, 2, 1 in any order. What is the probability of winning in the first, the second, and the third dice roll? What is the number of rolls required to have at least a 50% chance of winning?

**1.3** (B): A lotto game has 50 numbered balls, out of which six are picked at random. What is the probability of winning by betting on any six-number combination?

**1.4** (B): Three competing car companies, A, B, and C, have market shares of 60%, 30%, and 10%, respectively. The probabilities that the cars will show some construction defects are 5% for A, 7% for B, and 15% for C. What is the probability for any car bought at random to show some construction defect?

**1.5** (M): A bag contains ten billiard balls numbered from one to ten. If two balls are picked at random from the bag, what is the probability of getting:
(a) Two balls with even numbers?
(b) At least one ball with an odd number?
(c) Ball number eight in the selection?
You must propose two different methods of solving the exercise.

[7] The summing rule can be generalized, assuming $X = \{x_1, x_2, \ldots x_N\}$ and $Y = \{y_1, y_2, \ldots y_M\}$, into the following:

$$p(y_i) = p(y_i|x_1)p(x_1) + p(y_i|x_2)p(x_2) + \cdots p(y_i|x_N)p(x_N) = \sum_{j=1}^{N} p(y_i|x_j)p(x_j),$$

which can also be expressed as a matrix–vector relation, $P_Y = \tilde{U} P_X$ with $P_X = \{p(x_1), \ldots, p(x_N)\}$ and $P_Y = \{p(y_1), \ldots, p(y_M)\}$ being column vectors, and $U$ the $M \times N$ "conditional" or "transition" matrix:

$$\tilde{U} = \begin{pmatrix} p(y_1|x_1) & p(y_1|x_2) & \cdots & p(y_1|x_N) \\ p(y_2|x_1) & \cdots & \cdots & \vdots \\ \vdots & \cdots & \cdots & \vdots \\ p(y_M|x_1) & \cdots & \cdots & p(y_M|x_N) \end{pmatrix}.$$

**1.6** (M): The American roulette has 36 spots numbered 1, 2, . . . , 36, which are alternatively divided into 18 red spots (odd numbers) and 18 black spots (even numbers), plus two green spots, called 0 and 00. The roulette's outcome is any single number. It is possible to bet on any number (single or combination), or on red, black, odd, even, the first 12 (numbers 1–12), the second 12 (numbers 13–24), and the third 12 (numbers 25–36). The roulette payoffs are 35 to 1 for a winning number, 2 to 1 for a winning red/black/even/odd number and 3 to 1 for a winning first/second/third 12 number. What are the odds of winning any of these bets? Comments? What is the probability of winning at least once after playing single numbers 36 times successively?

**1.7** (T): In the lotto game of Exercise 1.3, what are the probabilities of having either exactly one or exactly two winning numbers in the six-number combination?

**1.8** (T): This is a television game show where the contestant may win a car. The car is standing behind one of three doors. The payer first designates one of the three doors. The host then opens another door, which reveals . . . a goat! Then the host asks the contestant: "Do you maintain your choice?" The question here is: what is in the contestant's interest? To maintain or to switch his or her choice? Clue: The answer is not intuitive!

# 2 Probability distributions

Chapter 1 enabled us to familiarize ourselves (say to revisit, or to brush up?) the concept of *probability*. As we have seen, any probability is associated with a given event $x_i$ from a given event space $S = \{x_1, x_2, \ldots, x_N\}$. The discrete set $\{p(x_1), p(x_2), \ldots, p(x_N)\}$ represents the *probability distribution function* or PDF, which will be the focus of this second chapter.

So far, we have considered single events that can be numbered. These are called *discrete events*, which correspond to event spaces having a finite size $N$ (no matter how big $N$ may be!). At this stage, we are ready to expand our perspective in order to consider event spaces having unbounded or infinite sizes ($N \to \infty$). In this case, we can still allocate an integer number to each discrete event, while the PDF, $p(x_i)$, remains a function of the *discrete* variable $x_i$. But we can conceive as well that the event corresponds to a *real* number, for instance, in the physical measurement of a quantity, such as length, angle, speed, or mass. This is another infinity of events that can be tagged by a real number $x$. In this case, the PDF, $p(x)$, is a function of the *continuous* variable $x$.

This chapter is an opportunity to look at the properties of both discrete and continuous PDFs, as well as to acquire a wealth of new conceptual tools!

## 2.1 Mean and variance

In this section, I shall introduce the notions of *mean* and *variance*. I will consider first discrete then continuous PDFs, with illustrative examples from the physical world. I will then show that in the limit of a large number of events, the discrete case can be approximated by the continuous case (the so-called *asymptotic* limit).

Consider then the *discrete* case, as defined by the sample/event space $S = \{x_1, x_2, \ldots, x_N\}$, where $N$ can be infinite (notation being $N \to \infty$). The associated PDF is called $p(x)$, which is a function of the random variable $x$, which takes the discrete values $x_i$ ($i = 1, \ldots, N$). When writing $p(x_i)$, this conceptually means "the probability that event $x$ takes the value $x_i$." The *mean*, which is noted $<x>$, or also $\bar{x}$, or $E(x)$, is given by the weighted sum

$$<x> = \sum_{i=1}^{N} x_i p(x_i). \tag{2.1}$$

The mean represents what is also called the *expectation value*, hence the notation $E(x)$. Contrary to what the word "expectation" would suggest, $E(x)$ is not the value one should expect for the outcome of event $x$ after some sufficient number of trials. Rather, it is the mean value of $x$ that one should expect to become, independent of the number of trials $N$, as this number indefinitely increases, and assuming that the PDF is constant over time.

As an illustration, take the event space $S = \{1, 2, 3, 4, 5, 6\}$ corresponding to the outcomes of rolling a single die. As we know, the PDF is $p(x) = 1/6$ for all events $x$. The mean value is, therefore,

$$<x> = \sum_{i=1}^{6} x_i p(x_i) = \sum_{i=1}^{6} i\frac{1}{6} = \frac{1}{6}(1+2+3+4+5+6) = 3.5. \tag{2.2}$$

This example illustrates what should be understood by "expected value." In no way would a die fall on $x = 3.5$, and nobody would "expect" such a result! The number simply represents the average count of spots, were we to perform an infinite (say, sufficiently large) number of die rolls. Looking back at Fig. 1.2 (counting spots from two dice), we observe that the PDF is centered about $x = 7$, which represents the mean value of all possibilities. The mean value should, therefore, not be interpreted as representing the event of highest probability, i.e., where the PDF exhibits a peak value, although this might be true in several cases, like the PDF in Fig. 1.2.

But the mean does not tell the whole story about a PDF. I introduce a second parameter called PDF *variance*. The variance, typically noted $\sigma^2$, is defined as the *expected value* of $(x - <x>)^2$. Thus, using basic algebra:

$$\begin{aligned}
\sigma^2 &= <(x - <x>)^2> \\
&= <x^2 - 2x<x> + <x>^2> \\
&= <x^2> - 2<x><x> + <<x>^2> \\
&= <x^2> - 2<x>^2 + <x>^2 \\
&= <x^2> - <x>^2
\end{aligned} \tag{2.3}$$

(noting that $<<x>^2> = <x>^2$, since $<x>^2$ is a number, not a variable). The variance is the difference between the expected value of $x^2$ and the square of the mean. It is always a positive number. We can thus take its square root, which is called the PDF *standard deviation*:

$$\sigma = \sqrt{\sigma^2} \equiv \sqrt{<x^2> - <x>^2}. \tag{2.4}$$

How does one calculate the mean square $<x^2>$? Consistent with the earlier definition in Eq. (2.1), we have

$$<x^2> = \sum_{i=1}^{N} x_i^2 p(x_i). \tag{2.5}$$

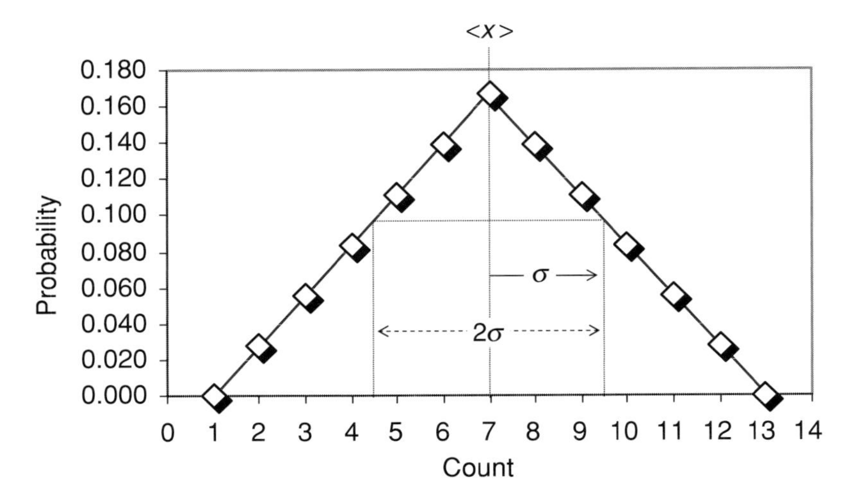


**Figure 2.1** Same probability distribution function (PDF) as in Fig. 1.2(b), illustrating the concepts of mean ($<x> = 7$) and standard deviation ($\sigma = 2.415$).

As an illustration, let us then calculate the standard deviations of the one-die and two-dice PDF. In the one-die case, we have

$$<x^2> = \sum_{i=1}^{6} i^2 \frac{1}{6} = \frac{1}{6}(1^2 + 2^2 + 3^2 + 4^2 + 5^2 + 6^2) = 15.166. \tag{2.6}$$

The one-die PDF variance is, thus, $\sigma^2 = <x^2> - <x>^2 = 15.166 - 3.5^2 = 2.916$, and the standard deviation is $\sigma = \sqrt{2.916} = 1.707$.

We can get a better sense of the meaning of such a "deviation" with the two-dice example. It takes a computer spreadsheet (or pocket-calculator patience) to compute $<x^2>$ with the PDF shown in Fig. 1.2(b), but the reader will easily check that the result is $<x^2> = 54.833$, hence $\sigma^2 = <x^2> - <x>^2 = 54.833 - 7^2 = 5.833$, and $\sigma = \sqrt{5.833} = 2.415$.

Reproducing, for convenience, the same PDF plot in Fig. 2.1, we see that the interval $[<x> - \sigma, <x> + \sigma] \approx [4.5, 9.5]$, of width $2\sigma$, defines the PDF region where the probabilities are the highest, i.e., about greater or equal to half the peak value. In other words, the outcome of the two-dice roll is most highly likely to fall within the $2\sigma$ region.

Statistics experts familiarly state that the event to be expected is $<x>$ with a $\pm\sigma$ accuracy or *trust interval*. However, such a conclusion does not apply with other types of PDF, for instance, with the *uniform distribution*, as I will show further below.

## 2.2 Exponential, Poisson, and binomial distributions

I shall now consider three basic examples of discrete PDFs. As we shall see, these PDFs govern many physical phenomena and even human society! They are the

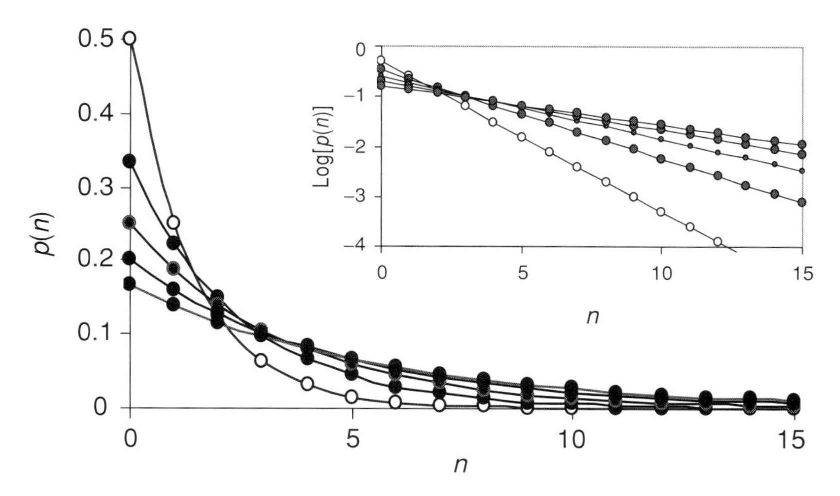


**Figure 2.2** Plots of *discrete-exponential* or *Bose–Einstein* probability distribution corresponding to mean values $<n> = N = 1, 2, 3, 4, 5$ (open symbols corresponding to the case $<n> = 1$). The inset shows the same plots in decimal logarithmic scale.

*discrete-exponential* distribution, the *Poisson* distribution, and the *binomial* distribution. To recall that these three PDFs concern discrete random variables, I will use the notations $<n>$ and $p(n)$, instead of $<x>$ and $p(x)$, which we reserve for continuous distributions.

The *discrete-exponential distribution*, also referred to as the *Bose–Einstein* (BE) distribution, is defined according to

$$p(n) = \frac{1}{N+1}\left(\frac{N}{N+1}\right)^n, \tag{2.7}$$

where $<n> = N$ is the mean value. This PDF variance is simply $\sigma^2 = N + N^2$.

Figure 2.2 shows plots of the exponential distribution for various values of $N$, in both linear and logarithmic scales. The continuous lines linking the data are only shown to guide the eye, recalling that the PDF applies to the discrete variable $n$. The PDF is seen to be linear in the logarithmic plot, which allows a better visualization of the evanescent tail at high $n$. As the figure also illustrates, the peak value of the exponential/BE distribution is reached at the origin $n = 0$, with $p(0) = 1/(N + 1)$.

In the physics world, such a distribution is representative of *thermal processes*. Such processes concern, for instance, the emission of photons (the electromagnetic energy quanta) by hot sources, such as an incandescent light bulb or a star, like the sun, or the distribution of electrons in atomic energy levels, and many other physical phenomena. As we shall see in the forthcoming chapters, the discrete-exponential distribution is important in information theory. We will also see that the exponential PDF is commonly found in human society, for instance, concerning the distribution of alphabetic characters in Western languages.

Let us consider next another PDF of interest, which is the *Poisson distribution*. This PDF is used to predict the number of occurrences of a discrete event over a fixed

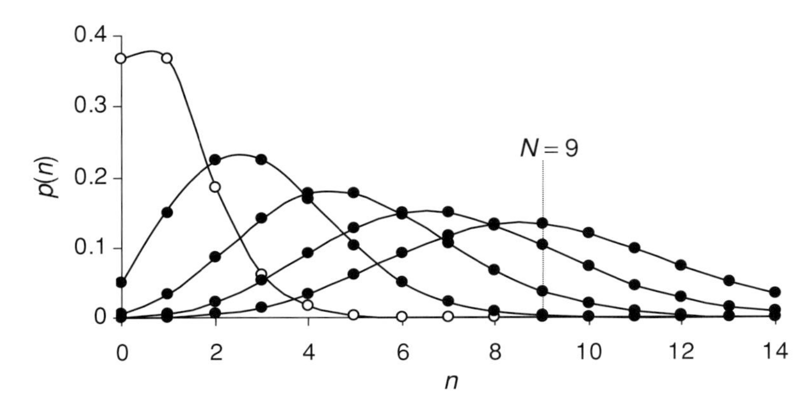


**Figure 2.3** Plots of the *Poisson* distribution corresponding to mean values $<n> = N = 1, 3, 5, 7, 9$ (open symbols corresponding to the case $<n> = 1$).

time interval. If $N$ is the expected number of occurrences over that time interval, the probability that the count is *exactly* $n$ is

$$p(n) = \mathrm{e}^{-N} \frac{N^n}{n!}. \tag{2.8}$$

As a key property, the Poisson PDF variance is equal to the PDF mean, i.e., $\sigma^2 = N$. Figure 2.3 shows plots of the Poisson PDF for various values of the mean $N$.

It is seen from the figure that as $N$ increases, the distribution widens (but slowly, according to $\sigma = \sqrt{N}$). Also, the line joining the discrete points progressively takes the shape of a symmetric bell curve centered about the mean, as illustrated for $N = 9$ (the mean coinciding with the PDF peak only for $N \to \infty$). This property will be discussed further on, when considering continuous distributions.

There exist numerous examples in the physical world of the Poisson distribution, referred to as *Poisson processes*. In atomic physics, for instance, the Poisson PDF defines the count of nuclei decaying by radioactivity over a period $t$. Given the decay rate $\lambda$ (number of decays per second), the mean count is $N = \lambda t$, and the Poisson PDF $p(n)$ gives the probability of counting $n$ decays over that period. In laser physics, the Poisson PDF corresponds to the count of photons emitted by a coherent light source, or laser. Poisson processes are also found at macroscopic and human scales: for instance, the number of cars passing under a highway bridge over a single lane, the number of phone calls handled by a central office, the number of Internet hits received by a website, the number of bonds traded in the stock exchange, or the number of raindrops falling on a roof window. Each of these counts, being made within a given amount of time, from seconds to minutes to hours to days, obeys Poisson statistics.

The *binomial distribution*, also called the *Bernoulli distribution*, describes the number of successes recorded in a discrete sequence of independent trials. Here, let's call $k$ the random variable, which is the number of successes, and $n$ the fixed number of trials, with $k \leq n$. If the probability of success for a single independent trial is $q$, the binomial

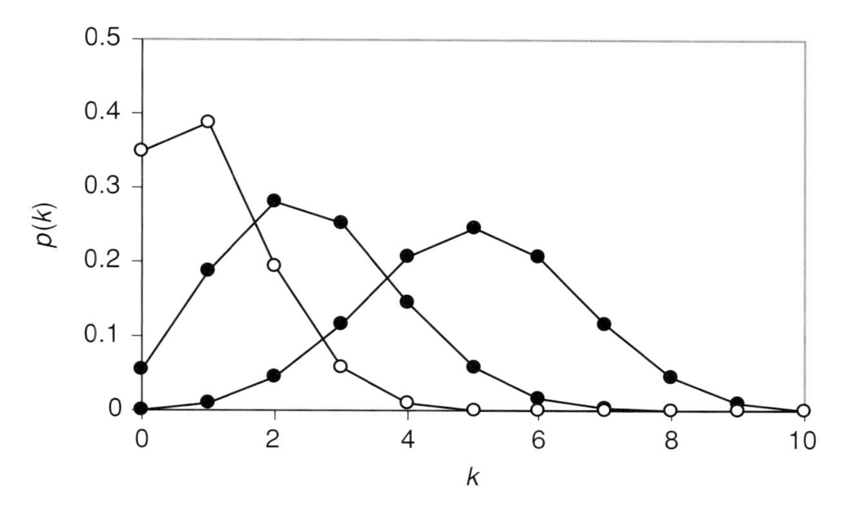


**Figure 2.4** Plots of the *binomial* or *Bernoulli* distribution corresponding to parameters $q = 0.1, 0.25, 0.5$ with $n = 10$ mean (open symbols corresponding to the case $q = 0.1$).

PDF is defined according to:

$$p(k) = C_n^k q^k (1-q)^{n-k} \equiv \frac{n!}{k!(n-k)!} q^k (1-q)^{n-k}. \tag{2.9}$$

Such a definition makes sense, considering that the probability of obtaining $k$ independent successes is $q^k$ and the probability of failure for the remaining $n - k$ events is $(1-q)^{n-k}$. The product of the two is the joint probability of the group of measurements, as observed in any specific order. The combinatorial factor $C_n^k$ is the number of possible orders for successive success or failure measurements. The mean and variance of the binomial distribution is $N = nq$ and $\sigma^2 = N(1-q)$, respectively. Figure 2.4 shows plots of the binomial PDF for $n = 10$ and different values of $q$. We observe that the PDF is asymmetric except when $q = 0.5$, in which case the peak value is exactly centered about $N$.

A physical illustration of the Bernoulli distribution is the passing, one at a time, of a stream of individual particles (like light quanta or photons) through a piece of absorbing material. These particles may either successfully pass through the material (success, probability $q$), or absorbed by it (failure, probability $1 - q$). Assuming an ideal particle counter, the probability of counting $k$ particles after the stream has passed though the material is effectively$p(k)$, as defined by Eq. (2.9). See also Section 2.5 in the bean machine example.

We note that the binomial distribution is defined over the finite random-variable interval $0 \leq k \leq n$. Since there is no restriction on the size of the sampling space defined by $n$, the PDF can be extended to the limit $n \to \infty$. It can be shown that in this infinite limit, the binomial PDF is equal to the Poisson PDF previously described. It can also be shown that, for large mean numbers $N$, the Poisson (or binomial) PDF asymptotically converges towards the *Gaussian* or *normal* distribution, which is described in the next section.

## 2.3 Continuous distributions

At the beginning of this chapter, I discussed the possibility that the events form a continuous and infinite suite of real numbers, which, in the physical world, represent the unbounded set of measurements of a physical quantity. The likelihood of the measurement then corresponds mathematically to a *continuous* probability distribution. Like discrete PDF, a continuous PDF $p(x)$ must satisfy the property

$$0 \leq p(x) \leq 1 \tag{2.10}$$

for all values $x$ belonging to the event space $\lfloor x_{\min}, x_{\max} \rfloor$ within which the events are defined. By definition, the sum of all probabilities must be equal to unity. With a continuous variable, such a sum is an *integral*, i.e.,

$$\int_{x=x_{\min}}^{x=x_{\max}} p(x)\mathrm{d}x = 1. \tag{2.11}$$

Consistent with the above definition, the PDF mean or expected value $<x>$ and the mean square are defined as follows:

$$<x> = \int_{x=x_{\min}}^{x=x_{\max}} x p(x)\mathrm{d}x \tag{2.12}$$

$$<x^2> = \int_{x=x_{\min}}^{x=x_{\max}} x^2 p(x)\mathrm{d}x, \tag{2.13}$$

with the *variance* being $\sigma^2 = <x^2> - <x>^2$, just as in the discrete-PDF case. Note that for any continuous PDF one can replace the interval $[x_{\min}, x_{\max}]$ by $[-\infty, +\infty]$, with the convention that $p(x) = 0$ for $x \notin [x_{\min}, x_{\max}]$.

## 2.4 Uniform, exponential, and Gaussian (normal) distributions

In the following, we shall consider three basic types of continuous distribution: *uniform*, *exponential*, and *Gaussian (normal)*.

The most elementary type of continuous PDF is the *uniform distribution*. It gives a good pretext to open up our integral-calculus toolbox. Such a distribution is defined as $p(x) = \text{const.}$ over a certain interval $[x_{\min}, x_{\max}]$ and $p(x) = 0$ elsewhere. The constant is found by applying Eq. (2.11), i.e.,

$$\int_{x=x_{\min}}^{x=x_{\max}} p(x)\mathrm{d}x = \text{const.} \int_{x=x_{\min}}^{x=x_{\max}} \mathrm{d}x = \text{const.}(x_{\max} - x_{\min}) \equiv \text{const.}\Delta x = 1, \tag{2.14}$$

where $\Delta x = x_{\max} - x_{\min}$ is the sample-space width, which gives const. $= 1/\Delta x$. Figure 2.5 shows a plot of the uniform distribution thus defined. The figure also illustrates that the PDF is actually defined over the infinite space $[-\infty, +\infty]$, while being

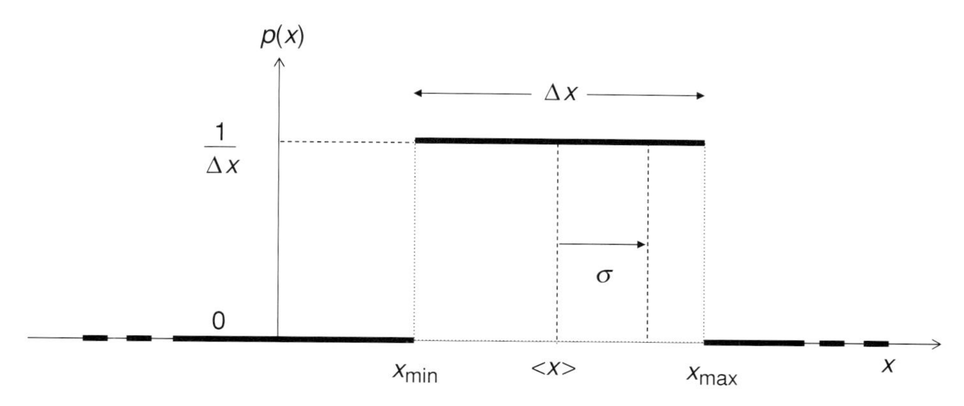


**Figure 2.5** Uniform probability distribution of the continuous variable $x$, see text for description.

identically zero outside the relevant sample interval $[x_{\min}, x_{\max}]$. It is left as an exercise for the reader to show that the mean and standard deviation of a continuous, uniform PDF are:

$$<x> == \frac{x_{\max} + x_{\min}}{2}, \tag{2.15}$$

$$\sigma = \frac{\Delta x}{\sqrt{12}}. \tag{2.16}$$

We see from the above results that for a uniform PDF the mean $<x>$ corresponds to the mid-point $(x_{\max} + x_{\min})/2$ of the interval $[x_{\min}, x_{\max}]$. This was expected, since all events $x$ in this interval are equally probable. However, the deviation $\sigma$ is different from the half width of the distribution, $\Delta x/2$, which was not intuitive! This is because (as defined earlier, and to recall) the variance is the mean value of the *square* of the difference $x - <x>$, i.e., $\sigma^2 = <(x - <x>)^2>$. Thus, the quantity $2\sigma = \Delta x/\sqrt{3}$ *does not* define a specific interval where the continuous events $x$ are more likely to be observed, although the contrary is true for most other types of nonuniform PDF.

Uniform continuous distributions, which take the shape of square or step functions, are not generally found in the physical world. One remarkable example, however, is provided by the so-called Fermi function in semiconductors while under absolute-zero temperature ($T = 0$ kelvin). Such a function is the probability $f(E)$ of having an electron at a given energy $E$. The physics shows that, at zero temperature, $f(E) = 1/E_{\mathrm{F}}$ for $0 \leq E \leq E_{\mathrm{F}}$ and $f(E) = 0$ for $E \geq E_{\mathrm{F}}$, where $E_{\mathrm{F}}$ is referred to as the Fermi energy level. To take a simpler example, assume that one is able to measure some physical parameter within an absolutely defined range. For instance, the distribution of frequencies $F$ from microwave or sunlight radiation, as analyzed through a perfect, square-shaped bandpass filter (suppose that measurements outside that filter range are not observed or irrelevant to the test). If the filter is not too large (or is sufficiently narrow!), and under some circumstances, then one may, indeed, observe that the measurement probability $p(F)$ is uniformly distributed. In any case, this would not mean that the probability is uniform in the absolute definition, but in a local-domain sense, as defined by the observation window of measurement apparatus.

Next, we consider the *continuous-exponential distribution*. The PDF is defined for $x \geq 0$ as follows:

$$p(x) = \lambda \mathrm{e}^{-\lambda x}, \tag{2.17}$$

where $\lambda$ is a strictly positive constant called the *rate parameter*. The mean and variance of the exponential PDF are $<x> = 1/\lambda$ and $\sigma^2 = 1/\lambda^2$, respectively (this is left as an exercise).

Between the discrete-exponential and the continuous-exponential PDFs, there is no one-to-one correspondence, except in the limit of large means. Indeed, setting $N = 1/\lambda$ in Eq. (2.17), we can redefine the discrete-exponential PDF ($x = n$ integer) under the form

$$p(n) = \frac{\lambda}{\lambda + 1} \mathrm{e}^{-n \log(1+\lambda)}. \tag{2.18}$$

Such a definition corresponds to that of the continuous PDF (Eq. 2.17) only in the limit $\lambda \to 0$ (for which $\log(1 + \lambda) \to \lambda$), which corresponds to the limit $N \to \infty$.

The continuous-exponential PDF is used to model the timing of physical events that happen at a constant average rate. If $x = t$ is a time variable, the event occurs at exact time $t$ with probability $p(t) = \lambda \exp(-\lambda t)$. We note that in this case, $\lambda$ has the dimension of an inverse time (e.g., inverse seconds or inverse days), meaning that the PDF is a probability rate (in /s or /day units) rather than being dimensionless. The exponential PDF is a maximum for $t = 0$ and decays rapidly as time increases. On average, the events occur at time $<t> = 1/\lambda \equiv \tau$, where $\tau$ is a characteristic *time constant*. In atomic physics, $\tau$ is called the decay constant, or also the 1/e lifetime. For radioactive atoms or for atoms used in laser materials, this means that the disintegration or photon emission occurs in average at the time $t = \tau$. The probability that any atom decays at a time *after* $t$ is given by

$$\begin{aligned} p(T > t) &= \int_t^{+\infty} p(t')\mathrm{d}t' = \int_t^{+\infty} \tau \exp(-t'/\tau)\mathrm{d}t' \\ &= \tau \left[ -\frac{\exp(-t/\tau)}{\tau} \right]_t^{+\infty} = \exp(-t/\tau). \end{aligned} \tag{2.19}$$

The integration expresses the fact that the probability $p(T > t)$ is given by the continuous sum of all probabilities $p(t')$ where $t'$ belongs to the time interval $[t, +\infty]$. The result shows that the probability that any atom decays after $t = \tau$ is $p(T > \tau) = \mathrm{e}^{-1} = 1/\mathrm{e} \approx 0.36$, hence the name 1/e lifetime. The probabilities that atoms decay after times $t = 2\tau, 3\tau, \ldots$, etc., are 13%, 5%, etc., illustrating that there is always a finite number of "surviving" atoms remaining in their original state and awaiting decay, but their number decreases exponentially over time.

Other applications of the exponential PDF can be found in daily life. For instance, if a person regularly drives above the speed limit and if the highway patrol makes regular controls, the probability of getting a speeding ticket only *after* a given time $t$, i.e., $p(T > t) = \exp(-t/\tau)$, is rapidly vanishing, just like the drivers' luck. The probability that he or she will get a ticket *before* time $t$, i.e., $p(T < t) = 1 - \exp(-t/\tau)$, is rapidly increasing towards unity, representing a situation reaching 100% likelihood. If $\tau = 1$ year represents the mean time for bad drivers to get a speeding ticket, the probability of only getting a ticket 2 to 3 years after that time is only 13% to 5%, meaning that there is an 87% to 95% chance of getting it well before then!

The exponential distribution is also used to characterize reliability and failure in manufactured products or systems, such as TV sets or car engines. Given the mean time to observe a given failure, $\tau$ (now called *mean time to failure* or MTTF), the probabilities that the failure will be observed *before* or *after* time $t$ are $p(T<t) = 1 - \exp(-t/\tau)$ or $p(T>t) = \exp(-t/\tau)$, respectively. The function $p(T>t)$ is generally referred to as the *reliability function*. Its complement, $p(T<t) = 1 - p(T>t)$ is referred to as the *failure function*. For instance, if $\tau = 5$ years represents the mean time to get a car-engine problem, the probability of getting the problem within one year is $p(T<1 \text{ year}) = 1 - \exp(-1/5) = 0.18$, and after one year $p(T>1 \text{ year}) = 1 - p(T < 1 \text{ year}) = 0.82$. This means that there is close to a 20% chance of having the problem before one year, even if the car engine (or driving safety) is supposed to be problem-free for a mean period of 5 years. On the other hand, the odds of having a problem after one year are 82%, but this prediction covers an infinite amount of time. It is possible to make a more detailed failure prediction for a given period spanning times $t_1$ to $t_2 > t_1$. Indeed, the probability of getting a failure between these two times is:

$$\begin{aligned} p(t_1 < T < t_2) &= 1 - [p(T < t_1) + p(T > t_2)] \\ &= 1 - [1 - \exp(-\lambda t_1) + \exp(-\lambda t_2)] \\ &= \exp(-\lambda t_1) - \exp(-\lambda t_2). \end{aligned} \tag{2.20}$$

With the above formula, one can determine the failure probabilities concerning any specific periods defined by $[t_1, t_2]$.

Next, we consider as a last but key example, another continuous PDF, which is the *Gaussian or normal distribution*. With a mean $<x> = N$ and a variance $\sigma^2$, it is formally defined according to:

$$p(x) = \frac{1}{\sigma\sqrt{2\pi}} \exp\left[-\frac{(x-N)^2}{2\sigma^2}\right]. \tag{2.21}$$

Since the function $\exp(-u^2)$ is symmetrical with a peak value centered at $u = 0$, we see that the Gaussian PDF is centered about its mean, $<x> = N$, with a peak value of $p(N) = p_{\text{peak}} = 1/(\sigma\sqrt{2\pi})$. For values $x = N \pm \sigma\sqrt{2}$, we observe from the definition that the probability drops to $e^{-1}p_{\text{peak}} = p_{\text{peak}}/e \approx 0.367 p_{\text{peak}}$. Figure 2.6 shows plots of Gaussian PDFs with mean $N = 0$ and different standard deviations.

The characteristic bell shape has justified over time the popular name of bell distribution, which is well known to a large public. The surface $S$ under the curve, which is defined by two points $x_1, x_2$, i.e.,

$$S = p(x_1 < x < x_2) = \int_{x1}^{x2} p(x)dx, \tag{2.22}$$

represents the probability of event $x$ taking a value in the interval $[x_1, x_2]$. It can be shown by integration in Eq. (2.22) that $p(N - \sigma < x < N + \sigma) \approx 0.682$, meaning that

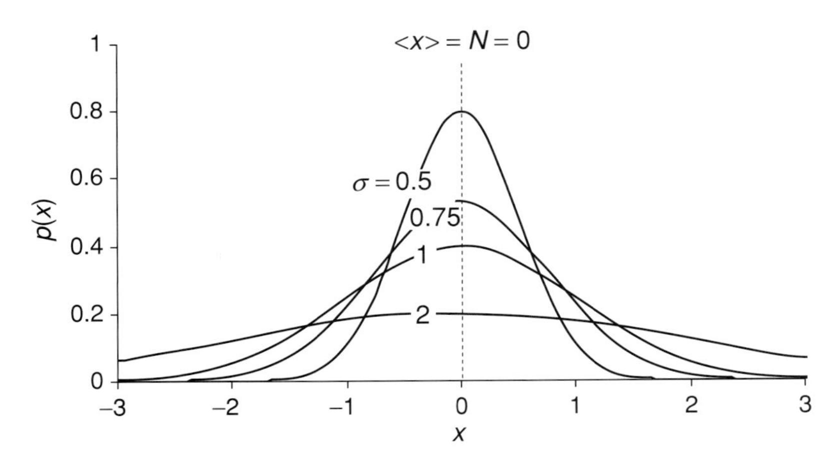


**Figure 2.6** Gaussian probability distribution with mean $<x> = N = 0$ and standard deviations $\sigma = 0.5, 0.75, 1$, and 2.

68.27% of the bell surface concerns events falling within two standard deviations ($\pm\sigma$) of the mean ($N$). Likewise, for the intervals $2\sigma$ and $3\sigma$ about the mean, the surfaces represent 95.4% and 99.7 of the total bell surface, respectively.

A physical parameter obeying a Gaussian PDF, for instance, electrical noise in radio or TV signals, is experimentally characterized through discrete sampling, even if the measurement apparatus (e.g., an analog oscilloscope) provides a continuous signal. It is interesting to see what a succession of such sampling measurements looks like in the real world. Figure 2.7 shows the plot of a series of 200 samplings of a random variable $x$ following a Gaussian distribution with $<x> = 0$ and $\sigma = 0.1$, as generated by a computer program. We observe that, as expected, the values of $x$ are randomly distributed above and below the $x$ axis. The sampling points form a cloud that is denser near the axis, defining a region of width $2\sigma$. The figure also includes the corresponding plot of $x^2$, which shows that most of the sampling points are found between $x^2 = 0$ and $x^2 = <x^2> = \sigma$. It is important to distinguish the discrete sampling points (here numbering 200) from the continuous, Gaussian PDF. To compare the two, we can draw a *histogram* of the sampling points, as shown in Fig. 2.8. The histogram represents the counts of points corresponding to the different values of $x$ in Fig. 2.7. For clarity, I have multiplied the sampled variable $x$ by a hundredfold and truncated the result ($y = 100x$) to an integer, which gives values ranging from $y = -23$ to $y = +29$.

As seen from Fig. 2.8, the envelope of the histogram is quite different from that of the actual Gaussian distribution, also plotted in the figure for comparison (with $\sigma_y = 100 \times \sigma_x = 10$). The reason for this discrepancy is twofold. First, once gathered into a histogram, the discrete samplings do not have a sufficient number (here 200) to reproduce the smooth and continuous features of the Gaussian PDF. Second, the data were arbitrarily arranged into truncated integer bins, which enhances the discontinuity of the histogram's envelope. To show how this truncation changes the envelope, a second histogram was made with $z = 50x$ (see inset in Fig. 2.8). This second histogram has a smoother envelope, because there are more data in each of the integer bins. For this

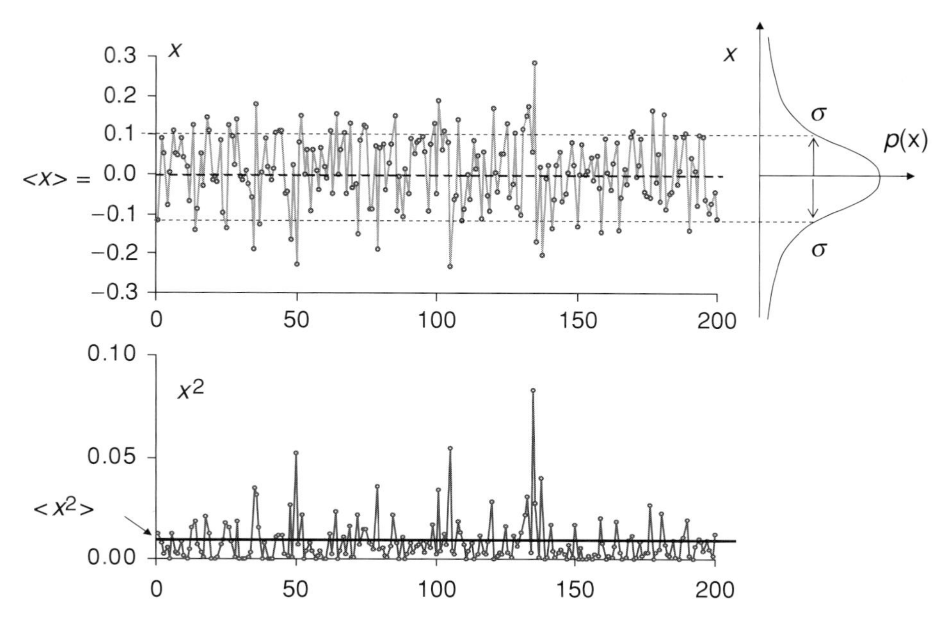


**Figure 2.7** Example of discrete samplings (200 events) of a Gaussian distribution ($<x> = 0$, $\sigma = 0.1$), showing the outcome for random variables $x$ (top) and $x^2$ (bottom).

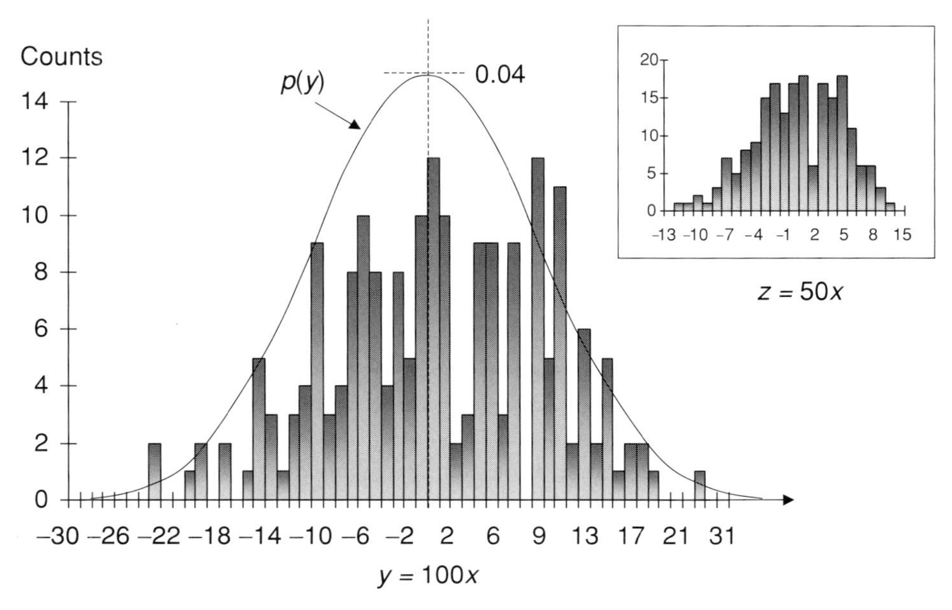


**Figure 2.8** Histogram of the 200 sampling points $x$ shown in Fig. 2.7, as converted into $y = 100x$, and corresponding Gaussian distribution envelope $p(y)$. The inset shows a denser histogram corresponding to the same $x$ data with $z = 50x$.

reason, the envelope shape is closer to that of a bell. To obtain a smooth histogram envelope that would closely fit the Gaussian bell curve, one would need to acquire $10^4$–$10^5$ sampling points and arrange them into hundreds of histogram bins.

The lesson learnt is that experimental statistics require large numbers of samplings in order to reflect a given probability law, for instance, the Gaussian PDF (but it is not limited to this case). We have previously reached a similar conclusion from our earlier coin-flipping experiment, for which the associated probability distribution is discrete and quite elementary ($p(\text{heads}) = p(\text{tails}) = 1/2$), see Chapter 1 and Fig. 1.3. To recall, it took no less than 700 samples to approach the expected uniform distribution with reasonable accuracy.

The Gaussian or normal probability distribution characterizes a large variety of random processes found in physics, in engineering, and in many other domains of science. In most random processes indeed, the uncertainty associated with *continuous* parameters, which is also referred to as *noise*, obeys a Gaussian (normal) PDF. Here is a nonlimitative list of Gaussian (normal) processes:

- Experimental measurement errors (the mean $<x>$ being taken as the value to be retained);
- Manufacturing, in the distribution of production yields and quality scoring;
- Telecommunications, in the distribution of 1/0 bit errors in digital receivers;
- Photonics, to approximate the transverse or spatial distribution of light intensity in optical fibers or in laser beams;
- Education and training, in the distribution of intelligence (IQ) test scores, or professional qualifications and performance ratings;
- Information theory, which will be developed in Chapter 4 when analyzing *continuous channels* with noise.
- Medicine, in the distribution of blood pressure and hair length, or of the *logarithm* of body weight or height;
- Economics and finance, in the distribution of the *logarithm* of interest rates, exchange rates, stock returns, and inflation.

Processes that are associated with a Gaussian (normal) distribution are said to respond to *normality*. If normality is satisfied only with the logarithm of the variable $x$ (as seen in the last two above examples), the process is said to be *log-normal*.[1] By way of a simplified explanation, normality comes from the *additive* effect of independent random factors, while log-normality comes from their *multiplicative* effect.

Generally, the Gaussian (normal) law represents a good approximation of most continuous distributions, provided the number of events or samples is relatively large. This will be shown in the next section. Furthermore, the Gaussian (normal) law represents the asymptotic limit of most discrete probability distributions with large mean $<k>$, two key examples being the *binomial* and the *Poisson* PDFs described in the previous section. As we have seen, the binomial distribution $p_k = C_n^k q^k (1-q)^{n-k}$, which is

[1] For advanced reference, the *log–normal distribution* is defined as: $p(x) = \frac{1}{x\sigma\sqrt{2\pi}} \exp\left[-\frac{(\log x - \mu)^2}{2s^2}\right]$. Its mean and variance are $N = \exp(\mu + s^2/2)$ and $\sigma^2 = \left[\exp\left(s^2\right) - 1\right]\exp\left(2\mu + s^2\right)$, respectively.

defined for integers $k = 0, \ldots, n$, Eq. (2.9) converges towards the Poisson distribution, Eq. (2.8) in the limit of large $n$. In the same limit, it can be shown that both distributions converge towards a Gaussian (normal) PDF of same mean $<k>$ and variance $\sigma^2 = nq(1-q)$.

## 2.5 Central-limit theorem

In probability theory, there exist several central-limit theorems, which show that *the sum of large numbers of independent random variables, each having a different probability distribution, asymptotically converges towards some kind of limiting PDF*.

Remarkably, this limiting PDF is the same, regardless of the initial distribution of these variables. The most well known of these theorems is referred to as the *central-limit theorem* (CLT). The CLT states that if the variance of the initial distribution is finite, the limiting PDF is the Gaussian (normal) distribution. The CLT thus explains why the Gaussian (normal) distribution is found in so many random processes: such processes usually stem from the *additive effect* of several independent or uncorrelated random variables, which individually obey any PDF type.

A simplified formulation of the CLT is as follows. Let $x$ be a random variable of a given parent distribution $p_{\text{parent}}(x)$. The parent distribution is characterized by a mean $<x> = N$ and a finite variance $\sigma^2$. Let $x_1, x_2, \ldots, x_k$ be a series of $k$ independent samples from this distribution. Define the sum

$$S_k = x_1 + x_2 +, \ldots, x_k. \tag{2.23}$$

Since the random variables (or samples) are independent, the mean and the variance of the sum in Eq. (2.23) are $<S_K> = kN$ and $\sigma_k^2 = k\sigma^2$, respectively.

The CLT simply states that *the probability distribution of $S_k$ asymptotically becomes Gaussian (normal) as the number of samples k increases*, or $k \to \infty$. A more general formulation of the CLT assumes a set of independent random variables $X_1 X_2, \ldots, X_k$. Each of these variables $X_i$ has a different probability distribution $p_i(x)$. Each distribution $p_i(x)$ has a mean $N_i$ and a variance $\sigma_i^2$. Define the sum $S_k = X_1 + X_2 +, \ldots, X_k$. Since the variables $X_1 X_2, \ldots, X_k$ are independent, the mean and variance of the sum are $<S_K> = N_1 + N_2 +, \ldots, N_k$ and $\sigma_k^2 = \sigma_1^2 + \sigma_2^2 +, \ldots, \sigma_k^2$, respectively. Just as in the previous formulation, the CLT simply states that the probability distribution of the sum $S_k$ is asymptotically Gaussian (normal).

I will not expand on the formal proof of the CLT, which is beyond the scope of this book.[2] However, for both clarification and fun purposes, the reader might check up this proof through some nicely illustrated online experiments using interactive Java

[2] Formal demonstrations of the CLT can be found in many academic books and in some websites. See, for instance: http://mathworld.wolfram.com/CentralLimitTheorem.html.

applets.[3] I consider here an experimental example of CLT proof, using one such web tool.[4]

## Rolling dice and adding spots

This is an experiment similar to that described earlier in Chapter 1 and illustrated in Figs. 1.1 and 1.2. As we know, each individual die has a uniform discrete distribution defined by $p(x) = 1/6$ with $x_i = 1, 2, 3, 4, 5, 6$ being the event space. If we roll two dice and sum up the spots, the event space is $x_i = 2, 3, 4, 5, 6, 7, 8, 9, 10, 11, 12$ and the probability distribution has a triangular (or witch's hat) shape centered about $<x> = 7$, see Fig. 1.2. Interestingly, the CLT does *not* apply to the two-dice case, the limiting PDF being still the witch hat, as can be easily verified. As a simplified explanation, this is because the event space is too limited. But when using three dice or more, the CLT is observed to apply, as illustrated in Fig. 2.9. The figure shows results obtained with five dice for $k = 10$, 100, 1000, 10 000, and 100 000. It is seen that as $k$ increases, the resulting histogram envelope progressively takes a bell shape. Ultimately, the histogram takes a symmetrical (discretized) bell shape limited to the $26 = 30 - 5 + 1$ integer bins of the event space. A 1000-dice experiment with an adequate number of trials $k$ would yield a similar histogram with $5001 = 6000 - 1000 + 1$ discrete bins, which is much closer to the idea of a smooth envelope, albeit the resulting PDF is discrete, not continuous. Only an infinite number of dice and an infinite number of rolls could provide a histogram match of the limiting Gaussian (normal) envelope.

To complete the illustration of the CLT, consider the school game of a pegboard matrix, also known as a pinball machine, bean machine, quincunx, or Galton box, and whose principle is at the root of the Japanese gambling parlors called Pachinko. At each step of the game, a ball bounces on a peg (or a nail) to choose a left or right path randomly, according to a uniform, two-valued distribution. The triangular arrangement of pegs or nails makes it possible to repeat the ball's choice as many times as there are rows ($n$) in the matrix. At the bottom and after the final row, the ball rests in a single bin, which is associated with some reward or gain. It is easily established that the probability for the ball to be found in a given bin $k$ follows the binomial distribution, Eq. (2.9). If the number $n$ of rows becomes large, and for a sufficiently large number of such trials, the histogram distribution of balls into the bottom bins takes a Gaussian-like envelope, which represents a nice, mechanical schoolroom illustration of the CLT.

This concludes the second chapter on probability basics. Most of the mathematical tools that are required to approach information theory have been described in these two chapters.

[3] See, for instance:
www.stat.sc.edu/~west/javahtml/CLT.html,
www.rand.org/statistics/applets/clt.html,
www.math.csusb.edu/faculty/stanton/m262/central_limit_theorem/clt_old.html,
www.ruf.rice.edu/%7Elane/stat_sim/sampling_dist/index.html,
www.vias.org/simulations/simusoft_cenlimit.html.

[4] I am grateful to Professor Todd Ogden for permission to reproduce the simulation results obtained from his web tool in www.stat.sc.edu/~west/javahtml/CLT.html.

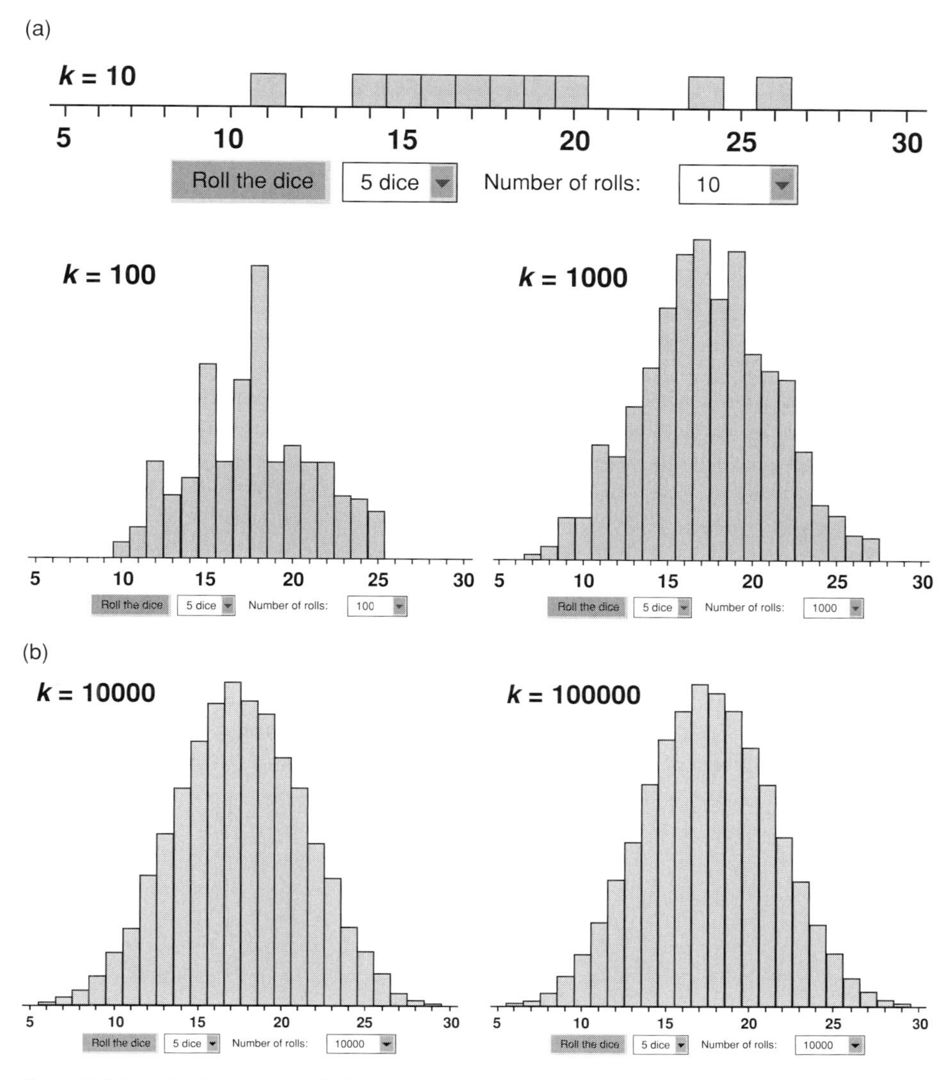


**Figure 2.9** Proof of the central-limit theorem, through five-dice rolling experiment, (a) from $k = 10$ to $k = 1000$ trials, (b) from $k = 10\ 000$ to $k = 100\ 000$ trials, showing asymptotic convergence of histogram shape towards a Gaussian or normal distribution envelope.

## 2.6 Exercises

**2.1** (B): Calculate the mean and standard deviation of a discrete uniform distribution defined over the integer interval [1, 5].

**2.2** (B): A radioactive substance has a decay rate of 0.1 events per second. Taking a group of ten atoms, what is the probability that half of the atoms have decayed after one minute?

**2.3** (B): What is the probability of counting five heads after tossing a coin ten times?

**2.4** (B): Assume that 10% of the population contracted the flu. Taking 20 people at random, what is the probability of finding three people with the flu? What is the probability of finding not more than three?

**2.5** (M): Show that the mean and standard deviation of a uniform, continuous PDF defined over the real interval $[x_{\min}, x_{\max}]$ of finite width $\Delta x$ are $<x> = (x_{\max} + x_{\min})/2$ and $\sigma = \Delta x/\sqrt{12}$, respectively.

**2.6** (B): Calculate the mean and variance of the exponential distribution $p(x) = \lambda e^{-\lambda x}$ $(x \geq 0, \lambda > 0)$.

**2.7** (B): A brand of TV set has a mean time to failure (MTFF) of five years with a warranty of one year. What is the probability that the TV set will fail before the warranty expires? What is the probability of a failure happening between one and five years?

**2.8** (B): Assume that, every year, there are 50 car accidents per 1000 new drivers, following an exponential law. What is the probability that a new driver will:
(a) Have one car accident after 2 years?
(b) Have no accident over 5 years?
(c) Have at least one accident between 2 and 5 years?

**2.9** (T): Prove that the mean and variance of the Bose–Einstein distribution are $<n> = N$ and $\sigma^2 = N + N^2$, respectively.

**2.10** (T): Prove that the mean and variance of the Poisson distribution satisfy $<n> = N = \sigma^2$.

# 3 Measuring information

In effect, the concept of information is obvious to anyone in life. Yet the word captures so much that we may doubt that any real definition satisfactory to a large majority of either educated or lay people may ever exist. Etymology may then help to give the word some skeleton. Information comes from the Latin *informatio* and the related verb *informare* meaning: to conceive, to explain, to sketch, to make something understood or known, to get someone knowledgeable about something. Thus, *informatio* is the action and art of shaping or packaging this piece of knowledge into some sensible form, hopefully complete, intelligible, and unambiguous to the recipient.

With this background in mind, we can conceive of information as taking different forms: a sensory input, an identification pattern, a game or process rule, a set of facts or instructions meant to guide choices or actions, a record for future reference, a message for immediate feedback. So information is diversified and conceptually intractable. Let us clarify here from the inception and quite frankly: a theory of information is unable to tell what information actually is or may represent in terms of objective value to any of its recipients! As we shall learn through this series of chapters, however, it is possible to measure information scientifically. The information measure does not concern value or usefulness of information, which remains the ultimate recipient's paradigm. Rather, information theory provides mathematical methods for *informatio*, in the ancestral Latin sense, but, since the twentieth century, using powerful computer technologies, which make it possible to handle phenomenal amounts of raw information messages.

This chapter will help us to conceive the "measure of information" in the Shannon or "classical" definition, which represents the central axiom in information theory. We will do this in two steps. First, we will analyze in what ways information might "make sense," but without any philosophical pretence. It is important in our view to go through this preliminary reflection before approaching information theory. No doubt, it has the potential to open up many interesting discussions and stimulate a brainstorming session, which could eventually get everyone lost in the class. However, the benefit of the reflection is to acquire the right focus, and safely engage us on the information-theory journey, which is a more practical and useful goal than that of philosophically mastering the information concept.

## 3.1 Making sense of information

In ordinary life and circumstances, *information* represents any piece of knowledge concerning facts, data, descriptions, characterizations, rules, or means to perform tasks, which can be *transmitted*, *exchanged*, or *acquired* in many ways. Such information can be transmitted through sounds, images, signals, notices, messages, announcements, voice, mails, memos, documents, books, encyclopedias, tutorials, chapters, training, and news media. There exist a seemingly boundless number of possible channels for information.

According to the above description, information has no intrinsic *meaning* in itself. The notion of meaningfulness is tangible only to the person or the entity to which the information is destined. For instance, long-band radio stations or foreign newspapers are meaningful only to those who speak the tongue. Within such channels, information makes sense only to the populations who have an interest in the acquisition or the use of the corresponding information.

What one usually calls *intelligence* is the faculty to relate together separate pieces of information,[1] to make sense of them in a specific context or with respect to a higher level of information knowledge. One may refer to intelligence as the means coherently to integrate information into a vital, self-preserving, or self-evolving way.

Another (but somewhat very specific) definition of *intelligence* is the systematic collection of information on target subjects by states, agencies, or private consultants. This activity is made in the purpose of gaining some strategic or economic advantage, of gaining a higher perspective on a business, a market, a competitor, or a country. The goal is to acquire or to maintain a competitive edge, for reasons of security and superior decision-making. The collected information can also represent a product for sale, as in the proliferating consultancy businesses.

Most people are concerned with the first type of intelligence, a skill whose acquisition and development begins in early school through the apprenticeship of language, then writing, math, literature, geography, history, and other knowledge. At a later stage, we eventually specialize into some form of integrated knowledge, which defines our situation, responsibility, wealth, and fate in society. Yet, the intelligent acquisition and processing of information never ceases, whether in private or in professional life. It is a commonplace to say that we are *controlled* by information. From a materialistic viewpoint, the primary goal of life is to collect the information that is useful for various personal reasons, such as societal compliance, mind growth, satisfaction, self-image, empowerment, standard of living, or material survival.

Fortunately, intelligence does act as an information filter. It is able to control the flow of information and to select what is relevant to any of the above needs. The rest represents "useless" information, like spam mail. Its fate is to vanish into oblivion. On the other hand, "useful" information is what intelligence looks after, with various degrees of expectation and priorities. The last notion of priority implies a personal investment to acquire and integrate information. If information must guide decisions and actions and, overall, future growth, this effort is of vital importance. The need to deploy effort in the

[1] The word *intelligence* comes from the Latin *inter-ligere*, literally meaning "linking things together."

acquisition and integration of information is too often overlooked, because of the fallacy that one could get informed through the news media, or just by browsing the Internet from a cell phone. To get informed is not only to access raw information, but also to make sense of it; this is where the effort comes. Intelligence may be a gift, but it takes some effort to make intelligent use of information.

As we all know, the principle of so-called computers is to process information, as reduced into data.[2] Computers perform this data processing at higher speeds and with larger amounts than the human brain can handle, which reflects a man-made form of higher intelligence. But this intelligence is not that of a machine. It is that of the designers who have conceived the machine and developed the code that is operated by the machine. Hence, the old-fashioned appellation of artificial intelligence, which, despite its phenomenal information-processing power, only emulates human intelligence. The man–machine intelligence challenge is well illustrated by chess contests, where the latter often wins but only for reasons of memory (having registered all champions' strategies in chess history) and computing power (having the possibility to analyze several moves in advance). The chances to win over the machine are higher if the player adopts some tactics that the machine has not prerecorded or cannot figure out the rationale thereof. One may recall Arthur C. Clarke's *2001: A Space Odyssey* (of which S. Kubrick made the famous movie), in which HAL,[3] the mightiest supercomputer ever designed in history, is eventually defeated by Dave, the more intelligent astronaut. For those who do not know the story, HAL could not compute the possibility that Dave would take a life chance by forcing his way back to the space station without wearing a helmet. The computer's intelligence may be virtually unbounded, but it is also exposed to some gaps in which human beings may easily compete. Computer hackers are here to illustrate that the most sophisticated firewalls can be broken into by means of exploiting such gaps. But that is another debate.

Since information is not about intelligence, what would be the use of an information theory? Could information theory ever be a science, as information science is to computers? Let us look at this question more closely. In accurate terminology, *information science* (or *informatics*), is a branch of *computer science*. Corresponding definitions go as follows:[4]

- *Information science* is primarily concerned with the structure, creation, management, storage, *retrieval*, dissemination, and transfer of information. It also includes studying the application of information in *organizations*, on its usage and the interaction between people, organizations, and *information systems*.
- *Computer science* encompasses a variety of topics that relate to computation, such as abstract analysis of algorithms, (. . . ) *programming languages*, *program* design, *software* and computer hardware. It is knowledge accumulated through scientific methodology by computation or by the use of a computer.

[2] The word *compute* comes from the Latin *com* (together) and *putare* (to root, reckon, number, appraise, estimate).

[3] The name HAL stems from a one-letter shift of the IBM (Information Business Machines) company acronym.

[4] http://en.wikipedia.org/wiki/information_science, http://en.wikipedia.org/wiki/Computer_science.

We observe from the above definitions that information science, a branch of computer science, is not concerned in assessing the value within information, other than taking care of it for some user benefit (acquiring, storing, transferring, retrieving, sharing . . .). In this view, the Internet is a perfect application of information science. It is left to the user to find the value and derive the benefits of information through intelligent use.

In contrast with the above, *information theory* is a branch of applied mathematics. It is concerned with the issues of coding, transmitting, and retrieving *information*. The three key applications of information theory are the transmission of data through special codes that enable *error correction* (transmission accuracy), *compression* (transmission and storage conciseness), *and encryption* (secrecy and security of transmission). Again, we observe that information theory is not meant to assess the value of information, or its relevance in terms of intelligent use. Internet communications depend heavily on these three applications: error correction, to preserve data integrity after multiple switching or routing and global transport; data compression, to accelerate uploading and downloading of web pages, text, and image files; and cryptography, to secure transactions or to protect privacy. Yet the intelligent use of the Internet information is left to the end users, discussion groups, web-content managers, and developers of websites. In this respect, information theory could be viewed as a misnomer, just like information science, because there is neither theory nor science that can assess how one can properly deal with information and its contents, in the broadest sense of the recipient's benefit. The future may reveal new theories of information, which could empower the user with some automatic assessment of the information usefulness or value in terms of scientific, societal, personal, and global impact. Without a shadow of a doubt, this would lead to a worrisome information computer, with Internet pervasiveness in both professional and private areas, taking care of everything in order for us not to worry about anything, just like the space companion HAL. This might also change the face of the Internet, with more and more intelligent browsers and search engines to present us the information the way we may appreciate it the most. Maybe we are not that far from such a frightening reality. Nevertheless, this only remains a matter of philosophical debate and speculation.

To close this preliminary discussion, we may simply conclude that information theory is neutral about information content and its subjective value. Its purpose is to derive mathematical algorithms for measuring, coding, and processing information, as reduced to immaterial data. The rest of this chapter is dedicated to the issue of *measuring information*, which is central to information theory. The nice and exciting thing indeed, is that information can be measured, and this will be our starting point in the information-theory journey.

## 3.2 Measuring information

Haven't we experienced rushing to the mailbox, the newsstand, or the TV? This exceptional interest in the mail or the news has many possible reasons: appointment confirmation, contract signature, conclusion of litigation, events of nationwide impact, such as disasters, elections, sports, but also house hunting, job classifieds, favorite rubrics, or, for children, Sunday cartoons! We have all also noticed that the value attributed to the mail or the newspaper decreases relatively quickly. Once read, both usually go to the recycling bin. This indicates that the information in these types of news has a special meaning in terms of being new or unexpected, which sharpens our interest.

In a scale of information interest, unexpected facts are what give information its highest scores. In contrast, there is decreasing interest in hearing or reading again the same news, except for the purpose of memorizing the information details. Jokes provide a good example here. The real information hidden in the joke is not the story but the conclusion or punch line. The joke may be a good one for various complex reasons, but an important common element is that the meaning of the punch line must be completely unexpected. A joke for which everyone in the audience already knows the conclusion is of little interest. This observation leads to a sense that *the interest of information is the highest when it is fully unexpected to the recipient*. The intrinsic importance of information is thus intimately associated with some degree of uncertainty. If an event is very likely, we gain little information in learning that it finally occurred. The converse is also true: unlikely or surprising events are those containing the highest information potential. Thus, information and event probabilities are connected, which bring us on track towards a measure definition. Consider next a practical example.

## A lotto surprise

Assume that I play the lotto, and the information I am looking after is the lotto's results. The official information finally comes as the following message:

$$x = \text{The winning combination is 7, 13, 17, 18.}$$

The meaningful part of this information is the four-number set. After I compare these numbers with the ones printed on my ticket, the information reduces to either $x =$ I have won, or $x =$ I have lost. Concisely expressed, the two possible information events are $x = W$ or $x = L$, respectively, with $X = \{W, L\}$ representing the event set. To each of these events ($x = W$ or $x = L$) is associated a probability, noted $p(x)$. For simplicity, we shall note $p(x = W) = p(W)$ and $p(x = L) = p(L)$.

To recall from Chapter 1, a probability is a positive number taking any value between 0 (impossible event) and 1 (absolutely certain event). The sum of all probabilities, which represent the combined likelihood of all possible events in $X$, is equal to unity.

With the lotto, we know from experience that the odds of winning are quite small, and the odds of losing are, therefore, substantially high. For instance, we may assume:

$$p(W) = 0.000001, \tag{3.1}$$

$$p(L) = 1 - p(W) = 0.999999, \tag{3.2}$$

meaning that the odds of winning are $1/1\,000\,000$ or 1 in a million.

How much information do I get when hearing the lotto result? Obviously, my expectations of winning being not very high, I won't learn much upon realizing that I have, indeed, lost. But in the contrary case, if I win, this is quite a phenomenal *surprise*! There is a lot of information in the winning event, to the same extent there is little information in the losing event.

Now we are looking for a formal way of defining the amount of information, which we could use as an objective measure reference $I(x)$ for each possible event $x$. We may postulate that $I(x)$ should approach zero for events close to absolute certainty ($p(x) \to 1$), and infinity for events reaching impossibility ($p(x) \to 0$). We must also require that for two events $x_1, x_2$ such that $p(x_1) < p(x_2)$, we have $I(x_1) > I(x_2)$, meaning that the less likely event is associated with greater information. An information measure $I(x)$ that satisfies all these properties is the following:

$$I(x) = \log\left(\frac{1}{p(x)}\right) = -\log[p(x)]. \tag{3.3}$$

In the above, the function log represents the natural logarithm, with the unit of $I(x)$ being called a *nat*. Note the minus sign in the definition's right-hand side, which ensures that the measure of information $I(x)$ is a positive number. This definition only applies to events of nonzero probability ($p(x) > 0$), meaning that the definition does not apply to impossible events ($p(x) = 0$), or only in the sense of a limit.

In our lotto example, we have

$$I(W) = -\log(0.000001) = 13.8 \text{ nats}, \tag{3.4}$$

$$I(L) = -\log(0.999999) = 0.000001 \text{ nat}. \tag{3.5}$$

Thus, to win this lotto game represents some *14 nats* of information, while losing means negligible information, namely, orders of magnitude below ($10^{-6}$ nat).

In the definition of $I(x)$, it is also possible to choose base-two or base-ten logarithms. This changes the information scale, but the qualitative result remains the same: information continuously increases as the likelihood of events decrease. I will come back to this issue later.

Next, let us explore some more properties of the proposed information measure. Consider two unrelated or independent events $A$ and $B$, with probabilities $p(A)$ and $p(B)$. We define the joint probability $p(A, B)$ as the probability that both events $A$ and $B$ occur. As we have seen in Chapter 1, if $A$ and $B$ are independent events, then $p(A, B) = p(A)p(B)$. The information associated with the joint events is, therefore,[5]

$$\begin{aligned} I(A, B) &= -\log[p(A, B)] \\ &= -\log[p(A)p(B)] \\ &= -\log[p(A)] - \log[p(B)] \\ &\equiv I(A) + I(B). \end{aligned} \tag{3.6}$$

The result shows that the information measure of the joint event is the sum of the information measures associated with each event considered separately. Information is, thus, an *additive* quantity. Such a property holds true only if the events are independent. The more complex case where events are not independent will be addressed later on.

[5] Using the rule for logarithms: $\log(XY) = \log X + \log Y$ ($X$ and $Y$ being positive).

In the following, I will refer to information measure simply as information, in the IT meaning of the word.

What if the two events $A$ and $B$ are not independent? Then we know from Bayes's theorem that $p(A, B) = p(A \mid B)p(B) = p(B \mid A)p(A)$. The information associated with the joint event is, therefore,

$$\begin{aligned} I(A, B) &= -\log[p(A, B)] \\ &= -\log[p(A)p(B \mid A)] \\ &= -\log[p(A)] - \log[p(B \mid A)] \\ &\equiv I(A) + I[p(B \mid A)]. \end{aligned} \quad (3.7)$$

The result shows that the information of the joint event is the sum of the information concerning event $A$ and the information concerning event $B$ – knowing that event $A$ occurred. The term $I[p(B \mid A)]$ is referred to as *conditional information*.

## 3.3 Information bits

Since we have established that information can be measured according to its degree of surprise, it is interesting to find the minimal information measurement one can make. This concept is encapsulated in the familiar coin-tossing game.

Since a coin has two sides (heads and tails), the probabilities that it falls on either side are equal (overlooking here any other material possibility). This feature is translated mathematically into the relation $p(x = \text{heads}) = p(x = \text{tails}) = 1/2$. Consistent with the definition in Eq. (3.3), the information that one gets from tossing a coin is $I(x) = -\log[p(x)]$ or, in this case, $I(\text{heads}) = I(\text{tails}) = -\log(1/2) = 0.693$ nat. We may choose logarithms in base 2 to make this information value an integer, namely, defining[6]

$$I(x) = -\log_2 p(x), \quad (3.8)$$

which gives $I(\text{heads}) = I(\text{tails}) = -\log_2(2^{-1}) = -(-1)\log_2(2) = 1$. The unit of this new measure of information is the *bit*, as short for *binary digit*.

What a bit represents is, thus, the exact amount of elementary information necessary to describe the outcome of a coin tossing. The property $I = 1$ *bit* means that the message used to transmit the information only requires a single symbol, out of a source of $2^I = 2$ possible symbols. We can thus use any character pair, for instance, *tails* = *0* and *heads* = *1*, for simplicity. The conclusion is that *the smallest information message is a bit*, namely, containing either 0 or 1 as a symbol.[7] The information regarding the outcome of the coin tossing can thus be communicated through a single message bit, which represents the minimum amount of measurable information.

[6] The conversion between natural and base-2 logarithms is given by the relation $\log_2 x = \log x / \log 2 \approx 1.442 \log x$.

[7] We could also have used base-3 or base-4 logarithms, defining the "trit" or the "quad," as representing three-valued or four-valued units of information. But the bit is the most elementary unit that cannot be sliced down into a smaller dimension.

If we toss the coin several times in a row, the information concerning the succession of outcomes is described by a string of as many single bits. For instance, the message consisting in the string 1101001101 ... means that the first two outcomes are heads, the third is tails, and so on.

To illustrate the concept of minimal information measure further, consider the case of a die. As we know, a die is a cube with six spotted faces, corresponding to the event source $X = \{1, 2, 3, 4, 5, 6\}$. The faces have equal probabilities of showing up as the die rolls out and stops (assuming no other physical possibility). The equiprobability translates into the relation $p(x = 1) = p(x = 2) = \ldots = p(x = 6) = 1/6$. The information contained in any die-roll measurement is, therefore,

$$I(X) = -\log_2(1/6) = \log_2(6) \approx 2.584 \text{ bits}. \tag{3.9}$$

The above result is puzzling because it shows that the *bit* information can be a noninteger, i.e., any positive real number. But how can one form a message string with *2.584 bits*? The explanation is not to confuse between the information measure (the bit) and the number $n$ of 1/0 symbols that is actually required to form the message. What the above result tells us is that the die-roll information can be coded by any string of $n$ symbols, satisfying $n \geq 2.584$. This means that the minimum string length is $n = 3$. It does, indeed, take three binary bits to represent all numbers from one to six, using the same symbol or block length.[8]

Next, we are going to show that binary coding is not the only way to encapsulate this information. Indeed, the full information of a single die roll can also be coded through a block of three YES or NO symbols, which provide the answer to *three* independent questions Q1, Q2, and Q3. Here is an example of the three questions:

Q1: Is the result $x$ even?

$$\text{If YES, then } x \in \{2, 4, 6\} = U_1, \text{ if NO, then } x \in \{1, 3, 5\} = U_2.$$

Q2: Is the result $x$ strictly greater than 3?

$$\text{If YES, then } x \in \{4, 5, 6\} = V_1; \text{ if NO, then } x \in \{1, 2, 3\} = V_2.$$

Q3: Is the result $x$ divided by 3?

$$\text{If YES, then } x \in \{3, 6\} = W_1; \text{if NO, then } x \in \{1, 2, 4, 5\} = W_2.$$

The combined answers to the three questions yield the value of $x$ (in ensemble language, one writes $x = U_i \cap V_j \cap W_k$, with $i, j, k = 1$ or 2). For instance, we have $5 = U_2 \cap V_1 \cap W_2$. It is straightforward to verify that the code correspondence is:

1 = NO/NO/NO,

2 = YES/NO/NO,

[8] The choice of code being arbitrary; we can use the binary representation, 1 = 001, 2 = 010, 3 = 011, 4 = 100, 5 = 101, 6 = 110, noting that the blocks 000 and 111 are unused.

3 = NO/NO/YES,

4 = YES/YES/NO,

5 = NO/YES/NO,

6 = YES/YES/YES,

while the two messages YES/NO/YES and NO/YES/YES are unused (in ensemble language, $U_1 \cap V_2 \cap W_1 = U_2 \cap V_1 \cap W_1 = \emptyset$, meaning that these ensembles are empty).

Regardless of the code, we see that the symbol blocks require three symbols of binary value bits (YES/NO or 1/0), while the information carried by the message is only 2.584 bits. Why do we have an extra 0.416 bit of information in the message? The explanation for this mystery is in fact quite simple. Assume a source that has eight equiprobable symbols, i.e., $Y = \{1, 2, 3, 4, 5, 6, 7, 8\}$, which can also be written $Y = X \cup \{7, 8\}$. The information of $Y$ is $I(Y) = -\log_2(1/8) = \log_2 2^3 = 3\log_2 2 =$ 3 bits. Since our previous messages describe the outcome of a die roll, the symbols (7, 8), which have a probability of zero, are never used. The consequence of never using these two symbols is that the actual probabilities of the six other symbols are raised by a factor of 8/6. This increase in likelihood corresponds to a relative information decrease (or less surprise!) of $\Delta I = -\log(8/6) = -0.416$ bit. Thus, the net information in the message is $I_{\text{net}} = I(Y) + \Delta I = 3 - 0.416 = 2.584$ bits.

One may justly argue that we get noninteger information because of the arbitrary choice of a base-2 logarithm. For the die roll, we could equally well choose a base-6 logarithm, i.e., $I(X) = -\log_6(x) = \log_6(6) = 1$ *sit* (for "six-ary digit"). Then the information of the die roll can be given by a single sit symbol, instead of three bits. However, the sit is six-valued, which requires six different symbols per elementary information, as opposed to the two-valued bit, which only requires two symbols. Which of the two message blocks is better: one made of a single six-valued symbol, or one made of three two-valued symbols (bits)? Obviously, the first type is the shortest possible, but the symbol is more complex to identify. The second type is longer, but the symbol interpretation the most straightforward. Actually, the choice of symbol representation, by the use of multivalued symbol alphabets, is fully arbitrary. However, for computer implementation, the bit remains the most practical way of coding information.

The lesson learnt is that information messages always require an *integer* number of symbols, the 1/0 bit being the simplest or most elementary type. The bit-measure of information ($I$), however, is any *real* positive number. Since the number of required message bits $n$ cannot be less than the information $I$ to be conveyed ($n \geq I$); it is, therefore, given by the integer equal to or immediately greater than $I$.

## 3.4 Rényi's fake coin

The "fake-coin" determination problem was originally described by the mathematician A. Rényi.[9] It brilliantly illustrates for the purposes of this entry-level chapter how

[9] A. Rényi, *A Diary on Information Theory* (New York: John Wiley & Sons, 1984), p. 10.

the information measure can be applied to find the solution of complex optimization problems, in particular, where the outcomes are ternary or multivalued. Let's move directly onto the fun stuff.

Assume that a medieval jeweler was presented with 27 gold coins, all looking strictly identical, but with one of them being fake, being made of a lighter metal. The jeweler only has a scale with two pans. What is the minimum number of weight measurements needed for the jeweler to figure out which coin is the fake?

*Answer*: A first hunch comes from realizing that the problem is wholly similar to the 2-faced coin or the 6-faced die examples. All coins presented have equal probabilities, $p(1/27)$, of being the fake one. According to our information-theory knowledge, we conclude that the number of information bits required to identify one out of 27 coins is $I = \log_2 27 = \log_2 3^3 = 3 \log_2 3$ bit. We just need to find a technique to acquire the information $I$ through a *minimum* number of trials or weighting measurements.

A painstaking approach but not the smartest, as we shall see, consists in comparing the weights of all coins with respect to a same reference coin, which the jeweler would initially pick up at random. If he happens to pick up for reference the fake coin, he gets the answer right away with the first measurement. The probability for this lucky conclusion is $1/27$, or 3.7%. In the worst case, where all of his successive measurements balance out, it would take 25 operations to conclude the test.[10] Yet, we may observe that this is not a bad method if the scale is too delicate and cannot take more than one coin in each pan. Let us assume that the scale is robust and sufficiently accurate to weigh groups of ten coins or so. Consider then the outcome of any scale measurement with groups of several coins with equal numbers, as selected at random. There are three possible, and equiprobable outcomes:

(a) Left pan heavier;
(b) Right pan heavier;
(c) Pans balanced.

The outcome is a ternary answer (YES/NO/NEUTRAL) with equiprobable outcomes. Each of these single measurements (a), (b), (c), thus, yields an information of $I_m = -\log_2(1/3) = \log_2 3$ bit, and $n$ such independent measurements yield the information $nI_m = n \log_2 3 \equiv nI/3$. This result shows that the minimum number of measurements required to sort the fake coin from the group is $n = 3$, since $3I_m = I$. This is a second hunch towards the problem solution. However, we don't know anything yet about how to proceed, in order to get the answer through this "theoretical" minimum of three operations.

The solution provided by Rényi is illustrated in Fig. 3.1. It consists in making a succession of selective measurements, first with two groups of nine coins, then two groups of three coins, then with two groups of one coin, each one identifying where the fake coin is located. This solution proves that three measurements indeed are sufficient to get the answer!

[10] Indeed, if the first 25 measurements with coin numbers 2–26 balance out, the reference coin (number 1) cannot be fake, so coin number 27 is the fake one.

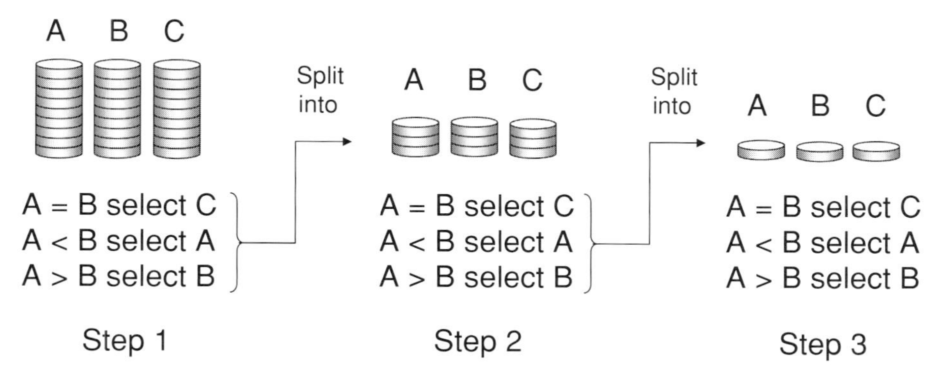


**Figure 3.1** Solution of Rényi's fake-coin problem: the 27 coins are split into three groups (A, B, C). The scale first compares weights of groups (A, B), which identifies the group X = A, B, or C containing the fake (lighter) coin. The same process is repeated with the selected group, X being split into three groups of three coins (step 2), and three groups of one coin (step 3). The final selection is the fake coin.

There exists at least one other way to solve Rényi's fake-coin determination problem, which to the best of my knowledge I believe is original.[11] The solution consists in making the three measurements only using groups of nine coins. The idea is to assign to each coin a position in space, forming a 3 × 3 coin cube, as illustrated in Fig. 3.2.

The coins are thus identified or labeled with a coefficient $c_{ijk}$, where each of the indices $i, j, k$ indicates a plane to which the coin belongs (coin $c_{111}$ is located at bottom-left on the front side and coin $c_{333}$ is located at top-right on the back side). To describe the measurement algorithm, one needs to define nine groups corresponding to all possible planes. For the planes defined by index $i$ = const., the three groups are:

$$\begin{aligned} P_{i=1} &= (c_{111}, c_{112}, c_{113}, c_{121}, c_{122}, c_{123}, c_{131}, c_{132}, c_{133}) \equiv \{c_{1jk}\} \\ P_{i=2} &= (c_{211}, c_{212}, c_{213}, c_{221}, c_{222}, c_{223}, c_{231}, c_{232}, c_{233}) \equiv \{c_{2jk}\} \\ P_{i=3} &= (c_{311}, c_{312}, c_{313}, c_{321}, c_{322}, c_{323}, c_{331}, c_{332}, c_{333}) \equiv \{c_{3jk}\}. \end{aligned} \tag{3.10}$$

The other definitions concerning planes $j$ = const. and $k$ = const. are straightforward. By convention, we mark with an asterisk any group containing the fake coin. If we put then $P_{i=1}$ and $P_{i=2}$ on the scale (step 1), we get three possible measurement outcomes (the signs meaning group weights being equal, lower, or greater):

$$\begin{aligned} P_{i=1} &= P_{i=2} \to P^*_{i=3} \\ P_{i=1} &< P_{i=2} \to P^*_{i=1} \\ P_{i=1} &> P_{i=2} \to P^*_{i=2}. \end{aligned} \tag{3.11}$$

We then proceed with step two, taking, for instance, $P_{j=1}$ and $P_{j=2}$, which yields either

[11] As proposed in 2004 by J.-P. Blondel of Alcatel (private discussion), which I have reformulated here in algorithmic form.

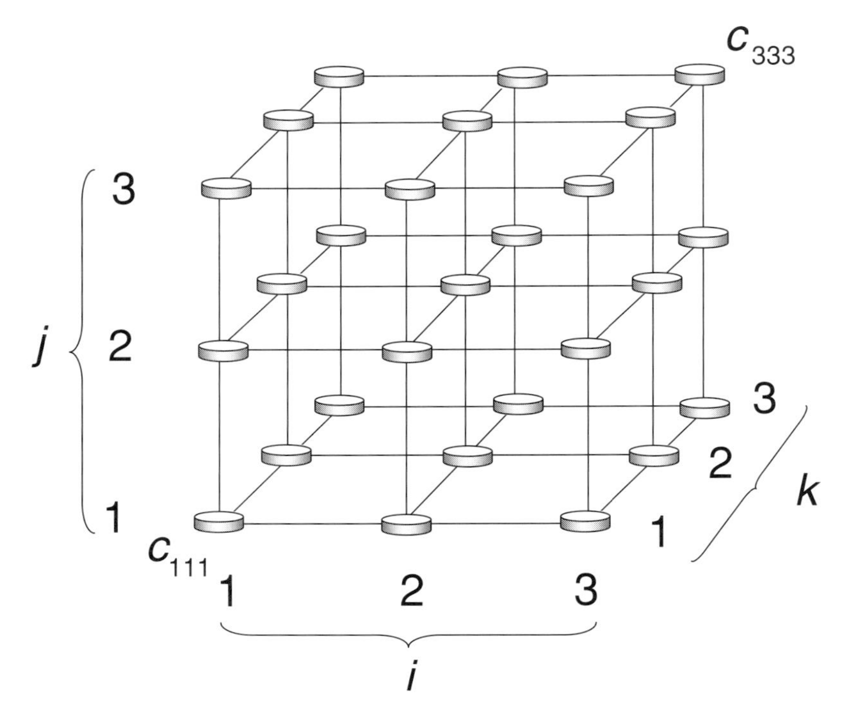


**Figure 3.2** Alternative solution of Rényi's fake-coin problem: each of the 27 coins is assigned a coefficient $c_{ijk}$, corresponding to its position within intersecting planes ($i, j, k = 1, 2, 3$) forming a $3 \times 3$ cube. The coefficients $c_{111}$ and $c_{333}$, corresponding to the coins located in front-bottom-left and back-top-right sides (respectively) are explicitly shown.

of the outcomes:

$$\begin{aligned} P_{j=1} &= P_{j=2} \rightarrow P^*_{j=3} \\ P_{j=1} &< P_{j=2} \rightarrow P^*_{j=1} \\ P_{j=1} &> P_{j=2} \rightarrow P^*_{j=2}. \end{aligned} \tag{3.12}$$

We then finally proceed with step three, taking, for instance, $P_{k=1}$ and $P_{k=2}$, which yields either of the outcomes:

$$\begin{aligned} P_{k=1} &= P_{k=2} \rightarrow P^*_{k=3} \\ P_{k=1} &< P_{k=2} \rightarrow P^*_{k=1} \\ P_{k=1} &> P_{k=2} \rightarrow P^*_{k=2}. \end{aligned} \tag{3.13}$$

We observe that the three operations lead one to identify three different marked groups. For instance, assume that the three marked groups are $P^*_{i=3}, P^*_{j=1}, P^*_{k=2}$. This immediately tells us that the fake-coin coefficient $c^*_{ijk}$ is of the form $c_{3jk}$ *and* $c_{i1k}$ *and* $c_{ij2}$, or in mathematical notation, using the Kronecker symbol,[12] $c^*_{ijk} = c_{ijk}\delta_{i3}\delta_{j1}\delta_{k2} \equiv c_{312}$.

[12] By definition, $\delta_{ij} = \delta_{ji} = 1$ for $i = j$ and $\delta_{ij} = 0$ otherwise.

In ensemble-theory language, this solution is given by the ensemble intersection $c^*_{ijk} = \{P^*_{i=3} \cap P^*_{j=1} \cap P^*_{k=2}\}$. Note that this solution can be generalized with $N = 3^p$ coins ($p \geq 3$), where $p$ is the dimension of a *hyper-cube* with three coins per side.

The alternative solution to the Rényi fake-coin problem does not seem too practical to implement physically, considering the difficulty in individually labeling the coins, and the hassle of regrouping them successively into the proposed arrangements. (Does it really save time and simplify the experiment?) It is yet interesting to note that the problem accepts more than one mathematically optimal solution. The interest of the hyper-cube algorithm is its capacity for handling problems of the type $N = m^p$, where both $m$ and $p$ can have arbitrary large sizes and the measuring device is $m$-ary (or gives $m$ possible outcomes). While computers can routinely solve the issue in the general case, we can observe that the information measure provides a hunch of the minimum computing operations, and eventually boils down to greater speed and time savings.

## 3.5 Exercises

**3.1** (B): Picking a single card out a 32-card deck, what is the information on the outcome?

**3.2** (B): Summing up the spots of a two-dice roll, how many message bits are required to provide the information on any possible outcome?

**3.3** (B): A strand of DNA has four possible nucleotides, named A, T, C, and G. Assume that for a given insect species, the probabilities of having each nucleotide in a sequence of eight nucleotides are: $p(A) = 1/4$, $p(T) = 1/16$, $p(C) = 5/16$, and $p(G) = 3/8$. What is the information associated with each nucleotide within the sequence (according to the information-theory definition)?

**3.4** (M): Two cards are simultaneously picked up from a 32-card deck and placed face down on a table. What is the information related to any of the events:

(a) One of the two cards is the Queen of Hearts?
(b) One of the two cards is the King of Hearts?
(c) The two cards are the King and Queen of Hearts?
(d) Knowing that one of the cards is the Queen of Hearts, the second is the King of Hearts?

Conclusions?

**3.5** (T): You must guess an integer number between 1 and 64, by asking as many questions as you want, which are answered by YES or NO. What is the minimal number of questions required to guess the number with 100% certainty? Provide an example of such a minimal list of questions.

# 4 Entropy

The concept of *entropy* is central to information theory (IT). The name, of Greek origin (*entropia, tropos*), means *turning point* or *transformation*. It was first coined in 1864 by the physicist R. Clausius, who postulated the second law of thermodynamics.[1] Among other implications, this law establishes the impossibility of perpetual motion, and also that the entropy of a thermally isolated system (such as our Universe) can only increase.[2] Because of its universal implications and its conceptual subtlety, the word entropy has always been enshrouded in some mystery, even, as today, to large and educated audiences.

The subsequent works of L. Boltzmann, which set the grounds of statistical mechanics, made it possible to provide further clarifications of the definition of entropy, as a *natural measure of disorder*. The precursors and founders of the later information theory (L. Szilárd, H. Nyquist, R. Hartley, J. von Neumann, C. Shannon, E. Jaynes, and L. Brillouin) drew as many parallels between the *measure of information* (the uncertainty in communication-source messages) and *physical entropy* (the disorder or chaos within material systems). Comparing information with disorder is not at all intuitive. This is because information (as we conceive it) is pretty much the conceptual opposite of disorder! Even more striking is the fact that the respective formulations for entropy that have been successively made in physics and IT happen to match exactly. A legend has it that Shannon chose the word "entropy" from the following advice of his colleague von Neumann: "Call it entropy. No one knows what entropy is, so if you call it that you will win any argument."

This chapter will give us the opportunity to familiarize ourselves with the concept of entropy and its multiple variants. So as not to miss the nice parallel with physics, we will start first with Boltzmann's precursor definition, then move to Shannon's definition, and develop the concept from there.

## 4.1 From Boltzmann to Shannon

The derivation of physical entropy is based on Boltzmann's work on statistical mechanics. Put simply, statistical mechanics is the study of physical systems made of large groups of

[1] This choice could also be attributed to the phonetic similarity with the German *Energie* (energy, or *energia* in Greek), so the word can also be interpreted as "energy turning point" or "point of energy transformation."

[2] For a basic definition of the second law of thermodynamics, see, for instance, http://en.wikipedia.org/wiki/Entropy.

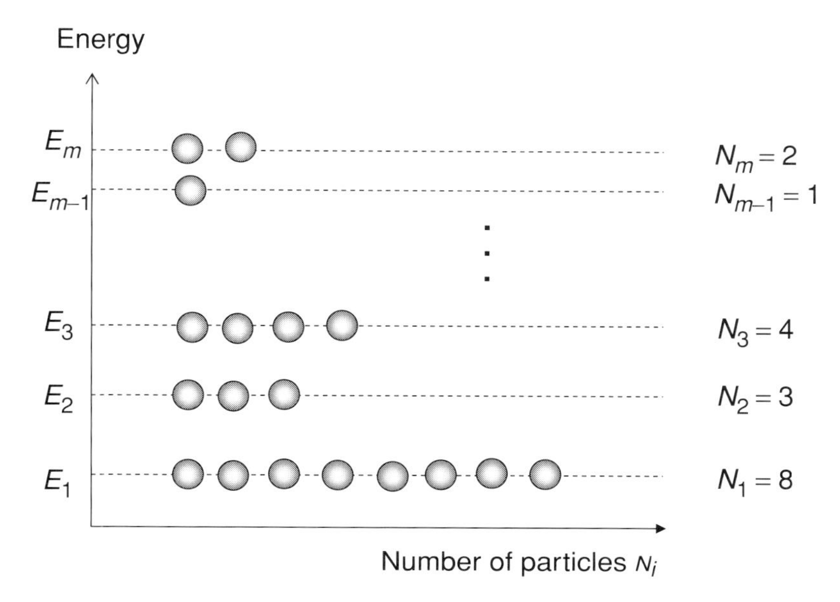


**Figure 4.1** Energy-level diagram showing how a set of $N$ identical particles in a physical macrosystem can be distributed to occupy, by subsets of number $N_i$, different microstates of energy $E_i$ $(i = 1 \dots m)$.

particles, for which it is possible to assign both *microscopic* (individual) and *macroscopic* (collective) properties.

Consider a macroscopic physical system, which we refer to for short as a *macrosystem*. Assume that it is made of $N$ particles. Each individual particle is allowed to occupy one out of $m$ possible microscopic states, or microstates,[3] which are characterized by an energy $E_i$ $(i = 1 \dots m)$, as illustrated in Fig. 4.1. Calling $N_i$ the number of particles occupying, or "populating," the microstate of energy $E_i$, the total number of particles in the macroscopic system is

$$N = \sum_{i=1}^{m} N_i, \tag{4.1}$$

and the total macrosystem energy is

$$E = \sum_{i=1}^{m} N_i E_i. \tag{4.2}$$

Now let's perform some combinatorics. We have $N$ particles each with $m$ possible energy states, with each state having a population $N_i$. The number of ways $W$ to arrange the $N$ particles into these $m$ boxes of populations $N_i$ is given by:

$$W = \frac{N!}{N_1! N_2! \dots N_m!} \tag{4.3}$$

[3] A microscopic state, or microstate, is defined by a unique position to be occupied by a particle at atomic scale, out of several possibilities, namely, constituting a discrete set of energy levels.

(see Appendix A for a detailed demonstration). When the number of particles $N$ is large, we obtain the following limit (Appendix A):

$$H = \lim_{N\to\infty} \frac{\log W}{N} = -\sum_{i=1}^{m} p_i \log p_i, \tag{4.4}$$

where $p_i = N_i/N$ represents the probability of finding the particle in the microstate of energy $E_i$. As formulated, the limit $H$ can thus be interpreted as representing the *average value* (or "expectation value" or "statistical mean") of the quantity $-\log p_i$, namely, $H = -\langle \log p \rangle$. This result became the *Boltzmann theorem*, whose author called $H$ "entropy."

Now let us move to Shannon's definition of entropy. In his landmark paper,[4] Shannon seeks for an improved and comprehensive definition of *information measure*, which he called $H$. In the following, I summarize the essential steps of the demonstration leading to Shannon's definition.

Assuming a random source with an event space, $X$, comprising $N$ elements or symbols with probabilities $p_i$ ($i = 1 \ldots N$), the unknown function $H$ should meet three conditions:

(1) $H = H(p_1, p_2, \ldots, p_N)$ is a continuous function of the probability set $p_i$;
(2) If all probabilities were equal (namely, $p_i = 1/N$), the function $H$ should be monotonously increasing with $N$;[5]
(3) If any occurrence breaks down into two successive possibilities, the original $H$ should break down into a weighed sum of the corresponding individual values of $H$.[6]

Shannon's formal demonstration (see Appendix B) shows that the unique function satisfying the three requirements (1)–(3) is the following:

$$H = -K \sum_{i=1}^{N} p_i \log p_i, \tag{4.5}$$

where $K$ is an arbitrary positive constant, which we can set to $K = 1$, since the logarithm definition applies to any choice of base ($K = \log_p x / \log_q x$ with $p \neq q$ being positive real numbers). It is clear that an equivalent notation of Eq. (4.5) is

$$H(X) = -\sum_{x\in X} p(x) \log p(x) \equiv \sum_{x\in X} p(x) I(x), \tag{4.6}$$

where $x$ is a symbol from the source $X$ and $I(x)$ is the associated information measure (as defined in Chapter 3).

[4] C. E. Shannon, A mathematical theory of communication. *Bell Syst. Tech. J.*, **27** (1948), 379–423, 623–56. This paper can be freely downloaded from http://cm.bell-labs.com/cm/ms/what/shannonday/paper.html.

[5] Based on the fact that equally likely occurrences provide more choice, then higher uncertainty.

[6] Say there initially exist two equiprobable events ($a$, $b$) with $p(a) = p(b) = 1/2$, so $H$ is the function $H(1/2, 1/2)$; assume next that event ($b$) corresponds to two possibilities of probabilities $1/3$ and $2/3$, respectively. According to requirement (3), we should have $H = H(1/2, 1/2) + (1/2)H(1/3, 2/3)$ with the coefficient $1/2$ in the second right-hand-side term being justified by the fact that the event ($b$) occurs only half of the time, on average.

The function $H$, which Shannon called *entropy*, is seen to be formally identical to the function of entropy defined by Boltzmann. However, we should note that Shannon's entropy *is not an asymptotic limit*. Rather, *it is exactly defined for any source having a finite number of symbols $N$*.

We observe then from the definition that Shannon's entropy is the *average value* of the quantity $I(x) = -\log p(x)$, namely, $H = \langle I \rangle = -\langle \log p \rangle$, where $I(x)$ is the *information* measure associated with a symbol $x$ of probability $p(x)$. *The entropy of a source is, therefore, the average amount of information per source symbol*, which we may also call the *source information*.

If all symbols are equiprobable ($p(x) = 1/N$), the source entropy is given by

$$H = -\sum_{x \in X} p(x) \log p(x) = -\sum_{i=1}^{N} \frac{1}{N} \log \frac{1}{N} \equiv \log N, \tag{4.7}$$

which is equal to the information $I = \log N$ of all individual symbols. We will have several occasions in the future to use such a property.

What is the unit of entropy? Were we to choose the natural logarithm, the unit of $H$ would be *nat/symbol*. However, as seen in Chapter 3, it is more sensible to use the base-2 logarithm, which gives entropy the unit of *bit/symbol*. This choice is also consistent with the fact that $N = 2^q$ equiprobable symbols, i.e., with probability $1/N = 1/2^q$, can be represented by $\log_2 2^q = q$ bits, meaning that all symbols from this source are made of exactly $q$ bits. In this case, there is no difference between the source entropy, $H = q$, the symbol information, $I = q$, and the symbol bit length, $l = q$. In the rest of this book, it will be implicitly assumed that the logarithm is of base two.

Let us illustrate next the source-entropy concept and its properties through practical examples based on various dice games and even more interestingly, on our language.

## 4.2 Entropy in dice

Here we will consider dice games, and see through different examples the relation between entropy and information. For a single die roll, the six outcomes are equiprobable with probability $p(x) = 1/6$. As a straightforward application of the definition, or Eq. (4.7), the source entropy is thus (base-2 logarithm implicit):

$$H = -\sum_{i=1}^{6} \frac{1}{6} \log \frac{1}{6} \equiv \log 6 = 2.584 \text{ bit/symbol}. \tag{4.8}$$

We also have for the information: $I = \log 6 = 2.584$ bits. This result means that it takes 3 bits (as the nearest upper integer) to describe any of the die-roll outcomes with the same symbol length, i.e., in binary representation:

$$x = 1 \rightarrow x100 \qquad x = 2 \rightarrow x010 \qquad x = 3 \rightarrow x110$$
$$x = 4 \rightarrow x001, \qquad x = 5 \rightarrow x101, \qquad x = 6 \rightarrow x011,$$

**Table 4.1** Calculation of entropy associated with the result of rolling two dice. The columns show the different possibilities of obtaining values from 2 to 12, the corresponding probability $p$ and the product $-p\log_2 p$ whose summation (bottom) is the source entropy $H$, which is equal here to 3.274 bit/symbol.

| Sum of dice numbers | Probability ($p$) | $-p\log_2(p)$ |
|---|---|---|
| $\mathbf{2} = 1 + 1$ | 0.027 777 778 | 0.143 609 03 |
| $\mathbf{3} = 1 + 2 = 2 + 1$ | 0.055 555 556 | 0.231 662 5 |
| $\mathbf{4} = 2 + 2 = 3 + 1 = 1 + 3$ | 0.083 333 333 | 0.298 746 88 |
| $\mathbf{5} = 4 + 1 = 1 + 4 = 3 + 2 = 2 + 3$ | 0.111 111 111 | 0.352 213 89 |
| $\mathbf{6} = 5 + 1 = 1 + 5 = 4 + 2 = 2 + 4 = 3 + 3$ | 0.138 888 889 | 0.395 555 13 |
| $\mathbf{7} = 6 + 1 = 1 + 6 = 5 + 2 = 2 + 5 = 4 + 3 = 3 + 4$ | 0.166 666 667 | 0.430 827 08 |
| $\mathbf{8} = 6 + 2 = 2 + 6 = 5 + 3 = 3 + 5 = 4 + 4$ | 0.138 888 889 | 0.395 555 13 |
| $\mathbf{9} = 6 + 3 = 3 + 6 = 5 + 4 = 4 + 5$ | 0.111 111 111 | 0.352 213 89 |
| $\mathbf{10} = 6 + 4 = 4 + 6 = 5 + 5$ | 0.083 333 333 | 0.298 746 88 |
| $\mathbf{11} = 6 + 5 = 5 + 6$ | 0.055 555 556 | 0.231 662 5 |
| $\mathbf{12} = 6 + 6$ | 0.027 777 778 | 0.143 609 03 |

$\sum = 1.$
Source entropy $\sum = 3.274.$

where the first bit $x$ is zero for all six outcomes. Nothing obliges us to attribute to each outcome its corresponding binary value. We might as well adopt any arbitrary 3-bit mapping such as:

$$x = 1 \rightarrow 100 \qquad x = 2 \rightarrow 010 \qquad x = 3 \rightarrow 110$$
$$x = 4 \rightarrow 001, \qquad x = 5 \rightarrow 101, \qquad x = 6 \rightarrow 011.$$

The above example illustrates a case where entropy and symbol information are equal, owing to the equiprobability property. The following examples illustrate the more general case, highlighting the difference between entropy and information.

## Two-dice roll

The game consists in adding the spots obtained from rolling two dice. The minimum result is $x = 2$ $(= 1 + 1)$ and the maximum is $x = 12$ $(= 6 + 6)$, corresponding to the event space $X = \{2, 3, 4, 5, 6, 7, 8, 9, 10, 11, 12\}$. The probability distribution $p(x)$ was described in Chapter 1, see Figs. 1.1 and 1.2. Table 4.1 details the 11 different event possibilities and their respective probabilities $p(x)$. The table illustrates that there exist 36 equiprobable dice-arrangement outcomes, giving first $p(1) = p(12) = 1/36 = 0.027$. The probability increases for all other arrangements up to a maximum corresponding to the event $x = 7$ with $p(7) = 6/36 = 0.166$. Summing up the values of $-p(x)\log p(x)$, the source entropy is found to be $H = 3.274$ bit/symbol. This result shows that, *on average*, the event can be described through a number of bits between 3 and 4. This was expected since the source has 11 elements, which requires a maximum of $2^4 = 4$ bits, while most of the events (namely, $x = 2, 3, 4, 5, 6, 7, 8, 9$) can be coded in principle with only $2^3 = 3$ bits. The issue of finding the best code to attribute a symbol of minimal length to each of the events will be addressed later.

### The 421

The dice game called 421 for short was popular in last century's French cafés. The game uses three dice and the winning roll is where the numbers 4, 2, and 1 show up, regardless of order. The interest of this example is to illustrate Shannon's property (3), which is described in the previous section and also in Appendix B. The probability of obtaining $x = 4, 2, 1$, like any specific combination where the three numbers are different, is $p(421) = (1/6)(1/6)(1/6) \times 3! = 1/36 = 0.0277$ (each die face has 1/6 chance and there are 3! possible dice permutations). The probability of winning is, thus, close to 3%, which (interestingly enough) is strictly equal to that of the double six winner ($p(66) = (1/6)(1/6) = 1/36$) in many other games using dice. The odds on missing a 4, 2, 1 roll are $p(\text{other}) = 1 - p(421) = 35/36 = 0.972$. A straightforward calculation of the source entropy gives:

$$\begin{aligned} H(421, \text{other}) &= -p(421)\log_2 p(421) - [1 - p(421)]\log_2[1 - p(421)] \\ &= \frac{1}{36}\log_2 36 + \frac{35}{36}\log_2\frac{36}{35} \equiv 0.183 \text{ bit/symbol.} \end{aligned} \tag{4.9}$$

The result shows that for certain sources, the average information is not only a real number involving "fractions" of bits (as we have seen), but also a number that can be *substantially smaller than a single bit*! This intriguing feature will be clarified in a following chapter describing coding and coding optimality.

Next, I shall illustrate Shannon's property (3) based on this example. What we did consisted in partitioning all possible events (result of dice rolls) into two subcategories, namely, winning ($x = 4, 2, 1$) and losing ($x = \text{other}$), which led to the entropy $H = 0.183$ bit/symbol. Consider now all dice-roll possibilities, which are equiprobable. The total number of possibilities (regardless of degeneracy) is $N = 6 \times 6 \times 6 = 216$, each of which is associated with a probability $p = 1/216$. The corresponding information is, by definition, $I(216) = \log 216 = 7.7548$ bits.

We now make a partition between the winning and the losing events. This gives $n(421) = 3! = 6$, and $n(\text{other}) = 216 - 6 = 210$. According to Shannon's rule (3), and following Eq. (B17) of Appendix B, we have:

$$\begin{aligned} I(N) &= H(p_1, p_2) + p_1 I(n_1) + p_2 I(n_2) \\ &= H(421, \text{other}) + p(421)I[n(421)] + p(\text{other})I[n(\text{other})] \\ &= 0.1831 + (1/36)I(6) + (35/36)I(210) = 0.1831 + 0.0718 + 7.4999 \\ &\equiv 7.7548 \text{ bits,} \end{aligned} \tag{4.10}$$

which is the expected result.

One could argue that there was no real point in making the above verification, since property (3) is ingrained in the entropy definition. This observation is correct: the exercise was simply meant to illustrate that the property works through a practical example. Yet, as we shall see, we can take advantage of property (3) to make the rules more complex and exciting, when introducing further "winning" subgroups of interest. For instance, we could keep "4, 2, 1" as the top winner of the 421 game, but attribute 10 bonus points for any different dice-roll result $mnp$ in which $m + n + p = 7$. Such a rule modification creates a new partition within the subgroup we initially called "other." Let's then decompose "other" into "bonus" and "null," which gives the source information decomposition:

$$\begin{aligned} I(N) = {} & H(421, \text{bonus}, \text{null}) + p(421)I[n(421)] \\ & + p(\text{bonus})I[n(\text{bonus})] + p(\text{null})I[n(\text{null})]. \end{aligned} \tag{4.11}$$

The reader can easily establish that there exists only $n(\text{bonus}) = 9$ possibilities for the "bonus" subgroup (which, to recall, excludes any of the "421" cases), thus $p(\text{bonus}) = 9/216$, and $p(\text{null}) = 1 - p(421) - p(\text{bonus}) = 1 - 1/36 - 9/216 = 0.9305$ (this alternative rule of the 421 game gives about 7% chances of winning something, which more than doubles the earlier 3% and increases the excitement). We obtain the corresponding source entropy;

$$\begin{aligned} & H(421, \text{bonus}, \text{null}) \\ & \quad = -p(421)\log_2[p(421)] - p(\text{bonus})\log_2[p(\text{bonus})] - p(\text{null})\log_2[p(\text{null})] \\ & \quad \equiv 0.4313 \text{ bit/symbol}. \end{aligned} \tag{4.12}$$

We observe that the entropy $H(421, \text{bonus}, \text{null}) \equiv 0.4313$ bit/symbol is more than the double of $H(421, \text{other}) = 0.183$ bit/symbol, a feature that indicates that the new game has more diversity in outcomes, which corresponds to a greater number of "exciting" possibilities, while the information of the game, $I(N)$, remains unchanged. Thus, entropy can be viewed as representing a measure of game "excitement," while in contrast information is a global measure of game "surprise," which is not the same notion. Consider, indeed, two extreme possibilities for games:

Game A: the probability of winning is very high, e.g., $p_A(\text{win}) = 90\%$;

Game B: the probability of winning is very low, e.g., $p_B(\text{win}) = 0.000\,01\%$.

One easily computes the game entropies and information:

$H(A) = 0.468$ bit/symbol $\quad I_{\text{win}}(A) = 0.152$ bit,

$H(B) = 0.000\,002\,5$ bit/symbol $\quad I_{\text{win}}(B) = 23.2$ bit.

We observe that, comparatively, game A has significant entropy and low information, while the reverse applies to game B. Game A is more exciting to play because the player wins much more often, hence a high entropy (but low information). Game B has more of a surprise potential because of the low chances of winning, hence a high information (but low entropy).

## 4.3 Language entropy

Here, we shall see how Shannon's entropy can be applied to analyze *languages*, as primary sources of word symbols, and to make interesting comparisons between different types of language.

As opposed to dialects, human languages through history have always possessed some written counterparts. Most of these language "scripts" are based on a unique, finite-size alphabet of symbols, which one has been trained in early age to recognize and manipulate (and sometimes to learn later the hard way!).[7] Here, I will conventionally call symbols "characters," and their event set the "alphabet." This is not to be confused with the "language source," which represents the set of all words that can be constructed from said alphabet. As experienced right from early school, not all characters and words are born equal. Some characters and words are more likely to be used; some are more rarely seen. In European tongues, the use of characters such as X or Z is relatively seldom, while A or E are comparatively more frequent, a fact that we will analyze further down.

However, the statistics of characters and words are also pretty much context-dependent. Indeed, it is clear that a political speech, a financial report, a mortgage contract, an inventory of botanical species, a thesis on biochemistry, or a submarine's operating manual (etc.), may exhibit statistics quite different from ordinary texts! This observation does not diminish the fact that, within a *language source*, (the set of all possible words, or character arrangements therein), words and characters are not all treated equal. To reach the fundamentals through a high-level analysis of language, let us consider just the basic *character* statistics.

A first observation is that in any language, the probability distribution of characters (PDF), as based on any literature survey, is not strictly unique. Indeed, the PDF depends not only on the type of literature surveyed (e.g., newspapers, novels, dictionaries, technical reports) but also on the contextual epoch. Additionally, in any given language practiced worldwide, one may expect significant qualitative differences. The Continental and North-American variations of English, or the French used in Belgium, Quebec, or Africa, are not strictly the same, owing to the rich variety of local words, expressions, idioms, and literature.

A possible PDF for English alphabetical characters, which was realized in 1942,[8] is shown in Fig. 4.2. We first observe from the figure that the discrete PDF nearly obeys an *exponential law*. As expected, the space character (sp) is the most frequent (18.7%). It is followed by the letters E, T, A, O, and N, whose occurrence probabilities decrease from 10.7% and 5.8%. The entropy calculation for this source is $H = 4.065$ bit/symbol. If we remove the most likely occurring space character (whose frequency is not meaningful) from the source alphabet, the entropy increases to $H = 4.140$ bit/symbol.

[7] The "alphabet" of symbols, as meaning here the list of distinct ways of forming characters, or voice sounds, or word prefixes, roots, and suffixes, or even full words, may yet be quite large, as the phenomenally rich Chinese and Japanese languages illustrate.

[8] F. Pratt, *Secret and Urgent* (Indianapolis: The Bobbs-Merrill Book Company, 1942). Cited in J. C. Hancock, *An Introduction to the Principles of Communication Theory* (New York: McGraw Hill, 1961).

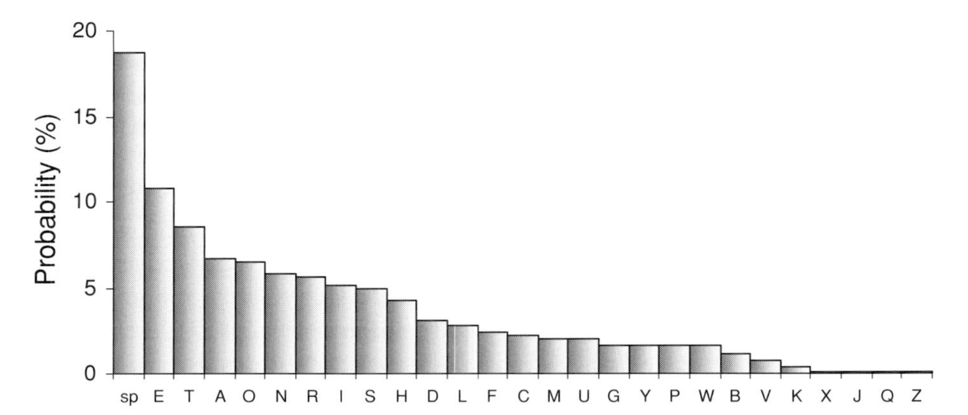


**Figure 4.2** Probability distribution of English alphabetical characters, as inventoried in 1942. The source entropy with or without the space character is $H = 4.065$ bit/symbol and $H = 4.140$ bit/symbol, respectively.

Figure 4.3 shows the probability distributions for English, German, French, Italian, Spanish, and Portuguese, after inventories realized in 1956 or earlier.[9] For comparison purposes, the data were plotted in the decreasing-probability sequence corresponding to English (Fig. 4.2). We observe that the various distributions exhibit a fair amount of mutual correlation, together or by subgroups. Such a correlation is more apparent if we plot in ordinates the different symbol probabilities against the English data, as shown in Fig. 4.4. The two conclusions are:

(a) Languages make uneven use of symbol characters (corresponding to discrete-exponential PDF);
(b) The PDFs are unique to each language, albeit showing a certain degree of mutual correlation.

Consider next *language entropy*. The source entropies, as calculated from the data in Fig. 4.3, together with the 10 most frequent characters, are listed in Table 4.2. The English source is seen to have the highest entropy (4.147 bit/symbol), and the Italian one the lowest (3.984 bit/symbol), corresponding to a small difference of 0.16 bit/symbol. *A higher entropy corresponds to a fuller use of the alphabet "spectrum,"* which creates more uncertainty among the different symbols. Referring back to Fig. 4.3, we observe that compared with the other languages, English is richer in H, F, and Y, while German is richer in E, N, G, B, and Z, which can intuitively explain that their source entropies are the highest. Compared with the earlier survey of 1942, which gave $H(1942) = 4.140$ bit/symbol, this new survey gives $H(1956) = 4.147$ bit/symbol. This difference is not really meaningful, especially because we do not have a basis for comparison between the type and magnitude of the two samples that were used. We can merely infer that the introduction of new words or *neologisms*, especially technical neologisms, contribute

[9] H. Fouché-Gaines, *Cryptanalysis, a study of ciphers and their solutions* (New York: Dover Publications, 1956).

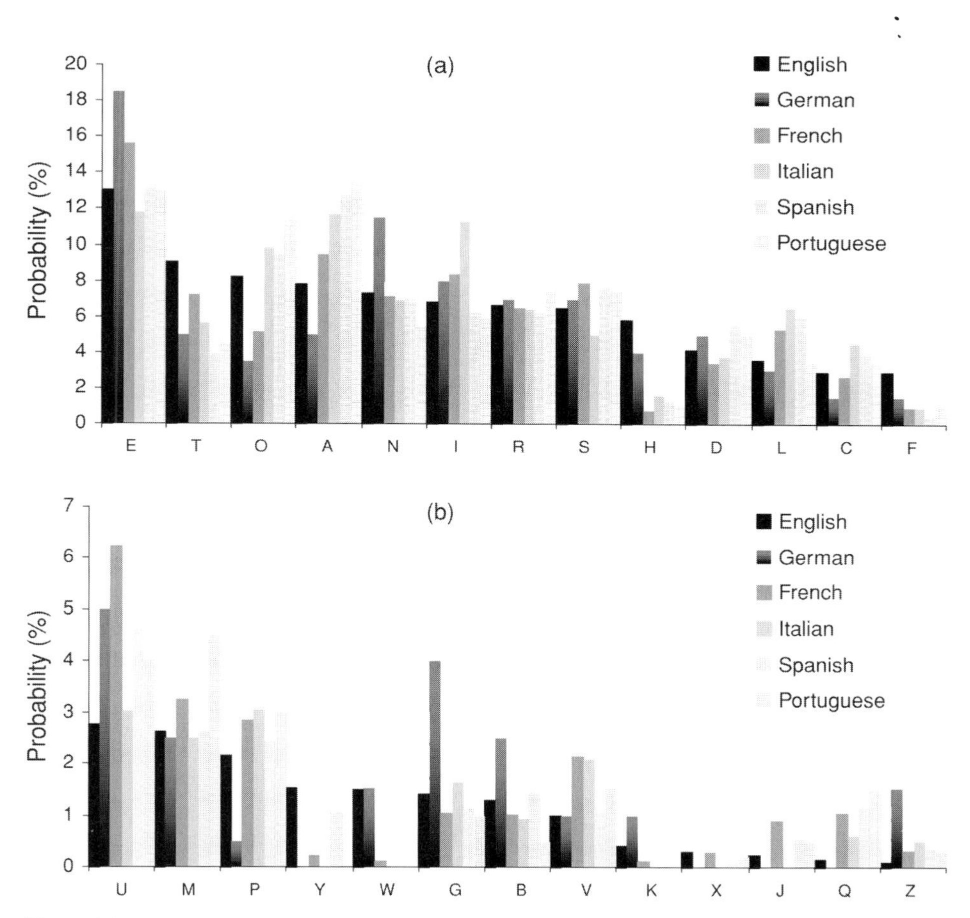


**Figure 4.3** Probability distributions of character symbols for English, German, French, Italian, Spanish and Portuguese, as inventoried in 1956 or earlier and as ordered according to English. The corresponding source entropy and 10 most frequent letters are shown in Table 4.2.

over the years to increase entropy. Note that the abandonment of old-fashioned words does not necessarily counterbalance this effect. Indeed, the old words are most likely to have a "classical" alphabet structure, while the new ones are more likely to be unusual, bringing a flurry of new symbol-character patterns, which mix roots from different languages and technical jargon.

What does an entropy ranging from $H = 3.991$ to 4.147 bit/symbol actually mean for any language? We must compare this result with an absolute reference. Assume, for the time being, that the reference is provided by the maximum source entropy, which one can get from a uniformly distributed source. The maximum entropy of a 26-character source, which cannot be surpassed, is thus $H_{\max} = I(1/26) = \log 26 = 4.700$ bit/symbol. We see from the result that *our European languages are not too far from the maximum entropy limit*, namely, within 85% to 88% of this absolute reference. As described in forthcoming chapters, there exist different possibilities of coding language alphabets

**Table 4.2** Alphabetical source entropies $H$ of English, German, French, Italian, Spanish, and Portuguese, and the corresponding ten most frequent characters.

| | $H$ (bit/symbol) | Ten most frequent characters |
|---|---|---|
| English | 4.147 | E T O A N I R S H D |
| German | 4.030 | E N I R S T A D U H |
| French | 4.046 | E A I S T N R U L O |
| Italian | 3.984 | E A I O N L R T S C |
| Spanish | 4.038 | E A O S N I R L D U |
| Portuguese | 3.991 | A E O S R I N D T M |

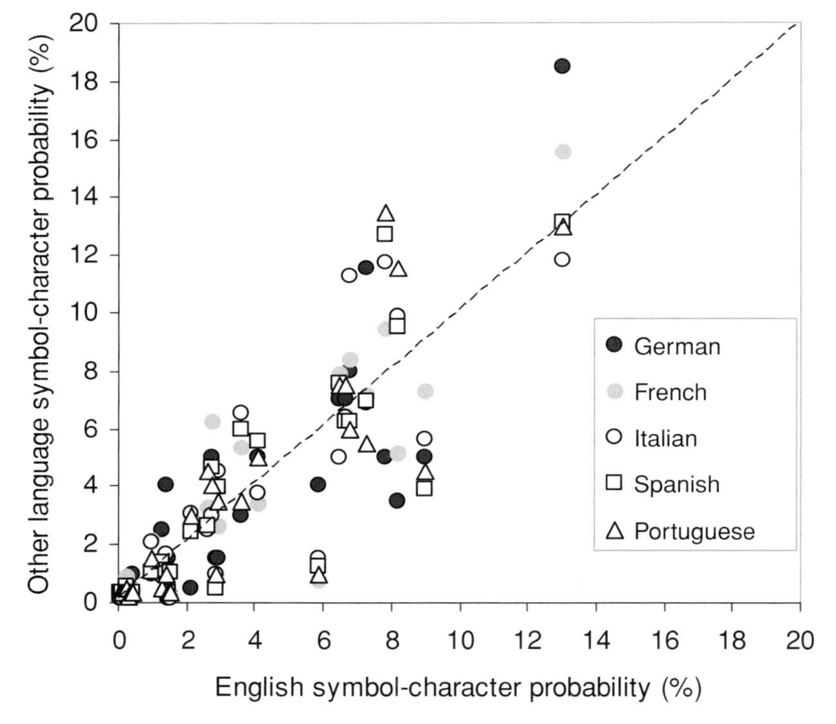


**Figure 4.4** Correlation between English character symbols with that of other European languages, according to data shown in Fig. 4.3.

with more compact symbols, or codewords, in such a way that the coded language source may approach this entropy limit.

Table 4.2 also reveals that there exist substantial differences between the ten most frequent symbol characters. Note that this (1956) survey yields for English the sequence ET*OA*N*IR*SHD, while the earlier data (1941) shown in Fig. 4.2 yields the sequence ET*AO*N*RI*SHD. This discrepancy is, however, not significant considering that the probability differences between A and O and between R and I are between 0.1% and 0.4%, which can be considered as an effect of statistical noise. The most interesting side of these character hierarchies is that they can provide information as to which language is used in a given document, in particular if the document has been encrypted according to certain coding rules (this is referred to as a ciphertext). The task of decryption, recovering what is referred to as the *plaintext*, is facilitated by the analysis of the frequencies (or probabilities) at which certain coded symbols or groups of coded symbols are observed to appear.[10] To illustrate the effectiveness of frequency analysis, let us perform a basic experiment. The following paragraph, which includes 1004 alphabetic characters, has been written without any preparation, or concern for contextual meaning (the reader may just skip it):

During last winter, the weather has been unusually cold, with records in temperatures, rainfalls and snow levels throughout the country. According to national meteorology data, such extreme conditions have not been observed since at least two centuries. In some areas, the populations of entire towns and counties have been obliged to stay confined into their homes, being unable to take their cars even to the nearest train station, and in some case, to the nearest food and utility stores. The persistent ice formation and accumulation due to the strong winds caused power and telephone wires to break in many regions, particularly the mountain ones of more difficult road access. Such incidents could not be rapidly repaired, not only because these adverse conditions settled without showing any sign of improvement, but also because of the shortage of local intervention teams, which were generally overwhelmed by basic maintenance and security tasks, and in some case because of the lack of repair equipment or adequate training. The combined effects of fog, snow and icing hazards in airports have also caused a majority of them to shut down all domestic traffic without advanced notice. According to all expectations, the President declared the status of national emergency, involving the full mobilization of police and army forces.

A rapid character count of the above paragraph (with a home computer[11]) provides the corresponding probability distribution. For comparison purposes, we shall use this time an English-language probability distribution based on a more recent (1982) survey.[12] This survey was based on an analysis of newspapers and novels with a total sample of 100 362 alphabetic characters. The results are shown in Figs. 4.5 and 4.6 for the distribution plot and the correlation plot, respectively. We observe from these two plots a remarkable resemblance between the two distributions and a strong correlation between them.[13] This result indicates that any large and random sample of English text, provided it does not include specialized words or acronyms, closely complies with the symbol-distribution statistics. Such compliance is not as surprising as the fact that it is so good considering the relatively limited size of the sample paragraph.

[10] It is noteworthy that frequency analysis was invented as early as the ninth century, by an Arab scientist and linguist, Al-Kindi.

[11] This experiment is easy to perform with a home computer. First write or copy and paste a paragraph on any subject, which should include at least $N = 1000$ alphabetical characters. Then use the find and replace command to substitute letter A with a blank, and so on until Z. Each time, the computer will indicate how many substitutions were effected, which directly provides the corresponding letter count. The data can then be tabulated at each step and the character counts changed into probabilities by dividing them by $N$. The whole measurement and tabulating operation should take less than ten minutes.

[12] S. Singh, *The Code Book: The Science of Secrecy from Ancient Egypt to Quantum Cryptography* (New York: Anchor Books, 1999).

[13] The reader may trust that this was a single-shot experiment, the plots being realized *after* having written the sample paragraph, with no attempt to modify it retroactively in view of improving the agreement.

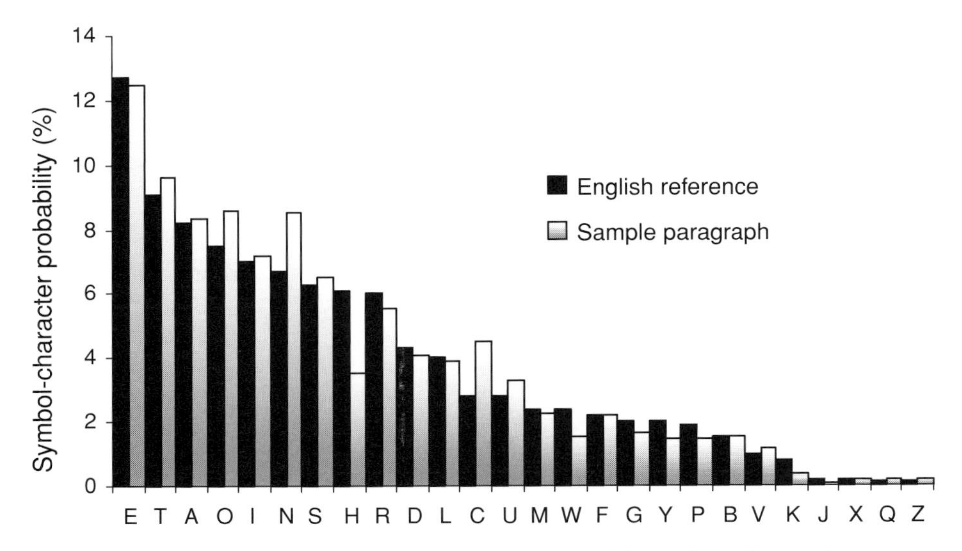


**Figure 4.5** Probability distributions of symbol-characters used in English reference (as per a 1982 survey) and as computed from the author's sample paragraph shown in the text.

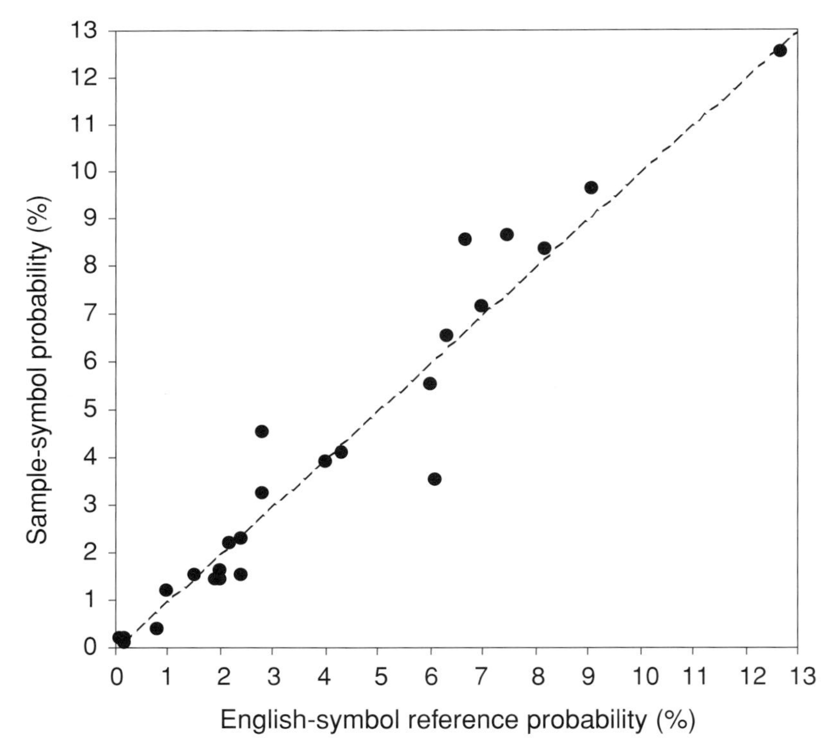


**Figure 4.6** Correlation between English reference and author's sample probability distributions.

As an interesting feature, the entropy of the 1982 English symbol-character source is computed as $H(1982) = 4.185$ bit/symbol.[14] This is larger than the 1956 data ($H(1956) = 4.147$ bit/symbol) and the 1942 data ($H(1942) = 4.140$ bit/symbol). If we were to attribute equal reference value to these data, a rapid calculation shows that the English source entropy grew by 0.17% in 1942–1956, and by approximately 0.9% in 1956–1982. This growth also corresponds to 1.1% over 40 years, which can be linearly extrapolated to 2.75% in a century, say a growth rate of 3%/century. Applying this rate up to the year of this book writing (2005), the entropy should be at least $H(2005) = 4.20$ bit/symbol. One century later (2105), it should be at least $H(2105) = 4.32$ bit/symbol. In reference to the absolute A–Z source limit ($H_{\max} = 4.700$ bit/symbol), this would represent an efficiency of alphabet use of $H(2105)/H_{\max} = 92\%$, to compare with today's efficiency, i.e., $H(2005)/H_{\max} = 89\%$. We can only speculate that a 100% efficiency may never be reached by any language unless cultural influence and mixing makes it eventually lose its peculiar linguistic and root structures.

## 4.4 Maximum entropy (discrete source)

We have seen that the information related to any event $x$ having probability $p(x)$ is defined as $I(x) = -\log p(x)$. Thus, information increases as the probability decreases or as the event becomes less likely. The information eventually becomes infinite ($I(x) \to \infty$) in the limit where the event becomes "impossible" ($p(x) \to 0$). Let $y$ be the complementary event of $x$, with probability $p(y) = 1 - p(x)$. In the previous limit, the event $y$ becomes "absolutely certain" ($p(y) \to 1$), and consistently, its information vanishes ($I(y) = -\log p(y) \to 0$).

The above shows that *information is unbounded, but its infinite limit is reached only for impossible events that cannot be observed*. What about entropy? We know that entropy is the measure of the average information concerning a set of events (or a system described by these events). Is this average information bounded? Can it be maximized, and to which event would the maximum correspond?

To answer these questions, I shall proceed from the simple to the general, then to the more complex. I consider first a system with two events, then with $k$ events, then with an infinite number of discrete events. Finally, I introduce some constraints in the entropy maximization problem.

Assume first two complementary events $x_1, x_2$ with probabilities $p(x_1) = q$ and $p(x_2) = 1 - q$, respectively. By definition, the entropy of the source $X = \{x_1, x_2\}$ is given by

$$\begin{aligned} H(X) &= -\sum_{x \in X} p(x) \log p(x) \\ &= -x_1 \log p(x_1) - x_2 \log p(x_2) \\ &= -q \log q - (1-q) \log(1-q) \equiv f(q). \end{aligned} \tag{4.13}$$

[14] The entropy of the sample paragraph is $H = 4.137$ bit/symbol, which indicates that the sample is reasonably close to the reference ($H(1982) = 4.185$ bit/symbol), meaning that there is no parasitic effect due to the author's choice of the subject or his own use of words.

Note the introduction of the function $f(q)$:

$$f(q) = -q\log q - (1-q)\log(1-q) \equiv f(q), \tag{4.14}$$

which will be used several times through these chapters.

A first observation from the definition of $H(X) = f(q)$ is that if one of the two events becomes "impossible," i.e., $q = \varepsilon \to 0$ or $1 - q = \varepsilon \to 0$, the entropy remains bounded. Indeed, the corresponding term vanishes, since $\varepsilon \log \varepsilon \to 0$ when $\varepsilon \to 0$. In such a limit, however, the other term corresponding to the complementary event, which becomes "absolutely certain," also vanishes (since $u \log u \to 0$ when $u \to 1$). Thus, in this limit the entropy also vanishes, or $H(X) \to 0$, meaning that the source's information is identical to zero as a statistical average. This situation of zero entropy corresponds to a fictitious system, which would be frozen in a state of either "impossibility" or "absolute certainty."

We assume next that the two events are equiprobable, i.e., $q = 1/2$. We then obtain from Eq. (4.13):

$$H(X) = f\left(\frac{1}{2}\right) = -\frac{1}{2}\log\frac{1}{2} - \left(1-\frac{1}{2}\right)\log\left(1-\frac{1}{2}\right) = -\log\frac{1}{2} = 1. \tag{4.15}$$

The result is that the source's average information (its entropy) is *exactly one bit*. This means that it takes a single bit to define the system: either event $x_1$ or event $x_2$ is observed, with equal likelihood. The source information is, thus, given by a simple YES/NO answer, which requires a single bit to formulate. In conditions of equiprobability, the uncertainty is evenly distributed between the two events. Such an observation intuitively conveys the sense that the entropy is a maximum in this case. We can immediately verify this by plotting the function $H(X) \equiv f(q) = -q\log q - (1-q)\log(1-q)$ defined in Eq. (4.14), from $q = 0$($x_1$ impossible) to $q = 1$($x_1$ absolutely certain). The plot is shown in Fig. 4.7. As expected, the maximum entropy $H_{\max} = 1$ bit is reached for $q = 1/2$, corresponding to the case of equiprobable events.

Formally, the property of entropy maximization can be proved by taking the derivative $\mathrm{d}H/\mathrm{d}q$ and finding the root, i.e., $\mathrm{d}H/\mathrm{d}q = \log[(1-q)/q] = 0$, which yields $q = 1/2$.

We now extend the demonstration to the case of a source with $k$ discrete events, $X = \{x_1, x_2, \ldots, x_k\}$ with associated probabilities $p(x_1), p(x_2), \ldots, p(x_k)$. This is a problem of multidimensional optimization, which requires advanced analytical methods (here, namely, the *Lagrange multipliers* method).

The solution is demonstrated in Appendix C. As expected, the entropy is maximum when all the $k$ source events are equiprobable, i.e., $p(x_1) = p(x_2) = \cdots = p(x_k) = 1/k$, which yields $H_{\max} = \log k$. If we assume that $k$ is a power of two, i.e., $k = 2^M$, then $H_{\max} = \log 2^M = M$ bits. For instance, a source of $2^{10} = 1024$ equiprobable events has an entropy of $H = 10$ bits.

The rest of this chapter, and Appendix C, concerns the issue of PDF optimization for entropy maximization, under parameter constraints. This topic is a bit more advanced than the preceding material. It may be skipped, without compromising the understanding of the following chapters.

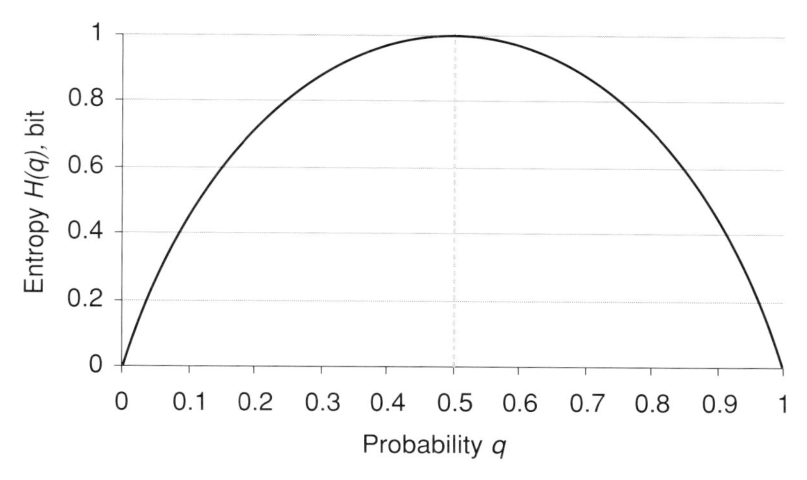


**Figure 4.7** Plot of the entropy $H(X) = f(q)$ for a source with two complementary events with probabilities $q$ and $1 - q$ (Eq. (4.14)), showing maximum point $H_{\max} = H(0.5) = 1$ where events are equiprobable.

In Appendix C, I also analyze the problem of entropy maximization through PDF optimization with the introduction of *constraints*.

A first type of constraint is to require that the PDF have a predefined mean value $\langle x \rangle = N$. In this case, the following conclusions can be reached:

- If the event space $X = \{x_1, x_2, \ldots, x_k\}$ is the *infinite* set of integer numbers $(x_1 = 0, x_2 = 1, x_3 = 2 \ldots)$, $k \to \infty$, the optimal PDF is the *discrete-exponential distribution*, also called the *Bose–Einstein* distribution (see Chapter 1);
- If the event space $X = \{x_1, x_2, \ldots, x_k\}$ is a *finite* set of nonnegative real numbers, the optimal PDF is the (continuous) *Boltzmann* distribution.

The Bose–Einstein distribution characterizes chaotic processes, such as the emission of light photons by thermal sources (e.g., candle, light bulb, Sun, star) or the spontaneous emission of photons in laser media. The Boltzmann distribution describes the random arrangement of electrons in discrete atomic energy levels, when the atomic systems are observed at thermal equilibrium.

As shown in Appendix C, the (maximum) entropy corresponding to the Boltzmann distribution, as defined in *nats*, is:

$$H_{\max} = \langle m \rangle \frac{h\nu}{k_B T}, \tag{4.16}$$

where $h\nu$ is the quantum of light energy (*photon*), $k_B T$ is the quantum of thermal energy (*phonon*) and $\langle m \rangle$ is the mean number of phonons at absolute temperature $T$ and oscillation frequency $\nu$ ($h$ = Planck's constant, $k_B$ = Boltzmann's constant). The quantity $\Delta E = \langle m \rangle h\nu$ is, thus, the mean electromagnetic energy or heat that can be radiated by the atomic system. The ratio $\Delta E/(k_B T)$ is the number of phonons required to keep the system in such a state. The maximal entropy $H_{\max}$, which represents the average information to define the system, is just equal to this simple ratio! This feature establishes a nice connection between information theory and atomic physics. Letting $S = k_B H_{max}$, which is consistent with the physics definition of entropy, we note that $S = \Delta E/T$, which corresponds to the well known Clausius relation between system entropy ($S$), heat contents ($\Delta E$), and absolute temperature ($T$).[15]

Consider next the possibility of imposing an arbitrary number of constraints on the probability distribution function (PDF) *moments*, $\langle x\rangle, \langle x^2\rangle, \ldots, \langle x^n\rangle$.[16] The general PDF solution for which entropy is maximized takes the nice analytical form (Appendix C):

$$p_j = \exp\left(\lambda_0 - 1 + \lambda_1 x_j + \lambda_2 x_j^2 + \cdots + \lambda_n x_j^n\right), \tag{4.17}$$

where $\lambda_0, \lambda_1, \lambda_2, \ldots, \lambda_n$ are the Lagrange multipliers, which must be computed numerically. The corresponding (maximum) entropy is simply given by:

$$H_{max} = 1 - \left(\lambda_0 + \lambda_1\langle x\rangle + \lambda_2\langle x^2\rangle + \cdots + \lambda_n\langle x^n\rangle\right) = 1 - \sum_{i=0}^{n} \lambda_i \langle x^i\rangle. \tag{4.18}$$

Maximization of entropy is not limited to constraining PDF moments. Indeed, any set of known functions $g_k(x)$ and their mean $\langle g_k\rangle$ ($k = 1, \ldots, n$) can be used to define and impose as many constraints. It is easily established that, in this case, the optimal PDF and maximum entropy takes the form:

$$p_j = \exp[-1 + \lambda_0 g_0(x_j) + \lambda_1 g_1(x_j) + \cdots + \lambda_n g_n(x_j)], \tag{4.19}$$

$$H_{max} = 1 - \sum_{i=0}^{n} \lambda_i \langle g_i\rangle. \tag{4.20}$$

From the observation in the real world of the functions or parameters $g_k(x)$, $\langle g_k\rangle$, it is thus possible heuristically to infer a PDF that best models reality (maximum entropy at macroscopic scale), assuming that a large number of independent microstates govern the process. This is discussed later.

Entropy maximization leads to numerically defined PDF solutions, which are essentially nonphysical. Searching for such solutions, however, is not a matter of pure academic interest. It can lead to new insights as to how the physical reality tends to exist in states of maximum entropy.

For instance, I showed in previous work[17] that given constraints in $\langle x\rangle, \sigma^2 = \langle x^2\rangle - \langle x\rangle$, the entropy of *amplified coherent light* is fairly close to that of the optimal-numerical PDF giving maximal entropy. For increasing photon numbers $\langle x\rangle$, the physical photon-statistics PDF and the optimal-numerical PDF are observed to converge asymptotically towards the same Gaussian distribution, as a consequence of the central-limit theorem (Chapter 1).

[15] It is beyond the scope of these chapters to discuss the parallels between entropy in Shannon's information-theory and entropy in physics. Further and accessible considerations concerning this (however complex) subject can be found, for instance, in: www.tim-thompson.com/entropy1.html, www.panspermia.org/seconlaw.htm, http://en.wikipedia.org/wiki/Entropy.

[16] A moment of order $k$ is, by definition, the mean value of $x^k$, namely, $\langle x^k\rangle = \sum x_i^k p_i$.

[17] E. Desurvire, How close to maximum entropy is amplified coherent light? *Opt. Fiber Technol.*, **6** (2000), 357. See also, E. Desurvire, D. Bayart, B. Desthieux, and S. Bigo, *Erbium-Doped Fiber Amplifiers: Device and System Developments* (New York: John Wiley & Sons, 2002), p. 202.

The fact that several discrete (and continuous) random physical processes are ruled by exact or near-maximum entropy distributions can be explained by the following: if a system $X$ offers a large number of possible random arrangements $x_i$ (called *microstates*), and if such arrangements are equally probable (all independent of each other or past microstate history), then the system's entropy $H(X)$ is a maximum, as I have shown earlier. In some physical processes (like amplification of coherent light), there is no reason for microstates to be equiprobable, which explains the discrepancy between observed and maximal entropies. The discrepancy, however, vanishes when the number of possible microstates becomes infinite. In some other physical processes (such as the electron occupation of atomic energy levels, or the spontaneous emission of photons by single atoms), the microstates are strictly equiprobable and the system's entropy is maximum, as we have seen.

The so-called *maximum entropy principle*,[18] is used in several domains of statistical sciences, from engineering and physics to computer vision, image processing and reconstruction, language analysis, urban design, marketing, elections, business, economics, and finance! The underlying motivation of *maximum-entropy models* (MEM) is to derive, heuristically, the PDF of a complex random process by using available data samples observed from reality. These experimental data are then used as constraints to compute the maximum-entropy PDF, elegantly referred to as *epistemic*. The philosophy and rationale of the maximum-entropy principle approach is: "Given facts or relations concerning events, which are verified in the physical world, what is the best statistical model available to predict the rest?"

The issue of maximizing entropy is revisited in Chapter 5 when considering *continuous* sources.

## 4.5 Exercises

**4.1** (B): What is the entropy in the action of picking, at random, one card out of a 32-card deck? What is the entropy for picking a hand of four cards?

**4.2** (B): A bag contains eight balls, including one red, two blue, two green, and three yellow ones. The events consist in picking, at random, a ball from the bag, seeing the color, and then replacing the ball. What is the entropy of the source?

**4.3** (M): Download from any website the text of the US Declaration of Independence,

www.ushistory.org/declaration/document/index.htm,

www.law.indiana.edu/uslawdocs/declaration.html,

www.loc.gov/rr/program/bib/ourdocs/DeclarInd.html,

www.archives.gov/exhibits/charters/declaration.html,

[18] See, for instance: www.answers.com/topic/principle-of-maximum-entropy?cat=technology&hl?function=hl?derivation=&hl?partition=, www.answers.com/maximum+entropy+probability+distribution?cat=technology.

from, "When in the Course of human events ..." to "... our Fortunes and our sacred Honor." What is the entropy of the selected text, as viewed as a source of 26 + 1 characters (including space)?

What is the entropy of the selected text, without the space character? (Computing method clue: see Chapter 4, note 11.)

**4.4** (M): Copy into a text file a "very big" prime number, for instance, a large "Mersenne," which you can download from the websites:

www.mersenne.org/,

www.isthe.com/chongo/tech/math/prime/mersenne.html#M32582657,

http://primes.utm.edu/largest.html.

For expediency, reduce the text source to only a few ten thousand digits. Compute the entropy of the source. Could the result have been guessed directly? (Computing method clue: see Chapter 4, note 11.)

# 5 Mutual information and more entropies

This chapter marks a key turning point in our journey in information-theory land. Heretofore, we have just covered some very basic notions of IT, which have led us, nonetheless, to grasp the subtle concepts of *information* and *entropy*. Here, we are going to make significant steps into the depths of Shannon's theory, and hopefully begin to appreciate its power and elegance. This chapter is going to be somewhat more mathematically demanding, but it is guaranteed to be not significantly more complex than the preceding materials. Let's say that there is more ink involved in the equations and the derivation of the key results. But this light investment will turn out well worth it to appreciate the forthcoming chapters!

I will first introduce two more entropy definitions: *joint* and *conditional entropies*, just as there are joint and conditional probabilities. This leads to a new fundamental notion, that of *mutual information*, which is central to IT and the various Shannon's laws. Then I introduce *relative entropy*, based on the concept of "distance" between two PDFs. Relative entropy broadens the perspective beyond this chapter, in particular with an (optional) appendix exploration of the second law of thermodynamics, as analyzed in the light of information theory.

## 5.1 Joint and conditional entropies

So far, in this book, the notions of *probability distribution* and *entropy* have been associated with single, independent events $x$, as selected from a discrete source $X = \{x\}$. With the example of written language (Chapter 4), we have seen that the occurrence of single characters from the alphabetical-character source $X = \{A, \ldots, Z\}$ is bound to a quasi-exponential PDF, which varies according to language, context, and time.

Considering the fact that language is made out of words, and not single characters, we realize that the previous analysis is incomplete. For instance, we could make an inventory of all words made of two, three, or more characters, and derive a new set of statistics. We could then say how often in English the letter E appears next to the letter A, or what the probability is that a given English word of five characters simultaneously contains the letters A, T, and N, and, given the knowledge that the first two letters are TH, what the probability is that the third one is E, to form the ubiquitous word THE.

This observation leads us to defining the entropy associated with *joint events*, on one hand, or *conditional events* on the other hand. To do this, let us briefly recall the properties of *joint* and *conditional probabilities*, which have been outlined in Chapter 1, and give them further attention here.

Consider two events $x \in X$ and $y \in Y$, where $X, Y$ are two random sources. The two events can occur simultaneously, regardless of any sequence order: we just observe $x$ *and* $y$, or $y$ *and* $x$, which conveys the same information. Alternatively, we can observe $x$ *then* $y$, but for some reason we seem never to get $y$ *then* $x$. Our observation could also be that whenever $x$ happens, we are likely to see $y$ happen as well. Alternatively, the fact that $x$ happens could have no incidence on the outcome of $y$. For instance, define the following event sources:

$$X = \{\text{quarterly results of a major telecom company}\},$$
$$Y = \{\text{stock market}\},$$
$$Z = \{\text{weather forecast}\},$$
$$W = \{\text{outbound city traffic}\}.$$

It is safe to say (scientifically speaking!) that events $x$ from $X$ and have no impact whatsoever on events $z$ or $w$ from $Z$ or $W$. Thus events $x$ and $z$ (or $x$ and $w$) are *independent*. The probabilities $p(x)$ that $x =$ excellent, and $p(z)$ that $z =$ cold and rainy also have different and unrelated PDFs. As we know from Chapter 1, the joint probability that we observe both events simultaneously is given by the mere product:

$$p(x, y) = p(\text{excellent, cold and rainy}) = p(x)p(y). \tag{5.1}$$

In Eq. (5.1), the function $p(x, y)$ is called the *joint probability* or *joint distribution* of events $(x, y)$. The meaning of the "," in the argument is that of the logical AND.[1] The joint distribution is, thus, always symmetrical, i.e., $p(x, y) = p(y, x)$, because the joint events $x$ AND $y$ and $y$ AND $x$ are strictly same. Since $p(x, y)$ is a probability, we have the summing properties

$$\begin{cases} \sum\limits_{y \in Y} p(x, y) = p(x) \\ \sum\limits_{x \in X} p(x, y) = p(y) \\ \sum\limits_{x \in X} \sum\limits_{y \in Y} p(x, y) = 1. \end{cases} \tag{5.2}$$

But the relation concerning any event pairs $(x, y)$ or $(z, w)$ of the above-defined sets is not at all this straightforward. This is because we should expect that these events are *not* independent: good or bad quarterly results do affect the stock market, good or bad weather forecasts do affect the outbound city traffic.

One then defines the *conditional probability* $p(a|b)$ as representing the probability of observing event $a$ given the observation or knowledge of event $b$. If $p(a|b) = p(a)$, this means that event $a$ occurs regardless of $b$, or that the two are independent. Of

[1] Joint probabilities $p(x, y)$ can also be written in the logical form $p(x \wedge y)$ where the sign $\wedge$ stands for the logical or Boolean operation AND.

course, the same conclusion applies if $p(b|a) = p(b)$. As we have seen in Chapter 1, the fundamental relation between joint and conditional probabilities, is given by *Bayes's theorem*:

$$p(a, b) = p(a|b)p(b) = p(b|a)p(a). \tag{5.3}$$

Using the summing properties in Eq. (5.2), we also have

$$\begin{cases} \sum_{b \in B} p(a|b)p(b) = p(a) \\ \sum_{a \in A} p(b|a)p(a) = p(b) \\ \sum_{a \in A} \sum_{b \in B} p(b|a)p(a) = \sum_{a \in A} \sum_{b \in B} p(a|b)p(b) = 1, \end{cases} \tag{5.4}$$

noting that conditional probabilities $p(a|x)$ *do not* sum over $x$ up to unity, unlike the probabilities $p(x)$ associated with single events.[2] It is straightforward to verify from the above relations that if the events $a$ and $b$ are independent, then $p(a|b) = p(a)$ and $p(b|a) = p(b)$. I provide next a numerical example of joint and conditional probabilities, which we will also use further on to introduce new entropy concepts.

## Stock exchange

Let the first event source be three possible conclusions for the quarterly sales report from a public company, namely:

$$x \in X = \{\text{good, same, bad}\} \equiv \{x_1, x_2, x_3\},$$

meaning that the results are in excess of the predictions (good), or on target (same), or under target (bad). The second event source is the company's stock value, as reflected by the stock exchange, with

$$y \in Y = \{\text{up, steady, down}\} \equiv \{y_1, y_2, y_3\}.$$

Then we assume that there exists some form of correlation between the company's results ($x \in X$) and its stock value ($y \in Y$). A numerical example of joint and conditional probability data, $p(x_i, y_j)$, $p(x_i|y_j)$ and $p(y_j|x_i)$, is shown in Table 5.1. The first group of numerical data, shown at the top of Table 5.1 corresponds to the joint probabilities $p(y_j, x_i) \equiv p(x_i, y_j)$. Summing up the data by rows ($i$) or by columns ($j$) yields the probabilities $p(x_i)$ or $p(y_j)$, respectively. The double checksum (bottom right), which yields unity through summing by row or by column, is also shown for consistency. The two other groups of numerical data in Table 5.1 correspond to the conditional probabilities $p(x_i|y_j)$ and $p(y_j|x_i)$. These are calculated through Bayes's theorem $p(x_i|y_j) = p(x_i, y_j)/p(y_j)$ and $p(y_j|x_i) = p(x_i, y_j)/p(x_i)$. The intermediate columns providing the data $p(x_i|y_j)p(y_j) = p(y_j|x_i)p(x_i) \equiv p(x_i, y_j)$ and their checksums by column are shown in the table for consistency.

[2] Yet we have the property $\sum_{a \in A} p(a|b) = 1$ for any event $b \in B$, summing over all possible events $a \in A$.

**Table 5.1** Example of joint and conditional probability distributions constructed from the two event sources $x \in X = \{\text{good, same, bad}\} \equiv \{x_1, x_2, x_3\}$ for a company's quarterly sales results and $y \in Y = \{\text{up, steady, down}\} \equiv \{y_1, y_2, y_3\}$ for its stock value. The table on top shows the joint probability $p(x_i, y_j)$, consistent with the probabilities $p(x_i)$ and $p(y_j)$ of the individual events $x_i$ and $y_j$ (values given in column or line $p$, respectively). The right column and bottom line provide the different checksums. The two other tables (middle and bottom) show the conditional probabilities $p(y_j|x_i)$ and $p(x_i|y_j)$, along with their different checksums.

| $p(x_i, y_j)$ | $p$ | $y_1$ (up)<br>0.300 | $y_2$ (steady)<br>0.500 | $y_3$ (down)<br>0.200 | $\sum =$ |
|---|---|---|---|---|---|
| $x_1$ (good) | **0.160** | 0.075 | 0.075 | 0.010 | **0.160** |
| $x_2$ (same) | **0.750** | 0.210 | 0.400 | 0.140 | **0.750** |
| $x_3$ (bad) | **0.090** | 0.015 | 0.025 | 0.050 | **0.090** |
| | $\sum =$ | **0.300** | **0.500** | **0.200** | **1.000** |

| $p(y_j, x_i)$ | $p$ | $y_1$ (up)<br>0.300 | $p(y_1, x_i)$<br>$= p(y_1\|x_i)p(x_i)$ | $y_2$ (steady)<br>0.500 | $p(y_2, x_i)$<br>$= p(y_2\|x_i)p(x_i)$ | $y_3$ (down)<br>0.200 | $p(y_3, x_i)$<br>$= p(y_3\|x_i)p(x_i)$ | |
|---|---|---|---|---|---|---|---|---|
| $x_1$ (good) | **0.160** | 0.469 | 0.075 | 0.469 | 0.075 | 0.063 | 0.010 | |
| $x_2$ (same) | **0.750** | 0.280 | 0.210 | 0.533 | 0.400 | 0.187 | 0.140 | |
| $x_3$ (bad) | **0.090** | 0.167 | 0.015 | 0.278 | 0.025 | 0.556 | 0.050 | |
| | $\sum =$ | | **0.300** | | **0.500** | | **0.200** | **1.000** |

| $p(x_i, y_j)$ | $p$ | $x_1$ (good)<br>0.160 | $p(x_1 y_j)p(y_j)$<br>$= p(x_1\|y_j)p(y_j)$ | $x_2$ (same)<br>0.750 | $p(x_2, y_j)$<br>$= p(x_2\|y_j)p(y_j)$ | $x_3$ (bad)<br>0.090 | $p(x_3, y_j)$<br>$= p(x_3\|y_j)p(y_j)$ | |
|---|---|---|---|---|---|---|---|---|
| $y_1$ (up) | **0.300** | 0.250 | 0.075 | 0.700 | 0.210 | 0.050 | 0.015 | |
| $y_2$ (steady) | **0.500** | 0.150 | 0.210 | 0.800 | 0.400 | 0.050 | 0.025 | |
| $y_3$ (down) | **0.200** | 0.010 | 0.015 | 0.700 | 0.1 | 0.250 | 0.050 | |
| | $\sum =$ | | **0.300** | | **0.750** | | **0.090** | **1.000** |

We define the *joint entropy*, or the entropy $H(X, Y)$ associated with the joint distribution $p(x, y)$ with $x \in X$ and $y \in Y$, namely:[3]

$$H(X, Y) = -\sum_{x \in X} \sum_{y \in Y} p(x, y) \log_2 p(x, y). \tag{5.5}$$

This joint entropy represents the average information derived from joint events occurring from two sources $X$ and $Y$. The unit of $H(X, Y)$ is *bit/symbol*.

We then define the *conditional entropy* $H(X|Y)$ through:

$$H(X|Y) = -\sum_{x \in X} \sum_{y \in Y} p(x, y) \log_2 p(x|y). \tag{5.6}$$

The conditional entropy $H(X|Y)$ corresponds to *the average information conveyed by the conditional PDF,* $p(x|y)$. Put simply, $H(X|Y)$ represents *the information we learn from source* $X$*, given the information we have from source* $Y$. Its unit is also *bit/symbol*.

[3] Or, equivalently, $H(X, Y) = -\sum_i \sum_j p(x_i, y_j) \log_2 p(x_i, y_j)$.

In the joint and conditional entropy definitions, note that the two-dimensional averaging over the event space $\{X, Y\}$ consistently involves the *joint distribution* $p(x, y)$.

As a property, the conditional entropy $H(X|Y)$ is generally different from $H(Y|X)$, which is defined as

$$H(Y|X) = -\sum_{x\in X}\sum_{y\in Y} p(x, y)\log_2 p(y|x), \tag{5.7}$$

since in the general case, $p(x|y) \neq p(y|x)$. The conditional entropy $H(X|Y)$ or $H(Y|X)$ is sometimes referred to by the elegant term *equivocation*.

It is easily verified that if the sources $X$, $Y$ represent independent events, then

$$\text{(a) } H(X, Y) = H(X) + H(Y), \tag{5.8}$$

$$\text{(b) } H(X|Y) = H(X)m \text{ and } H(Y|X) = H(Y). \tag{5.9}$$

As it can also be easily established, the joint and conditional entropy are related to each other through the *chain rule*:

$$\begin{cases} H(X, Y) = H(X|Y) + H(Y) \\ H(X, Y) = H(Y|X) + H(X). \end{cases} \tag{5.10}$$

We may find the chain rule easier to memorize under the form

$$\begin{cases} H(X|Y) = H(X, Y) - H(Y) \\ H(Y|X) = H(X, Y) - H(X), \end{cases} \tag{5.11}$$

which states that, given a source $X$ or $Y$, any advance knowledge (or "conditioning") from the other source $Y$ or $X$ reduces the joint entropy "reserve" $H(X, Y)$ by the net amount $H(Y)$ or $H(X)$, respectively; these are positive quantities. In other words, the prior information one may gain from a given source is made at the expense of the information available from the other source, unless the two are independent.

We can illustrate the above properties through our stock-exchange PDF data from Table 5.1. We want, here, to determine how the average information from the company's sales, $H(X)$, is affected from that concerning the stocks, $H(Y)$, and the reverse. The computations of $H(X)$, $H(Y)$, $H(X, Y)$, $H(Y|X)$ and $H(X|Y)$ are detailed in Table 5.2. It is seen from the table that the results and stock entropies compute to $H(X) = 1.046$ and $H(Y) = 1.485$, respectively (for easier reading, I omit here the bit/symbol units). The joint entropy is found to be $H(X, Y) = 2.466$, which is lower than the sum $H(X) + H(Y) = 2.532$. This proves that the two sources are not independent, namely, that they have some information in common. We find indeed that the conditional entropies satisfy:

$$H(Y|X) = 1.418 < H(Y) = 1.485,$$

$$H(X|Y) = 0.980 < H(X) = 1.046,$$

or, equivalently, (using four decimal places, for accuracy (see Table 5.2):

$$H(Y) - H(Y|X) = 1.4855 - 1.4188 = 0.0667,$$

$$H(X) - H(X|Y) = 1.0469 - 0.9802 = 0.0667.$$

**Table 5.2** From top to bottom: calculation of entropies $H(X)$, $H(Y)$, $H(X, Y)$, $H(Y|X)$ and $H(X|Y)$, as based on the numerical example of Table 5.1. The two equations shown at the bottom of the table prove the fundamental relations between single-event entropies, conditional entropies, and the joint entropy.

$u_i = p(x_i)$
$v_j = p(y_j)$

| | $u_i$ | $-u_i \log u_i$ | | $v_j$ | $-v_j \log v_j$ |
|---|---|---|---|---|---|
| $x_1$**(good)** | 0.160 | 0.423 | $y_1$**(up)** | 0.300 | 0.521 |
| $x_2$**(same)** | 0.750 | 0.311 | $y_2$**(steady)** | 0.500 | 0.500 |
| $x_3$**(bad)** | 0.090 | 0.313 | $y_3$**(down)** | 0.200 | 0.464 |
| | $H(X) =$ | 1.0469 | | $H(Y) =$ | 1.4855 |

$H(X) + H(Y) = 2.5324$
$u_{ij} = p(x_i, y_j)$

| | $u_{i1}$ | $-u_{i1} \log u_{i1}$ | $u_{i2}$ | $-u_{i2} \log u_{i2}$ | $u_{i3}$ | $-u_{i3} \log u_{i3}$ |
|---|---|---|---|---|---|---|
| $x_1$ | 0.075 | 0.280 | 0.075 | 0.280 | 0.010 | 0.066 |
| $x_2$ | 0.210 | 0.473 | 0.400 | 0.529 | 0.140 | 0.397 |
| $x_3$ | 0.015 | 0.091 | 0.025 | 0.133 | 0.050 | 0.216 |
| | $\sum =$ | 0.844 | $\sum =$ | 0.942 | $\sum =$ | 0.680 |

$H(X, Y) = 2.4657$
$v_{ji} = p(y_j|x_i)$

| | $v_{1i}$ | $u_{i1}$ | $-u_{i1} \log v_{1i}$ | $v_{2i}$ | $u_{i2}$ | $-u_{i2} \log v_{2i}$ | $v_{3i}$ | $u_{i3}$ | $-u_{i3} \log v_{3i}$ |
|---|---|---|---|---|---|---|---|---|---|
| $x_1$ | 0.469 | 0.075 | 0.082 | 0.469 | 0.075 | 0.082 | 0.063 | 0.010 | 0.040 |
| $x_2$ | 0.280 | 0.210 | 0.386 | 0.533 | 0.400 | 0.363 | 0.187 | 0.140 | 0.339 |
| $x_3$ | 0.167 | 0.015 | 0.039 | 0.278 | 0.025 | 0.046 | 0.556 | 0.050 | 0.042 |
| | | $\sum =$ | 0.506 | | $\sum =$ | 0.491 | | $\sum =$ | 0.421 |

$H(Y|X) = 1.4188$
$w_{ij} = p(x_i|y_j)$

| | $w_{1j}$ | $u_{1j}$ | $-u_{i1} \log w_{1j}$ | $w_{2j}$ | $u_{2j}$ | $-u_{i2} \log w_{2j}$ | $w_{3j}$ | $u_{3j}$ | $-u_{i3} \log w_{3j}$ |
|---|---|---|---|---|---|---|---|---|---|
| $y_1$ | 0.250 | 0.075 | 0.150 | 0.700 | 0.210 | 0.108 | 0.050 | 0.015 | 0.065 |
| $y_2$ | 0.150 | 0.075 | 0.205 | 0.800 | 0.400 | 0.129 | 0.050 | 0.025 | 0.108 |
| $y_3$ | 0.050 | 0.010 | 0.043 | 0.700 | 0.140 | 0.072 | 0.250 | 0.050 | 0.100 |
| | | $\sum =$ | 0.398 | | $\sum =$ | 0.309 | | $\sum =$ | 0.421 |

$H(X|Y) = 0.9802$
$H(X, Y) = H(Y|X) + H(X)$
$2.4657 = 1.4188 + 1.0469$
$= H(X|Y) + H(Y)$
$= 0.9802 + 1.4855.$

These two results mean that the prior knowledge of the company's stocks contains an average of 0.0667 bit/symbol of information on the company's quarterly result, and the reverse is also true. As we shall see in the next section, the two differences above are always equal and they are called *mutual information*. Simply put, *the mutual information is the average information that two sources share in common.*

## 5.2 Mutual information

I introduce yet another type of entropy definition, which will bring us to some closing point and our final reward. We can define the *mutual information* of two sources $X$ and $Y$ as the bit/symbol quantity:

$$H(X;Y) = \sum_{x \in X} \sum_{y \in Y} p(x, y) \log \frac{p(x, y)}{p(x)p(y)}. \tag{5.12}$$

We may note the *absence of a minus sign* in the above definition, unlike in $H(X, Y)$, $H(Y|X)$, and $H(X|Y)$. Also note the ";" separator, which distinguishes mutual information from joint entropy $H(X, Y)$. Mutual information is also often referred to in some textbooks as $I(X; Y)$ instead of $H(X; Y)$.

Since the logarithm argument is unity when the two sources are independent ($p(x, y) = p(x)p(y)$), we immediately observe that the mutual information is equal to zero in this case. This reflects the fact that independent sources do not have any information in common.

It takes a bit of painstaking but straightforward computation to show the following three equalities:

$$\begin{aligned} H(X;Y) &= H(X) - H(X|Y) \\ &= H(Y) - H(Y|X) \\ &= H(X) + H(Y) - H(X, Y). \end{aligned} \tag{5.13}$$

The first two equalities above confirm the observation derived from our previous numerical example. They can be interpreted according to the following statement: *mutual information is the reduction of uncertainty in $X$ that we get from the knowledge of $Y$* (and the reverse).

The last equality, as rewritten under the form

$$H(X, Y) = H(X) + H(Y) - H(X;Y), \tag{5.14}$$

shows that *the joint entropy of two sources is generally less than the sum of the source entropies*. The difference is the mutual information that the sources have in common, which reduces the net uncertainty or joint entropy.

Finally, we note from the three relations in Eq. (5.13) that the mutual information is symmetrical in the arguments, namely, $H(X;Y) = H(Y;X)$, as is expected from its very meaning.

The different entropy definitions introduced up to this point may seem a bit abstract and their different relations apparently not very practical to memorize! But the situation becomes different after we draw an analogy with the property of ensembles.

Consider, indeed, two ensembles, called $A$ and $B$. The two ensembles may be united to form a whole, which is noted $F = A \cup B$ ($A$ union $B$). The two ensembles may or may not have elements in common. The set of common elements is called $G = A \cap B$ ($A$ intersection $B$). The same definitions of union and intersection apply to any three ensembles $A$, $B$, and $C$. Figure 5.1 shows *Venn diagram* representations of such ensemble combinations. While Venn diagrams were introduced in Chapter 1, I shall provide further

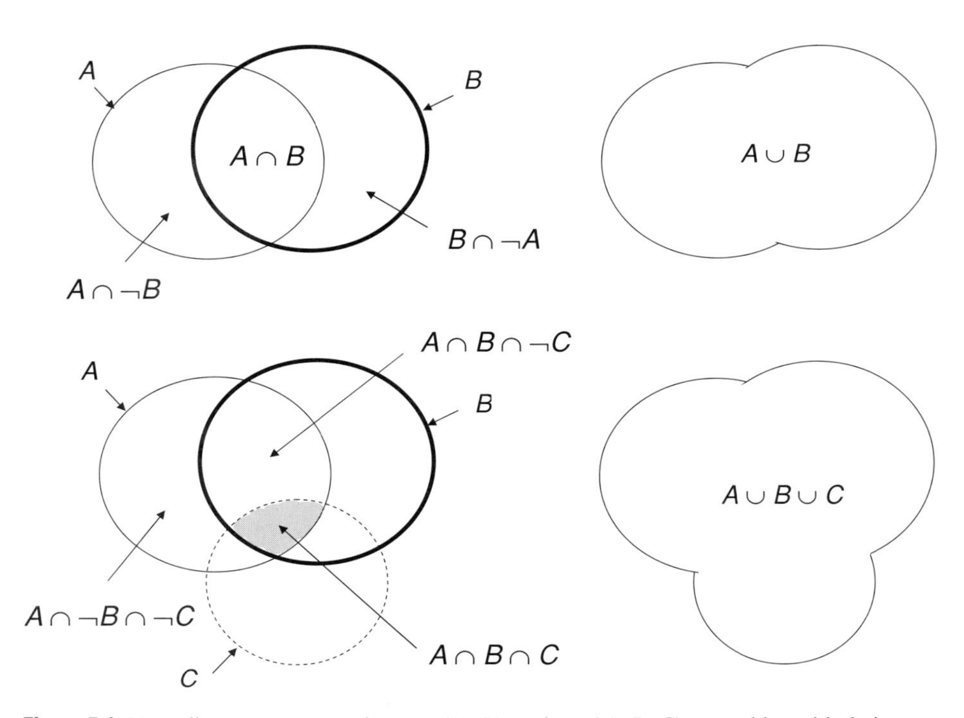


**Figure 5.1** Venn diagrams representing two $(A, B)$ or three $(A, B, C)$ ensembles with their unions $(A \cup B$ or $A \cup B \cup C)$ and their intersections $(A \cap B$ or $A \cap B \cap C)$. The intersections defined by $A \cap \neg B$, $B \cap \neg A$, and $A \cap \neg B \cap \neg C$, $A \cap B \cap \neg C$ are also shown.

related concepts here. In the case of two ensembles, there exist four subset possibilities, as defined by their elements's properties:

Elements common to $A$ *or* $B$: $A \cup B$,

Elements common to $A$ *and* $B$: $A \cap B$,

Elements from $A$ *and not* $B$: $A \cap \neg B$,

Elements from $B$ *and not* $A$: $B \cap \neg A$.

In the conventional notations shown at right, we see that the symbol $\cup$ stands for a logical OR, the symbol $\cap$ stands for a logical AND, and the symbol $\neg$ stands for a logical NO. These three different symbols, which are also called *Boolean operators*,[4] make it possible to perform various mathematical computations in the field called *Boolean logic*. In the case of three ensembles $A, B, C$, we observe that there exist many more subset possibilities (e.g., $A \cap B \cap C$, $A \cap B \cap \neg C$); it is left as an exercise to the reader to enumerate and formalize them in terms of Boolean expressions. The interest of the above visual description with the Venn diagrams is the straightforward correspondence with the various entropy definitions that have been introduced. Indeed, it can be shown

[4] To be accurate, Boolean logic uses instead the symbol $\vee$ for "or" (the equivalent of $\cup$ in ensemble language), and the symbol $\wedge$ for "and" (the equivalent of $\cap$ in ensemble language).

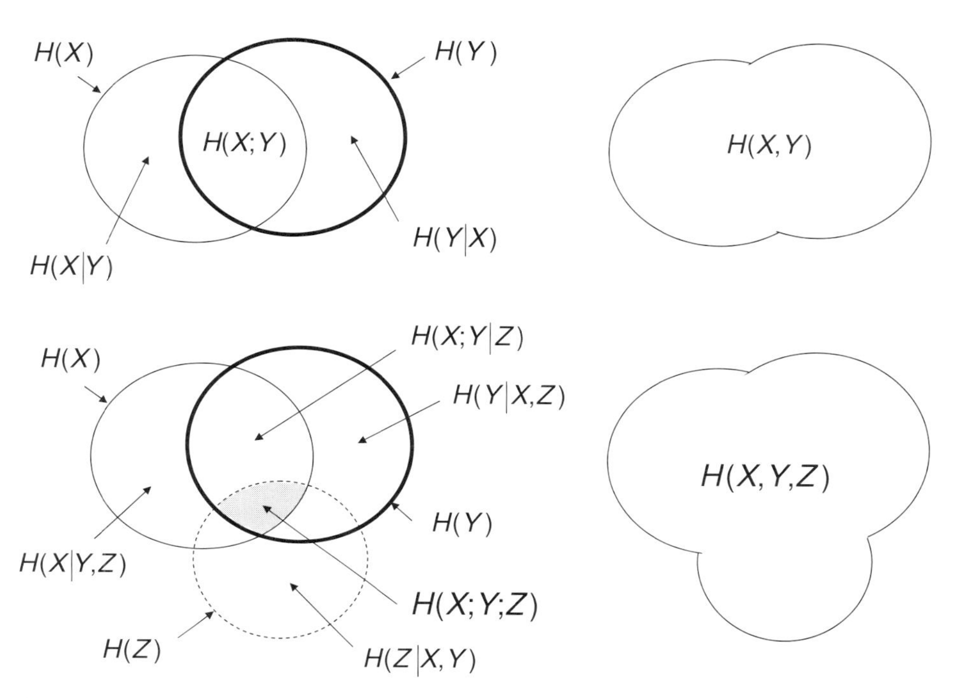


**Figure 5.2** Venn diagram representation of entropy $H(U)$, joint entropy $H(U, V)$, conditional entropy $H(U|V)$, and mutual information $H(U; V)$, for two ($U = X, Y$) or three ($U, V = X, Y, Z$) sources.

that the following equivalences hold:

$$\begin{cases} H(X, Y) \leftrightarrow H(X \cup Y) \\ H(X; Y) \leftrightarrow H(X \cap Y) \\ H(X|Y) \leftrightarrow H(X \cap \neg Y) \\ H(Y|X) \leftrightarrow H(Y \cap \neg X). \end{cases} \tag{5.15}$$

The first equivalence in Eq. (5.15) means that the joint entropy of two sources is the entropy of the source defined by their combined events.

The second equivalence in Eq. (5.15) means that the mutual information of two sources is the entropy of the source containing the events they have in common.

The last two equivalences in Eq. (5.15) provide the definition of the conditional entropy of a source $U$ given the information on a source $V$. The conditional entropy is given by the contributions of all the events belonging to $U$ but *not* to $V$. This property is far from obvious, unless we can visualize it. Figure 5.2 illustrates all the above logical equivalences through Venn diagrams, using up to three sources.

Considering the two-source case, we can immediately visualize from Fig. 5.2 to which subsets the differences $H(X) - H(X; Y)$ and $H(Y) - H(X; Y)$ actually correspond. Given the identities listed in Eq. (5.15), we can call these two subsets $H(X|Y)$ and $H(Y|X)$, respectively, which proves the previous point. We also observe from the Venn diagram that $H(X|Y) \leq H(X)$ and $H(Y|X) \leq H(Y)$, with equality if the sources are independent.

The above property can be summarized by the statement according to which *conditioning reduces entropy*. A formal demonstration, using the concept of "relative entropy," is provided later.

The three-source case, as illustrated in Fig. 5.2, is somewhat more tricky because it generates more complex entropy definitions with three arguments $X$, $Y$, and $Z$. Conceptually, defining joint or conditional entropies and mutual information with three (or more) sources is not this difficult. Considering the joint probability $p(x, y, z)$ for the three sources, we can indeed generalize the previous two-source definitions according to the following:

$$H(X, Y, Z) = -\sum_{x\in X}\sum_{y\in Y}\sum_{z\in Z} p(x, y, z) \log p(x, y, z), \tag{5.16}$$

$$H(X; Y; Z) = +\sum_{x\in X}\sum_{y\in Y}\sum_{z\in Z} p(x, y, z) \log \frac{p(x, y, z)}{p(x)p(y)p(z)}, \tag{5.17}$$

$$H(Z|X, Y) = -\sum_{x\in X}\sum_{y\in Y}\sum_{z\in Z} p(x, y, z) \log p(z|x, y), \tag{5.18}$$

$$H(X, Y|Z) = -\sum_{x\in X}\sum_{y\in Y}\sum_{z\in Z} p(x, y, z) \log p(x, y|z). \tag{5.19}$$

These four definitions correspond to the joint entropy of the three sources $X$, $Y$, $Z$ (Eq. (5.16)), the mutual information of the three sources $X$, $Y$, $Z$ (Eq. (5.17)), the entropy of source $Z$ given the known entropy of $X$, $Y$ (Eq. (5.18)), and the joint entropy of $X$, $Y$ given the known entropy of $Z$ (Eq. (5.19)). The last two definitions are seen to involve conditional probabilities of higher orders, namely, $p(z|x, y)$ and $p(x, y|z)$, which are easily determined from the generalization of Bayes's theorem.[5] Other entropies of the type $H(X; Y|Z)$ and $H(X|Y; Z)$ are more tricky to determine from the above definitions. But we can resort in all confidence to the equivalence relations and the corresponding two-source or three-source Venn diagrams shown in Fig. 5.2. Indeed, a straightforward observation of the diagrams leads to the following correspondences:

$$H(X; Y|Z) = H(X; Y) - H(Z), \tag{5.20}$$

$$H(X|Y; Z) = H(X) - H(Y; Z). \tag{5.21}$$

Finally, the Venn diagrams (with the help of Eq. (5.15) make it possible to establish the following properties for $H(X, Y|Z)$ and $H(X|Y, Z)$. The first chain rule is

$$H(X, Y|Z) = H(X|Z) + H(Y|X, Z), \tag{5.22}$$

[5] As we have

$$p(x, y, z) = p(z|x, y)p(x)p(y) \to p(z|x, y) \equiv p(x, y, z)/[p(x)p(y)]$$
$$\text{and } p(x, y, z) = p(x, y|z)p(z) \to p(x, y|z) =\equiv p(x, y, z)/p(z).$$

which is easy to memorize if a condition $|z$ is applied to both sides of the definition of joint entropy, $H(X, Y) = H(X) + H(Y|X)$. The second chain rule,

$$H(X, Y|Z) = H(Y|Z) + H(X|Y, Z), \tag{5.23}$$

comes from the permutation in Eq. (5.22) of the sources $X, Y$, since the joint entropy $H(X, Y)$ is symmetrical with respect to the arguments.

The lesson learnt from using Venn diagrams is that there is, in fact, little to memorize, as long as we are allowed to make drawings! The only general rule to remember is:

$H(U|Z)$ is equal to the entropy $H(U)$ defined by the source $U$ (for instance, $U = X, Y$ or $U = X; Y$) minus the entropy $H(Z)$ defined by the source $Z$, the reverse being true for $H(Z|U)$. But the use of Venn diagrams require us not to forget the unique correspondence between the ensemble or Boolean operators ($\cup \cap \neg$) and the separators (, ; |) in the entropy-function arguments.

## 5.3 Relative entropy

In this section, I introduce the notion of *distance* between two event sources and the associated concept of *relative entropy*.

The mathematical concept of *distance* between two real variables $x, y$ is familiarly known as the quantity $d = |x - y|$. For two points $A, B$ in the plane, with coordinates $(x_A, y_A)$ and $(x_B, y_B)$, respectively, the distance is defined as $d = \sqrt{(x_A - x_B)^2 + (y_A - y_B)^2}$.

More generally, any definition of distance $d(X, Y)$ between two entities $X, Y$ must obey four axiomatic principles:

(a) *Positivity*, $d(X, Y) \geq 0$;
(b) *Symmetry*, $d(X, Y) = d(Y, X)$;
(c) *Nullity for self*, $d(X, X) = 0$;
(d) *Triangle inequality*, $d(X, Z) \leq d(Y, X) + d(Y, Z)$.

Consider now the quantity $D(X, Y)$, which we define as

$$D(X, Y) = H(X, Y) - H(X; Y). \tag{5.24}$$

From the visual reference of the Venn diagrams in Fig. 5.2 (top), it is readily verified that $D(X, Y)$ satisfies at least the first three above distance axioms (a), (b), and (c). The last axiom, (d), or the triangle inequality, can also be proven through the Venn diagrams when considering three ensembles $X, Y, Z$, which I leave here as an exercise. Therefore, $D(X, Y)$ represents a distance between the two sources $X, Y$.

It is straightforward to visualize from the Venn diagrams in Fig. 5.2 (top) that

$$D(X, Y) = H(X|Y) + H(Y|X), \tag{5.25}$$

or using the definition for the conditional entropies, Eq. (5.6), and grouping them together,

$$D(X, Y) = -\sum_{x\in X}\sum_{y\in Y} p(x, y)\log[p(x|y)p(y|x)]. \tag{5.26}$$

We note from the above definition of distance that the weighted sum involves the joint distribution $p(x, y)$.

The concept of distance can now be further refined. I shall introduce the *relative entropy* between two PDFs, which is also called the *Kullback–Leibler (KL) distance* or the *discrimination*. Consider two PDFs, $p(x)$ and $q(x)$, where the argument $x$ belongs to a single source $X$. The *relative entropy*, or *KL distance*, is noted $D[p(x)||q(x)]$ and is defined as follows:

$$D[p(x)\|q(x)] = \left\langle \log\frac{p(x)}{q(x)}\right\rangle_p = \sum_{x\in X} p(x)\log\frac{p(x)}{q(x)}. \tag{5.27}$$

In this definition, the continuity limit $\varepsilon\log(\varepsilon) \equiv 0\,(\varepsilon \to 0)$ applies, while, by convention, we must set $\varepsilon\log(\varepsilon/\varepsilon') \equiv 0(\varepsilon, \varepsilon' \to 0)$.

The relative entropy is not strictly a distance, since it is generally not symmetric ($D(p\|q) \neq D(q\|p)$, as the averaging is based on the PDF $p$ or $q$ in the first argument) and, furthermore, it does not satisfy the triangle inequality. It is, however, zero for $p = q$ ($D(q\|q) = 0$), and it can be verified as an exercise that it is *always nonnegative* ($D[p\|q] \geq 0$).

An important case of relative entropy is where $q(x)$ is a *uniform distribution*. If the source $X$ has $N$ events, the uniform PDF is thus defined as $q(x) \equiv 1/N$. Replacing this definition in Eq. (5.27) yields:

$$\begin{aligned} D[p(x)\|q(x)] &= \sum_{x\in X} p(x)\log\frac{p(x)}{1/N} = \log N\sum_{x\in X} p(x) + \sum_{x\in X} p(x)\log p(x) \\ &\equiv \log N - H(X). \end{aligned} \tag{5.28}$$

Since the distance $D(p\|q)$ is always nonnegative, it follows from the above that $H(X) \leq \log N$. This result shows that the entropy of a source $X$ with $N$ elements has $\log N$ for its upper bound, which (in the absence of any other constraint) represents the *entropy maximum*. This is consistent with the conclusion reached earlier in Chapter 4, where I addressed the issue of maximizing entropy for discrete sources.

Assume next that $p$ and $q$ are *joint distributions* of two variables $x, y$. Similarly to the definition in Eq. (5.27), the *relative entropy* between the two joint distributions is:

$$\begin{aligned} D[p(x, y)\|q(x, y)] &= \left\langle \log\frac{p(x, y)}{q(x, y)}\right\rangle_p \\ &= \sum_{x\in X}\sum_{y\in Y} p(x, y)\log\frac{p(x, y)}{q(x, y)}. \end{aligned} \tag{5.29}$$

We note that the expectation value is computed over the distribution $p(x, y)$, and, therefore, $D[p(x, y)\|q(x, y)] \neq D[q(x, y)\|p(x, y)]$ in the general case.

The relative entropy is also related to the *mutual information*. Indeed, recalling the definition of mutual information, Eq. (2.37), we get

$$H(X;Y) = \sum_{x \in X} \sum_{y \in Y} p(x, y) \log \frac{p(x, y)}{p(x)p(y)} \equiv D[p(x, y) \| p(x)p(y)], \tag{5.30}$$

which shows that the mutual information between two sources $X, Y$ is the relative entropy (or KL distance) between the joint distribution $p(x, y)$ and the distribution product $p(x)p(y)$. Since the relative entropy (or KL distance) is always nonnegative ($D(.\|.) \geq 0$), it follows that *mutual information is always nonnegative* ($H(X;Y) \geq 0$).

This is consistent with results obtained in Section 5.2. Indeed, recalling the chain rules in Eq. (5.13) and combining them with the property of nonnegativity (obtained in section herein) yields:

$$H(X;Y) = H(X) - H(X|Y) = H(Y) - H(Y|X) \geq 0, \tag{5.31}$$

which thus implies the two inequalities

$$\begin{cases} H(X|Y) \leq H(X) \\ H(Y|X) \leq H(Y). \end{cases} \tag{5.32}$$

The above result can be summarized under the fundamental conclusion, which has already been established: *conditioning reduces entropy*.

Thus, given two sources $X, Y$, the information we obtain from source $X$ given the prior knowledge of the information from $Y$ is less than or equal to that available from $X$ alone, meaning that entropy has been reduced by the fact of conditioning. The strict inequality applies in the case where the two sources have nonzero mutual information ($H(X;Y) > 0$). If the two sources are disjoint, or made of independent events, then the equality applies, and conditioning from $Y$ has no effect on the information of $X$.

Next, I shall introduce another definition, which is that of *conditional relative entropy*, given the *joint distributions* $p$ and $q$ over the space $\{X, Y\}$:

$$\begin{aligned} D[p(y|x) \| q(y|x)] &= \left\langle \log \frac{p(y|x)}{q(y|x)} \right\rangle_p \\ &= \sum_{x \in X} \sum_{y \in Y} p(x, y) \log \frac{p(y|x)}{q(y|x)}, \end{aligned} \tag{5.33}$$

noting that averaging is made through the joint distribution, $p(x, y)$. This definition is similar to that of the relative entropy for joint distributions, Eq. (5.29), except that it applies here to *conditional* joint distributions.

From Eqs. (5.27), (5.29), and (5.33), we can derive the following relation, or *chain rule*, between the relative entropies of two single-variate distributions $p(x), q(x)$ and their two-variate or joint distributions $p(x, y), q(x, y)$:

$$D[p(x, y) \| q(x, y)] = D[p(x) \| q(x)] + D[p(y|x) \| q(y|x)]. \tag{5.34}$$

This chain rule can be memorized if one decomposes $p(x, y)$ and $q(x, y)$ through Bayes's theorem, i.e., $p(x, y) = p(y|x)p(x)$ and $q(x, y) = q(y|x)q(x)$, and rearranges the four factors into two distance terms $D[*\|*]$. A similar chain rule thus also applies with $x$ and $y$ being switched in the right-hand side.

The usefulness of relative entropy and conditional-relative entropy can only be appreciated at a more advanced IT level, which is beyond the scope of these chapters. To satisfy a demanding student's curiosity, however, an illustrative example concerning the *second law of thermodynamics* is provided. For this, one needs to model the time evolution of probability distributions, which involves *Markov chains*. The concept of Markov chains and its application to the second law of thermodynamics are described in Appendix D, which is to be regarded as a tractable, bur somewhat advanced topic.

## 5.4 Exercises

**5.1** (B): Given all possible events $a \in A$, and the conditional probability $p(a|b)$ for any event $b \in B$, demonstrate the summation property:

$$\sum_{a \in A} p(a|b) = 1.$$

**5.2** (M): If two sources $X, Y$ represent independent events, then prove that
(a) $H(X, Y) = H(X) + H(Y)$,
(b) $H(X|Y) = H(X)$ and $H(Y|X) = H(Y)$.

**5.3** (M): Prove the entropy chain rules, which apply to the most general case:

$$\begin{cases} H(X, Y) = H(Y|X) + H(X) \\ H(X, Y) = H(X|Y) + H(Y). \end{cases}$$

**5.4** (M): Prove that for any single-event source $X$, $H(X|X) = 0$.

**5.5** (M): Prove that for any single-event source $X$, $H(X; X) = H(X)$.

**5.6** (M): Establish the following three equivalent properties for mutual information:

$$\begin{aligned} H(X; Y) &= H(X) - H(X|Y) \\ &= H(Y) - H(Y|X) \\ &= H(X) + H(Y) - H(X, Y). \end{aligned}$$

**5.7** (M): Two sources, $A = \{a_1, a_2\}$ and $B = \{b_1, b_2\}$, have a joint probability distribution defined as follows:

$$\begin{aligned} p(a_1, b_1) = 0.3,\ p(a_1, b_2) = 0.4, \\ p(a_2, b_1) = 0.1,\ p(a_2, b_2) = 0.2. \end{aligned}$$

Calculate the joint entropy $H(A, B)$, the conditional entropy $H(A|B)$ and $H(B|A)$, and the mutual information $H(A; B)$.

**5.8** (B): Show that $D(X, Y)$, as defined by

$$D(X, Y) = H(X, Y) - H(X; Y),$$

can also be expressed as:

$$D(X, Y) = H(X|Y)H(Y|X).$$

**5.9** (T): With the use of the Venn diagrams, prove that

$$D(X, Y) = H(X, Y) - H(X; Y)$$

satisfies the triangle inequality,

$$d(X, Z) \leq d(Y, X) + d(Y, Z),$$

for any sources $X, Y, Z$.

**5.10** (M): Prove that the Kullback–Leibler distance between two PDFs, $p$ and $q$, is always nonnegative, or $D[p\|q] \geq 0$. Clue: assume that $f = p/q$ satisfies Jensen's inequality,

$$\langle f(u) \rangle \leq f(\langle u \rangle).$$

# 6 Differential entropy

So far, we have assumed that the source of random symbols or events, $X$, is *discrete*, meaning that the source is made of a set (finite or infinite) of discrete elements, $x_i$. To such a discrete source is associated a PDF of discrete variable, $p(x = x_i)$, which I have called $p(x)$, for convenience. In Chapter 4, I have defined the source's *entropy* according to Shannon as $H(X) = -\sum_i p(x_i) \log p(x_i)$, and described several other entropy variants for multiple discrete sources, such as *joint entropy*, $H(X, Y)$, *conditional entropy*, $H(X|Y)$, *mutual information*, $H(X; Y)$, *relative entropy* (or *Kullback–Leibler [KL] distance*) for discrete single or multivariate PDFs, $D[p\|q]$, and *conditional relative entropy*, $D[p(y|x)\|q(y|x)]$. In this chapter, we shall expand our conceptual horizons by considering the entropy of *continuous* sources, to which are associated PDFs of continuous variables. It is referred to as *differential entropy*, and we shall analyze here its properties as well as those of all of its above-listed variants. This will require the use of some integral calculus, but only at a relatively basic level. An interesting issue, which is often overlooked, concerns the comparison between discrete and differential entropies, which is nontrivial. We will review different examples of differential entropy, along with illustrative applications of the KL distance. Then we will address the issue of maximizing differential entropy (finding the optimal PDF corresponding to the upper entropy bound under a set of constraints), as was done in Chapter 4 in the discrete-PDF case.

## 6.1 Entropy of continuous sources

A continuous source of random events is characterized by a real variable $x$ and its associate PDF, $p(x)$, which is a continuous "density function" over a certain domain $X$. Several examples of continuous PDFs were described in Chapter 2, including the uniform, exponential, and Gaussian (normal) PDFs.

By analogy with the discrete case, one defines the Shannon entropy of a continuous source $X$ according to:

$$H(X) = -\int_X p(x) \log p(x) \mathrm{d}x. \tag{6.1}$$

As in the discrete case, the logarithm in the integrand is conventionally chosen in base two, the unit of entropy being *bit/symbol*. In some specific cases, the definition could preferably involve the natural logarithm, which defines entropy in units of *nat/symbol*. The entropy of a continuous source, as defined above, is referred to as *differential entropy*.[1] Similar definitions apply to multivariate PDFs, yielding the *joint entropy*, $H(X, Y)$, the *conditional entropy*, $H(X|Y)$, the *mutual information*, $H(X; Y)$, and the *relative entropy* or *KL distance*, $D(p\|q)$ with following definitions:

$$H(X, Y) = -\int_X \int_Y p(x, y) \log p(x, y) \mathrm{d}x \mathrm{d}y, \tag{6.2}$$

$$H(X|Y) = -\int_X \int_Y p(x, y) \log p(x|y) \mathrm{d}x \mathrm{d}y, \tag{6.3}$$

$$H(X; Y) = \int_X \int_Y p(x, y) \log \frac{p(x, y)}{p(x)p(y)} \mathrm{d}x \mathrm{d}y, \tag{6.4}$$

$$D[p(x)\|q(x)] = \int_X p(x) \log \frac{p(x)}{q(x)} \mathrm{d}x, \tag{6.5}$$

$$D[p(y|x)\|q(y|x)] = \int_X \int_Y p(x, y) \log \frac{p(y|x)}{q(y|x)} \mathrm{d}x \mathrm{d}y. \tag{6.6}$$

The above definitions of differential entropies appear to come naturally as the generalization of the discrete-source case (Chapter 5). However, such a generalization is nontrivial, and far from being mathematically straightforward, as we shall see!

A first argument that revealed the above issue was provided by Shannon in his seminal paper.[2] Our attention is first brought to the fact that, unlike discrete sums, integral sums are defined within a given coordinate system. Considering single integrals, such as in Eq. (6.1), and using the relation $p(x)\mathrm{d}x = p(y)\mathrm{d}y$, we have

$$\begin{aligned} H(Y) &= -\int_X p(y) \log[p(y)] \mathrm{d}y \\ &= -\int_X p(x) \frac{\mathrm{d}x}{\mathrm{d}y} \log\left[p(x) \frac{\mathrm{d}x}{\mathrm{d}y}\right] \mathrm{d}y \\ &= -\int_X p(x) \left[\log p(x) + \log \frac{\mathrm{d}x}{\mathrm{d}y}\right] \mathrm{d}x \\ &= H(X) - \int_X p(x) \log\left[\frac{\mathrm{d}x}{\mathrm{d}y}\right] \mathrm{d}x \\ &\equiv H(X) - C. \end{aligned} \tag{6.7}$$

[1] The term "differential" comes from the fact that the probability $P(x \leq y)$ is defined as $P(x \leq y) = \int_{x_{\min}}^{y} p(x)\mathrm{d}x$, meaning that the PDF $p(x)$, if it exists, is positive and is integrable over the interval considered, $[x_{\min}, y]$, and is also the derivative of $P(x \leq y)$ with respect to $y$, i.e.,

$$p(x) = \frac{\mathrm{d}}{\mathrm{d}y} \int_{x_{\min}}^{y} p(x)\mathrm{d}x.$$

[2] C. E. Shannon, A mathematical theory of communication. *Bell Syst. Tech. J.*, **27** (1948), 79–423, 623–56. This paper can be freely downloaded from http://cm.bell-labs.com/cm/ms/what/shannonday/paper.html.

This result establishes that the entropy of a continuous source is defined within some arbitrary constant shift $C$, as appearing in the last RHS term.[3] For instance, it is easy to show that the change of variable

$$y = ax,$$

($a$ being a positive constant) translates into the entropy shift

$$H(Y) = H(X) + \log a.$$

The shift can be positive or negative, depending on whether $a$ is greater or less than two (or than e, for the natural logarithm), respectively.

A second observation is that differential entropy can be *nonpositive* ($H(X) \leq 0$) unlike in the discrete case where entropy is always strictly positive ($H(X) > 0$).[4] A straightforward illustration of this upsetting feature is provided by the continuous uniform distribution over the real or finite interval $x \in [a, b]$;

$$p(x) = \frac{1}{b-a}. \tag{6.8}$$

From the definition in Eq. (6.1), it is easily computed that the differential entropy is $H(X) = \log_2(b-a)$. The entropy is zero or negative if $b - a \leq 2$ (or $b - a \leq$ e, for the natural logarithm).

The difference between discrete and differential entropy becomes trickier if one attempts to connect the two definitions. This issue is analyzed in Appendix E. The comparison consists in sampling, or "discretizing" a continuous source, and calculating the corresponding discrete entropy, $H''$. The discrete entropy is then compared with the differential entropy, $H'$. As the Appendix shows, the two entropies are shifted from each other by a constant $n = -\log \Delta$, or $H'' = H' + n$. The constant $n$ corresponds to the number of extra bits required to discretize the continuous distribution with bins of size $\Delta = 1/2^n$. The key issue is that in the integral limit $n \to \infty$ or $\Delta \to 0$, this constant is infinite! The conclusion is that an infinite discrete sum, corresponding to the integral limit $\Delta \to 0$, is associated with an infinite number of degrees of freedom, and hence, with *infinite entropy*.

The differential entropy defined in Eq. (6.1), as applying to any integration domain $X$, is always finite, however. This proof is left as an (advanced) exercise. In the experimental domain, calculations are always made in discrete steps. If the source is continuous, the experimentally calculated entropy is not $H'$, but its discretized version $H''$, which assumes a heuristic sampling bin $\Delta$. The constant $n = -\log \Delta$ should then be subtracted from the discrete entropy $H''$ in order to obtain the differential entropy $H'$, thus ensuring reconciliation between the two concepts.

[3] For multivariate expressions, e.g., two-dimensional, this constant is

$$C = \int_X \int_Y p(x, y) \log\left[J\left(\frac{x}{y}\right)\right] \mathrm{d}x\mathrm{d}y,$$

where $J(x/y)$ is the Jacobian matrix.

[4] Excluding the case of a discrete distribution where $p(x_i) = \delta_{ij}$ (Kronecker symbol).

The various relations and properties that were obtained in the discrete case between entropy, conditional entropy, joint entropy, relative entropy, and mutual information also apply to the continuous case. To recall, for convenience, these relations and properties are:

$$\begin{cases} H(X,Y) = H(Y|X) + H(X) \\ H(X,Y) = H(X|Y) + H(Y), \end{cases} \tag{6.9}$$

$$\begin{aligned} H(X;Y) &= H(X) - H(X|Y) \\ &= H(Y) - H(Y|X) \\ &= H(X) + H(Y) - H(X,Y), \end{aligned} \tag{6.10}$$

$$\begin{aligned} D(X,Y) &= H(X,Y) - H(X;Y) \\ &= H(X|Y) + H(Y|X), \end{aligned} \tag{6.11}$$

$$D(p\|q) \geq 0, \tag{6.12}$$

$$H(X;Y) = D[p(x,y)\|p(x)p(y)] \geq 0, \tag{6.13}$$

$$D[p(x,y)\|q(x,y)] = D[p(x)\|q(x)] + D[p(y|x)\|q(y|x)]. \tag{6.14}$$

In particular, it follows from Eqs. (6.13) and (6.10) that $H(X|Y) \leq H(X)$ and $H(Y|X) \leq H(Y)$, with equality if the sources are independent. Thus for continuous sources, *conditioning reduces differential entropy*, just as in the discrete case.

I shall describe next a few examples of PDFs lending themselves to closed-form, or analytical definitions of differential entropy, with some illustrations regarding relative entropy and KL distance.

Consider first the *continuous uniform distribution*, defined over the real interval of width $u = b - a$, Eq. (6.8). As we have seen, the corresponding bit/symbol entropy is $H_{\text{uniform}} = \log_2(b-a)$. We note that the entropy $H_{\text{uniform}}$ is nonpositive if $u \leq 1$. The result shows that the entropy of a continuous uniform distribution of width $u$ increases as the logarithm of $u$. In the particular cases where $u = 2^N$, with $N$ being integer, then $H_{\text{uniform}} = N$ bit/symbol. In the limit $u, N \to \infty$, or $p(x) = 1/u = 2^{-N} \to 0$, corresponding to a uniform distribution of infinite width, the entropy is infinite, corresponding to an infinite number of degrees of freedom for source events having themselves an infinite information, $I(x) = -\log[p(x)] = N$ bits. We thus observe that, short of any constraints on the definition interval, or PDF mean, the entropy is unbounded, or $H_{\text{uniform}} \to +\infty$ as $N \to \infty$.

We may compute the relative entropy, or KL distance, between any continuous PDF $p(x)$ defined over the domain $X = [a,b]$ with $u = (b-a) = 2^N$ and the corresponding uniform PDF, which we now call $q(x) = 1/u = 2^{-N}$, according to Eq. (6.5):

$$\begin{aligned} D[p(x)\|q(x)] &= \int_X p(x) \log \frac{p(x)}{q(x)} \mathrm{d}x \\ &= \int_X p(x) \log\left[2^N p(x)\right] \mathrm{d}x \\ &= \int_X p(x) \left[\log(2^N) + \log p(x)\right] \mathrm{d}x \end{aligned} \tag{6.15}$$

$$= N \int_X p(x)\mathrm{d}x + \int_X p(x) \log p(x)\mathrm{d}x$$
$$\equiv N - H_p(X),$$

where $H_p(X)$ is the entropy of the source $X$ and $p(x)$ is the associated PDF. Since the KL distance is nonnegative, $D(p\|q) \geq 0$, the above result implies that $H_p(X) \leq N$, meaning that $N = \log u$ represents the upper bound, or maximum entropy for all PDF defined over $X$.

Consider next the *continuous-exponential distribution*, which is defined as $p(x) = \lambda e^{-\lambda x}$ with $x \geq 0$ (nonnegative real), $\lambda = 1/\langle x \rangle \equiv 1/\tau$ a strictly positive constant, and $\tau = \langle x \rangle$ the PDF mean, also called the $1/\mathrm{e}$ lifetime (Chapter 2). An elementary calculation (see Exercises) while taking the natural logarithm in the entropy definition yields the *nat/symbol* entropy:

$$H_{\mathrm{exp}} = \ln\left(\frac{\mathrm{e}}{\lambda}\right) = \ln(\mathrm{e}\tau) = 1 + \ln \tau. \tag{6.16}$$

Thus, the entropy of a continuous-exponential distribution $H_{\mathrm{exp}}$ can be negative or null if $\tau \leq 1/\mathrm{e}$ and increases as the logarithm of the lifetime $\tau$. Short of any constraints on the mean or lifetime, the entropy is unbounded, or $H_{\mathrm{exp}} \to +\infty$ as $\tau \to \infty$.

Assuming $\tau = \mathrm{e}^{N-1}$, where $N$ is an integer, we obtain $H_{\mathrm{exp}} = N$ nat/symbol, which compares with the entropy of the uniform distribution, $H_{\mathrm{uniform}} = N$ bit/symbol, for a source defined over any real interval $x \in [a, b]$ having the width $u = (b - a) = 2^N$. We may compute the relative entropy, or KL distance, between any continuous PDF $p(x)$ defined over the domain $x \geq 0$ and the exponential PDF, which we now call $q(x) = \lambda \mathrm{e}^{-\lambda x}$, according to Eq. (6.5), and using natural logarithms:

$$\begin{aligned}
D[p(x)\|q(x)] &= \int_X p(x) \log \frac{p(x)}{q(x)} \mathrm{d}x \\
&= \int_X p(x) \ln\left[\frac{p(x)}{\lambda \mathrm{e}^{-\lambda x}}\right] \mathrm{d}x \\
&= \int_X p(x) \{\ln[p(x)] - \ln(\lambda) + \lambda x\}\, \mathrm{d}x \\
&= \int_X p(x) \ln[p(x)]\mathrm{d}x - \ln(\lambda) \int_X p(x)\mathrm{d}x + \lambda \int_X x p(x)\mathrm{d}x \\
&\equiv \lambda \langle x \rangle_p - \ln(\lambda) - H_p(X),
\end{aligned} \tag{6.17}$$

where $H_p(X)$ is the entropy of $X$ with PDF $p(x)$ and $\langle x \rangle_p$ is the PDF mean value. From this result, we can determine the relative entropy or KL distance between two exponential PDFs, i.e., $p_{\mathrm{exp}}(x) = \mu \mathrm{e}^{-\mu x} (\mu > 0)$ and $q_{\mathrm{exp}}(x) = \lambda \mathrm{e}^{-\lambda x}$, which gives

$$D[p_{\mathrm{exp}}(x)\|q_{\mathrm{exp}}(x)] = \frac{\lambda}{\mu} - \ln(\lambda) - \ln\left(\frac{\mathrm{e}}{\mu}\right) \equiv \frac{\lambda}{\mu} - 1 + \ln\left(\frac{\mu}{\lambda}\right). \tag{6.18}$$

Setting $u = \lambda/\mu$, we have $D = u - \ln(\mathrm{e}u)$. It can easily be checked that $D$ reaches a minimum of zero for $u = 1$ ($\lambda = \mu$, or $p(x) = q(x)$), as expected. In the limits $\lambda \to 0$ ($p(x) \to \phi(x)$) or $\mu \to 0$ ($p(x) \to \phi(x)$), the KL distance $D$ is infinite, but $\phi(x)$, which emulates a step or uniform distribution over the interval $x \in [0, +\infty]$, is not a valid PDF. It is left as an exercise to study an exponential PDF variant that is defined over a finite interval $x \in [0, m]$, and to reconcile the infinite limit with the uniform-distribution case.

Consider next the case of the *normal* or *Gaussian distribution*, $p(x)$, as defined in Eq. (2.21). Using first natural logarithms, the differential entropy calculation yields:

$$
\begin{aligned}
H_{\text{normal}} &= -\int p(x) \ln\left\{\frac{1}{\sqrt{2\pi\sigma^2}} \exp\left[-\frac{(x-x_0)^2}{2\sigma^2}\right]\right\} \mathrm{d}x \\
&= \int p(x) \frac{(x-x_0)^2}{2\sigma^2} \mathrm{d}x - \int p(x) \ln\left(\frac{1}{\sqrt{2\pi\sigma^2}}\right) \mathrm{d}x \\
&= \frac{1}{2\sigma^2} \langle (x-x_0)^2 \rangle + \frac{1}{2} \ln(2\pi\sigma^2) \int p(x) \mathrm{d}x \\
&= \frac{1}{2\sigma^2} \sigma^2 + \frac{1}{2} \ln(2\pi\sigma^2) \equiv \frac{1}{2} \ln(2\pi e \sigma^2),
\end{aligned}
\tag{6.19}
$$

where we used the property $\langle (x-x_0)^2 \rangle = \langle x^2 \rangle - \langle x_0^2 \rangle = \sigma^2$. In bit/symbol units, this result gives

$$
H_{\text{normal}} \approx 2.047 + 0.72 \log \sigma^2 = 2.047 + 1.44 \log \sigma. \tag{6.20}
$$

We thus find that the entropy of the normal or Gaussian distribution can be negative or zero if $\sigma \leq 1/\sqrt{2\pi \mathrm{e}}$, and increases as the logarithm of its variance or deviation. Short of any constraints on the variance $\sigma$, the entropy is unbounded, or $H_{\text{normal}} \to +\infty$ as $\sigma \to \infty$.

We may compute the relative entropy, or KL distance, between any continuous PDF $p(x)$ and the normal or Gaussian PDF, $q_{\text{normal}}(x)$, according to Eq. (6.5) with mean and variance $(x_0, \sigma_0^2)$ and using natural logarithms:

$$
\begin{aligned}
D[p(x) \| q_{\text{normal}}(x)] &= \int_X p(x) \ln\left[\frac{p(x)}{q_{\text{normal}}(x)}\right] \mathrm{d}x \\
&\equiv -H_p(X) + \frac{1}{2} \text{in}(2\pi\sigma_0^2) + \frac{1}{2\sigma_0^2} \left[\langle x^2 \rangle_p - 2x_0 \langle x \rangle_p + x_0^2\right],
\end{aligned}
\tag{6.21}
$$

where $H_p(X)$ is the entropy of the source $X$ with associated PDF $p(x)$. We may apply the above definition to the case where $p(x)$ is also a normal or Gaussian PDF with mean and variance $(x_1, \sigma_1^2)$ and obtain (see Exercises):

$$
D\left[p_{\text{normal}}(x) \| q_{\text{normal}}(x)\right] = \frac{1}{2}\left[\frac{\sigma_1^2}{\sigma_0^2} + \left(\frac{x_0 - x_1}{\sigma_0}\right) - 1 - \ln\left(\frac{\sigma_1^2}{\sigma_0^2}\right)\right]. \tag{6.22}
$$

As expected, the relative entropy or KL distance $D$ vanishes if the two PDFs are identical, $(x_0, \sigma_0^2) = (x_1, \sigma_1^2)$. Setting $u = \sigma_1^2/\sigma_0^2$ and $v = [(x_0 - x_1)/\sigma_0]^2$, we have $D = (u^2 - \ln u - 1 + v)/2$. It is easily checked that $D$ reaches a minimum for $u = 1$, namely, $D_{\min} = v/2$, which is zero for $v = 0$. Also, $D$ is infinite in the two limiting cases $u \to 0(\sigma_1 \to 0)$ or $u \to +\infty(\sigma_0 \to 0)$, which correspond to $p(x) \to \delta(x - x_1)$, or $q(x) \to \delta(x - x_0)$, respectively, where $\delta(x)$ is the *Dirac* or *delta distribution*.[5]

There are plenty of continuous-PDF types for which differential entropy comes out in closed-form or analytical expressions. A list of the most important PDFs comprises:[6]

- Uniform,
- Normal or Gaussian,
- Exponential,
- Rayleigh,
- Beta,
- Cauchy,
- Chi,
- Chi-squared,
- Erlang,
- F,
- Gamma,
- Laplace,
- Logistic,
- Log-normal,
- Maxwell–Boltzmann,
- Generalized normal,
- Pareto,
- Student's $t$,
- Triangular,
- Weibull,
- Multivariate normal.

## 6.2 Maximum entropy (continuous source)

In Chapter 4, we addressed the problem of finding the maximum entropy. For discrete sources, this problem can be resolved with or without making restricting assumptions regarding the PDF, i.e., with or without assuming constraints. Without constraints, the straightforward result is that maximum entropy is reached with the *uniform (discrete)* PDF with $N$ equiprobable events, i.e., $p = 1/N$ yielding $H_{\max} = \log N$. Introducing

[5] The Dirac or delta distribution, which is not a function (except to the physicists!), satisfies the properties

$$\delta(x-u) = 0 \text{ for } x \neq u, \lim_{x \to u} \delta(x-u) = +\infty, \int_{-\infty}^{+\infty} \delta(x)\mathrm{d}x = 1, \text{ and } \int_{-\infty}^{+\infty} \delta(x-u) f(x)\mathrm{d}x = f(u).$$

[6] See table and PDF definition links at bottom of web page http://en.wikipedia.org/wiki/Differential_entropy. See also T. M. Cover and J. A. Thomas, *Elements of Information Theory* (New York: John Wiley & Sons, 1991), Table 16.1, pp. 486–7.

constraints, such as the PDF mean, the entropy is found to be maximized by the *discrete-exponential*, or *Bose–Einstein* distribution. We are now facing a similar problem, this time with continuous distributions and associated differential entropy. The previous examples in this chapter have illustrated that without constraints (such as the PDF mean), entropy is, in this case, virtually unbounded. It just takes a normal (Gaussian), a uniform, or an exponential PDF with the appropriate limits to reach entropy infinity!

In the continuous-PDF case, the issue of maximizing entropy is, therefore, bounded to the introduction of constraints. Given such constraints, the problem consists in determining the *optimal* PDF for which entropy is maximized, this maximum entropy representing the upper bound for all possible PDFs under the same constraints. Albeit seemingly abstract or fuzzy, this last statement should become crystal clear after we have gone through the optimization problem.

The continuous PDF, which maximizes entropy under certain given constraints, can be found through the *Lagrange-multipliers method*, as previously used in Chapter 4, and described in Appendix C for the discrete PDF case. Here we are going to explore the method more extensively, but let us not be scared, this development should remain relatively basic and accessible, if the reader or student has made it this far!

First of all, let us see how we can find the PDF, which may obey some set of assumed constraints. Let us also make the problem the most general possible. Let $g_0(x), g_1(x), \ldots, g_n(x)$ be the assumed sets of $n+1$ constraint functions of the variable $x$ and their mean values and $\langle g_0\rangle, \langle g_1\rangle, \ldots, \langle g_n\rangle$ defined by

$$\langle g_k\rangle = \int_X g_k(x)p(x)\mathrm{d}x. \tag{6.23}$$

It is implicitly assumed that $\langle g_0\rangle, \langle g_1\rangle, \ldots, \langle g_n\rangle$ are all finite, which is an issue to be discussed later on. The first constraint $g_0(x) = 1$ is there to impose the PDF normalization condition, or $\int_X p(x)\mathrm{d}x = 1$. Using the entropy definition in Eq. (6.1), the method consists in minimizing the "functional" $f$ as defined as:

$$f = \int_X \left[ -p(x)\log p(x) + \sum_{k=0}^{n} \lambda_k g_k(x)p(x) \right] \mathrm{d}x, \tag{6.24}$$

where $\lambda_0, \lambda_1, \ldots, \lambda_n$ are the unknown Lagrange multipliers. Minimizing $f$ yields:

$$\begin{aligned}
\frac{\mathrm{d}f}{\mathrm{d}p} &= \frac{\mathrm{d}}{\mathrm{d}p}\int_X \left[ -p(x)\log p(x) + \sum_{k=0}^{n} \lambda_k g_k(x)p(x) \right] \mathrm{d}x \\
&= \int_X \frac{\mathrm{d}}{\mathrm{d}p}\left[ -p(x)\log p(x) + \sum_{k=0}^{n} \lambda_k g_k(x)p(x) \right] \mathrm{d}x \\
&= \int_X \left[ -\log p(x) - 1 + \lambda_0 + \sum_{k=1}^{n} \lambda_k g_k(x) \right] \mathrm{d}x = 0 \qquad (6.25) \\
&\Rightarrow -\log p(x) - 1 + \lambda_0 + \sum_{k=1}^{n} \lambda_k g_k(x) = 0,
\end{aligned}$$

which yields the general, but unique, PDF solution:

$$p(x) = \exp\left[\lambda_0 - 1 + \sum_{k=1}^{n} \lambda_k g_k(x)\right]. \tag{6.26}$$

Nicely, the solution is exclusively defined by the discrete set of parameters $\lambda_0, \lambda_1, \ldots, \lambda_n$, which we must now find by using the $n+1$ constraints defined in Eq. (6.23). Integrating in Eq. (6.24), and equating the result with $f = 0$, yields the corresponding expression of maximum entropy:

$$H_{\max} = 1 - (\lambda_0 + \lambda_1 \langle g_1 \rangle + \lambda_2 \langle g_2 \rangle +, \ldots, \lambda_n \langle g_n \rangle). \tag{6.27}$$

As previously mentioned, however, an implicit condition is that the means $\langle g_0 \rangle, \langle g_1 \rangle, \ldots, \langle g_n \rangle$ be finite. This condition is of consequence in the possible values for the parameters $\lambda_0, \lambda_1, \ldots, \lambda_n$, as we see next.

Assume, for instance, that our constraint functions are defined by $g_k(x) = x^k$ for all $k$, or $\langle g_k \rangle = \langle x^k \rangle$, which corresponds to the PDF moments of order $k$. The PDF solution takes then the form:

$$p(x) = A \exp(\lambda_1 x + \lambda_2 x^2 + \lambda_3 x^3 + \lambda_4 x^4 + \lambda_5 x^5 + \cdots + \lambda_n x^n), \tag{6.28}$$

with $A = \exp(\lambda_0 - 1)$. Substituting the PDF solution into the constraints in Eq. (6.23) yields:

$$\begin{cases} \int_X p(x)\mathrm{d}x = A \int_X \exp(\lambda_1 x + \lambda_2 x^2 + \lambda_3 x^3 + \lambda_4 x^4 + \lambda_5 x^5 + \cdots + \lambda_n x^n) x = 1 \\ \int_X x^k p(x)\mathrm{d}x = A \int_X x^k \exp(\lambda_1 x + \lambda_2 x^2 + \lambda_3 x^3 + \lambda_4 x^4 + \lambda_5 x^5 + \cdots + \lambda_n x^n)\mathrm{d}x = \langle x^k \rangle \\ (k = 1, \ldots, n). \end{cases} \tag{6.29}$$

For any of the integrals involved in Eq. (6.29) to converge, a certain number of additional conditions should be met, depending on the integration domain $X$ and the number of constraints $n + 1$:

(a) $X$ is finite, or $X = [x_{\min}, x_{\max}] \Rightarrow$ no extra conditions;
(b) $X$ has an infinite upper or lower bound:
  (i) $X = [x_{\min}, +\infty] \Rightarrow$ condition is $\lambda_n < 0$,
  (ii) $X = [-\infty, x_{\max}] \Rightarrow$ condition is $\lambda_n < 0$.
(c) $X$ is unbounded, or $X = [-\infty, +\infty]$:
  (i) $n$ is even $\Rightarrow$ condition is $\lambda_n < 0$,
  (ii) $n$ is odd $\Rightarrow$ condition are $\lambda_n = 0$ and $\lambda_{n-1} < 0$.

The above shows that in the situations corresponding to (a), (b), and (c(ii)), it is possible to find the set of Lagrange parameters $\lambda_0, \lambda_1, \ldots, \lambda_n$ that define the optimal PDF and its maximal entropy. Except for $n \leq 2$ (as we shall see), this takes an iterative numerical resolution in which the above assumptions are explicitly introduced from the start in order to ensure convergence. The situation (c(ii)) poses a specific problem. Indeed, if we must set $\lambda_n = 0$, we are left with $n + 1$ equations (Eq. (6.29)) to determine the

$n$ Lagrange parameters $\lambda_0, \lambda_1, \ldots, \lambda_{n-1}$. In the general case, this problem cannot be solved. However, numerical computation makes it possible to find an upper bound to the entropy under said constraints, and to determine an ad-hoc PDF whose entropy approaches it with arbitrary accuracy. While the exact optimal PDF solution does not exist, it can yet be approximated with very high numerical precision.

Consider next two simple cases of interest which correspond to $g_k(x) = x^k$ in the two situations (b(i)) with $n = 1$ and $x_{\min} = 0$, and (c(i)) with $n = 2$, respectively.

In the first case, the PDF and maximum entropy solutions are:

$$p(x) = \exp(\lambda_0 - 1 + \lambda_1 x), \tag{6.30}$$

$$H_{\max} = 1 - (\lambda_0 + \lambda_1 \langle x \rangle), \tag{6.31}$$

with the convergence condition $\lambda_1 < 0$.

The solution defined in Eq. (6.30) corresponds to the *continuous-exponential distribution*. It is easily established from the constraints in Eq. (6.23) that $\lambda_0 = 1 - \ln\langle x \rangle$ and $\lambda_1 = -1/\langle x \rangle$, which gives $p(x) = (1/\langle x \rangle)\exp(-x/\langle x \rangle)$ and $H_{\max} = \ln(\mathrm{e}\langle x \rangle)$, as the expected result, see Eq. (6.16).

In the second case ($n = 2$), the PDF and maximum entropy solutions are:

$$p(x) = \exp(\lambda_0 - 1 + \lambda_1 x + \lambda_2 x^2), \tag{6.32}$$

$$H_{\max} = 1 - \left(\lambda_0 + \lambda_1 \langle x \rangle + \lambda_2 \langle x^2 \rangle\right), \tag{6.33}$$

where $\lambda_2 < 0$.

Compared to the previous case, the analytical computation of the three Lagrange parameters $\lambda_0, \lambda_1, \ldots, \lambda_2$ from the constraints in Eq. (6.23) is a bit tedious, albeit elementary, which I leave as a good math-training exercise. We eventually find

$$\begin{cases} \exp(\lambda_0 - 1) = \dfrac{1}{\sqrt{2\pi\sigma^2}} \exp\left[-\dfrac{\langle x \rangle^2}{2\sigma^2}\right] \\ \lambda_1 = \dfrac{\langle x \rangle}{\sigma^2} \\ \lambda_2 = -\dfrac{1}{2\sigma^2}. \end{cases} \tag{6.34}$$

Substitution of the above Lagrange parameters in Eqs. (6.32)–(6.33) yields:

$$p(x) = \frac{1}{\sqrt{2\pi\sigma^2}} \exp\left[-\frac{(x - \langle x \rangle)^2}{2\sigma^2}\right] \tag{6.35}$$

and with $\sigma^2 = \langle x^2 \rangle - \langle x \rangle^2$, as defined in *nat/symbol*:

$$H_{\max} = \ln\sqrt{2\pi e \sigma^2}. \tag{6.36}$$

We recognize in Eq. (6.35) the definition of the Gaussian (normal) distribution. It is a fundamental result in information theory that the optimal continuous PDF for which entropy is maximized, under a constraint in the first two moments $(\langle x \rangle, \sigma^2)$, is precisely the Gaussian (normal) distribution. The entropy of any continuous PDF (as defined over the event space $X = [-\infty, +\infty]$), therefore, has $H_{\max} = \ln\sqrt{2\pi \mathrm{e}\sigma^2}$ as an upper bound.

The issue of *entropy maximization* is key to the solution of several problems in statistical physics and engineering (and many other fields as well) where the PDF is unknown, while a certain set of constraints $g_k(x)$, $\langle g_k \rangle$ are known from real-life or experimental observation. See previous discussion in Chapter 4 regarding the *maximum entropy principle*.

## 6.3 Exercises

**6.1** (B): Calculate the differential entropy of the continuous source defined over the real interval $X = [a, b]$, with $a \neq b$, assuming a uniform PDF.

**6.2** (B): Calculate the differential entropy associated with an exponential PDF.

**6.3** (T): Show that the differential entropy

$$H(X) = -\int_X p(x) \log p(x) \mathrm{d}x$$

with $0 < p(x) \leq 1$, is always finite.

**6.4** (T): Assume the probability distribution

$$p(x) = \alpha \mathrm{e}^{-\lambda x},$$

which is defined for $x \in [0, m], m, \alpha, \lambda$ being nonnegative real.

(a) Determine the PDF constant $\alpha$, the PDF mean $\langle x \rangle$ and the PDF entropy $H$.

(b) Show that in the limit $\lambda \to 0$ the PDF becomes uniformly distributed with $p(x) \to 1/m$.

(c) Calculate the relative entropy, or KL distance $D[p\|q]$ between the two PDFs,

$$q(x) = \alpha \mathrm{e}^{-\lambda x}$$

and

$$p(x) = \beta \mathrm{e}^{-\mu x}.$$

(d) Determine the KL distance $D[p\|q]$ in the limit $\lambda \to 0$ or $q(x) \to 1/m$.

(e) Show that in the limit $\lambda, \mu \to 0$ the distance $D[p\|q]$ vanishes.

**6.5** (M): Show that the relative entropy, or KL distance, between any continuous PDF $p(x)$ and the normal (Gaussian PDF),

$$q(x) = \frac{1}{\sqrt{2\pi\sigma^2}} \exp\left[-\frac{(x - x_0)^2}{2\sigma^2}\right]$$

is given by

$$D[p(x)\|q(x)] = -H_p(X) + \frac{1}{2}\ln(2\pi\sigma^2) + \frac{1}{2\sigma^2}\left[\langle x^2 \rangle_p - 2x_0\langle x \rangle_p + x_0^2\right],$$

where $H_p(X)$ is the source entropy associated with the distribution $p(x)$.

**6.6** (B): Determine the two parameters $\lambda_0, \lambda_1$ and the domain $x \in X$ to make the function

$$p(x) = \exp(\lambda_0 - 1 + \lambda_1 x)$$

a probability distribution.

**6.7** (T): determine the three parameters $\lambda_0, \lambda_1, \lambda_2$ to make the function

$$p(x) = \exp(\lambda_0 - 1 + \lambda_1 x + \lambda_2 x^2)$$

a probability distribution over the domain $x \in X = [-\infty, +\infty]$.

*Clues*: where appropriate, effect the variable substitution $x = y - \lambda_1/(2\lambda_2)$, set $\lambda_2 = -\alpha(\alpha > 0)$, and use the result

$$\int_{-\infty}^{+\infty} \exp(-\alpha x^2) \equiv \sqrt{\pi/\alpha}.$$

# 7 Algorithmic entropy and Kolmogorov complexity

This chapter will take us into a world very different from all that we have seen so far concerning Shannon's information theory. As we shall see, it is a strange world made of virtual computers (universal Turing machines) and abstract axioms that can be demonstrated without mathematics merely by the force of logic, as well as relatively involved formalism. If the mere evocation of Shannon, of information theory, or of entropy may raise eyebrows in one's professional circle, how much more so that of Kolmogorov complexity! This chapter will remove some of the mystery surrounding "complexity," also called "algorithmic entropy," without pretending to uncover it all. Why address such a subject right here, in the middle of our description of Shannon's information theory? Because, as we shall see, algorithmic entropy and Shannon entropy meet conceptually at some point, to the extent of being asymptotically bounded, even if they come from totally uncorrelated basic assumptions! This remarkable convergence between fields must make integral part of our IT culture, even if this chapter will only provide a flavor. It may be perceived as being somewhat more difficult or demanding than the preceding chapters, but the extra investment, as we believe, is well worth it. In any case, this chapter can be revisited later on, should the reader prefer to keep focused on Shannon's theory and move directly to the next stage, without venturing into the intriguing sidetracks of *algorithmic information theory*.

## 7.1 Defining algorithmic entropy

The concept of *information*, which has been described extensively in Chapter 3, also evolved beyond Shannon's view, and independently of his classical theory. An alternative definition, which is referred to as *algorithmic information*, is attributed to G. Chaitin, R. Solomonoff, and A. Kolmogorov.[1] Algorithmic information opened the way to the field of *algorithmic information theory* (AIT).

From AIT's perspective, any source event $x$ is treated as an object variable. The event may consist in any symbol sequence, whether random or deterministic. The focus of AIT is *not* on the source $X$, the statistical ensemble of all possible events or symbolic sequences, but on this particular sequence $x$.

[1] See, for instance: http://home.mira.net/~reynella/debate/informat.htm and useful links therein.

Entering the AIT domain simply requires one to acknowledge the following three basic definitions:

(a) The *complexity* $K(x)$ is defined as *the smallest size of a program* $q(x)$ *necessary to generate the sequence* $x$.
(b) Such a program is a finite set of binary instructions with a length of $|q(x)|$ bits.
(c) The program can be implemented by a *Turing machine* (TM).

The smallest program size, $\min|q(x)| = K(x)$, which is called the *complexity* of $x$, is equivalently referred to as *algorithmic information content*, *algorithmic complexity*, *algorithmic entropy*, or *Kolmogorov complexity* (also called *Kolmogorov–Chaitin* complexity and sometimes noted $KC(x)$ instead).[2] The Turing machine, which is named after its inventor, A. Turing,[3] can be viewed as the most elementary and ideal implementation of a computer, and is, as we shall see, of infinite computation power (due to speed, not memory size).

To clarify and develop all of the above, we ought first to understand what a Turing machine looks like and how it works; this is addressed in the next section. Then we will investigate Kolmogorov complexity and its properties. Interestingly, I will show that Kolmogorov complexity is in fact incomputable! Finally, I will show that Kolmogorov complexity and Shannon's entropy are symptotically bounded, a most remarkable feature considering the previous property.

## 7.2 The Turing machine

The Turing machine (TM) is an abstract, idealized, or paper version of the simplest and most elementary computing device. As Fig. 7.1 illustrates, it consists of a *tape* and a *read/write head*. The tape is of indefinite length and it contains a succession of memory *cells*, into which are written the bit symbols 0 or 1.[4] By convention, cells that were never written or were left blank are read as containing the 0 bit. The tape can be made to move left or right by one cell at a time.

The operations of the head and tape are defined by a *table of instructions* $\{I_1, I_2, \ldots, I_N\}$ of finite size $N$, also called an *action table*. The action table is not a program to be read sequentially. Rather, it is a set of instructions corresponding to different possibilities to be considered by the machine, as I shall clarify. At each instruction step, the machine's head is initially positioned at a single tape cell.

[2] See, for instance: http://en.wikipedia.org/wiki/Kolmogorov_complexity, http://szabo.best.vwh.net/kolmogorov.html.

[3] See, for instance: http://plato.stanford.edu/archives/spr2002/entries/turing-machine/, http://en.wikipedia.org/wiki/Universal_Turing_machine, www.turing.org.uk/turing/, www.alanturing.net/, www.cs.usfca.edu/www.AlanTuring.net/turing_archive/pages/Reference%20Articles/What%20is%20a%20Turing20Machine.html.

[4] More generally, the symbols that can be put into the cells, including a conventional blank symbol, could be selected from any finite alphabet.

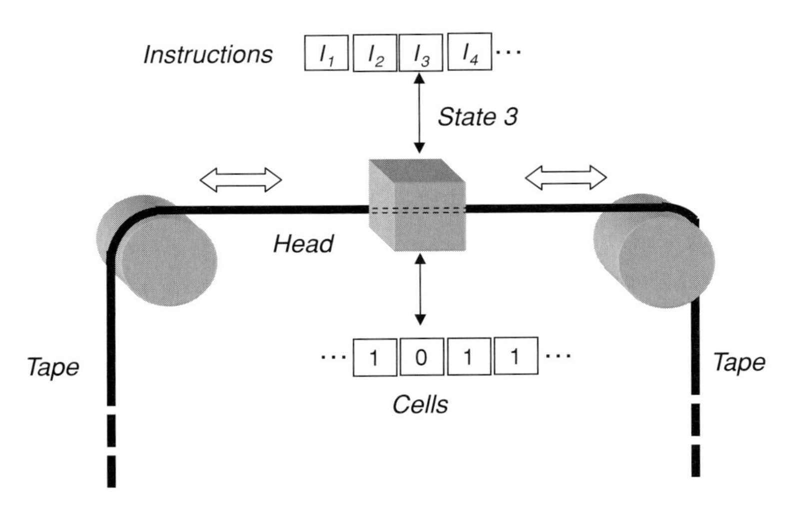


**Figure 7.1** Schematic representation of Turing machine.

The machine is said to be in an *input state* $s_i$, which corresponds to a specific instruction in the action table. This instruction tells the machine to perform three basic operations altogether:

(a) Given the cell's content, what new content (namely, 1 or 0) is to be written into the cell;
(b) In what direction the tape should be moved, namely, left or right;
(c) Into what new state $s_j$ the machine should be moved.

For instance, given the input state $s_1$ the corresponding instruction could be:

(a) If reading 0 then write 1,
(b) Move tape to the left,
(c) Go into state $s_3$.

If the cell reading is 0, the machine changes the contents, then moves according to the two other actions, (b) and (c). If the cell reading is 1, then it halts. To cover this other possibility, a second instruction can be introduced into the action table, for instance;

(a) If reading 1 then write 0,
(b) Move tape right,
(c) Go into state $s_2$.

These two sets of instructions in the action table can be summarized as follows:

$$\begin{aligned} s_1; 0 &\to 1; \mathrm{L}; s_3 \\ s_1; 1 &\to 0; \mathrm{R}; s_2. \end{aligned} \tag{7.1}$$

Since the instructions move the TM into new states $s_2, s_3$, the instructions corresponding to input states $s_2, s_3$ should also be found in the table. There are no restrictions concerning

**Table 7.1** Example of action table for Turing machine.

| Input state | Change of cell contents | Move tape (R, right; L, left) | Output state |
|---|---|---|---|
| $s_1$ | $0 \to 1$ | R | $s_2$ |
| $s_2$ | $0 \to 1$ | R | $s_3$ |
| $s_3$ | $0 \to 0$ | R | $s_4$ |
| $s_4$ | $0 \to 1$ | R | $s_5$ |
| $s_5$ | $0 \to 1$ | L | $s_1$ |

**Table 7.2** Tape changes during Turing machine computation, using the action table defined in Table 7.1. The input tape is blank, representing a string of 0 bits or 0000000 ... The initial position of the head is on the leftmost zero bit. At each step, the position of the head on the tape is shown by the bold, underlined bit. The machine halts at step 6, leaving the codeword 1101100... on the output tape.

| Step | Input state | Tape contents | Output state |
|---|---|---|---|
| **1** | $s_1$ | $\underline{\mathbf{0}}\,0\,0\,0\,0\,0\,0\ldots$ | $s_2$ |
| **2** | $s_2$ | $1\,\underline{\mathbf{0}}\,0\,0\,0\,0\,0\ldots$ | $s_3$ |
| **3** | $s_3$ | $1\,1\,\underline{\mathbf{0}}\,0\,0\,0\,0\ldots$ | $s_4$ |
| **4** | $s_4$ | $1\,1\,0\,\underline{\mathbf{0}}\,0\,0\,0\ldots$ | $s_5$ |
| **5** | $s_5$ | $1\,1\,0\,1\,\underline{\mathbf{0}}\,0\,0\ldots$ | $s_1$ |
| **6** | $s_1$ | $1\,1\,0\,\underline{\mathbf{1}}\,1\,0\,0\ldots$ | ? |
| Halt | | | |

the possibility of returning to a state previously used as an input (e.g., $s_1; 0 \to 1; \mathrm{R}; s_1$). If there is no output state corresponding to the instruction, the machine halts. The same happens if the cell reading is not a case covered by the action table given a new input state.

To summarize, the tape and machine keep on moving, as long as the combination of initial state and cell reading have a matching definition in the action table. The final output of the program is the bit sequence written on the tape, which is left at the point where the machine halted. To illustrate how the TM works, consider next a few program examples including the basic operations of addition, subtraction, multiplication, and division.

### Example 7.1

This is a TM program that creates the 5-bit sequence 11011 out of an initially blank tape with equivalent sequence $\underline{0}0000\ldots$ Table 7.1 shows the action table instructions and Table 7.2 shows the step-by-step implementation on the tape. It is assumed that the first initial state is $s_1$ and that the head is located at the leftmost cell, as underlined. As Table 7.2 illustrates, each of the computation steps is characterized by a left-to-right tape move (or equivalently, right-to-left head move, as viewed from the tape) with a possible change in the cell contents. It is seen that the machine halts at step 6, because there is no instruction in the action table concerning an input state $s_1$ with a cell reading of 1.

**Table 7.3** Example of action table for Turing machine defining the addition of two numbers in the unary system.

| Input state | Change of cell contents | Move tape | Output state |
|---|---|---|---|
| $s_1$ | $1 \to 1$ | R | $s_2$ |
| $s_2$ | $0 \to 0$ | R | $s_3$ |
| | $1 \to 1$ | R | $s_2$ |
| $s_3$ | $0 \to 0$ | L | $s_5$ |
| | $1 \to 0$ | L | $s_4$ |
| $s_4$ | $0 \to 1$ | R | $s_2$ |

### Example 7.2

This is a TM program that *adds* two integers. Such an operation requires one to use the *unary* system. In the unary system, the only symbol is 1 (for instance) and an integer number $n$ is represented by a string of $n$ consecutive 1s. The decimal numbers 3 and 7 are, thus, represented in unary as 111 and 1111111, respectively. Adding unary numbers is just a matter of concatenating the two strings, namely, $3 + 7 \equiv 111 + 1111111 = 1111111111 \equiv 10$. Because we are manipulating two different numbers, we need an extra blank symbol, 0, to use as a delimiter to show where each of the numbers ends. Thus, the two numbers, 2 and 3, must be noted as the unary string 1101110. We use this string as the TM input tape.

The action table and the step-by-step implementation of its instructions are shown in Table 7.3 and 7.4, respectively. It is seen that the TM halts at step 14. The output tape then contains the string $1111100 \equiv 5$, which is the expected result. It is left to the reader as an exercise to show that the program also works with other integer choices, meaning that this is a true adding machine! We note that the program only requires four states with two symbols, which illustrates its simplicity. The output state $s_5$, which is undefined, forces the TM to halt. Analyzing the tape moves line by line in Table 7.4, we can better understand how the algorithm works.

- Basically, each time the head sees a 1 in the cell, the only action is to look at the next cell to the right, and so on, until a 0 input is hit (here, at step 3). The head then inspects the next cell with input state $s_3$.
- If the result is another 0, then the output state is $s_5$, meaning that the machine must halt at the next step, the output tape remaining as it is. This is because reading two successive 0 means that either (a) the second operand is equal to zero, or (b) that the addition is completed.
- If the result is a 1, then the program flips the two cell values from 01 to 10, and the same inspection as before is resumed, until another 0 is hit (here, at step 12). If the sequence 00 is identified from there, the TM halts, meaning that the addition is completed.

**Table 7.4** Tape changes during Turing machine computation, using the action table in Table 7.3, which defines the addition of two numbers. The input tape is set to 1101110, which in unary corresponds to the two numbers $2 \equiv 11$ and $3 \equiv 111$, as delimited by the blank symbol 0. The initial position of the head is on the leftmost bit. The output tape is seen to contain the string $1111100 \equiv 5$.

| Step | Input state | Tape contents | Output state |
|---|---|---|---|
| **1** | $s_1$ | **1**101110 | $s_2$ |
| **2** | $s_2$ | 1**1**01110 | $s_2$ |
| **3** | $s_2$ | 11**0**1110 | $s_3$ |
| **4** | $s_3$ | 110**1**110 | $s_4$ |
| **5** | $s_4$ | 11**0**0110 | $s_2$ |
| **6** | $s_2$ | 111**0**110 | $s_3$ |
| **7** | $s_3$ | 1110**1**10 | $s_4$ |
| **8** | $s_4$ | 111**0**010 | $s_2$ |
| **9** | $s_2$ | 1111**0**10 | $s_3$ |
| **10** | $s_3$ | 11110**1**0 | $s_4$ |
| **11** | $s_4$ | 1111**0**00 | $s_2$ |
| **12** | $s_2$ | 11111**0**0 | $s_3$ |
| **13** | $s_3$ | 111110**0** | $s_5$ |
| **14** | $s_5$ | 11111**0**0 | Halt |

Performing the addition of large numbers only increases the number of computation steps and the overall computation time. But since the tape is infinite, there is no problem of overflow. The TM is, however, not capable of handling infinite numbers, or real numbers with an infinity of decimal places, as I shall discuss later.

### Example 7.3

This concerns the operation of *subtraction*. It is performed in a way similar to addition, e.g., $7 - 3 \equiv 1111111 - 111 = 1111 \equiv 4$. Assume first that the first operand $(i)$ is greater than or equal to the second operand $(j)$. Table 7.5 shows an action table performing the subtraction and outputting $k = i - j$. It is left as an exercise to verify how the proposed algorithm (call it *Sub*0) effectively works.

If the first operand $(i)$ is strictly less than the second $(j)$, or $i < j$, we must use a subtraction algorithm different from *Sub*0; call it *Sub*1. The problem is now to determine which out of *Sub*0 or *Sub*1 is relevant and must be assigned to the TM. We must then define a new algorithm, call it *Comp*, which compares the two operands $(i, j)$ and determines which of the two cases, $i \geq j$ or $i < j$, applies. It is left as an exercise to determine an action table for implementing *Comp*. Based on the output of *Comp*, the TM can be then assigned to perform either *Sub*0 or *Sub*1, with the additional information from *Comp* that the result of the subtraction is nonnegative $(k = i - j \geq 0)$ or strictly negative $(k = i - j \leq 0)$. Note that the case $i = j$ $(k = 0)$ is implicit in the output of *Sub0*.

**Table 7.5** Example of action table for Turing machine defining the subtraction of two numbers $i$, $j$ ($i \geq j$) in unary system.

| Input state | Change of cell contents | Move tape | Output state |
|---|---|---|---|
| $s_1$ | $0 \to 0$ | R | $s_2$ |
| | $1 \to 1$ | R | $s_1$ |
| $s_2$ | $0 \to 0$ | R | $s_2$ |
| | $1 \to 0$ | L | $s_3$ |
| | $B \to B$ | R | $s_5$ |
| $s_3$ | $0 \to 0$ | L | $s_3$ |
| | $1 \to 0$ | R | $s_2$ |

I will not develop the analysis further here, or a full TM program for subtraction. Suffice it to realize that *Sub*1 can be implemented as a mirror algorithm of *Sub*0, proceeding from right to left, or performing the operation $k' = j - i$. The information printed by *Comp* on the tape tells whether the output of the subtraction is positive or negative (noting that the case $i = j$ ($k = 0$) is implicit in the output of *Sub*0). The lesson learnt is that the TM is capable of performing subtraction in the general case, should the algorithm and corresponding action table cover all possible cases, with the TM halting on completion of the task. It is left as an exercise to analyze how such a program can be implemented. This exercise, and the preceding ones, should generate a feeling of intimacy with the basic TM. As we will see, the TM reserves some surprises, and this is why we should enjoy this exploration.

---

### Example 7.4

This concerns the operation of *multiplication*. A TM program performing the multiplication of two integers,[5] is shown in Table 7.6.

The directions of the move are either > (right) or < (left). As in previous examples, the two numbers to multiply and written in the input tape must be expressed in the unary system, with the convention that integer $n$ is represented by a string of $n + 1$ 1s ($0 \equiv 1$, $1 \equiv 11$, $2 \equiv 111$, $3 \equiv 1111$, etc.). Here, I have introduced a supplemental start instruction (1,_,1,_,>) to position the head to the second cell to the right and the TM in input state 1. This instruction ensures that the program can be run with various simulators available on the Internet.[6] It is seen from Table 7.6 that the multiplication program requires as much as 15 states, plus the undefined "halt" state. Also, the TM must be able to handle (or read and write) six different symbols including the "blank" underscore, which corresponds to an absence of read/write action (and is not to be confused with

[5] See www.ams.org/featurecolumn/archive/turing_multiply_code.html and the line-by-line comments to analyze the corresponding algorithm. This program can be tested in Turing machine simulators available on the Internet, see for instance: http://ironphoenix.org/tril/tm/, www.turing.org.uk/turing/scrapbook/tmjava.html, www.cheransoft.com/vturing/index.html.

[6] See, for instance: http://ironphoenix.org/tril/tm/, www.turing.org.uk/turing/scrapbook/tmjava.html, www.cheransoft.com/vturing/index.html.

**Table 7.6** Action table for Turing machine multiplication program. The table contains 38 instructions using 15 states numbered 1, ..., 15, plus the halt state (H), and six symbols noted 1, X, Y, Z, W plus the underscore _, which stands for blank. The instruction nomenclature is "input state, read symbol, output state, write symbol, move right (>) or left (<)." The numbers $m$, $n$ to be multiplied are entered in the input tape under the string form _M_N_, where $M$ and $N$ are strings of $m+1$ and $n+1$ successive 1s, respectively. The head is initially located to the leftmost (blank) symbol. The output tape takes the form _M_N_P_, where $P$ is the unary representation of $mn = p$ ($P$ is a string of $p+1$ successive 1s).

| | | | |
|---|---|---|---|
| **1** | 1, _, 1, _, > | **20** | 10, _, 11, _, < |
| **2** | 1, 1, 2, W, > | **21** | 10, 1, 12, Z, > |
| **3** | 2, 1, 2, 1, > | **22** | 12, 1, 12, 1, > |
| **4** | 2, _, 3, _, > | **23** | 12, _, 13, _, > |
| **5** | 3, 1, 4, Y, > | **24** | 13, 1, 13, 1, > |
| **6** | 4, 1, 4, 1, > | **25** | 13, _, 14, 1, < |
| **7** | 4, _, 5, _, > | **26** | 14, 1, 14, 1, < |
| **8** | 5, _, 6, 1, < | **27** | 14, _, 14, _, < |
| **9** | 6, _, 6, _, < | **28** | 14, Z, 10, Z, > |
| **10** | 6, 1, 6, 1, < | **29** | 14, Y, 10, Y, > |
| **11** | 6, Z, 6, Z, < | **30** | 11, 1, 11, 1, < |
| **12** | 6, Y, 6, Y, < | **31** | 11, _, 11, _, < |
| **13** | 6, X, 7, X, > | **32** | 11, Z, 11, 1, < |
| **14** | 6, W, 7, W, > | **33** | 11, Y, 6, Y, < |
| **15** | 7, _, 8, _, > | **34** | 8, Y, 8, 1, < |
| **16** | 7, 1, 9, X, > | **35** | 8, _, 8, _, < |
| **17** | 9, 1, 9, 1, > | **36** | 8, X, 8, 1, < |
| **18** | 9, _, 9, _, > | **37** | 8, W, 15, 1, < |
| **19** | 9, Y, 10, Y, > | **38** | 15, _, H, _, > |

an erase/blank character). The table also shows that 38 different instructions, or *TM transitions*, are necessary to complete the multiplication of the two input numbers. To recall, such an operation can be performed with any numbers of any size, without limitation in size, since the TM tape is theoretically infinite in length. I will come back to this point later.

---

### Example 7.5

This concerns the operation of *division*, which is also relatively simple to implement with a TM. Given two integers $m, n$, the task is to find the quotient $[m/n]$ and the remainder $r = m - n[m/n]$, where the brackets indicate the integer part of the ratio $m/n$. A first test consists in verifying that $n > 0$, i.e., that the field containing $n$ has at least two 1s (with the convention $1_{\text{decimal}} \equiv 110_{\text{unary}}$). This test can be performed through the above-described *Comp* program, which is also derived in an exercise. If $n = 0$, a special character meaning "zero divide" must be output to the tape and the TM must be put into the halt state. A second test consists in checking whether $m \geq n$ or $m < n$, which is again performed by the *Comp* program. In the second case ($m < n$), the TM must output $[m/n] = 0$ and $r = 0$, and then halt. In the first case ($m \geq n$), the TM must compute the value of the quotient $[m/n] \geq 1$. Computing $[m/n]$ is a matter of iteratively removing a string copy of $n$ from $m$, while keeping count of the number of removals, $k$. This is equivalent to performing the subtraction $m - n$ through the above-described program *Sub*0, iterated as many times ($k$) as the result of *Sub*0 is nonnegative. The remainder $r$ is the number of 1s remaining in the original $m$ field after this iteration. See Exercise 7.5 for a practical illustration of this algorithm.

---

A key application of division concerns data conversion from unary to binary (or to decimal). It is a simple exercise to determine the succession of divisions by powers of two (or powers of ten) required to retrieve the digits (or decimals) of a given unary number. Similarly, multiplication and addition can be used to convert data from binary (or decimal) into unary. It is given as an exercise to determine an algorithm involving a succession of multiplications and additions necessary to perform such a conversion, given a binary (or decimal) number. The key conclusion is that both input and output TM tapes can be formatted into any arbitrary $M$-ary number representation (assuming that the head has the capability to read/write the corresponding $M$ symbol characters), while the unary system is used for the TM computations. The TM is, thus, able to manage all number representations for input and output, which illustrates that it is a truly fundamental computing machine, as we may realize increasingly through the rest of this chapter.

It is most recommended to explore the previously referenced Internet sites, which present live TM simulators, and thus make it possible to visualize that Turing machine at work. Some of these simulators include Java applets, from which ready-made TM programs may be selected through pop-up menus and then executed. Some of these sites make it possible to create and test one's own programs, with the option to run them either step by step or at some preselected speed. It is quite fascinating to observe in real-time the execution of a TM program, as it conveys the impression of an elementary intelligence. But let us not be mistaken, the intelligence is not that of the machine, but of the human logic that built the action table!

We may wonder how long it takes the TM to perform big operations, or operations involving big operands. Actually, we ought to measure it not by mere execution time, but by the number of required TM transitions from start to halt states. Such transitions are the elementary tape moves from one TM state to the next TM state, passing through one cell at a time. Recalling that the TM is not a physical machine, it can be made with arbitrarily small size and arbitrarily high speed, so that the execution time for any single transition is not a relevant parameter. The number of states defined in the action table provides a measure of the complexity of the algorithm (the word "complexity" is not chosen here by chance, as we shall see later on). But as we know, the TM can use any of these states several times before reaching the halt state, therefore, a simple algorithm may have a long execution length. Also, the size of the input and of the output data affects the execution length. For instance, a simple character-erasure algorithm may take as many transitions as there are symbol characters in the input data. Therefore, the relevant parameter to measure the execution length is the number of transitions required

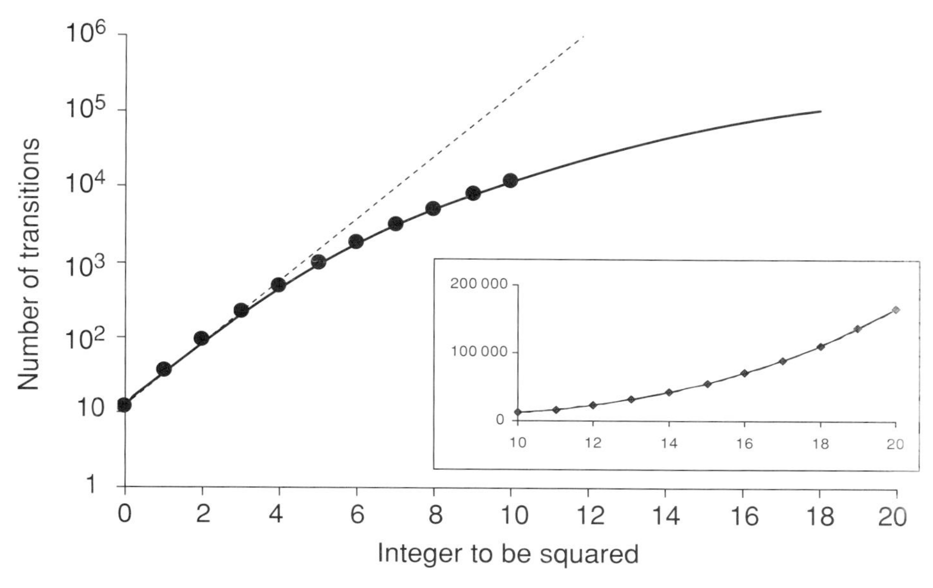


**Figure 7.2** Number of Turing-machine transitions to compute the square of small integers, according to action-table program defined in Table 7.6. The dots correspond to results obtained with a Turing-machine simulator up to $n = 10$, the smooth line corresponds to a sixth-order fitting and extrapolation polynomial up to $n = 20$. The inset shows the same data from $n = 10$ to $n = 20$ in linear plot with a fourth-order fitting.

to execute the full algorithmic computation, and depends on the sizes of the input and of the output.

To provide a sense of computation length, let us look at a practical example. Consider, for instance, the following question: "Given two numbers of maximum size $n$, how many TM transitions are required to perform their multiplication?" We may answer this question by analyzing the operation of squaring integers, as ranging from $n = 0$ to $n = 20$, and each time count the number of transitions. The action table corresponds to that of Table 7.6, with $n, n$ for the two input operands. Figure 7.2 shows a plot of the transition counts, as obtained from a TM simulator up to a maximum computation power of $n = 10$. Polynomial extrapolation makes it possible to evaluate the result up to $n = 20$, either in logarithmic or linear scales. It is observed from the figure that the number of transitions to compute $n^2$ is subexponential to the integer size $n$. For $n = 0 - 1$, the computation requires 12–35 transitions; for $n = 10$, it requires 11 825 transitions, and for $n = 20$ the result is evaluated to be 170 851. The fact that the trend is subexponential can be simply explained by the following argument: for sufficiently large $n$, the TM is essentially found in the states that move the machine towards (state 13) and back to (state 14) the right-hand side of the tape in order to store the 1 symbols defining the result. Each time the trip is one cell longer. With the relation $2\sum_{k=1}^{N} k = N(N+1)$, where $k$ is the number of cell moves to the result field to write 1 and $N = n^2$ the total number of round-trips, we can evaluate the number of transitions as asymptotically $\approx N^2 = n^4$. This is precisely why the linear curve in Fig. 7.2 is very well fitted by a fourth-order polynomial. We can thus conclude that for squaring an integer $n$, the number of TM computations (or transitions) is asymptotically $n^4$: the result of squaring integer $n$ is a string of $n^2 + 1$ symbols, but it takes $n^4$ head moves to write the result onto the tape.

In the preceding five application examples, I have illustrated how the TM can perform various types of basic computations. In all cases, integers were used. But what about *real* numbers? Can Turing machines also perform computations with real numbers? These are the questions we address next.

Considering real numbers, we ought first to distinguish two cases. First, a real number $n$ is associated with a size $|n|$, which ranges from zero to infinity. Second, some real numbers are nonintegers, and, therefore, are defined through a certain suite of decimal places. The suite can be of finite size, such as $5/4 = 1.25$ (the other decimal places being zero), or of infinite size, such as $1/3 = 0.333333\ldots$ or $13/11 = 1.181818\ldots$ (decimals, or patterns of decimals, being infinitely repeated). We will consider later on the case of numbers whose infinite decimal suites do not exhibit patterns (*irrational* and *transcendental* numbers, whose decimals may or may not be computable through a polynomial equation).

Just as with our physical computers, real numbers with finite size and finite numbers of decimal places can be handled by the TM through the *exponent format* $n = p_n \times 10^{q_n}$ or more generally through any type of *floating-point* representation. For instance, $203.5 = 2.0350 \times 10^2 = 20.350 \times 10^1 = 2035.0 \times 10^{-1}$ are three possibilities for the same real number 203.5. The case where the number of decimal places is infinite can also be processed by a TM, but only up to some truncation level, just as with any physical computer. For instance:

$$13/11 = 1.1181818\ldots \approx 1.11818 \times 10^1 \approx 1.1181 \times 10^1$$
$$\approx 1.118 \times 10^1 \ldots, \text{ and so on.}$$

Concerning *transcendental* numbers,[7] like $\pi$ or e, we may think that because of its infinite memory size, the TM could have an advantage over a physical computer. Alas, the number of decimals for such numbers being infinite, the TM would also never halt in the task of performing its infinite computation, and this points to the same problem of truncation. It is then possible to devise a TM program that outputs the decimal places of $\pi$ or e, and stores them on the TM tape, but the machine would never halt unless the program included an algorithmic counter to set a limit to the output's size. The same restriction applies to physical computers, with the difference that the limit is intrinsically defined by memory size and execution time.

It would appear from the above analysis and conclusions that *any* real number having an infinite number of decimal places, including transcendental or irrational numbers, such as $\pi$, e, $\sqrt{2}\ldots$, is TM computable, at least up to some truncation level. But it turns out to our surprise that this is not the case!

[7] Meaning numbers that are not a solution of any polynomial equation; not to be confused with *irrational numbers* (such as $\sqrt{2}$) which apply to this case (namely, here, the solution of $x^2 - 2 = 0$). See for instance http://en.wikipedia.org/wiki/Transcendental_number.

## 7.3 Universal Turing machine

Indeed, Turing showed that for some numbers there exists no table of instructions that is capable of generating the $n$th digit of its representation for any arbitrary $n$. Such numbers are said to be *incomputable numbers*. Now we are entering the core of the subject of "algorithmic complexity," where our mind is going to be further challenged!

For starters, we shall note the following:

- The number of *computable* numbers is relatively small compared with the number of *incomputable* ones;
- *Computable* numbers with an *infinite* set of decimal places can be computed by a TM program of *finite size*.

See further for a discussion regarding *computable numbers*.

We have seen that the TM needs an action table, call it a program, to perform computations. But where would such a program be physically located? The answer is that we can use the input tape as a way to store this program. It may be written in the form of a symbol string of finite length, followed by a delimiter symbol. The rest of the tape thus contains the input data string to be processed by this program. Thus, the first string in the input tape instructs the TM on how to process the second string in the input tape. Let us refer to this implementation as an enhanced TM. Nicely enough, Turing has shown that not only can such an enhanced TM be implemented, but it also has the capability to simulate the behavior of *any other TM*. For the second reason, it is called a *universal Turing machine* or UTM.

Basically, a UTM is a TM that executes a program stored in its input tape. Simple UTMs can be built with a surprisingly small number of states and small alphabet sizes. The smallest known UTM uses two states and 18 symbols, and is noted $UTM\ 2 \times 18$, or is conventionally referred to as $UTM(2, 18)$. Other well known small UTMs are $UTM(3, 10)$, $UTM(4, 6)$, $UTM(5, 5)$, $UTM(7, 4)$, $UTM(10, 3)$, and $UTM(22, 2)$, the latter being the smallest known binary UTM.

As we have seen in the previous section, and its related exercises, the TM is able to emulate all basic operators ($+, -, \times, \div, \leq, >, \ldots$) and, thus, to execute any algorithm that a real computer can perform. The internal TM variables are identified by specific symbol markers, as first initialized or assigned from the input tape data, and then processed by means of these operators. All features of a high-level computer program can be implemented by a TM: loops with conditional blocks (IF . . . , THEN . . . , ELSE . . . , GOTO), function calling, variable retrieval, or procedures or subroutines (which we can conceive as representing a subset of instructions within the action table and having specific enter and exit states).

The so-called *Church–Turing thesis* states that a UTM is capable to solve all problems that have a solution under an "algorithm" or an "effective computation method," as these terms are usually understood, provided sufficient storage space and computing time are provided.[8] The Church–Turing thesis can be formulated as follows:

Any computation process or algorithm that can be devised by a mathematician can be effectively implemented on a Turing machine.

Or, equivalently:

Anything a real computer can compute, a Turing machine can compute.[9]

As mentioned, there are, however, a few qualitative differences between a TM and a real computer. A first key difference is that the TM's memory is theoretically infinite. A real computer can be equipped with whatever amount of memory space is required. But the need to manage this finite-size memory space affects the program structure and algorithms. A real computer also efficiently accesses and manages its memory space by indexing and virtual caches. But this is difficult, if not impractical, to implement in a TM. A TM can be built with several parallel tapes (and read/write heads), but it can be shown that this only partly alleviates the problem of memory access. A second key difference is that the TM algorithms are usually more general or universal than those implemented in real computers. The main reason is that they are not bounded to data type or format limits and, therefore, do not encounter unexpected program failures or crashes. The only peculiar situation to be encountered with a TM is that where it would run forever or never halt. Given a table of instructions, and a specific input tape or all possible input tapes, the question of whether or not a TM will eventually halt is known as the *halting problem*.

Turing showed that, in the general case, *the halting problem cannot be solved*. In algorithmic information-theory jargon, the problem is said in this case to be *undecidable*. The term "undecidable" means that the YES/NO answer to the question, "Can it be solved?" is neither formally provable nor unprovable. Given all possible instructions and input data there is no universal algorithm enabling one to determine whether a TM would halt. It has been shown, however, that the halting problem is *solvable* for TMs having *fewer than four states*, the four-state case being still an open issue. The five-state case has been shown to be *almost certainly undecidable*.

It is not in the scope of this chapter to venture into the details of this rather complex and abstract domain of mathematical science. However, it is possible to provide here a

[8] See: http://en.wikipedia.org/wiki/Church-Turing_thesis, http://www.cs.princeton.edu/introcs/76universality/. In particular (as abridged from http://en.wikipedia.org/wiki/Church-Turing_thesis), an *algorithm* must satisfy the following requirements:

(a) It must consist of a finite set of simple and precise instructions described with a finite number of symbols;
(b) It must always produce the result in a finite number of steps;
(c) In principle, it could be carried out by a human being with paper and pencil;
(d) Its execution requires no intelligence of the human being except that which is needed to understand and correctly execute the instructions.

[9] http://en.wikipedia.org/wiki/Universal_Turing_machine.

simple sketch of proof of the insolvability of the halting problem in the general case, as the following illustrates.[10]

As we know, any TM is uniquely defined by its action table or program. Such a program can be encoded into a string of symbols, for instance, in the binary system with 0 and 1 bits. According to the coding rules, each individual instruction $k$ in the table takes the form of a binary number $s_k$ of defined size. Concatenating the coded instructions gives the string $s = s_1 s_2 s_3 \ldots s_k \ldots s_N\#$, where $N$ is the total number of instructions in the table, and # is an end delimiter, e.g., $s = 110101110\ldots 1000010\#$. Because of the coding rules, not every binary string picked at random defines a TM, but the contrary is true: *for any TM there is a corresponding unique binary string* $s$. We can then index each TM according to its unique string value and put these TMs into an ordered list. Although there exists an infinity of such strings, it is said that they are *countable*, just like integer numbers,[11] but unlike real numbers. Let this counter index be $n$, and call $T(n)$ the TM that uses this program. We now have an infinite catalog of TM, which can be listed as $T(1), T(2), T(3), \ldots$

Consider next the input data on the TM tape. The data also form a string of finite size, which corresponds to any binary number. Likewise, we can index and order these data into a list, forming the infinite, but countable, set of all possible input data strings, with $p$ as the index. Call $T(n, p)$ the unique Turing machine $T(n)$, which has the data string $p$ on its input tape. If $T(n, p)$ halts, call the output data $T^*(n, p)$. The halting problem is summarized by the question: "*Given a Turing machine* $T(n, p)$, *is there any rule to determine whether it will halt?*" With the previous definitions, we can now prove the undecidability of the halting problem. I shall do this by assuming that such a rule (call it the halting rule) does exist; then I will show that this assumption leads to an intractable contradiction!

With the halting rule, we know for sure whether or not a given machine $T(n, p)$ does halt. In all the favorable cases (halting TMs), we can safely run the computations to a conclusion, and tabulate the output data $T^*(n, p)$. We can, thus, fill in a two-dimensional array of $T^*(n, p)$ such as that in Table 7.7. In the unfavorable cases (nonhalting TMs), there is no point in running the machines, since the halting rule tells us that they never halt. In this case, we just enter a $\times$ mark in the corresponding array cell. This array is infinite in two dimensions, so it can never be physically completed, but this does not matter for the demonstration. The halting rule yields a Boolean answer to the statement "$T(n, p)$ halts," under the form $q =$ true or $q =$ false. We can, therefore, design another universal Turing machine $H(t)$ with a program $t$ that does the following:

- In the favorable case ($q =$ true), emulate $T(p, p)$ to compute $T^*(p, p)$; the output is then $T^*(p, p) + 1$ (to differentiate from the output of machine $T(p, p)$).
- In the unfavorable case ($q =$ false), skip calculation and output 0.

[10] See http://pass.maths.org.uk/issue5/turing/, and also http://en.wikipedia.org/wiki/Halting_problem.

[11] Integers are countable, because given any two integer bounds ($n_1, n_2 \geq n_1$), there exists a finite number $n_2 - n_1 - 1$ of integers between the two bounds.

**Table 7.7** Array representing the infinite sets of Turing machines $T(n)$ having $p$ for input data, with examples of corresponding output data $T^*(n, p)$ returned when the machine $T(n, p)$ halts, expressed as decimal numbers. The arrangements known by the rule *not* to halt are indicated with a cross. The bottom row defines a hypothetical machine $H(t)$, which always halts, returns $T^*(p, p) + 1$ if $T(p)$ halts and 0 otherwise.

| | 1 | 2 | 3 | 4 | ... | $p$ | ... |
|---|---|---|---|---|---|---|---|
| $T(1)$ | 17 | 640 | 25 | × | ... | 201 | |
| $T(2)$ | 28 | × | 33 | 11 | ... | 44 | ... |
| $T(3)$ | × | 8 | × | 73 | ... | 3000 | ... |
| $T(4)$ | 4 | × | 3 | 54 | ... | × | ... |
| ⋮ | ⋮ | ⋮ | ⋮ | ⋮ | | ⋮ | |
| $T(n)$ | 650 | 13 | × | 27 | ... | 5 | ... |
| ⋮ | ⋮ | ⋮ | ⋮ | ⋮ | | ⋮ | |
| $H(t)$ | 18 | 0 | 0 | 55 | ... | $T^*(p, p) + 1$ or 0 | ... |

The various outputs of $H(t)$, based on the values of $T(p, p)$, are shown in the bottom row of Table 7.7. The new TM $H(t)$, thus, *always* halts, since in either case ($q$ = true or $q$ = false), and over the infinite set of possible input data $p$, it has a definite output. We have, thus, identified a TM that always halts, regardless of the input. Since it is a TM, it should appear somewhere in the array. But $H(t)$ is nowhere to be found. Indeed, the machine $H(t, t)$ has an output $H^*(t, t)$, which, by definition, is different from the outputs $T^*(t, t)$ of any known machines $T(t, t)$. The paradox can be summarized as follows:

If a halting rule existed for any Turing machine, one could build another Turing machine that always halts for any possible input. But such a machine appears nowhere in the infinite list of possible Turing machines.

It can be concluded, therefore, (but only as a proof sketch) that, in the general case, *the halting rule does not exist, and that the halting problem is generally undecidable.* The fact that there is no solution to the halting problem in the general case does not preclude an intuitive, if not formal solution in any given particular case. It is trivial, indeed, to define Turing machines that for any data input (a) either certainly halt or (b) certainly never halt.[12]

At this point, we shall leave the halting problem and the puzzling universe of Turing machines, in order to resume our focus in information theory. With what we have learnt about universal Turing machines, we are now able to broaden our appreciation of information theory with the new concept of *algorithmic entropy* or *algorithmic complexity*, or *Kolmogorov complexity*.

[12] It is sufficient to include in the TM program an instruction that is certain to be called regardless the input data, and in which the output state is undefined (TM certain to halt), or an instruction creating a nested loop (TM certain never to halt).

## 7.4 Kolmogorov complexity

In Section 7.1, I introduced, without discussion, the definition of *complexity*, which I equivalently referred to as *algorithmic information content*, *algorithmic complexity*, *algorithmic entropy*, or *Kolmogorov–Chaitin complexity*.

In what is called *algorithmic information theory*, the object variable $x$ represents any symbol sequence from an events source $X$, whether random or deterministic. The underlying concept is that there must exist one, several, or an infinity of universal Turing-machine (UTM) programs, $q$, which are capable of outputting the sequence $x$, and then halt. We shall note any such programs as $q(x)$. As mentioned earlier, the *Kolmogorov complexity* of $x$, noted $K(x)$, is defined as the *minimum size* of $q(x)$, i.e.:

$$K(x) = \min |q(x)|, \tag{7.2}$$

where the symbol $|\cdot|$ stands for the *program length*, i.e., the number of bits defining the program.

The concept of Kolmogorov *complexity* nicely parallels that of *information* in Shannon's theory. Recall from Chapter 3 that, given a source event $x \in X$, the corresponding information is

$$I(x) = -\log p(x). \tag{7.3}$$

Shannon's information provides the *minimum number of bits required to describe this event*. More accurately, the minimum length of the description is given by the integer $\lceil I(x) \rceil$, where the symbol means the closest integer greater than or equal to $I(x)$. As we have seen in Chapter 4, Shannon's entropy $H(X)$ represents the statistical average of the source information, i.e.,

$$H(X) = \langle I(x) \rangle_X = \sum_{x \in X} p(x) I(x) = -\sum_{x \in X} p(x) \log p(x). \tag{7.4}$$

Thus, the entropy (or strictly $\lceil H(x) \rceil$ for a rounded integer) provides the minimum number of bits required *on average* to describe the source events. In algorithmic information theory, the Kolmogorov complexity (or algorithmic entropy) represents the minimum UTM program size required to describe the source event $x$, without any knowledge of its probability $p(x)$.

Thus, *complexity* measures the absolute information contents of an individual event from a source, in contrast to Shannon *entropy*, which measures the average information contents of the whole event source. In the following, I shall establish that *the average minimum size of programs describing a random source, $\langle K(x) \rangle_X$, is approximately equal to the source's entropy $H(X)$*. As can be authoritatively stated:[13]

> It is an amazing fact that the expected length of the shortest binary computer description of a random variable is approximately equal to its [source] entropy. Thus, the shortest computer description acts as a universal code, which is uniformly good for all probability distributions. In this sense, algorithmic complexity is a conceptual precursor of entropy.

[13] T. M. Cover and J. A. Thomas, *Elements of information theory* (New York: John Wiley & Sons, 1991).

The following examples will convey a practical understanding of Kolmogorov complexity and its close relation to Shannon entropy, as events become increasingly random:

---

**Example 7.6**

Consider the event sequence $x = ABABABABAB$ from the source alphabet $X^n$ with $X = \{A, B, C, \ldots, Z\}$. If we want to save this sequence for a permanent archive, we might just write down $ABABABABAB$ on a piece of paper. But if the sequence had 1000 characters with 500 repeats of $AB$, we would not do that. Instead, we would write down "repeat $AB$ 500 times." It then takes only 19 alphanumeric characters (including spaces) instead of 1000, to describe $x$ exactly. The statement

$$q(x) = \text{repeat } AB \text{ 500 times} \tag{7.5}$$

is thus a Turing machine program. As encoded into ASCII, which takes seven bits per alphanumeric character, the program length is $|q| = 19 \times 7 = 133$ bits, as opposed to 7000 bits if the sequence was crudely encoded "as is." But this is not the shortest possible program. For instance, we could also define it under the more concise form $q(x) = \text{A} = 1, \text{B} = 0, \text{R } 10 \times 500$, where R stands for "repeat" and $\times$ for "as many times as." The length of $q(x)$ is now 15 characters or $15 \times 7 = 105$ bits. If this program represent the absolutely minimal way to describe such a sequence, the sequence complexity is $K(x) = 105$ bits.

---

**Example 7.7**

Consider the following binary sequence:

$$\begin{aligned} x = 11.00100100\,00111111\,01101010\,10001000\,10000101\,10100011\,00001000\,11010 \\ 011\,00010011\,00011001\,10001010\,00101110\,00000011\,01110000\,01110011\,0100010 \\ 0\,10100100\,00001001\,00111000\,00100010\,00101001\,10011111\,00110001\,11010000 \\ 00001000\,00101110\,11111010\,10011000\,11101100\,01001110\,01101100\,10001001, \end{aligned}$$

which is made of 258 bits, i.e., 149 0s and 109 1s, plus a radix point (the equivalent of the decimal point in the binary system). Such a sequence has no recognizable pattern and is close to random. But, as it happens, it corresponds to the exact binary expansion of the number $\pi = 3.141592653589\ldots$ up to 256 digital places[14] (or eight decimal places). Since there are various algorithms to calculate $\pi$ up to any arbitrary precision, one could replace the above string by the shortest such algorithm. For instance, one definition of $\pi$ is given by the series

$$\pi_n = 4 \sum_{k=1}^{n} \frac{(-1)^{k+1}}{2k - 1}, \tag{7.6}$$

[14] See binary expansion of $\pi$ over 30 000 digits in www.befria.nu/elias/pi/binpi.html.

with $\pi_n \to \pi$ as $n \to \infty$. It can be shown that the error $|\pi - \pi_n|$ is of the order of $1/(2n)$, so it takes about one billion terms ($n = 10^9$) to achieve an accuracy of $1/(2 \times 10^9) = 5 \times 10^{-10}$, which is better than eight decimal places. Based on the above formula, the TM algorithm could then be packaged under the form (operand symbols being straightforward to interpret):

$$q(x) = 4 \times S[k, 1, n; P(-1; k+1)/(2 \times k - 1)], \tag{7.7}$$

which, in this general form, takes only 27 ASCII characters or 189 bits, plus the number of bits to define the truncation parameter, $n$. As the number of bits required to represent $n$ sufficiently large is approximately $\log_2 n$, we conclude that $|q| \approx 189 + \log_2(n)$. For $n = 10^9$, we have $\log_2(n) \approx 30$, and the program length is approximately $|q| \approx 189 + 30 \approx 220$ bits, which is about 40 bits shorter than the length of the sequence $x$, i.e., $|x| = 258$ bits. The TM is capable of outputting $x$ with a program length that is somewhat shorter. As it happens, the series formula in Eq. (7.6) converges relatively slowly, in contrast to several alternative definitions. With a definition having a more rapid convergence (e.g., $\approx 1/n^3$),[15] we may thus obtain the condition $|q| \ll |x|$, illustrating that the algorithmic definition must rapidly pay off for strings with relatively large sizes. It is quite a remarkable feature that long random sequences from transcendental numbers, such as $\pi$, e, $\log 2$, $\sqrt{2}, \ldots$, may be so defined in only a few bits of TM instructions, which conveys a flavor of the elegance of algorithmic information theory.

---

### Example 7.8

Consider the following string of 150 decimal symbols:

$x =$ 4390646281 3640989838 1872779754 1099387485 5579862843 0145920705
9431329445 6125451990 7325732423 7580094766 7581012661 2285404850
7226973202 5731849141 4938800048.

As in the previous example, the above string has no recognizable pattern, but it is closer to a truly random sequence (the frequencies corresponding to the $0, 1, 2, \ldots, 9$ symbols being 16, 13, 16, 13, 19, 15, 10, 15, 17, 16, respectively, which is close to 150/10). The string actually corresponds to a sampling of the decimal expansion of $\sqrt{2}$, starting from its millionth decimal.[16] While we observe that the sequence is nearly random, we know, because of its finite size, that it is computable, no matter how complex the algorithm. Therefore, there must exist a TM program that outputs $x$, given the information that $x$

[15] For instance,

$$\pi_n = 3 + 4\sum_{k=1}^{n} \frac{(-1)^{k+1}}{2k(2k+1)(2k+2)}.$$

See also: http://mathworld.wolfram.com/PiFormulas.html, www.geom.uiuc.edu/~huberty/math5337/groupe/expresspi.html, http://en.wikipedia.org/wiki/Pi.

[16] See expansion of $\sqrt{2}$ to 1 000 000 decimals in http://antwrp.gsfc.nasa.gov/htmltest/gifcity/sqrt2.1mil.

represents the decimals of $\sqrt{2}$ from $n = 1\,000\,000$ to $1\,000\,149$. It is, thus, possible to shrink the string information to less than the string's bit length, or using a four-bit code for decimals, to less than $4^*(150 + \lceil \log_2 n \rceil) = 680$ bits. This number can be viewed as representing an upper bound for the minimum program length required to output $x$.

---

**Example 7.9**

Consider the following 150-decimal string:

$$x = 1406859655\;4459325754\;8112159013\;6078068662\;8894754577\;4091431997$$
$$5387666328\;2313491092\;3281754384\;6809379687\;2005607612\;0145807590$$
$$2895743612\;9022633078\;1424279313.$$

This string is very close to a purely random sequence. It was generated from the previous example's sequence after performing two operations: (a), replacing, at random, symbols appearing more often by symbols appearing less often, so that they all appear exactly 15 times in the sequence; and (b), several successive random shufflings of blocks of symbols of decreasing size. Just considering the second operation, the number of possible random rearrangements is calculated to be $n = 150!/(10!)^{10}$, which, from Stirling's formula, yields $n \approx \mathrm{e}^{120.2} \approx 10^{52.2}$. It would take, therefore, up to some $10^{52}$ trials for a TM to find the right computing parameters to reproduce the above string, short of knowing the shuffling algorithm that was used.[17] To imagine the size of $10^{52}$, assume that a TM is capable of computing at the amazing speed of 1 billion trials per second. It is easy to find that the computation time is about $10^{35}$ years, or $10^{26}$ *billion years*! Any hint of the operations (a) and (b) could reduce the computing time to the same order as in the previous example. But since the author has chosen to destroy irremediably the records of his (b) operations, one may consider that $x$ is definitely incomputable and taken as a random sequence for which no algorithm can be found within reasonable computing time. To define the above sequence, the only alternative for the TM is to describe it symbol by symbol, or bit by bit. By definition, a truly random sequence of 150 bits should have a *descriptive complexity* of, at the very least, 150 bits. It represents the minimal length of a TM program capable of outputting this sequence.

---

The lesson learnt from these examples is that a complexity $K(x)$ can be associated to any symbol string $x$, which is defined as the minimum program length for a *universal* Turing machine to output it. What happens if one uses a *nonuniversal* Turing machine? The answer is that the complexity of $x$ *increases*. It is possible to show this property

[17] Yet, sophisticated computer algorithms may be able to reconstruct the sequence, based on the hint that it represents a shuffling of a selection of the decimals of some computable number. What we did here is similar to cryptographic coding. With such a hint, cryptographic analysis can perform various "attacks" on the encrypted sequence and eventually output the original plain sequence. It is simply a matter of computation time and memory, but Turing machines are not limited in this respect. Therefore, the proposed example is not an "incomputable" string, even if it may take a maximum of $10^{52} \approx 2^{173}$ trials to output the original sequence and TM code. Consistently, cryptographists refer to the *complexity* of an attack problem as being of size $2^n$, where $n$ is the space of possibilities to try.

through a relatively simple demonstration (by "simple," I mean "simple to understand," as long as the delicate reasoning is carefully followed).

Call $U$ a universal TM and $O$ any other machine. The programs to output $x$ are $q_U(x)$ and $q_O(x)$, respectively, which yields the outputs $U[q_U(x)] = x$ and $O[q_O(x)] = x$. The complexity of $x$ is, thus, dependent on which machine is used. By definition, we have $K_U(x) = \min_U |q_U(x)|$ and $K_O(x) = \min_O |q_O(x)|$, where the minima $\min_{U,O}$ correspond to the two different machine types.

Now comes the argument whereby the two complexities can be related. Since $U$ is universal, by definition it has the capability of simulating the behavior of $O$. Therefore, there must exist a program $s_O$, independent of $x$, which we can input to $U$ to instruct $U$ how to simulate $O$. The $U$ tape can thus be loaded with the program $s_O$, followed by the program $q_O$, so that we obtain the "simulation" output $U[s_O q_O(x)] \equiv U[q_U^*(x)] = x$. The length of the simulation program $q_U^*$ is simply $|q_U^*| = |s_O| + |q_O(x)|$, which we can write $|q_U^*| = c_O + |q_O(x)|$, where $c_O$ is a positive constant. The associated complexity is given by

$$\begin{aligned} K_U(x) &= \min_U |q_U^*(x)| = \min_U |q_U(x)| \leq \min_O |q_U^*(x)| \\ &= \min_O |q_O(x) + c_O| = \min_O |q_O(x)| + c_O \equiv K_O(x) + c_O. \end{aligned} \tag{7.8}$$

In the above, I have introduced the property $\min_U |q_U(x)| \leq \min_O |q_U^*(x)|$, which reflects the fact that the program $q_U^*$ includes the additional features of the $O$ machine simulation that $q_U(x)$ does not have. So its minimum length in $O$ is expected to be somewhat greater than the minimum length of $q_U$ in $U$. The above results thus establish the inequality:

$$K_U(x) \leq K_O(x) + c_O, \tag{7.9}$$

which shows that the minimum complexity is obtained from the universal machine $U$.

The result in Eq. (7.9) applies to all machines $O$ different from $U$ (otherwise, equality stands, along with $c_O = 0$). But what if $O$ was *also* a universal machine? The result in Eq. (7.9) still applies since there are an infinite number of possible universal machines other than $U$. Then let us use $O$ to simulate each $U$ in turn. The same reasoning as above leads to the inequality

$$K_O(x) \leq K_U(x) + c_U, \tag{7.10}$$

where $c_U$ is a positive constant independent of $x$. From Eqs. (7.9) and (7.10) it is seen that the case $K_U = K_O$ implies $c_O = c_U = 0$, meaning that $U$ and $O$ are identical machines. Assume, in the general case, that $K_U \neq K_O$. This condition also implies $c_O \neq c_U$ and $\sup(c_O, c_U) > 0$. Combining Eqs. (7.9) and (7.10) and omitting the variable $x$ for simplicity, yields:

$$c_O \geq K_U - K_O \geq -c_U, \tag{7.11}$$

which gives a final condition on the absolute complexity difference,[18]

$$0 < |K_U - K_O| \le \sup(c_O, c_U) \equiv c, \tag{7.12}$$

with $c = \sup(c_O, c_U) > 0$. Since $K_U, K_O, c_O, c_U$ are program description *lengths*, their minimum value is unity (one bit), so the double inequality

$$1 < |K_U - K_O| \le c \tag{7.13}$$

also holds. The conclusion is that it is possible to find two different *universal* Turing machines capable of outputting $x$, and the minimum difference in the complexities is at least one bit.

We consider next the issue of determining the complexity of a string $x$ with a *known length* $l(x)$. In Example 7.9, we have seen that if the string $x$ is incomputable, but has a finite length $l(x)$, one can devise a halting program $q_0$ that defines $x$ as a mere symbol-by-symbol or bit-by-bit description. The length of such a program (or descriptive length) is $|q(x)| = l(x) + c$, where $c$ is a constant taking care of TM instructions, such as "print" or "stop on string delimiter #." We conclude, therefore, that the associated complexity is $K(x) = \min_U |q(x)| = l(x) + \min_U c \equiv l(x) + c'$ (in the following, I will use $c$ or $c'$ to mean any nonzero positive constants). The fact that the string length $l(x)$ is known a priori is an important piece of information in the definition of the program $q(x)$. In this condition, it is sensible to call this complexity the *conditional complexity* of $x$, which we note $K[x \mid l(x)]$. Conditional complexity literally means

The descriptive length of the program that outputs $x$ with knowledge of the length of $x$.

Based on the above, we have established the following property for *any string of known length* $l(x)$:

$$K[x \mid l(x)] \le l(x) + c \tag{7.14}$$

(the sign $\le$ being introduced instead of $=$ to signify any arbitrary constant $c$), which provides an upper bound for conditional complexity.

Assume next the case of strings whose lengths are unknown a priori but can be defined by some algorithm (or through any computable analytical or iterative formula). Thus (because there is an algorithm), there must exist a TM program $q$ such that $|q(x)| \le l(x) + c$. Since an algorithm is generally not able to tell the string length $l(x)$, therefore, the upper bound in Eq. (7.14) does not apply for $K(x)$. If we want a machine to output a string of definite length $l(x) = n$, the program must contain this information in the form of an instruction. A possible way of encoding this instruction is:

- Convert $n$ into a binary number of $\log n$ bits (strictly, $\lceil \log_2 n \rceil$);
- Form a string $m$, where the bits are repeated twice;
- Append a delimiter such as 01 or 10.

[18] This is shown as follows:

(a) If $K_U > K_O$, then $-c_U \le 0 < |K_U - K_O| \le c_O \le \sup(c_O, c_U)$;
(b) If $K_U < K_O$, then $-c_O \le 0 < |K_U - K_O| \le c_U \le \sup(c_O, c_U)$.

For instance, $n = 11$, or 1011 in binary, is encoded as $m = 1100111101$. This is not a minimal-length code (as compression algorithms exist), but it surely defines $n$. We see that this code length is $2 \log n + 2 = 2 \log l(x) + 2$. The basic program that describes $x$ and that can tell the TM to halt after outputting $n$ bits is the same as that which prints $x$ bit by bit ($q_0$), but with an additional instruction giving the value of $n$. Based on Eq. (7.14) and the above result, the length of such a program is, therefore, $|q_0(x)| = K[x \mid l(x)] + c + 2 \log l(x) + 2 \approx K[x \mid l(x)] + 2 \log l(x) + c$ for sufficiently long sequences. But as we know, if $x$ can be described by some algorithm, the corresponding program length is such that $|q(x)| \leq |q_0(x)|$, meaning that the complexity of $x$ has the upper bound

$$K(x) \leq K[x \mid l(x)] + 2 \log l(x) + c. \tag{7.15}$$

We have established that the complexity $K(x)$ of any string $x$ has an upper bound defined by either Eq. (7.14) or Eq. (7.15).

The knowledge of an upper bound for complexity is a valuable piece of information, but it does not tell us in general how to measure it. An important and most puzzling theorem of algorithmic information theory is that *complexity cannot be computed*. This theorem can be stated as follows:

There exists no known algorithm or formula that, given any sequence $x$, a Turing machine can output $K(x)$.

This leads one to conclude that the problem of determining the Kolmogorov complexity of any $x$ is *undecidable*! The proof of this surprising theorem turns out to be relatively simple,[19] as shown below.

Assume that a program $q$ exists such that a UTM can output the result $U[q(x)] = K(x)$ for *any* $x$. We can then make up a simple program which, given $x$, could find at least one string having at least the same complexity as $x$, which we call $K(x) \equiv K$. Such a program $r$ is algorithmically defined as follows:

***Input*** $K$,

***For*** $n = 1$ to infinity,

***Define*** all strings $s$ of length $n$,

***If*** $K(s) \geq K$ ***print*** $s$ ***then halt***,

***Continue.***

This program has a length $|r| = |q| + 2 \log l(K) + c$. Since the program length grows as the logarithm of $K$, there exists a value $K$ for which $|r(K)| < K$. The program eventually outputs a string $s$ of complexity $K(s) \geq K > |r|$, which is strictly greater than the program size! By definition, the complexity $K(s)$ of $s$ is its minimum descriptive length. By definition, however, there is no TM program that can output $s$ with a length shorter that $K(s)$. Therefore, the hypothetical program $q$, which computes $K(x)$ given $x$, simply cannot exist. Complexity cannot be computed by Turing machines (or, for that matter, by *any* computing machine).

[19] As adapted from http://en.wikipedia.org/wiki/Kolmogorov_complexity.

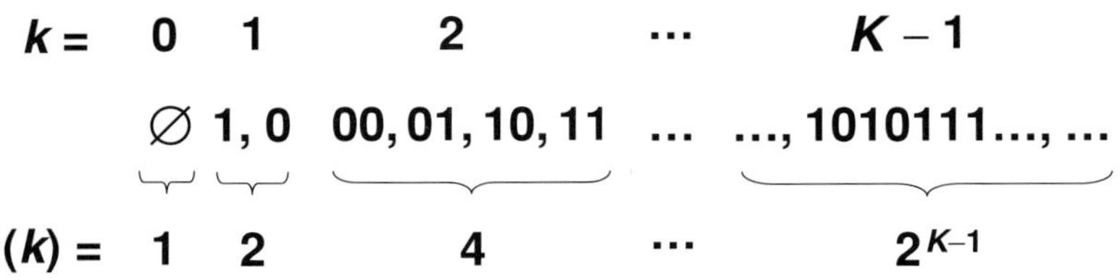

$k =$ 0 1 2 ... $K-1$

∅ 1, 0 00, 01, 10, 11 ... ..., 1010111..., ...

$n(k) =$ 1 2 4 ... $2^{K-1}$

**Figure 7.3** Enumeration of bit strings of size $k = 0$ to $k = K - 1$.

While *complexity cannot be computed*, we can show that it is at least possible to achieve the following:

(a) Given a complexity $K$, to tell how many strings $x$ have a complexity less than $K$, i.e., satisfying $K(x) < K$;
(b) For a string of given length $n$, to define the upper bound of its complexity, namely, to find an analytical formula for the upper bound of $K[x \mid l(x)]$ in Eq. (7.15).

To prove the first statement, (a), is a matter of finding how many strings $x$ have a complexity below a given value $K$, i.e., satisfying $K(x) < K$. This can be proven by the following enumeration argument. Figure 7.3 shows the count of all possible binary strings of length $k$, from $k = 0$ (empty string) to $k = K - 1$. The sum of all possibilities illustrated in the figure yields

$$\sum_{k=0}^{K-1} n(k) = \sum_{k=0}^{K-1} 2^k = \frac{1 - 2^K}{1 - 2} = 2^K - 1 < 2^K. \tag{7.16}$$

Each of these possible strings represents a TM program, to which a unique output $x$ corresponds. There are fewer than $2^K$ programs smaller than $K$, therefore *there exists fewer than* $2^K$ *strings of complexity* $< K$.

The statement in (b) can be proven as follows. Given any binary string $x$ of length $n$, we must find the minimal-length program $q$ that outputs it. Assume first that the string has $k$ bits of value 1, thus $n - k$ bits of value 0, with $0 < k < n$. We can then create a catalog of all possible strings containing $k$ 1s, and index them in some arbitrary order. For instance, with $n = 4$ and $k = 2$, we have $x_1 = 1100$, $x_2 = 1010$, $x_3 = 1001$, $x_4 = 0101$, $x_5 = 0011$, $x_6 = 0110$. Having this catalog, we can simply define any string $x_i$ according to some index $i$. Given $n$ and $k$, the range of index values $i$ is $C_n^k = n!/[k!(n-k)!]$, which represents the number of ways to assign $k$ 1s into $n$ bit positions. The program $q$ defining $x$ must contain information on both $k$ and $i$. As we have seen, it takes $2 \log k + 2$ bits to define $k$ by repeating its bits, with a two-bit delimiter at the end (noting that fancier compression algorithms of shorter lengths also exist). The second number $i$ can be defined after the delimiter, but without repeating bits, which occupies a size of $\log C_n^k$ bits. We must also include a generic program, which, given $k$, generates the catalog. Assume that its length is $c$, which is independent of $k$. The total length of the program $q$ is, therefore,

$$|q(k, n)| = 2 \log k + 2 + \log C_n^k + c \equiv 2 \log k + \log C_n^k + c'. \tag{7.17}$$

The next task is to evaluate $\log C_n^k$. Using Stirling's formula,[20] it is possible to show that for sufficiently large $n$:[21]

$$\log C_n^k \approx \log \frac{1}{\sqrt{2\pi}} + nf\left(\frac{k}{n}\right), \tag{7.18}$$

where the function $f$ is defined by $f(u) = -u \log u - (1-u)\log(1-u)$, and, as usual, all logarithms are in base 2. The function $f(k/n)$ is defined over the interval $0 < k/n < 1$. To include the special case where all bits are identical ($k/n = 0$ or $k/n = 1$) we can elongate the function by setting $f(0) = f(1) = 0$, which represents the analytical limit of $f(y)$ for real $y$. We note that $f(u)$ is the same function as defined in Eq. (4.14) for the entropy of two complementary events. Its graph is plotted in Fig. 4.7, showing a maximum of $f(1/2) = 1$ for $u = 1/2$.

Substituting Eq. (7.18) into Eq. (7.17), we, thus, obtain:

$$|q(k,n)| \approx 2\log k + nf\left(\frac{k}{n}\right) + c'', \tag{7.19}$$

where $c''$ is a constant. Consistently, this program length represents an upper bound to the complexity of string $x$ with $k$ 1s, i.e.,

$$K(x \mid k_{\text{ones}}) \leq 2\log k + nf\left(\frac{k}{n}\right) + c''. \tag{7.20}$$

In the general case, a string $x$ of length $n$ can have any number $k$ of 1 bits, with $0 < k/n < 1$. We first observe that the integer $k$ is defined by the sum of the $a_j$ bits forming the string, namely, if $x = a_n a_{n-1} \ldots a_2 a_1$ ($a_j = 0$ or 1) we have $k = \sum_{j=1}^{n} a_j$. Second, we observe that since $k < n$ we have $\log k < \log n$. Based on these two observations, the general approximation formula giving the upper bound for the complexity $K(x \mid n)$ of *any* binary string of length $n$ defined by $x = a_n a_{n-1} \ldots a_2 a_1$ ($a_j = 0$ or 1), is:

$$K(x \mid n) \leq 2\log n + nf\left(\frac{1}{n}\sum_{j=1}^{n} a_j\right) + c''. \tag{7.21}$$

To recall, Eq. (7.21) represents the Stirling approximation of the exact definition:

$$K(x \mid n) \leq 2\log n + \log C_n^k + c. \tag{7.22}$$

[20] See Eq. (A9) in Appendix A.

[21] Applying Stirling's formula yields, after some algebra:

$$C_n^k \approx \frac{1}{\sqrt{2\pi}} \exp\left\{n\left[\left(1+\frac{1}{2n}\right)\ln(n) - \left(\frac{k}{n}+\frac{1}{2n}\right)\ln(k) - \left(1-\frac{k}{n}+\frac{1}{2n}\right)\ln\left[n\left(1-\frac{k}{n}\right)\right]\right]\right\}.$$

In the limit $n \gg 1$, and after regrouping the terms, the formula reduces to:

$$C_n^k \approx \frac{1}{\sqrt{2\pi}} \exp\left\{n\left[-\frac{k}{n}\ln\left(\frac{k}{n}\right) - \left(1-\frac{k}{n}\right)\ln\left(1-\frac{k}{n}\right)\right]\right\} = \frac{1}{\sqrt{2\pi}} 2^{nf(k/n)},$$

where

$$f(u) = -u\frac{\ln u}{\ln 2} - (1-u)\frac{\ln u}{\ln 2}(1-u) \equiv -u\log_2 u - (1-u)\log_2(1-u).$$

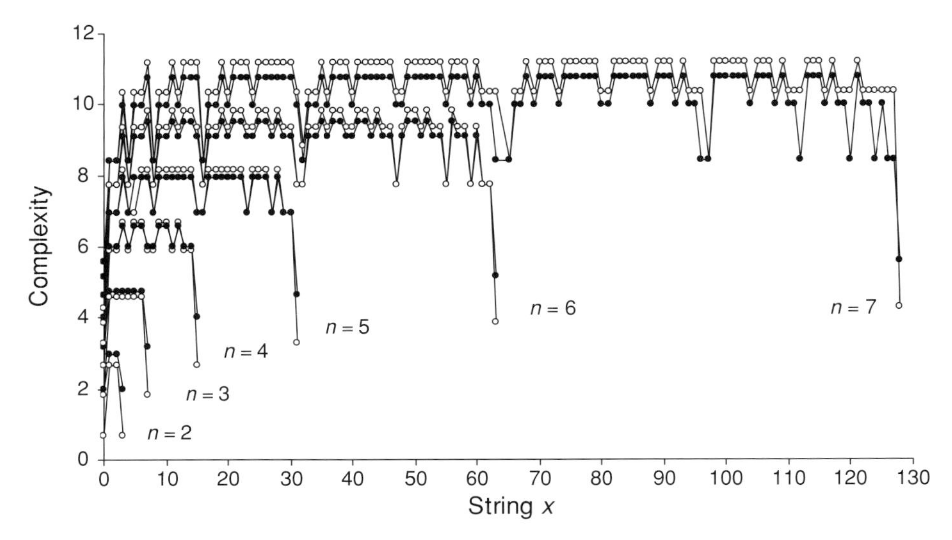


**Figure 7.4** Upper bound of conditional complexity $K(x \mid n)$ for each binary string $x$ of exactly $n$ bits ($n = 2$ to $n = 7$), as defined by Eq. (7.21) with the Stirling approximation (open symbols), and as defined by Eq. (7.22) without approximation (dark symbols). In each series of size $n$, the strings $x$ are ordered according to their equivalent decimal value.

Figure 7.4 shows plots of the upper bound of $K(x \mid n)$ for each binary string of exactly $n$ bits, according to (a) the Stirling approximation in Eq. (7.21) with $c'' = \log(1/\sqrt{2\pi}) + c$ (taking $c = 0$), and (b) the corresponding exact definition in Eq. (7.22), also taking $c = 0$. For each series of length $n$, the strings $x$ are ordered according to their equivalent decimal value (e.g., $x = 11$ corresponds to $x = 1011$ in the series $n = 4$, $x = 01011$ in the series $n = 5$, $x = 001011$ in the series $n = 6$, etc.).

We first observe from the figure that the upper bound of $K(x \mid n)$ oscillates between different values. For even bit sequences ($n$ even), the absolute minima are obtained for $f(u) = 0$ or $C_n^k = 1$, and correspond to the cases where all bits are identical. The absolute maxima are obtained for $f(u) = 1$ or $C_n^k = 0.5$, which corresponds to the cases where there is an equal number of 0 and 1 bits in the string. For odd bit sequences ($n$ odd), the conclusions are similar with all bits identical but one (minima) or with an approximately equal number of 0 and 1 bits in the string.

Second, we observe that the approximated definition (Eq. (7.21)) and the exact definition (Eq. (7.22)) provide nearly similar results. It is expected that the difference rapidly vanishes for string lengths $n$ sufficiently large.

Third, we observe that the complexity is generally greater than the string length $n$, which appears to be in contradiction with the result obtained in Eq. (7.14), i.e., $K[x \mid l(x) = n] \leq n + c$. Such a contradiction is lifted if we rewrite Eq. (7.21) in the form:

$$K(x \mid n) \leq n \left[ \frac{2 \log n}{n} + f\left(\frac{k}{n}\right) + \frac{c''}{n} \right] \tag{7.23}$$

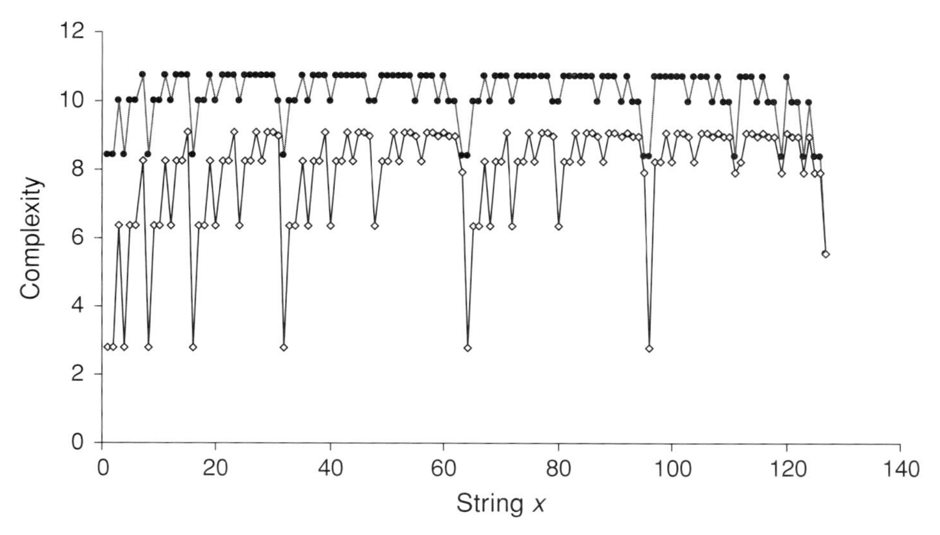


**Figure 7.5** Upper bound of conditional complexity for each binary strings $x$ of length $n = 7$: $K(x \mid n)$, as defined by Eq. (7.22) (dark symbols) and $K(x \mid n, k)$, as defined by Eq. (7.24) (open symbols) with $k$ being the number of 1s in the string.

and take the limit for large $n$, which gives $K(x \mid n) \leq nf(k/n) \leq n = l(x)$. The upper bound $n = l(x)$ stems from the fact that $f(u)$ varies from zero to unity. It is a maximum for $u = k/n = 0.5$, which corresponds to the case where there is an equal number of 0 or 1 bits in the string. In the case where the number of 1 bits in the string is known, we have also established from the above that

$$K(x \mid n, k_{\text{ones}}) \leq 2\log k + \log C_n^k + c. \tag{7.24}$$

It is exactly the same result as in Eq. (7.22) with $\log n$ replaced by $\log k$ ($k \geq 1$). Since $k \leq n$, the upper bound of $K(x \mid n, k_{\text{ones}})$ is, therefore, lower than that of $K(x \mid n)$. The comparison between the two definitions is shown in Fig. 7.5 for $n = 7$. We observe from the figure that $K(x \mid n, k_{\text{ones}})$ has a finer structure than $K(x \mid n)$ as we scan the string catalog from $x = 0000000$ to $x = 1111111$, which reflects the periodic changes in the number of 1 bits within each of the strings.

Consider next the following problem: given two strings $x$ and $y$, what is the size of the smallest program that can output both $x$ and $y$ *simultaneously*?

A first possibility is that $x$ and $y$ are *algorithmically independent*, i.e., there is no algorithm $q$ that is capable of computing both $x$ *and* $y$. Let $q_x$ and $q_y$ ($q_x \neq q_y$) be the two programs describing $x$ and $y$, respectively, from the same universal Turing machine $U$ (i.e., $U[q_x] = x$, $U[q_y] = y$). We can then chain the two programs to form $q_{xy} = q_x q_y$, a program that computes $x$ *then* $y$. There is no need for any additional instruction to the machine. Therefore, the program length is simply $|q_{xy}| = |q_x| + |q_y|$. The *minimal length* of such a program is:

$$K(x, y) = \min_U |q_{xy}| = \min_U |q_x| + \min_U |q_y| \equiv K(x) + K(y), \tag{7.25}$$

which shows that if two strings are algorithmically independent their *joint complexity* $K(x, y)$ is given by the sum of their individual complexities. It is clear that $K(x, y) = K(y, x)$. What if $x$ and $y$ are *not* algorithmically independent? This means that the computation of $x$ provides some clue as to the computation of $y$. A program calculating $y$ could be $q_y = q_x q_{y|x}$. The machine $U$ first computes $x$ then uses the program $q_{y|x}$ to compute $y$.

Next, we shall define the *conditional complexity* $K(y \mid q_x)$, which represents the minimal size of a program describing $y$ given the program describing $x$. It is also noted $K(y \mid x*)$, with $x* = q_x$, or, for simplicity, $K(y \mid x)$. This last notation should be used with the awareness that $\mid x$ is a condition on the program $q_x$, not on the string $x$.

The issue of finding the minimal size of $q_y = q_x q_{y|x}$ is far from trivial. Chaitin showed[22] that

$$\begin{aligned} K(x, y) &\leq K(x) + K(y \mid x) + c \\ &\leftrightarrow \\ K(y \mid x) &= K(x, y) - K(x) + c', \end{aligned} \tag{7.26}$$

where $c$ represents a small overhead constant, which is one bit for sufficiently long strings. The second inequality stems from the first, with $c' \geq 0$ being a nonnegative constant.[23] Since the joint complexity $K(x, y)$ is symmetrical in the arguments $x, y$, we also have

$$K(x \mid y) = K(x, y) - K(y) + c'. \tag{7.27}$$

If $x$ and $y$ are algorithmically independent, it is clear that $q_{y|x} = q_y$ (there is no clue from $x$ to compute $y$), or equivalently $K(y \mid x) = K(y)$, and likewise, $K(x \mid y) = K(x)$. In this case, $K(x, y) = K(x) + K(y) + c'$.

We can now define the *mutual complexity* $K(x; y)$ of $x$ and $y$ (note the delimiter ";") according to either of the following:

$$\begin{cases} K(x; y) = K(x) + K(y) - K(x, y) \\ K(x; y) = K(x) - K(x \mid y) + c' \\ K(x; y) = K(y) - K(y \mid x) + c', \end{cases} \tag{7.28}$$

where $c'$ is a nonnegative constant. In Eq. (7.28), the last two definitions stem from the first one and the properties in Eqs. (7.26) and (7.27).

The above results represent various relations of algorithmic complexity between two strings $x, y$. We immediately note that such relations bear a striking resemblance with that concerning the joint or conditional *entropies* and *mutual information* of two *random-event sources* $X, Y$ according to classical IT.

[22] See: G. J. Chaitin, A theory of program size formally identical to information theory. *J ACM*, **22** (1975), 329–40, www.cs.auckland.ac.nz/CDMTCS/chaitin/acm75.pdf. See also: G. J. Chaitin, Algorithmic information theory. *IBM J. Res. Dev.*, **21** (1977), 350–9, 496, www.cs.auckland.ac.nz/CDMTCS/chaitin/ibm.pdf.

[23] From the first definition in Eq. (7.26) we obtain $K(y \mid x) \geq K(x) - K(x, y) - c$, therefore, there exists a constant $c' \geq 0$ for which $K(y \mid x) = K(x) - K(x, y) + c'$.

Indeed, to recall from Chapter 5, the conditional and joint entropies are related through

$$H(X \mid Y) = H(X, Y) - H(Y) \tag{7.29}$$

and

$$H(Y \mid X) = H(X, Y) - H(X), \tag{7.30}$$

which are definitions similar to those in Eqs. (7.26) and (7.27) for the conditional complexity. From Chapter 5, we also have, for the mutual information,

$$\begin{cases} H(x; y) = H(X) + H(Y) - H(X, Y) \\ H(x; y) = H(X) - H(X \mid Y) \\ H(x; y) = H(Y) - H(Y \mid X), \end{cases} \tag{7.31}$$

which are definitions similar to that in Eq. (7.28) for the mutual complexity. It is quite remarkable that the chaining relations of conditional or joint complexities and conditional or joint entropies should be so similar (except in algorithmic IT for the finite constant $c'$, which is nonzero in the general case), given the conceptual differences between Kolmogorov complexity and Shannon entropy! As a matter of fact, such a resemblance between algorithmic and classical IT is not at all fortuitous. As stated at the beginning of this chapter, complexity and entropy are approximately equal when it comes to random events or sequences: the average size of a minimal-length program describing *random* events or sequences $x$ from a source $X$ is, indeed, approximately equal to the source entropy, or $\langle K(x)\rangle_X \approx H(X)$. The conceptual convergence between algorithmic entropy and Shannon entropy is formalized in the next section.

## 7.5 Kolmogorov complexity vs. Shannon's entropy

As we have seen, complexity can be viewed as a measure of information, just like Shannon's entropy. The key difference is that complexity $K(x)$ measures the information from an individual event $x$, while entropy $H(X)$ measures the average information from an event source $X$. Despite this important conceptual difference, I have shown in the previous section the remarkable similarity existing between chain rules governing the two information measures. In this section, I shall formally establish the actual (however approximate) relation between Shannon's entropy and Kolmogorov complexity.

Consider a source $X$ of random events $x_i$ with associated probabilities $p(x_i)$. For simplicity, we will first assume that the source is binary, i.e., the only two possible events are $x_1 = 0$ or $x_2 = 1$, thus $p(x_2) = 1 - p(x_1)$. We can record the succession of $n$ such events under the form of a binary string of length $n$, which we define as $x = x_i^{(1)} x_i^{(2)} x_i^{(3)} \ldots x_i^{(n)}$ with $i = 1, 2$.

We can then estimate the upper bound of the conditional complexity of $K(x \mid n)$ according to Eq. (7.21), which I repeat here with the new notations:

$$K(x \mid n) \leq 2\log n + nf\left(\frac{1}{n}\sum_{j=1}^{n} x_i^{(j)}\right) + c, \tag{7.32}$$

where (to recall) $f(u) = -u\log_2 u - (1-u)\log_2(1-u)$.

Next, we take the expectation value[24] of both sides in Eq. (7.32) to obtain:

$$\begin{aligned}
\langle K(x \mid n)\rangle &\leq \left\langle 2\log n + nf\left(\frac{1}{n}\sum_{j=1}^{n} x_i^{(j)}\right) + c\right\rangle \\
&= 2\log n + c + n\left\langle f\left(\frac{1}{n}\sum_{j=1}^{n} x_i^{(j)}\right)\right\rangle \\
&\leq 2\log n + c + nf\left(\frac{1}{n}\sum_{j=1}^{n} \left\langle x_i^{(j)}\right\rangle\right) \\
&= 2\log n + c + nf(q) \\
&\equiv 2\log n + c + nH(X).
\end{aligned} \tag{7.33}$$

To get the above result, we have made use of three properties.

- First, we have applied *Jensen's inequality*, which states that for any concave[25] function $F$, we have $\langle F(u)\rangle \leq F(\langle u\rangle)$;
- Second, we made the substitution $\langle x_i^{(j)}\rangle = \langle x_i\rangle = x_1 p(x_1) + x_2 p(x_2) \equiv q$;
- Third, we have used the property $F(q) = H(X)$, which is the definition of the entropy of a binary, random-event source $X$, see Eq. (4.13).

In the limit of large $n$, the result in Eq. (7.33) yields:

$$\frac{\langle K(x \mid n)\rangle}{n} \leq \frac{2\log n}{n} + \frac{c}{n} + H(X) \approx H(X). \tag{7.34}$$

This result means that for strings of sufficiently long length $n$, the average "per bit" complexity $\langle K(x \mid n)\rangle / n$ has the source entropy $H(X)$ as an upper bound. This can be equivalently stated: *the average complexity of a random bit string, $\langle K(x \mid n)\rangle$, is upper-bounded by the entropy $nH(X)$ of the source that generates it.* Note that the same conclusion is reached concerning nonbinary sources having $M$-ary symbols.[26]

[24] Since all events are independent, the probability of obtaining the string $x = x_i^{(1)} x_i^{(2)} x_i^{(3)} \ldots x_i^{(n)}$ is $p(x) = \prod_{j=1}^{n} p(x_i^{(j)})$. The expectation value $\langle K(x)\rangle$ thus means $\sum_x p(x)K(x)$, or the statistical average over all possibilities of strings $x$.

[25] Jensen's inequality applies to *concave functions*, which have the property that they always lie below any chord (such as $\sqrt{x}$, $-x^2$, or $\log(x)$).

[26] T. M. Cover and J. A. Thomas, *Elements of information theory* (New York: John Wiley & Sons, 1991).

Next, we try to find a lower bound to $\langle K(x \mid n)\rangle /n$. To each string $x$ corresponds a minimal-length *program* $q$, which is able to output $x$ from a universal Turing machine $U$, i.e., $U(q, n) = x$. It will be shown in Chapter 8 that the average length $L = \langle l(q)\rangle$ of such programs *cannot exceed* the source entropy, $nH(X)$. Equivalently stated, *the source's entropy is a lower bound of the mean program length*, i.e., $nH(X) \leq L$. It is also shown there that if the program length for each $q$ is chosen such that $l(q) = -\log p(x)$, then the equality stands, i.e., $L = nH(X)$. Here, I shall conveniently use this property to complete the demonstration. As we know, the conditional complexity $K(x \mid n)$ is precisely the shortest program length that can compute $x$. Thus, we have $\langle K(x \mid n)\rangle = \langle l(q)\rangle = L \geq nH(X)$, and

$$H(X) \leq \frac{\langle K(x \mid n)\rangle}{n}. \tag{7.35}$$

Combining the results in Eqs. (7.34) and (7.35), we obtain the double inequality

$$H(X) \leq \frac{\langle K(x \mid n)\rangle}{n} \leq \frac{2\log n}{n} + \frac{c}{n} + H(X). \tag{7.36}$$

It is seen from the final result in Eq. (7.36) that as the string length $n$ increases, the two boundaries converge to $H(X)$, and thus the per-bit complexity $\langle K(x \mid n)\rangle /n$ and the source entropy $H(X)$ become identical. For the purpose of the demonstration, we needed to consider "bits" and "strings." But we could also consider $X^*$ as a random source of strings $x$ with probability $p(x)$ and entropy $H(X^*)$. Thus, we have $\langle K(x)\rangle \approx H(X^*)$ as the asymptotic limit, which eliminates the need to refer to a "per-bit" average complexity.

The above result, thus, establishes the truly amazing and quite elegant property according to which *Kolmogorov complexity and Shannon's entropy give very similar measures of information*. Such an asymptotic relationship holds despite the profound conceptual difference existing between algorithmic and Shannon information theories.

## 7.6 Exercises

**7.1** (M): Use a Turing machine to add the two numbers $i = 4$ and $j = 2$, using the action table in Table 7.3.

**7.2** (B): Use a Turing machine to subtract the two numbers $i = 4$ and $j = 3$, using the action table in Table 7.5.

**7.3** (T): Solve Exercise 7.2 first. Then complete the subtraction algorithm by introducing a new TM state aiming to clean up the useless 0 in the output tape sequence, in the general case with $i \geq j$.

**7.4** (T): Define, for a Turing machine, an algorithm *Comp*, and a corresponding action table, whose task is to compare two integers $i, j$, and whose output is either *Comp*($i \geq j$) $= 0$ or *Comp*($i < j$) $= 1$. Clue: begin the analysis by solving Exercise 7.2 first.

**7.5** (T): Determine the number of division or subtraction operations required to convert the unary number

$$N_1 = 11\_1111\_1111\_1111\_0$$

into its decimal ($N_{10}$) representation. Notes: (a) separators _ have been introduced in the definition of $N_1$ for the sake of reading clarity; (b) the convention of the unary representation chosen here is $1_{10} = 10$.

**7.6** (T): Show how a Turing machine can convert the binary number

$$M_2 = 1001$$

into its unary ($M_1$) representation. Note: use the convention for the unary representation $1_{10} = 10$.

# 8 Information coding

This chapter is about *coding information*, which is the art of packaging and formatting information into meaningful *codewords*. Such codewords are meant to be recognized by computing machines for efficient processing or by human beings for practical understanding. The number of possible codes and corresponding codewords is infinite, just like the number of events to which information can be associated, in Shannon's meaning. This is the point where information theory will start revealing its elegance and power. We will learn that codes can be characterized by a certain *efficiency*, which implies that some codes are more efficient than others. This will lead us to a description of *the first of Shannon's theorems*, concerning *source coding*. As we shall see, coding is a rich subject, with many practical consequences and applications; in particular in the way we efficiently communicate information. We will first start our exploration of information coding with numbers and then with language, which conveys some background and flavor as a preparation to approach the more formal theory leading to the abstract concept of *code optimality*.

## 8.1 Coding numbers

Consider a source made of $N$ different events. We can label the events through a set of numbers ranging from 1 to $N$, which constitute a basic source *code*. This code represents one out of $N!$ different possibilities. In the code, each of the numbers represents a *codeword*. One refers to the list of codewords, here $\{1, 2, 3, 4, 5, 6, 7, 8, 9, 10, 11, 12, \ldots, N\}$ as a *dictionary*.

We notice here that our $N$ codewords use decimal numbers. In fact, these codewords are generated by some unique combinations of characters, as selected from a smaller dictionary, here $\{1, 2, 3, 4, 5, 6, 7, 8, 9, 0\}$. The smallest dictionary is referred to as a codeword *alphabet*. In Roman antiquity, one would have used instead the alphabet

$$\{\mathrm{I, II, III, IV, V, VI, VII, VIII, IX, X, L, C, D, M}\},$$

which corresponds to Roman numerals. As we know from school, the corresponding codewords are formed according to certain rules[1] (can we recognize in MDCCLXXXIX

[1] The correspondence being $\{1, 2, 3, 4, 5, 6, 7, 8, 9, 10, 50, 100, 500, 1000\} \equiv$ {I, II, III, IV, V, VI, VII, VIII, IX, X, L, C, D, M}, which does not include any character for zero. Note the subtractive rule $9 \equiv$ IX and

the date of the French Revolution?). Despite the oddity of their code rules, it is noteworthy that Roman numerals are still in use, for instance to represent the hours on clock dials, number pages in book prefaces, express copyright dates, enumerate cases in mathematical descriptions, or count series in games, such as US football. It is also interesting to note that the Roman-numeral system was in fact inspired by the Greek system in use in 400 BC.[2]

The ten-character dictionary of our decimal system, {0, 1, 2, 3, 4, 5, 6, 7, 8, 9} was first used by the Hindus in 400 BC, and then transmitted later to the West by the Arabs, hence the misnomer *Arabic numerals*, which should rather be *Hindu-Arabic numerals*. The introduction of a 0 character made it possible to greatly simplify numerals. Indeed, with only two- or three-character codewords (00–99 or 000–999), up to 100 or 1000 numerals can be generated. The *hexadecimal system* is based on the 16-character alphabet

$$\{0, 1, 2, 3, 4, 5, 6, 7, 8, 9, \mathrm{A}, \mathrm{B}, \mathrm{C}, \mathrm{D}, \mathrm{E}, \mathrm{F}\}.$$

The advantage of the hexadecimal system is that the codeword length is shorter than in the decimal case, at the expense of using a greater number of alphabet characters. The drawback of both decimal and hexadecimal codes is that they are based on relatively large alphabets. In practice, such characters may not be simple to generate, write, or faithfully interpret, as we all know from handwriting experience. Such characters need, in turn, to be coded through a simpler alphabet. The most basic code (and number representation) corresponds to the *binary system,* which uses the two-character {0, 1} alphabet.[3] A single-character, binary codeword is referred to as a *bit*, short for binary digit. In the following, we will equivalently refer to codewords as "numbers."

Binary and decimal numbers are conventionally written as the ordered character sequences $B = \ldots b_3b_2b_1b_0$ and $D = \ldots d_3d_2d_1d_0$, respectively. The conversion between decimal and binary numbers is given by the following power expansion:

$$\begin{aligned} \ldots d_3d_2d_1d_0 &\equiv \ldots d_3 \times 10^3 + d_2 \times 10^2 + d_1 \times 10^1 + d_0 \times 10^0 \\ &= \ldots b_3 \times 2^3 + b_2 \times 2^2 + b_1 \times 2^1 + b_0 \times 2^0 \\ &\equiv \ldots b_3b_2b_1b_0. \end{aligned} \tag{8.1}$$

not 9 ≡ VIII. To generate numbers greater than 10 ≡ X, the rule is as follows: 11 ≡ XI, 12 ≡ XII, . . . , 18 ≡ XVIII, 19 ≡ XIX, and 20 ≡ XX, and. Then 30 ≡ XXX, 31 ≡ XXXI, . . . , up to 39 ≡ XXXIX. Because of the absence of a zero, the powers of ten would be character-consuming should they have to be repeated by as many X. To make numbers more compact, the Romans chose to represent 50, 100, 500, and 1000 by the symbols L, C, D, and M, respectively, with the subtractive rule 40 ≡ XL, 400 ≡ CD, 90 ≡ XC, 900 ≡ CM. Thus, 2006 is represented by MMVI, while 1999 is represented by MCMXCIX. There are additional rules for representing greater numbers, see for instance:
http://en.wikipedia.org/wiki/Roman_numerals#IIII_or_IV.3 F, and
http://ostermiller.org/calc/roman.html.
For more on the numeral systems used by different civilizations through history, see
http://en.wikipedia.org/wiki/Arabic_numerals.

2 See: http://en.wikipedia.org/wiki/Greek_numerals.

3 The *unary* system uses a single-character alphabet {1}. Numbers are represented by this character's repetition, starting from zero: 0 = 1, 1 = 11, 2 = 111, 3 = 1111, etc. Albeit not practical, such a code can have interesting applications, for instance, in Turing machines (see Chapter 7).

Thus, the decimal number $D = 3 = 1 \times 2^1 + 1 \times 2^0$ corresponds to the binary number $B = 11$, and the decimal

$$\begin{aligned} D &= 1539 \\ &= 1 \times 2^{10} + 1 \times 2^9 + 0 \times 2^8 + 0 \times 2^7 + 0 \times 2^6 + 0 \times 2^5 \\ &\quad + 0 \times 2^4 + 0 \times 2^3 + 0 \times 2^2 + 1 \times 2^1 + 1 \times 2^0 \end{aligned}$$

corresponds to the binary $B = 11\,000\,000\,011$, for instance. It is easy to establish that the maximum decimal value for an $n$-bit binary number is $D = 2^n - 1$. For instance, $D = 7 = 2^3 - 1$ corresponds to $B = 111$, and $D = 15 = 2^4 - 1$ corresponds to $B = 1111$, which represent the maximum decimal values for 3-bit and 4-bits binary numbers, respectively.

For practical handling, long binary numbers are usually split into subgroups of eight bits, which are called *bytes*. The byte itself can be divided into two subgroups of four bits. Since $B = 1111$ is equal to $D = 15$, one can represent a byte through a two-character codeword based on the alphabet {0, 1, 2, 3, 4, 5, 6, 7, 8, 9, A, B, C, D, E, F}, which corresponds to the *hexadecimal* representation. For instance, the hexadecimal $H = 7\text{A}$ corresponds to the binary $B = 01111010$ and the decimal $D = 122$. The hexadecimal system is just a convenient way of representing binary numbers through a 16-character alphabet. It is also a base-16 representation, which immediately translates into the binary system by blocks of four bits. A byte thus covers the decimal-number range $2^8 - 1 = 255$, which is also conveniently represented by hexadecimal numbers from 00 to FF. Note that with the zero, the number of symbols that can be represented by $n$ bits is actually $2^n$.

## 8.2 Coding language

In Chapter 4, we have analyzed the *entropy in language*. Languages, especially the ones we can't read, can be viewed as random sources of alphabet characters. As we have seen, alphabet characters do not have the same probability of occurrence, because words make preferential use of certain letters, like E, T, A, O, N... in English. The probability distribution of most languages' alphabets is exponential (see Fig. 4.3). The distribution varies according to geographic derivatives, and the type of communication (e.g., informal, technical, literary). The language dictionary slowly evolves through generations, as new words are introduced, and older ones are abandoned. Even the alphabet somewhat evolves, for instance with the new character @, which became much more common with the Internet generation. In this section, we shall consider the issue of coding language by substituting the conventional A–Z alphabet with a decimal, and then a binary coding system.

A first coding approach could consist in attributing a decimal number to each alphabet character: for instance, A = 1, B = 2, and so on, down to Z = 26. This requires 26 codewords with two decimal symbols varying from 0 to 9, i.e., going from 00 to 99. This makes 100 coding possibilities, leaving extra room for $100 - 26 = 74$ other alphabetical or alphanumerical symbols. This reserve would be adequate to code other symbols, such as lower-case a–z characters, characters with accents and alterations (é, è, à, ä, ç, ù, ü, ñ . . . ), space and punctuation characters, parentheses and quotes, math operands and symbols $(+, -, \times, \div, =, >, <)$, the decimals 0–9, and other special characters (*, %, #, §, &, ‘, ~, ∧, ¨, °, _, |, /, \, [, ], {, }, $, €, £, @ . . . ). This whole character bank forms our computer-keyboard alphabet, to be completed with several other computer commands. It can, thus, be theoretically coded through 100 decimal numbers (00 to 99). But such a decimal coding has never been of any practical use, except in the early times of cryptography, with the unique character–decimal correspondence representing the "secret code."[4]

A second and more powerful coding approach consists in converting the whole keyboard alphabet into *binary codewords*. The advantage is that the latter can be processed by computers, without any other form of encoding. Since we have $2^7 = 128$, we observe that codewords of only seven-bit length are more than sufficient to cover a full keyboard-symbol alphabet. This is the reason why most computers use *ASCII*,[5] a standard code invented in 1961 and based on *seven-bit codewords*. Note that ASCII is not the only possible code for computers: indeed, *EBCDIC*[6] is another standard based on *eight-bit codewords*, or bytes. The extra bit of EBCDIC makes it possible to code twice as many alphabet characters as in ASCII, namely $2^8 = 256$ codewords. In the 1970s, ASCII was, in fact, extended to eight-bit codewords, which, in particular, makes it possible to include all language-specific characters. The correspondence table between ASCII and keyboard characters can be found on the Internet.[7] Table 8.1 illustrates the correspondence for the most commonly used keyboard characters. For instance, the letter A is coded as 1000001, the letter b is coded as 1100010, and the character @ is coded as 1000000. We observe from the table that all alphabetical characters or letters (lower or upper case) begin with 1 as the leftmost (highest-weight or seventh) bit. We also observe that all upper-case letters have 0 in the second leftmost (or sixth) bit position, while lower-case letters have 1 instead. Thus, the $2 \times 26 = 52$ upper and lower-case letter alphabet is actually using six-bit codewords, while the 26-letter alphabet is using five-bit codewords.

A five-bit codeword can cover $2^5 = 32$ characters, which is apparently not sufficient for representing the 26 letters and the 10 numbers, unless numbers can be written as words (e.g. 3 = three ). To increase the possibilities of a five-bit code, a trick is to introduce two shift characters, one to announce a shift from *letters* to *figures* (concerning codewords to follow) and the other for the reverse operation. The introduction of these two shift characters, thus, virtually permits one to re-use the set of 26 letter codewords as another set of 26 figure codewords, including numbers and punctuation, leaving four extra symbols to use for space, carriage return, line feed, and blank. This is the principle

[4] For a detailed introduction to early secret codes and cryptography, see for instance E. Desurvire, *Wiley Survival Guide in Global Telecommunications, Broadband Access, Optical Components and Networks, and Cryptography,* (New York: J. Wiley & Sons, 2004).

[5] American Standard Code for Information Interchange.

[6] Extended Binary Coded Decimal Interchange Code.

[7] See, for instance, http://–wikipedia.org/wiki/ASCII.

**Table 8.1** ASCII code table for common keyboard characters (extract).

| ← Bit word | | | | 7 | 0 | 0 | 0 | 0 | 1 | 1 | 1 | 1 |
|---|---|---|---|---|---|---|---|---|---|---|---|---|
| | | | | 6 | 0 | 0 | 1 | 1 | 0 | 0 | 1 | 1 |
| 4 | 3 | 2 | 1 | 5 | 0 | 1 | 0 | 1 | 0 | 1 | 0 | 1 |
| 0 | 0 | 0 | 0 | | | | space | 0 | @ | P | \ | p |
| 0 | 0 | 0 | 1 | | | | ! | 1 | A | Q | a | q |
| 0 | 0 | 1 | 0 | | | | " | 2 | B | R | b | r |
| 0 | 0 | 1 | 1 | | | | # | 3 | C | S | c | s |
| 0 | 1 | 0 | 0 | | | | $ | 4 | D | T | d | t |
| 0 | 1 | 0 | 1 | | | | % | 5 | E | U | e | u |
| 0 | 1 | 1 | 0 | | | | & | 6 | F | V | f | v |
| 0 | 1 | 1 | 1 | | | | ' | 7 | G | W | g | w |
| 1 | 0 | 0 | 0 | | | | ( | 8 | H | X | h | x |
| 1 | 0 | 0 | 1 | | | | ) | 9 | I | Y | i | y |
| 1 | 0 | 1 | 0 | | | | * | : | J | Z | j | z |
| 1 | 0 | 1 | 1 | | | | + | ; | K | [ | k | { |
| 1 | 1 | 0 | 0 | | | | , | < | L | \ | l | \| |
| 1 | 1 | 0 | 1 | | | | - | = | M | ] | m | } |
| 1 | 1 | 1 | 0 | | | | . | > | N | ^ | n | ~ |
| 1 | 1 | 1 | 1 | | | | / | ? | O | — | o | |

of the *Baudot code*, now better known as *International Alphabet IA2* and still in use in telex machines.

Consider now the code-source *entropy*. If all ASCII codewords were equally likely (having a uniform probability distribution), the corresponding entropy would be $H = \log 128 = \log 2^7 \equiv 7$ *bit/symbol*, which is precisely the codeword length. In the case where the codeword length matches the entropy of the code source, it is said that the code is *optimal*. This important concept of *code optimality* will be met repeatedly throughout this chapter.

But, as we are aware, language characters do not have a uniform probability, and, therefore, the source entropy must be less than the above maximum (7 bit/symbol). For ordinary text files, the most likely symbols are spaces and lower-case letters, which follow an exponential distribution. In Chapter 4, we established that the plain 26-character English alphabet (A–Z) has an entropy of *4.185 bit/symbol* (1982 survey). We would then expect the entropy of the ASCII source to be somewhat greater than this value, considering the greater diversity of characters, but substantially less than 7 bit/symbol, because of the nonuniformity of the code source. Thus, as applied to language, ASCII can be regarded as being a nonoptimal code, since its codeword length is greater than the actual source entropy! For computer files, like tabulated data, source programs, or HTML Internet pages, however, the character statistics are quite different from that of the English language, and the corresponding probability distribution is somewhat closer to uniform. In this respect, ASCII is closer to code optimality.

Once a *codeword length* has been fixed, and is, by definition, the same for all codewords, there is no possibility of further optimizing the code. On the other hand, *code optimization* seeks for optimal codes which are based on *variable-length* codewords.

We will look at the issue of code optimization in the next two sections, starting with an analysis of the Morse code.

## 8.3 The Morse code

Another approach for coding language is to use codewords with variable lengths. The rationale is to make the length of the most frequently-used characters the shortest possible, and the reverse for the least frequently used ones. With this approach, the average codeword length is shorter than that of a fixed-length code, and this will bring us closer to coding optimality.

The *Morse code* is a historical illustration of the above concept. Such a code has been widely used in pre-computer ages for military communications (from the American Civil War to the First World War), for the early beginnings of the public telegraph, which, as a true revolution of the time, brought the telegram,[8] and for maritime communications and safety.[9] Today, its use is only restricted to nostalgic amateur groups.[10] The Morse code is a binary-like or pseudo-binary code based on the two character values $dit = \bullet$ and $da = —$ (or *dot* and *dash*, respectively). While anybody knows the meaning of SOS, fewer people know that it actually means "Save Our Souls," and maybe even fewer people know the Morse transcription:

$$\bullet\bullet\bullet/---/\bullet\bullet\bullet \text{ (dit dit dit/da da da/dit dit dit)},$$

as repeated several times.

As symbolized above by the slash, each Morse codeword must in fact be separated by short pauses. Such pauses are meant for unambiguous identification of the codewords. This is because Morse messages are meant to be generated, to be written, and to be read *in real time* by human operators, and not by a machine.

Table 8.2 shows the "Continental" international correspondence of the 44 Morse symbols with alphanumerical characters and various punctuation signs. The list is completed with another five symbols for messaging commands. Note that there is no Morse symbol for "space," for consideration of economy. Morse messages make sense without spaces, just like HELLOHOWAREYOU or HAPPYBIRTHDAYTOYOU. It does not preclude that the sending operator may use the "wait" symbol once in a while, to take a breath or if he is accidentally interrupted whilst broadcasting a message. Yet full texts can be coded with all punctuation marks (except the exclamation point, !), which makes the Morse code very complete as a communication means. As one observes from Table 8.2, the shortest Morse symbols are attributed to the most common letters in

[8] The first telegram was sent from Baltimore to Washington, DC over electrical wires by Morse in 1844, see http://en.wikipedia.org/wiki/Electrical_telegraph.

[9] According to international maritime safety regulations, ships at sea no longer need to be equipped with Morse-based alarm systems with SOS signaling as in the past. Since 1999, indeed, the regulatory alternative is now the Global Maritime Distress and Safety System (GMDSS), which uses satellite and other communication principles.

[10] The Morse code still has fans world-wide, who collect and use old machine and even organize High Speed Telegraphy Championships.

**Table 8.2** Correspondence of the Continental International Morse code with alphanumerical characters, punctuation, and other command characters. The nine letters most frequently used in European languages are placed at the left.

| | | | | | | | | | |
|---|---|---|---|---|---|---|---|---|---|
| E | • | B | –••• | . | •–•–•– | 0 | ––––– | call T | –•– |
| T | – | C | –•–• | , | ––••–– | 1 | •–––– | error | ••••••• |
| A | •– | D | –•• | ? | ••––•• | 2 | ••––– | wait | •–••• |
| N | –• | F | ••–• | : | –––••• | 3 | •••–– | end M | •–•–• |
| I | •• | G | ––• | ; | –•–•–• | 4 | ••••– | end B | •••–•– |
| M | –– | H | •••• | - | –•••– | 5 | ••••• | | |
| S | ••• | J | •––– | / | –••–• | 6 | –•••• | | |
| O | ––– | K | –•– | " | •–••–• | 7 | ––••• | | |
| R | •–• | L | •–•• | | | 8 | –––•• | | |
| | | P | •––• | | | 9 | ––––• | | |
| | | Q | ––•– | | | | | | |
| | | U | ••– | | | | | | |
| | | V | •••– | | | | | | |
| | | W | •–– | | | | | | |
| | | X | –••– | | | | | | |
| | | Y | –•–– | | | | | | |
| | | Z | ––•• | | | | | | |

call T = call to transmit, end M = end message, end B = end broadcasting.

European languages, i.e., E, T, A, N, I . . . while the longest symbols are attributed to the least frequent letters, such as J, Q, X, Y, Z. In this way, operators save lots of time when generating or writing down Morse codewords.

Another trick in the Morse code is that *letter symbols* take a maximum of four dit/da characters, while *number symbols* are exactly five characters long. This makes it easier for operators to distinguish between letters and numbers, and avoids any risk of confusion for numbers (mistakes in numbers having potentially more important consequences, unlike with letters, which can be intuitively corrected, or whose mistakes are immediately noticeable). The Morse code has proven quite efficient for rapid messaging between "human entities" having limited telecommunications equipment. Certain civilian and military boats still carry on-board Morse machines as *light guns*: in adverse conditions when the radio is down because of power failure or enemy scrambling, a point-to-point and "radio-silent" Morse communication by day or by night may be the only solution. And even a small piece of mirror with the sun or a flashlight works very well to communicate over distances of kilometers, and can be included as part of any survivor's equipment, for vital SOS messaging.

The Morse is, thus, a first example of a variable-length code. Since the codeword length is decreased in proportion to the symbol frequency, we should expect that the entropy of Morse-code source is quite smaller than that of an ASCII code reduced to the same A–Z letters. In fact, the entropy analysis of the Morse code is not as straightforward as it may first appear. Earlier, I referred to the code as being *pseudo-binary*, even if it uses only two characters, which provided a hint. Indeed, the code makes use of short pauses or blanks between two codewords, without which the code would be unintelligible. For instance, the beginning message HELLO

$$\bullet\bullet\bullet\bullet / \bullet / \bullet - \bullet\bullet / \bullet - \bullet\bullet / - - -$$

transmitted without blanks would look like

$$\bullet\bullet\bullet\bullet\bullet\bullet - \bullet\bullet\bullet - \bullet\bullet - - -,$$

which from Table 8.2 can be interpreted in several different ways (e.g., 5ELRJ or SVEFAM or EEEEEETIA2, etc.). This illustrates the property that, without such blanks, the Morse code is not *uniquely decodable* and is useless (say, except for mere SOS purposes). This notion of unique decodability will be further addressed in the next section. Here, we shall analyze what these information-less, but indispensable blanks represent in terms of code entropy.[11]

The idea to begin our analysis is to look at the blanks (/) as representing an extra symbol character in the Morse code, which is systematically present at the end of any codeword. Thus dit/ and da/ actually form *digrams* (two-character symbols) as opposed to *monograms* (single-character symbols). Two possibilities exist for introducing this extra blank symbol.

The first possibility is to set the blank to a binary-code (dit/da) value, which must meet two requirements: (a) it is not already taken by any Morse code symbol, and (b) the concatenation of the blank to any Morse codeword, forming the new "digram" codeword, should be *uniquely decodable*. Referring to Table 8.2, the smallest binary symbol for "blank" should be $- - - - - -$. With such a convention, whenever one hears six *das*, it is definitely a blank without ambiguity or error, and one knows for sure which other conventional Morse codeword precedes or follows. The detection of $- - - - - -$ as a new symbol is also an indication that blanks are now being coded! But we now have a tax to pay: the minimum symbol size of this new Morse code is 7 (digram symbols "E-blank" and "T-blank"), and the maximum size is 13 (digram symbol "error-blank"). We shall therefore discard this effective, but poorly economical approach.

The second possibility is to convert the pseudo-binary Morse code into a *ternary* one. In base three, the characters are 0, 1, and 2, which are called *trits* for ternary digits).[12] A single trit, thus, codes three numbers, and $n$ trits make up $3^n$ coding possibilities. Our extended Morse code having $49 + 1 = 50$ symbols, we see that $n = 4$ trits are required, although $3^4 = 81$ is far in excess of what we actually need. Here, we shall not attempt to optimize the length of this ternary coding system, but only to use the property offered by a third alphabet character to represent blanks uniquely. Setting the convention da $= 0$, dit $= 1$, we can only have blank $= 2$. The ternary codeword is simply generated by appending 2 at the end of the binary Morse codeword. The first two columns in Table 8.3 show the ternary codeword correspondence with the A–Z letters of the Morse code. For instance $R = \bullet - \bullet$ becomes 1012 in the proposed ternary representation. Actually, this alternative Morse code is not different from the conventional one, if one conceives of

[11] To my knowledge, the following (including in the next section) constitutes an original information-theory analysis of the Morse code.

[12] Likewise, in the base-4 or quaternary system, the characters 0, 1, 2, and 3 are called *quads*.

**Table 8.3** Ternary representation of the Morse-code letters into *trit* codewords (CW), with the introduction of a character, 2, to signal the blank immediately following conventional Morse symbols ($\bullet = 1$, $- = 0$, blank $= 2$), as shown in the first two columns. Column 3 shows the source probability distribution $p(x)$, which is the same as used in Fig. 4.5 for English-language reference (1982 survey). Columns 4 and 5 show the detailed calculation of the bit/symbol ($H_2$) and trit/symbol ($H_3$) entropy, using base-2 and base-3 logarithms, respectively. Column 6 shows the codeword length $l(x)$ associated with each trit symbol, and Column 7 shows the calculation of the mean codeword length $L$ (*effective code entropy*). The last two columns represent the same as Columns 6 and 7, but with a different coding solution with codeword length $l'(x)$ yielding the mean $L'$ (see text for description). The calculation results (source entropy, effective code entropy, and coding efficiency) are shown at bottom.

| Morse symbol | Morse trit CW $(x)$ | $p(x)$ | $p\log_2(p)$ | $p\log_3(p)$ | CW length $l(x)$ | Mean CW $l(x)p(x)$ | Other CW $l'(x)$ | Mean CW length $l'(x)p(x)$ |
|---|---|---|---|---|---|---|---|---|
| E | 12 | 0.127 | 0.378 | 0.239 | 2 | 0.254 | 2 | 0.254 |
| T | 02 | 0.091 | 0.314 | 0.198 | 2 | 0.181 | 2 | 0.181 |
| A | 102 | 0.082 | 0.295 | 0.186 | 3 | 0.245 | 2 | 0.64 |
| O | 0012 | 0.075 | 0.280 | 0.177 | 4 | 0.299 | 2 | 0.150 |
| I | 112 | 0.070 | 0.268 | 0.169 | 3 | 0.209 | 2 | 0.140 |
| N | 012 | 0.067 | 0.261 | 0.165 | 3 | 0.200 | 2 | 0.134 |
| S | 1112 | 0.063 | 0.251 | 0.158 | 4 | 0.51 | 3 | 0.188 |
| H | 11112 | 0.061 | 0.246 | 0.155 | 5 | 0.304 | 3 | 0.182 |
| R | 1012 | 0.060 | 0.243 | 0.153 | 4 | 0.239 | 3 | 0.179 |
| D | 0112 | 0.043 | 0.195 | 0.123 | 4 | 0.171 | 3 | 0.129 |
| L | 10112 | 0.040 | 0.185 | 0.117 | 5 | 0.199 | 3 | 0.120 |
| C | 01012 | 0.028 | 0.144 | 0.091 | 5 | 0.140 | 3 | 0.084 |
| U | 1102 | 0.028 | 0.144 | 0.091 | 4 | 0.112 | 4 | 0.112 |
| M | 002 | 0.024 | 0.129 | 0.081 | 3 | 0.072 | 4 | 0.096 |
| W | 1002 | 0.024 | 0.129 | 0.081 | 4 | 0.096 | 4 | 0.096 |
| F | 11012 | 0.022 | 0.121 | 0.076 | 5 | 0.110 | 4 | 0.088 |
| G | 0012 | 0.020 | 0.113 | 0.071 | 4 | 0.080 | 4 | 0.080 |
| Y | 11102 | 0.020 | 0.113 | 0.071 | 5 | 0.100 | 4 | 0.080 |
| P | 10112 | 0.019 | 0.108 | 0.068 | 5 | 0.095 | 5 | 0.095 |
| B | 01112 | 0.015 | 0.091 | 0.057 | 5 | 0.075 | 5 | 0.075 |
| V | 11102 | 0.010 | 0.066 | 0.042 | 5 | 0.050 | 5 | 0.050 |
| K | 0102 | 0.008 | 0.056 | 0.035 | 4 | 0.032 | 5 | 0.040 |
| J | 10002 | 0.002 | 0.018 | 0.011 | 5 | 0.010 | 5 | 0.010 |
| X | 01102 | 0.002 | 0.018 | 0.011 | 5 | 0.010 | 5 | 0.010 |
| Q | 00102 | 0.001 | 0.010 | 0.006 | 5 | 0.005 | 5 | 0.005 |
| Z | 00112 | 0.001 | 0.010 | 0.006 | 5 | 0.005 | 5 | 0.005 |
| $\sum$ | | 1.000 | | | | | | |
| Entropy | | | $H_2 = 4.185$ bit/symbol (source) | $H_3 = 2.640$ trit/symbol (source) | | $L = 3.544$ trit/symbol (code) | $L' = 2.744$ trit/symbol (code) | |
| Coding efficiency | | | | | | 74.49% | 96.23% | |

the last character 2 in each codeword as another sound, which is different from dit or da (say, do or du). For the purposes of entropy analysis, we shall consider here only the symbols $x$ corresponding to A–Z, for which we know the probability distribution, $p(x)$, taking the English-language PDF described in Chapter 4. It is easily established that the PDF of the trit codewords is the same as that of the bit (or conventional Morse) codewords.[13] Table 8.3 also shows the source entropy in either base-2 (bit/symbol) or base-3 (trit/symbol) logarithms. By convention, entropy in logarithm base $M$ will be called here $H_M$. Consistently, it is defined according to:

$$H_M = -\sum_x p(x) \log_M p(x). \tag{8.2}$$

Recalling that $\log_M x = \ln x / \ln M$, where ln is the natural logarithm, the relation between base-$M$ entropy and conventional (base-2) entropy is the following:

$$\ln(M) H_M = \ln(2) H_2. \tag{8.3}$$

Looking at Table 8.3, the calculations for base 2 and base 3 source entropies yield $H_2 = 4.185$ bit/symbol (English-language entropy) and $H_3 = 2.640$ trit/symbol, respectively. The next step in our analysis is to define a way to measure how efficient a given code is in using the most concise codewords, regardless of the logarithmic base. This issue is addressed in the next section, which will also use our new Morse code by way of an illustrative example.

## 8.4 Mean code length and coding efficiency

Let's introduce the *mean codeword length* (also called *expected length*) according to the definition

$$L(X) = \langle l \rangle_X = \sum_{x \in X} p(x) l(x), \tag{8.4}$$

where $l(x)$ is the codeword length corresponding to symbol $x$ from source $X$. With a binary code, the unit of $L$ is bit/symbol. This defines the mean or expected codeword length as *effective code entropy*.[14]

We shall, again, use the ternary Morse code as an illustrative example. In this case, $L$ is in units of *trit/symbol*. Columns 6 and 7 in Table 8.3 detail the calculation of the mean codeword length, as based on the above definition, the ternary codewords previously introduced (Column 2) and the distribution $p(x)$. As Table 8.3 shows, the mean codeword length is $L = 3.544$ trit/symbol. Since $L$ has the dimensions of entropy, we can compare it with the codeword source, using the ratio $\eta = H_3/L$, with

[13] This is because the joint and conditional probabilities of the Morse/blank digrams $x/y$ satisfy $p(y = \text{blank} \mid x) = p(x)$ and $p(y = \text{blank}, x) = p(x)$, with $p(y = \text{blank}) = 1$.

[14] More accurately the unit of the mean codeword length is *bit/codeword* or *trit/codeword*, but, for simplicity and clarity, I shall use here the names bit/symbol or trit/symbol, it being understood that there is a one-to-one correspondence between codewords and symbols.

$H_3 = 2.640$ trit/symbol. We, thus, find that $\eta = 74.5\%$. As we shall see further in this chapter, such a ratio defines *coding efficiency,* a parameter that cannot exceed unity. Put simply:

The mean codeword length (or effective code entropy) cannot exceed the source entropy.

This is a fundamental property of codes, which was originally demonstrated by Shannon.

We can thus conclude that with a 74% coding efficiency, the Morse code (analyzed as a ternary code) is a reasonably good choice. Yet, despite its popularity and usefulness in the past (and until as recently as 1999), the Morse code is quite far from optimal. The main reason for this is the use of the blank, which is first required for the code to be uniquely decodable, and second for being usable by human operators. Such a blank, however, does not carry any information whatsoever, and it takes precious codeword resources! This observation shows that we could obtain significantly greater coding efficiencies if the Morse code was uniquely decodable without making use of blanks. This improved code could be transmitted as uninterrupted bit or trit sequences. But it would be only intelligible to machines, because human beings would be too slow to recognize the unique symbol patterns in such sequences. Considering then both binary and ternary codings (with source entropies $H_{\text{source}}$ shown in Table 8.3), and either fixed or variable symbol or codeword sizes, there are four basic possibilities:

(i) *Fixed-length binary* codewords with $l = 5$ *bit* ($2^5 = 32$), giving $L \equiv 5.000$ bit/symbol, or $H_{\text{source}}/L = 83.7\%$;
(ii) *Fixed-length ternary* codewords with $l = 3$ *trit* ($3^3 = 27$), giving $L \equiv 3.000$ trit/symbol, or $H_{\text{source}}/L = 88.0\%$;
(iii) *Variable-length binary* codewords with lengths between $l = 1$ and $l = 10$ bits, giving $L \equiv 4.212$ bit/symbol, or $H_{\text{source}}/L = 99.33\%$ (with optimal codes using $3 \leq l \leq 9$ bits);
(iv) *Variable-length ternary* codewords with lengths between $l = 1$ and $l = 3$ trits, giving $L \equiv 2.744$ trit/symbol or $H_{\text{source}}/L = 96.2\%$.

The result shown in case (iii) is derived from an optimal coding approach (Huffmann coding) to be described in Chapter 9. The result shown in case (iv) will be demonstrated in the next section. At this stage, we can just observe that all the alternative solutions (i)–(iv) have coding efficiencies significantly greater than the Morse code. We also note that the efficiency seems to be greater for the multi-level ($M$-ary) codes with $M > 2$, but I will show next that it is not always true.

Consider, for simplicity, the case of fixed-length $M$-ary codes, for which it is straightforward to calculate the coding efficiency. For instance, quaternary codewords, made with $n$ quad characters, called 0, 1, 2, and 3, can generate $4^n$ symbol possibilities. Since $4^2 < 26 < 4^3 = 64$, codewords with 3 quads are required for the A–Z alphabet. The mean codeword length is, thus, $L = 3$ quad/symbol. The source entropy is $H_4 = (\ln 2/\ln 4)\, H_2 = 2.092$ quad/symbol. Thus, the coding efficiency is $\eta = H_4/L = 69.7\%$, which is lower than that of ternary, fixed-length coding (88.0%), as seen in case (ii). The reason is that going from ternary to quaternary coding does not reduce the codeword length, which remains equal to three. The situation would be quite different if we used variable-length codewords, since a single quad can represent four possibilities, instead of three for the trit. Let us look at the extreme case of a 26-ary coding, i.e., a coding made of 26 alphabet characters (called *26-its*) which can be represented, for instance, by an electrical current with 26 intensity levels. Each codeword is, thus, made of a single *26-it*, and the mean codeword length is $L = 1$ *26-it/symbol*. The source entropy is $H_{26} = (\ln 2/\ln 26)H_2 = 0.890$ *26-it/symbol*. The coding efficiency is, therefore, $\eta = H_{26}/L = 89.0\%$, which represents the highest efficiency for a fixed-length coding (one character/symbol) of the A–Z source. However, if we relate this last result to the efficiency of a *variable-length* binary or ternary code, cases (iii) and (iv) above, we see that the second yields higher performance.

The main conclusions of the above analysis are:

(1) The Morse code can be viewed as being a variable-length ternary code; it is quite efficient for human-operator use, but not for binary or ternary machines;
(2) One can code symbols through either fixed-length or variable-length codewords, using any $M$-ary representation ($M =$ number of different code characters or code alphabet size);
(3) The mean codeword length ($L$) represents the average length of all possible source codewords; it cannot be smaller than the source entropy ($H$), which defines a coding efficiency ($L/H$) with a maximum of 100%;
(4) For fixed-length codewords, one can increase the coding efficiency by moving from binary to $M$-ary codes (e.g., ternary, quaternary, . . . ), the efficiency increasing with $M$, continuously or by steps;
(5) The coding efficiency can be increased by moving from fixed-length to variable-length codewords (the shortest codewords being used for the most likely source symbols and the reverse for the longest codewords).

The property (3), known as *Shannon's source coding theorem*, and the last property (5) will be demonstrated next.

## 8.5 Optimizing coding efficiency

In this chapter so far, we have learnt through heuristic arguments that coding efficiency is substantially improved when the code has a *variable length*. It makes economical sense, indeed, to use short codewords for the most frequently used symbols, and keep the longer codewords for the least frequently used ones, which is globally (but not strictly) the principle of the old Morse code. We must now establish some rules that assign the most adequate codeword length to each of the source symbols, with the purpose of achieving a code with optimal efficiency.

Consider a basic example. Assume a source with five symbols, called A, B, C, D, and E, with the associated probabilities shown below:

| $x =$ A | B | C | D | E |
|---|---|---|---|---|
| $p(x) = 0.05$ | 0.2 | 0.05 | 0.4 | 0.3 |

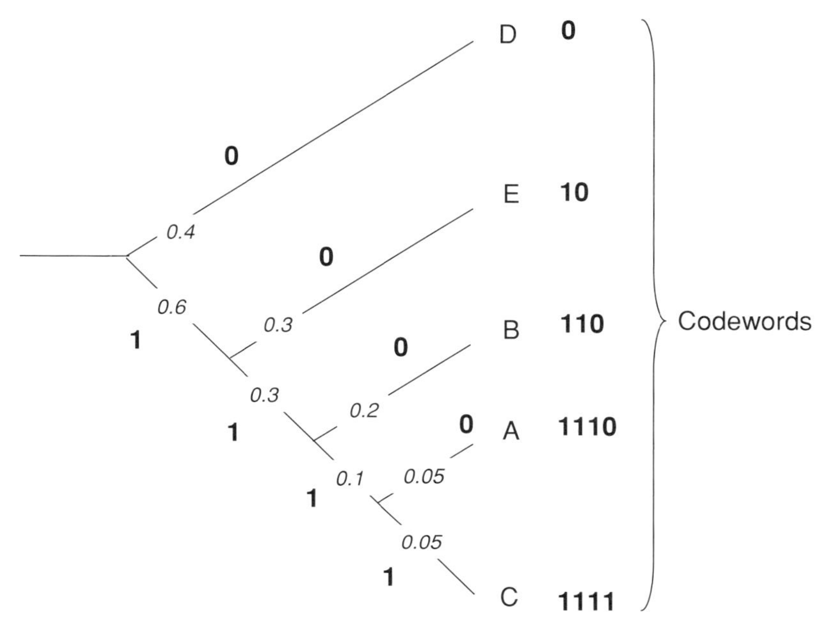


**Figure 8.1** Coding tree for the source $x = \{A, B, C, D, E\}$ with associated probability distribution $p(x) = \{0.05, 0.2, 0.05, 0.4, 0.3\}$.

This source entropy is easily calculated to be $H = 1.946$ bit/symbol. One method of assigning a code to each of these symbols is to draw a logical *coding tree*, as shown in Figs. 8.1 and 8.2.

The two figures show that each leaf of the coding tree represents a unique codeword. In both figures, the symbols shown at the right are placed in decreasing order of probability. The coding tree thus assigns each codeword according to successive choices between 0 and 1 as the branches split up. In Fig. 8.1, for instance, symbol D is assigned the codeword 0, meaning that the other source symbols, A, C, B, and E, should be assigned codewords beginning with a 1, and so on. Alternatively (Fig. 8.2), we can assign 0 as the leftmost codeword bit of symbols D, and E, and 1 to all others, and so on. It is straightforward to calculate the mean codeword length from Eq. (8.4), which gives $L = 2.0$ bit/symbol and $L = 2.3$ bit/symbol, for the two coding trees, respectively. The coding efficiencies are found to be 97.3% and 84.6%, respectively, showing that the first code (Fig. 8.1) is much closer to optimality that the second code (Fig. 8.2). In the last case, the efficiency is relatively low because we have chosen a minimum codeword length of 2 bits, as opposed to one bit in the first case.

In these examples, the choice of codewords (not codeword lengths) seemed to be arbitrary, but in fact such a choice obeyed some implicit rules.

First, different symbols should be assigned to strictly different codewords. In this case, the code is said to be *nonsingular*.

Second (and as previously seen), the code should be *uniquely decodable*, meaning that an *uninterrupted string of codewords* leads to only one and strictly one symbol-sequence

**Table 8.4** Different types of codes for a 5-ary source example made of a binary codeword alphabet.

| Source | Singular | Nonsingular, not uniquely decodable | Uniquely decodable, not prefix | Prefix |
|---|---|---|---|---|
| A | 00 | 0 | 101 | 101 |
| B | 01 | 1 | 1011 | 100 |
| C | 10 | 01 | 110 | 110 |
| D | 00 | 10 | 1101 | 00 |
| E | 10 | 11 | 10110 | 01 |

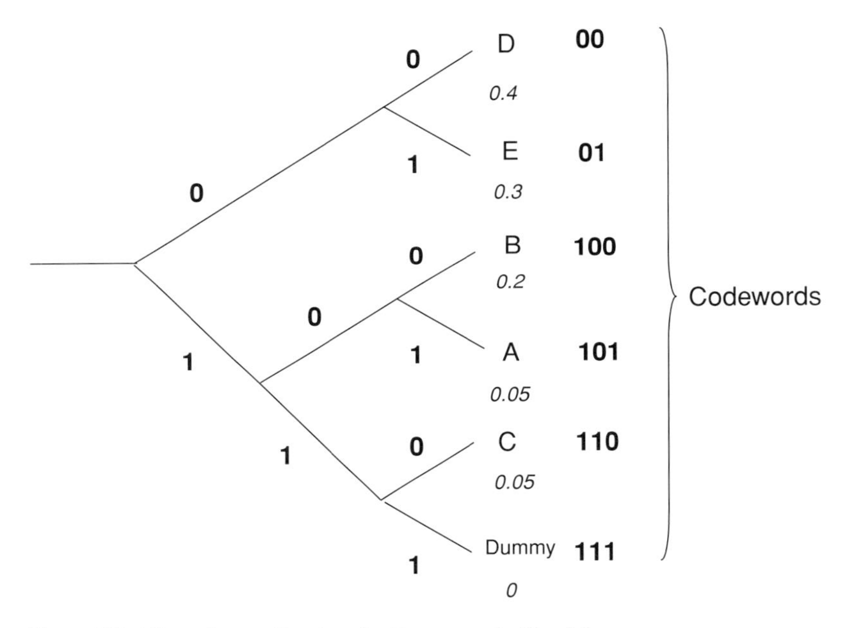


**Figure 8.2** Alternative coding tree for the source in Fig. 8.1.

interpretation. Indeed, consider the code sequence 1101110010... made from a string of codewords from the code in Fig. 8.1. Scanning from left to right, since 1 and 11 are not codewords, the first matching codeword is 110, or symbol B. The same analysis continues until we find the meaning BADE. The code is, thus, uniquely decodable. We also note that under this last condition, no codeword represents the beginning bits or *prefix* of another codeword. This means that the codeword can be instantaneously interpreted after reading a certain number of bits (this number being known from the identified prefix), without having to look at the codeword coming next in the sequence. For this reason, such codes are called *prefix codes* or *instantaneous codes*. Note that a code may be uniquely decodable without being of the prefix or instantaneous type. Table 8.4 shows examples of singular codes, nonsingular but nonuniquely decodable, uniquely

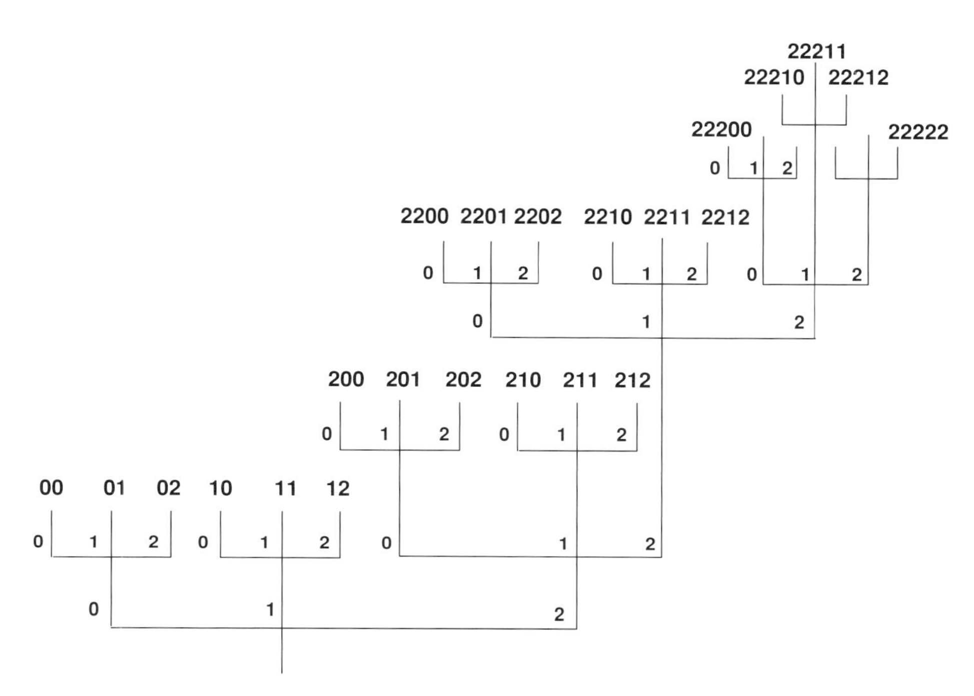


**Figure 8.3** Coding tree for A–Z letter source, using ternary codewords (see also Table 8.3).

decodable but not prefix, and prefix, in the case of a 5-ary code alphabet made of five different binary codewords.

We shall now use the coding-tree method to see if we might improve the coding efficiency of the A–Z English-character source, using a ternary alphabet with variable size (case (iv) in Section 8.4). Figure 8.3 shows a possible coding tree for a 27-symbol source (26 letters, plus a dummy). The method consists of assigning the first group of 2-bit codewords (00, 01, 02, 10, 11, 12) to the first six most frequent characters, then the second group of 3-bit codewords (200, 201, 202, 210, 211, 212) to the following six most frequent characters, and so on, until all attributions are exhausted (except for a dummy 22222 codeword, which is never used). The codeword length corresponding to this character–codeword mapping is shown in Table 8.3 in Column 8 as $l'(x)$. Column 9 of Table 8.3 details the calculation of the mean codeword length $L'$. The result is $L' = 2.744$ trit/symbol, which corresponds to a coding efficiency of $\eta = H_3/L = 96.23\%$. This coding efficiency is far greater than any efficiency obtained so far with fixed and variable-length codes for the English A–Z character source.

What we have learnt from the above examples is that there exist many possibilities for assigning variable-length codewords to a given symbol source. Also, we have seen that the *coding tree* can be used as a tool towards generating codes with shorter mean codeword length. We must now formalize our approach, in order to determine the shortest mean codeword length achievable for any given source, and a method of finding the corresponding optimal code, which is described next.

## 8.6 Shannon's source-coding theorem

In this section, we address the following question:

Given a symbol source, what is the smallest mean codeword length achievable?

The answer is provided by a relatively simple demonstration. Let $H_M(X)$ be the entropy of a $M$-ary source $X = \{x\}$ with codeword lengths $l(x)$ and corresponding probability $p(x)$. The mean codeword length is $L(X) = \sum l(x)p(x)$. We then estimate the difference $\Delta = L(X) - H_M(X)$ according to:

$$\begin{aligned}\Delta &= L(X) - H_M(X) \\ &= \sum_{x\in X} l(x)p(x) + \sum_{x\in X} p(x)\log_M p(x) \\ &= \sum_{x\in X} p(x)[l(x) + \log_M p(x)] \\ &\equiv \langle l + \log_M p\rangle_X.\end{aligned} \tag{8.5}$$

A sufficient (but not necessary) condition for the difference $\Delta$ to vanish is that each $l(x)$ takes the specific value $l^*(x)$, as defined by

$$l^*(x) = -\log_M p(x). \tag{8.6}$$

The above result means that if the codeword length $l(x)$ of symbol $x$ is chosen to be specifically equal to $l^*(x) = -\log_M p(x)$, then $\Delta = 0$ and the mean codeword length $L$ equals the source entropy $H_M$. We note that this optimal length $l^*(x)$ represents the *information* contained in the symbol $x$, which was defined in Chapter 3 as being $I(x) = -\log_M p(x)$, and which is a measure of the symbol *uncertainty*. Thus, the codeword lengths are short for symbols with less information (more frequent, or less uncertain) and are long for symbols with more information (less frequent, or more uncertain). Since the codeword lengths must be integer numbers, the optimal condition $l^*(x) = -\log_M p(x)$ can only be approximately satisfied in the general case. The codeword lengths must then be chosen as the integer number greater than but closest to $l^*(x)$.

The above demonstration does *not* prove that the choice of codeword length $l^*(x)$ *minimizes* the mean codeword length $L$. We don't know, either, if the difference $\Delta$ is positive or negative in the general case. To establish both properties, we must use the so-called *Kraft–McMillan inequality*, which is demonstrated in Appendix F. This inequality, which applies to *prefix codes*,[15] states

$$\sum_{x\in X} M^{-l(x)} \leq 1. \tag{8.7}$$

Let's prove first that $l^*(x)$ is the minimum mean codeword length. For this, we apply the *Lagrange-multipliers* method, which we have already used in Chapters 4 and 6 to

[15] As it turns out, the Kraft inequality also applies to the greater class of uniquely decodable codes (McMillan inequality), hence the name *Kraft–McMillan inequality*.

maximize the entropy of discrete and continuous sources. To recall, this method makes it possible to find the function $l_{\mathrm{opt}}(x)$ for which any function $L(l)$ is minimized (or maximized), given the constraint that $C(l) = u$, where $L, C$ are any derivable functions of $l$ and $u$ is a constant. Formally, one first defines:

$$J(l) = L(l) + \lambda C(l), \tag{8.8}$$

where $L(l) = \sum_x p(x)l(x)$ and $C(l) = \sum_x M^{-l(x)} \equiv \sum_x \mathrm{e}^{-l(x)\ln M} = u$. The derivation of Eq. (8.8) with respect to $l(x)$ leads to

$$\frac{\partial J}{\partial l(x)} = p(x) - \lambda l M^{-l(x)} l(x) \ln M. \tag{8.9}$$

Setting the result in Eq. (8.9) to zero, we obtain the parameter $\lambda$, which, with the constraint, yields the optimum function $l_{\mathrm{opt}}(x)$ :[16]

$$l_{\mathrm{opt}}(x) = -\log_M p(x) - \log_M u. \tag{8.10}$$

Since by definition $u \leq 1$, the smallest codeword-length for any $x$ is given by

$$l^*(x) = -\log_M p(x), \tag{8.11}$$

which we had previously found in Eq. (8.6) to be the function for which $L(X) = H(X)$.

The above demonstration proves that the codeword length distribution $l^*(x)$ yields the *shortest mean codeword length* $L(X)$. We can now prove that the source entropy $H(X)$ represents an absolute lower bound for $L(X)$. Indeed,

$$\begin{aligned}
\Delta &= L(X) - H_M(X) \\
&= \sum_{x \in X} p(x)[l(x) + \log_M p(x)] \\
&= \sum_{x \in X} p(x)[\log_M p(x) - \log_M M^{-l(x)}] \\
&= \sum_{x \in X} p(x)[\log_M p(x) - \log_M(AM^{-l(x)}) + \log_M A] \\
&= \sum_{x \in X} p(x) \log_M \left[\frac{p(x)}{AM^{-l(x)}}\right] + \sum_{x \in X} p(x) \log_M A \\
&\equiv D(p \| AM^{-l(x)}) + \log_M A.
\end{aligned} \tag{8.12}$$

In the above development, we have introduced the constant $A = 1/\sum_x M^{-l(x)}$, which makes the function $AM^{-l(x)}$ a probability distribution. Thus $D(p\|AM^{-l(x)})$ is a *Kullback–Leibler distance*, which, from Chapter 5, we know to be always nonnegative. Because of the Kraft–McMillan inequality, we also have $A \geq 1$, or $\log_M A \geq 0$. The conclusion is that we always have $\Delta \geq 0$, which leads to *Shannon's source-coding*

[16] Namely, $\lambda = 1/(u \ln M)$, which comes from substituting the result $M^{-l(x)} = p(x)/(\lambda \ln M)$ into the constraint $C(l) = \sum_x M^{-l(x)} = u$, and using $\sum_x p(x) = 1$. This leads to $p(x) = M^{-l(x)}/u$, from which we obtain $l(x) \equiv l_{\mathrm{opt}}(x) = -\log_M p(x) - \log_M u$.

*theorem*:

$$L(X) \geq H_M(X). \tag{8.13}$$

The theorem translates into the following:

The mean codeword length cannot be shorter than the source entropy: equality is achieved when the codeword lengths are chosen such that $l(x) = l^*(x) = -\log_M p(x)$.

In the general case, there is no reason for $l^*(x)$ to be an integer. In this case, a compromise is to take the *smallest integer value greater than or equal to* $l^*(x)$, which is noted $\lceil l^*(x) \rceil$. Such a realistic code assignment is called a *Shannon code*, which is also known as *Shannon–Fano code*. We, thus, have $l^*(x) = \lceil l^*(x) \rceil - \varepsilon(x)$, with $0 \leq \varepsilon(x) < 1$. This relation gives the realistic minimal value of the mean codeword length, which we shall call here $L^{**}$, and define according to:

$$\begin{aligned} L^{**}(X) &= \sum_{x \in X} p(x) \lceil l^*(x) \rceil \\ &= \sum_{x \in X} p(x) \left[ l^*(x) + \varepsilon(x) \right] \\ &\equiv H_M(X) + \langle \varepsilon \rangle_X . \end{aligned} \tag{8.14}$$

Since $0 \leq \langle \varepsilon \rangle_X < 1$, it immediately follows that

$$H_M(X) \leq L^{**}(X) < H_M(X) + 1, \tag{8.15}$$

which represents another fundamental property of minimal-length codes (or Shannon–Fano codes). The result can be restated as follows:

There exists a code whose mean codeword length falls into the interval $[H_M, H_M + 1]$; it is equal to the source entropy $H_M$ if the distribution $l^* = -\log_M p$ has integer values (dyadic source).

This *source-coding theorem* was the first to be established by Shannon, which explains its other appellation of *Shannon's first theorem*.

By definition, we call *redundancy* the difference $\rho = L - H$. We call $\rho_{\text{bound}}$ the upper bound of the redundancy, i.e., $\rho < \rho_{\text{bound}}$ is always satisfied. The Shannon–Fano code redundancy *bound* is therefore $\rho_{\text{bound}} = 1$ bit/symbol, which means that in the general case, $L^{**} < H + 1$, consistently with Eq. (8.15). In the specific case of dyadic sources ($l^* = -\log_M p$ is an integer for all $x$), we have $L^* = H$, and $\rho = 0$ bit/symbol.

We have seen from the preceding demonstration of Shannon's source-coding theorem, that there exists an optimum codeword-length assignment ($l^*(x)$, or more generally, $l^{**}(x)$) for which the mean codeword length is minimized. What if we were to use a different assignment? Assume that this assignment takes the form $\hat{l}(x) = \lceil -\log_M q(x) \rceil$, where $q(x)$ is an arbitrary distribution. It follows from this assignment that the mean codeword length becomes:

$$\begin{aligned} \hat{L}(X) &= \sum_{x \in X} p(x) \hat{l}(x) \\ &= \sum_{x \in X} p(x) \left\lceil \log_M \frac{1}{q(x)} \right\rceil \end{aligned}$$

$$
\begin{aligned}
&= \sum_{x \in X} p(x) \left[ \log_M \frac{1}{q(x)} + \varepsilon(x) \right] \\
&= \sum_{x \in X} p(x) \left[ \log_M \frac{p(x)}{q(x)p(x)} + \varepsilon(x) \right] \\
&= \sum_{x \in X} p(x) \left[ \log_M \frac{p(x)}{q(x)} \right] - \sum_{x \in X} p(x) [\log_M p(x)] + \langle \varepsilon \rangle_X \\
&\equiv D(p \| q) + H_M(X) + \langle \varepsilon \rangle_X.
\end{aligned} \tag{8.16}
$$

Since $0 \leq \langle \varepsilon \rangle_X < 1$, it immediately follows from the above result that

$$
\begin{aligned}
H_M(X) + D(p \parallel q) &\leq \hat{L}(X) \\
&< H_M(X) + D(p \parallel q) + 1.
\end{aligned} \tag{8.17}
$$

The result illustrates that a "wrong" choice of probability distribution results in a "penalty" of $D(p \parallel q)$ for the smallest mean codeword length, which shifts the boundary interval according to $\hat{L}(X) \in [H_M(X) + D, H_M(x) + D + 1]$. Such a penalty can, however, be small, and a "wrong" probability distribution can, in fact, yield a mean codeword length $\hat{L}(X)$ smaller than $L^{**}(X)$. This fact will be proven in Chapter 9, which concerns optimal coding, and specifically *Huffman codes*.

Since the above properties apply to any type of $M$-ary coding, we may wonder (independently of technological considerations for representing, acquiring, and displaying symbols) whether a specific choice of $M$ could also lead to an absolute minimum for the expected length $L(X)$. Could the binary system be optimal in this respect, considering the fact that the codewords have the longest possible lengths? A simple example with the English-character source ($X = \{A - Z\}$, 1982 survey, see Chapter 4) provides a first clue. Table 8.5 shows the mean codeword lengths $L^{**}(X)$ for various multi-level codes: *binary*, *ternary*, *quaternary*, and *26-ary*, with the details of their respective code-length assignments.

We observe from Table 8.5 that the binary code has the smallest mean codeword length relative to the source entropy, i.e., $L^{**}(x) = 4.577$ bit/symbol, yielding the highest coding efficiency of $\eta = 91.42\%$. This result is better than the value of $\eta = 83.70\%$ found for a fixed-length binary code with 5 bit/symbol (such as ASCII reduced to lower-case or upper-case letters). Moving from binary to ternary is seen to decrease the coding efficiency to $\eta = 84.37\%$. The efficiency increases somewhat with the quaternary code ($\eta = 84.63\%$.), but apart from such small irregularities due to the integer truncation effect, it decreases with increasing $M$. For $M = 26$, the efficiency has dropped to $\eta = 72.63\%$. This last result is substantially lower than the value of $\eta = 89.0\%$, which we found earlier for fixed-length, 26-ary coding having one character/symbol. The conclusion of this exercise is that the *Shannon code does not necessary yield better efficiencies than a fixed-length code*, against all expectations. In the example of the English-character source, the Shannon code is better with binary codewords, but it is

**Table 8.5** Shannon codes for the English-character source, using binary, ternary, quaternary, and 26-ary codewords.

| $x$ | $p(x)$ | $p\log_2(p)$ | Binary coding: Ideal CW length $l^*(x)$ | Binary coding: Real CW length $l^{**}(x)$ | Binary coding: Mean CW length $l^{**}(x)p(x)$ | $p\log_3(x)$ | Ternary coding: Ideal CW length $l^*(x)$ | Ternary coding: Real CW length $l^{**}(x)$ | Ternary coding: Mean CW length $l^{**}(x)p(x)$ |
|---|---|---|---|---|---|---|---|---|---|
| E | 0.127 | 0.378 | 2.9771 | 3 | 0.381 | 0.239 | 1.8783 | 2 | 0.254 |
| T | 0.091 | 0.314 | 3.4623 | 4 | 0.363 | 0.198 | 2.1845 | 3 | 0.272 |
| A | 0.082 | 0.295 | 3.6126 | 4 | 0.327 | 0.186 | 2.2793 | 3 | 0.245 |
| O | 0.075 | 0.280 | 3.7413 | 4 | 0.299 | 0.177 | 2.3605 | 3 | 0.224 |
| I | 0.070 | 0.268 | 3.8408 | 4 | 0.279 | 0.169 | 2.4233 | 3 | 0.209 |
| N | 0.067 | 0.261 | 3.904 | 4 | 0.267 | 0.165 | 2.4632 | 3 | 0.200 |
| S | 0.063 | 0.251 | 3.9928 | 4 | 0.251 | 0.158 | 2.5192 | 3 | 0.188 |
| H | 0.061 | 0.246 | 4.0394 | 5 | 0.304 | 0.155 | 2.5486 | 3 | 0.182 |
| R | 0.060 | 0.243 | 4.0632 | 5 | 0.299 | 0.153 | 2.5636 | 3 | 0.179 |
| D | 0.043 | 0.195 | 4.5438 | 5 | 0.214 | 0.123 | 2.8668 | 3 | 0.129 |
| L | 0.040 | 0.185 | 4.6482 | 5 | 0.199 | 0.117 | 2.9327 | 3 | 0.120 |
| C | 0.028 | 0.144 | 5.1628 | 6 | 0.167 | 0.091 | 3.2573 | 4 | 0.112 |
| U | 0.028 | 0.144 | 5.1628 | 6 | 0.167 | 0.091 | 3.2573 | 4 | 0.112 |
| M | 0.024 | 0.129 | 5.3851 | 6 | 0.144 | 0.081 | 3.3976 | 4 | 0.096 |
| W | 0.024 | 0.129 | 5.3851 | 6 | 0.144 | 0.081 | 3.3976 | 4 | 0.096 |
| F | 0.022 | 0.121 | 5.5107 | 6 | 0.132 | 0.076 | 3.4768 | 4 | 0.088 |
| G | 0.020 | 0.113 | 5.6482 | 6 | 0.120 | 0.071 | 3.5636 | 4 | 0.080 |
| Y | 0.020 | 0.113 | 5.6482 | 6 | 0.120 | 0.071 | 3.5636 | 4 | 0.080 |
| P | 0.019 | 0.108 | 5.7222 | 6 | 0.114 | 0.068 | 3.6103 | 4 | 0.076 |
| B | 0.015 | 0.091 | 6.0632 | 7 | 0.105 | 0.057 | 3.8255 | 4 | 0.060 |
| V | 0.010 | 0.066 | 6.6482 | 7 | 0.070 | 0.042 | 4.1945 | 5 | 0.050 |
| K | 0.008 | 0.056 | 6.9701 | 7 | 0.056 | 0.035 | 4.3976 | 5 | 0.040 |
| J | 0.002 | 0.018 | 8.9701 | 9 | 0.018 | 0.011 | 5.6595 | 6 | 0.012 |
| X | 0.002 | 0.018 | 8.9701 | 9 | 0.018 | 0.011 | 5.6595 | 6 | 0.012 |
| Q | 0.001 | 0.010 | 9.9701 | 10 | 0.010 | 0.006 | 6.2904 | 7 | 0.007 |
| Z | 0.001 | 0.010 | 9.9701 | 10 | 0.010 | 0.006 | 6.2904 | 7 | 0.007 |
| $\sum$ | 1.000 | | | | | | | | |
| Entropy | | $H_2 = 4.185$ bit/symbol (source) | | | $L^{**} = 4.577$ bit/symbol (code) | $H_3 = 2.640$ trit/symbol (source) | | | $L^{**} = 3.19$ trit/symbol (code) |
| Coding efficiency | | | | | 91.42% | | | | 84.37% |

| $x$ | $p(x)$ | $p\log_4(p)$ | Quaternary coding: Ideal CW length $l^*(x)$ | Quaternary coding: Real CW length $l^{**}(x)$ | Quaternary coding: Mean CW length $l^{**}(x)p(x)$ | $p\log_{26}(x)$ | 26-ary coding: Ideal CW length $l^*(x)$ | 26-ary coding: Real CW length $l^{**}(x)$ | 26-ary coding: Mean CW length $l^{**}(x)p(x)$ |
|---|---|---|---|---|---|---|---|---|---|
| E | 0.127 | 0.189 | 1.4885 | 2 | 0.254 | 0.080 | 0.6334 | 1 | 0.127 |
| T | 0.091 | 0.157 | 1.7312 | 2 | 0.181 | 0.067 | 0.7366 | 1 | 0.091 |
| A | 0.082 | 0.148 | 1.8063 | 2 | 0.164 | 0.063 | 0.7686 | 1 | 0.082 |

*(cont.)*

**Table 8.5** *(cont.)*

| | | Quaternary coding | | | | 26-ary coding | | | |
|---|---|---|---|---|---|---|---|---|---|
| $x$ | $p(x)$ | $p\log_4(p)$ | Ideal CW length $l^*(x)$ | Real CW length $l^{**}(x)$ | Mean CW length $l^{**}(x)p(x)$ | $p\log_{26}(x)$ | Ideal CW length $l^*(x)$ | Real CW length $l^{**}(x)$ | Mean CW length $l^{**}(x)p(x)$ |
| O | 0.075 | 0.140 | 1.8706 | 2 | 0.150 | 0.060 | 0.7959 | 1 | 0.075 |
| I | 0.070 | 0.134 | 1.9204 | 2 | 0.140 | 0.057 | 0.8171 | 1 | 0.070 |
| N | 0.067 | 0.130 | 1.952 | 2 | 0.134 | 0.055 | 0.8306 | 1 | 0.067 |
| S | 0.063 | 0.125 | 1.9964 | 2 | 0.126 | 0.053 | 0.8495 | 1 | 0.063 |
| H | 0.061 | 0.123 | 2.0197 | 3 | 0.182 | 0.052 | 0.8594 | 1 | 0.061 |
| R | 0.060 | 0.122 | 2.0316 | 3 | 0.179 | 0.052 | 0.8644 | 1 | 0.060 |
| D | 0.043 | 0.097 | 2.2719 | 3 | 0.129 | 0.041 | 0.9667 | 1 | 0.043 |
| L | 0.040 | 0.093 | 2.3241 | 3 | 0.120 | 0.039 | 0.9889 | 1 | 0.040 |
| C | 0.028 | 0.072 | 2.5814 | 3 | 0.084 | 0.031 | 1.0984 | 2 | 0.056 |
| U | 0.028 | 0.072 | 2.5814 | 3 | 0.084 | 0.031 | 1.0984 | 2 | 0.056 |
| M | 0.024 | 0.064 | 2.6926 | 3 | 0.072 | 0.027 | 1.1457 | 2 | 0.048 |
| W | 0.024 | 0.064 | 2.6926 | 3 | 0.072 | 0.027 | 1.1457 | 2 | 0.048 |
| F | 0.022 | 0.060 | 2.7553 | 3 | 0.066 | 0.026 | 1.1724 | 2 | 0.044 |
| G | 0.020 | 0.056 | 2.8241 | 3 | 0.060 | 0.024 | 1.2016 | 2 | 0.040 |
| Y | 0.020 | 0.056 | 2.8241 | 3 | 0.060 | 0.024 | 1.2016 | 2 | 0.040 |
| P | 0.019 | 0.054 | 2.8611 | 3 | 0.057 | 0.023 | 1.2174 | 2 | 0.038 |
| B | 0.015 | 0.045 | 3.0316 | 4 | 0.060 | 0.019 | 1.2899 | 2 | 0.030 |
| V | 0.010 | 0.033 | 3.3241 | 4 | 0.040 | 0.014 | 1.4144 | 2 | 0.020 |
| K | 0.008 | 0.028 | 3.4851 | 4 | 0.032 | 0.012 | 1.4829 | 2 | 0.016 |
| J | 0.002 | 0.009 | 4.4851 | 5 | 0.010 | 0.004 | 1.9084 | 2 | 0.004 |
| X | 0.002 | 0.009 | 4.4851 | 5 | 0.010 | 0.004 | 1.9084 | 2 | 0.004 |
| Q | 0.001 | 0.005 | 4.9851 | 5 | 0.005 | 0.002 | 2.1211 | 3 | 0.003 |
| Z | 0.001 | 0.005 | 4.9851 | 5 | 0.005 | 0.002 | 2.1211 | 3 | 0.003 |
| $\sum$ | 1.000 | | | | | | | | |
| Entropy | | $H_4 = 2.092$ quad/symbol (source) | | | $L^{**} = 2.472$ quad/symbol (code) | $H_{26} = 0.890$ 26-it/symbol (source) | | | $L^{**} = 1.226$ 26-it/symbol (code) |
| Coding efficiency | | | | | 84.63% | | | | 72.63% |

worse for any other $M$-ary codewords. This property[17] can easily be explained through the definition of coding efficiency:

$$\frac{1}{\eta} = \frac{L^{**}}{H_M}$$

$$= \frac{\sum_{x \in X} p(x)\lceil -\log_M x \rceil}{H_M}$$

[17] To my knowledge, this is not found in any textbooks.

$$= \frac{\sum_{x \in X} p(x)[-\log_M x + \varepsilon_M(x)]}{H_M} \tag{8.18}$$

$$= \frac{-\sum_{x \in X} p(x) \log_M x + \sum_{x \in X} p(x)\varepsilon_M(x)}{H_M}$$

$$\equiv 1 + \frac{\langle \varepsilon_M \rangle_X}{H_M},$$

which gives, using the relation $H_M \ln M = H_2 \ln 2$,

$$\eta \equiv \frac{1}{1 + \dfrac{\ln M}{H_2 \ln 2}\langle \varepsilon_M \rangle_X}. \tag{8.19}$$

It is seen that the coding efficiency decreases in inverse proportion to $\ln M$, which formally confirms our previous numerical results. However, this decrease is also a function of $\langle \varepsilon_M \rangle_X$, which represents the mean value of the truncations of $-\log_M p(x)$. Both depend on $M$ and the distribution $p(x)$, therefore, there is no general rule for estimating it. We may, however, conservatively assume that for most probability distributions of interest we should have $0.25 \leq \langle \varepsilon_M \rangle_X \leq 0.5$, which is well satisfied in the example in Table 8.5,[18] and which yields the coding efficiency boundaries:

$$\frac{1}{1 + \dfrac{\ln M}{2H_2 \ln 2}} \leq \eta \leq \frac{1}{1 + \dfrac{\ln M}{4H_2 \ln 2}}. \tag{8.20}$$

This last result confirms that the efficiency of Shannon codes is highest for binary codewords and decreases as the reciprocal of the number of alphabet characters used for the codewords ($\ln M/\ln 2$).

I will show next that the Shannon code is generally optimal when considering *individual* (and not average) codeword lengths, which we have noted $l^{**}(x)$. Consider, indeed, any different code for which the individual codeword length is $l(x)$. A first theorem states that the probability $P$ for which $l^{**}(x) \geq l(x) + n$ ($n$ = number of excess bits between the two codes) satisfies $P \leq 1/2^{n-1}$.[19] Thus, the probability that the Shannon code yields an individual codeword length in excess of two bits (or more) is $1/2$ at maximum. Using the property $P(l^{**} \geq l) = 1 - P(l^{**} < l)$, we can express this theorem under the alternative form:

$$P\left[l^{**}(x) < l(x) + n\right] \geq 1 - \frac{2}{2^n}, \tag{8.21}$$

which shows that the probability of the complementary event ($l^{**} < l + n$) becomes closer to unity as the number of excess bits $n$ increases. Such a condition, however, is not strong enough to guarantee with high probability that $l^{**}(x) < l(x)$ in the majority of cases and for any code associated with $l(x)$.

[18] The detailed calculation yields $\langle \varepsilon_2 \rangle_X \approx 0.39$, $\langle \varepsilon_3 \rangle_X \approx 0.48$, $\langle \varepsilon_4 \rangle_X \approx 0.38$, $\langle \varepsilon_5 \rangle_X \approx 0.47$, and $\langle \varepsilon_{26} \rangle_X \approx 0.33$.

[19] See T. M. Cover and J. A. Thomas, *Elements of Information Theory* (New York: John Wiley & Sons, 1991).

A second theorem of Shannon codes yields an interesting and quite unexpected property. It applies to probability distributions for which $l^{**} = l^*$, meaning that $-\log p(x)$ is an integer for all $x$. As we have seen earlier, such a distribution is said to be *dyadic*. This theorem states that for a *dyadic* source distribution (see proof in previous note):[20]

$$P\left[l^*(x) < l(x)\right] \geq P\left[l^*(x) > l(x)\right], \tag{8.22}$$

where (to recall) $l^*(x)$ is the individual codeword length given by Shannon coding and $l(x)$ is that given by any other code. It is, therefore, *always more likely* for dyadic sources that the Shannon code yields *strictly shorter* individual codeword lengths than the opposite. This is indeed a stronger statement than in the previous case, concerning nondyadic sources. It should be emphasized that these two theorems of Shannon-coding optimality concern *individual*, and not *mean* codeword lengths. As we shall see in the next chapter, there exist codes for which the mean codeword length is actually shorter than that given by the Shannon code.

The different properties that I have established in this section provide a first and vivid illustration of the predictive power of information theory. It allows one to know beforehand the expected minimal code sizes, regardless of the source type and the coding technique (within the essential constraint that the code be a prefix one). This knowledge makes it possible to estimate the efficiency of any coding algorithm out of an infinite number of possibilities, and how close the algorithm is to the ideal or optimal case. The following chapter will take us one step further into the issue of *coding optimality*, which stems from the introduction of *Huffman codes*.

## 8.7 Exercises

**8.1** (B): Consider the source with the symbols, codewords and associated probabilities shown in Table 8.6:

**Table 8.6** Data for Exercise 8.1.

| Symbol | Codeword $x$ | Probability $p(x)$ |
|---|---|---|
| A | 1110 | 0.05 |
| B | 110 | 0.2 |
| C | 1111 | 0.05 |
| D | 0 | 0.4 |
| E | 10 | 0.3 |

Calculate the mean codeword length, the source entropy and the coding efficiency.

**8.2** (M): Provide an example of a dyadic source for a dictionary of four binary codewords, and illustrate Shannon's source-coding theorem.

[20] See T. M. Cover and J. A. Thomas, *Elements of Information Theory* (New York: John Wiley & Sons, 1991).

**8.3** (M): Assign a uniquely decodable code to the symbol-source distribution given in Table 8.7.

**Table 8.7** Data for Exercise 8.3.

| Symbol $x$ | Probability $p(x)$ |
|---|---|
| A | 0.302 |
| B | 0.105 |
| C | 0.125 |
| D | 0.025 |
| E | 0.177 |
| F | 0.016 |
| G | 0.250 |
| $\sum$ | 1.000 |

and determine the corresponding coding efficiency.

**8.4** (M): Assign a Shannon–Fano code to the source defined in Exercise 8.3, and determine the corresponding coding efficiency $\eta = H(X)/L^{**}(X)$ and code redundancy $L^{**}(X) - H(X)$.

# 9 Optimal coding and compression

The previous chapter introduced the concept of *coding optimality*, as based on variable-length codewords. As we have learnt, an optimal code is one for which the mean codeword length closely approaches or is equal to the source entropy. There exist several families of codes that can be called optimal, as based on various types of algorithms. This chapter, and the following, will provide an overview of this rich subject, which finds many applications in communications, in particular in the domain of *data compression*. In this chapter, I will introduce *Huffman codes*, and then I will describe how they can be used to perform data compression to the limits predicted by Shannon. I will then introduce the principle of *block codes*, which also enable data compression.

## 9.1 Huffman codes

As we have learnt earlier, variable-length codes are in the general case more efficient than fixed-length ones. The most frequent source symbols are assigned the shortest codewords, and the reverse for the less frequent ones. The *coding-tree method* makes it possible to find some heuristic codeword assignment, according to the above rule. Despite the lack of further guidance, the result proved effective, considering that we obtained $\eta = 96.23\%$ with a ternary coding of the English-character source (see Fig. 8.3, Table 8.3). But we have no clue as to whether other coding trees with greater coding efficiencies may ever exist, unless we try out all the possibilities, which is impractical.

The *Huffman coding* algorithm provides a near-final answer to the above code-optimality issue. The coding algorithm consists in four steps (here in binary implementation, which is easy to generalize to the $M$-ary case[1]):

(i) List the symbols in decreasing order of frequency/probability;
(ii) Attribute a 0 and a 1 bit to the last two symbols of the list;
(iii) Add up their probabilities, make of the pair a single symbol, and reorder the list (in the event of equal probabilities, always move the pair to the highest position);
(iv) Restart from step one, until there is only one symbol pair left.

[1] In ternary coding, for instance, the symbols must be grouped together by three; in quaternary coding, they should be grouped by four, and so on. If there are not enough symbols to complete the $M$ groups, dummy symbols having zero probabilities can be introduced at the end of the list to complete the tree.

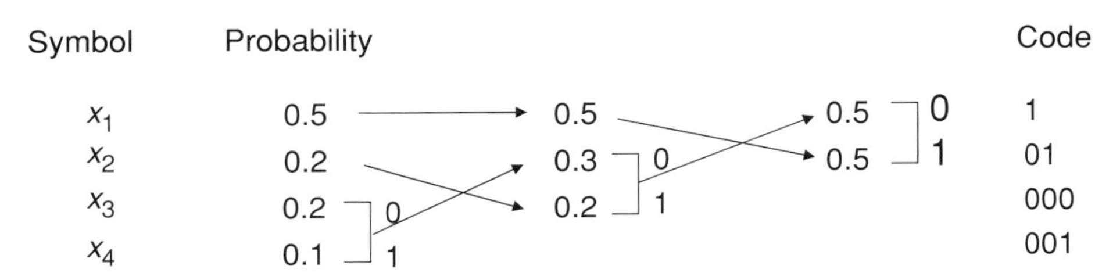


**Figure 9.1** Implementation of Huffman coding for the source $x = \{x_1, x_2, x_3, x_4\}$ with associated probability distribution $p(x) = \{0.5, 0.2, 0.2, 0.1\}$.

An example of Huffman-coding implementation is provided in Fig. 9.1. The source is $X = \{x_1, x_2, x_3, x_4\}$, with respective probabilities $p(x) = \{0.5, 0.2, 0.2, 0.1\}$. The figure shows how the above-described pairs are formed and their probability sums replaced in the next list. We see that the last two symbols $(x_3, x_4)$ represent a probability of 0.3, which comes in second position, after $x_1$ and before $x_2$. Then the pair $[x_2, (x_3, x_4)]$ represents the same probability as $x_1$, but we place it on top of the next list, according to the above rule. The groundwork is now complete. The code assignment simply consists in following the arrows and writing down the 0 or 1 bits encountered in the path. The codeword is the same word but with the bits written in the reverse order. For instance, we find for $x_4$ the bit path 100, so the codeword is 001. For $x_2$, we find 10, so the codeword is 01, and so on.

How efficient is Huffman coding? It is easy to calculate that in the above example, the mean codeword length $L$ is 1.80 bit/word. The entropy $H$ is 1.76 bit/symbol. The coding efficiency is, therefore, $\eta = 1.76/1.8 = 97.8\%$, which is nearly ideal. For comparison, if we had attributed to each of the symbols the same codeword length of two bits, we would have $L = 2$ and $\eta = 1.76/2 = 88\%$. If we used the Shannon–Fano code, it is easily verified that the mean length $L^{**}$ is 2.1 bit/word, and the efficiency drops to $\eta = 1.76/2.1 = 83\%$.

To become further convinced of the merits of Huffman coding, consider again the English-character source, for which we have already explored several coding possibilities (namely, 5-bit ASCII, Morse, and Shannon–Fano). It is a patient exercise (although not without fun, as the reader should experience) to proceed by hand through the successive steps of the Huffman algorithm on a large piece of paper or a blackboard.[2] The results are shown in Table 9.1 (codeword assignment) and Fig. 9.2 (coding tree). As the table shows, the minimum codeword length is three bits (two symbols) and the maximum is nine bits (four symbols). The mean codeword length is 4.212 bit/symbol, yielding a remarkably high efficiency of $\eta = 4.184/4.212 = 99.33\%$. A comparison between Table 9.1 (Huffman code) and Table 8.5 (Shannon code, binary) only reveals apparently small differences in codeword length assignments. The remarkable feature is that such apparently small differences actually make it possible to increase the coding efficiency from 91.42% (Shannon code) to 99.33% (Huffman code).

[2] A recommended class project is to write a computer program aiming to perform a Huffman-coding assignment, with team competition for the fastest and most compact algorithm, while taking sources with a large number of symbols for test examples.

**Table 9.1** Binary Huffman code for English-letter source, showing assigned codeword and calculation of mean codeword length, leading to a coding efficiency of $\eta = 99.33\%$. The two columns at far right show the new probability distribution $q(x) = -\log(2^{-l(x)})$ and the Kullback–Leibler distance $D(p\|q)$.

| $x$ | $p(x)$ | $-p(x)\log_2(x)$ | Codeword | Length $l(x)$ | $p(x)l(x)$ | $q(x)$ | $D(p\|q)$ |
|---|---|---|---|---|---|---|---|
| E | 0.127 | 0.378 | 011 | 3 | 0.380 | 0.125 | 0.002 |
| T | 0.091 | 0.314 | 111 | 3 | 0.272 | 0.125 | −0.042 |
| A | 0.082 | 0.295 | 0001 | 4 | 0.327 | 0.063 | 0.032 |
| O | 0.075 | 0.280 | 0010 | 4 | 0.299 | 0.063 | 0.019 |
| I | 0.070 | 0.268 | 0100 | 4 | 0.279 | 0.063 | 0.011 |
| N | 0.067 | 0.261 | 0101 | 4 | 0.267 | 0.063 | 0.006 |
| S | 0.063 | 0.251 | 1000 | 4 | 0.251 | 0.063 | 0.000 |
| H | 0.061 | 0.246 | 1001 | 4 | 0.243 | 0.063 | −0.002 |
| R | 0.060 | 0.243 | 1010 | 4 | 0.239 | 0.063 | −0.004 |
| D | 0.043 | 0.195 | 00000 | 5 | 0.214 | 0.031 | 0.020 |
| L | 0.040 | 0.185 | 00110 | 5 | 0.199 | 0.031 | 0.014 |
| C | 0.028 | 0.144 | 10110 | 5 | 0.140 | 0.031 | −0.005 |
| U | 0.028 | 0.144 | 10111 | 5 | 0.140 | 0.031 | −0.005 |
| M | 0.024 | 0.129 | 11001 | 5 | 0.120 | 0.031 | −0.009 |
| W | 0.024 | 0.129 | 11010 | 5 | 0.120 | 0.031 | −0.009 |
| F | 0.022 | 0.121 | 11011 | 5 | 0.110 | 0.031 | −0.011 |
| G | 0.020 | 0.113 | 000010 | 6 | 0.120 | 0.016 | 0.007 |
| Y | 0.020 | 0.113 | 000011 | 6 | 0.120 | 0.016 | 0.007 |
| P | 0.019 | 0.108 | 001110 | 6 | 0.114 | 0.016 | 0.005 |
| B | 0.015 | 0.091 | 001111 | 6 | 0.090 | 0.016 | −0.001 |
| V | 0.010 | 0.066 | 110001 | 6 | 0.060 | 0.016 | −0.006 |
| K | 0.008 | 0.056 | 1100000 | 7 | 0.056 | 0.008 | 0.000 |
| J | 0.002 | 0.018 | 110000100 | 9 | 0.018 | 0.002 | 0.000 |
| X | 0.002 | 0.018 | 110000101 | 9 | 0.018 | 0.002 | 0.000 |
| Q | 0.001 | 0.010 | 110000110 | 9 | 0.009 | 0.002 | −0.001 |
| Z | 0.001 | 0.010 | 110000111 | 9 | 0.009 | 0.002 | −0.001 |
| $\sum$ | 1.000 | | | | | 1.000 | |
| Source entropy $H_2$ | | 4.184 bit/symbol | | | | | |
| Coding efficiency $\eta$ | 99.33% | | | | | | |
| Mean codeword length $L$ | | | | | 4.212 | | |
| KL distance | | | | | | | 0.028 |

Looking next at Fig. 9.2, the coding tree is seen to be of order nine, meaning that there are nine branch splits from the root, corresponding to the maximum codeword size. A complete tree of order 9 has $2^9 = 512$ terminal nodes, and $2^{10} - 1 = 1023$ nodes in total. Our coding tree, thus, represents a trimmed version of the complete order-9 tree. One says that the first is "embedded" into the second. Note that the symbol/codeword assignment we have derived here is not unique. For instance, the codewords of letters E and T can be swapped, and any codeword of length $l$ can be swapped within the group of

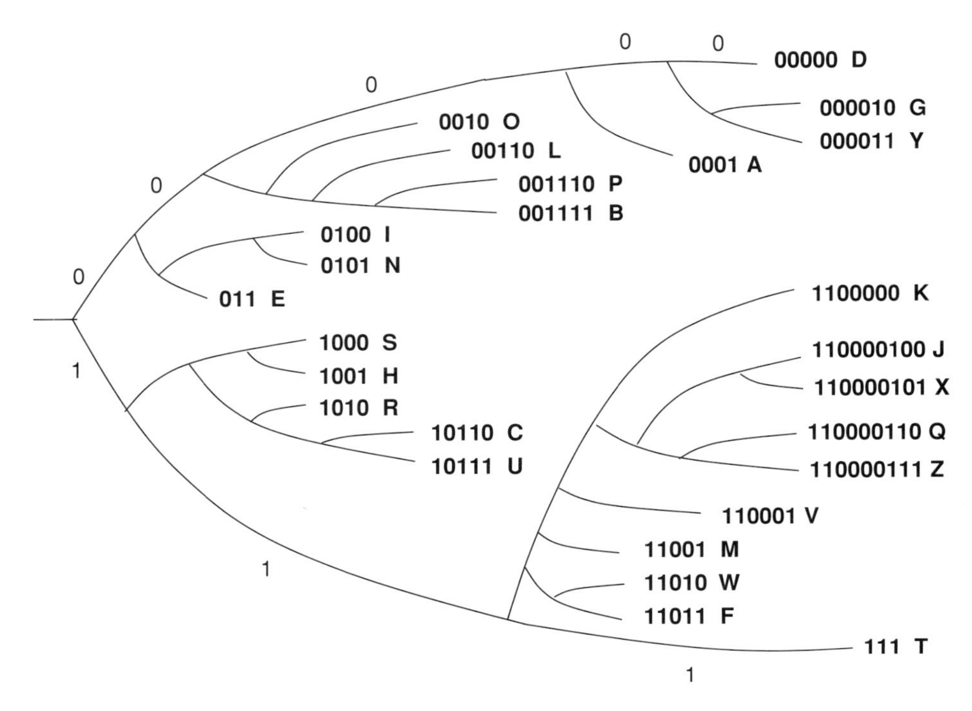


**Figure 9.2** Coding tree for A–Z English-character source and binary codeword assignment (see also Table 8.5). The convention is that branches pointing upwards are labeled 0 and branches pointing downwards are labeled 1.

$m$ symbols assigned the same length (representing $m!$ possible substitutions, for instance $7! = 5040$ for symbol groups numbering $m = 7$). Additionally, all bits can be logically inverted, which generates a duplicated set of codewords.

From the coding tree in Fig. 9.2, one calculates that the number of possible symbol/codeword assignments yielding the same coding efficiency is, in fact:

$$N = 2 \times (2! \times 7! \times 7! \times 5! \times 1! \times 4!) = 292\,626\,432\,000 = 2.92\,10^{11}, \quad (9.1)$$

which represents no less than some *300 billion* possibilities!

The two previous examples illustrate that Huffman coding is an optimal code that brings coding efficiency relatively close to the 100% limit. However, such a conclusion appears to be in contradiction with a key result obtained in Chapter 8 concerning the Shannon–Fano code. Indeed, we have seen that any codeword-length assignment $\hat{l}(x) = \lceil -\log q(x) \rceil$ that is different from that of the Shannon–Fano code, namely, $l^{**}(x) = \lceil -\log p(x) \rceil$, results in a "penalty" for the mean codeword length, $\hat{L}(X)$. Such a penalty is equal to the Kullback–Leibler (KL) distance $D \equiv D(p\|q)$, or, after Eq. (8.17), $\hat{L}(X) \in [H_M(X) + D,\ H_M(x) + D + 1]$. Yet, we should not conclude from this result that $\hat{L}(X) > L^{**}(X)$. As a matter of fact, the situation is exactly opposite for our English-letter coding example. Indeed, we find from Table 9.1 that $D(p\|q) = 0.028$ (and, incidentally, $\langle \varepsilon \rangle_X \equiv 0$ since $q(x)$ is of the form $2^{-m}$). Referring to the results in both Table 8.5 and Table 9.1, we obtain from Eq. (8.15) and Eq. (8.17):

$$L^{**}(X) = 4.577 \in [4.184, 4.184 + 1] \equiv [4.184, 5.184]$$
$$\hat{L}(X) = 4.212 \in [4.184 + 0.028, 4.184 + 0.028 + 1] \equiv [4.212, 5.212],$$

which shows not only that the Huffman code is more efficient (or closer to optimal) than the Shannon code ($\hat{L}(X) < L^{**}(X)$) but also that in this example the penalty $D(p\|q)$ is very close to the difference $\Delta = \hat{L}(X) - H_2(X)$. Incidentally, it appears here that $\Delta = 0$, but this is only an effect of number accuracy.[3]

How does the Huffman code perform with *dyadic* sources? It is easy to establish from any basic example with source $p(x) = 2^{-n(x)}$ ($n(x)$ = integer) that, in such a case, the Huffman-code algorithm assigns the individual codeword length $l^*(x) = -\log 2^{-n(x)} = n(x)$. This is the same result as for the Shannon–Fano code and, as we have seen earlier, the resulting mean length is the absolute minimum, or $\hat{L}(X) = L^{**}(X) \equiv H(X)$. It is a nice exercise to prove directly the general property according to which Huffman coding of dyadic sources is 100% efficient.

The important lesson and general conclusion to retain from the above analysis is that the mean codeword length of Huffman codes $\hat{L}(X)$ is the smallest achievable, regardless of the source distribution. One can also state that Huffman coding is optimal with respect to any other code, meaning that the latter always yields longer or equal codeword lengths. Such a property of optimality can be stated as follows:

If $L(X)$ is the mean codeword length of any code different from the Huffman code, then $L(X) \geq \hat{L}(X)$.

The reader may refer to most IT textbooks for a formal proof of Huffman codes optimality. As we have seen in Chapter 8, however, the *individual* codeword lengths, $l^*(x) \equiv \lceil -\log p(x) \rceil$ assigned by the Shannon–Fano code are *generally shorter* than that provided by any other codes, including Huffman codes. Formally, this property can be written under the form

$$\begin{aligned} l^*(x) &\leq \hat{l}(x), \quad \textit{most often} \\ L^{**}(X) &\geq \hat{L}(x), \quad \textit{always}. \end{aligned} \tag{9.2}$$

The secondary conclusion is that for *individual* codeword lengths, the Shannon code is optimal in *most* cases, while for *mean* codeword lengths, the Huffman code is optimal in *all* cases.[4] This represents a most important property for *data compression* applications, to be described in the next section.

As we have seen earlier, the redundancy bound of Shannon–Fano codes is $\rho_{\text{bound}} = 1$, meaning that in the general case, $\rho = L^{**} - H < 1$. It can be shown that for Huffman codes,[5] the redundancy bound is $\rho_{\text{bound}} = p_{\min} + \log_2[(2\log_2 \text{e})/\text{e}] \approx p_{\min} + 0.08607$,

[3] The detailed calculation to five decimal places shows that $H_2(X) = 4.18396$, $L(X) = 4.21236$, $D(p\|q) = 0.02840$, and $\Delta = 0.05680$, which represents a small nonzero difference.

[4] Note that the English-letter example (see Tables 8.5 and 9.1) corresponds to the case where the Huffman code is also optimal for individual codeword lengths.

[5] See: D. J. C. MacKay, *Information Theory, Inference and Learning Algorithms* (Cambridge, UK: Cambridge University Press, 2003).

where $p_{\min}$ is the smallest source-symbol probability. The Huffman coding efficiency, $\eta = H/\hat{L}$, is, thus, guaranteed to be strictly greater than $1 - \rho/\hat{L}$. In the example of Table 9.1, for instance, we find $\hat{L} - H = 4.212 - 4.184 = 0.028$, which is lower than $\rho_{\text{bound}} = p_{\min} + 0.08607 \equiv 0.001 + 0.08607 = 0.0877$, and $\eta = 99.33\%$, which is greater than $1 - 0.0877/4.212 = 97.9\%$. Even tighter redundancy bounds can also be found for specific probability distributions.[6]

## 9.2 Data compression

Any information source, such as texts, messages, recordings, data files, voice or music recordings, still or motion pictures, can be coded into sequences of binary codewords. As we have learnt, the encoding conversion can use either fixed-length or variable-length codes. In the first case, the bit size of the resulting sequence length is a multiple of the codeword length. In the second case, the sequence can be made shorter through an adequate choice of code, or through a second coding conversion, from a first fixed-length code to a better, variable-length code. Converting a binary codeword sequence into a shorter one is called *data compression*.

Since the advent of the Internet and electronic mail, we know from direct personal experience what data compression means: if compression is not used, it may seem to take forever to upload or download files like text, slides, or pictures. Put simply, compression is the art of *packing a maximum of data into a smallest number of bits*. It can be achieved by the proper and optimal choice of *variable-length codes*. As we have learnt, Shannon–Fano and Huffman codes are both optimal, since, in the general case, their mean codeword length is shorter than the fixed-length value given by $L = \log_2 N$ ($N$ = number of source symbols to encode).

The *compression rate* is defined by taking a fixed-length code as a reference, for instance the ASCII code with its 7 bit/symbol codewords. If, with a given symbol source, a variable-length code yields a shorter mean codeword length $L < 7$, the compression rate is defined as $r = 1 - L/7$. The compression rate should not be confused with *coding efficiency*. The latter measures how close the mean codeword length is to the source entropy, which we know to be the ultimate lower bound. This observation intuitively suggests that *efficient codes do not lend themselves to efficient compression*. Yet, this correct conclusion does not mean that further compression towards the limits given by the source entropy is impossible, as we will see in the last section in this chapter, which is concerned with *block coding*.

The previous examples with the A–Z English-character source shown in Table 8.5 (Shannon–Fano coding) and Table 9.1 (Huffman coding) represent good illustrations of the data compression concept. Here, we shall take as a comparative reference the codeword length of *five bits* used by ASCII to designate the restricted set of lower case or upper case letters. With the results from the two aforementioned examples, namely

[6] See: R. G. Gallager, Variations on a theme by Huffman. *IEEE Trans. Inform. Theory*, **24** (1978), 668–74.

$L^{**} = 4.577$ bit/symbol and $\tilde{L} = 4.212$ bit/symbol, respectively, we obtain:

$$r^{**} = 1 - \frac{L^{**}}{5} = 1 - \frac{4.577}{5} = 8.46\% \quad \text{(Shannon–Fano)},$$

$$\tilde{r} = 1 - \frac{\tilde{L}}{5} = 1 - \frac{4.212}{5} = 15.76\% \quad \text{(Huffman)}.$$

The above compression rates may look modest, and indeed they are when considering the specific example of the English language. Recall that the mean codeword length cannot be lower than the source entropy $H$. Thus, there exists no code for which $L < H$. Since, in this example, the source entropy is $H = 4.185$ bit/symbol, we see that the compression rate is intrinsically limited to $r = 1 - 4.185/5 = 16.3\%$. But this limit cannot be reached, since no code can beat the Huffman code, and $\tilde{r} = 15.76\%$, thus, represents the really achievable compression limit for this source example.

Here we come to a fine point in the issue of data compression. It could have been intuitively concluded that, given a source, the compression rate is forever fixed by the coding algorithm that achieves compression. But as it turns out, the compression rate also depends on the source distribution itself. To illustrate this point, consider four different ASCII text sources and check how Huffman compression works out for each of them. Here are three English sentences, or "datafiles," which have been made (ridiculously so) nontypical:

Datafile 1: "Zanzibar zoo zebra Zazie has a zest for Jazz"

Datafile 2: "Alex fixed the xenon tube with wax"

Datafile 3: "The sass of Sierra snakes in sunny season"

The above sentences are nontypical in the fact that certain letters are unusually redundant for English. Therefore, the frequency distributions associated with the symbol-characters used are very different from average English text sources. For comparison purposes, we introduce a fourth example representing an ordinary English sentence of similar character length:

Datafile 4: "There is a parking lot two blocks from here"

We shall now analyze how much these four sources (or datafiles) can be compressed through Huffman coding. Table 9.2 shows the corresponding symbol-character frequencies and distributions (overlooking spaces and upper or lower case differences), codeword assignment, mean codeword length, and compression rate, taking for reference the fixed-codeword length $L_{\text{ref}} = \lceil \log_2 26 \rceil = 5$, corresponding to a restricted, five-bit ASCII code.

It is seen from the results in Table 9.2 that the compression ratio varies from 20% (datafile 4) to about 35% (datafile 3). The difference in compression ratio cannot be attributed to the datafile length: the best and the worst results are given by datafiles having the same length of 34–35 characters. With only 28 characters, the shortest datafile (datafile 2) has a compression ratio of only 27%. The number of different characters is about the same for datafiles 1–3, but the compression ratio varies from 27%

**Table 9.2** Examples of file compression through Huffman coding based on the four sentences shown in the text. For each file, the tables list the characters ($x$), their frequency ($f$), their probability ( $p(x)$), the codeword assignment (CW), and the codeword length ($l$ ($x$)).

| Datafile 1 (zebra) | | | | | | Datafile 2 (xenon) | | | | | | Datafile 3 (snakes) | | | | | | Datafile 4 (parking) | | | | | |
|---|---|---|---|---|---|---|---|---|---|---|---|---|---|---|---|---|---|---|---|---|---|---|---|
| x | f | $p(x)$ | CW | $l(x)$ | $l(x)p(x)$ | x | f | $p(x)$ | CW | $l(x)$ | $l(x)p(x)$ | x | f | $p(x)$ | CW | $l(x)$ | $l(x)p(x)$ | x | f | $p(x)$ | CW | $l(x)$ | $l(x)p(x)$ |
| Z | 9 | 0.2500 | 10 | 2 | 0.500 | E | 5 | 0.1786 | 11 | 2 | 0.357 | S | 9 | 0.2647 | 01 | 2 | 0.529 | E | 4 | 0.1143 | 100 | 3 | 0.343 |
| A | 7 | 0.1944 | 11 | 2 | 0.389 | X | 4 | 0.1429 | 000 | 3 | 0.429 | N | 5 | 0.1471 | 000 | 3 | 0.441 | O | 4 | 0.1143 | 011 | 3 | 0.343 |
| E | 3 | 0.0833 | 0001 | 4 | 0.333 | T | 3 | 0.1071 | 0010 | 4 | 0.429 | A | 4 | 0.1176 | 110 | 3 | 0.353 | R | 4 | 0.1143 | 010 | 3 | 0.343 |
| O | 3 | 0.0833 | 0010 | 4 | 0.333 | A | 2 | 0.0714 | 0101 | 4 | 0.286 | E | 4 | 0.1176 | 111 | 3 | 0.353 | T | 3 | 0.0857 | 1110 | 4 | 0.343 |
| R | 3 | 0.0833 | 0011 | 4 | 0.333 | H | 2 | 0.0714 | 0110 | 4 | 0.286 | I | 2 | 0.0588 | 1001 | 4 | 0.235 | A | 2 | 0.0571 | 0011 | 4 | 0.229 |
| B | 2 | 0.0556 | 1010 | 4 | 0.222 | I | 2 | 0.0714 | 0111 | 4 | 0.286 | O | 2 | 0.0588 | 1010 | 4 | 0.235 | H | 2 | 0.0571 | 0010 | 4 | 0.229 |
| I | 2 | 0.0556 | 1011 | 4 | 0.222 | N | 2 | 0.0714 | 1000 | 4 | 0.286 | R | 2 | 0.0588 | 1011 | 4 | 0.235 | I | 2 | 0.0571 | 0001 | 4 | 0.229 |
| S | 2 | 0.0556 | 00000 | 5 | 0.278 | W | 2 | 0.0714 | 1001 | 4 | 0.286 | F | 1 | 0.0294 | 00100 | 5 | 0.147 | K | 2 | 0.0571 | 0000 | 4 | 0.229 |
| F | 1 | 0.0278 | 00001 | 5 | 0.139 | B | 1 | 0.0357 | 1010 | 4 | 0.143 | H | 1 | 0.0294 | 00101 | 5 | 0.147 | L | 2 | 0.0571 | 11111 | 5 | 0.286 |
| H | 1 | 0.0278 | 10000 | 5 | 0.139 | D | 1 | 0.0357 | 1011 | 4 | 0.143 | K | 1 | 0.0294 | 00110 | 5 | 0.147 | S | 2 | 0.0571 | 11110 | 5 | 0.286 |
| J | 1 | 0.0278 | 10001 | 5 | 0.139 | F | 1 | 0.0357 | 00110 | 5 | 0.179 | T | 1 | 0.0294 | 00111 | 5 | 0.147 | B | 1 | 0.0286 | 11011 | 5 | 0.143 |
| N | 1 | 0.0278 | 10010 | 5 | 0.139 | L | 1 | 0.0357 | 00111 | 5 | 0.179 | U | 1 | 0.0294 | 10000 | 5 | 0.147 | C | 1 | 0.0286 | 11010 | 5 | 0.143 |
| T | 1 | 0.0278 | 10011 | 5 | 0.139 | O | 1 | 0.0357 | 01111 | 5 | 0.179 | Y | 1 | 0.0294 | 10001 | 5 | 0.147 | F | 1 | 0.0286 | 11001 | 5 | 0.143 |
| C | 0 | 0.0000 | | | | U | 1 | 0.0357 | 01001 | 5 | 0.179 | B | 0 | 0.0000 | | | | G | 1 | 0.0286 | 11000 | 5 | 0.143 |
| D | 0 | 0.0000 | | | | C | 0 | 0.0000 | | | | C | 0 | 0.0000 | | | | M | 1 | 0.0286 | 10111 | 5 | 0.143 |
| G | 0 | 0.0000 | | | | G | 0 | 0.0000 | | | | D | 0 | 0.0000 | | | | N | 1 | 0.0286 | 10110 | 5 | 0.143 |
| K | 0 | 0.0000 | | | | J | 0 | 0.0000 | | | | G | 0 | 0.0000 | | | | P | 1 | 0.0286 | 10101 | 5 | 0.143 |
| L | 0 | 0.0000 | | | | K | 0 | 0.0000 | | | | J | 0 | 0.0000 | | | | W | 1 | 0.0286 | 10100 | 5 | 0.143 |
| M | 0 | 0.0000 | | | | M | 0 | 0.0000 | | | | L | 0 | 0.0000 | | | | D | 0 | 0.0000 | | | |
| P | 0 | 0.0000 | | | | P | 0 | 0.0000 | | | | M | 0 | 0.0000 | | | | J | 0 | 0.0000 | | | |
| Q | 0 | 0.0000 | | | | Q | 0 | 0.0000 | | | | P | 0 | 0.0000 | | | | Q | 0 | 0.0000 | | | |
| U | 0 | 0.0000 | | | | R | 0 | 0.0000 | | | | Q | 0 | 0.0000 | | | | U | 0 | 0.0000 | | | |
| V | 0 | 0.0000 | | | | S | 0 | 0.0000 | | | | V | 0 | 0.0000 | | | | V | 0 | 0.0000 | | | |
| W | 0 | 0.0000 | | | | V | 0 | 0.0000 | | | | W | 0 | 0.0000 | | | | X | 0 | 0.0000 | | | |
| X | 0 | 0.0000 | | | | Y | 0 | 0.0000 | | | | X | 0 | 0.0000 | | | | Y | 0 | 0.0000 | | | |
| Y | 0 | 0.0000 | | | | Z | 0 | 0.0000 | | | | Z | 0 | 0.0000 | | | | Z | 0 | 0.0000 | | | |
| $\sum$ | 36 | 1.0000 | | | | | 28 | 1.0000 | | | | | 34 | 1.0000 | | | | | 35 | 1.0000 | | | |
| Mean CW length (bit/symbol) | | | | | **3.306** | | | | | | **3.643** | | | | | | **3.265** | | | | | | **4.000** |
| Compression ratio | | | | | **33.89%** | | | | | | **27.14%** | | | | | | **34.71%** | | | | | | **20.00%** |

to 34.7%. A closer look at the table data shows that the factor that appears to increase the compression ratio is the frequency spread in the top group of most frequent characters. If the most frequent characters have dissimilar frequencies, then shorter codewords can be assigned to a larger number of symbols. We observe that the first three datafiles, corresponding to nontypical English texts, lend themselves to greater compression than the fourth datafile, corresponding to ordinary English. There is no need to go through tedious statistics to conclude beforehand that increasing the length of such English-text datafiles would give compression ratios increasingly closer to the limit of $\tilde{r} = 15.76\%$. Clearly, this is because the probability distribution of long English-text sequences will duplicate with increasing fidelity the standard distribution for which we have found this compression limit. On the other hand, shorter sequences of only a few characters might have significantly higher compression ratios. To take an extreme example, the datafile "AAA" (for American Automobile Association) takes 1 bit/symbol and, thus, has a compression ratio of $r = 1 - 1/5 = 80\%$. If we take the full ASCII code for reference (7 bit/character), the compression becomes $r = 1 - 1/7 = 85.7\%$.[7]

The above examples have shown that for any given datafile, there exists an optimal (Huffman) code that achieves maximum data compression. As we have seen, the codeword assignment is different for each datafile to be compressed. Therefore, one needs to keep track of which code is used for compression in order to be able to recover the original, uncompressed data. This information, which we refer to as *overhead*, must then be transmitted along with the compressed data, which we refer to as *payload*. Since the overhead bits reduce the effective compression rate, it is clear that the overhead size should be the smallest possible relatively to the payload. In the previous examples, the overhead is simply the one-to-one correspondence table between codewords and symbols. Using five-bit (ASCII) codewords to designate each of the character symbols, and a five-bit field to designate the corresponding compressed codewords makes a ten-bit overhead per datafile symbol. Taking, for instance, datafile 3 (Table 9.2), there are 13 symbols, which produces 130 bits of overhead. It is easily calculated that the payload represents 111 bits, which leads to a total of 130 + 111 = 241 bits for the complete compressed file (overhead + payload). In contrast, a five-bit ASCII code for the same uncompressed datafile would represent only 170 bits, as can also be easily verified. The compressed file thus turns out to be 40% bigger than the uncompressed one! The conclusion is that

[7] This consideration illustrates the interest of *acronyms*. Their primary use is to save text space, easing up reading and avoiding burdensome redundancies. This is particularly true with technical papers, where the publication space is usually limited. An equally important use of acronyms is to capture abstract concepts into small groups of characters, for instance ADSL (*asymmetric digital subscriber line)* or HTML (*hypertext markup language)*. The most popular acronyms are the ones that are easy to remember, such as FAQ (*frequently asked questions*), IMHO (*in my humble opinion*), WYSIWYG (*what you see is what you get*), NIMBY (*not in my backyard*), and the champion VERONICA (*very easy rodent-oriented netwide index to computerized archives*), for instance. The repeated use of acronyms makes them progressively accepted as true English words or generic brand names, to the point that their original character-to-word correspondence is eventually forgotten by their users, for instance: PC for *personal computer*, GSM for *global system for mobile* [*communications*], LASER for *light amplification by stimulated emission of radiation*, NASDAQ for *National Association of Securities Dealers Automated Quotations*, etc. Language may thus act a natural self-compression machine, which uses the human mind as a convenient dictionary. In practice, this dictionary is only rarely referred to, since the acronym gains its own meaning by repeated use.

compression can be efficient with a given datafile, but it is a worthless operation if the resulting overhead (needed to decompress the data) is significantly larger than the payload. But the "overhead tax" is substantially reduced when significantly longer datafiles are compressed. The case of plain English-text datafiles is not the best example, because as their size increases the symbol probability distribution becomes closer to that of the standard English-source reference, for which (as we have shown) the compression ratio is limited to $\tilde{r} = 15.76\%$.[8]

One can alleviate the overhead tax represented by the coding-tree information by using a standard common reference, which is called the *codebook*. Such a codebook contains different optimal coding trees for generic sources as varied as standard languages (English, French, German, etc.), programming-language source codes ($C^{++}$, Pascal, FORTRAN, HTML, Java, etc.), tabulated records (students, payroll, company statistics, accounting, etc.), or just raw binary datafiles, for instance. An optimal coding tree devised for *all* inventoried source types can also be included in the codebook. Choosing a specific coding tree from the codebook is called *semantic-dependent coding*. This choice means that one has prior knowledge of the type of source or source semantics considered, and this knowledge guides the choice of the most appropriate code to select from the codebook menu.

If the codebook contains $N$ coding trees for these different sources, the overhead only represents $\log_2 N$ bits. This overhead must be included at the beginning of the compressed file, to indicate which coding tree (or source *mapping*) has been used for compression. The coding efficiency (or compression ratio) obtained with a codebook is never greater than that obtained with a case-specific Huffman coding. However, if one includes the overhead bits in the compressed file, the conclusion could be the opposite. Indeed, a 1985 experiment consisted in comparing results obtained by compressing different types of programming-language source codes (out of 530 programs), using either a codebook or case-by-case Huffman coding, the corresponding overheads being included into each computation.[9] The result was that the codebook approach always produced higher compression ratios. The conclusion is that, taking into account the overhead, a nonoptimal coding tree picked from a generic codebook may yield a better compression performance than a case-specific Huffman coding tree. Other studies have sought to generate a universal codebook tree, which could apply indifferently to English, French, German, Italian, Spanish, or Portuguese, for instance. The spirit of the approach and procedure can be described as follows. First, an optimal coding tree is computed with all sources combined (taking for database reference the full contents of a number of recent newspapers, magazines, books, and so on from each language). Second, optimal coding trees are computed for each individual source. The task then is to identify what the different trees share in common and devise the universal tree accordingly, subject to certain constraints to be respected for each language source.

[8] Under the simplifying assumption of a single A–Z alphabet without spaces, punctuation, or any other symbols.

[9] See R. M. Capocelli, R. Giancarlo, and I. J. Taneja, Bounds on the redundancy of Huffman codes. *IEEE Trans. Infor. Theory*, **32** (1986), 854–7.

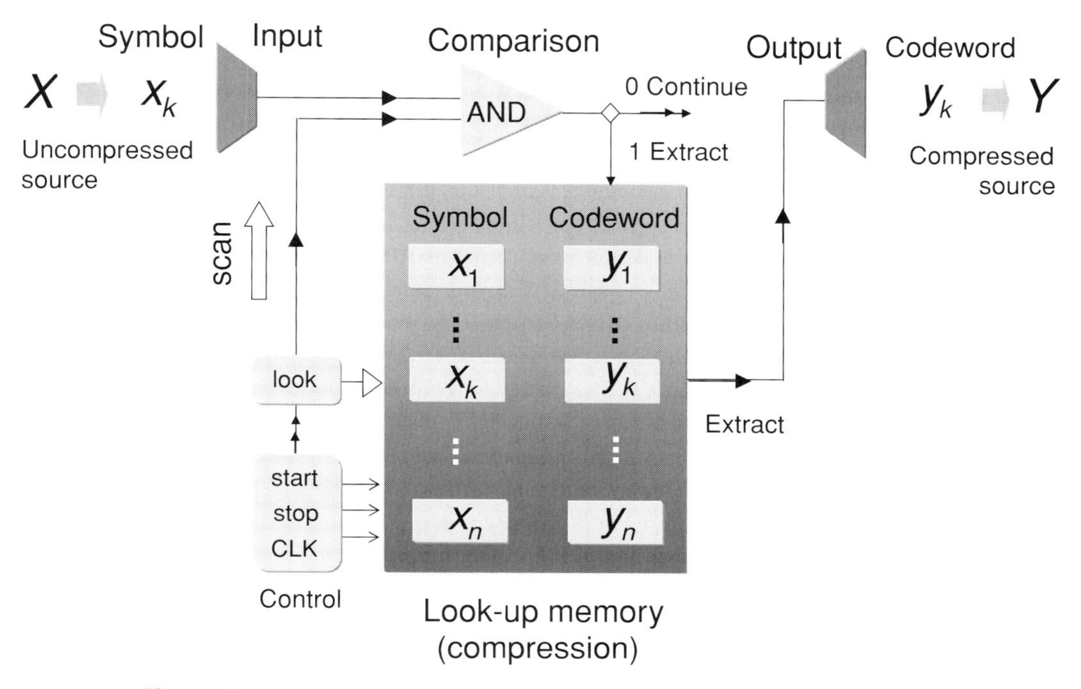


**Figure 9.3** Generic implementation of data compression from source $X$ (uncompressed source symbols $x_k$) into source $Y$ (compressed codewords $y_k$) through a memory look-up system. The operation of decompression is the same, with the roles of $Y$ and $X$ being interchanged.

The same code-compression techniques and conclusions apply to data sources far more complex than text, such as digital *sound* and still or motion *pictures*. Such sources contain large amounts and varieties of random symbols, whose frequency distributions significantly depart from the uniform and exponential generic types. As we have seen, this makes compression algorithms all the more efficient. Since the files can be made arbitrarily large, the overhead tax can become relatively insignificant. Case-specific compression based on Huffman coding (referred to as *Huffman compression*) is widely used in telecom devices such as faxes and modems, and in multimedia applications with well known standards for music, images, and video, known as *MP3*, *JPEG*, and *MPEG*, for instance. A brief overview of common compression standards is given in Appendix G. The task of compressing sound and pictures is, however, made more complex because the source's probability distribution is inherently unknown, and no generic codebook may exist or prove effective. A solution to this problem is referred to as *adaptive coding*, where optimal coding trees are devised "on the fly" as new source symbols come in the sequence. This will be described in Chapter 10.

The reverse operation of back-translating a coded datafile or a codeword sequence, is called *decompression*. Figure 9.3 illustrates the generic system layout used for code compression or decompression. As the figure shows, a dedicated memory provides the one-to-one correspondence between source symbols and codewords (compression) or the reverse (decompression). The operation of extracting the information stored in the memory is called *table look-up*. The memory is scanned address by address, until the symbol (or codeword) is found, in which case the corresponding codeword (or symbol) is output from the memory. It is, basically, the same operation as if we look at Table 9.1 (the memory) to find the codeword for the source character U (the output or answer being 10111), or to find the source character for the codeword 110000100 (the output or answer being J). Note that decompression with a Huffman code is easier than with any other code, because we know from the codeword length and prefix where to look at in the memory. For a machine, it takes a simple program to determine at what memory location to start the look-up search, given the codeword size and prefix bits. For block codes (see next section), the compression/decompression operation is similar, but compression first requires arranging the input symbols into sequences of predefined size (e.g., bytes for binary data). As Fig. 9.3. illustrates, the compression/decompression apparatus is completed with a control subsystem ensuring a certain number of generic functions, including synchronization between the input symbol or codewords and the memory.

Assigning individual codewords to each source symbol, in a way that is *uniquely decodable*, corresponds to what is called *lossless compression*. The idea conveyed by this adjective is that no information is lost through the compression. Assuming that the compressed source remains unaltered by storage or transmission conditions, the reverse operation of code decompression results in a perfect restitution of the original symbol sequence, with no alteration or error. In contrast, lossy compression algorithms cause some information loss, which translates into errors upon decompression. This information loss, and the restitution errors, is sufficiently minor to remain unnoticed, as in a sound track or a movie picture. Such lossy compression algorithms are inherently not applicable to data files, where absolute exactness is a prerequisite, as in any text document (press, literature, technical) or tabulated data (services, finance, banking, records, computer programs).

## 9.3 Block codes

So far in this chapter, I have described different codes with either fixed-length or variable-length codewords. In this section, we shall consider yet another coding strategy, which consists in encoding the symbols by *blocks*. The result is a *block code,* a generic term, which applies to any code that manipulates groups of codewords, either by concatenation or by the attribution of new codewords for specific groups of source symbols.

Most generally, block codes can be attributed different "sub-blocks" or *fields*. For instance, a given field can be reserved for the *payload* (the sequence of codewords to be transmitted) and another field to the *overhead* (the information describing how to handle and decompress the payload). Block codes are used, for instance, in *error-correction*, which will be described in Chapter 11. Here, we shall focus on the simplest type of block codes, which make it possible to encode source symbols with higher bit/codeword efficiency. As we shall see, block coding results in significantly longer codewords, and the codeword dictionary is also significantly greater. The key advantage is that the mean bit/codeword is considerably reduced, which enables much more data to be packed per

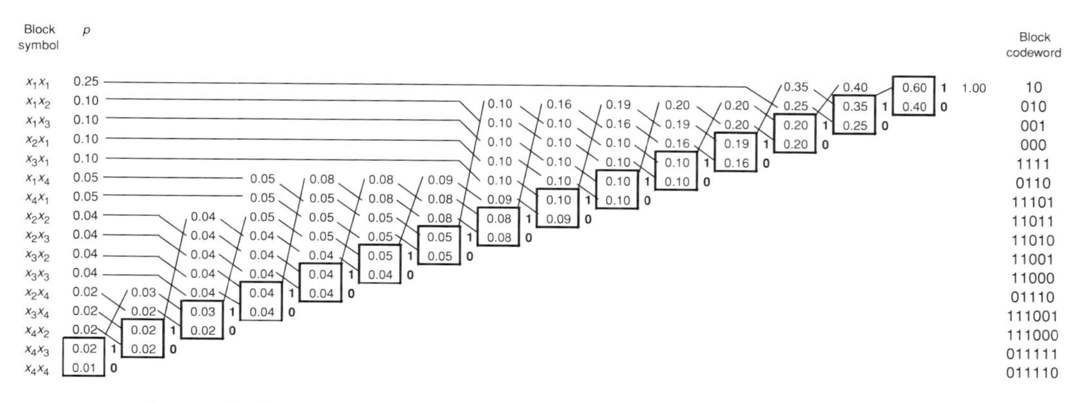


**Figure 9.4** Block coding by symbol pairs $x_ix_j$ of the source $X = \{x_1, x_2, x_3, x_4\}$, with respective probabilities $p(x) = \{0.5, 0.2, 0.2, 0.1\}$ through the Huffman-coding algorithm. The column at left shows the block codeword probabilities in decreasing order. The column at right shows the block codeword assignment resulting from the algorithm implementation.

transmitted bit. I shall illustrate this property with three examples, which have also been designed to make the subject entertaining.

---

**Example 9.1:** ***Four-event source***

Consider the four-event source $X = \{x_1, x_2, x_3, x_4\}$ previously analyzed in Fig. 9.1, which is defined by the distribution $p(x) = \{0.5, 0.2, 0.2, 0.1\}$. As we have seen, its entropy is $H(X) = 1.76$ bit/symbol, and its optimal mean length (Huffman coding) is $L(X) = 1.80$ bit/symbol. The corresponding coding efficiency is $\eta = 1.76/1.80 = 97.8\%$. We shall now implement block-coding by grouping the symbols in ordered pairs $(x_ix_j)$, forming the new and extended 16-event source $X' = \{x_1x_1, x_1x_2, x_1x_3, \ldots, x_4x_2, x_4x_3, x_4x_4\}$ with associated probabilities $p(x_ix_j) = p(x_i)p(x_j)$, the events being assumed to be independent. It is a patient exercise to determine the Huffman code for $X'$. The result is shown in Fig. 9.4, with details and results summarized in Table 9.3. We find that the extended source entropy is $H(X') = 3.522$ bit/symbol, which is equal to $2H(X)$, as expected for a source of joint independent events. Then we find that the mean codeword length is $L(X') = 3.570$ bit/word. The coding efficiency is, therefore, $\eta = 3.522/3.570 = 98.65\%$, which is an improvement on the previous efficiency $\eta = 97.77\%$ of the single-codeword Huffman code. Since each block code represents a pair of symbols, the actual mean symbol length is $L(X')/2 = 1.78$ bit/symbol. This reduction in symbol length may look small with respect to $L(X) = 1.80$ bit/symbol for the single-symbol code but, as we have seen, it has a noticeable impact on coding efficiency.

We can continue to improve the coding efficiency by grouping symbols in ordered triplets $(x_ix_jx_k)$, quadruplets $(x_ix_jx_kx_l)$, or $n$-tuplets $(x_ix_jx_k, \ldots, x_\#)$ of arbitrary large sizes, corresponding to a source $X^{(n)}$ assigned with $4^n$ individual codewords. The source $X^{(n)}$ is referred to as the $n$th *extension* of the source $X$.

As we know well from *Shannon's source-coding theorem*, (Chapter 8), the mean codeword length is $\hat{L}(X^{(n)}) \geq H(X^{(n)}) = nH(X)$. For large $n$, we have $\hat{L}(X^{(n)}) \approx nH(X)$,

**Table 9.3** Block coding by symbol pairs $x_i x_j$ of the source $X = \{x_1, x_2, x_3, x_4\}$, with respective probabilities $p(x) = \{0.5, 0.2, 0.2, 0.1\}$ through the Huffman coding algorithm. The table at left shows the block codeword probabilities and the corresponding entropy of the extended source $X' = \{x_1x_1, x_1x_2, x_1x_3, \ldots x_4x_2, x_4x_3, x_4x_4\}$. The table at right shows the block symbols ordered in decreasing probabilities with their Huffman block-codeword assignment.

| Symbol $i$ | Symbol $j$ | $p(x_i)$ | $p(x_j)$ | Block symbol | $p_{ij} = p(x_i)p(x_j)$ | $-p_i p_j \log_2 p_{ij}$ | Block symbol | $p_{ij}$ | Block codeword | $l(x_i x_j)$ | $p_{ij} l(x_i x_j)$ |
|---|---|---|---|---|---|---|---|---|---|---|---|
| $x_1$ | $x_1$ | 0.5 | 0.5 | $x_1x_1$ | 0.250 | 0.500 | $x_1x_1$ | 0.250 | 10 | 2 | 0.500 |
| | $x_2$ | 0.5 | 0.2 | $x_1x_2$ | 0.100 | 0.332 | $x_1x_2$ | 0.100 | 010 | 3 | 0.300 |
| | $x_3$ | 0.5 | 0.2 | $x_1x_3$ | 0.100 | 0.332 | $x_1x_3$ | 0.100 | 001 | 3 | 0.300 |
| | $x_4$ | 0.5 | 0.1 | $x_1x_4$ | 0.050 | 0.216 | $x_2x_1$ | 0.100 | 000 | 3 | 0.300 |
| $x_2$ | $x_1$ | 0.2 | 0.5 | $x_2x_1$ | 0.100 | 0.332 | $x_3x_1$ | 0.050 | 1111 | 4 | 0.400 |
| | $x_2$ | 0.2 | 0.2 | $x_2x_2$ | 0.040 | 0.186 | $x_1x_4$ | 0.040 | 0110 | 4 | 0.200 |
| | $x_3$ | 0.2 | 0.2 | $x_2x_3$ | 0.040 | 0.186 | $x_4x_1$ | 0.040 | 11101 | 5 | 0.250 |
| | $x_4$ | 0.2 | 0.1 | $x_2x_4$ | 0.020 | 0.113 | $x_2x_2$ | 0.020 | 11011 | 5 | 0.200 |
| $x_3$ | $x_1$ | 0.2 | 0.5 | $x_3x_1$ | 0.100 | 0.332 | $x_2x_3$ | 0.100 | 11010 | 5 | 0.200 |
| | $x_2$ | 0.2 | 0.2 | $x_3x_2$ | 0.040 | 0.186 | $x_3x_2$ | 0.040 | 11001 | 5 | 0.200 |
| | $x_3$ | 0.2 | 0.2 | $x_3x_3$ | 0.040 | 0.186 | $x_3x_3$ | 0.040 | 11000 | 5 | 0.200 |
| | $x_4$ | 0.2 | 0.1 | $x_3x_4$ | 0.020 | 0.113 | $x_2x_4$ | 0.020 | 01110 | 5 | 0.100 |
| $x_4$ | $x_1$ | 0.1 | 0.5 | $x_4x_1$ | 0.050 | 0.216 | $x_3x_4$ | 0.050 | 111001 | 6 | 0.120 |
| | $x_2$ | 0.1 | 0.2 | $x_4x_2$ | 0.020 | 0.113 | $x_4x_2$ | 0.020 | 111000 | 6 | 0.120 |
| | $x_3$ | 0.1 | 0.2 | $x_4x_3$ | 0.020 | 0.113 | $x_4x_3$ | 0.020 | 011111 | 6 | 0.120 |
| | $x_4$ | 0.1 | 0.1 | $x_4x_4$ | 0.010 | 0.066 | $x_4x_4$ | 0.010 | 011110 | 6 | 0.060 |
| $\sum$ | | | | | 1.000 | | | | 1.000 | | |
| Extended source entropy | | | | | | 3.522 | | | | | |
| Mean block length | | | | | | | | | | | 3.570 bit/word |
| Efficiency | | | | | | | | | | | 98.65% |

or $\hat{L}(X^{(n)})/n \approx H(X) = 1.76$ bit/symbol and for the coding efficiency, $\eta \approx 100\%$. This example shows that one can reach the theoretical limit of 100% coding efficiency with arbitrary accuracy, but at the price of using an extended dictionary of codewords, most with relatively long lengths.

---

**Example 9.2:** ***26-event source; the English-language characters***

Block coding may not be so practical when applied to sources having more than two events. This is because Huffman coding is very close to being the most efficient, and the extra complexity introduced by using an extended source with a long list of variable-length codewords is not so much worth it. To illustrate this point, consider the case of the English language. For simplicity, it can be viewed as a 26-event source, namely producing the A–Z symbol characters, which we call $X^{(1)}$. The key question we want to address here is: *"Could one use a different alphabet and its associated block code to convey more information in any length of text?"* Indeed, is it possible to squeeze a piece of English text by means of a *super-alphabet*? One may conceive of such a super-alphabet as being made from character pairs (also called digrams), representing altogether $26 \times 26 = 676$ new symbol characters. Thus, all English books and written materials using this super-alphabet would be twice as short as the originals! This would reduce their production costs, their price and weight, and possibly they could be faster to read. But such improvements would be at the expense of having to learn and master the use of 676 different characters.[10] Here, the point is not to propose changing the English alphabet, but rather to analyze how text information could be compressed for saving memory space and speeding up transmission between computers. To build this new code, we ought to assign the shortest codewords to the most frequently used digrams,

[10] Two illustrative examples of "super-alphabets" are the symbolic/ideographic/logogram *kanjis* of the Chinese and Japanese languages. In Japanese, children must learn 1006 kanjis (*Gakushuu*) over a six-year elementary school cycle. To read Japanese newspapers one must master 1945 official kanjis (*Jōyō*). The Jōyō extends to 2928 kanjis when including people's names. Note that written Japanese is completed with two syllabary alphabets (*Hiragana* and *Katakana*), each having 46 different characters. In modern Chinese, literacy requires mastering about 3000 kanjis, while educated people may know between 4000 and 5000 kanjis. Comprehensive Chinese dictionaries include between 40 000 and 80 000 kanjis. These impressive figures should not obscure the fact that Western languages also have phenomenal inventories of dictionary words, despite their limited 26-character (or so) alphabets. Apart from technical literature, a few thousand words are required to master reading and writing English, with up to 10 000 for the most educated people. The inventories of French and German come to 100 000 to 185 000 words, while English is credited with a whopping 500 000 words (750 000 if old English is included). As with oriental languages, it is not clear, however, if such a profusion of language "codewords" is truly representative of any current or relevant use. Another consideration is that words in Western languages are usually recognized "at once" by the educated human brain, without detailed character-by-character analysis, which in fact makes alphabetical codewords similar to super-alphabetical symbols, just like Chinese or Japanese kanjis. The latter might be more complex to draw and to memorize (especially concerning their individual phonetics!), but they have a more compact form than the alphabet-based Western "codewords." However, the key advantage of languages using limited-size alphabets (such as based on 26 characters, and 28 or 29 for Swedish or Norwegian) is the easiness to learn, read, and pronounce, and write words, especially in view of handwriting skills and adult personalization. With super-alphabets, these different tasks are made far more difficult (without considering complex spelling and pronunciation rules).

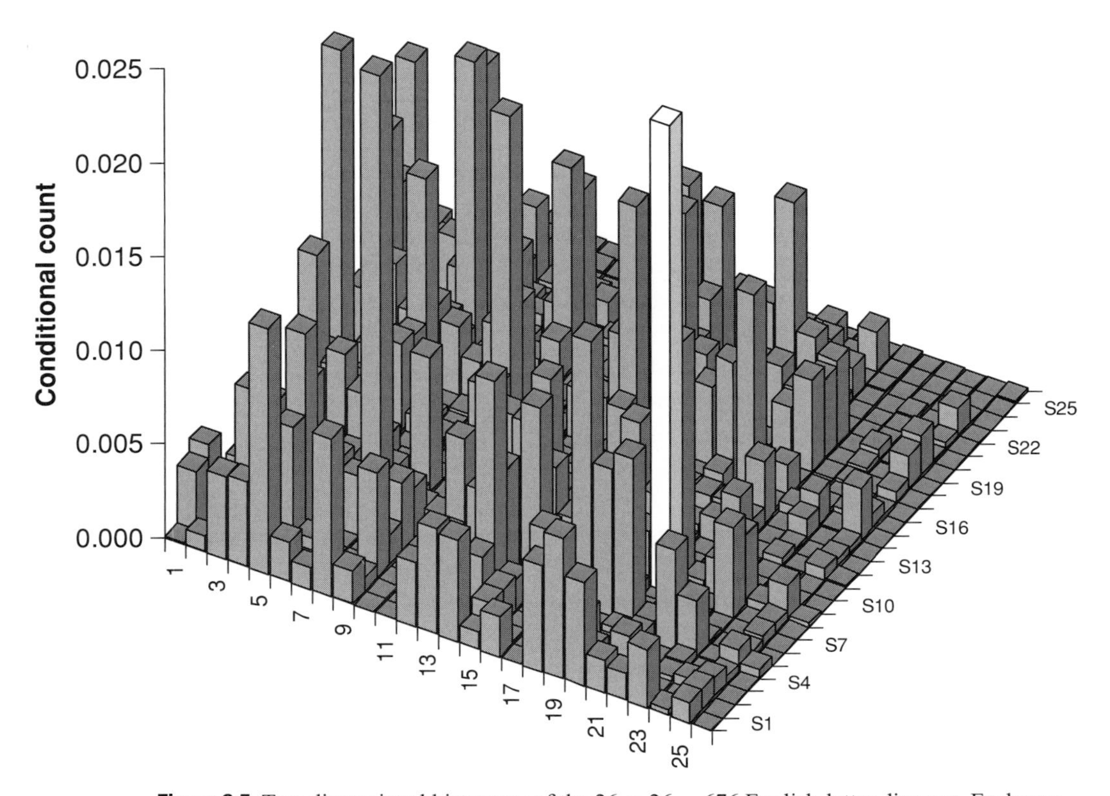


**Figure 9.5** Two-dimensional histogram of the $26 \times 26 = 676$ English-letter digrams. Each row, corresponds to the distribution of counts $c(y|x)$, where $x$ is the first letter of the digram. The front row corresponds to the distribution $c(A|x)$. The most frequent English digram, TH, corresponding to $c(H|T)$ is highlighted.

and the longest codewords to the less frequently used ones. A two-dimensional histogram obtained by counting the frequency of English digrams $xy$ out of a 10 000-letter text is shown in Fig. 9.5.[11] Each of these counts is noted $c(y|x)$, see note.[12] We can view these digrams as forming an extended language source $X^{(2)}$, with virtually independent super-alphabet symbols of probability $p(xy) = c(y|x)/10\,000$.[13] We can then rearrange them in order of decreasing frequency, as shown in Fig. 9.6 for the leading group ($c(y|x) \geq 100$).[14] From the figure, we observe that the three most frequent digrams are TH, HE, and AN. It immediately comes to the mind that one could readily change these

[11] Plotted after analyzing raw data from: H. Fouché-Gaines, *Cryptanalysis, a Study of Ciphers and Their Solutions* (New York: Dover Publications, 1956).

[12] The conversion of the histogram data $c(y|x)$ into conditional probabilities $p(y|x)$ is given by the property $N\sum_x c(y|x)p(x) \equiv p(y) \equiv \sum_x p(y|x)p(x)$, which gives $p(y|x) = Nc(y|x) = c(y|x)p(y)/\sum_x c(y|x)p(x)$.

[13] For simplicity, the 27th "space" character was omitted in this count, as reflecting a tradition of old cryptography; single letters are, therefore, not included, but this does not change the generality of the analysis, since they can be coded as digrams with "space" as a second character.

[14] It is no surprise that we find TH, HE, AN, and IN as the most frequent digrams in English, suggesting an inflation of words and word prefixes, such as THE, THEN, THERE, HE, HERE, AN, AND, IN, and so on.

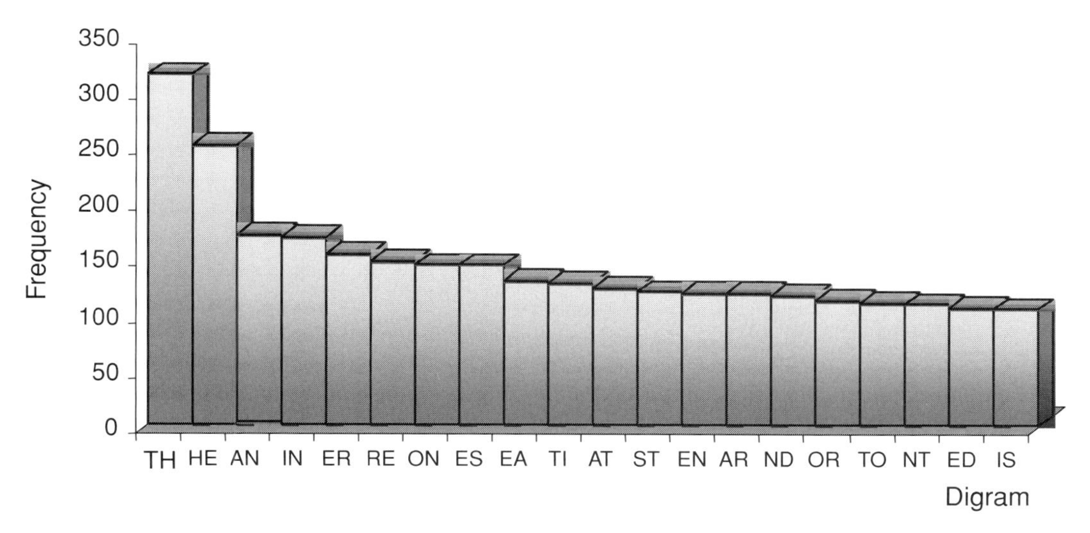


**Figure 9.6** Arranging the leading English-letter digrams in order of decreasing frequency out of a 10 000-letter count. The corresponding probability distribution is obtained by normalizing this histogram by 10 000 (from data in Fig. 9.5 with $c(y|x) \geq 100$).

into single characters, calling them, for instance, $\dot{T}$, $\hat{H}$, or $\tilde{A}$, respectively.[15] Ancient scriptures include lots of such symbolic digram contractions, also called *ligatures*, which helped the early writers, sculptors, or printers to produce their works more effectively or artistically.[16] Figure 9.7 shows a logarithmic-scale plot of the 676 digrams in decreasing order of frequency. The distribution is seen to be very nearly exponential, as the numerical fit indicates. The digram counts that were found to be zero in the reference were changed to 0.5 and 0.1, to allow and to optimize this exponential fit. The fact that the distribution is exponential suggests that Shannon–Fano coding should be a good approximation of an optimal (Huffman) code, considering the relatively large size of the source.

The digram-source entropy is calculated to be $H(X^{(2)}) = 7.748$ bit/digram. Interestingly, it is smaller than $H'' = 2 \times H(X^{(1)}) = 8.368$, where $H(X^{(1)}) = 4.185$ bit/character is the entropy of the English-language source. This fact is a positive indication that the

[15] These are not to be confused with *phonetics*, which uses other super-alphabet codes in an attempt to emulate and classify foreign-language sounds and vocal spelling.

[16] A surviving contemporary example of digram ligatures is the symbol character &, called *ampersand* (or *esperluette* in French). It is in all computer keyboards, and is now used internationally as the word contraction of "and"; see for reference: *http://en.wikipedia.org/wiki/Ligature_(typography), www.adobe.fr/type/topics/theampersand.html*. The graphics of the ampersand character actually come from the Latin word *et*. While the French use the same word as Latin, *et*, the Italian and Spanish use the shorter words *e* and *y*, respectively, which do not call for symbol contraction, in contrast with most other languages. A second example of a ligature, also in all computer keyboards, is the "at sign" character @, a contraction for *at*, as used in e-mail addresses to indicate the domain name. Interestingly, its origins can be traced back to the Middle Ages to designate weight or liquid capacity (in Spanish and Portuguese, *amphora* is *arroba*, which yields the current French name of *arobase*). The character has also been commonly used in English as a contraction of *at* to designate prices. See http://en.wikipedia.org/wiki/At_sign.

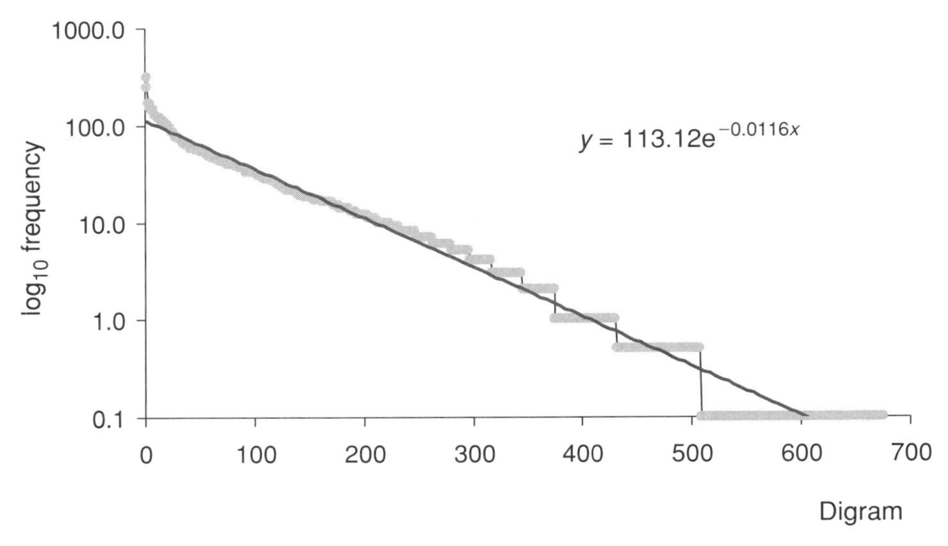


**Figure 9.7** Plot of the 676 English-letter digram frequencies (logarithmic scale) out of a 10 000-letter count, showing that the distribution is very nearly exponential (from data in Fig. 9.5).

English digrams contain *less* information, or uncertainty, than the mere concatenation of letters by pairs. For the probability distribution, this implies that $p(xy) \geq p(x)p(y)$, most generally for any symbol pair $x$, $y$, meaning that English letters are most often correlated to a certain extent, and correlation means less information. For instance, we find for the first four leading digrams ($xy =$ TH, HE, AN, IN) in Fig. 9.6 (dividing counts by 10 000), the following probabilities:

$$\begin{aligned} p(\mathrm{TH}) &= 0.0313 > p(\mathrm{T})p(\mathrm{H}) = 0.0055, \\ p(\mathrm{HE}) &= 0.0249 > p(\mathrm{H})p(\mathrm{E}) = 0.0077, \\ p(\mathrm{AN}) &= 0.0171 > p(\mathrm{A})p(\mathrm{N}) = 0.0054, \\ p(\mathrm{IN}) &= 0.0168 > p(\mathrm{I})p(\mathrm{N}) = 0.0055, \end{aligned}$$

which indeed illustrates a strong correlation attached to the four digrams TH, HE, AN, and IN.

We shall consider next the coding of English digrams through binary codewords. This will provide a good test for comparing the efficiencies of Shannon–Fano and Huffman codes.

Owing to the size of the source, the calculation of the Huffman coding-tree must be performed with a computer.[17] From my own computer program and test runs I found that the Shannon–Fano and Huffman codeword lengths for the English-digram source vary from $l = 5$ to $l = 17$ (but obviously with different assignments), leading to the

[17] Different algorithms and source programs can be found in the Internet, see for instance, www.compressconsult.com/huffman/#codelengths and www.cs.mu.oz.au/~alistair/inplace.c.

following results:

$$\text{Shannon–Fano code: } L^{**} = 8.262 \text{ bit/digram} \qquad \eta = 93.78\%,$$
$$\text{Huffman code: } \tilde{L} = 7.778 \text{ bit/digram} \qquad \eta = 99.617\%,$$

where for the source entropy the reference used is $H = 7.748$ bit/digram.

As expected, the Huffman code is found to be more efficient than the Shannon–Fano code. It is close to 0.3% of the source entropy, which represents a record value for the examples described so far in these chapters! For comparison, we found earlier (Table 9.1) that the efficiency of the English monograms is $\eta = 99.33\%$ ($L^{**} =$ 4.212 bit/monogram), which is close to 0.6% of the source entropy. The efficiency of monogram coding is, therefore, not significantly different from that of digram coding. Were we to compress an English book with Huffman codes, while assuming *five bits per character* in the original book, we would get the following compression rates:

$$\tilde{r} = 1 - \frac{\tilde{L}}{5} = 1 - \frac{4.212}{5} = 15.76\% \quad \text{(monogram code)},$$
$$\tilde{r} = 1 - \frac{\tilde{L}}{10} = 1 - \frac{8.262}{10} = 17.38\% \quad \text{(digram code)}.$$

The above figures are, in fact, quite conservative, because we took five-bit coding for arbitrary reference. As we know, uncompressed texts use seven-bit ASCII. Overlooking capitals, punctuation, and numbers (these being less frequent than lower-case letters), and neglecting spaces, we can make a rough estimate of the compression rates with respect to ASCII, as follows:

$$\tilde{r} = 1 - \frac{\tilde{L}}{7} = 1 - \frac{4.212}{7} = 39.82\% \quad \text{(monogram code)},$$
$$\tilde{r} = 1 - \frac{\tilde{L}}{14} = 1 - \frac{8.262}{14} = 40.98\% \quad \text{(digram code)},$$

which are the values typically found in standard data-compression software based on Huffman coding.[18]

---

The first lesson learnt from this whole exercise is that block-coding makes it possible to improve the compression rate. In the case of English, however, the improvement from using digram blocks instead of monogram characters is not very significant. Note that the performance could be further improved by using longer blocks (trigrams, quadrigrams, etc.), but at the expense of manipulating enormous dictionaries with codewords having relatively long average lengths. The second lesson is that block codes also make it possible to increase the coding efficiency when the events in the original source exhibit a fair amount of mutual correlation, as we have seen with English digrams. However, we should not conclude that block codes are more efficient with such sources, and are less efficient with sources of mutually uncorrelated events. The last example shall prove this point.

[18] See, for instance, www.ics.uci.edu/~dan/pubs/DC-Sec4.html.

**Example 9.3:** ***Two-event source; the roulette game***

We analyze the *roulette game*, as inspired by previous work,[19] but which is further developed here. This game originated in France in the seventeenth century, and was later imported to America with minor modifications, hence, the alternative expressions of *French wheel* or *American wheel*. To recall, the principle of the roulette is to make various types of bets on the ball landing at random in one of 36 spots, numbered from 1 to 36, of the rotating wheel. If the ball stops on the right spot number, the gain is 36 times the amount that was bet. To increase the odds on winning anything, bets can also be made on one, two, or four numbers at once, but this reduces the gain in the same proportion. To increase the variety of betting possibilities further, the wheel spots are also divided into families: (a) numbers 1–18 and 18–36, (b) odd and even numbers, (c) red or black, and (d) numbers 1–12, 13–24, and 25–36.[20] A key feature of the roulette, which is not friendly to the players, is the existence of an extra green number, called 0 in the French wheel. The American wheel also has this number 0 plus a second one, called 00. When the ball lands on either 0 or 00, all bet proceeds go directly to the "bank." It is easy to establish that the odds on winning in the roulette game are relatively low, no matter how hard one may try with any playing strategy.[21]

Here, we shall simplify the game by assuming that the gambling exclusively concerns a single number selected from 1 to 36. The corresponding probability of winning is $p$ (win) $= p$ (no 0 or 00) $\times p$ (number selected) $= (1 - 2/38) \times (1/36) = 1/38 = 0.0263$, and the probability of losing is $p$ (lose) $= 1 - p$ (win).[22] This is a two-event source

[19] See B. Osgood, *Mathematics of the Information Age (2004)*, p. 64, at www-ee.stanford.edu/~osgood/Sophomore%20College/Math%20of%20Info.htm.

[20] See interactive example, while safely taking bets with "free money" at (for instance) www.mondo-casinos.com/gratuit/roulette/index.php (*note*: this reference is for study purposes and does not constitute in any way a recommendation of gambling).

[21] If the roulette outcome is strictly random (can one always be sure of this?), the odds on the bank to pocket all the bets with the 0 or 00 outcomes are $1/37 = 2.7\%$ (French) and $2/38 = 5.4\%$ (American). For the individual player, the odds on winning any single-number bet are $(1 - 1/37)/36 = 1/36 = 2.7\%$ (French) and $(1 - 2/38)/36 = 1/38 = 2.6\%$. This means that (with respect to this player) the bank wins with probabilities 97.3% (French) and 97.4% (American). The odds on winning are obviously greater with the other bets (a)–(c), namely $(1 - 1/37)/2 = 48.6\%$ (French) and $(1 - 2/38)/2 = 47.3\%$ (American) for (a)–(c) and $(1 - 1/37)/3 = 32.4\%$ (French) and $(1 - 2/38)/3 = 31.5\%$ (American) for (d), but we note that they are lower than 50% (a)–(c) or 33% (d), which, on average, is always in favor of the bank.

[22] We note that in the real game, the odds on winning are less than 1/36, which makes the player's situation "unfair." It is possible to play with a high probability of making zero gains, but this comes with a low probability of making maximal loss. Indeed, if we place one token on each of the 36 numbers, two events can happen:

(a) One of the 1–36 numbers comes out: we win 36 tokens, which is the exact amount we bet, and the net gain is zero; this has a high probability of $36/38 = 94.75\ \% < 100\%$.

(b) The 0 or 00 comes out, the bank wins the 36 tokens (our loss is maximum); this has a low probability of $2/38 = 5.25\% > 0\%$.

Similar conclusions apply when tokens are placed in equal numbers in the "odd/even," "red/black," or "1–12, 13–24, 25–36" fields. The first lesson learnt is that even the least risky gambling options, which have zero gain, most likely come with a small chance of maximal loss. Even in this extreme gambling option, the game definitely remains favorable to the bank! Consider now the odds of winning if one plays the same number 36 times in a row. The probability to lose all games is $(1 - 1/38)^{36} = 38.3\%$, so there

with entropy defined by $H(X) \equiv f(p) = -p \log p - (1-p)\log(1-p)$, as described in Chapter 4. Substituting $p$ (win) in this definition yields $H(X) = 0.17557$ bit/symbol. As we shall see later, we will need such a high accuracy to be able to analyze the code performance.

Call the two source events W and L for the outcomes "win" and "lose." A succession of $n$ roulette bets can, thus, be described by the sequence LLLLLWLL..., which is made of $n$ symbols of the L/W type. As we have learnt so far in this chapter, it would be a waste to represent such a sequence with an $n$-bit codeword. Rather, we should seek for a variable-length codeword assignment, which uses the shortest codewords for the most likely sequences, and the longest codewords for the least likely ones. We should expect the mean codeword length to approach the extended source entropy $H(X^{(n)}) = nH(X) \equiv 0.17557n = n/5.69$, which is indeed significantly shorter than $n$. To find such a code, we may proceed step by step, considering sequences with increasing lengths, and determining a corresponding block code. We will compare the block code efficiency with that of a Huffman code.

Consider first a sequence of four games ($n = 4$). Overlooking the order of the individual outcomes, there exist five *types* of possible sequences, in decreasing likelihood: LLLL (complete loss), LLLW (one win), LLWW (two wins), LWWW (three wins), and WWWW (four wins). We must then take into account the exact order of the individual W/L associated to a given sequence type. For instance, the sequence types we called LLLW and LWWW actually correspond to four possible unique outcomes:

LLLW, LLWL, LWLLL, WLLL;
LWWW, WLWW, WWLW, WWWL.

The sequence type LLWW is associated with six unique outcomes:

LLWW, LWLW, LWWL, WLWL, WWLL, WLLW,

while the sequences LLLL and WWWW have only one unique possibility. Most generally, the number of ways of selecting $k$ slots from a sequence of $n$ slots is given by the *combinatorial coefficient* $C_n^k$ (see Chapter 1):

$$C_n^k = \frac{n!}{k!(n-k)!}, \tag{9.3}$$

which is also written $\binom{k}{n}$, and where $q! = 1 \times 2 \times 3 \times \cdots \times (q-1) \times q$ is called the *factorial* of the integer $q$ (by convention, $0! = 1$). Consistently, we find for LLLW or LWWW the number $C_4^1 = 4!/(1!3!) = 4$, and for LLWW the number $C_4^2 = 4!/(2!2!) = 6$. Calling $p$ the probability of winning a single game in the sequence, we can associate

is a 61.7% chance of winning one way or another, which looks a reasonable bet. But this number assumes that we play 36 games, regardless of their outcome. The most likely possibility to win is that our number comes once (and only once), which has the probability $C_{36}^1(1/38)^1(1-1/38)^{35} = 37.2\%$. In this event, we win 36 tokens, but the price to pay is to lose 35 tokens for the other 35 games. The net result is that, with 61.7% chances and in the most favorable case, we have not gained anything at all! Furthermore, there are 38.3% chances that we lose 36 tokens. The second lesson learnt is the same as the first lesson!

**Table 9.4** Construction of variable-length block code for the American roulette, considering a succession of $n = 4$ independent games for which the outcome is lose (L) or win (W). The sequence LLLL (type A) corresponds to four wins, the sequence LLLW (type B) to three losses and one win, and so on. Each sequence type has a probability $p_1$ and a number of ordered possibilities $N$, yielding a net probability $p = Np_1$. Each ordered possibility is labeled by a block code made of a header and a trailer block. The header is coded with the uniquely decodable word of length $HB$, which indicates the sequence type. The trailer field has the minimum number of bits $TB$ required to code $N$. The resulting codeword length is $M = HB + TB$. At bottom are shown the total number of extended-source events, the mean codeword length $L = \langle M \rangle$, the extended-source entropy $nH$ and the coding efficiency $\eta = H/L$.

$N = 4$ games

| Type | Sequence type | Probability, $p_1$ | Possibilities, $N$ | No of trailer bits, $TB$ | Probability, $p = Np^1$ | Header | No of header bits, $HB$ | No of bits required, $M = HB + TB$ | $Mp$ |
|---|---|---|---|---|---|---|---|---|---|
| A | LLLL | 0.89881955 | 1 | 0 | 0.89881955 | 1 | 1 | 1 | 0.89882 |
| B | LLLW | 0.02429242 | 4 | 2 | 0.09716968 | 01 | 2 | 4 | 0.38868 |
| C | LLWW | 0.00065655 | 6 | 3 | 0.00393931 | 001 | 3 | 6 | 0.02364 |
| D | LWWW | $1.7745 \times 10^{-5}$ | 4 | 2 | $7.0979 \times 10^{-5}$ | 0001 | 4 | 6 | 0.00043 |
| E | WWWW | $4.7959 \times 10^{-7}$ | 1 | 0 | $4.7959 \times 10^{-7}$ | 0000 | 4 | 4 | 0.00000 |
| $\sum$ | | | **16** | | $1.0000 \times 10^{00}$ | | | | |
| $L$ | | | | | | | | | **1.31156** |

Extended source entropy $nH = 0.70226$.
Efficiency $\eta = 53.544\%$.

each of the above sequence types with the corresponding probabilities:

$$\begin{cases} p(LLLL) = C_4^0(1-p)^4 \\ p(WWWW) = C_4^0 p^4 \\ p(LLLW) = C_4^1 p(1-p)^3 \\ p(WWWL) = C_4^1 p^3(1-p) \\ p(LLWW) = C_4^2 p^2(1-p)^2, \end{cases} \tag{9.4}$$

recalling that each of these types is associated with $N = C_4^k$ unique outcomes. Table 9.4 lists the data computed with $p = 1/38 = 0.0263$, the resulting probabilities being shown in decreasing order, with the sequence types referred to as A, B, C, D, and E. To label each of the sequence outcomes, we shall construct a variable-length block code made with the concatenation of a *header block* with a *trailer block*. The header block is coded with a uniquely decodable word of length $HB$, which indicates the sequence type (A, B, C, D, or E). The trailer block has the minimum number of bits $TB$ required to code $N$. For instance, the outcome LLLL (type A) has the header 1 and an empty trailer. The outcome LLLW (type B) has the header 01 and a trailer of two bits (namely 00, 01, 10, and 11) which labels each of the four unique possibilities in the type-B sequence (namely LLLW, LLWL, LWLLL, and WLLL). The resulting codeword length is, therefore, $M = HB + TB$. Table 9.4 shows the mean codeword length $L = \langle M \rangle$ and the coding efficiency $\eta = nH/L$, where $nH$ is the entropy of the extended source corresponding to the $n$ independent outcomes of the $n$ successive games. The results indicate that our block code takes $L = 1.311$ bit/word to describe any game sequence uniquely, corresponding to a coding efficiency of $\eta = 53.544\%$.

The efficiency of our block code ($\eta = 53.544\%$) is not outstanding, indeed, but we can measure the progress made by comparing it to that given by a fixed-length code. Indeed, if we use a four-bit codeword to describe any of the game sequences (e.g., 1010 for WLWL), the efficiency drops to $\eta = 4H/4 = 0.1755$ or 17.5%. The use of our variable-length block code has, in fact, more than doubled the coding efficiency! With respect to four-bit codewords, the compression rate obtained with the block code is $r = 1 - 1.311/4 = 67.2\%$.

This first example with sequences of $n = 4$ games illustrates that the efficiency of the block code increases with the length of the sequence. Indeed, long sequences of events contain *more information* than short sequences of events, because their associated probabilities are *lower*.[23] Therefore, we can infer that our block coding becomes increasingly efficient with longer sequences, which we shall now verify. Table 9.5 shows the results obtained with sequences of length $n = 32$. As the table first shows, the number of events is dramatically increased to a whopping 4 294 967 296 or about 4.3 billion possibilities! The code is seen to have an average length of $L = 5.771$ bit/word, corresponding to a coding efficiency of $\eta = 97.34\%$. The corresponding compression rate is $r = 1 - 5.771/32 = 81.9\%$.

The results obtained for $n = 32$ games seem to confirm the previous observation, according to which the code becomes increasingly efficient with ever-longer sequences. As it turns out, however, this is not the case! Figure 9.8 shows a plot of the coding efficiency $\eta$, as computed with sequence lengths from $n = 1$ to $n = 129$. From the figure, we observe that the efficiency increase exhibits some irregularities, with local peaks appearing whenever $n$ is a power of two ($n = 2^k$), but also for other intermediate values. The irregularity observed between the peaks is explained by the fact that the trailer bits, which label all outcome possibilities within a sequence type, are generally not fully used. The use is optimal whenever $n = 2^k$, which yields a local peak in efficiency. Looking at Table 9.4 and Table 9.5, we observe indeed that it takes exactly two trailer bits to describe the four possibilities of the sequence LLLW, and it takes exactly four trailer bits to describe the 16 possibilities of the sequence LLLL LLLL LLLL LLLW. This is not chance, since the number of possibilities is $C_n^1 = n = 2^k$. Since these sequences always come in second position in the distribution, this feature has a major impact in reducing the mean codeword length. Other sequences with $2^k < n < 2^{k+1}$ require an extra trailer bit, which explains the efficiency drops observed in Fig. 9.8 each time where $n = 2^k + 1$ ($n \geq 8$). The figure also shows that, unexpectedly, the efficiency peaks decrease past a maximum point obtained for $n = 32$. This effect can be attributed to the fact that the number of possibilities $C_n^q$ ($1 < q < n$) grows exponentially with length.[24] Such a growth dramatically increases the number of codewords with inefficient use

[23] Isn't this observation counterintuitive? We would, indeed, expect that there is more uncertainty (or entropy) in shorter game series than in longer ones. But there is more information contained in the succession of repeated events than in any single event taken separately.

[24] It can easily be verified that for even values of $n$, the maximum of $C_n^k$ is reached at $k = n/2$. Using Stirling's formula, we obtain $C_n^{n/2} \approx 2^{n+1}/\sqrt{2\pi n}$.

**Table 9.5** As for Table 9.4 but with $n = 32$.

| Case | Events | Probability, $p_1$ | Possibilities, $N$ | No of trailer bits, $TB$ | Probability, $p = Np_1$ | Header | No of header bits, $HB$ | No of bits required, $M = HB + TB$ | $Mp$ |
|---|---|---|---|---|---|---|---|---|---|
| A | LLLL LLLL LLLL LLLL LLLL LLLL LLLL LLLL | 0.425971044 | 1 | 0 | 0.42597104 | 1 | 1 | 1 | 0.425971044 |
| B | LLLL LLLL LLLL LLLL LLLL LLLL LLLL LLLW | 0.011512731 | 32 | 5 | 0.36840739 | 01 | 2 | 7 | 2.578851726 |
| C | LLLL LLLL LLLL LLLL LLLL LLLL LLLL LLWW | 0.000311155 | 496 | 9 | 0.15433283 | 001 | 3 | 12 | 1.851993904 |
| D | LLLL LLLL LLLL LLLL LLLL LLLL LLLL LWWW | $8.40959 \times 10^{-6}$ | 4 960 | 13 | 0.04171157 | 0001 | 4 | 17 | 0.709096765 |
| E | LLLL LLLL LLLL LLLL LLLL LLLL LLLL WWWW | $2.27286 \times 10^{-7}$ | 35 960 | 16 | 0.00817321 | 0000 1 | 5 | 21 | 0.171637492 |
| F | LLLL LLLL LLLL LLLL LLLL LLLL LLLW WWWW | $6.14287 \times 10^{-9}$ | 201 376 | 18 | 0.00123703 | 0000 01 | 6 | 24 | 0.029688647 |
| G | LLLL LLLL LLLL LLLL LLLL LLLL LLWW WWWW | $1.66024 \times 10^{-10}$ | 906 192 | 20 | 0.00015045 | etc. | 7 | 27 | 0.004062129 |
| H | LLLL LLLL LLLL LLLL LLLL LLLL LWWW WWWW | $4.48712 \times 10^{-12}$ | 3 365 856 | 22 | $1.5103 \times 10^{-5}$ | | 8 | 30 | 0.00045309 |
| I | LLLL LLLL LLLL LLLL LLLL LLLL WWWW WWWW | $1.21274 \times 10^{-13}$ | 10 518 300 | 24 | $1.2756 \times 10^{-6}$ | | 9 | 33 | $4.20945 \times 10^{-5}$ |
| J | LLLL LLLL LLLL LLLL LLLL LLLW WWWW WWWW | $3.27766 \times 10^{-15}$ | 28 048 800 | 25 | $9.1935 \times 10^{-8}$ | | 10 | 35 | $3.21771 \times 10^{-6}$ |
| K | LLLL LLLL LLLL LLLL LLLL LLWW WWWW WWWW | $8.85855 \times 10^{-17}$ | 64 512 240 | 26 | $5.7149 \times 10^{-9}$ | | 11 | 37 | $2.1145 \times 10^{-7}$ |
| L | LLLL LLLL LLLL LLLL LLLL LWWW WWWW WWWW | $2.3942 \times 10^{-18}$ | 129 024 480 | 27 | $3.0891 \times 10^{-10}$ | | 12 | 39 | $1.20475 \times 10^{-8}$ |
| M | LLLL LLLL LLLL LLLL LLLL WWWW WWWW WWWW | $6.47082 \times 10^{-20}$ | 225 792 840 | 28 | $1.4611 \times 10^{-11}$ | | 13 | 41 | $5.99037 \times 10^{-10}$ |
| N | LLLL LLLL LLLL LLLL LLLW WWWW WWWW WWWW | $1.74887 \times 10^{-21}$ | 347 373 600 | 29 | $6.0751 \times 10^{-13}$ | | 14 | 43 | $2.6123 \times 10^{-11}$ |
| O | LLLL LLLL LLLL LLLL LLWW WWWW WWWW WWWW | $4.72668 \times 10^{-23}$ | 471 435 600 | 29 | $2.2283 \times 10^{-14}$ | | 15 | 44 | $9.80463 \times 10^{-13}$ |
| P | LLLL LLLL LLLL LLLL LWWW WWWW WWWW WWWW | $1.27748 \times 10^{-24}$ | 565 722 720 | 30 | $7.227 \times 10^{-16}$ | | 16 | 46 | $3.32442 \times 10^{-14}$ |
| Q | LLLL LLLL LLLL LLLL WWWW WWWW WWWW WWWW | $3.45265 \times 10^{-26}$ | 601 080 390 | 30 | $2.0753 \times 10^{-17}$ | | 17 | 47 | $9.754 \times 10^{-16}$ |
| R | LLLL LLLL LLLL LLLW WWWW WWWW WWWW WWWW | $9.33149 \times 10^{-28}$ | 565 722 720 | 30 | $5.279 \times 10^{-19}$ | | 18 | 48 | $2.53394 \times 10^{-17}$ |
| S | LLLL LLLL LLLL LLWW WWWW WWWW WWWW WWWW | $2.52202 \times 10^{-29}$ | 471 435 600 | 29 | $1.189 \times 10^{-20}$ | | 19 | 48 | $5.70706 \times 10^{-19}$ |
| T | LLLL LLLL LLLL LWWW WWWW WWWW WWWW WWWW | $6.81628 \times 10^{-31}$ | 347 373 600 | 29 | $2.3678 \times 10^{-22}$ | | 20 | 49 | $1.16022 \times 10^{-20}$ |
| U | LLLL LLLL LLLL WWWW WWWW WWWW WWWW WWWW | $1.84224 \times 10^{-32}$ | 225 792 840 | 28 | $4.1596 \times 10^{-24}$ | | 21 | 49 | $2.03822 \times 10^{-22}$ |
| V | LLLL LLLL LLLW WWWW WWWW WWWW WWWW WWWW | $4.97902 \times 10^{-34}$ | 129 024 480 | 27 | $6.4242 \times 10^{-26}$ | | 22 | 49 | $3.14784 \times 10^{-24}$ |
| W | LLLL LLLL LLWW WWWW WWWW WWWW WWWW WWWW | $1.34568 \times 10^{-35}$ | 64 512 240 | 26 | $8.6813 \times 10^{-28}$ | | 23 | 49 | $4.25383 \times 10^{-26}$ |
| X | LLLL LLLL LWWW WWWW WWWW WWWW WWWW WWWW | $3.63698 \times 10^{-37}$ | 28 048 800 | 25 | $1.0201 \times 10^{-29}$ | | 24 | 49 | $4.99863 \times 10^{-28}$ |
| Y | LLLL LLLL WWWW WWWW WWWW WWWW WWWW WWWW | $9.82966 \times 10^{-39}$ | 10 518 300 | 24 | $1.0339 \times 10^{-31}$ | | 25 | 49 | $5.06618 \times 10^{-30}$ |
| Z | LLLL LLLW WWWW WWWW WWWW WWWW WWWW WWWW | $2.65667 \times 10^{-40}$ | 3 365 856 | 22 | $8.942 \times 10^{-34}$ | | 26 | 48 | $4.29214 \times 10^{-32}$ |
| AA | LLLL LLWW WWWW WWWW WWWW WWWW WWWW WWWW | $7.18018 \times 10^{-42}$ | 906 192 | 20 | $6.5066 \times 10^{-36}$ | | 27 | 47 | $3.05811 \times 10^{-34}$ |
| AB | LLLL LWWW WWWW WWWW WWWW WWWW WWWW WWWW | $1.94059 \times 10^{-43}$ | 201 376 | 18 | $3.9079 \times 10^{-38}$ | | 28 | 46 | $1.79762 \times 10^{-36}$ |
| AC | LLLL WWWW WWWW WWWW WWWW WWWW WWWW WWWW | $5.24483 \times 10^{-45}$ | 35 960 | 16 | $1.886 \times 10^{-40}$ | | 29 | 45 | $8.4719 \times 10^{-39}$ |
| AD | LLLW WWWW WWWW WWWW WWWW WWWW WWWW WWWW | $1.41752 \times 10^{-46}$ | 4 960 | 13 | $7.0309 \times 10^{-43}$ | | 30 | 43 | $3.02329 \times 10^{-41}$ |
| AE | LLWW WWWW WWWW WWWW WWWW WWWW WWWW WWWW | $3.83114 \times 10^{-48}$ | 496 | 9 | $1.9002 \times 10^{-45}$ | | 31 | 40 | $7.60099 \times 10^{-44}$ |
| AF | LWWW WWWW WWWW WWWW WWWW WWWW WWWW WWWW | $1.03544 \times 10^{-49}$ | 32 | 5 | $3.3134 \times 10^{-48}$ | | 32 | 37 | $1.22597 \times 10^{-46}$ |
| AG | WWWW WWWW WWWW WWWW WWWW WWWW WWWW WWWW | $2.7985 \times 10^{-51}$ | 1 | 0 | $2.7985 \times 10^{-51}$ | | 32 | 32 | $8.95519 \times 10^{-50}$ |
| $\sum$ | | | **4 294 967 296** | | $1.0000 \times 10^{00}$ | | | | |
| $L$ | | | | | | | | | **5.771800335** |

Extended source entropy $nH = 5.618081$.
Efficiency $\eta = 97.34\%$.

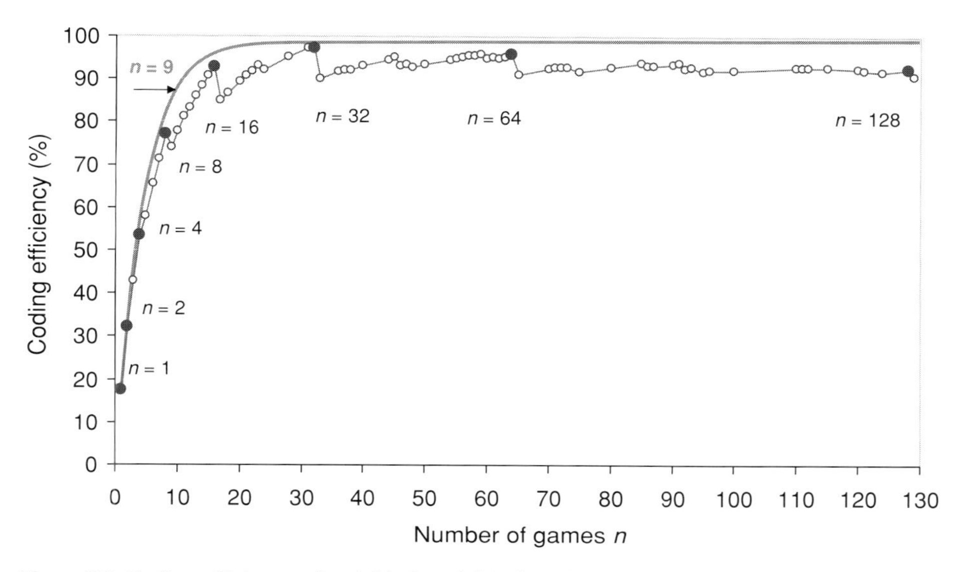


**Figure 9.8** Coding efficiency of variable-length block code describing a sequence of $n$ roulette game outcomes. The data corresponding to $n = 2^k$ (where $k$ is an integer) are shown in dark. The continuous curve corresponds to the envelope of ideal Huffman coding, as computed up to $n = 9$ and extrapolated beyond $n = 9$.

of trailer bits, which progressively reduces the maximum compression effect obtained each time that $n = 2^k$. Concerning the curve envelope ($n = 9$), which is also plotted in Fig. 9.8, see further.

The above example illustrates that block coding can be quite efficient and yield high data compression, but care should be given to find the optimum sequence length, if it exists. A relevant question is: *how does block coding compare with Huffman coding*? The second solution, which we know to be optimal in terms of compression power, involves coding all events of the extended source $X^{(n)}$ through the coding-tree procedure described earlier. If the number of extended-source events is reasonably small, the procedure is easily and rapidly implemented. But it becomes impractical or prohibitive when the number of source events grows exponentially with the block size $n$. In the case where $n = 16$, for instance, the number of events is $2^{16} = 4\ 294\ 967\ 296$ or 4 billion. In the case where $n = 64$, the number of events becomes $2^{64} = 18\,446\,744\,073\,709\,500\,000$, or 18 billion billion! It does not make any sense to implement Huffman coding in such a case. Yet, we have been able here to determine a block code and to calculate its performance up to $2^{129} = 6.8 \times 10^{38} = 680 \times 10^9 \times 10^9 \times 10^9 \times 10^9$. The explanation is that the block-code assignment only takes the analysis of $n + 1$ block types, as opposed to $2^n$ individual events. With a home computer, we can easily compute Huffman codes up to a few thousand elements. What about Shannon–Fano coding? As we know, the codeword assignment is straightforward in this case. As with block codes, we do not have to consider the $2^n$ events individually to compute the mean codeword length.

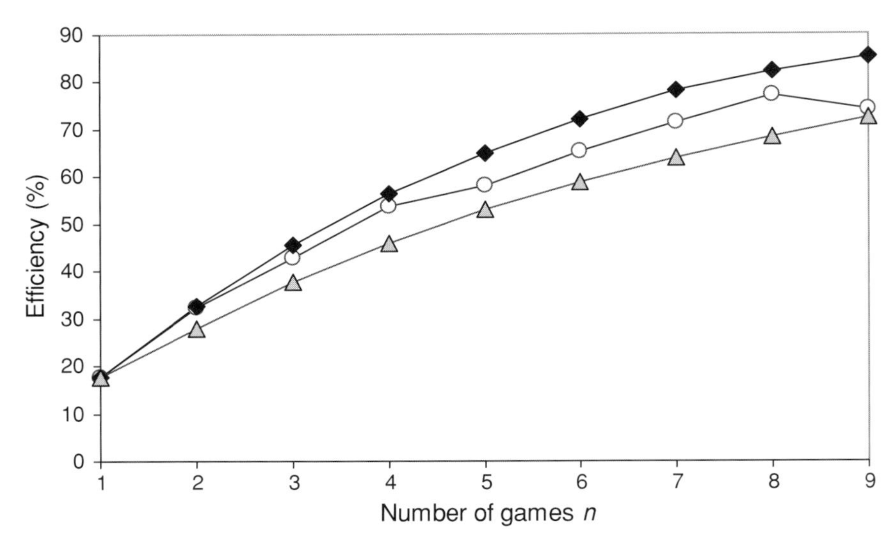


**Figure 9.9** Comparison between Huffman coding (diamonds), block-coding (open circles), and Shannon–Fano coding (triangles) in the roulette-game example of Fig. 9.8, with data calculated up to $n = 9$.

The results for both Huffman coding and Shannon–Fano coding, as applied to the roulette-game example, are shown in Fig. 9.9 for game sequence lengths $n \leq 9$. We observe from the figure that, as expected, the efficiency of the Huffman code is greater than that of any other codes. Unlike the block code, both Huffman and Shannon–Fano codes grow smoothly with the sequence length $n$. Interestingly, the block-code performance is observed to be bounded by the two envelopes made by the Huffman code (upper boundary) and the Shannon–Fano code (lower boundary). The Huffman-code data in Fig. 9.9 have also been plotted in Fig. 9.8. In this previous figure, an arbitrary, yet conservative extrapolation of what the Huffman-code performance should look like for sequences beyond $n = 9$ is also shown. We expect that there is no optimal sequence length, and know that the coding efficiency at $n = 32$ is greater than that of our proposed block code ($\eta = 97.34\%$). It would take the power of a workstation or mainframe computer to determine the exact curve and find the values of $n$ for which the Huffman-code efficiency approaches $\eta = 99\%$, 99.9%, 99.99%, 99.999%, etc.

The Shannon–Fano code efficiency, computed up to $n = 90$, is plotted in Fig. 9.10, along with the previous data from Fig. 9.8. We observe that the initially smooth behavior of the Shannon code breaks near $N = 16$, leading to aperiodic oscillations similar to that of the block code. Interestingly, the efficiency of the Shannon–Fano code is seen to increase globally towards the upper limit set by the Huffman code. As expected, the Shannon–Fano code does not exhibit any optimum sequence length, apart from the existence of local maxima.

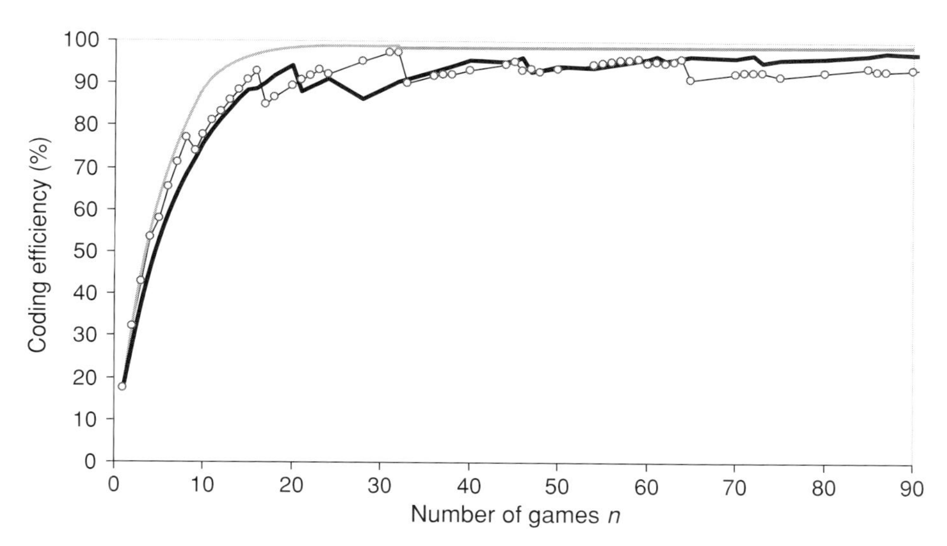


**Figure 9.10** As in Fig. 9.9, with additional results from Shannon–Fano coding up to $n = 90$ (bold line), showing a global convergence trend towards Huffman-coding efficiency.

In this example, we pushed the investigation relatively far, but this was for the sake of mathematical curiosity. In practice, there is little interest nor anything practical to implement Huffman or Shannon–Fano codings when the source has a number of events significantly greater that $2^{16} = 65\,536$. Should we take this number as a reference, this means that compression codes may not be applied to binary sources with lengths greater that $n = 16$, for which the uncompressed representation is 16 bits (two bytes). In contrast, a block code such as used in this roulette example can be reasonably and efficiently applied to sources up to $2^{32} = 4\,294\,967\,296$ elements, which requires the incredibly shorter length of 5.77 bit/word on average to encode, as opposed to 32 bit/word! How this last solution can be effectively implemented is a question of memory space (i.e., $2^{32}$ addresses for table look-up) and considerations of computer speed, both having economical impacts in practical applications. Block coding remains advantageous because of the possibility of splitting events into many categories and types, which separates the tasks of code assignment between a *header* and a *trailer*. Each of these code sub-blocks has fewer bits to handle, and this is what makes the approach more practical when dealing with sources with relatively large numbers of events.

---

## 9.4 Exercises

**9.1** (M): Assign a Huffman code to the two-dice roll distribution described in Chapter 1 (as listed in Table 9.6 overleaf), with the results of the roll $\{2, 3, 4, \ldots, 12\}$ being symbolized by the characters $\{A, B, C, \ldots, K\}$ and calculate the coding efficiency.

**Table 9.6** Data for Exercise 9.1.

| Symbol $x$ | Probability $p(x)$ |
|---|---|
| A = 2 | 0.028 |
| B = 3 | 0.056 |
| C = 4 | 0.083 |
| D = 5 | 0.111 |
| E = 6 | 0.139 |
| F = 7 | 0.167 |
| G = 8 | 0.139 |
| H = 9 | 0.111 |
| I = 10 | 0.083 |
| J = 11 | 0.056 |
| K = 12 | 0.028 |
| $\sum$ | 1.000 |

**9.2** (T): Prove that Huffman coding for uniformly distributed sources of $N = 2^n$ symbols ($n$ an integer) yield a mean codeword length of $l(x) = n$.

**9.3** (T): Show that for dyadic sources, the Huffman code is 100% efficient.
*Clue*: Prove this first using two-element and three-element sources, then conclude in the general case.

**9.4** (M): Find a block code to describe the outcome of five successive coin tosses, and determine the corresponding coding efficiency.

# 10 Integer, arithmetic, and adaptive coding

This second chapter concludes our exploration tour of coding and data compression. We shall first consider *integer coding*, which represents another family branch of optimal codes (next to Shannon–Fano and Huffman coding). Integer coding applies to the case where the source symbols are fully known, but the probability distribution is only partially known (thus, the previous optimal codes cannot be implemented). Three main integer codes, called *Elias*, *Fibonacci*, and *Golomb–Rice*, will then be described. Together with the previous chapter, this description will complete our inventory of *static codes*, namely codes that apply to cases where the source symbols are known, and the matter is to assign the optimal code type. In the most general case, the source symbols and their distribution are unknown, or the distribution may change according to the amount of symbols being collected. Then, we must find new algorithms to assign optimal codes without such knowledge; this is referred to as *dynamic coding*. The three main algorithms for dynamic coding to be considered here are referred to as *arithmetic coding*, *adaptive Huffman coding*, and *Lempel–Ziv coding*.

## 10.1 Integer coding

The principle of *integer coding* is to assign an optimal (and predefined) codeword to a list of $n$ known symbols, which we may call $\{1, 2, 3, \ldots, n\}$. In such a list, the symbols are ranked in order of decreasing frequency or probability, or mathematically speaking, in order of "nonincreasing" frequency or probability. This ranking assumes that at least the ranks of the most likely symbols are known beforehand; the remaining less-likely symbols being arranged in the list in any order. In this case, neither Shannon–Fano coding nor Huffman coding can be implemented, and we must then be looking for new types of "heuristic" code, which will exhibit minimal redundancy (the difference between the obtained mean codeword length and the source entropy). Such is the rationale for integer coding. The most frequently used algorithms are given by the *Elias*, *Fibonacci*, and *Golomb–Rice codes*, which I shall describe next.

*Elias codes* come in two different types, which are named *Elias-gamma* and *Elias-delta*. The correspondence between the first 32 integers, their uncompressed binary representation, and the Elias-gamma or delta codewords is shown in Table 10.1. An explanation of the codeword assignment follows.

**Table 10.1** Various types of integer coding: (a) nonparameterized with *Elias* codes (*gamma* and *delta*) and (b) parameterized with *Fibonacci* code ($m = 2$) and *Golomb* code (simple with $m = 8$ and actual with $m = 6$).

| $i$ | $\lfloor \log_2 i \rfloor$ | Uncompressed | Elias codes | | Fibonacci $m = 2$ | Golomb | |
|---|---|---|---|---|---|---|---|
| | | | Gamma | Delta | | Simple $m = 8$ | Actual $m = 6$ |
| 1 | 0 | 0000 0001 | 1 | 1 | 11 | 1 000 | 1 00 |
| 2 | 1 | 0000 0010 | 0 10 | 0 100 | 011 | 1 001 | 1 01 |
| 3 | 1 | 0000 0011 | 0 11 | 0 101 | 0011 | 1 010 | 1 100 |
| 4 | 2 | 0000 0100 | 00 100 | 0 1100 | 1011 | 1 011 | 1 101 |
| 5 | 2 | 0000 0101 | 00 101 | 0 1101 | 00011 | 1 100 | 1 110 |
| 6 | 2 | 0000 0110 | 00 110 | 0 1110 | 10011 | 1 101 | 1 111 |
| 7 | 2 | 0000 0111 | 00 111 | 0 1111 | 01011 | 1 110 | 01 00 |
| 8 | 3 | 0000 1000 | 000 1000 | 00 100000 | 000011 | 1 111 | 01 01 |
| 9 | 3 | 0000 1001 | 000 1001 | 00 100001 | 100011 | 01 000 | 01 100 |
| 10 | 3 | 0000 1010 | 000 1010 | 00 100010 | 010011 | 01 001 | 01 101 |
| 11 | 3 | 0000 1011 | 000 1011 | 00 100011 | 001011 | 01 010 | 01 110 |
| 12 | 3 | 0000 1100 | 000 1100 | 00 100100 | 101011 | 01 011 | 01 111 |
| 13 | 3 | 0000 1101 | 000 1101 | 00 100101 | 0000011 | 01 100 | 001 00 |
| 14 | 3 | 0000 1110 | 000 1110 | 00 100110 | 1000011 | 01 101 | 001 01 |
| 15 | 3 | 0000 1111 | 000 1111 | 00 100111 | 0100011 | 01 110 | 001 100 |
| 16 | 4 | 0001 0000 | 0000 10000 | 00 1010000 | 0010011 | 01 111 | 001 101 |
| 17 | 4 | 0001 0001 | 0000 10001 | 00 1010001 | 0001011 | 001 000 | 001 110 |
| 18 | 4 | 0001 0010 | 0000 10010 | 00 1010010 | 1001011 | 001 001 | 001 111 |
| 19 | 4 | 0001 0011 | 0000 10011 | 00 1010011 | 0101011 | 001 010 | 0001 00 |
| 20 | 4 | 0001 0100 | 0000 10100 | 00 1010100 | 1101011 | 001 011 | 0001 01 |
| 21 | 4 | 0001 0101 | 0000 10101 | 00 1010101 | 00000011 | 001 100 | 0001 100 |
| 22 | 4 | 0001 0110 | 0000 10110 | 00 1010110 | 10000011 | 001 101 | 0001 101 |
| 23 | 4 | 0001 0111 | 0000 10111 | 00 1010111 | 01000011 | 001 110 | 0001 110 |
| 24 | 4 | 0001 1000 | 0000 11000 | 00 1011000 | 00100011 | 001 111 | 0001 111 |
| 25 | 4 | 0001 1001 | 0000 11001 | 00 1011001 | 00010011 | 0001 000 | 00001 00 |
| 26 | 4 | 0001 1010 | 0000 11010 | 00 1011010 | 00001011 | 0001 001 | 00001 01 |
| 27 | 4 | 0001 1011 | 0000 11011 | 00 1011011 | 10001011 | 0001 010 | 00001 100 |
| 28 | 4 | 0001 1100 | 0000 11100 | 00 1011100 | 01001011 | 0001 011 | 00001 101 |
| 29 | 4 | 0001 1101 | 0000 11101 | 00 1011101 | 11001011 | 0001 100 | 00001 110 |
| 30 | 4 | 0001 1110 | 0000 11110 | 00 1011110 | 00101011 | 0001 101 | 00001 111 |
| 31 | 4 | 0001 1111 | 0000 11111 | 00 1011111 | 10101011 | 0001 110 | 000001 00 |
| 32 | 5 | 0010 0000 | 00000 100000 | 00 11000000 | 11101011 | 0001 111 | 000001 01 |

The Elias-gamma codeword of an integer $i$ is given by its binary representation up to the 1 bit of highest weight, prefaced by a number of zeros equal to $\lfloor \log_2 i \rfloor$. The expression $\lfloor x \rfloor$ (*floor*($x$)) means the smallest integer near or equal to $x$. Thus, we have $\lfloor \log_2 1 \rfloor = 0$, $\lfloor \log_2 2 \rfloor = 1$, $\lfloor \log_2 3 \rfloor = 1$, $\lfloor \log_2 4 \rfloor = 2$, and so on. According to the above rules, the Elias-gamma codewords for $i = 4$ and $i = 13$ are gamma(4) = 00_100 and gamma(13) = 000_1101, respectively (the underscore _ being introduced for clarity).

The Elias-*delta* codeword is defined by gamma($\lfloor \log_2 i \rfloor + 1$), followed by the minimal binary representation of $i$ with the most significant 1 bit being removed. With $i = 4$, for instance, we have $i = 100\_2$ and gamma($\lfloor \log_2 4 \rfloor + 1$) = 011, which, with this rule, yields delta(4) = 011_00. With $i = 14$, we have $i = 1110\_2$ and gamma($\lfloor \log_2 14 \rfloor + 1$) = gamma(4) = 00100, which, with this rule, gives delta(14) = 00100_110. Table 10.1 shows that for small integers (except $i = 1$) the Elias-delta codewords are longer than the gamma codewords. The lengths become equal for $i = 16$ to $i = 31$. The situation is then reversed for $i \geq 32$. This shows that Elias-delta coding is preferable for sources with $i \leq 31$, while Elias-gamma coding is preferable for larger sources. The important difference between the two codes is their asymptotic limit when the source entropy goes to infinity. It can be checked, using a short tabulating program, that the asymptotic limit of the coding efficiency, $\eta = H/L$, is equal to 50% for the Elias-gamma code, while it is equal to 100% for the Elias-delta code.[1] Therefore, the Elias-delta code is asymptotically optimal.

The fact that the two Elias codes are not optimal (except asymptotically for Elias-delta) does not preclude their use for data compression. For instance, taking the English-language source with the distribution listed in Table 8.3 and the corresponding entropy $H = 4.185$ bit/symbol, we find that both Elias-gamma and Elias-delta codings have a mean codeword length of $L = 5.241$ bit/word, corresponding to an efficiency of $\eta = H/L = 79.83\%$. For comparison, Huffman coding yields $\hat{L} = 4.212$ bit/word and $\eta = 99.33\%$. The Elias codes, thus, make it possible to achieve a nonoptimal but acceptable coding performance on limited-size sources. The same conclusion applies to Elias-delta coding for large sources, with the advantage of being straightforward to implement, in contrast with Huffman coding. An Elias-delta variant, known as *Elias-omega* or *recursive Elias coding* makes it possible to shorten the codeword lengths, but with limited benefits as the source size increases.[2]

The Elias coding approach is referred to as *nonparameterized*. This means that the symbol–codeword correspondence is fixed with the code choice (gamma, delta, recursive). In *parameterized* coding, an integer parameter $m$ is introduced to create another degree of freedom in the choice and optimization of codeword lengths. This is the case of the *Golomb codes* and the *Fibonacci codes*, which I describe next.

The Fibonacci codes are based on the *Fibonacci numbers of order* $m \geq 2$. Fibonacci numbers form a suite of integer numbers $F(-m+1)$, $F(-m+2)$, ..., $F(0)$, $F(1) \ldots F(k)$, which are defined as follows:

(a) The number $F(k)$ with $k \geq 1$ is equal to the sum of all preceding $m$ numbers;
(b) The numbers $F(-m+1)$ to $F(0)$ are all equal to unity.

Taking, for instance, $m = 2$, we have $F(-1) = F(0) = 1$, thus, $F(1) = F(-1) + F(0) = 2$, $F(2) = F(1) + F(0) = 3$, and so on, which yields the Fibonacci-number suite 1, 1, 2, 3, 5, 8, 13, 21, 34, ... The construction of a Fibonacci code based on the $m = 2$ parameter for the integer set $\{1, 2, \ldots, 34\}$ is illustrated by the example shown in Table 10.2. As the table indicates, the integer set $\{1, 2, \ldots, 34\}$ is first listed

[1] As intermediate values, we find for a 64-element, uniformly distributed source ($H = \log 64 = 6$), the coding efficiencies of $\eta$(gamma) = 64.8% and $\eta$(delta) = 68.8%.

[2] See, for instance: http://en.wikipedia.org/wiki/Elias_omega_coding.

**Table 10.2** Construction of a Fibonacci code of order $m = 2$ from the suite of Fibonacci numbers shown at the bottom.

| $i$ | | | | | | | | $F(i)$ |
|---|---|---|---|---|---|---|---|---|
| 1 | | | | | | | 1 | 11 |
| 2 | | | | | | 1 | 0 | 011 |
| 3 | | | | | 1 | 0 | 0 | 0011 |
| 4 | | | | | 1 | 0 | 1 | 1011 |
| 5 | | | | 1 | 0 | 0 | 0 | 00011 |
| 6 | | | | 1 | 0 | 0 | 1 | 10011 |
| 7 | | | | 1 | 0 | 1 | 0 | 01011 |
| 8 | | | 1 | 0 | 0 | 0 | 0 | 000011 |
| 9 | | | 1 | 0 | 0 | 0 | 1 | 100011 |
| 10 | | | 1 | 0 | 0 | 1 | 0 | 010011 |
| 11 | | | 1 | 0 | 1 | 0 | 0 | 001011 |
| 12 | | | 1 | 0 | 1 | 0 | 1 | 101011 |
| 13 | | 1 | 0 | 0 | 0 | 0 | 0 | 0000011 |
| 14 | | 1 | 0 | 0 | 0 | 0 | 1 | 1000011 |
| 15 | | 1 | 0 | 0 | 0 | 1 | 0 | 0100011 |
| 16 | | 1 | 0 | 0 | 1 | 0 | 0 | 0010011 |
| 17 | | 1 | 0 | 0 | 1 | 0 | 1 | 0001011 |
| 18 | | 1 | 0 | 1 | 0 | 0 | 0 | 1001011 |
| 19 | | 1 | 0 | 1 | 0 | 0 | 1 | 0101011 |
| 20 | | 1 | 0 | 1 | 0 | 1 | 0 | 1101011 |
| 21 | 1 | 0 | 0 | 0 | 0 | 0 | 0 | 00000011 |
| 22 | 1 | 0 | 0 | 0 | 0 | 0 | 1 | 10000011 |
| 23 | 1 | 0 | 0 | 0 | 0 | 1 | 0 | 01000011 |
| 24 | 1 | 0 | 0 | 0 | 1 | 0 | 0 | 00100011 |
| 25 | 1 | 0 | 0 | 0 | 1 | 0 | 1 | 00010011 |
| 26 | 1 | 0 | 0 | 1 | 0 | 0 | 0 | 00001011 |
| 27 | 1 | 0 | 0 | 1 | 0 | 0 | 1 | 10001011 |
| 28 | 1 | 0 | 0 | 1 | 0 | 1 | 0 | 01001011 |
| 29 | 1 | 0 | 1 | 0 | 0 | 0 | 0 | 11001011 |
| 30 | 1 | 0 | 1 | 0 | 0 | 0 | 1 | 00101011 |
| 31 | 1 | 0 | 1 | 0 | 0 | 1 | 0 | 10101011 |
| 32 | 1 | 0 | 1 | 0 | 1 | 0 | 0 | 11101011 |
| Fibonacci: | 21 | 13 | 8 | 5 | 3 | 2 | 1 | |

in increasing order. The suite of Fibonacci numbers, $\{1, 2, 3, 5, 8, 13, 21\}$, starting with $F(0) = F(1) = 1$ up to $F(7) = 21$, is written at bottom, from right to left, defining seven columns. It is easily checked that all integer numbers are given by a sum of the Fibonacci numbers, for instance:

$$7 = 5 + 2 = 1 \times F(4) + 0 \times F(3) + 1 \times F(2) + 0 \times F(1),$$

$$12 = 8 + 3 + 1 = 1 \times F(5) + 0 \times F(4) + 1 \times F(3) + 0 \times F(2) + 1 \times F(1),$$

which can be coded as $7_{10} \equiv 1010_{\text{Fibonacci}}$ and $12_{10} \equiv 10101_{\text{Fibonacci}}$, respectively.[3] The second column in Table 10.2 shows the codewords obtained according to such

[3] It is left as an exercise to show that a Fibonacci code of order $m = 3$ requires level-three coding.

a decomposition into Fibonacci numbers. The actual Fibonacci code is obtained by taking the mirror image of this initial codeword and appending a 1 postfix, as seen from the last column at right: thus $f(i = 1) = 11$, $f(i = 2) = 011$, $f(i = 3) = 0011$, $f(i = 5) = 00011$, and so on. The result of this operation is a prefix code, i.e., a code for which no codeword is the prefix of another codeword. This code example is also listed in Table 10.1, for comparison with the Elias codes. The comparison shows that the Fibonacci codewords are significantly shorter. Using the English-language source (Table 8.3), we find that our Fibonacci code has a mean codeword length of $L = 4.928$ bit/word, which corresponds to an efficiency of $\eta = H/L = 4.184/4.928 = 84.90\%$, and represents an improvement on the previous Elias-gamma and delta codes ($\eta = 79.83\%$).

It can be shown that Fibonacci codes are not asymptotically optimal, like Elias-gamma codes but unlike Elias-delta codes. Higher-order Fibonacci codes ($m > 2$) have better compression rates, provided that the source size is large and the probability distribution nearly uniform. Even if the Elias-delta codes are asymptotically optimal, Fibonacci codes of order two perform better with any source of size up to $10^6/2$ (precisely, $n = 514\,228$).[4] This fact illustrates that *asymptotic code optimality is not the only criterion for selecting the most efficient code*. Such a feature was also illustrated in the previous example. The comparative code performance also depends on the type of source *distribution*. Recalling that the Fibonacci code is parameterized by an integer $m \geq 2$, one must find an optimal value of $m$ for each source distribution under consideration. Such an optimization is advantageous, but case-specific. This represents both an advantage and a drawback, in comparison with *nonparameterized* codes (such as Elias codes), which are fixed once and for all.

The *Golomb codes* constitute a second important category of parameterized codes. For a Golomb code with parameter $m$, the codeword $G(i)$ of integer $i$ is made of two parts:

(a) A *prefix*, which is made of *1 preceded by $q$ zero bits*, with $q$ being the quotient $q = \lfloor (i-1)/m \rfloor$,
(b) A *suffix*, which is the *binary representation in $\lceil \log_2 m \rceil$ bits of the remainder* $r = i - 1 - qm$.

The first rule (a) can also be changed with the definition $q = \lfloor i/m \rfloor$ if, by convention, the list of integers $i$ is made to start from zero. The second rule (b) represents a simplified version and is not the one actually used in Golomb codes. However, we shall use it as a first step for easily introducing the concept. To provide a practical example, assume $m = 4$. We have, for the first 12 integers:

$$i = 1, \ldots, 4 \to q = 0 \to \text{prefix} = 1,$$
$$i = 5, \ldots, 8 \to q = 1 \to \text{prefix} = 01,$$
$$i = 9, \ldots, 12 \to q = 2 \to \text{prefix} = 001, \text{ etc.},$$

[4] D. A. Lelewer and D. S. Hirschberg, Data compression. *Computing Surveys*, **19** (1987), 261–97, see www.ics.uci.edu/~dan/pubs/DataCompression.html.

and

$$i = 1, \ldots, 4 \rightarrow r = 0, 1, 2, 3 \rightarrow \text{suffix} = 00, 01, 10, 11,$$
$$i = 5, \ldots, 8 \rightarrow r = 0, 1, 2, 3 \rightarrow \text{suffix} = 00, 01, 10, 11,$$
$$i = 9, \ldots, 12 \rightarrow r = 0, 1, 2, 3 \rightarrow \text{suffix} = 00, 01, 10, 11.$$

We, thus, observe that the prefix increases by one bit at each multiple of $m = 4$ and that the suffix has a constant length and changes with a periodicity of $m = 4$. From the above rules and with this example we obtain $G(3) = 1\_10$, $G(5) = 01\_00$ and $G(12) = 001\_11$, for instance (the underscore _ being introduced for clarity). Note the prefix rule of *1 preceded by q zero bits* is only conventional.[5] We can also define the prefix as *0 preceded by q one bits*, which represents the complement of the previous prefix (e.g., $G(3) = 0\_10$, $G(5) = 10\_00$, and $G(12) = 110\_11$). Another convention for the suffix is to take the smallest number of bits for the binary representation of $r$, which only changes 00 into 0. Thus, we have $G(3) = 1\_0$ instead of $1\_00$, $G(5) = 01\_0$ instead of $01\_00$, and so on. This convention reduces the code length by one bit each time $i = km + 1$ ($k$ an integer). With their prefix increase by blocks of $m$ and their suffix $m$ periodicity, the Golomb codes are straightforward to generate. Table 10.1 shows the nonoptimized Golomb code for $m = 8$.

Consider next the actual Golomb code, which uses a more complicated rule (b) for defining the suffix. This rule consists of coding the suffix with $c = \lfloor \log_2 m \rfloor$ bits for the first $c$ values of $r$ (with 0 as the leading bit), and with $c + 1$ bits for the other values (with 1 as the leading bit). Consider, for instance, $m = 5$. We have $c = \lfloor \log_2 5 \rfloor = 2$. Thus, we have, for the first 15 integers:

$$i = 1, \ldots, 5 \rightarrow q = 0 \rightarrow \text{prefix} = 1,$$
$$i = 6, \ldots, 10 \rightarrow q = 1 \rightarrow \text{prefix} = 01,$$
$$i = 11, \ldots, 15 \rightarrow q = 2 \rightarrow \text{prefix} = 001, \text{ etc.},$$

and

$$i = 1, \ldots, 5 \rightarrow r = 0, 1, 2, 3, 4 \rightarrow \text{suffix} = 00, 01, 100, 101, 110,$$
$$i = 6, \ldots, 10 \rightarrow r = 0, 1, 2, 3, 4 \rightarrow \text{suffix} = 00, 01, 100, 101, 110,$$
$$i = 11, \ldots, 15 \rightarrow r = 0, 1, 2, 3, 4 \rightarrow \text{suffix} = 00, 01, 100, 101, 110,$$

showing that in each period, the first two suffixes are coded with two bits with a leading 0, and the other suffixes are coded with three bits with a leading 1. We note that the three-bit suffix codes do not correspond to a binary representation of $r$, unlike with our previous definition. Table 10.1 shows the actual Golomb code corresponding to the case $m = 6$. We observe that the second definition makes it possible to shorten the length of

[5] Coding a number $n$ by $n - 1$ zeros followed by a one bit is referred to as *unary coding*. For instance, the numbers $n = 2$, $n = 5$, and $n = 7$ are represented in unary coding as 01, 00001, and 0000001, respectively. This definition should not be confused with the "unary number representation," which uses only one symbol character (e.g., 1) and is defined according to the rule $0_{\text{decimal}} \equiv 1_{\text{unary}}$, $1_{\text{decimal}} \equiv 11_{\text{unary}}$, $2_{\text{decimal}} \equiv 111_{\text{unary}}$, etc.

most codewords in the list. If we apply the Golomb code to the English-symbol source (Table 8.3), we obtain mean codeword lengths of

$$L = 4.465 \text{ bit/word (first, simple definition with } m = 8),$$
$$L = 4.316 \text{ bit/word (second, actual definition with } m = 6),$$

corresponding to efficiencies ($\eta = H/L \equiv 4.184/L$) of $\eta = 93.69\%$ and $\eta = 96.94\%$, respectively. These two results represent a significant improvement on the previous Elias (gamma or delta) codes ($\eta = 79.83\%$) and second-order Fibonacci codes ($\eta = 84.90\%$). It is clear that the Golomb-code parameter $m$ must be optimized according to the source size and distribution type. Golomb codes with low $m$ have relatively small codewords for the first few integers (owing to the short suffix of length $\log_2 m$), but the length rapidly increases because of the fast prefix increment. On the contrary, Golomb codes with high $m$ have relatively large codewords for the first few integers (owing to the long suffix), but the length increases slowly because of the slow prefix increment. Golomb codes with $m = 2^k$ ($k$ an integer) have also been known as *Rice codes*. For this reason one generally refers to *Golomb–Rice codes* to designate them altogether (with $m = 2^k$ corresponding to Rice codes). For sources of specific distribution types, it is possible to determine the optimal exponent parameter $k$ that minimizes the mean codeword length. In the general case, this parameter can also be determined through heuristic methods.

Integer coding based on various Elias or Golomb–Rice derivatives finds many applications in the field of *database management*, ensuring rapid access to and optimal indexing of library files. An illustrative application is the indexing of very large databases, such as the inventories of *nucleotide sequences* in biology. Golomb–Rice codes are also used in *sound compression* standards (see Appendix G).

## 10.2 Arithmetic coding

In static codes (Shannon–Fano, Huffman, block, and integer codings), it is implicitly assumed that the source characteristics (events and probabilities) are known, with the exception of integer coding, which only requires knowledge of the most likely events. In any case, these codes are fixed and optimized once and for all, so this approach is called *defined-word coding*. In the general case, one may not have such a prior knowledge. It is also possible that the source properties change from time to time (nonstationary source), in which case static codes lose their optimality or become inadequate. Nontypical English texts, such as those analyzed in Chapter 9 (Table 9.2), provide an illustrative example of deviation from the stationary source reference.[6] As we saw in that chapter, we

[6] A famous example of a very unusual English-text source is the 1939 novel *Gadsby* (E. V. Wright), which, in over 50 000 words, does not contain any character E whatsoever! Here is an extract from page 1:

If youth, throughout all history, had a champion to stand up for it; to show a doubting world that a child can think; and, possibly, do it practically; you wouldn't constantly run across folks today who claim that "a child don't know anything." A child's brain starts functioning at birth; and has, amongst its many infant

can use *universal codebooks*, which contain libraries of best codes for a variety of sources, such as text, programming-language codes, or datafiles, or a statistical mix of all possible combinations thereof. But the codebook approach is inadequate if the source under investigation escapes any of these known types. Static coding is, therefore, intrinsically limited, although it is most convenient because the symbol or codeword correspondence only requires a one-time calculation and optimization. While Huffman coding is ultimately optimal (overlooking overhead information), it is computationally intensive for large sources. Block codes have been shown in Chapter 9 to offer some simplification advantage by coding symbols into "super-symbol" groups, but with the drawback that the number of codewords rapidly becomes intractable with increasing group sizes, for basic considerations of memory space and read–write times.

In the general situation, where the source characteristics are unknown, one must implement what is equivalently referred to as *dynamic* or *adaptive* or *stream coding*, which evokes the time-changing character of the codeword assignment according to the evolving source characteristics. The basic philosophy of dynamic coding is to devise an optimal code "on the fly" for any sequence of incoming symbols, while minimizing the number of operations required at each intermediate step. One does not need to know the distribution of the single symbols forming the sequence, or the length of the sequence to encode, which represents a significant advantage over static coding. The symbol alphabet and the distribution may also radically change from one input sequence to the next, and the code is able to dynamically adapt to this. *Arithmetic coding*, which is described in this section, represents an intermediate case where the code is dynamically configured from the source, but the resulting codeword dictionary is then kept for extensive use, just as in a static code. Two other approaches, referred to as *adaptive Huffman coding*, and *Lempel–Ziv* (LZ) *coding*, are truly dynamically adaptive coding algorithms. These are described in the following two sections.

In *arithmetic coding*, it is assumed that both encoding and decoding machines have identical *programs,* which make it possible to compute the source's probability distribution and associated joint probabilities of all orders (see further). A specificity of arithmetic coding is that a specific "end-of-sequence" symbol is always required, as I shall illustrate.

convolutions, thousands of dormant atoms, into which God has put a mystic possibility for noticing an adults act, and figuring out its purport.

Also well known in this genre is the later 1969 French novel *La Disparition* (G. Perec), translated into English in 1995 as *A Void*, while fully respecting the author's spirit and using no E. Other translations of Perec's novel also exist in German and Danish, although the stunt was too hard in this last case for a full translation. To complete the story of such literary oddities, an early pioneer of the genre is reportedly H. Holland, who wrote a short 1928 novel called *Eve's Legend*, which uses no vowels *other* than E. In the same style, the author Perec also published *Les Revenentes* (1972). Here is a sample of *Eve's Legend*:

Men were never perfect, yet the three brethren Veres were ever esteemed, respected, revered, even when the rest, whether the select few, whether the mere herd, were left neglected . . .

See:
www.webrary.org/Maillist/msg/2001/2/Re.missingletterquotEquot.html,
www.ling.ed.ac.uk/linguist//issues/11/11-1701.html#?CFID=18397914&CFTOKEN=46891046.

Consider the basic example of a source $X = \{a, b, c\}$ with probabilities $p_X(x_i) = \{0.4, 0.4, 0.2\}$. We assume that symbol $c$ is exclusively used to signal the end of message. The interval containing all real values $p$ such that $0 \le p < 1$ is noted $[0, 1)$. Our encoder's program then proceeds as follows (Fig. 10.1):

- *Step 1*: The interval $[0, 1)$ is first divided into three subintervals $[0.0, 0.4)$, $[0.4, 0.8)$, and $[0.8, 1.0)$, corresponding to the symbol events $a$, $b$, and $c$, respectively. The interval $[0,1)$ is also divided into two equal regions, labeled in binary with prefixes 0 and 1, as shown in the right-hand side. We observe from the figure that the prefix 1 so far corresponds to either $a$ or $b$, and the prefix 0 to either $b$ or $c$.
- *Step 2*: Each of the previous subintervals, except the last one $[0.8, 1.0)$, corresponding to event $c$, is divided into three subintervals, the widths of which correspond to the joint probabilities $p(x_1, x_2) = p(x_2|x_1)p(x_1)$ of either joint events $aa$, $ab$, $ac$, or $ba$, $bb$, $bc$. The regions corresponding to labels 0 and 1 are also divided into two equal parts, which are labeled with prefixes 00, 01, 10, and 11. We observe from the figure that:
  - The prefix 11 corresponds to either $aa$ or $ab$;
  - The prefix 10 corresponds to either $ab$, $ac$, or $ba$;
  - The prefix 01 corresponds to either $ba$, $bb$, or $bc$;
  - The prefix 00 corresponds to either $bc$ or $c$;
- *Step 3*: Each of the previous subintervals, except the last ones corresponding to final events $c$, is divided again into three subintervals, the widths of which correspond to the joint probabilities $p(x_1, x_2, x_3) = p(x_3 | x_1, x_2)p(x_3)$ of joint events $aaa$, $aab$, $aac$, $aba$, $abb$, $abc$, $baa$, $bab$, $bac$, $bba$, $bbb$, $bbc$. The region corresponding to prefixes 00, 01, 10, and 11 are also divided into two equal parts, which are labeled 000 to 111 and correspond to different joint-event possibilities, except for 000 which is exclusively attached to event $c$.

The encoder is capable of executing the above steps an arbitrary number of times in order to find the prefix attached to a string of any length and ending in $c$. To clarify this point, assume that the string (or joint event) to encode is $bc$. We observe from Fig. 10.1(a) that the unique subinterval which is fully contained in the region defined by string $bc$ has the prefix 00111. We can then use this prefix as the unique codeword for string $bc$. The same observation applies, for instance, to the string $aac$, which gets 11001 as a unique prefix and codeword. Assume next that the machine must encode the string $babc$. The magnification in Fig. 10.1(b) shows that the subinterval corresponding to prefix 1000001 is the only one to be fully contained in the region defined by string $babc$, and, therefore, it should be used as the unique codeword. In summary, the encoder assigns a unique codeword to any symbol string (or joint event) $x_1x_2, \ldots, x_nc$ by slicing down the interval $[0, 1)$ into as many subintervals of widths $p(x_1, x_2, \ldots, x_n, c) = p(c|x_1, x_2, \ldots, x_n)p(c)$. The algorithm to perform this operation is described in detail in Appendix H. Our description represents a simplification of the more general algorithm described in MacKay (2003).[7]

[7] D. J. C. MacKay, *A Short Course in Information Theory* (Cambridge, UK: Cambridge University Press, 2003).

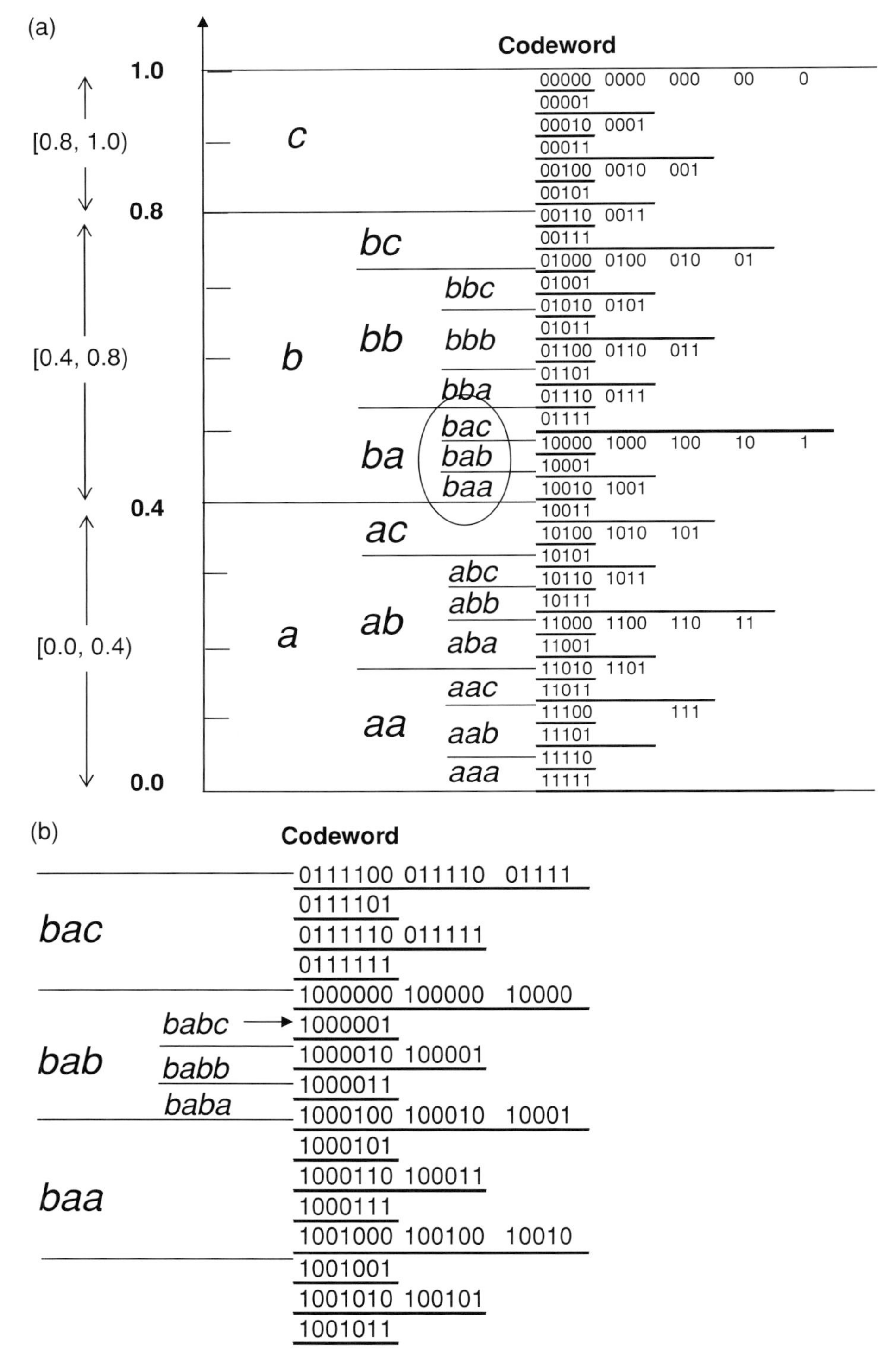


**Figure 10.1** Principle of arithmetic coding for the symbol sequence *babc* from event source $X = \{a, b, c\}$, showing coding of joint probabilities: (a) up to three events with five-bit codewords; (b) up to four events with seven-bit codewords.

Given a source $X = \{x_1, x_2, \ldots, x_n\}$ and its probability characteristics, the coding algorithm enables the *encoding machine* (encoder) to calculate rapidly the subinterval $[u_N, v_N)$ corresponding to any sequence $a_1 a_2, \ldots, a_{N-1} x_n$, where the symbols $a_k$ can take any value, save $x_n$. If the message source and characteristics are unknown, the encoder must acquire the entire sequence $a_1, a_2, \ldots, a_{N-1}$, identify the different symbols used ($x_i$), and calculate the corresponding probability distribution $p(x_i)$ attached to each of the message symbols specifically used in this sequence. We can, however, assume that the incoming messages always use the same type of source (e.g., English text, programming-language source code, bank datafiles, etc.), in which case, such an operation simply consists of prior calibration work. In this sense, the code is adaptive, but it is supposed to be extensively used once it has been thus defined, just like any other variable-length static code. The key difference here with a purely static code is that the codewords are not calculated symbol by symbol (or by symbol blocks of predefined length), but by symbol sequences having any arbitrary length. The maximum sequence length is defined by the precision of the coding arithmetic, namely the maximum number of bits that the machine can handle for any given codeword. As we have seen, the use of $m$-bit codewords makes it possible to represent probabilities with an accuracy of $1 \times 2^{-m}$. Referring to Fig. 10.1, for instance, the five-bit codewords define subintervals having a minimum width of $2^{-5} = 1/32 = 0.03125$, exactly. This number defines the minimum difference allowed between symbol-sequence probabilities.

As previously explained, the source's probability distribution, including joint probabilities up to some maximum order, must be calculated at first by the encoding machine. As soon as this initial calibration round is complete, encoding can then be performed on the fly, meaning that the codeword assignment (encoding) is performed almost at the same rate as the message-sequence symbols $a_1, a_2, \ldots, a_{N-1}$ are input to the encoder. Such a construction consists of building up the right codeword prefix corresponding to the beginning of the received sequence. Referring to Fig. 10.1, for instance, the reception of $a_1 = b$ as the first sequence symbol does not make it possible to conclude between 00, 01, or 10 as possible two-bit prefixes. If the next symbol is $a_2 = b$, the machine can make the choice of 01, since the interval of the sequence $bb$ is described by this prefix, as the figure indicates. If the next symbol is $a_3 = a$, the prefix is changed to 011, and so on. This shows that the encoder builds the codeword prefix practically at the same rate as symbols come in, because more than one symbol may be required to make the choice of the next prefix bit. If the choice is unique, this means that the codeword is complete, and basically, that the sequence is ended. For instance, the three-symbol sequence $bbc$ in Fig. 10.1 yields the final five-bit codeword 01001. In contrast, the three-symbol sequence $bbb$ is still at a two-bit 01 prefix level, and the encoding machine is waiting for a fourth symbol to choose between the 010 or 011 prefixes.

As we have seen, on reaching the maximum codeword size, the termination symbol $c = x_n$ is used by the encoder to close the message sequence, or to complete the codeword. A nice feature is that the encoding process does not need to halt at this point. Indeed, encoding can resume with the next incoming sequence, without any interruption.

Therefore, the encoding process may continue indefinitely over time. As another feature, the encoder may append *two* termination symbols $cc = x_n x_n$ at the end of a codeword, to instruct the decoder that the probability distribution characteristics are being maintained in the following message. Special terminations, or $cc'c'' \ldots = x_n x_{n+1} x_{n+2} \ldots$, whose symbols are not used by the original message source, may also be appended to the sequence for other signaling purposes, such as instructing a change in probability distribution or in the source alphabet.

Having identical and exact knowledge of the message-source characteristics, the *decoding machine* (decoder) can compute all possible subintervals $[u_N, v_N)$ to arbitrary small widths, as permitted by the arithmetic resolution, and store them in its memory. How this knowledge can be extracted from a program, and without communicating with the encoder, is a complex issue. As with the encoding process, the determination of the probability distribution is a matter of initial calibration, using the same program as the encoder (see later). The successive message bits input to the decoder are interpreted as codeword prefixes. The decoder performs the same task as reading Fig. 10.1 from right to left. Each codeword prefix points to a group of subintervals stored in the decoder's memory. As soon as a full codeword is identified (e.g., 01001 in the figure), the memory outputs the corresponding string (e.g., *bbc*). Therefore, decoding is also performed *on the fly*, since the memory pointer can move as fast as the message bits are received. Like encoding, but as the reverse operation, retrieving the next symbol in the sequence is not done on a bit-by-bit basis but out of progressive choices according to the prefix patterns.

As we have seen, arithmetic coding and decoding need to compute the source's probability distribution, with conditional probabilities of arbitrary order. A possibility is that both encoder and decoder use the same *codebook* reference, but such a solution is generally not optimal, and also lacks any flexibility. Rather, the system should be adaptable to any source type, for which the characteristics may change over time, including the size and definition of the symbol alphabet (e.g., changing from English-text language to computer datafiles or digital images). The initial encoder and decoder calibration, which provides the source's probability distribution and conditional probabilities, requires some initial computation steps. Let me describe here how such a computation works. From the encoder side, the first process consists of identifying the symbols $x_i$ and their distribution $p(x_i)$. The idea is to monitor a sufficient number of "symbol events," in order to convert the raw frequency histogram into an actual PDF. Because the encoder introduces the extra termination symbol $c$, which is not part of the message source, the corresponding probability $p(c)$ must be fixed to some arbitrary value, for instance, $p(c) = 0.1$. The first incoming symbol identified, say $x_1$, is assigned the initial probability $p(x_1) = 0.9$ (which satisfies $p(x_1) + p(c) = 1$). If the second incoming symbol is different, say $x_2$, the probability distribution becomes $p(x_1) = p(x_2) = 0.45$. This calibration process continues until a full distribution $p(x_i)$ is obtained, with $\sum_i p(x_i) = 1$ and $c \equiv x_n$. As for the conditional probabilities, $p(a_k = x_i | a_1 a_2 \ldots a_{k-1})$, they can be assigned according to the *Laplace* or the *Dirichlet* models. To explain the Laplace model, consider two events, $x$ and $y$. Let $F_x$ be the number of times that the event $x$ has been counted in the sequence $a_1 a_2 \ldots a_{k-1}$, and $F_y$ the count for event $y$. The Laplace rule defines the conditional probability as

$$p(a_k = x|a_1a_2 \ldots a_{k-1}) = \frac{F_x + 1}{F_x + F_y + 2}, \tag{10.1}$$

with the same relation applying for $p(a_k = y|a_1a_2 \ldots a_{k-1})$, being obtained by interchanging $x$ and $y$ (we note that consistently, the sum of the two conditional probabilities is equal to unity). For instance, considering the sequence $xxyxxxy$, we have $F_x = 5$ and $F_y = 2$, thus $p(a_8 = x|a_1a_2 \ldots a_7) = 6/9 = 2/3$ and $p(a_8 = y|a_1a_2 \ldots a_7) = 3/9 = 1/3$. For an $n$-event source $X = \{x_1, x_2, \ldots, x_n\}$, Laplace's rule is:

$$p(a_k = x_i|a_1a_2 \ldots a_{k-1}) = \frac{F_{x_i} + 1}{\sum_{x_i}(F_{x_i} + 1)}. \tag{10.2}$$

Note that the Laplace rule only represents a *model* to determine the conditional probabilities heuristically. Such an assignment is arbitrary and does not need to be exact or accurate. What matters is that both encoder and decoder use the same definition. It can yet be refined using the *Dirichlet* model:

$$p(a_k = x_i|a_1a_2, \ldots, a_{k-1}) = \frac{F_{x_i} + \alpha}{\sum_{x_i}(F_{x_i} + \alpha)}, \tag{10.3}$$

where $\alpha$ is an adjustable constant, for instance $\alpha = 0.05 - 0.01$. With the knowledge of the distributions $p(x_i)$ and $p(x_i|a_1a_2, \ldots, a_{k-1})$, the encoder can then implement the arithmetic-coding algorithm described in Appendix H. From the decoder's side, the calibration process is similar, except that it operates in the opposite way. The decoder, which has the same resolution as the encoder, identifies the different message codewords received and assigns a probability interval to each possible prefix, as Fig. 10.1 illustrates, reading from right to left. The correspondence between the identified codewords and the original source symbols is only a matter of convention, like the A–Z sequence of characters in the English alphabet.

It can be shown that arithmetic coding is near optimal, as the codeword length, $l(s)$, for a given symbol sequence, $s$, closely approaches the Shannon limit $-\log p(s)$, which represents the information contents of the sequence. Owing to its versatility with respect to source types and its capability of coding and decoding "on the fly," arithmetic coding is used in many *still or motion image-compression standards*, such as JPEG and MPEG (see Appendix G).

Another interesting application of arithmetic coding concerns the *generation of random numbers*. Indeed, random-bit strings can be generated by feeding an arithmetic *decoder* with uniformly distributed bit streams, such as produced by a pseudo-random word generator.[8] The decoder then outputs what it interprets to be a suite of symbol sequences picked up within the [0, 1) probability interval. The symbol sequences form random bit streams having probability-distribution characteristics departing from uniformity, i.e., $p(x_1 = 0) \neq p(x_2 = 1)$.

[8] A pseudo-random word can be a pre-established bit pattern with uniform 1/0 bit distribution, which is cyclically repeated by bit translation or permutation.

Arithmetic coding can also be used for *fast data-entry devices*. The principle is for a human operator to acquire information and produce a maximum of information bits through a minimal number of body gestures. A computer keyboard only provides the one-to-one correspondence between alphanumerical characters and their ASCII codewords. The keyboard is designed to have the most frequently used letters in specific locations that the ten fingers learn to reach automatically, without searching. Let us imagine instead a fancy dynamic keyboard, where the most frequently used letters and *most likely letter groups* would always be found near the last character that was entered, like finding *he* immediately after typing *t* or *nd* after *a*, corresponding to the words *the* and *and*, respectively. Ready-made word terminations, like *cept*, *dition*, *stitute*, *tinue*, *vey*, and *vention*, would show up as soon as the text *con* had been input to the keyboard, for instance. These word terminations could also be arranged according to their frequent use in the specific message context. Such a dynamic keyboard would make it possible to achieve rapid text acquisition using a single-click, perhaps with a mouse or an optical or eye pointer or tracker. A representative application of this principle is provided by the project *Dasher* (European languages), also named *Daishoya* (Japanese).[9] It consists of a text-entry interface, which can be driven by natural pointing gestures, using a joystick, a touch-screen, a tracker or roller ball, a mouse, or even an eye tracker. Experienced readers can perform text acquisition with a single finger or eye motion at rates of 20 to 40 words per minute, which is nearly as fast as the normal writing rate and even faster in the last case. Practical device applications concern palmtop computers, wearable computers, one-handed computers, and hands-free computers for various working environments and for the disabled.

## 10.3 Adaptive Huffman coding

*Adaptive Huffman coding* is also known as *FGK*, after *Faller, Gallager* and *Knuth*, and as *algorithm V*, after improvements of FGK from *Vitter*.[10]

The FGK principle represents a dynamic implementation of Huffman *coding trees*, which is based on a running estimate of the symbol probability distribution. The code is *optimal* but only within the context of a given source message. Both encoder and decoder adapt themselves to the changing probability distribution, which makes the method suitable to encode or decode time-evolving or nonstationary sources.

The key advantage of adaptive vs. static Huffman coding is that the data encoding and decoding is, indeed, performed "on the fly," through a *single-pass* conversion process. In contrast, the static scheme requires two passes: the first one for the coding-tree determination, the second for the coding. However, if the source's characteristics are time-invariant, this operation only needs to be performed once, and the other incoming message sequences are coded and decoded through a single pass. If the source's

[9] See details with animated screenshot demonstrations in www.inference.phy.cam.ac.uk/dasher; the software is freely available.

[10] See: D. A. Lelewer and D. S. Hirschberg, Data compression. *Computing Surveys*, **19** (1987), 261–97, www.ics.uci.edu/~dan/pubs/DataCompression.html.

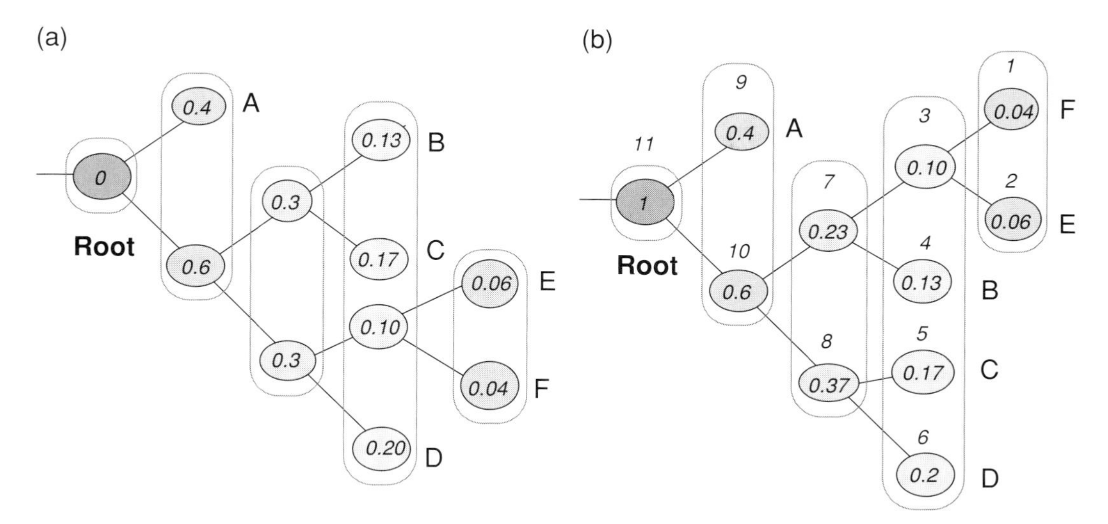


**Figure 10.2** Example of source coding tree (a) without sibling property, and (b) with sibling property (Huffman tree). The number shown inside each node's circle is the corresponding probability, and in case (b) the nodes are labeled 0 to 11.

characteristics evolve rapidly, however, then the coding tree must be re-evaluated for each message sequence, which justifies the interest of the single-pass adaptive method.

The FGK algorithm uses what is called the *sibling property* of coding trees, as introduced by Gallager.[11] This algorithm is defined as follows:

A coding tree has a sibling property in the case where all nodes, except the root and the terminal (leaf) nodes, have a sibling node and can be listed in order of nondecreasing weights.

To grasp the meaning of this seemingly obscure definition, and understand the concept of "sibling property," consider the two illustrative examples shown in Fig. 10.2. The two coding trees shown in Fig. 10.2 are associated with the same source $X = \{A, B, C, D, E, F\}$, whose probability distribution is $p(x_i) = \{0.4, 0.2, 0.17, 0.13, 0.06, 0.04\}$. The weights (or combined probabilities) of each of the nodes are indicated. Except for the root node at left, and the leaf nodes at right, intermediate nodes are seen to come with a *sibling* of equal or lower weight. Looking at the tree (a) in the figure, we observe that there are five sibling pairs responding to this description. However, the group formed by the pairs (0.2–0.1) and (0.17–0.13) is not ordered according to increasing or decreasing weights. The above-stated "sibling property" rule requires that the nodes be arranged by successive pairs of nondecreasing weights, (0.1–0.13) then (0.17–0.2), and this is precisely the case of the second tree (b) shown in Fig. 10.2. Not surprisingly, this rule "compliant" tree is a *Huffman tree*, as one may easily check. Also, Gallager showed another powerful property, according to which

A binary prefix code is a Huffman code if and only if its coding tree has the sibling property.

[11] See: R. G. Gallager, Variations on a theme by Huffman. *IEEE Trans. Inform. Theory*, **24** (1978), 668–74.

The implementation of the FGK algorithm proceeds as follows. At the start, the tree is a single leaf node, which is referred to as the *Ø-node*. Assume that there are $n$ symbols in the message sequence to be encoded. The encoder does not need to know what $n$ might be. The idea is always to keep the Ø-node for the $n - m$ symbols that have not been observed yet by the encoder, and to compute the Huffman coding tree for the other $m$ symbols, out of an observed sequence of $k$ symbols. The resulting coding tree is a Huffman tree $h(k)$, which has $k + 1$ leaves: one leaf is the Ø-node (probability or weight zero) and the other $k$ leaves represent the $k$ symbols (nonzero probabilities or weights), ordered by siblings of nondecreasing weights, and labeled in that order. Whenever a new or previously unobserved symbol is identified, the Ø-node is split into a new Ø-node and a new leaf node is created for this symbol. The coding tree is also reconfigured. Figure 10.3 illustrates the evolution of the coding tree, as recomputed at each step for the 12-symbol sequence example GOODTOSEEYOU, from step $k = 1$(G) to step $k = 6$ (GOODTO).

From the orderly sequence of node weights, we observe that all trees in Fig. 10.3 have the sibling property. The evolving codeword assignment for each different symbol (G, O, D, T, S, E, Y, U, Ø), as determined by both encoder and decoder from the same static Huffman algorithm, is also shown in the figure. The symbol codewords up to step $k = 12$ are shown in Table 10.3. We observe from the table that, as expected, the codeword assignment changes at each computational step $k$. The table also shows, for each step $k$, the value of entropy $H$, the mean codeword length $L$, and the corresponding coding efficiency $\eta$. The entropy and coding efficiency are plotted in Fig. 10.4. We observe from the figure that the entropy increases with the message length, following a sawtooth pattern. The entropy's slope progressively decreases while remaining globally positive. The occasional downward slope changes correspond to the accidental occurrence of repeated symbols, which decreases the mean uncertainty or Shannon information. Clearly, the drops observed at $k = 6$ and $k = 11$ can be attributed to the repeated occurrences of O in the GOODTOSEEYOU message sequence. For a sufficiently long message sequence, it is expected that the entropy reaches the limit of that of the English-language source, which was shown in Chapter 4 (1982 poll) to be $H = 4.185$ bit/symbol. Such a convergence must be quite rapid, since a message as short as GOODTOSEEYOU has the entropy of $H = 2.751$ bit/symbol (Table 10.3), which represents 66% of this limit. We also observe from Fig. 10.4 that the coding efficiency rapidly converges to 100% as new symbols come in, although with a similar saw-tooth progression for entropy. With this message-sequence example, the efficiency reached at $k = 12$ is $\eta = 91.72\%$ (Table 10.3), a relatively high performance due to the optimality of the Huffman coding (the mean codeword length $L$ always remaining within one bit of the source entropy $H$, as the table data also indicate).

As we have seen, the encoder dynamically updates its coding tree at each step $k$, starting with a single Ø-node leaf. From the receiving side, the *decoder* just has to perform the same operation. However, for the decoder to update its coding tree, it needs the following basic information:

(a) For a new symbol: the Ø-node's current codeword and the new symbol definition;
(b) For a symbol previously observed: the symbol's current codeword.

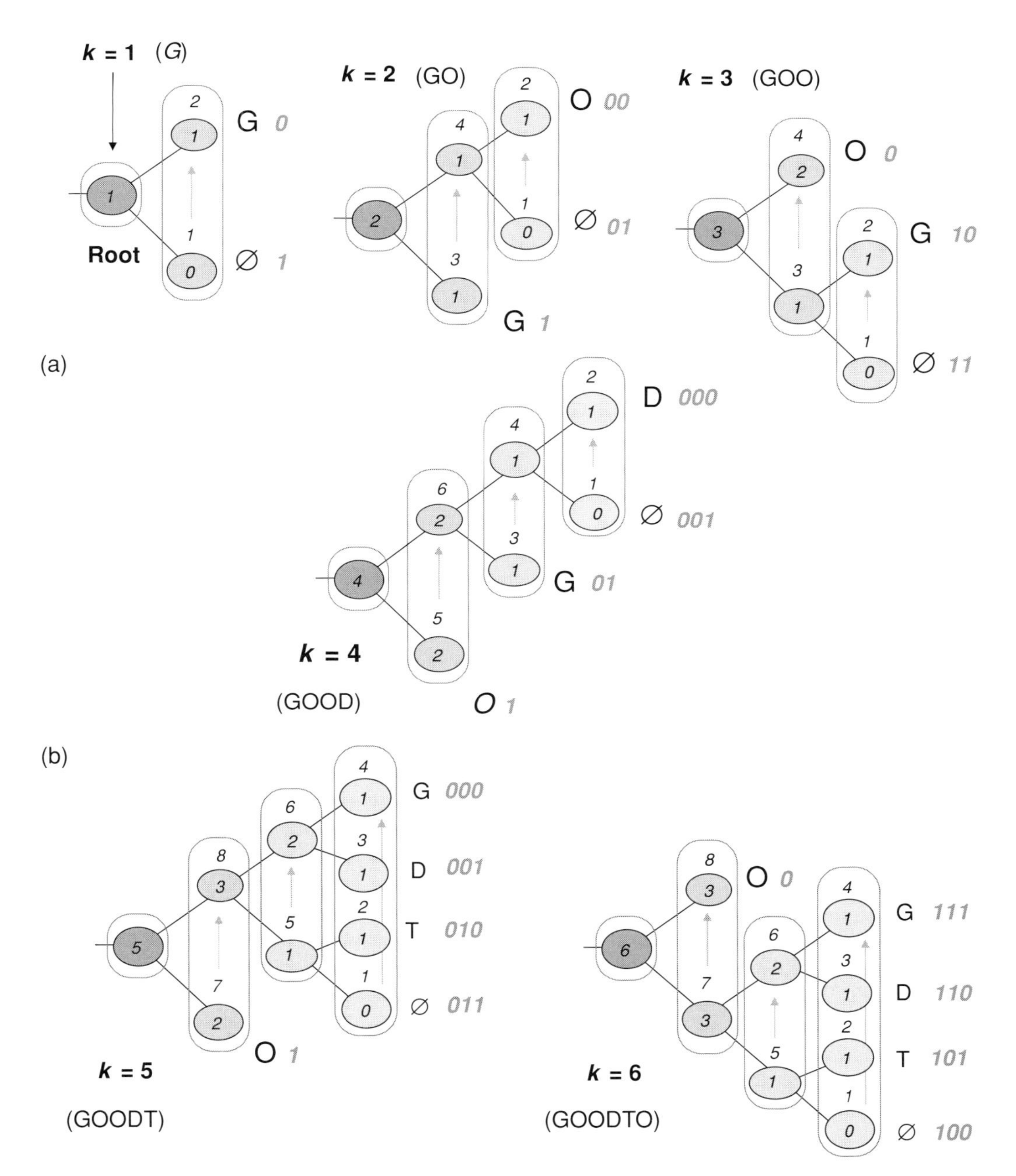


**Figure 10.3** Evolution of the adaptive Huffman coding tree with the message example GOODTOSEEYOU. The index $k$ is the number of symbols received: (a) $k = 1$–4; (b) $k = 5$–6. Nodes are relabeled according to the *sibling-property rule*. The number inside each node's circle is the corresponding weight. The steady or changing codewords associated with each symbol are shown at the right. The evolution of the coding tree and codeword assignment up to $k = 12$ is shown in Table 10.3.

**Table 10.3** Evolution of coding tree according to the *adaptive Huffman coding algorithm* with the 12-symbol message example GOODTOSEEYOU. The index $k$ is the number of symbols received. A possible codeword assignment to each symbol, including that of the $\varnothing$-node, is shown for each step. See Fig. 10.3 for the corresponding coding trees up to step $k = 6$. The entropy and coding efficiency are plotted in Fig. 10.4.

| Symbol | $k=1$ | $k=2$ | $k=3$ | $k=4$ | $k=5$ | $k=6$ | $k=7$ | $k=8$ | $k=9$ | $k=10$ | $k=11$ | $k=12$ |
|---|---|---|---|---|---|---|---|---|---|---|---|---|
| G | 0 | 1 | 10 | 01 | 1 | 111 | 001 | 010 | 011 | 100 | 100 | 101 |
| O | | 00 | 0 | 1 | 000 | 0 | 1 | 00 | 00 | 01 | 00 | 01 |
| D | | | | 000 | 001 | 110 | 010 | 011 | 100 | 101 | 101 | 10 |
| T | | | | | 010 | 101 | 011 | 100 | 101 | 110 | 110 | 111 |
| S | | | | | | | 0000 | 101 | 010 | 111 | 111 | 0000 |
| E | | | | | | | | 110 | 11 | 000 | 010 | 001 |
| Y | | | | | | | | | | 0010 | 0110 | 0001 |
| U | | | | | | | | | | | | 1000 |
| Ø | 1 | 01 | 11 | 001 | 011 | 100 | 0001 | 111 | 0101 | 0011 | 0011 | 1001 |
| Source entropy $H$ (bit/symbol) | 0 | 0.5 | 0.918 | 1.5 | 1.921 | 1.792 | 2.128 | 2.405 | 2.419 | 2.646 | 2.550 | 2.751 |
| mean codeword length $L$ (bit/word) | 1 | 1.5 | 1.333 | 1.75 | 2.200 | 2.000 | 2.285 | 2.625 | 2.777 | 2.800 | 2.727 | 3.000 |
| Coding efficiency $\eta$ (%) | 0.00 | 33.33 | 68.87 | 85.71 | 87.36 | 89.62 | 93.10 | 91.64 | 87.10 | 94.52 | 93.51 | 91.72 |

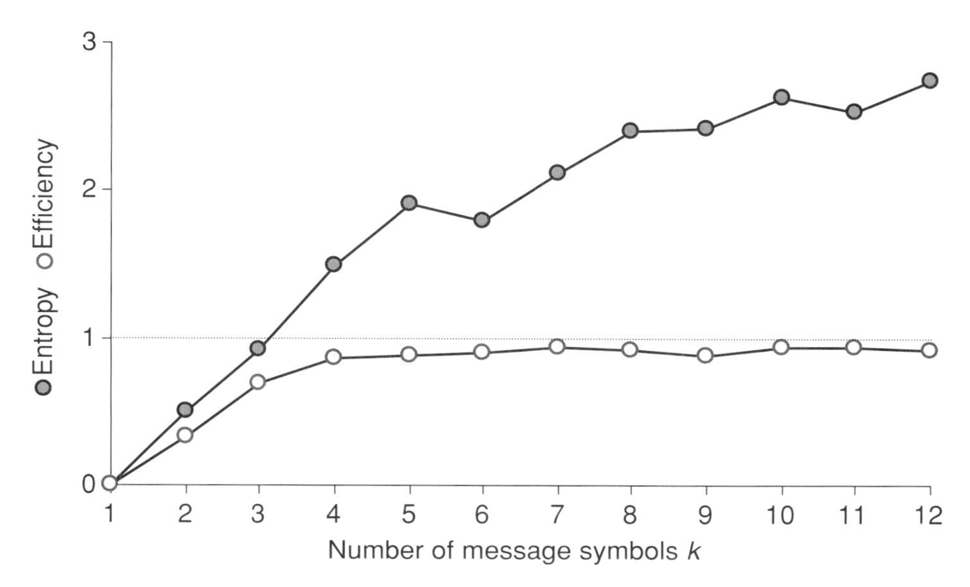


**Figure 10.4** Evolution of entropy $H$ and coding efficiency $\eta$ in adaptive Huffman coding as a function of the number of symbols received $k$, which corresponds to the example in Fig. 10.3 and Tables 10.3 and 10.4.

In case (a), the ∅-node's current codeword acts as a signal to the decoder that a new symbol must be entered in the tree (initially, the ∅-node codeword is set to zero). The previous ∅-node is split into a new ∅-node and a new leaf for this symbol. The other information concerns the symbol description, for instance its ASCII value. In case (b), the weight of the symbol's terminal node (leaf) simply needs to be incremented by one unit, which is sufficient to recompute the coding tree. Note that the information in case (b) is pure payload (previously-seen symbol transmission), while that in case (a) is both payload and overhead (new symbol transmission with definition). The data listed in Table 10.4 represent the information provided by the encoder to the decoder for the message string GOODTOSEEYOU, as based on the codewords of Table 10.3, to be sequentially used by the decoder for coding-tree updating and decoding. The table also shows the number of information bits that the encoder outputs at each step. Assume that a fixed-length codeword with $n$ bits is required to define a new symbol, for instance $n = 7$ for the reduced ASCII code. As this table indicates, the seven-bit codes for symbols G and O are $\langle 1000111 \rangle$ and $\langle 1001111 \rangle$, respectively. So in Table 10.4, the encoder, thus, needs to output

$$\langle 0 \rangle \langle 1000111 \rangle = \langle 01000111 \rangle \text{ at stage } k = 1 \text{ (new symbol G)},$$
$$\langle 1001111 \rangle = \langle 11001111 \rangle \text{ at stage } k = 2 \text{ (new symbol O)},$$
$$\langle 00 \rangle \text{ at stage } k = 3 \text{ (old symbol O)},$$

and so on. Since the first part of each of these codewords is a prefix code known by the decoder (from the previous stage) and the second part is a standard fixed-length code, the decoder can make sense of the codewords as an uninterrupted stream, without

**Table 10.4** (Top) information provided by the encoder at each coding step $k$ for the GOODTOSEEYOU example described in Table 10.3, with corresponding codewords and symbol definitions $x$. For clarity, the last two lines indicate whether the incoming symbols in the message string are new or old, and the message status; (Bottom) number of bits required to convey the information at each step, where $n$ is the fixed codeword length used to define the symbols.

| $k = 1$ | $k = 2$ | $k = 3$ | $k = 4$ | $k = 5$ | $k = 6$ | $k = 7$ | $k = 8$ | $k = 9$ | $k = 10$ | $k = 11$ | $k = 12$ |
|---|---|---|---|---|---|---|---|---|---|---|---|
| 0 | 1 | 00 | 11 | 011 | 000 | 100 | 0001 | 110 | 0101 | 01 | 0011 |
| G | O | | D | T | | S | E | | Y | | U |
| New | New | Old | New | New | Old | New | New | Old | New | Old | New |
| G | GO | GOO | GOOD | GOODT | GOODTO | GOODTOS | GOODTOSE | GOODTOSEE | GOODTOSEEY | GOODTOSEE YO | GOODTOSEEYOU |
| $1 + n$ | $1 + n$ | 2 | $2 + n$ | $3 + n$ | 3 | $3 + n$ | $4 + n$ | 3 | $4 + n$ | 2 | $4 + n$ |

needing any codeword delimiters or blanks. The data shown at the bottom of Table 10.4 indicate that the total information required to encode GOODTOSEEYOU takes $32 + 8n$ bits, which, for $n = 7$, comes to 88 bits.

For comparison purposes, consider the case of *static* Huffman coding. In this case, the encoder first reads the full message GOODTOSEEYOU and then computes the Huffman coding tree, which yields the same codeword assignment as shown in Table 10.3, except that the symbol Ø is not used. The encoder must then define the coding tree as overhead information. From the table data, and assuming $n$ bits to define each of the eight symbols (G, O, D, T, S, E, Y, U), the overhead size comes to $2x(2 + n) + 3x(3 + n) + 3x(4+ n) = 29 + 8n$. With $n = 7$, the overhead size is, therefore, 85 bits. On the other hand, the GOODTOSEEYOU message payload represents $2x(2) + 3x(3) + 3x(4) = 25$ bits. The total message length (overhead + payload) is, therefore, $85 + 25 =$ *110* bits. This result compares with the *88-bit* full message length of the adaptive FGK coding, which is 20% shorter. The FGK performance can be even further improved by decreasing the size of the overhead information, namely the definition of the $N$ identified source symbols. Such a definition requires $N \log_2 N$ bits. Using seven-bit ASCII, the source alphabet size is $2^7 = 128$, which covers more than the ensemble of computer-keyboard characters. If the message to be encoded uses fewer than 128 symbols, the overhead can be reduced to fewer than seven bits. Regardless of the source size, it is possible to make a list of all possible symbols and to attribute to each one a code number of variable length, for instance defined by $\lceil -\log p(x_i) \rceil$, where $p(x_i)$ is a conservative estimate of the symbol probability distribution with long messages. With this approach, the *average* overhead size is minimal, the most frequent symbols having shorter code-number definitions, and the reverse for the least frequent ones. Both encoder and decoder must share this standard symbol codebook.

We conclude from this tedious but meaningful exercise that when taking into account the overhead, *adaptive coding may yield a performance significantly greater than that of static coding*. Such an advantage must be combined with the fact that encoding and decoding can be performed dynamically "on the fly," unlike in the static case. Furthermore, changes in the source's characteristics (symbol alphabet and distribution) do not affect the optimality of the code, which, as its name indicates, is adaptive. The price to pay for these benefits is the extensive computations required (updating the coding tree for each symbol received). In contrast, static Huffman coding is advantageous as the message-source characteristics are fixed, or only slowly evolving in time. In this case, significantly longer messages can be optimally encoded with the same coding tree, without needing updates. Static-coding-tree updates can, however, be forwarded periodically, representing a negligible loss of coding performance, due to the large information-payload size transmitted in between.[12] As an example of an application, FGK is used for dynamic data compression and file archival in a UNIX-environment utility known as *compact*.

[12] It is an interesting class project to perform the comparison between static and adaptive Huffman codings, based on the same English-text message but considering extracts of different sizes. The goal is to find the break-even point (message size) where the performance of adaptive coding, taking into account the overhead of transmitting the 26-letter alphabet codeword, becomes superior to that of static coding.

An improvement on FGK is provided by the *Vitter algorithm*, or "*algorithm V*." The difference with FGK is that the coding tree recomputations are subject to certain constraints. The nodes are still numbered according to the sibling rule, but in the numbering all leaves of weight $w$ are put under the internal node of the same weight. The ensemble of nodes of the same weight is said to form a *block*, always with an internal node as the leader with the highest number. The coding-tree update can be seen as moving certain nodes from one block to another with greater weight. The rule is that moving nodes automatically take the place of the leading node of the target block. A second constraint concerns the codeword assignment. The algorithm seeks to *minimize* both functions defined by $s = \sum l(x_i)$ (called *external path length*) and $m = \max[l(x_i)]$ (called *tree height*), where $l(x_i)$ is the codeword length assigned to symbol $x_i$. Note that the decoder is equipped with the same program, and is, therefore, able to reconfigure the coding tree and codeword assignment in a way strictly identical to that of the encoder. The above constraints guarantee a coding tree with minimal height and the minimal number of codeword bits. This *Vitter tree* is then best suited to the next update, but only under the assumption that the next symbol in the message sequence is not new (or is not already a tree leaf) and that all symbols are nearly equally probable. For these reasons, algorithm V generally outperforms FGK in terms of number of transmitted bits, and it can be shown that this number is, at worse, one extra bit greater than that of the static Huffman code payload.

## 10.4 Lempel–Ziv coding

The adaptive coding devised in 1977 by Lempel and Ziv, now widely referred to as *Lempel–Ziv* (LZ), or *Ziv–Lempel,* or more commonly LZ77, is fundamentally different from the previous Huffman/FGK/algorithm-V approach. In a way that recalls the principle of static arithmetic coding, the LZ codewords are generated by symbol *blocks*. As previously described, arithmetic coding is a *defined-word* scheme: it maps all possible source-symbol blocks into a final codeword set, as allowed by the maximum codeword size (or arithmetic resolution). With LZ, symbol blocks are dynamically analyzed and corresponding codewords are generated on the fly. One sometimes refers to LZ as a *free-parse* algorithm, meaning that codewords are generated as the incoming message sequence is parsed.

Here, the word *parsing* defines the action of breaking up a message string into different substring patterns, which are called *phrases*. The term *distinct parsing* refers to the case where no two phrases are identical.

Basically, the LZ algorithm consists of analyzing the source-message string by blocks of variable size (substrings). The maximum block size is defined by some prescribed integer $L_1$, for instance $L_1 = 16$ bits. The substring analysis makes it possible to identify previously observed patterns or phrases. These phrases are assigned a codeword whose length, called $L_2$, is *fixed*, for instance $L_2 = 8$ bits. The phrases are defined so that their probabilities of occurrence are nearly equal. As a result, the most frequently occurring symbols are grouped into longer phrases, and the reverse for the least frequently occurring ones. Therefore, the same codeword length is used indifferently to represent long or short phrases (and even single symbols), all of them having nearly equal occurrence probability. This should not be confused with the principle of block codes (Chapter 9) where the codeword length is fixed but the corresponding distribution is nonuniform. Another key difference is that LZ is an *adaptive* algorithm, which is capable of learning from the source's characteristics (the most frequently used symbol-sequence patterns, like English words or their fractions) and to generate optimal codewords on the fly, without having to parse the entire message sequence, or to assume any probabilistic model for the source.

We shall now analyze the details of the LZ algorithm by considering a binary sequence and parsing it into distinct phrases. In the following, we shall use binary message strings with 1 and 0 bits as the symbol alphabet, but this does not remove the generality of the LZ algorithm in terms of alphabet size. As a working example, assume the following 25-bit sequence:

$$0101101101001000101101100.$$

Parsing the sequence consists of generating a list of all *distinct* phrases that can be identified in the sequence, each one being different from any one previously observed. Each phrase is then assigned an address number $k = 1, 2, 3, \ldots,$ which is called a *pointer*. Using the above sequence, this parsing action gives:

*Phrase:*

0 1 01 10 11 010 0100 0101 101 100;

*Pointer (decimal):*

1 2 3 4 5 6 7 8 9 10;

*Pointer (binary):*

0001 0010 0011 0100 0101 0110 0111 1000 1001 1010.

We note that the 10 identified phrases have equal probability, namely $p = 1/10$. We also note that each phrase is the *prefix* of some other phrase coming up in the list: for instance, phrase $k = 3$ (01) is the prefix of phrase $k = 6$ (010); in turn, phrase $k = 6$ is the prefix of phrases $k = 7$ (0100) and $k = 8$ (0101), and so on. We can, therefore, associate a prefix phrase with the two other phrases that differ only by the last bit, for instance $7 \leftrightarrow (6, 0)$ and $8 \leftrightarrow (6, 1)$, where the first number is the prefix pointer and the second number is the differing bit (referred to as *extension character*). The LZ codeword is given by concatenating the pointer and the extension character, with the pointer expressed in binary. The first two phrases, which are made of a single bit, are concatenated with pointer $k = 0$, which corresponds to the empty phrase. From these rules, we obtain from our example:

*Phrase:*

0 1 01 10 11 010 0100 0101 101 100;

*Pointer (decimal):*

1 2 3 4 5 6 7 8 9 10;

*(Prefix pointer, bit), decimal:*

$(0, 0)(0, 1)(1, 1)(2, 0)(2, 1)(3, 0)(6, 0)(6, 1)(4, 1)(4, 0).$

**Table 10.5** *Lempel–Ziv* (LZ) codeword dictionary generated from the message example *0101101101001000101101100*. The last bit of the LZ codewords is highlighted for reading clarity.

| Pointer | Phrase | (Prefix pointer, bit) | Codeword |
|---|---|---|---|
| 0 | ∅ | – | – |
| 1 | 0 | (0, 0) | 0000*0* |
| 2 | 1 | (0, 1) | 0000*1* |
| 3 | 01 | (1, 1) | 0001*1* |
| 4 | 10 | (2, 0) | 0010*0* |
| 5 | 11 | (2, 1) | 0010*1* |
| 6 | 010 | (3, 0) | 0011*0* |
| 7 | 0100 | (6, 0) | 0110*0* |
| 8 | 0101 | (6, 1) | 0110*1* |
| 9 | 101 | (4, 1) | 0100*1* |
| 10 | 100 | (4, 0) | 0100*0* |

The sequence is, thus, coded into the following decimal representation:

$$(0, 0)\ (0, 1)\ (1, 1)\ (2, 0)\ (2, 1)\ (3, 0)\ (6, 0)\ (6, 1)\ (4, 1)\ (4, 0),$$

which must be converted into a string of binary codewords, as listed in Table 10.5. The set of LZ codewords is referred to as a *dictionary*. Given the fixed codeword length $L_2$, which is the number of pointer or address bits plus one, the maximum dictionary size is, therefore, $2^{L(2)-1}$. In this example, we allot four bits to the pointer, which limits the dictionary size to $2^4 = 16$ codewords of length $L_2 = 5$ bits. Based on Table 10.5, the initial *25-bit* message is thus converted into the following coded string (underscores being introduced for reading clarity):

$$00000\_00001\_00011\_00100\_00101\_00110\_01100\_01101\_01001\_01000,$$

which is *50 bits* long. The LZ code, thus, realizes a twofold expansion of the source message, which is a characteristic feature of the initialization stage. As the message length increases, the bit phrases become more redundant, and the benefits of LZ coding for data compression begin to appear, as discussed later.

Decoding an LZ sequence is straightforward. A nice feature is that the decoder does not need to know the codeword size. As we have seen, the LZ algorithm imposes the constraint that the first two codewords in the sequence be of the form 000 . . . 00 or 000 . . . 01, namely a 0 or 1 bit preceded by a prefix of size $L_2 - 1$ bits. This information automatically provides the codeword length $L_2$. The decoder then registers successive codewords up to rank $2^{L_2-1}$, which represent the LZ code alphabet, and puts them into a dictionary along with their corresponding source-message contents. For instance, the codeword 10001 at line 10 of the decoder's memory (pointer = 10) means 1 preceded by the message contents of line $1000_2 \equiv 8_{10}$ (pointer = 8). According to the same rule, this content has been established to be 0101, so the content of line 10 is 0101 + 1 = 01011. The whole process is strictly equivalent to reconstructing the equivalent of Table 10.5 from top to bottom, but starting from each new codeword received up to $2^{L_2-1}$. It is a remarkable feature that the decoder is able to interpret the LZ code without any dictionary being ever transmitted.

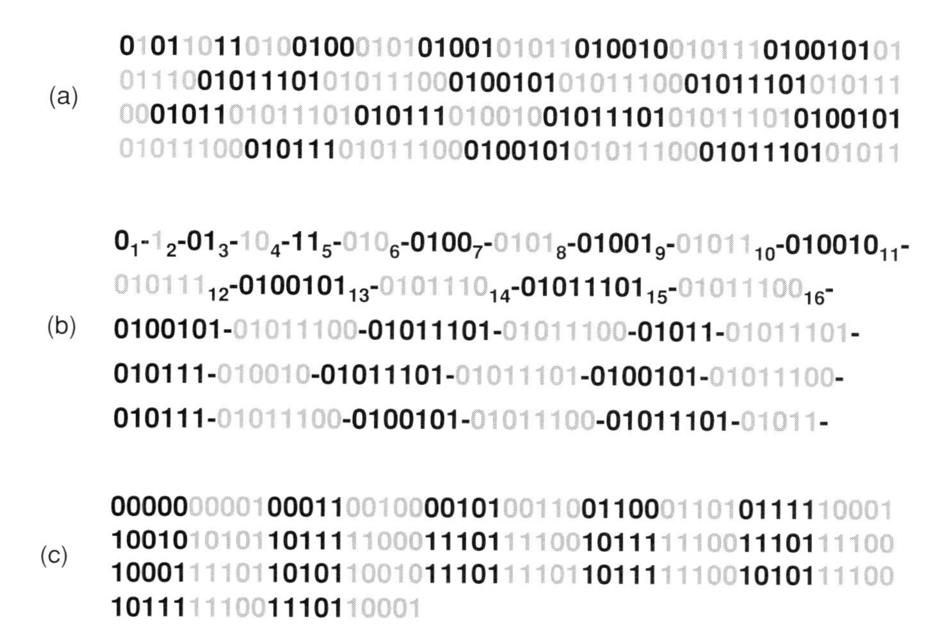

**Figure 10.5** Coding a message bit sequence with the *Lempel–Ziv* (LZ) algorithm:

(a) Input message (200 bits including 95 0s and 105 1s);
(b) Parsing of message (a) into distinct phrases, as shown in alternative black and gray colors; the first 16 phrases generate the LZ codewords (see correspondence in Table 10.6); the 16 subscripts correspond to pointer addresses;
(c) LZ code sequence (five-bit codewords) representing 25% compressed version of source message (a).

If the LZ code must be reconfigured, the encoder can send either of the two codewords 000 . . . 00 or 000 . . . 01, recalling that these two have only appeared once at the very beginning of the coded sequence. In our example (Table 10.5), note that only 00001 can be used for this purpose.[13] Receiving an unexpected codeword is an anomaly, which instructs the LZ decoder that a new dictionary has been used in the sequence bits to follow. The decoder then reads the next codeword to check out the (possibly new) codeword size, and the process of reconstructing the dictionary from scratch starts again. Because the dictionary construction is sequential and as rapid as the codeword acquisition, the flow of message or code data remains uninterrupted, even in the process of dictionary reconstruction. This feature makes LZ coding a truly *dynamic*, *adaptive*, and "*free-parse*" algorithm.

A puzzling question remains: how is a code with fixed-length codewords ever able to achieve any data compression? Before going through the details of a formal

[13] Indeed, 00000 may appear several times in the sequence, since we don't have a codeword for the contents 00, and 00 is the prefix of no codeword in the dictionary. In the event that any of the 16 source-message phrases is followed by 00$x$, the code necessarily uses 00000 to represent the first and single 0. The next codeword depends on the third bit $x$. If $x = 0$, the code uses again $\langle 00000 \rangle$ to represent the second and single 0. If $x = 1$, the second and third bits, 01 form the prefix of several content possibilities in the dictionary, including the stand-alone 01 (Table 10.5).

**Table 10.6** *Lempel–Ziv* (LZ) codeword dictionary generated from distinct parsing of the 200-bit message example shown in Fig. 10.5. The codeword frequency is also shown.

| Pointer | Phrase | (Prefix pointer, bit) | Codeword | Frequency |
|---|---|---|---|---|
| 0 | Ø | – | – | 0 |
| 1 | 0 | (0, 0) | 0000***0*** | 1 |
| 2 | 1 | (0, 1) | 0000***1*** | 1 |
| 3 | 01 | (1, 1) | 0001***1*** | 1 |
| 4 | 10 | (2, 0) | 0010***0*** | 1 |
| 5 | 11 | (2, 1) | 0010***1*** | 1 |
| 6 | 010 | (3, 0) | 0011***0*** | 1 |
| 7 | 0100 | (6, 0) | 0110***0*** | 1 |
| 8 | 0101 | (6, 1) | 0110***1*** | 1 |
| 9 | 01001 | (7, 1) | 0111***1*** | 1 |
| 10 | 01011 | (8, 1) | 1000***1*** | 3 |
| 11 | 010010 | (9, 0) | 1001***0*** | 2 |
| 12 | 010111 | (10, 1) | 1010***1*** | 3 |
| 13 | 0100101 | (11, 1) | 1011***1*** | 4 |
| 14 | 0101110 | (12, 0) | 1100***0*** | 1 |
| 15 | 01011101 | (14, 1) | 1110***1*** | 6 |
| 16 | 01011100 | (14, 0) | 1110***0*** | 6 |

demonstration, let us consider a significantly longer message, for instance the 200-bit-long sequence shown in Fig. 10.5. In this second example, I have purposefully overemphasized the bit-pattern redundancy at the beginning of the message to obtain 16 individual phrases of rapidly increasing size. The corresponding LZ codeword dictionary, as based on four-bit pointer addresses, is shown in Table 10.6. I have also continued the rest of the message by preferentially using the longest phrases previously identified and repeating them once in a while, again to emphasize redundancy. As a result, we obtain 34 phrases. Since each phrase corresponds to a five-bit LZ codeword (Table 10.6), the total LZ sequence length is $34 \times 5 = 170$ bits, as shown in Fig. 10.5. Thus, the LZ code has achieved a compression equal to $1 - 170/200 = 25\%$ of the initial, 200-bit source message. From the codeword frequency analysis (Table 10.6), we can also calculate the entropy of the compressed-message source. As easily calculated from the table data, the result turns out to be $H = 3.601$ bit/word. Since the codeword length is $L_2 = 5$ bits, this result corresponds to a coding efficiency of $\eta = 3.601/5 = 70.03\%$.

In this example, we have first generated a LZ codeword dictionary until the number of available pointer addresses was exhausted. We have then re-used the dictionary as a "static code" to process the rest of the message, which provided the compression effect. In reality, compression occurs even as the dictionary is generated and before memory-size exhaustion, but this effect is only observed in relatively long message sequences (see further). To emphasize, the LZ algorithm is *not* meant to be implemented as a static code past the point of memory saturation (although this remains a valid option), but to work as a dynamic code, regardless of the sequence length. Since the number of codeword entries increases indefinitely, the process of updating the dictionary should be periodically reinitiated. A new dictionary can be reconstructed from scratch, or cleaned up from the earlier entries to make up new space.

Given the maximum codeword size $L_2$, the LZ algorithm can identify as many as $2^{L_2-1}$ individual phrases, for instance $2^{15} = 32\,768$ phrases with $L_2 = 16$. Such a number is more than sufficient to cover all possible patterns in a redundant source message and, therefore, ensure effective code compression. This remains true in the case where the source characteristics evolve over time. If the memory size is assumed to be sufficiently large to fit a virtually unlimited number of entries, a key question is whether or not the coded sequence can ever be shorter than the original message. To answer this question, we need to know the number of possible phrases, $c(n)$, and the LZ codeword size, $L_2$, given a message length of $n$ bits. The Lempel and Ziv analysis of distinct parsing, which provides the answer, is detailed in Appendix I, and we shall use the result here. Given a message of $n$ bits, the operation of *distinct parsing* yields $c(n)$ phrases to which $c(n)$ unique codewords are associated. The number of pointer-address bits required to define such a parsing is $\log c(n)$, which gives $L_2(n) = \log c(n) + 1$ for the codeword size.[14] Therefore, the coded version of the message has a full sequence length of:

$$L_s(n) = c(n)[\log c(n) + 1]. \tag{10.4}$$

The code compression (or expansion) is given by the ratio:

$$R(n) = \frac{L_s(n)}{n} = \frac{c(n)}{n}[\log c(n) + 1]. \tag{10.5}$$

The issue now is to determine the upper bound of $\eta_n$ in the limit of infinitely long messages ($n \to \infty$). In Appendix I, it is shown that the number of distinct-parsing phrases $c(n)$ is bounded according to:

$$c(n) \le \frac{n}{(1 - \varepsilon_n)\log n}, \tag{10.6}$$

where $\varepsilon_n$ vanishes as $n \to \infty$. We, thus, observe that the growth of the number of phrases with message size is *sublinear*, with an absolute upper bound of $n/\log n$. Another useful information is that the codeword size is bounded according to

$$L_2(n) = \log c(n) + 1 \le \log\left[\frac{n}{(1 - \varepsilon_n)\log n}\right] + 1 \le \log(2n). \tag{10.7}$$

The codeword length, thus, increases somewhat slower than the logarithm (base 2) of the message length $n$. Both properties concerning the growth of $c(n)$ and $L_2(n)$ are advantageous for memory size considerations. Replacing next the property in Eq. (10.6) in Eq. (10.5) yields for the compression ratio:

$$\begin{aligned} R(n) &\le \frac{1}{(1 - \varepsilon_n)\log n}\left\{\log\left[\frac{n}{(1 - \varepsilon_n)\log n}\right] + 1\right\} \\ &\equiv \frac{1}{(1 - \varepsilon_n)}\left\{1 + \frac{1 - \log\log n - \log(1 - \varepsilon_n)}{\log n}\right\}. \end{aligned} \tag{10.8}$$

Taking the infinite limit in Eq. (10.8) leads to the conclusion $R(n \to \infty) \le 1$, which proves that compression is possible (the *compression rate* being defined as $r = 1 - R(n)$). Yet, we don't know exactly how much compression can be achieved.

[14] In the example of Fig. 10.5, for instance, we have $n = 71$ bits, $c(n) = 16$ and $L_2 = 5$.

A more general and finer-grain analysis, which requires a very involved demonstration, is provided in.[15] Here, I shall just provide the conclusion, which takes the form

$$\limsup_{n\to\infty} R(n) \leq H(X), \tag{10.9}$$

where $H(X)$ is the entropy of the binary source $X = \{0, 1\}$ from which the message string is generated.[16] This property shows that the LZ code is *asymptotically optimal.*

In Section 3.2, indeed, we have seen that a code is *optimal* if, given the source entropy $H(X)$, the mean codeword length $\bar{L}$ satisfies:

$$H(X) \leq \bar{L} < H(X) + 1. \tag{10.10}$$

This inequality also applies to any optimal block code with fixed length $L_s(n)$, as applied to an $n$-bit message from extended source $X^{(n)}$:

$$H(X^{(n)}) \leq L_s(n) < H(X^{(n)}) + 1. \tag{10.11}$$

Dividing Eq. (10.11) by $n$ and taking the limit $n \to \infty$ yields:

$$\lim_{n\to\infty} \frac{H(X^{(n)})}{n} \leq \lim_{n\to\infty} \frac{L_s(n)}{n} < \lim_{n\to\infty} \frac{H(X^{(n)})}{n} + \lim_{n\to\infty} \frac{1}{n}, \tag{10.12}$$

or, equivalently,

$$\lim_{n\to\infty} \frac{L_s(n)}{n} = \lim_{n\to\infty} \frac{H(X^{(n)})}{n} = H(X). \tag{10.13}$$

The last equality in Eq. (10.13) is made under the assumption that, in the infinite limit, the extended source becomes *memoryless*. This concept means that the source events become asymptotically independent as their number increases, which justifies the limit.

Returning to the earlier issue, we can, thus, conclude from the result in Eq. (10.9) that LZ coding is *asymptotically optimal*, with an average codeword length (bit/source-symbol) ultimately *no greater* than the source entropy.

The original LZ77 algorithm (also called LZ1) was further developed into a LZ78 version (also called LZ2). The latter version corresponds to the *dictionary-based* approach that I have described. The original LZ77 uses, instead, a *sliding-window* approach. The LZ77 algorithm checks out the input symbol sequence, and searches for any match in a sliding-window buffer (the dictionary equivalent). The algorithm output, called a *token*, is made of three numbers: the offset of the sliding window, where the matched sequence is found to begin; the length of the matched sequence; and the unique code of the last unmatched symbol. Both LZ77 and LZ78 have generated a prolific family of variants (and related patents), which can be listed as:

- (For LZ77): *LZR*, *LZSS* (Lempel–Ziv–Storer–Szymanski), *LZB*, and *LZH* (Lampel–Ziv–Haruyasu);
- (For LZ78): *LZW* (Lempel–Ziv–Welch), *LZC*, *LZT*, *LZMW*, *LZJ*, and *LZFG*.

[15] T. M. Cover and J. A. Thomas *Elements of Information Theory* (New York: John Wiley & Sons, 1991).

[16] Recall that the entropy of a binary source is bounded to the maximum $H_{\max}(X) = 1$, which corresponds to a uniformly distributed source.

The LZ algorithm family has led to several data-compression standards and file-archival utilities known as *Zip* (from LZC), *GNU zip* or *gzip* (from LZ77), *PKZIP* (from LZW), *PKARK*, *Bzip/Bzip2*, *ppmz*, *pack*, *lzexe* (a MS-DOS utility), and *compress* (from LZC and LZW, a UNIX utility), for instance. Such programs make it possible to compress most ASCII-source files by a factor of two, and sometimes up to three, but with different performance and computing speeds.[17] The algorithm LZW is also used for image compression in the well-known formats called *GIF* (*graphic interchange format*) and *TIFF* (*tagged image file format*); for instance, GIF has become a leading standard for the exchange of graphics on the World Wide Web. The original LZ77 algorithm is used in the patent-free image format *PNG* (*portable network graphics*). See Appendix G for a more detailed description of data and image compression standards.

## 10.5 Exercises

**10.1** (B): Determine the coding efficiency corresponding to the symbol source defined in Table 10.7, using Elias-gamma and Elias-delta codes.

**Table 10.7** Data for Exercise 10.1.

| Symbol $x$ | $p(x)$ |
|---|---|
| A | 0.302 |
| B | 0.105 |
| C | 0.125 |
| D | 0.025 |
| E | 0.177 |
| F | 0.016 |
| G | 0.125 |
| H | 0.125 |

**10.2** (T): Determine the Fibonacci code of order $m = 3$ up to the first 31 integers (*Clue*: use level-three coding and be sure to obtain a prefix code).

**10.3** (M): Determine the Golomb code with parameter $m = 3$ for integers $i = 0$ to $i = 12$.

**10.4** (M): Determine the step-by-step coding assignment in dynamic Huffman coding for the message sequence AABCABB and the resulting coding efficiency.

**10.5** (M): Determine a Lempel–Ziv code for the 30-symbol sequence

ABBABBBABABAAABABBBAABAABAABAB

and the corresponding compression rate.

[17] See: D. A. Lelewer and D. S. Hirschberg, Data compression. *Computing Surveys*, **19** (1987), 261–97, www.ics.uci.edu/~dan/pubs/DataCompression.html and D. J. C. MacKay, *Information Theory, Inference and Learning Algorithms* (Cambridge: Cambridge University Press, 2003).

# 11 Error correction

This chapter is concerned with a remarkable type of code, whose purpose is to ensure that any errors occurring during the transmission of data can be *identified* and *automatically corrected*. These codes are referred to as *error-correcting codes* (ECC). The field of error-correcting codes is rather involved and diverse; therefore, this chapter will only constitute a first exposure and a basic introduction of the key principles and algorithms. The two main families of ECC, *linear block codes* and *cyclic codes*, will be considered. I will then describe in further detail some specifics concerning the most popular ECC types used in both telecommunications and information technology. The last section concerns the evaluation of *corrected bit-error-rates* (BER), or BER improvement, after information reception and ECC decoding.

## 11.1 Communication channel

The communication of information through a message sequence is made over what we shall now call a *communication channel* or, in Shannon's terminology, a *channel*. This channel first comprises a *source*, which generates the message symbols from some alphabet. Next to the source comes an *encoder*, which transforms the symbols or symbol arrangements into codewords, using one of the many possible coding algorithms reviewed in Chapters 9 and 10, whose purpose is to compress the information into the smallest number of bits. Next is a *transmitter*, which converts the codewords into physical waveforms or signals. These signals are then propagated through a physical *transmission pipe*, which can be made of vacuum, air, copper wire, coaxial wire, or optical fiber. At the end of the pipe is a *receiver*, whose function is to convert the received signals into the original codeword data. Next comes a *decoder*, which decodes and decompresses the data and restitutes the original symbol sequence to the message's recipient. In an *ideal* communications channel, there is no loss, nor is there any alteration of the message information thus communicated. But in *realistic* or *nonideal* channels, there is always a finite possibility that a part of this information may be lost or altered, for a variety of physical reasons, e.g., signal distortion and additive noise, which will not be analyzed here. The only feature to be considered here is that a fraction of the received codewords or symbols may differ from the ones that were transmitted. In this case, it is said that there exist *symbol errors*.

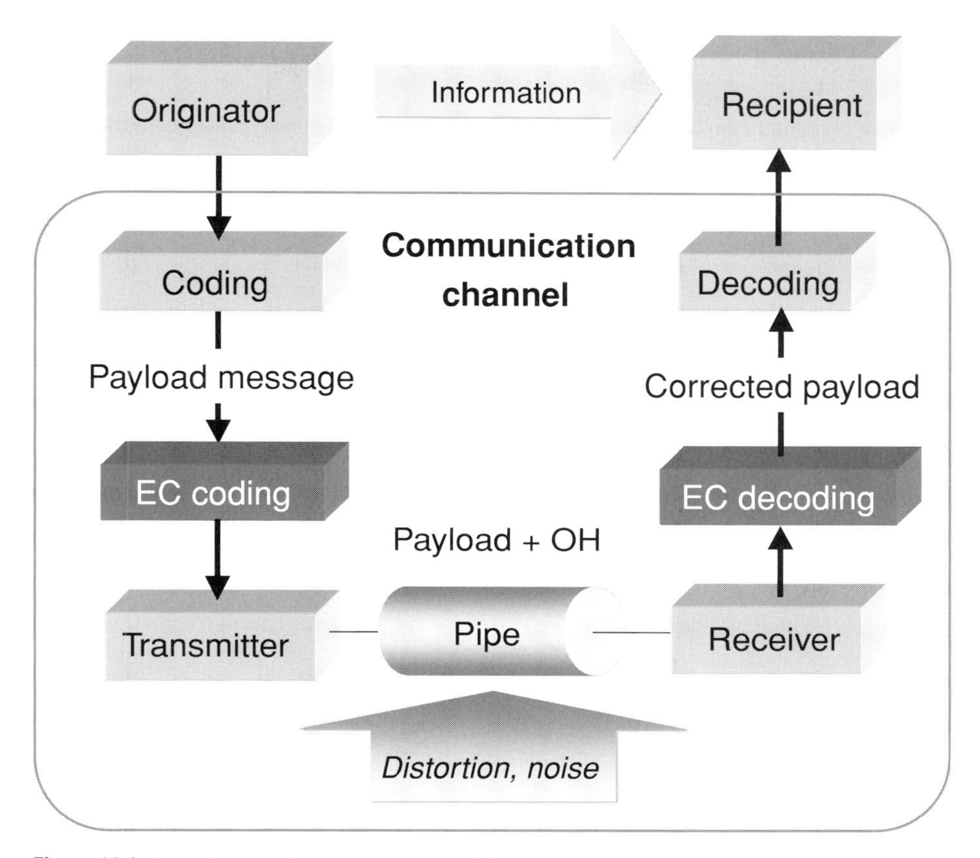


**Figure 11.1** Basic layout of error-correction (EC) coding and decoding within a communication channel (OH = overhead).

We may just see ECC as another type of coding step or "coding layer," which is to be inserted in the message transmission chain. More specifically, error-correction encoding should be performed *after* data encoding (prior to transmission) and decoding (error correction) should be performed *before* data decoding (after reception), as illustrated in Fig. 11.1. As the figure indicates, the coded message to be transmitted is now referred to as a *payload*. Let us refer to the originator's message as the data *payload*. The key purpose of ECC is to ensure that the payload arrives at the destination with the *smallest possible probability of symbol errors*. The tax to pay for ECC implementation is that some additional information must be included in the message code, referred to as *control bits*. Such control bits, which disappear after ECC decoding, correspond to what is called *overhead information*, or "overhead" for short.

How does ECC work? A first and most basic approach consists of appending a certain number of *redundancy* bits to the payload data. The decoder then uses these redundancy bits to decide whether (a) the payload bits have been properly detected, and (b) any resulting corrective action is required to restore the "errored" bit (as telecom jargon goes) to their correct initial value. An obvious redundancy scheme is to repeat the information a certain number of times. It is just like communicating a phone number or someone's name over a poor telephone connection or in noisy surroundings. By repeating the information two or three times, the hope is that the number or the name spelling is going to be unambiguously communicated. For instance, one can transmit 0000 instead of a single 0 symbol, and 1111 instead of a single 1 symbol. According to this code, the received sequence (underscores introduced for clarity)

$$0010\_1111\_0111\_1011$$

is interpreted as *very likely* to represent the payload message:

$$0000\_1111\_1111\_1111.$$

In the above guess of the correct message, we have followed an intuitive rule of *majority logic*. The bit errors were automatically corrected based on the fact that the three bit anomalies in the four-bit groups were easily identifiable. However, there is no absolute certainty that such an intuitive "majority-rule" decoding may yield the actual payload. Indeed, if two errors occur in any of the above four-bit words (however less likely the event), receiving 1100 or any other permutation thereof could be decoded as 1111 or 0000, while the correct codeword could have been either 0000 or 1111, which represents an undecidable problem. In such a case, majority logic fails to correct errors, unless more redundancy bits are used. Clearly, decoding reliability grows with increasing information redundancy and overhead, but at the expense of wasting the channel transmission capacity (a concept that we will have to develop in the next chapter).

The two following sections describe efficient and far smarter ways to correct bit errors with minimal redundancy or overhead, as based on the two main ECC families called *linear block codes* and *cyclic codes*.

## 11.2 Linear block codes

A first possible strategy for error correction is to transmit the information by slicing it into successive *blocks* of a fixed, $n$-bit length. By convention, the first segment of the block is made of $k$ *payload bits*, corresponding to $2^k$ possible (payload) message sequences. The second segment is made of $m = n - k$ bits, which we shall call *parity bits*, and which will be used for EC coding. These $m$ parity bits represent the *overhead information*. We thus have an $(n, k)$ block made of $n$ bits, which contains $k$ payload bits, and $n - k$ parity bits. This arrangement is most generally referred to as a *linear block code*.

By definition, the *bit rate* $B$ is the number of payload bits generated or received per unit time (e.g., $B = 1$ Mbit/s). It is also the rate at which payload bits are encoded by the transmitter and decoded by the receiver, in the absence of ECC. The *code rate* is defined as the ratio $R = k/n$, which represents the proportion of payload bits in the block ($1 - R$ representing the redundancy or overhead proportion). To provide decoded or corrected data at the initial payload-bit-rate $B$, the encoded-block bit rate should be increased by a factor $1/R$, corresponding to a percent additional bandwidth of $1/R - 1$, which is referred to as the percent *bandwidth expansion factor*. The rate at which bit errors are detected is called the *bit error rate* (BER). The key effect of ECC is, thus, to reduce the BER by orders of magnitude, literally! The intriguing fact, however, is that there is no such a thing as "absolute" error correction, which would provide BER $= 0 \times 10^0$ to all orders. The reality is that ECC may provide BER reduction to *any* arbitrary order, but there is a tax to pay for reaching this "absolute" by means of increasingly sophisticated ECC algorithms with increasingly longer overheads.

How do ECC linear block codes work? The task is to find out an appropriate coding algorithm for the parity bits, making it possible, on EC decoding, to detect and correct errors in the received block, with adequate or target precision. Because such an algorithm begins with the input data, the approach is called *forward error correction*, or FEC.[1] The following description requires some familiarity with basic matrix-vector formalism.

We focus now on how linear block codes are actually generated. Define first the input message/payload bits by the $k$-vector $X = (x_1 \ldots x_k)$, and the block code (EC encoder output) by the $n$-vector $Y = (y_1 \ldots y_n)$. The block-code bits $y_i$ are calculated from the linear combination:

$$y_i = g_{1i}x_1 + g_{2i}x_2 + \cdots + g_{ki}x_k, \tag{11.1}$$

where $g_{lm}$ ($l = 1 \ldots k$, $m = 1 \ldots n$) are binary coefficients. This definition can be put into a vector-matrix product, $Y = X\tilde{G}$, or, explicitly,

$$Y = (y_1 \ldots y_n) = (x_1 \quad x_2 \quad \ldots \quad x_k)\begin{pmatrix} g_{11} & g_{12} & \cdots & g_{1n} \\ g_{21} & g_{22} & & \vdots \\ \vdots & & & \vdots \\ g_{k1} & \cdots & \cdots & g_{kn} \end{pmatrix} \equiv X\tilde{G}. \tag{11.2}$$

The matrix $\tilde{G}$ is called the *generator matrix*. It is always possible to rewrite the generator matrix under the so-called *systematic form*:

$$\tilde{G} = [I_k|P] = \left(\begin{array}{cccc|cccc} 1 & \cdots & \cdots & 0 & p_{11} & p_{21} & \cdots & p_{1m} \\ 0 & 1 & & \vdots & p_{12} & p_{22} & & \vdots \\ \vdots & & & \vdots & \vdots & & & \vdots \\ 0 & \cdots & \cdots & 1 & p_{1k} & \cdots & \cdots & p_{km} \end{array}\right), \tag{11.3}$$

where $I_k$ is the $k \times k$ identity matrix (which leaves the $k$ message bits unchanged) and $P$ is an $m \times k$ matrix which determines the redundant parity bits. Thus the encoder output $Y = X\tilde{G} = X[I_k|P]$ is a payload block of $k$ bits followed by a parity block of $m = n - k$ bits.[2] This arrangement, where the payload bit-sequence is left unchanged by the EC encoder, is called a *systematic code*.

[1] There exist possibilities for *backward error correction* (BEC), but the approach means some form of backward-and-forward communication between the two ends of the channel, at the high expense of channel use or bandwidth waste.

[2] Note that in another possible convention for the systematic form/code, as described in some textbooks, the parity bits may instead precede the payload bits.

For future use, we define the $n \times m$ *parity-check matrix:*

$$\tilde{H} = [P^{\mathrm{T}}|I_m] = \left(\begin{array}{cccc|cccc} p_{11} & p_{12} & \dots & p_{1k} & 1 & \dots & \dots & 0 \\ p_{21} & p_{22} & & \vdots & 0 & 1 & & \vdots \\ \vdots & & & \vdots & \vdots & & & \vdots \\ p_{1m} & \dots & \dots & p_{mk} & 0 & \dots & \dots & 1 \end{array}\right), \tag{11.4}$$

where $P^{\mathrm{T}}$ is the transposed matrix of $P$ (note how their coefficients are symmetrically permuted about the diagonal elements). We note then the important property

$$HG^{\mathrm{T}} = GH^{\mathrm{T}} = 0, \tag{11.5}$$

which stems from

$$HG^{\mathrm{T}} = [P^{\mathrm{T}}|I_m] \cdot \left[\frac{I_k}{P^{\mathrm{T}}}\right] = P^{\mathrm{T}} + P^{\mathrm{T}} \equiv 0 = (HG^{\mathrm{T}})^{\mathrm{T}} = GH^{\mathrm{T}}. \tag{11.6}$$

In this expression, we have used the property that for binary numbers $x$, the sum $x + x$ is identical to zero ($0 + 0 = 1 + 1 \equiv 0$). From the property in Eq. (11.5), we have

$$Y\tilde{H}^{\mathrm{T}} = 0, \tag{11.7}$$

which stems from $Y\tilde{H}^{\mathrm{T}} \equiv X\tilde{G}\tilde{H}^{\mathrm{T}} = X(\tilde{G}\tilde{H}^{\mathrm{T}}) = 0$.

We consider now the received block code, which we can define as the $n$-vector $Z$. Since the received block code is a "contaminated" version of the original block code $Y$, we can write $Z$ in the form:

$$Z = Y + E, \tag{11.8}$$

where $E$ is an error vector whose $i$th coordinate is 0 if there is no error and 1 otherwise. Post-multiplying Eq. (11.8) by $\tilde{H}^{\mathrm{T}}$ and using the property in Eq. (11.7), we obtain

$$S = Z\tilde{H}^{\mathrm{T}} = (Y + E)\tilde{H}^{\mathrm{T}} = E\tilde{H}^{\mathrm{T}}. \tag{11.9}$$

The $m$-vector $S$ is called the *syndrome*. The term "syndrome" is used to designate the information helping to diagnose a "disease," which, here, is the error "contamination" of the signal. A zero-syndrome vector means that the block contains no errors ($E = \vec{0}$). As the example described next illustrates, single-error occurrences ($E$ has only one nonzero coordinate) are in one-to-one correspondence with a given syndrome $S$. Thus when computing the syndrome $S = Z\tilde{H}^{\mathrm{T}}$ at the receiver end, one immediately knows two things:

(a) Whether there are errors, as indicated by $S \neq \vec{0}$ ;
(b) In the assumption that it is a single-error occurrence, where it is located.

At this point, we should illustrate the process of error detection and error correction through the following basic example.

Consider the block code $(n, k) = (7, 4)$, which has $k = 4$ message bits and $m = 3$ parity bits. Such a block, which is of the form $n = 2^m - 1$ and $k = n - m$, with $m \geq 3$,

**Table 11.1** Message words and corresponding block codes in an example of a Hamming code (7.4).

| Message word $X$ | Block code $Y$ |
|---|---|
| **0000** | 0000 **000** |
| **0001** | 0001 **111** |
| **0010** | 0010 **011** |
| **0011** | 0011 **100** |
| **0100** | 0100 **101** |
| **0101** | 0101 **010** |
| **0110** | 0110 **110** |
| **0111** | 0111 **001** |
| **1000** | 1000 **110** |
| **1001** | 1001 **001** |
| **1010** | 1010 **101** |
| **1011** | 1011 **010** |
| **1100** | 1100 **011** |
| **1101** | 1101 **100** |
| **1110** | 1110 **000** |
| **1111** | 1111 **111** |

is referred to as a *Hamming code.* According to the definition in Eq. (11.3), we define the following generator matrix (out of many other possibilities)

$$\tilde{G} = \left(\begin{array}{cccc|ccc} 1 & 0 & 0 & 0 & 1 & 1 & 0 \\ 0 & 1 & 0 & 0 & 1 & 0 & 1 \\ 0 & 0 & 1 & 0 & 0 & 1 & 1 \\ 0 & 0 & 0 & 1 & 1 & 1 & 1 \end{array}\right), \tag{11.10}$$

which gives, from Eq. (11.4), the parity-check matrix

$$\tilde{H} = \left(\begin{array}{cccc|ccc} 1 & 1 & 0 & 1 & 1 & 0 & 0 \\ 1 & 0 & 1 & 1 & 0 & 1 & 0 \\ 0 & 1 & 1 & 1 & 0 & 0 & 1 \end{array}\right) \tag{11.11}$$

and its transposed version

$$\tilde{H}^{\mathrm{T}} = \left(\begin{array}{ccc} 1 & 1 & 0 \\ 1 & 0 & 1 \\ 0 & 1 & 1 \\ 1 & 1 & 1 \\ \hline 1 & 0 & 0 \\ 0 & 1 & 0 \\ 0 & 0 & 1 \end{array}\right). \tag{11.12}$$

The $2^k = 16$ possible message words $X$ and their corresponding block codes $Y$, as calculated from Eqs. (11.2) and (11.10), are listed in Table 11.1. For clarity, the parity bits are shown in bold numbers. As expected, the block codes are made of a first sequence

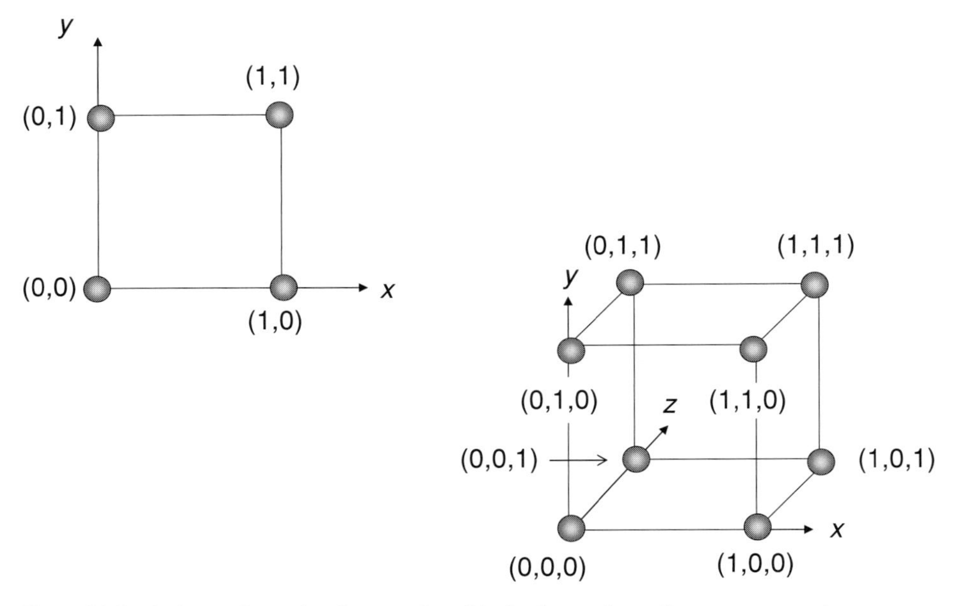


**Figure 11.2** Placing codewords of two or three bits in the vertices of a square or a cube, respectively, to illustrate Hamming distance.

of four (original) messages bits followed by a second sequence of three parity bits. For instance, the message codeword 0111 is encoded into the block code 0111 001, where the three bits on the right are the parity bits.

For future use, I introduce the following definitions in ECC formalism:

- *Hamming weight* $w(X)$: number of 1 bits in a given (*block*) codeword $X$.
- *Hamming distance* $d(X, Y)$ *between two codewords* $X, Y$: number of bit positions in which the two codewords differ; consequently, the Hamming weight $w(X)$ is the distance between a given codeword $X$ and the all-zero codeword; also note the property $d(X, Y) = w(X + Y) = w(X - Y)$.
- *Minimum Hamming distance* $d_{\min}$: minimum number of 1 bits in the block codewords (excluding the all-zero codeword). Consequently, $d_{\min}$ is also the minimum number of bit positions by which any codeword differ. It is readily checked from the example in Table 11.1 that $d_{\min} = 3$, which is a specific property of Hamming codes.

It should be noted that the Hamming distance is a mathematical *distance* in the strict definition of the term, as described in Chapter 5, when introducing the *Kullback–Leibler* distance between two probability distributions. To illustrate the meaning of Hamming distance, Fig. 11.2 shows that it is possible to place all the different two-bit or three-bit codewords ($(a_i a_j)$ or $(a_i a_j a_k)$) on the vertices of a square or a cube, following the Cartesian coordinate system $x = a_i, y = a_j, z = a_k$, which corresponds to the vectors $(x, y, z) = (a_i, a_j, a_k)$. According to definition, the Hamming distance between codeword $(0, 0)$ and $(1, 0)$ or $(1, 1)$ is $d = 1$ or $d = 2$, respectively. Likewise, the distance between $(0, 0, 0)$ and $(1, 0, 0)$ or $(1, 0, 1)$ or $(1, 1, 1)$ is $d = 1$ or $d = 2$ or

**Table 11.2** Syndrome vector $S = (s_1, s_2, s_3)$ associated with single-error patterns $E = (e_1, e_2, e_3, e_4, e_5, e_6, e_7)$ in the Hamming block code (7, 4) defined in Table 11.1.

| Error pattern $E$ | Syndrome $S$ |
|---|---|
| **0000000** | 000 |
| **1000000** | 110 |
| **0100000** | 101 |
| **0010000** | 011 |
| **0001000** | 111 |
| **0000100** | 100 |
| **0000010** | 010 |
| **0000001** | 001 |

$d = 3$, respectively. We see from the illustration that the Hamming distance does not correspond to the Euclidian distance, but rather to the smallest number of square or cube edges to traverse in order to move from one point to the other.

Going back to our $(n, k) = (7, 4)$ block code example, assume next that the received block $Z$ contains a *single* bit error, whose position in the block is unknown. If the error concerns the first bit of $Z = Y + E$, this means that the value of the first bit of $Y$ has been increased (or decreased) by 1, corresponding to the error vector $E = (1, 0, 0, 0, 0, 0, 0)$. For an error occurring in the second bit, we have $E = (0, 1, 0, 0, 0, 0, 0)$, and so on, to bit seven. For each single-error occurrence, we can then calculate the corresponding syndrome vector using the relation $S = E\tilde{H}^{\mathrm{T}}$ in Eq. (11.9). The result of the computation is shown in Table 11.2. It is seen that a unique syndrome corresponds to each error occurrence. For instance, if the syndrome is $S = (0, 1, 1)$, we know that a bit error occurred in the third position of the block sequence, and correction is made by adding 1 to the bit (equivalently, by switching its polarity, or by adding $E$ to $Z$). Thus any single-error occurrence (whether in payload or in parity bit fields) can be readily identified and corrected.

What happens if there is more than one error in the received block? Consider the case of double errors in our $(n, k) = (7, 4)$ example. The possibilities of a single bit error correspond to seven different error patterns. For double errors, the number of possibilities is $C_7^2 = 7!/[2!(7-2)!] = 21$, corresponding to 21 different error patterns. Since the syndrome is only three bits, it can only take $2^3 - 1 = 7$ configurations. Thus, the syndrome is no longer associated with a unique error pattern. Here, the seven syndrome configurations correspond to $7 + 21 = 28$ error patterns of either one- or two-bit errors. It is an easy exercise to determine the actual number of error patterns associated with each syndrome. For instance, the results show that $S = (0, 0, 0)$, $S = (0, 0, 1)$, $S = (0, 1, 1)$ and $S = (1, 0, 1)$ are associated with one, two, three and four patterns of two-bit errors, respectively. Thus, syndrome decoding can only detect the presence of errors, but cannot locate them with 100% accuracy, since they are associated with more than one error pattern. It is clear that with Hamming codes having greater numbers of parity bits, this imperfect correction improves in effectiveness, as the mapping $E \rightarrow S$ becomes less redundant. The approach is, however, costly in bandwidth use, and error correction is, as we have announced, not "absolutely" efficient. This situation illustrates the fact that in most cases, *syndrome decoding cannot determine the exact error pattern* $E$, *but only the one that is most likely to correspond to the syndrome*, $E_0$. The Hamming distance between $E$ and $E_0$ is minimized but is nonzero. Therefore, the substitution $Z \to Z + E_0$ represents a best-choice correction rather than an exact corrective operation. This is referred to as *maximum-likelihood decoding*.

As we have seen, the $2^{n-k}$ different syndromes allow the absolute identification of any *single* bit error in the $(n, k)$ block code. Yet, each of these vectors corresponds to a finite number of *multiple* bit errors, albeit such occurrences are associated with lower and rapidly decreasing probabilities. For a block code of length $n$, there actually exist $2^n$ possible error patterns (from zero to $n$ bit errors). Out of these sets of error possibilities, syndrome decoding is able to detect and correct $2^k$ *single bit* errors, as we have seen earlier. What about the remaining error patterns, which contain more than one bit error? These are of two kinds. Since the block code has $2^k$ block codewords, there exist: (a) $2^k$ error patterns that exactly match any of the block codewords or belong to the block codeword set, and (b) $2^n - 2^k$ error patterns that do not belong to the block codeword set. In case (a), the syndrome vector is identical to zero ($S = (0, 0, 0)$), which means that the errors remain *undetected* (and for that matter, uncorrected). In case (b), the syndrome vector is nonzero ($S \neq (0, 0, 0)$), which indicates the existence of errors, but, as we have seen, such multiple bit errors cannot be corrected. Consider the first case (a) where the code fails to detect any error, and let us estimate the corresponding probability, $q(E)$. It is easily established that this probability is given by

$$q(E) = \sum_{i=1}^{n} A_i p^i (1-p)^{n-i}, \tag{11.13}$$

where $p$ is the probability of a single bit error and $A_i$ is the number of block codewords having a Hamming weight of $i$ (to recall, having $i$ bits equal to one). We note that $A_i$ for $i = 1, \ldots, d_{\min} - 1$. It is left as an exercise that in our $(n, k) = (7, 4)$ code example, we have $q(E) < 10^{-8}$ for $p \leq 10^{-3}$, and $q(E) < 10^{-14}$ for $p \leq 10^{-5}$, which illustrates that the likelihood of the code to fail in detecting any errors is comparatively small, even at single-bit-error rates as high as $p = 10^{-5} - 10^{-3}$.

In view of the above, the key question coming to mind is: "Given a block code, what is the maximum number of errors that can be detected and corrected with absolute certainty?" The rigorous answer to this question is provided by a fundamental property of linear block codes. This property states that the code has the power to correct *any* error patterns of Hamming weight $w$ (or any number of $w$ errors in the block code), provided that

$$w \leq \left\{ \frac{d_{\min} - 1}{2} \right\}, \tag{11.14}$$

where $d_{\min}$ is the *minimum Hamming distance*, and where the brackets $\{x\}$ indicate the largest integer contained in the argument $x$, namely $\{(d_{\min} - 1)/2\} = \lfloor (d_{\min} - 1)/2 \rfloor$. A formal demonstration of this property is provided in Appendix J. What about error detection? It is simpler to show that the code has the capability to detect up to $d_{\min} - 1$ errors. Indeed, $d_{\min}$ represents the minimum number of bit positions by which any two codewords differ. If the distance $d(X, Y)$ between a received codeword $Y$ and a message code $X$ is such that $d(X, Y) < d_{\min}$, or equivalently, $d(X, Y) \le d_{\min} - 1$, then $Y$ differs from $X$ by fewer than $d_{\min}$ positions and, hence, does not belong to the code. The code is, thus, able to recognize that there exists a number of $d(X, Y)$ errors up to a maximum of $d_{\min} - 1$. As we have seen, the code is, however, able to correct up to $\{(d_{\min} - 1)/2\}$ of these errors. In the case of Hamming codes, we have seen that $d_{\min} = 3$. This gives for these codes an error-detection capability of up to $d_{\min} - 1 = 2$ errors and an error-correction capability of up to $w = 1$ errors, independently of the block-code size.

## 11.3 Cyclic codes

*Cyclic codes* represent a subset of linear block codes that obey two essential properties:

(a) The sum of two codewords is also a block codeword;
(b) Any cyclic permutation of the block codeword is also a block codeword.

Cyclic codes corresponding to $(n, k)$ block codes can be generated by polynomials $g(p)$ of degree $n - k$. Here, I shall briefly describe the principle of this encoding method.

Assume a block code $Y = (y_0 \dots y_{n-1})$ of $n$ bits $y_i$, which are now labeled from 0 to $n - 1$. From this codeword, we can generate a polynomial

$$Y(p) = y_0 + y_1 p + y_2 p^2 + \cdots + y_{n-1} p^{n-1}, \tag{11.15}$$

where $p$ is a real variable. We multiply $Y(p)$ by $p$ and perform the following term rearrangements:

$$\begin{aligned} pY(p) &= y_0 p + y_1 p^2 + y_2 p^3 + \cdots + y_{n-1} p^n \\ &= y_0 p + y_1 p^2 + y_2 p^3 + \cdots + y_{n-1}(p^n + 1) - y_{n-1} \\ &= \left[ \frac{y_{n-1} + y_0 p + y_1 p^2 + y_2 p^3 + \cdots + y_{n-2} p^{n-1}}{p^n + 1} + y_{n-1} \right] (p^n + 1). \end{aligned} \tag{11.16}$$

To derive the above result, we have used the property that for binary numbers, subtraction is the same as addition. Equation (11.16) can then be put in the form:

$$\frac{pY(p)}{p^n + 1} = y_{n-1} + \frac{Y_1(p)}{p^n + 1}, \tag{11.17}$$

where

$$Y_1(p) = y_{n-1} + y_0 p + y_1 p^2 + y_2 p^3 + \cdots + y_{n-2} p^{n-1} \tag{11.18}$$

is the reminder of the division of $pY(p)$ by $M = p^n + 1$, or

$$Y_1(p) = pY(p) \bmod [M], \tag{11.19}$$

where mod$[M]$ stands for "modulo $M$" (the same way one writes $6{:}3 = 0$ mod[3], or $8{:}3 = 2$ mod[3]). It is seen from the definition in Eq. (11.17) that $Y_1(p)$ is a codeword polynomial representing a cyclically shifted version of $Y(p)$ (Eq. (11.15)). Likewise, the polynomials $Y_m(p) = p^m Y(p)\,\mathrm{mod}[M]$ are all cyclically shifted versions of the code $Y(p)$. It is then possible to use the $Y(p)$ polynomials as a new way of encoding messages, as I show next.

Define $g(p)$ as a *generator polynomial* of degree $n - k$, which *divides* $p^n + 1$. As an example for $n = 7$, we have the irreducible polynomial factorization:[3]

$$\begin{aligned} p^7 + 1 &= (p+1)(p^3 + p^2 + 1)(p^3 + p + 1) \\ &\equiv (p+1)g_1(p)g_2(p), \end{aligned} \tag{11.20}$$

showing that $g_1(p) = p^3 + p^2 + 1$, $g_2(p) = p^3 + p + 1$, $(p+1)g_1(p)$ and $(p+1)g_2(p)$ are possible generator polynomials for $k = 4$ (first two) and $k = 3$ (last two). Define next the message polynomial $X(p)$ of degree $k - 1$:

$$X(p) = x_0 + x_1 p + x_2 p^2 + \cdots x_{n-1} p^{k-1}. \tag{11.21}$$

We can now construct a cyclic code for the $2^k$ possible messages $X(p)$ according to the definition:

$$Y(p) = X(p)g(p), \tag{11.22}$$

where $g(p)$ is a generator polynomial. Equation (11.21) represents a polynomial version of the previous matrix definition in Eq. (11.2), i.e.,$Y = X\tilde{G}$. Although we will not make use of it here, one can define the *parity-check polynomial* $h(p)$ from the relation

$$g(p)h(p) = 1 + p^n, \tag{11.23}$$

or equivalently, $g(p)h(p) = 0\,\mathrm{mod}[1 + p^n]$, which is the polynomial version of the matrix Eq. (11.5), i.e., $HG^{\mathrm{T}} = GH^{\mathrm{T}} = 0$.

I illustrate next the polynomial encoding through the example of the $(n, k) = (7, 4)$ Hamming block code. From Eq. (11.20), we can choose $g(p) = p^3 + p + 1$ as the generator polynomial (which divides by $p^7 + 1$). The coefficients of the polynomial $Y(p)$, as calculated from Eq. (11.22) for all possible message polynomials $X(p)$, are listed in Table 11.3.

It is seen from Table 11.3 that the obtained code representation is not *systematic*, i.e., the message bits no longer represent a separate word at the beginning or end of the code, unlike in the linear block codes previously seen (Table 11.1). For instance, as the table shows, the original four-bit message 0111 is encoded into 01100011.

Next, I show how error detection and correction is performed in cyclic codes. Assume that $Z(p)$ is the result of transmitting $Y(p)$ through a noisy channel. We divide the result by $g(p)$ and put it in the form:

$$Z(p) = q(p)g(p) + s(p), \tag{11.24}$$

[3] An irreducible polynomial, like a prime number, can be divided only by itself or by unity.

**Table 11.3** Message polynomial coefficients and corresponding cyclic codeword polynomial coefficients in the (7, 4) block code example, as corresponding to the generator polynomial $g(p) = p^3 + p + 1$.

| Message word $X(p)\ x_3x_2x_1x_0$ | Cyclic code $Y(p)\ y_6y_5y_4y_3y_2y_1y_0$ |
|---|---|
| **0000** | 0000000 |
| **0001** | 0001011 |
| **0010** | 0010110 |
| **0011** | 0011101 |
| **0100** | 0101100 |
| **0101** | 0100111 |
| **0110** | 0111010 |
| **0111** | 0110001 |
| **1000** | 1011000 |
| **1001** | 1010011 |
| **1010** | 1001110 |
| **1011** | 1000101 |
| **1100** | 1110100 |
| **1101** | 1111011 |
| **1110** | 1100010 |
| **1111** | 1101001 |

where $q(p)$ and $s(p)$ are the quotient and the remainder of the division, respectively. If there were no errors, we would have $Z(p) = Y(p) = X(p)g(p)$, meaning that the quotient would be the original message, $q(p) = X(p)$, and the remainder would be zero, $s(p) = 0$. As done previously, we call $s(p)$ the *syndrome polynomial*. A nonzero syndrome means that errors are present in the received code. As previously shown for the vectors $E, S$, it is possible to map the error polynomial $e(p)$ into $s(p)$, which makes it possible to associate syndrome and error patterns. Looking at Table 11.3, we observe that the minimum Hamming distance for this cyclic code is $d_{\min} = 3$. According to Eq. (11.14), the Hamming weight of error patterns that can be corrected is $w \leq \{(d_{\min} - 1)/2\} = 1$, meaning that only single errors can be corrected. As previously discussed, this is a general property of Hamming codes, which is not affected by the choice of cyclic coding. The capabilities of error detection and correction of various cyclic codes are discussed next.

## 11.4 Error-correction code types

So far, this chapter has provided the basic conceptual tools of ECC principles and coding or decoding algorithms. Here, I shall complete this introduction by reviewing different ECC types used in information technologies (data standards) and telecommunications (packet or frame standards), as well as their error-correction capabilities.

### Hamming codes

As we have seen in Section 11.2, these are linear block codes of the form $(n, k) = (2^m - 1, n - m)$, with $m \geq 3$. The minimum distance of a Hamming code was shown to be $d_{\min} = 3$, giving a correction capability of one-bit error.

### Hadamard codes

These are linear block codes of the form $(n, k) = (2^m, m + 1)$, whose $2^{m+1}$ codewords are generated by *Hadamard matrices*.[4] It is left as an exercise to verify this statement and property. The minimum distance is $d_{\min} = n/2 = 2^{m-1}$, yielding a correction capability of $\{(2^{m-1} - 1)/2\}$ or 3–7 bit errors for $m = 4 - 5$.

### Cyclic redundancy check (CRC) codes

This is the generic name given to any cyclic code used for error detection. Most data packet or framing standards include a "CRC" trailer field, which ensures error correction for the packet or frame payload. Binary CRC codes $(n, k)$ can *detect* error bursts of length $\leq n - k$, as well as various other error-burst patterns, such as combinations of errors up to a maximum of $d_{\min} - 1$. CRC codes can *correct* all error patterns of odd Hamming weight (odd numbers of bit errors) when the generator polynomial has an even number of nonzero coefficients.

### Golay code

This can be viewed equivalently as a (23, 12) linear block code, or a (23, 12) cyclic code. In the last case, the code is generated by either of the two polynomials

$$g(p) = 1 + p^2 + p^4 + p^5 + p^6 + p^{10} + p^{11},$$

[4] By definition, a *Hadamard matrix* $M_n$ is an $nxn$ binary matrix ($n = 2^m$), in which each row is different from all the others by exactly $n/2$ positions. One row must contain only 0 bits, thus all other rows have $n/2$ zeros and $n/2$ ones. The smallest Hadamard matrix (out of four possibilities) is:

$$M_2 = \begin{pmatrix} 0 & 0 \\ 0 & 1 \end{pmatrix}.$$

A property of the Hadamard matrices is that it is possible to construct $M_{2n}$ from $M_n$, as follows:

$$M_{2n} = \begin{pmatrix} M_n & M_n \\ M_n & \bar{M}_n \end{pmatrix},$$

where $\bar{M}_n$ is the complementary matrix of $M_n$ (meaning that all bits values are switched). Thus, for $M_4$ we obtain:

$$M_4 = \begin{pmatrix} M_2 & M_2 \\ M_2 & \bar{M}_2 \end{pmatrix} \equiv \begin{pmatrix} 0 & 0 & 0 & 0 \\ 0 & 1 & 0 & 1 \\ 0 & 0 & 1 & 1 \\ 0 & 1 & 1 & 0 \end{pmatrix}.$$

The key property is that the rows of $M_n$ and $\bar{M}_n$ form a complete set of $2n$ codewords, which correspond to a linear block code $(n, k)$ of length $n = 2^m$ with $k = m + 1$ and minimum distance $d_{\min} = n/2 = 2^{m-1}$.

or

$$g(p) = 1 + p + p^5 + p^6 + p^7 + p^9 + p^{11},$$

which are both dividers of $1 + p^{23}$. The corresponding minimum distance is $d_{\min} = 7$, corresponding to a correction capability of three-bit errors.

## Maximum-length shift-register codes

These are cyclic codes of the form $(2^m - 1, m)$ with $m \geq 3$, which are generated by polynomials of the form $g(p) = (1 + p^n)/h(p)$, where $h(p)$ is a primitive polynomial of degree $m$ (meaning an irreducible polynomial dividing $1 + p^q$ with $q = 2^m - 1$ being the smallest possible integer). The minimum distance for this code is $d_{\min} = 2^{m-1}$, indicating the capability of correcting a maximum of $w = \{2^{m-1} - 1/2\}$ simultaneous errors. For instance, the code $(15, 4)$ has a three-bit-error correction capability. The code rate (number of message-payload bits divided by the block length) is $R = 4/15 = 26\%$, which is relatively poor in terms of bandwidth use. The maximum-length code sequences are labeled as "pseudo-noise," owing to their auto-correlation properties, which closely emulate white noise.

## Bose–Chaudhuri–Hocquenghem (BCH) codes

These are cyclic codes of the form $(2^m - 1, k)$ with $m \geq 3$, $n - k \leq mt$, where $m$ is an arbitrary positive integer, and $t$ is the number of errors that the code can correct (not all $m, t$ values being yet eligible). The BCH codes are generated by irreducible polynomials dividing $1 + p^{2m-1}$. The first of these polynomials is $g(p) = 1 + p + p^2$, corresponding to the block $(7, 4)$ with $t = 1$. The minimum distance for BCH codes is $d_{\min} = 2t + 1 = 3, 5, 7, \ldots$ corresponding to error-correcting capabilities of $t = 1, 2, 3, \ldots$, respectively. For instance, the block $(31, 11)$ with $t = 5$ can be corrected for a number of errors corresponding to almost half of the 11 message bits, while the code rate is $R = 11/31 = 35.5\%$. Table 11.4 lists the possible parameter combinations $(m, n, k, t)$ for BCH codes up to $m = 6$ corresponding to block lengths $7 \leq n \leq 63$, and their generating polynomials, as expressed in *octal* form.[5] Extended lists of BCH codes and polynomials for $m \leq 8$ and $m \leq 34$ can be found in.[6]

[5] The *octal* representation of a polynomial is readily understood by the following two examples, showing first the representation of the polynomial coefficients in binary, then in decimal, then, finally, in octal:

$$\begin{aligned}
Q_1(p) &= p^3 + p + 1 \\
&= 1 \times p^3 + 0 \times p^2 + 1 \times p + 1 \equiv 1011_{\text{binary}} \\
&\equiv 19_{\text{decimal}} \\
&= 2 \times 8^1 + 3 \times 8^0 = 23_{\text{octal}}, \\
Q_2(p) &= p^9 + p^7 + p^6 + p^4 + 1 \equiv 101101001_{\text{binary}} \\
&\equiv 721_{\text{decimal}} \\
&= 3 \times 8^3 + 5 \times 8^2 + 2 \times 8^1 + 1 \times 8^0 \equiv 3521_{\text{octal}}.
\end{aligned}$$

[6] J. G. Proakis, *Digital Communications* (New York: McGraw Hill, 2001), pp. 438–9, and other related references, therein.

**Table 11.4** List of BCH codes with block lengths $7 \le n \le 63$, and their generating polynomials $g(p)$, expressed in octal.

| $m$ | $n$ | $k$ | $t$ | $g(p)$ | $m$ | $n$ | $k$ | $t$ | $g(p)$ |
|---|---|---|---|---|---|---|---|---|---|
| 3 | 7 | 4 | 1 | 13 | 6 | 63 | 57 | 1 | 103 |
| 4 | 15 | 11 | 1 | 23 | | | 51 | 2 | 12471 |
| | | 7 | 2 | 721 | | | 45 | 3 | 1701317 |
| | | 5 | 3 | 2467 | | | 39 | 4 | 166623567 |
| | | | | | | | 36 | 5 | 1033500423 |
| 5 | 31 | 26 | 1 | 45 | | | 30 | 6 | 157464165547 |
| | | 21 | 2 | 3551 | | | 24 | 7 | 17323260404441 |
| | | 16 | 3 | 107657 | | | 18 | 10 | 1363026512351725 |
| | | 11 | 5 | 5423325 | | | 16 | 11 | 6331141367235453 |
| | | 6 | 7 | 313365047 | | | 10 | 13 | 472622305527250155 |
| | | | | | | | 7 | 15 | 5231045543503271737 |

## Reed–Solomon (RS) codes

These represent a subfamily of BCH codes based on a specific arrangement, noted $RS(N, K)$. The code block is made of $N$ symbols comprising $K$ message symbols and $N - K$ parity symbols. The message or parity symbols of length $m$ are nonbinary (e.g., $m = 8$ for *byte* symbols). The total block length is, thus, $Nm$. The RS code format is $RS(N = 2^m - 1, K = N - 2t)$. Its minimum distance is $d_{\min} = N - K + 1$, corresponding to a *symbol* error-correction capability of $t = (N - K)/2$. The code rate is $R = (N - 2t)/N = 1 - 2t/N$. For instance, the $RS(255, 231)$ code having $N = 255$ symbols ($m = 8$) and $t = 12$ ($K = 231$) corresponds to a code rate of 90% (or a relatively small bandwidth expansion factor of $N/K - 1 = 10.4\%$. Thus, symbol errors can be corrected up to about 1/20 of the block length with only $N - K = 24$ symbols, representing an ECC overhead of nearly 10%. This example illustrates the power of RS codes and justifies their widespread use in telecommunications.

## Concatenated block codes

It is possible to concatenate, or use two different ECCs successively. This is usually done with a nonbinary ECC (outer code), which is labeled $(N, K)$ and a binary ECC (inner code), which is labeled $(n, k)$. The message coding begins with the outer code and ends with the inner code, yielding a block of the form $(nN, kK)$, meaning that each of the $K$ nonbinary symbols is encoded into $k$ binary symbols, and similarly for the parity bits. At the receiving end, the block successively passes through the inner decoder and then the outer decoder. The corresponding code rate and minimum Hamming distance are given by $R' = kK/nN$ and $d'_{\min} = d_{\min} D_{\min}$, meaning that both parameters are given by the products of their counterparts for each code. Since the code rates $k/n$ and $K/N$ are less than unity, the resulting code rate is substantially reduced. However, the error-correction capability $w = \{(d_{\min} D_{\min} - 1)/2\}$ is substantially increased, approximately as the square of that of the individual codes. Using for instance the concatenation of

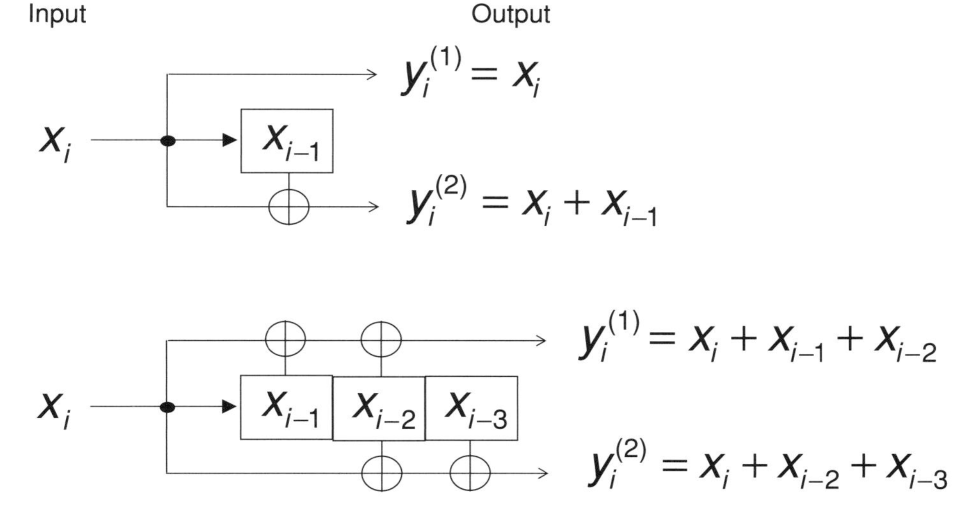


**Figure 11.3** Two examples of rate ½ convolutional encoders: (a) systematic, with constraint length $m = 2$; (b) nonsystematic, with constraint length $m = 4$.

two RS codes of the previous example, $RS(255, 231)$, the code rate is 81% (bandwidth expansion factor 22%) while the correction capability is 71 bit errors, representing 31% of the initial message block length.

## Convolutional codes

These are linear $(n, k)$ block codes with rate $R = k/n$, using a transformation rule, which is a function of the $m$ last transmitted bits. The number $m$ is called the code's *constraint length*. The $m - 1$ input bits are memorized into *shift registers*, which are all initially set to zero. Two examples of rate ½ convolutional encoders are provided in Fig. 11.3. The first example corresponds to an encoder with constraint length $m = 2$. It outputs one bit, which is identical to the input ($y_i^{(1)} = x_i$), and a second bit, which is the sum (modulo 2) of the last two input bits ($y_i^{(2)} = x_i + x_{i-1}$). Because the input message is reproduced in the output, this convolutional code is said to be *systematic*. The second example in Fig. 11.3 corresponds to a *nonsystematic*, $m = 4$ convolutional code. Another possibility is to feedback the output of one or several shift registers to the encoder input. This is referred to as a *recursive* convolutional encoder. The code examples shown in Fig. 11.3 are said to be *nonrecursive*. For relatively small values of $m$, decoding is performed through the *Viterbi algorithm*,[7] which is based on a principle of *optimal decision* for block error correction. It is beyond the scope of this introduction to enter the complexities of decoding algorithms for convolutional codes. Suffice it here to state that their error-correction capabilities can be tabulated according to their code rate $k/n$.[8] For $1/n$ codes with constraint length $m$, an upper bound for the

[7] See, for instance: http://en.wikipedia.org/wiki/Viterbi_algorithm.

[8] J. G. Proakis, *Digital Communications* (New York: McGraw Hill, 2001), pp. 492–6.

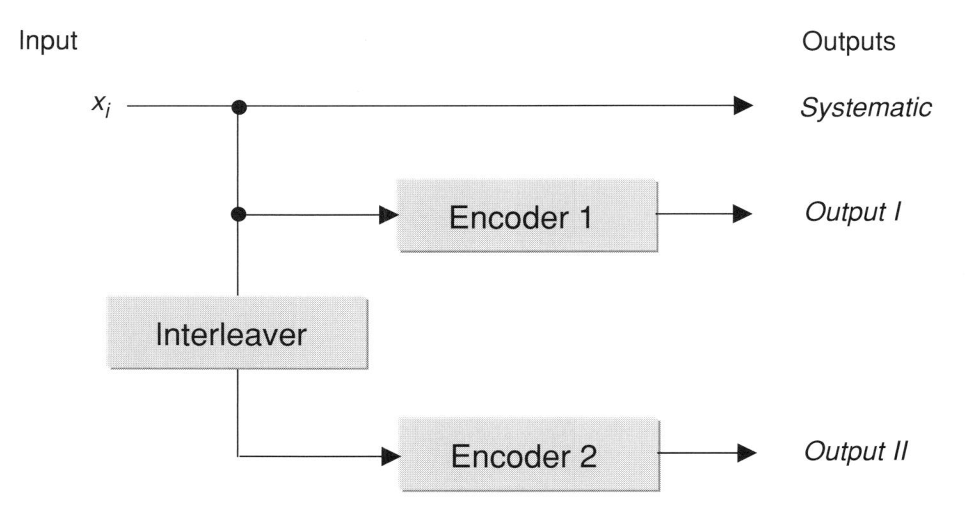


**Figure 11.4** Generic layout of turbo encoder.

minimum Hamming distance $d_{\min}$ is given by the following formula:

$$d_{\min} \leq \min_{p \geq 1} \left\{ \frac{2^{p-1}}{2^p - 1}(m + p - 1)n \right\}. \tag{11.25}$$

It is easily checked that with $n = 2$ and $m = 3$, for instance, the distance is bounded according to $d_{\min} \leq 5$, corresponding to a maximum of two-bit-error correction. Convolutional codes are commonly used in mobile-radio and satellite communications, and in particular in the wireless-connection standard called *Bluetooth*.

## Turbo codes

These high-performance codes represent a relatively recent (1993) and significant advance in the field of ECC. The principle is that the turbo encoder generates three sub-blocks, as illustrated in Fig. 11.4. As the figure shows, the first sub-block is the $k$-bit block of payload data that is output as uncoded and, thus, representing the *systematic output*. The second sub-block (output I) contains $m/2$ parity bits for the payload, computed using a convolutional code. The third sub-block (output II), also computed through a convolutional code, contains $m/2$ parity bits for a known or fixed pseudo-random permutation of the input payload bits. Such a permutation is effected through an *interleaver*, as seen in the figure. This pseudo-random interleaving makes the output I and output II sub-blocks substantially different from each other, at least in a large majority of input payload cases. The rationale in this approach is to ensure that one of these two sub-blocks has a high Hamming weight, or more ones than zeros, which (as it can be shown) makes the codeword identification and decoding easier than in the low-weight case. Thus, having two parallel encoders with substantially different or virtually uncorrelated outputs, corresponds to a "divide-and-conquer" ECC strategy. Here, we will not venture into the advanced principles of turbo decoding, but it is worth providing a high-level description of the corresponding strategy. The main task of the convolutional decoder is to compute an integer number $I$ for each received bit. This number, called a *soft bit*, which is output in the interval $[-127, 127]$, measures the likelihood for the received bit to be a 0 or a 1. The likelihood is measured as follows: $I = -127$ means certainly 0, $I = -100$ means very probably 0, $I = 0$ means equally likely to be either 0 or 1, $I = +100$ means very likely to be 1, and $I = +127$ means certainly likely to be 1. In the implementation, two parallel convolutional decoders are used to generate such likelihood measures for each of the $m/2$ bit parity sub-blocks and for the $k$-bit payload sub-block. Using their likelihood data, the two decoders can then construct two *hypotheses* for the $k$-bit *payload pattern*. These bit-pattern hypotheses are then compared. If they do match exactly, the bit pattern is output as representing the final decision. In the opposite case, the decoders exchange their likelihood measures. Each decoder includes the likelihood measure from the other decoder, which makes it possible for each of them to generate, once again, a new payload hypothesis. If the hypothesis comparison is not successful, the process is continued; it may take typically up to 15 to 18 iterations to converge! For turbo code designers, the matter and mission is to optimize the convolutional codes for efficient error correction, with maximal code rates (or minimal redundancy) and coding gains. In wireless-cellular applications, for instance, the two convolutional encoders are typically recursive, with a constraint length $m = 4$ and a code rate $1/3$, yielding overheads better than 20%. To quote some representative figures, the coding gain of turbo codes at BER $= 10^{-6}$ and after 10 iterations may be over 2 dB better than that of concatenated codes, 4.5 dB better than that of convolutional codes, and represent a 10 dB SNR improvement with respect to the uncoded BER case.[9] As two illustrative examples, suffice it to mention here the significant applications of turbo codes in the fields of aerospace and satellite communications[10] and 3G/cellular telephony (UMTS, cdma2000).[11]

This completes our "high-level" overview of the various ECC types and families. Since ECCs can only correct error patterns with up to $w = \{(d_{\min} - 1)/2\}$ bit errors, it is clear that, in the general case, the bit-error rate (BER) is never identical to zero. Bit errors are generated when the receiver makes the "wrong decision," namely outputs a 1 when the payload bit under consideration was a 0, and the reverse. Here, I shall not describe the different types of receiver architectures and decision techniques. Both are intimately dependent upon the type of signal *modulation formats*, the method of encoding bits into actual physical signal waveforms, whether electrical, radio, or optical. The ECC algorithms pervade all communication and information systems, and at practically all protocol layers: in point-to-point radio links (ground, aeronautical, satellite, and deep-space), in optical links (from fiber-to-the-home to metropolitan or core networks and undersea cable systems), in mobile cellular networks (GSM, GPRS, CDMA, UMTS/3G), in broadband wireless or wireline access (ADSL, 802.11/WiFi), in metropolitan, regional, and global voice or data transport (Ethernet, ATM, TCP/IP, or

[9] www.aero.org/publications/crosslink/winter2002/04.html.

[10] www.aero.org/publications/crosslink/winter2002/04.html.

[11] See, for instance: http://users.tkk.fi/~pat/coding/essays/turbo.pdf; http://www.csee.wvu.edu/~mvalenti/documents/valenti01.pdf.

Internet), and in global positioning systems (GPS, Galileo), to quote a few representative examples.

## 11.5 Corrected bit-error-rate

As we have seen in this chapter, ECCs can only correct error patterns having up to $w = \{(d_{\min} - 1)/2\}$ bit errors. It is clear, then, that in the general case, the bit error rate (BER) is always nonzero. This is despite the fact that it can be made arbitrarily close to zero through the appropriate ECC and under certain channel limiting conditions, as will be described in Chapter 12. Here, I will not detail the optimum receiver *decision* techniques, usually referred to *as hard-decision* and *soft-decision decoding*, respectively. Suffice it to state that in the soft-decision decoding approach, the receiver decision is optimized for each $1/0$ symbol type (e.g., use of different matching filters in multilevel signaling), which eventually minimizes symbol errors. On the other hand, hard-decision decoding consists of making a single choice between 1 and 0 values for each of the received symbols. Here, we shall first derive a simple expression for the hard-decision-decoding BER, which represents a general upper bound, regardless of the type of code used.

The uncorrected BER is defined by the function $p(x)$, which is the probability of misreading 1 or 0 symbols. In the modulation and detection format referred to as *intensity-modulation/direct-detection* (IM-DD) or, equivalently, *ON–OFF keying* (OOK), the parameter $x$ is the *Q-factor*.[12] Simply defined, the $Q$-factor is a function of the mean received signal powers, $\langle P_0 \rangle$ and $\langle P_1 \rangle$, which are associated with the 0 and 1 bit symbols, respectively, and of the corresponding Gaussian noise powers with standard deviations $\sigma_0 = \sqrt{\sigma_0^2}$, $\sigma_1 = \sqrt{\sigma_1^2}$, respectively. The $Q$-factor thus takes the simple form:[13]

$$Q = \frac{\langle P_1 \rangle - \langle P_0 \rangle}{\sigma_1 + \sigma_0}. \tag{11.26}$$

[12] See, for instance: E. Desurvire, *Survival Guides Series in Global Telecommunications, Signaling Principles, Network Protocols and Wireless Systems* (New York: J. Wiley & Sons, 2004), Ch. 1; E. Desurvire, *Erbium-Doped Fiber Amplifiers, Device and System Developments* (New York: J. Wiley & Sons, 2002), Ch. 3.

[13] One can also define the received *signal-to-noise ratio* (SNR) as follows:

$$\mathrm{SNR} = P_S/P_N,$$

where $P_S \equiv (P_1 + P_0)/2$ is the time-average signal power (assuming pseudo-random bit sequences), and $P_N$ is the total additive noise power. It can be shown (E. Desurvire, *Erbium-Doped Fiber Amplifiers, Device and System Developments* (New York: J. Wiley & Sons, 2002), Ch. 3.) that the SNR and the $Q$-factor are related through

$$\mathrm{SNR} = \frac{Q(Q+\sqrt{2})}{2},$$

$$\text{or, equivalently, } Q = 2\sqrt{2}\frac{\mathrm{SNR}}{1+\sqrt{1+4(\mathrm{SNR})}}.$$

For high SNR, the second formula can be approximated by $\mathrm{SNR} \approx Q^2/2$. In optical communications, only half of the noise power is taken into account, because noise exists in two electromagnetic-field polarizations. In this case, the above definition reduces to $\mathrm{SNR} \approx Q^2$.

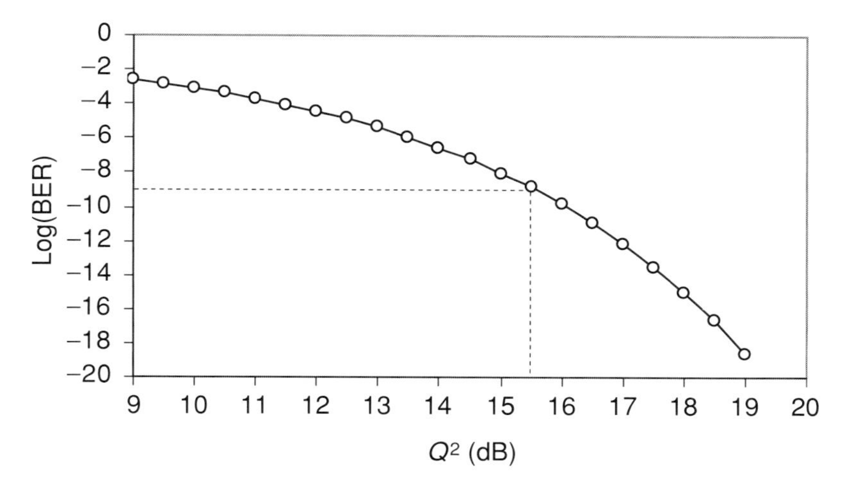


**Figure 11.5** Bit error rate (BER) as a function of the $Q$-factor expressed as $Q^2_{\mathrm{dB}}$, showing $\mathrm{BER} = 10^{-9}$ for $Q^2 = +15.5$ dB.

The probability of bit error (BER) is then given by the analytical expression and approximation:

$$p(Q) = \frac{1}{2}\operatorname{erfc}\left(\frac{Q}{\sqrt{2}}\right) \approx \frac{1}{Q\sqrt{2\pi}}\mathrm{e}^{-\frac{Q^2}{2}}, \tag{11.27}$$

where $\operatorname{erfc}(u)$ is the complementary error function, and where the approximation in the right-hand side is very accurate for $Q \geq 2$. It is easily checked that for $Q = 6$ we have $\mathrm{BER} = p(6) \approx 10^{-9}$, corresponding to one mistaken bit out of one billion received bits. One may also use the decibel definition for the $Q$-factor according to $Q^2_{\mathrm{dB}} = 20\log_{10} Q$, which gives $Q^2 = +15.5$ dB for $Q = 6$, or $\mathrm{BER} = 10^{-9}$. Figure 11.5 shows a plot of $\mathrm{BER} = p(Q) = f(Q^2_{\mathrm{dB}})$. Long before the massive development and ubiquitous implementation of ECC in optical telecommunications systems, the values of $\mathrm{BER} = 10^{-9}$ or $Q^2 = +15.5$ dB have been considered the de facto standard for "error-free" transmission.

Having, thus, defined the bit error probability, $p$, we can now analyze to what extent a BER can be corrected, or reduced, by ECC, using a simple demonstration.[14] For simplicity, we shall assume that the communication channel is *memoryless*. This means that there is no patterning effect within the possible bit sequences, or between codewords, namely that bit errors are uncorrelated. In a single transmission event, if $m$ errors occur in a block of $n$ bits, the corresponding error probability is, therefore, given by $p^m(1-p)^{n-m}$. The number of ways the $m$ errors can be arranged in the block is $C_n^m = n!/[m!(n-m)!]$. The total error probability is then $C_n^m p^m(1-p)^{n-m}$.

Consider now the corrected BER. To evaluate the upper bound for the corrected BER, we must take into account all error possibilities, but include the fact that all error patterns with a maximum of $w = \{(d_{\min} - 1)/2\}$ have been effectively corrected. The corresponding BER is then given by the summation over all error-pattern possibilities

[14] Such a simplified demonstration makes no pretence of being accurate or unique.

from $m = w + 1$ to $m = n$, representing the minimum and maximum numbers of incorrect bits, respectively, i.e.,

$$\mathrm{BER}_{\mathrm{corr}} \le \sum_{m=w+1}^{n} C_n^m p^m (1-p)^{n-m}. \tag{11.28}$$

It is seen from Eq. (11.28) that the BER upper-bound in the right-hand side can be expressed as the finite sum

$$\mathrm{BER}_{\mathrm{corr}} \le \mathrm{BER}_{w+1} + \mathrm{BER}_{w+2} + \cdots + \mathrm{BER}_n \tag{11.29}$$

with

$$\mathrm{BER}_m = C_n^m p^m (1-p)^{n-m}. \tag{11.30}$$

In such a sum, each term corresponds to events of increasing numbers of bit errors, which have fast-decreasing likelihoods. It is, therefore, possible to approximate the BER bound by the first few terms in the sum. The accuracy of such approximation wholly depends on the block length, the value of $w$ and the uncorrected bit-error probability, $p$. We shall consider a practical and realistic example, which also illustrates the powerful impact of ECC in BER correction.

Assume the block code $RS(n = 2^7 - 1 = 127, k = 119)$, which corresponds to the minimum Hamming distance $d_{\min} = (n-k)/2 = 4$. We are, therefore, interested in evaluating the BER contributions $\mathrm{BER}_5$, $\mathrm{BER}_6$, etc., in Eq. (11.29), which correspond to over four bits of noncorrectable errors. We choose a realistic-case situation where the bit error probability is $p = 10^{-4}$. From Eq. (11.30), we get, for the first three uncorrected BER contributions:

$$\begin{cases} \mathrm{BER}_5 = C_{127}^5 p^5 (1-p)^{122} \approx 2.5 \times 10^{-12} \\ \mathrm{BER}_6 = C_{127}^6 p^6 (1-p)^{121} \approx 4.9 \times 10^{-15} \\ \mathrm{BER}_7 = C_{127}^7 p^7 (1-p)^{120} \approx 8.5 \times 10^{-18}. \end{cases} \tag{11.31}$$

From the above evaluations, it is seen that the series $\mathrm{BER}_m$ is very rapidly converging, and that only the first contribution $\mathrm{BER}_5$ is actually significant. This means that for the corrected code, the primary source of errors is the "extra" bit error out of five error events, which cannot be corrected by the code. It is seen, however, that the ECC has reduced the BER from $p = 10^{-4}$ to $p = 2.5 \times 10^{-12}$, which represents quite a substantial improvement! Figure 11.6 shows plots of the uncorrected and corrected BER, using in the latter case the definition of $\mathrm{BER}_5$ in Eq. (11.31). which illustrates the BER improvement due to ECC.

Considering ON–OFF keying, and the above example, the uncorrected and corrected BER correspond to $Q_{\mathrm{unc}} \approx 3.73$ ($Q_{\mathrm{unc}}^2 = 11.4\,\mathrm{dB}$) and $Q_{\mathrm{corr}} \approx 6.91$ ($Q_{\mathrm{corr}}^2 = 16.8\,\mathrm{dB}$). One can then define the *coding gain* as the decibel ratio $\gamma = 20\log_{10}[Q_{\mathrm{corr}}/Q_{\mathrm{unc}}]$, or the decibel difference $\gamma = Q_{\mathrm{corr}}^2 - Q_{\mathrm{unc}}^2$, which, in this example, yields $\gamma = 20\log_{10}[6.91/3.73] = 16.4 - 11.8 = 5.4\,\mathrm{dB}$. This coding gain is indicated in Fig. 11.6 through the horizontal arrow. As a matter of fact, a 5.4 dB coding *gain* means that an identical BER can be achieved through ECC when the signal-to-noise ratio or SNR is decreased by 5.4 dB, as the figure illustrates. As mentioned in the previous section, the

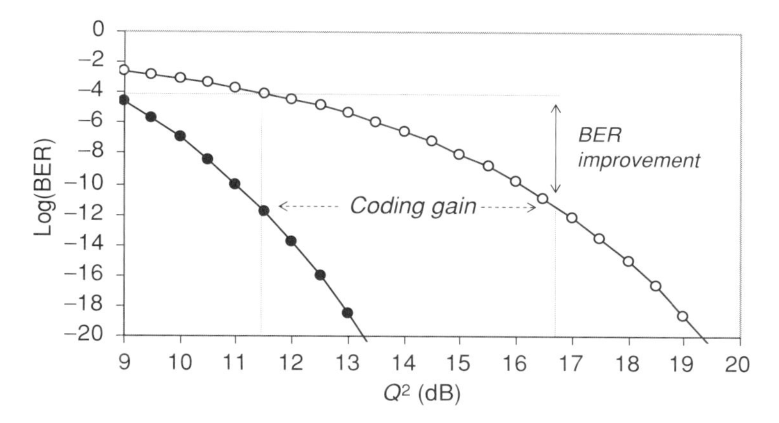


**Figure 11.6** Uncorrected BER (open circles) and corrected BER (closed circles) as a function of $Q^2_{\mathrm{dB}}$, showing BER improvement from $10^{-4}$ to $2.5 \times 10^{-12}$, and corresponding coding gain of 5.4 dB.

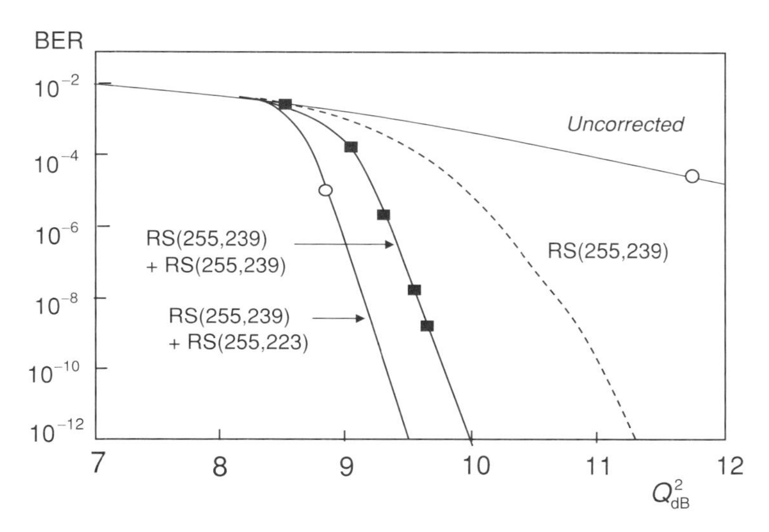


**Figure 11.7** Measured experimental data for uncorrected BER (open circle) and corrected BERs using $RS(255, 239)$ Reed–Solomon ECC (dashed line), and two concatenations $RS(255, 239) +$ $RS(255, 239)$ and $RS(255, 239) + RS(255, 223)$.

coding gain can be increased by the concatenation of two ECC codes, one serving as a first outer code, and the other as a second inner code. This feature is illustrated by the experimental data[15] shown in Fig. 11.7, where Reed–Solomon codes $RS(255, k)$ are used. As the figure shows, the implementation of a single $RS(255, 239)$ code, which has a redundancy of $255/239 - 1 = 6.7\%$, provides a coding gain of about 1.5 dB at $\mathrm{BER} = 10^{-4}$. The corrected BER is, however, less than $\mathrm{BER} = 10^{-12}$. When this RS

[15] Experimental data from: O. Ait Sab and J. Fang, Concatenated forward error correction schemes for long-haul DWDM optical transmission systems. In *Proc. European Conference on Optical Communications, ECOC'99*, Vol. II (1999), p. 290.

code is concatenated twice, i.e., using $RS(255, 239) + RS(255, 239)$ with a redundancy of $(255/239)^2 - 1 = 13.8\%$, the coding gain is 5.4 dB at BER $= 10^{-5}$. Alternatively, one can increase the number of parity bits in the inner code, i.e., using $RS(255, 223)$, which corresponds to the concatenated code $RS(255, 239) + RS(255, 223)$, which has a redundancy of $255^2/(239 \times 223) - 1 = 22.0\%$. The resulting coding gain is seen to improve with respect to the previous ECC concatenation, corresponding to 5.8 dB at BER $= 10^{-5}$. We also note from the figure that the BER is reduced by more than two orders of magnitude. Such data illustrate the powerful effect of ECC on error control. As I have mentioned earlier in this chapter, the BER is always nonzero, meaning that there is no such thing as an absolute error correction or BER $= 0.0 \times 10^0$ to infinite accuracy. Two key questions are: (a) how much a corrected BER can be made close to zero, and (b) whether there is a SNR (or $Q^2_{\text{dB}}$) limit under which ECCs are of limited effect in any BER correction? Such questions are answered in Chapter 13, through one of the most famous *Shannon theorems*.

## 11.6 Exercises

**11.1** (B): Tabulate the block codewords of the Hamming code $(n, k) = (7, 4)$ defined by the following generator matrix:

$$\tilde{G} = \left(\begin{array}{cccc|ccc} 1 & 0 & 0 & 0 & 1 & 0 & 1 \\ 0 & 1 & 0 & 0 & 1 & 1 & 1 \\ 0 & 0 & 1 & 0 & 0 & 1 & 0 \\ 0 & 0 & 0 & 1 & 1 & 1 & 0 \end{array}\right).$$

**11.2** (B): Assuming the Hamming code $(n, k) = (7, 4)$ with the parity-check matrix

$$\tilde{H} = \left(\begin{array}{cccc|ccc} 1 & 0 & 1 & 1 & 1 & 0 & 0 \\ 1 & 1 & 0 & 1 & 0 & 1 & 0 \\ 1 & 1 & 1 & 0 & 0 & 0 & 1 \end{array}\right),$$

determine the single errors and corrected codewords from the three received blocks that are defined as follows:

$$Z_1 = 1010111,$$
$$Z_2 = 0100001,$$
$$Z_3 = 0011110.$$

**11.3** (M): With the Hamming code $(n, k) = (7, 4)$ described in the text, whose block codewords are listed in Table 11.1 and syndrome vectors are listed in Table 11.2, assume *two* bit errors in the received blocks. To how many possible error patterns do each of the syndrome vectors correspond?

**11.4** (M): Calculate the probability that the Hamming code $(n, k) = (7, 4)$ described in the text (with block codewords listed in Table 11.1) fails to detect any multiple bit errors, while assuming single-bit-error probabilities of $p = 10^{-2}, 10^{-3}, 10^{-4}$, and $10^{-5}$.

**11.5** (T): Define a Hadamard code generated from a $4 \times 4$ matrix, then calculate the corresponding generator and parity-check matrices.

**11.6** (M): Construct a cyclic code from the generator polynomial

$$g(p) = p^3 + p^2 + 1$$

and tabulate the corresponding codewords.

**11.7** (M): Construct the systematic nonrecursive convolutional code $(n, k) = (3, 1)$ of constraint length $m = 3$ corresponding to the encoder

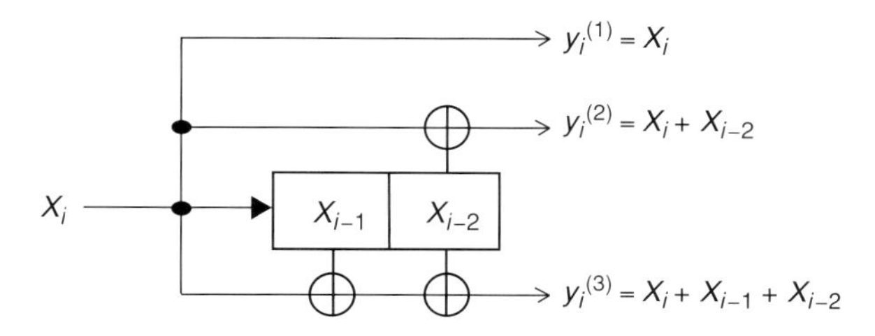


Draw the corresponding coding tree for up to four input bits $a, b, c, d$.

# 12 Channel entropy

This relatively short chapter on *channel entropy* describes the entropy properties of *communication channels*, of which I have given a generic description in Chapter 11 concerning error-correction coding. It will also serve to pave the way towards probably the most important of all Shannon's theorems, which concerns *channel coding*, as described in the more extensive Chapter 13. Here, we shall consider the different basic communication channels, starting with the *binary symmetric channel*, and continuing with nonbinary, asymmetric channel types. In each case, we analyze the channel's entropy characteristics and *mutual information*, given a discrete source transmitting symbols and information thereof, through the channel. This will lead us to define the *symbol error rate* (SER), which corresponds to the probability that symbols will be wrongly received or mistaken upon reception and decoding.

## 12.1 Binary symmetric channel

The concept of the *communication channel* was introduced in Chapter 11. To recall briefly, a communication channel is a transmission means for encoded information. Its constituents are an originator source (generating message symbols), an encoder, a transmitter, a physical transmission pipe, a receiver, a decoder, and a recipient source (restituting message symbols). The two sources (originator and recipient) may be discrete or continuous. The encoding and decoding scheme may include not only symbol-to-codeword conversion and the reverse, but also data compression and error correction, which we will not be concerned with in this chapter. Here, we shall consider *binary channels*. These include two binary sources $X = \{x_1, x_2\}$ as the originator and $Y = \{y_1, y_2\}$ as the recipient, with associated probability distributions $p(x_i)$ and $p(y_i)$, $i = 1, 2$. The events are discrete (as opposed to continuous); hence binary channels are one elementary variant of *discrete channels*.

A binary channel can be represented schematically through the diagram shown in Fig. 12.1. Each of the four paths corresponds to the event that, given a symbol $x_i$ from originator source $X$, a symbol $y_j$ is output from the recipient source $Y$. The conditional probability of this event is $p(y_j|x_i)$.

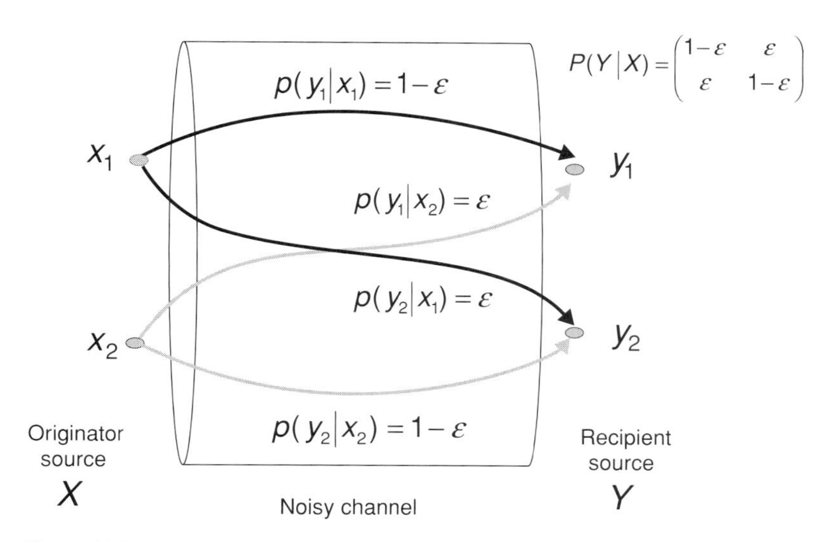


**Figure 12.1** Binary symmetric channel showing the originator source $X = \{x_1, x_2\}$, the recipient source $Y = \{y_1, y_2\}$, the conditional probabilities $p(y_j|x_i)$, and the corresponding transition matrix $P(Y|X)$.

In the ideal case of a *noiseless* channel, the two sources are 100% correlated,[1] i.e.,

$$\begin{aligned} p(y_j|x_i) &= 1 \quad \text{for } i = j, \\ p(y_j|x_i) &= 0 \quad \text{for } i \neq j, \end{aligned} \tag{12.1}$$

or $p(y_j|x_i) = \delta_{ij}$, with $\delta_{ij}$ being the Kronecker symbol. In this case, the channel is ideal indeed, since the transmission is 100% deterministic, which causes no corruption of information.

The set of conditional probabilities $p(y_j|x_i)$ associated with the channel defines a $j \times i$ *transition matrix*. This matrix, noted $P(Y|X)$, can be written as[2]

$$P(Y|X) = \begin{pmatrix} p(y_1|x_1) & p(y_1|x_2) \\ p(y_2|x_1) & p(y_2|x_2) \end{pmatrix}. \tag{12.2}$$

[1] Since there is no reason to have a one-to-one correspondence between the symbols or indices, we can also have $p(y_j|x_i) = 1$ for $i \neq j$ and $p(y_i|x_j) = 0$ for $i = j$. Thus, a perfect binary symmetric channel can equivalently have either of the transition matrixes (see definition in text):

$$P(Y|X) = \begin{pmatrix} 1 & 0 \\ 0 & 1 \end{pmatrix}$$

or

$$P(Y|X) = \begin{pmatrix} 0 & 1 \\ 1 & 0 \end{pmatrix}$$

Since in the binary system, bit parity (which of the 1 or 0 received bits represents the actual 0 in a given code) is a matter of convention, the noiseless channel can be seen as having an identity transition matrix.

[2] According to the transition-matrix representation, one can also define for any source $Z = X, Y$ the *probability vector* $P(Z) = [p(z_1), p(z_2)]$ to obtain the matrix-vector relation $P(Y) = P(Y|X)P(X)$, which stems from the property of conditional probabilities: $p(y_j) = \sum_i p(y_j|x_i)p(x_i)$.

To recall from Chapter 1, the properties of the transition-matrix elements are twofold:

(a) $p(y_j|x_1)p(x_1) + p(y_j|x_2)p(x_2) = p(y_j)$;
(b) $p(y_1|x_i) + p(y_2|x_i) = 1$ (column elements add to unity).

The channel is said to be *symmetric* if its transition matrix $P(Y|X)$ is symmetric, i.e., $[P(Y|X)]_{kl} = [P(Y|X)]_{lk}$, or the matrix elements are invariant by permutation of row and column indices, i.e., ${}^{\mathrm{T}}P(Y|X) \equiv P(Y|X)$.

In the case of the *ideal* or *noiseless channel*, we have, according to Eqs. (12.1) and (12.2):

$$P(Y|X) = \begin{pmatrix} 1 & 0 \\ 0 & 1 \end{pmatrix}, \tag{12.3}$$

which is the identity matrix. In the general, or *nonideal*, case, where the channel is corrupted by noise, we have $p(y_i|x_i) = 1 - \varepsilon$ and, hence, $p(y_{j\neq i}|x_i) = \varepsilon$, where $\varepsilon$ is a positive real number satisfying $0 \leq \varepsilon \leq 1$. The smaller $\varepsilon$, the closer the channel is to ideal or noiseless. The degree of noise "corruption" is, thus, defined by the parameter $\varepsilon$, which, as we shall see later in this chapter, defines the *symbol error probability*.

According to the definition in Eq. (4.2), in the most general case the transition matrix of the *binary symmetric channel* takes the form:

$$P(Y|X) = \begin{pmatrix} 1-\varepsilon & \varepsilon \\ \varepsilon & 1-\varepsilon \end{pmatrix}. \tag{12.4}$$

A limiting case is given by the parameter value $\varepsilon = 0.5$. In this case, the elements of the transition matrix in Eq. (12.4) are all equal to $p(y_j|x_i) = 0.5$. Thus, given the knowledge of the originator symbol $x_i$ (namely, $x_1$ or $x_2$), there is an equal probability that the recipient will output any of the symbols $y_j$ (namely, $y_1$ or $y_2$). It does not matter which originator symbol is fed into the channel, the recipient symbol output being indifferent, as in a perfect coin-flipping guess. In this case, the channel is said to be *useless*.

## 12.2 Nonbinary and asymmetric discrete channels

In this section, I provide a few illustrative examples of discrete communication channels, which, contrary to those described in the previous section, are *asymmetric*, and in some cases, *nonbinary*. These are referred to as the *Z channel*, the *binary erasure channel*, the *noisy typewriter*, and the *asymmetric channel with nonoverlapping outputs*.

---

**Example 12.1: *Z channel***

This is illustrated in Fig. 12.2. The Z channel is a binary channel in which one of the two originator symbols (say, $x_1$) is unaffected by noise and ideally converted into a unique recipient symbol (say, $y_1$), which implies $p(y_1|x_1) = 1$ and $p(y_2|x_1) = 0$. But the channel noise affects the other symbols, giving $p(y_1|x_2) = \varepsilon$ and $p(y_2|x_2) = 1 - \varepsilon$.

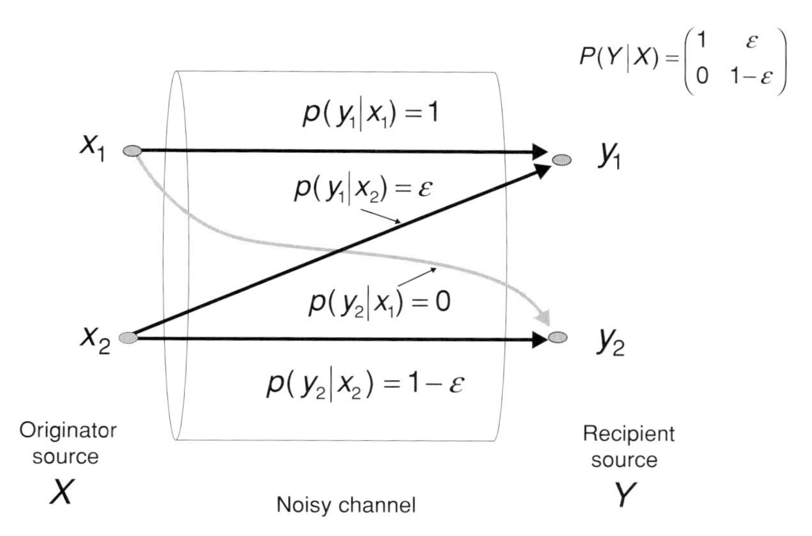


**Figure 12.2** Z channel showing the originator source $X = \{x_1, x_2\}$, the recipient source $Y = \{y_1, y_2\}$, the conditional probabilities $p(y_j|x_i)$, and the corresponding transition matrix $P(Y|X)$.

The corresponding transition matrix is, therefore:

$$P(Y|X) = \begin{pmatrix} 1 & \varepsilon \\ 0 & 1-\varepsilon \end{pmatrix}. \tag{12.5}$$

---

Note that stating that symbol $x_1$ is ideally transmitted through the channel and received as symbol $y_1$, does not mean that receiving symbol $y_1$ entails certainty of symbol $x_1$ being output by the originator.

---

**Example 12.2:** ***Binary erasure channel***

This is illustrated in Fig. 12.3. In the erasure channel, there exists a finite probability that the recipient source $Y$ outputs none of the symbols from the originator source $X$. This is equivalent to stating that $Y$ includes a supplemental symbol $y_3 =$ "void," which we shall call $\emptyset$. We, thus, define the probability of $Y$ having $y_3 = \emptyset$ for output as $p(\emptyset|x_1) = p(\emptyset|x_2) = \varepsilon$. It is easily established that the transition matrix corresponding to input $P(X) = [p(x_1), p(x_2)]$ and output $P(Y) = [p(y_1), p(\emptyset), p(y_2)]$ is defined as follows:

$$Y = \{y_1, y_2, y_3 = \emptyset\},$$

$$P(Y|X) = \begin{pmatrix} 1-\varepsilon & 0 \\ \varepsilon & \varepsilon \\ 0 & 1-\varepsilon \end{pmatrix}. \tag{12.6}$$

---

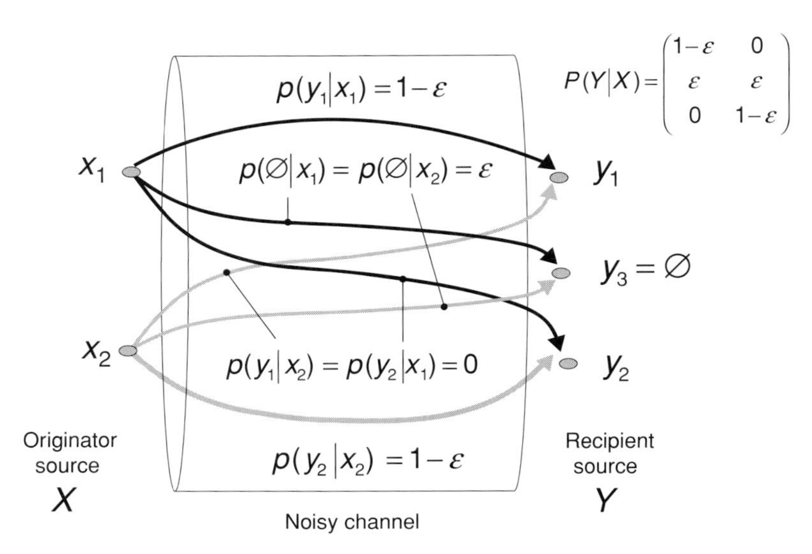


**Figure 12.3** Binary erasure channel, showing the originator source $X = \{x_1, x_2\}$, the recipient source $Y = \{y_1, y_2\}$, the conditional probabilities $p(y_j|x_i)$, and the corresponding transition matrix $P(Y|X)$, with the recipient source.

---

**Example 12.3:** ***Noisy typewriter***

This corresponds to a "typewriter" version of the binary symmetric channel, called the noisy typewriter, which is illustrated in Fig. 12.4. As seen from the figure, the source symbols are the characters {A, B, C, . . . , Z}, arranged in a linear keyboard sequence. When a given symbol character is input to the keyboard, there are equal chances of outputting the same character or one of its two keyboard neighbors. For instance, letter B as the input yields either A, B, or C as the output, with $p(\mathrm{A} \mid \mathrm{B}) = p(\mathrm{B} \mid \mathrm{B}) = p(\mathrm{C} \mid \mathrm{B}) = 1/3$, and so on for the 26 letters (input A yielding Z, A, or B), as illustrated in the figure. The corresponding transition matrix (reduced here for clarity to six input/output symbol characters, is the following:

$$P(Y|X) = \frac{1}{3}\begin{pmatrix} 1 & 1 & 0 & 0 & 0 & 1 \\ 1 & 1 & 1 & 0 & 0 & 0 \\ 0 & 1 & 1 & 1 & 0 & 0 \\ 0 & 0 & 1 & 1 & 1 & 0 \\ 0 & 0 & 0 & 1 & 1 & 1 \\ 1 & 0 & 0 & 0 & 1 & 1 \end{pmatrix}. \tag{12.7}$$

---

**Example 12.4:** ***Asymmetric channel with nonoverlapping outputs***

This corresponds to an asymmetric channel with nonoverlapping outputs, which maps source $X = \{x_1, x_2\}$ to source $Y = \{y_1, y_2, y_3, y_4\}$ according to the transition probabilities defined in Fig. 12.5. The property of this channel is that each of the outputs

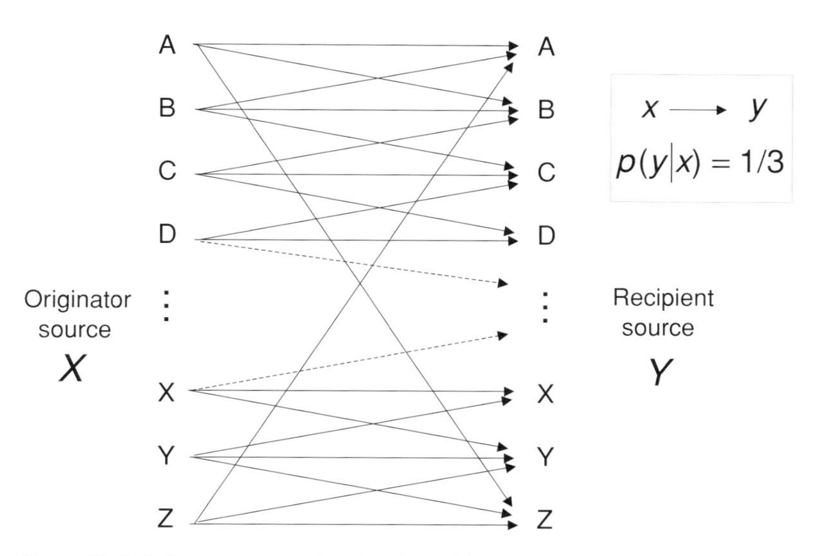


**Figure 12.4** Noisy typewriter, showing the originator source $X = \{x_1, x_2\}$, the recipient source $Y = \{y_1, y_2\}$, the conditional probabilities $p(y_j|x_i)$, and the corresponding transition matrix $P(Y|X)$, with $X, Y = \{A, B, C, \ldots, Z\}$.

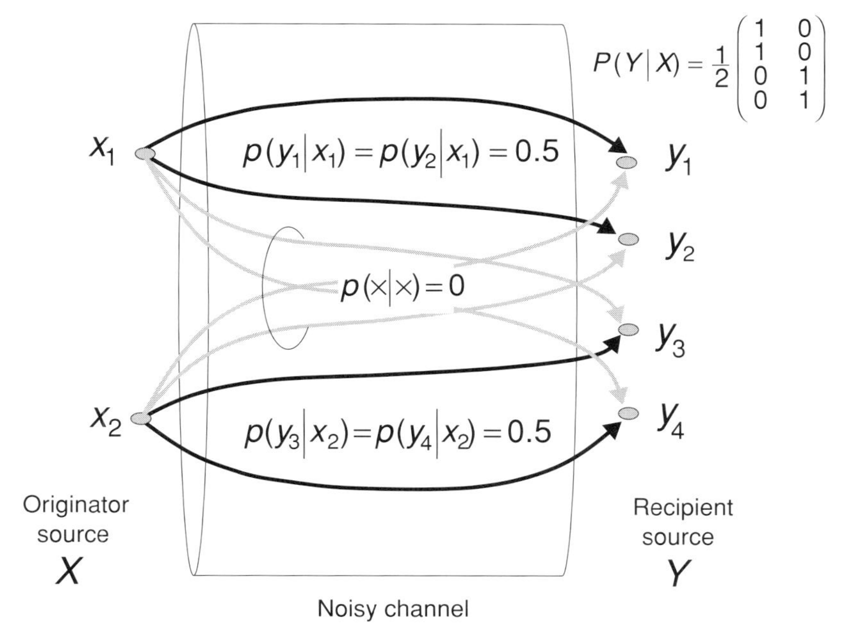


**Figure 12.5** Asymmetric channel with nonoverlapping outputs, showing the originator source $X = \{x_1, x_2\}$, the recipient source $Y = \{y_1, y_2\}$, the conditional probabilities $p(y_j|x_i)$, and the corresponding transition matrix $P(Y|X)$, with $X = \{x_1, x_2\}$ and $Y = \{y_1, y_2, y_3, y_4\}$.

$\{y_1, y_2, y_3, y_4\}$ has only one corresponding input $\{x_1, x_2\}$. While the output symbols are random, the input symbols are deterministic, which represents a special case of noisy channel where uncertainty does not increase, as discussed later in this chapter. It is easily established that the corresponding transition matrix is:

$$P(Y|X) = \frac{1}{2}\begin{pmatrix} 1 & 0 \\ 1 & 0 \\ 0 & 1 \\ 0 & 1 \end{pmatrix}. \tag{12.8}$$

---

The four above examples of asymmetric, nonbinary channels will be used again in Chapter 13 to illustrate the concept of channel capacity.

## 12.3 Channel entropy and mutual information

With the examples of the previous section, we can now make a practical use of the abstract definitions introduced in Chapter 5 concerning *conditional entropy* (also called *equivocation*), $H(Y|X)$, *joint entropy*, $H(X, Y)$, and *mutual information*, $H(Y; X)$, between two sources $X, Y$. It is useful to recall here the corresponding definitions and their relations:

$$H(Y|X) = -\sum_j \sum_i p(x_i, y_j) \log p(y_j|x_i), \tag{12.9}$$

$$\begin{aligned} H(X, Y) &= H(Y, X) \\ &= -\sum_j \sum_i p(x_i, y_j) \log p(x_i, y_j) \\ &= H(X) + H(Y|X) \\ &= H(Y) + H(X|Y), \end{aligned} \tag{12.10}$$

$$\begin{aligned} H(X; Y) &= H(Y) - H(Y|X) \\ &= H(X) - H(X|Y) \\ &= H(X) + H(Y) - H(X, Y). \end{aligned} \tag{12.11}$$

We consider first the elementary case of the *symmetric binary channel*, which was previously analyzed and illustrated in Fig. 12.1. The input probability distribution is, thus, defined by $p(x_1) = q$ and $p(x_2) = 1 - q$, where $q$ is a nonnegative real number such that $0 \leq q \leq 1$. According to definition, the entropy of the input source is $H(X) = -p(x_1) \log p(x_1) - p(x_2) \log p(x_2) \equiv f(q)$, where

$$f(q) = -q \log q - (1 - q) \log(1 - q). \tag{12.12}$$

The function $f(q)$, which was first introduced in Chapter 4, is plotted in Fig. 4.7. It is seen from the figure that it has a maximum of $f(q) = 1$ for $q = 0.5$, which corresponds to the case of maximal uncertainty in the source $X$. The output probabilities are calculated as follows:

$$\begin{aligned} p(y_1) &= p(y_1|x_1)p(x_1) + p(y_1|x_2)p(x_2) \\ &= (1-\varepsilon)q + \varepsilon(1-q) \\ &= q + \varepsilon - 2\varepsilon q \\ &\equiv r \\ p(y_2) &= 1 - r. \end{aligned} \tag{12.13}$$

We find, thus, the entropy of the output source:

$$H(Y) = f(q + \varepsilon - 2\varepsilon q). \tag{12.14}$$

Using next the transition matrix $P(Y|X)$ defined in Eq. (12.4) and *Bayes's theorem*, $p(x, y) = p(y|x)p(x)$, we find the joint probabilities $p(y_1, x_1) = (1-\varepsilon)q$, $p(y_1, x_2) = (1-q)\varepsilon$, $p(y_2, x_1) = \varepsilon q$ and $p(y_2, x_2) = (1-\varepsilon)(1-q)$. Replacing these results into Eq. (12.9), it is easily found that

$$H(Y|X) = -\varepsilon \log \varepsilon - (1-\varepsilon)\log(1-\varepsilon) \equiv f(\varepsilon), \tag{12.15}$$

and, hence,

$$\begin{aligned} H(X, Y) &= H(X) + H(Y|X) \\ &= f(q) + f(\varepsilon), \end{aligned} \tag{12.16}$$

$$\begin{aligned} H(X; Y) &= H(Y) - H(Y|X) \\ &= f(q + \varepsilon - 2\varepsilon q) - f(\varepsilon). \end{aligned} \tag{12.17}$$

These results can be commented on as follows.

First, we observe that the entropy of the output source $H(Y) = f(q + \varepsilon - 2\varepsilon q)$ is different from that of the input source $H(X) = f(q)$. Both reach the same maximum of $H = 1$ for $q = 0.5$. Furthermore, for any values of the noise parameter $\varepsilon$,[3] the output source entropy satisfies

$$H(Y) = f(q + \varepsilon - 2\varepsilon q) \geq f(q) = H(X), \tag{12.18}$$

which is illustrated by the family of curves shown in Fig. 12.6, for $\varepsilon = 0$ to $\varepsilon = 0.35$ (noting that the same curves are obtained with $\varepsilon \leftrightarrow 1 - \varepsilon$). As seen from the figure, the uncertainty introduced by channel noise ($\varepsilon$) increases the output source entropy $H(Y)$, which was expected. In the cases $q = 0$ or $q = 1$, which correspond to a deterministic source $X$ ($p(x_1) = 0$ and $p(x_2) = 1$, or $p(x_1) = 1$ and $p(x_2) = 0$), the output source entropy is nonzero, or $H(Y) = f(\varepsilon) = f(1-\varepsilon)$. The corresponding entropy can, thus, be seen as representing the exact measure of the channel's noise. It is inferred from the family of curves in Fig. 12.6 that in the limit $\varepsilon \to 0.5$, the output source entropy

[3] The parameter $\varepsilon$ defines the amount of noise in the binary symmetrical communications channel through the conditional probabilities $p(y_j|x_i)$. However, it should be noted that because of the symmetry of the transition matrix, the same amount of noise is associated with both $\varepsilon$ and $1-\varepsilon$. Thus, noise *increases* for increasing values of $\varepsilon$ in the interval $[0, 1/2]$, and *decreases* for increasing values of $\varepsilon$ in the interval $[1/2, 1]$.

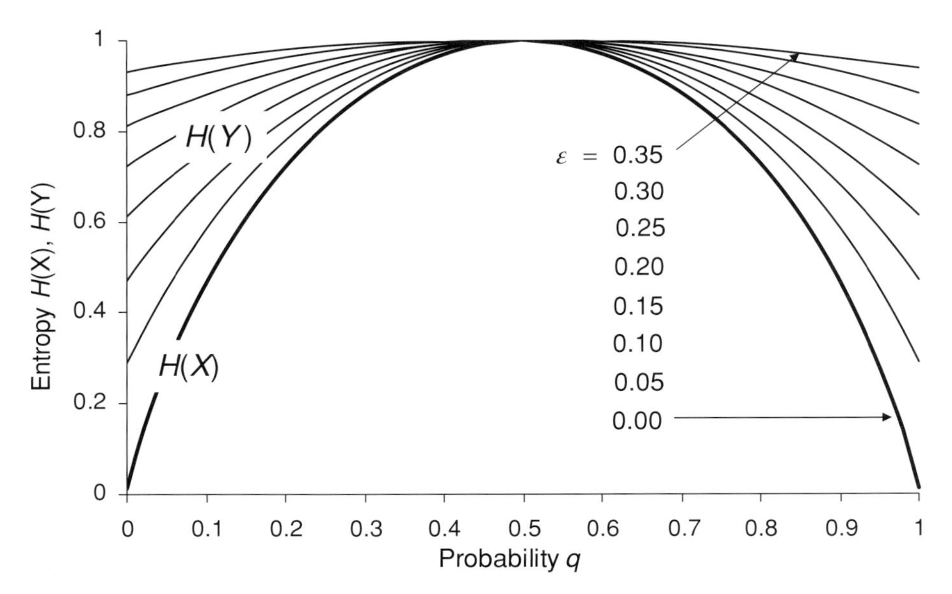


**Figure 12.6** Input $H(X)$ and output $H(Y)$ source entropies of binary symmetric channel, as functions of the probability $q = p(x_1)$ and channel noise parameter $\varepsilon$.

reaches the uniform limit $H(Y) = 1$, regardless of the input probability ($q$). The case $\varepsilon = 0.5$ corresponds to a maximum uncertainty in the output symbols, i.e., according to Eq. (12.13): $p(y_1) = q + \varepsilon - 2\varepsilon q \equiv 0.5 = p(y_2)$, which is independent of the input probability distribution. It is correct to call this communication channel *useless*, because the output symbols are 100% uncorrelated with the input symbols.

Second, we observe from the result in Eq. (12.15) that the equivocation $H(Y|X) = f(\varepsilon)$ is independent of the input probability distribution. This was expected, since *equivocation defines the uncertainty* of $Y$ *given the knowledge* of $X$. It represents a measure of the channel's noise, i.e., $H(Y|X) = f(\varepsilon)$. In the limit $\varepsilon \to 0$, we have $H(Y|X) \to 0$, which means that the communication channel (CC) is *ideal* or *noiseless*. The transition matrix of the ideal or noiseless channel is defined in Eq. (12.3). In this case, Eqs. (12.14)–(12.17) show that all entropy measures are equal, i.e., $H(X) = H(Y) = H(X, Y) = H(X; Y) = f(q)$.

Third, we consider the *mutual information* $H(X; Y)$ of the binary symmetric channel, which is defined in Eq. (12.17). Figure 12.7 shows plots of $H(Y; X)$ as a function of the input probability parameter $q$ and the channel noise parameter $\varepsilon$ for $\varepsilon = 0$ to $\varepsilon = 0.35$ (noting that the same curves are obtained with $\varepsilon \leftrightarrow 1 - \varepsilon$). The input source entropy $H(X)$ is also shown for reference.

We observe from Fig. 12.7 that the mutual information $H(X; Y)$ is never greater than the input source entropy $H(X)$. This is expected, since, by definition, $H(X; Y) = H(X) - H(X|Y) \leq H(X)$, Eq. (12.11), with the upper bound a result of the fact that $H(X|Y) \geq 0$. When the channel is noiseless or ideal ($\varepsilon = 0$), the mutual information equals the source entropy and, as we have seen earlier, all entropy measures are equal:

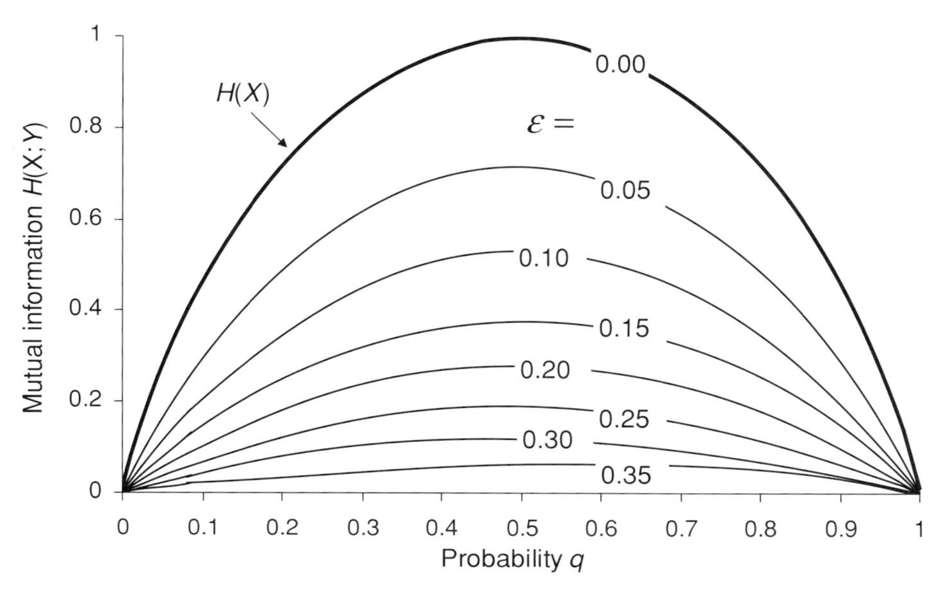


**Figure 12.7** Mutual information $H(X; Y)$ of binary symmetric channel, as a function of the probability $q = p(x_1)$ and channel noise parameter $\varepsilon$. The input source entropy $H(X)$ is also shown.

$H(X) = H(Y) = H(X, Y) = H(X; Y) = f(q)$. As the channel noise ($\varepsilon$) increases, the mutual information is seen to decrease with $0 \leq H(X; Y) < H(X)$. When the channel noise reaches the limit $\varepsilon \to 0.5$, the mutual information uniformly vanishes or $H(X; Y) \to 0$, regardless of the input probability ($q$). This is the case of the *useless channel*.

Finally, we observe from Fig. 12.7 that the mutual information is always maximum for $q = 0.5$, regardless of the channel noise $\varepsilon$. The conclusion is that *for noisy binary symmetric channels, the probability distribution that maximizes mutual information is the uniform distribution*.

The above demonstration, based on the example of the binary symmetric channel, illustrates that in noisy channels the entropy of the output source $H(Y)$ is *not* a correct measure of the information obtained from the recipient's side. This is because $H(Y)$ also measures the uncertainty introduced by the channel noise, which is not information. Thus, noise enhances the output source uncertainty without contributing any supplemental information according to the definition in Chapter 3. The right measure of the information available to the recipient is *mutual information*, i.e., the uncertainty of the output source $H(Y)$ minus the equivocation $H(Y|X)$. As we have seen, equivocation is the absolute measure of channel noise, and it is independent of the input probability distribution. The second important conclusion we have reached is that mutual information depends on the input probability distribution. Hence, given a noisy channel, *there must exist an input probability distribution for which the channel's mutual information is maximized*, meaning that the information obtained by the recipient is maximized. This conclusion anticipates Shannon's definition of *channel capacity,* which will be described in Chapter 13.

## 12.4 Symbol error rate

In this last section, we consider channels with sources $X$, $Y$ having equal sizes (i.e., numbers of symbol events), and analyze the effect of channel noise in the transmission of information from originator to recipient.

As previously established, the effect of channel noise is to increase the uncertainty in the recipient source $Y$, as translated by the entropy inequality $H(Y) \geq H(X)$ with respect to the input source entropy $H(X)$. As also observed, this increase of uncertainty does not correspond to additional information. Rather, it corresponds to an uncertainty in the effective transmission of information from originator to recipient. In noisy symmetric channels, the transition matrix $P(Y|X)$ is different from identity, which means that there is no deterministic (or one-to-one) correspondence between the input symbols $x_i$ and the output symbols $y_i$. Short of this unique correspondence between input and output symbols, any received message (i.e., sequence of symbols) is likely to be corrupted by a finite amount of symbol *errors*.

Given $x_i$ as input, a *symbol error* is defined as any discrepancy between the recipient output symbol $y_i$, which would have been expected with certainty in a noiseless channel, and the symbol $y_i^*$ actually output by the recipient. For messages of sufficient size (or number of symbols), the mean error count is called the *symbol error rate* (SER).

Let us now formalize the SER concept. Given the fact that the two sources $X$, $Y$ have equal size $N$, there exists a one-to-one symbol correspondence $x_i \leftrightarrow y_i$ between them. To each input message or input symbol sequence $x_i x_j x_k, \ldots$, thus, corresponds a unique output message or output symbol sequence $y_i y_j y_k, \ldots$, of matching size, in which each symbol is the faithful counterpart of the symbol of same rank in the input sequence. With the noisy channel, any error concerning the received symbol $y_i$ corresponds to the joint event $(y_j, x_{i\neq j})$. The probability associated with this specific symbol error is $p(y_j, x_{i\neq j})$.

The total channel SER is given by the sum of all possible symbol error probabilities according to:

$$\begin{aligned}\text{SER} &= \sum_{j=1}^{N} p(y_j, x_{i\neq j}) \\ &= \sum_{j=1}^{N} p(y_j|x_{i\neq j})p(x_i) \\ &= 1 - \sum_{i=1}^{N} p(y_i|x_i)p(x_i) \\ &= 1 - \sum_{i=1}^{N} p(y_i, x_i).\end{aligned} \tag{12.19}$$

The second line in Eq. (12.19) expresses the fact that the SER is the probability of the event complementary to that of the "absolutely zero error" event, the latter being defined as the sum of all possible "no error" events $(y_i, x_i)$ of associated probability $p(y_i, x_i)$.

According to Eq. (12.19), the SER of the binary symmetric channel is defined as

$$\begin{aligned} \text{SER} &= p(y_1|x_2)p(x_2) + p(y_2|x_1)p(x_1) \\ &= \varepsilon p(x_2) + \varepsilon p(x_1) \\ &= \varepsilon\left[p(x_1) + p(x_2)\right] \\ &= \varepsilon. \end{aligned} \tag{12.20}$$

The noise parameter $\varepsilon$, which we used in the earlier section to characterize the binary symmetric CC noise, thus, corresponds to the channel SER. Given any string of $N$ input symbols $x_i x_j x_k, \ldots, x_N$ and the corresponding string of $N$ output symbols $y_i y_j y_k, \ldots, y_N$, the SER gives the probability of any count of errors. For with $N$ sufficiently large, the average error count effectively measured is close to the SER. For instance, SER $= 0.001 = 10^{-3}$ corresponds to an average of one error in 1000 transmitted symbols, two errors in 2000 transmitted symbols, etc. This estimation becomes accurate for sufficiently long sequences, such that $N \gg 1/\text{SER}$.

More accurately, the SER should refer to a "mean error ratio" rather than an "error rate." The term comes from communication systems where symbols are transmitted at a certain symbol rate, i.e., the number of symbols transmitted per unit time. Given a symbol transmission rate of $N$ symbols per unit time, the corresponding *channel error rate* (or mean error count per unit time) is, thus, $N \times$ SER.

In binary channels, the SER is referred to as *bit error rate* or BER. For sufficiently long and random bit sequences, the bit probabilities become very nearly equal, i.e., $p(x_1 = \text{"0"}) \approx p(x_2 = \text{"1"}) = 1/2$. According to Eq. (12.20), we have, in this case,

$$\begin{aligned} \text{BER} &\approx \frac{1}{2}\left[p(y_1|x_2) + p(y_2|x_1)\right] \\ &\equiv \frac{1}{2}\left[p(0|1) + p(1|0)\right]. \end{aligned} \tag{12.21}$$

Most real-life binary channels are asymmetric, meaning that usually, $p(0|1) \neq p(1|0)$. Then the smallest BER is achieved when the sum $p(0|1) + p(1|0)$ is minimized.

As an easy illustration of SER (or BER) minimization, consider the case of the "Z channel," which was described earlier. To recall for convenience, the Z channel has the transition matrix

$$P(Y|X) = \begin{pmatrix} 1 & \varepsilon \\ 0 & 1-\varepsilon \end{pmatrix}. \tag{12.22}$$

According to Eq. (12.19), the corresponding symbol error rate is SER $= \varepsilon p(x_2) + 0 \times p(x_1) = \varepsilon p(x_2)$. The SER, thus, only depends on the probability $p(x_2)$. Minimizing the SER is, therefore, a matter of using an input probability distribution such that $p(x_2) \ll p(x_1) < 1$. In terms of coding, this means that the message sequences to be transmitted through the communications channel should contain the smallest possible number of symbols $x_2$, or a number that should not exceed some critical threshold. For instance, given the constraint $\varepsilon = p(y_1|x_2) = 0.25$ and a target error rate of SER $= 10^{-2}$, we should impose for the code $p(x_2) \leq \mathrm{SER}/\varepsilon = 10^{-2}/0.25 = 0.04$. The corresponding code should not have more than 4% of $x_2$ symbols in any input codeword sequence. Such a code has no reason to be optimal in terms of the efficient use of bits to transmit information. The point here is simply to illustrate that given a noisy communication channel, *the SER can be made arbitrarily small through an adequate choice of input source coding*. This conclusion anticipates Shannon's definition of *channel capacity*, which is described in Chapter 13.

Through the above example, we have seen that the channel SER can be minimized by the adequate choice of source code. The issue of SER minimization should not be confused with the principle of *error-correction codes* (ECC), which were described in Chapter 11. The principle of ECC codes is to reduce the channel SER to an arbitrarily small or negligibly small level. As we have learnt, error detection and correction are performed by means of extra redundancy or parity bits, which are included in block codewords. Such bits represent overhead information, which is meant to detect and correct errors automatically (up to some maximum), and which is removed once these tasks are accomplished. Error-correction codes, thus, make it possible to achieve transition matrices effectively approaching the identity matrix in Eq. (12.3) or Eq. (12.4) with $\varepsilon$ being made arbitrarily small.

## 12.5 Exercises

**12.1** (B): Show that any binary symmetric channel with $X$, $Y$ as input and output sources has an invariant probability vector $P(X)$ such that $P(Y) = P(X)$.

**12.2** (B): Determine the transition matrix of a channel obtained by cascading two identical binary symmetric channels, and show that the result is also a binary symmetric channel.

**12.3** (B): A binary channel successfully transmits bit 0 with a probability of 0.6 and bit 1 with a probability of 0.9. Assuming an input distribution $P(X) = (q, 1 - q)$ with $q = 0.2$, determine the output source entropy $H(Y)$.

**12.4** (M): A binary symmetric channel with noise parameter $\varepsilon$ is used to transmit codewords of five-bit length. Determine the probabilities of transmitting codewords with:

(a) One error;
(b) Two errors;
(c) At least two errors;
(d) One burst of two successive errors.

*Application*: provide in each case the numerical result, assuming that $\varepsilon = 0.1$.

# 13 Channel capacity and coding theorem

This relatively short but mathematically intense chapter brings us to the core of Shannon's information theory, with the definition of *channel capacity* and the subsequent, most famous *channel coding theorem* (CCT), the second most important theorem from Shannon (next to the *source coding theorem*, described in Chapter 8). The formal proof of the channel coding theorem is a bit tedious, and, therefore, does not lend itself to much oversimplification. I have sought, however, to guide the reader in as many steps as is necessary to reach the proof without hurdles. After defining *channel capacity*, we will consider the notion of *typical sequences* and *typical sets* (of such sequences) in codebooks, which will make it possible to tackle the said CCT. We will first proceed through a formal proof, as inspired from the original Shannon paper (but consistently with our notation, and with more explanation, where warranted); then with different, more intuitive or less formal approaches.

## 13.1 Channel capacity

In Chapter 12, I have shown that in a noisy channel, the *mutual information*, $H(X;Y) = H(Y) - H(Y|X)$, represents the measure of the true information contents in the output or recipient source $Y$, given the *equivocation* $H(Y|X)$, which measures the informationless channel noise. We have also shown that mutual information depends on the input probability distribution, $p(x)$. Shannon defines the *channel capacity* $C$ as representing the maximum achievable mutual information, as taken over all possible input probability distributions $p(x)$:

$$C = \max_{p(x)} H(X;Y). \tag{13.1}$$

As an illustrative example, consider first the case of the *binary symmetric channel* with noise parameter $\varepsilon$. As we have seen, the corresponding mutual information is given by Eq. (12.17), which I reproduce here for convenience:

$$H(X;Y) = f(q + \varepsilon - 2\varepsilon q) - f(\varepsilon), \tag{13.2}$$

where $q = p(x_1) = 1 - p(x_2)$, and $f(q) = -q \log q - (1 - q) \log(1 - q)$. According to Shannon's definition, the capacity of this channel corresponds to the maximum of

$H(X;Y)$ as defined above. It is straightforward to find this maximum by differentiating this definition with respect to the variable $q$ and finding the root:

$$\begin{aligned}\frac{\mathrm{d}H(X;Y)}{\mathrm{d}q} &= \frac{\mathrm{d}}{\mathrm{d}q}\left[f(q+\varepsilon-2\varepsilon q)-f(\varepsilon)\right] \\ &= 0,\end{aligned} \tag{13.3}$$

which, with $u = q + \varepsilon - 2\varepsilon q$ and the definition of the function $f(q)$ leads to

$$\begin{aligned}\frac{\mathrm{d}}{\mathrm{d}q}f(q+\varepsilon-2\varepsilon q) &= (1-2\varepsilon)\frac{\mathrm{d}f(u)}{\mathrm{d}u} \\ &= (1-2\varepsilon)\log\frac{1-u}{u}, \\ &= 0,\end{aligned} \tag{13.4}$$

or $u = 1/2$ and, hence, $q = 1/2 = p(x_1) = p(x_2)$.

This first result shows that *the mutual information of a binary symmetric channel is maximized with the uniform input distribution*. We have previously reached the same conclusion in Chapter 12, when plotting the mutual information $H(X;Y)$ as a function of the parameter $q$, see Fig. 12.7. With $q = 1/2$ we obtain $f(q + \varepsilon - 2\varepsilon q) = f(1/2) = 1$ and finally the capacity of the noisy binary channel:

$$\begin{aligned}C &= 1 - f(\varepsilon) \\ &= 1 + \varepsilon\log\varepsilon + (1-\varepsilon)\log(1-\varepsilon).\end{aligned} \tag{13.5}$$

With the binary symmetric channel, the "maximization problem" involved in the definition of channel capacity, Eq. (13.1), is seen to be relatively trivial. But in the general case, the maximization problem is less trivial, as I shall illustrate through a basic example.

Consider next the class of binary channels whose transition matrix is defined as:

$$P(Y|X) = \begin{pmatrix} a & 1-b \\ 1-a & b \end{pmatrix}, \tag{13.6}$$

where $a, b$ are real numbers belonging to the interval $[0, 1]$. Such a transition matrix corresponds to any binary channel, including the asymmetrical case, where $a \neq b$. It has been shown,[1] albeit without demonstration, that the channel capacity takes the form

$$C = \log(2^U + 2^V), \tag{13.7}$$

where

$$\begin{cases} U = \dfrac{(1-a)f(b) - bf(a)}{a+b-1} \\ V = \dfrac{(1-b)f(a) - af(b)}{a+b-1}. \end{cases} \tag{13.8}$$

[1] A. A. Bruen and M. A. Forcinito, *Cryptography, Information Theory and Error-Correction* (New York: John Wiley & Sons, 2005). Note that in this reference, the transition matrix is the transposed version of the one used here.

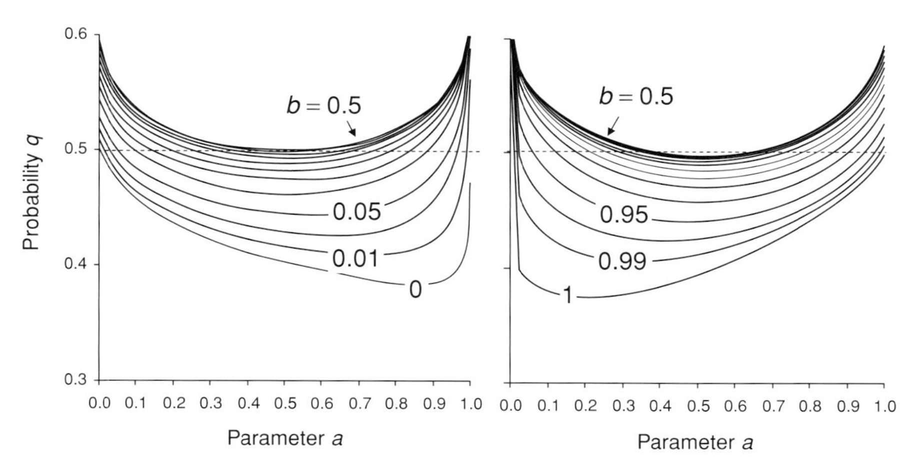


**Figure 13.1** Optimal input probability distribution $q = p(x_1) = 1 - p(x_2)$, corresponding to capacity $C$ of binary channel, plotted as a function of the two transition-matrix parameters $a$ and $b$ (step of 0.05).

The demonstration of the above result, which is elementary but far from straightforward, is provided in Appendix K. This Appendix also shows that the corresponding optimal distribution is given by $p(x_1) = q$, $p(x_2) = 1 - q$ with the parameter $q$ defined according to:

$$q = \frac{1}{a+b-1}\left(b - 1 + \frac{1}{1+2^W}\right), \tag{13.9}$$

with

$$W = \frac{f(a) - f(b)}{a+b-1} = V - U. \tag{13.10}$$

It is straightforward to verify that in the case $a = b = 1 - \varepsilon$ (binary symmetric channel), the channel capacity defined in Eqs. (13.7)–(13.8) reduces to $C = 1 - f(\varepsilon)$ and the optimal input distribution reduces to the uniform distribution $p(x_1) = p(x_2) = q = 1/2$.

The case $a + b = 1$, which seemingly corresponds to a pole in the above definitions, is trickier to analyze. Appendix K demonstrates that the functions $U$, $V$, $W$, and $q$ are all continuously defined in the limit $a + b \to 1$ (or for that matter, over the full plane $a, b \in [0, 1]$). In the limit $a + b \to 1$, it is shown that $H(Y; X) = C = 0$, which corresponds to the case of the *useless channel*. The fact that we also find, in this case, $q = 1/2$ (uniform input distribution) is only a consequence of the continuity of the function $q$. As a matter of fact, there is no optimal input distribution in useless channels, and all possible input distributions yield $C = 0$.

Figures 13.1 and 13.2 show 2D and 3D plots of the probability $q = p(x_1) = 1 - p(x_2)$ as a function of the transition-matrix parameters $a, b \in [0, 1]$ (sampled in steps of 0.05), as defined from Eqs. (13.9) and (13.10). It is seen from the figures that the optimal input probability distribution is typically *nonuniform*.[2] The 3D curve shows that the optimal

[2] Yet for each value of the parameter $b$, there exist two uniform distribution solutions, except at the point $a = b = 0.5$, where the solution is unique ($q = 1/2$).

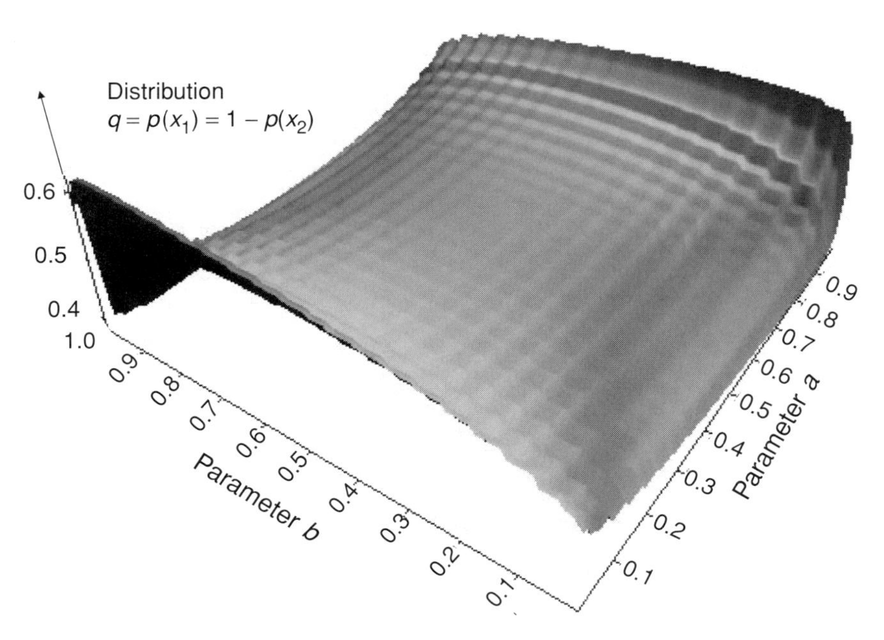


**Figure 13.2** 3D representation of the data shown in Fig. 3.1.

distribution has a saddle point at $a = b = 1/2$, which, as we have seen earlier, is one case of a useless channel ($a + b = 1$).

Figure 13.3 shows the corresponding 3D surface plot of the channel capacity $C$, as defined in Eqs. (13.7) and (13.8). It can be observed from the figure that the channel-capacity surface is symmetrically folded about the $a + b = 1$ axis, along which $C = 0$ (useless channel). The two maxima $C = 1$ located in the back and front of the figure correspond to the cases $a = b = 0$ and $a = b = 1$, which define the transition matrices:

$$P(Y|X) = \begin{pmatrix} 0 & 1 \\ 1 & 0 \end{pmatrix} \quad or \quad \begin{pmatrix} 1 & 0 \\ 0 & 1 \end{pmatrix}, \tag{13.11}$$

respectively. These two matrices define the two possible *noiseless* channels, for which there exists a one-to-one correspondence between the input and output symbols with 100% certainty.[3]

[3] Since there is no reason to have a one-to-one correspondence between the symbols or indices, we can also have $p(y_j|x_i) = 1$ for $i \neq j$ and $p(y_j|x_i) = 0$ for $i = j$. Thus, a perfect binary symmetric channel can equivalently have either of the transition matrixes (see definition in text):

$$P(Y|X) = \begin{pmatrix} 1 & 0 \\ 0 & 1 \end{pmatrix},$$

or

$$P(Y|X) = \begin{pmatrix} 0 & 1 \\ 1 & 0 \end{pmatrix}.$$

Since in the binary system, bit parity (which of the 1 or 0 received bits represents the actual 0 in a given code) is a matter of convention, the noiseless channel can be seen as having an identity transition matrix.

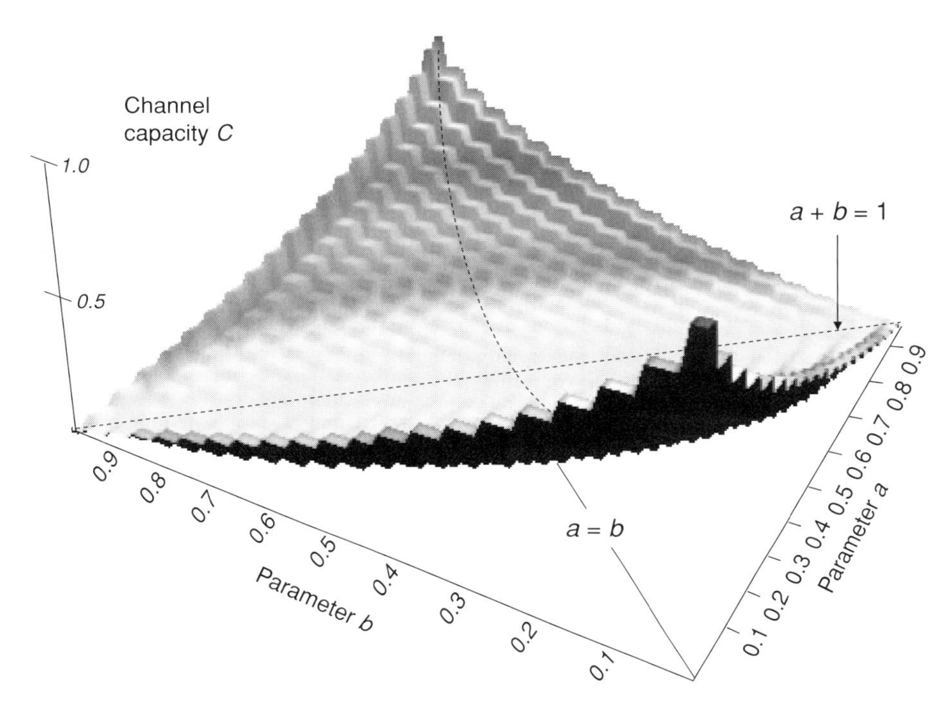


**Figure 13.3** Binary channel capacity $C$ as a function of the two transition-matrix parameters $a$ and $b$. The diagonal line $a + b = 1$ corresponds to the useless-channel capacity $C = 0$. The diagonal line $a = b$ corresponds to the binary symmetric-channel capacity $C = 1 - f(a)$.

The other diagonal line in Fig. 13.3 represents the channel capacities for which $a = b$ corresponds to the *binary symmetric channel*. Substitution of $b = a$ in Eq. (13.8) yields $U = V = -f(a)$ and, thus, $C = 1 - f(a)$ from Eq. (13.7), which is the result obtained earlier in Eq. (13.5). We also obtain from Eq. (13.10) $W = 0$, and, from Eq. (13.9), $q = 1/2$, which is the uniform distribution, as expected from the analysis made in Chapter 12.

To provide illustrations of channel capacity, we shall consider next the four discrete-channel examples described in Chapter 12. As usual, the input probability distribution is defined by the single parameter $q = p(x_1) = 1 - p(x_2)$.

---

**Example 13.1:** ***Z channel***

The transition matrix is shown in Eq. (12.5), to which the parameters $a = 1$ and $b = 1 - \varepsilon$ correspond. It is easily obtained from the above definitions that $U = 0$ and $V = W = -f(\varepsilon)/(1 - \varepsilon)$, which gives

$$C = \log\left(1 + 2^{-\frac{f(\varepsilon)}{1-\varepsilon}}\right), \tag{13.12}$$

$$q = \frac{1}{1-\varepsilon}\left(\varepsilon + \frac{1}{1 + 2^{-\frac{f(\varepsilon)}{1-\varepsilon}}}\right). \tag{13.13}$$

The case $\varepsilon = 0$, for which $a = b = 1$, corresponds to the noiseless channel with identity transition matrix, for which $C = 1$ and $q = 1/2$. The limiting case $\varepsilon \to 1$ ($a = 1$ and $b = 0$) falls into the category of useless channels for which $a + b = 1$, with $C = 0$ and $q = 1/2$ (the latter only representing a continuity solution,[4] but not the optimal distribution, as discussed at the end of Appendix K).

---

**Example 13.2:** ***Binary erasure channel***

The transition matrix is shown in Eq. (12.6). One must then calculate the entropies $H(Y)$ and $H(Y|X)$, according to the method illustrated in Appendix K (see Eqs. (K4)–(K5) and (K7)–(K8)). It is left to the reader as an easy exercise to show that $H(Y) = f(\varepsilon) + (1 - \varepsilon)f(q)$ and $H(Y|X) = f(\varepsilon)$. The mutual information is, thus, $H(X;Y) = H(Y) - H(Y|X) = (1 - \varepsilon)f(q)$ and it is maximal for $q = 1/2$ ($\mathrm{d}f/\mathrm{d}q = [\log(1 - q)/q]$). Substituting this result into $H(X;Y)$ yields the channel capacity $C = 1 - \varepsilon$. The number $1 - \varepsilon = p(y_1|x_1) = p(y_2|x_2)$ corresponds to the fraction of bits that are successfully transmitted through the channel (a fraction $\varepsilon$ being erased). The conclusion is that the binary-erasure channel capacity is reached with the uniform distribution as input, and it is equal to the fraction of nonerased bits.

---

**Example 13.3:** ***Noisy typewriter***

The transition matrix is shown in Eq. (12.7) for the simplified case of a six-character alphabet. Referring back to Fig. 12.4, we observe that each of the output characters, $y = \mathrm{A, B, C}, \ldots, \mathrm{Z}$, is characterized by three nonzero conditional probabilities, e.g., for $y = \mathrm{B}$: $p(\mathrm{B}|\mathrm{A}) = p(\mathrm{B}|\mathrm{B}) = p(\mathrm{B}|\mathrm{C}) = 1/3$. The corresponding probability is:

$$\begin{aligned} p(y = \mathrm{B}) &= p(\mathrm{B}|\mathrm{A})p(\mathrm{A}) + p(\mathrm{B}|\mathrm{B})p(\mathrm{B}) + p(\mathrm{B}|\mathrm{C})p(\mathrm{C}) \\ &= [p(\mathrm{A}) + p(\mathrm{B}) + p(\mathrm{C})]/3. \end{aligned} \tag{13.14}$$

The joint probabilities are

$$\begin{cases} p(\mathrm{B}, \mathrm{A}) = p(\mathrm{B}|\mathrm{A})p(\mathrm{A}) = p(\mathrm{A})/3 \\ p(\mathrm{B}, \mathrm{B}) = p(\mathrm{B}|\mathrm{B})p(\mathrm{B}) = p(\mathrm{B})/3 \\ p(\mathrm{B}, \mathrm{C}) = p(\mathrm{B}|\mathrm{C})p(\mathrm{C}) = p(\mathrm{C})/3. \end{cases} \tag{13.15}$$

Using the above definitions, we obtain

$$\begin{aligned} H(Y) &= -\sum_j p(y_j) \log p(y_j) \\ &= -\frac{p(\mathrm{A}) + p(\mathrm{B}) + p(\mathrm{C})}{3} \log \frac{p(\mathrm{A}) + p(\mathrm{B}) + p(\mathrm{C})}{3} + \leftrightarrow, \end{aligned} \tag{13.16}$$

[4] As shown in Appendix K, the continuity of the solutions $C$ and $q$ stems from the limit $1/(1 + 2^W) \approx a - \eta/2$, where $a + b = 1 - \eta$ and $\eta \to 0$. This limit yields $q = 1/2$, regardless of the value of $a$, and $C = -\log(a) \to 0$ for $a \to 1$

where the sign $\leftrightarrow$ means all possible circular permutations of the character triplets (i.e., BCD, DEF, ..., XYZ, YZA). The mutual information is given by:

$$\begin{aligned} H(Y|X) &= -\sum_i \sum_j p(x_i, y_j) \log p(y_j|x_j) \\ &= -[p(\mathrm{B,A}) \log p(\mathrm{B|A}) + p(\mathrm{B,B}) \log p(\mathrm{B|B}) + p(\mathrm{B,C}) \log p(\mathrm{B|C}) \leftrightarrow] \\ &= -\left[\frac{p(\mathrm{A}) + p(\mathrm{B}) + p(\mathrm{C})}{3} \log \frac{1}{3} + \leftrightarrow\right] \\ &= -3 \frac{p(\mathrm{A}) + p(\mathrm{B}) + \cdots + p(\mathrm{Z})}{3} \log \frac{1}{3} \\ &= \log 3. \end{aligned} \tag{13.17}$$

The channel capacity is, therefore:

$$\begin{aligned} C &= \max_{p(x)} \{H(Y) - H(Y|X)\} \\ &= \max_{p(x)} \left\{ -\frac{p(\mathrm{A}) + p(\mathrm{B}) + p(\mathrm{C})}{3} \log \frac{p(\mathrm{A}) + p(\mathrm{B}) + p(\mathrm{C})}{3} + \leftrightarrow \right\} - \log 3. \end{aligned} \tag{13.18}$$

The expression between brackets is maximized when for each character triplet, A, B, C, we have $p(\mathrm{A}) = p(\mathrm{B}) = p(\mathrm{C})$, which yields the optimal input distribution $p(x) = 1/26$. Thus, Eq. (13.18) becomes

$$\begin{aligned} C &= -\left\{\frac{1}{26} \log \frac{1}{26} + \leftrightarrow\right\} - \log 3 \\ &= \frac{26}{26} \log \frac{1}{26} - \log 3 \\ &= \log 26 - \log 3 \\ &= \log \frac{26}{3}. \end{aligned} \tag{13.19}$$

This result could have been obtained intuitively by considering that there exist 26/3 triplets of output characters that correspond to a unique input character.[5] Thus 26/3 symbols can effectively be transmitted without errors as if the channel were noiseless. The fact that 26/3 is not an integer does not change this conclusion (an alphabet of 27 characters would yield exactly nine error-free symbols).

---

**Example 13.4:** ***Asymmetric channel with nonoverlapping outputs***

The transition matrix is shown in Eq. (12.8). Since there is only one nonzero element in each row of the matrix, each output symbol corresponds to a single input symbol. This channel represents another case of the noisy typewriter (Example 3), but with

[5] Namely, outputs Z, A, B uniquely correspond to A as input, outputs C, D, E uniquely correspond to D as input, etc. Such outputs are referred to as "nonconfusable" subsets.

nonoverlapping outputs. Furthermore, we can write

$$\begin{aligned} p(y_1 \text{ or } y_2) &\equiv p(y_1) + p(y_2) \\ &= p(y_1|x_1)p(x_1) + p(y_1|x_2)p(x_2) + p(y_2|x_1)p(x_2) + p(y_2|x_2)p(x_2) \\ &= \frac{1}{2}[p(x_1) + p(x_2)] \equiv \frac{1}{2} = p(y_3 \text{ or } y_4). \end{aligned} \tag{13.20}$$

Define the new output symbols $z_1 = y_1$ or $y_2$, $z_2 = y_2$ or $y_3$. It is easily established that the conditional probabilities satisfy $p(z_1|x_1) = p(z_2|x_2) = 1/2$. The corresponding transition matrix is, thus, the identity matrix. This noisy channel is, therefore, equivalent to a *noiseless* channel, for which $C = 1$ with the uniform distribution as the optimal input distribution. This conclusion can also be reached by going through the same formal calculations and capacity optimization as in Example 3.

With the above examples, the *maximization problem* of obtaining the *channel capacity* and the corresponding *optimal input probability distribution* is seen to be relatively simple. But one should not hastily conclude that this applies to the general case! Rather, the solution of this problem, should such a solution exist and be unique, is generally complex and nontrivial. It must be found through numerical methods using *nonlinear optimization algorithms*.[6]

## 13.2 Typical sequences and the typical set

In this section, I introduce the two concepts of *typical sequences* and the *typical set*. This concept is central to the demonstration of Shannon's second theorem, known as the *channel coding theorem*, which is described in the next section.

*Typical sequences* can be defined according to the following. Assume an originator message source $X^n$ generating binary sequences of length $k$.[7] Any of the $2^k$ possible sequences generated by $X^n$ is of the form $x = x_1x_2x_3, \ldots, x_k$, where $x_i = 0$ or 1 are the message bits. We assume that the bit events $x_i$ are independent. The entropy of such a source is $H(X^k) = kH(X)$.[8] Assume next that the probability of any bit in the sequence

[6] A summary is provided in T. M. Cover and J. A. Thomas, *Elements of Information Theory* (New York: John Wiley & Sons, 1991), p. 191.

[7] The notation $X^k$ refers to the *extended source* corresponding to $k$ repeated observations or uses of the source $X$.

[8] Such a property comes from the definition of joint entropy, assuming two sources with independent events:

$$\begin{aligned} H(X, Y) &= -\sum_{xy} p(x, y) \log [p(x, y)] \\ &= -\sum_{xy} p(x)p(y) \log [p(x)p(y)] \\ &= -\sum_{x} p(x) \sum_{y} p(y) [\log p(x) + \log p(y)] \\ &= -\sum_{x} p(x) \log p(x) \sum_{y} p(y) - \sum_{x} p(x) \sum_{y} p(y) \log p(y) \\ &= -\sum_{x} p(x) \log p(x) - \sum_{y} p(y) \log p(y) \\ &\equiv H(X) + H(Y). \end{aligned}$$

With the extended source $X^2$, we obtain $H(X^2) = 2H(X)$, and consequently $H(X^k) = kH(X)$.

being one is $p(x_i = 1) = q$. We, thus, expect that any sequence roughly contains $kq$ bits equal to one, and $k(1-q)$ bits equal to zero.[9] This property becomes more accurately verified as the sequence length $k$ is sufficiently large, as we shall see later. The probability of a sequence $\theta$ containing *exactly* $kq$ 1 bits and $k(1-q)$ 0 bits is

$$p(\theta) = mq^{kq}(1-q)^{k(1-q)}, \tag{13.21}$$

where $m = C_k^{kq}$ is the number of possible $\theta$ sequences. Taking the minus logarithm (base 2) of both sides in Eq. (13.21) yields

$$\begin{aligned} -\log p(\theta) &= -\log m - \log q^{kq} - \log(1-q)^{k(1-q)} \\ &= \log\frac{1}{m} - kq\log q - k(1-q)\log(1-q) \\ &= \log\frac{1}{m} - k\left[-q\log q - (1-q)\log(1-q)\right] \\ &\equiv \log\frac{1}{m} + kf(q) \equiv \log\frac{1}{m} + kH(X) \\ &\equiv \log\frac{1}{m} + H(X^k). \end{aligned} \tag{13.22}$$

The result obtained in Eq. (13.22) shows that the probability of obtaining a sequence $\theta$ is

$$p(\theta) = m2^{-kH(X)} = m2^{-H(X^k)}. \tag{13.23}$$

Thus, all $\theta$ sequences are equiprobable, and each individual sequence in this set of size $m$ has the probability $p(x) \equiv p(\theta)/m = 2^{-kH(X)} = 2^{-H(X^k)}$. We shall now (tentatively) call any $\theta$ a *typical sequence*, and the set of such sequences $\theta$, the *typical set*.

An example of a typical set (as tentatively defined) is provided in Fig. 13.4. In this example, the parameters are chosen to be $k = 6$ and $q = 1/3$. For each sequence $\theta$ containing $j$ bits equal to one ($j = 0, 1, \ldots, 6$), the corresponding probability was calculated according to $p(x) = q^j(1-q)^{k-j}$. The source entropy is calculated to be $H(X^k) = kH(X) = kf(q) = 6f(1/3) = 6 \times 0.918 = 5.509$ bit. The typical set, thus, comprises $m = C_k^{kq} = C_6^2 = 15$ typical sequences with equal probabilities $p(x) = p(\theta)/m = 2^{-H(X^k)} = 2^{-5.509} = 0.022$. The figure shows the log probabilities of the entire set of $2^6 = 64$ possible sequences, along with the relative size of the typical set and associated probabilities. As seen in the figure, there exist two neighboring sets, A and B, with probabilities close to the typical set, namely $-\log p(x_{\mathrm{A}}) = 6.509$ and $-\log p(x_{\mathrm{B}}) = 4.509$, respectively. The two sets correspond to sequences similar to the $\theta$ sequences with one 1 bit either in excess (A) or in default (B). Their log probabilities, therefore, differ by $\pm 1$ with respect to that of the $\theta$ sequence. In particular,

[9] For simplicity, we shall assume that $kq$ is an integer number.

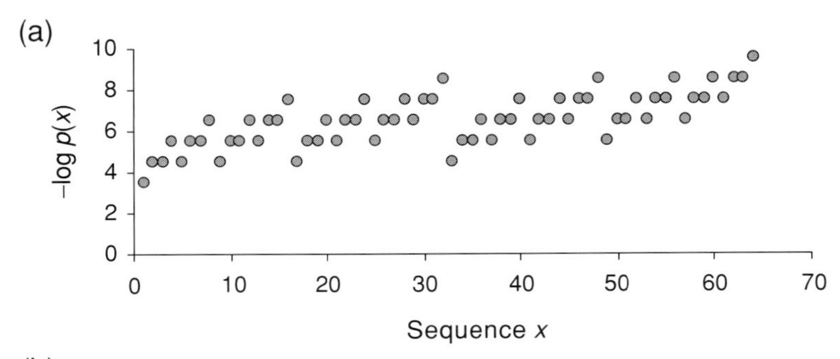


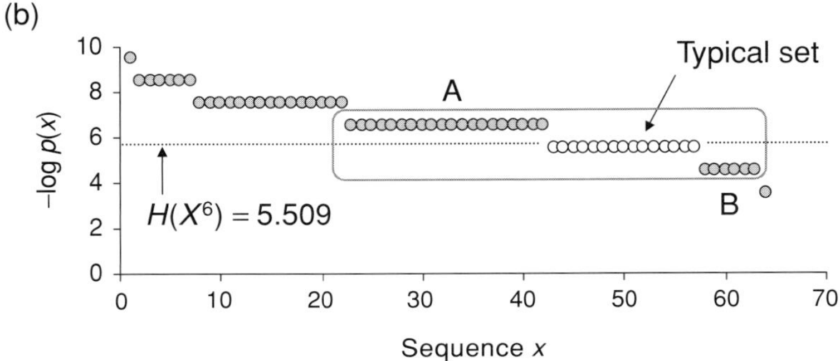


**Figure 13.4** Illustration of a tentative definition of the typical set for a binary sequence with length $k = 6$ and probability of 1 symbol $q = 1/3$ (source entropy $kH(X) = 5.509$ bit: (a) probabilities ($-\log_2 p(x)$) of each sequence $x$ as ordered from $x = 000000$ to $x = 111111$; (b) same as (a) with sequences ordered in increasing order of probability. The typical set corresponds to sequences having $kq = 2$ bits equal to 1, with uniform probability $-\log_2 p(x) = kH(X) = 5.509$. The two neighboring sets, A and B, with similar probabilities are indicated.

we have

$$\left| -\frac{\log p(x_{A\,or\,B})}{k} - H(X) \right| = \left| -\frac{kH(X) \pm 1}{k} - H(X) \right|$$
$$= \left| \pm \frac{1}{k} \right| \tag{13.24}$$
$$= \frac{1}{k} = \frac{1}{6} = 0.16.$$

Because the absolute differences defined by Eq. (13.24) are small, we can now extend our definition of the typical set to include the sequences from the sets A and B, which are roughly similar to $\theta$ sequences within one extra or missing 1 bit. We can state that:

- The sequences from A and B have roughly $kq$ 1 bits and $k(1-q)$ 0 bits;
- The corresponding probabilities are roughly equal to $2^{-H(X^k)}$.

According to this extended definition of the typical set, the total number of typical sequences is now $N = 15 + 6 + 20 = 41$. We notice that $N \approx 2^{H(X^k)} = 45.5$, which leads to a third statement:

- The number of typical sequences is roughly given by $2^{H(X^k)}$.

The typical set can be extended even further by including sequences differing from the $\theta$ sequences by a small number $u$ of extra or missing 1 bits. Their log probabilities, therefore, differ by $\pm u$ with respect to $H(X^k)$, and the absolute difference in Eq. (13.24) is equal to $u/k$. For long sequences ($k \gg u > 1$), the result can be made arbitrarily small.

The above analysis leads us to a most general definition of a *typical sequence*: *given a message source X with entropy $H(X) = f(q)$ and an arbitrary small number $\varepsilon$, any sequence x of length k, which satisfies*

$$\left| \frac{1}{k} \log \frac{1}{p(x)} - H(X) \right| < \varepsilon, \tag{13.25}$$

*is said to be "typical," within the error $\varepsilon$. Such typical sequences are roughly equiprobable with probability $2^{-kH(X)}$ and their number is roughly $2^{kH(X)}$.* The property of typical sequences, as defined in Eq. (13.25), is satisfied with arbitrary precision as the sequence length $k$ becomes large.[10] Alternatively, we can rewrite Eq. (13.25) in the form

$$\begin{aligned} &\left[ \frac{1}{k} \log \frac{1}{p(x)} - H(X) \right]^2 < \varepsilon^2 \\ &\leftrightarrow \\ &2^{-k[H(X)+\varepsilon]} < p(x) < 2^{-k[H(X)-\varepsilon]}, \end{aligned} \tag{13.26}$$

which shows the convergence between the upper and lower bounds of the probability $p(x)$ as $k$ increases.

Note that the typical set does not include the high-probability sequences (such as, in our example, the sequences with high numbers of 0 bits). However, any sequence $x$ selected at random is likely to belong to the typical set, since the typical set (roughly) has $2^{kH(X)}$ members out of (exactly) $2^k$ sequence possibilities. The probability $p$ that $x$ belongs to the typical set is, thus, (roughly):

$$p = 2^{k[H(X)-1]} = 2^{-k[1-H(X)]}, \tag{13.27}$$

which increases as $H(X)$ becomes closer to unity ($q \to 0.5$).[11]

## 13.3 Shannon's channel coding theorem

In this section, I describe probably the most famous theorem in information theory, which is referred to as the *channel coding theorem* (CCT), or *Shannon's second theorem*.[12] The

[10] This property is also known as the *asymptotic equipartition principle* (AEP). This principle states that given a source $X$ of entropy $H(X)$, any outcome $x$ of the extended source $X^k$ is most likely to fall into the typical set roughly defined by a uniform probability $p(x) = 2^{-kH(X)}$.

[11] In the limiting case $H(X) = 1, q = 0.5, p = 1$, all possible sequences belong to the typical set; they are all strictly equiprobable, but generally they do not have the same number $kq = k/2$ of 1 and 0 bits. Such a case corresponds to the roughest possible condition of typicality.

[12] To recall, the first theorem from Shannon, the *source-coding theorem*, was described in Chapter 8, see Eq. (8.15).

CCT can be stated in a number of different and equivalent ways. A possible definition, which reflects the original one from Shannon,[13] is:

Given a noisy communication channel with capacity $C$, and a symbol source $X$ with entropy $H(X)$, which is used at an information rate $R \leq C$, there exists a code for which message symbols can be transmitted through the channel with an arbitrary small error $\varepsilon$.

The demonstration of the CCT rests upon the subtle notion of "typical sets," as analyzed in the previous section. It proceeds according to the following steps:

- Assume the input source messages $x$ to have a length (number of symbols) $n$. With independent symbol outcomes, the corresponding extended-source entropy is $H(X^n) = nH(X)$. The typical set of $X^n$ roughly contains $2^{nH(X)}$ possible sequences, which represent the most probable input messages.
- Call $y$ the output message sequences received after transmission through the noisy channel. The set of $y$ sequences corresponds to a random source $Y^n$ of entropy $H(Y^n) = nH(Y)$.
- The channel capacity $C$ is the maximum of the mutual information $H(X;Y) = H(X) - H(X|Y)$. Assume that the source $X$ corresponds (or nearly corresponds) to this optimal condition.

Refer now to Fig. 13.5 and observe that:

- The typical set of $Y^n$ roughly contains $2^{nH(Y)}$ possible sequences, which represent the most probable output sequences $y$, other outputs having a comparatively small total probability.
- Given an output sequence $y_j$, there exist $2^{nH(X|Y)}$ most likely and equiprobable input sequences $x$ (also called "reasonable causes"), other inputs having comparatively small total probability.
- Given an input sequence $x_i$, there exist $2^{nH(Y|X)}$ most likely and equiprobable output sequences $y$ (also called "reasonable effects"), other outputs having comparatively small total probability.

Assume next that the originator is using the channel at an information (or code) rate $R < C$ per unit time, i.e., $R$ payload bits are generated per second, but the rate is strictly less than $C$ bits per second. Thus $nR$ payload bits are generated in the duration of each message sequence of length $n$ bits. To encode the payload information into message sequences, the originator chooses to use only the sequences belonging to the typical set of $X$. Therefore, $2^{nR}$ coded message sequences (or codewords) are randomly chosen from the set of $2^{nH(X)}$ typical sequences, and with a uniform probability. Accordingly, the probability that a given typical sequence $x_i$ will be selected for transmitting the coded message is:

$$p(x_i) = \frac{2^{nR}}{2^{nH(X)}} = 2^{n[R-H(X)]}. \tag{13.28}$$

[13] C. E. Shannon, A mathematical theory of communication. *Bell Syst. Tech. J.*, **27** (1948), 379–423, 623–56, http://cm.bell-labs.com/cm/ms/what/shannonday/shannon1948.pdf.

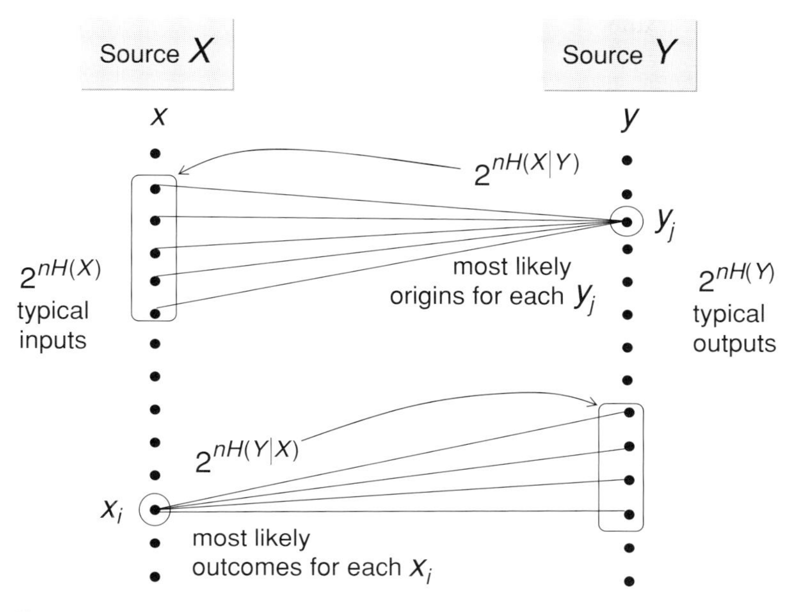


**Figure 13.5** Noisy communication channel relating input source $X$ (sequences $x$) and output source $Y$ (sequences $y$), showing the numbers of most likely inputs $x$ (typical set of $X$), of most likely outputs $y$ (typical set of $Y$), of most likely origins for given output $y_j$, and of most likely outcomes for given input $x_i$.

Assume next that a particular output sequence $y_j$ is received. What is the *no-error* probability, $\tilde{p}$, that $y_j$ exclusively corresponds to the message $x_i$? As we have seen, for any given output message $y_j$ there exist $\alpha = 2^{nH(X|Y)}$ most likely origins $x_i$. On the other hand, the probability that the originator message is *not* $x_i$ is $\bar{p} = 1 - p(x_i)$. Thus, we obtain from Eq. (13.28) the no-error probability:

$$\tilde{p} = \bar{p}^{\alpha} = [1 - p(x_i)]^{\alpha} = \{1 - 2^{n[R - H(X)]}\}^{2^{nH(X|Y)}}. \tag{13.29}$$

Since we assumed $R < C = H(X) - H(X|Y)$, we have $R - H(X) < -H(X|Y)$ or $R - H(X) = -H(X|Y) - \varepsilon$ where $\varepsilon$ is a positive nonzero number ($\varepsilon > 0$). Replacing this relation into Eq. (13.29) yields

$$\tilde{p} = \{1 - 2^{-n[H(X|Y) + \varepsilon]}\}^{2^{nH(X|Y)}}. \tag{13.30}$$

We now take the limit of the probability $\tilde{p}$ for large sequence lengths $n$. We substitute $u = 2^{nH(Y|X)}$ and $v = 2^{-n\varepsilon}$ into Eq. (13.30) and take the limit of large $n$ to obtain:

$$\begin{aligned} \tilde{p} &= \left(1 - \frac{v}{u}\right)^{u} = \mathrm{e}^{u \log(1 - \frac{v}{u})} \\ &\approx \mathrm{e}^{u(-\frac{v}{u})} = \mathrm{e}^{-v} \\ &\approx 1 - v = 1 - 2^{-n\varepsilon}. \end{aligned} \tag{13.31}$$

In the limit $n \to \infty$, we, thus, have $\tilde{p} \to 1$, which means that any output message $y_j$ corresponds exclusively to an input message $x_i$. The *error probability*, $p_e = 1 - \tilde{p}$, becomes identically zero as the sequence length $n$ becomes large, which means that, under the above prerequisites, *any original message can be transmitted through a noisy channel with asymptotically 100% accuracy or zero error*.

Concerning binary channels, the above conclusion can be readily reformulated in terms of the "block code" concept introduced in Chapter 11, according to the following:

Given a noisy binary communication channel with capacity $C$, there exists a block code $(n, k)$ of $2^k$ codewords of length $n$, and code rate $R = k/n \leq C$, for which message codewords can be transmitted through the channel with an arbitrary small error $\varepsilon$.

The above constitutes the formal demonstration of the CCT, as in accordance with the original Shannon paper. Having gone through this preliminary, but somewhat mathematically abstract, exercise, we can now revisit the CCT through simpler approaches, which are more immediately intuitive or concrete. The following, which includes three different approaches is inspired from reference.[14]

## Approach 1

Recall the example of the "noisy typewriter" channel described in Chapter 12 and illustrated in Fig. 12.4. The noisy typewriter is a channel translating single-character messages from keyboard to paper. But any input character $x_i$ (say, J) from the keyboard results in a printed character $y_j$, which has equal chances of being one of three possible characters (say, I, J, K). As we have seen earlier (Section 13.1, Example 3), it is possible to extract a subset of triplets from all the 26 possible output messages. If we add the space as a 27th character, there exist exactly $27/3 = 9$ such triplets. Such triplets are called *nonconfusable* outputs, because they correspond to a unique input, which we call *distinguishable*. The correspondence between this unique input and the unique output triplet is called a *fan*. Thus in the most general case, *inputs are said to be distinguishable if their fans are nonconfusable or do not overlap*.

We shall now use the concept of distinguishable inputs to revisit the CCT. Figure 13.5 illustrates the correspondence between the typical set of input sequences ($2^{nH(X)}$ elements) and the typical set of output sequences ($2^{nH(Y)}$ elements). It is seen that a fan of $2^{nH(Y|X)}$ *most likely* (and equiprobable) output sequences of the typical set of $Y^n$ corresponds to any input sequence $x_i$ of the typical set of $X^n$. Define $M$ as the number of nonconfusable fans, i.e., the fans that are attached to distinguishable input sequences. The number of output sequences corresponding to the $M$ distinguishable inputs is, thus, $M2^{nH(Y|X)}$. In the ideal case, where all inputs are distinguishable (or all fans are nonconfusable), we have $M2^{nH(Y|X)} = 2^{nH(Y)}$. If some of the fans overlap, we have $M2^{nH(Y|X)} < 2^{nH(Y)}$. The most general case, thus, corresponds to the condition

$$M2^{nH(Y|X)} \leq 2^{H(Y)}, \tag{13.32}$$

[14] A. A. Bruen and M. A. Forcinito, *Cryptography, Information Theory and Error-Correction* (New York: John Wiley & Sons, 2005), Ch. 12.

or, equivalently,

$$\begin{aligned} M \leq\ & 2^{nH(Y)-nH(Y|X)} = 2^{n[H(Y)-H(Y|X)]} \\ & \leftrightarrow \\ M \leq\ & 2^{nH(X;Y)} \leq 2^{nC}, \end{aligned} \tag{13.33}$$

where $C = \max H(X;Y)$ is the binary-channel capacity ($nC$ being the capacity of the binary channel's $n$th extension, i.e., using messages of $n$-bit length). If one only chooses distinguishable input sequences to transmit messages at the information or code rate $R$, one has $M = 2^{nR}$ possible messages, which gives $R = (\log M)/n$. Replacing this definition into Eq. (13.33) yields the condition for the channel rate:

$$R \leq C. \tag{13.34}$$

The fundamental property expressed in Eq. (13.34) is that *the channel capacity represents the maximum rate at which input messages can be made uniquely distinguishable*, corresponding to accurate transmission. We note here that the term "accurate" does not mean 100% error free. This is because the $M2^{nH(Y|X)}$ nonconfusable fans of the $M$ distinguishable input messages are only *most likely*. But by suitably increasing the message sequence length $n$, one can make such likelihood arbitrarily high, and thus the probability of error can be made arbitrarily small.

While this approach does not constitute a formal demonstration of the CCT (unlike the analysis that preceded), it provides a more intuitive and simple description thereof.

## Approach 2

This is similar to the first approach, but this time we look at the channel from the output end. The question we ask is: what is the condition for any *output* sequence to be distinguishable, i.e., to correspond to a unique (nonconfusable) fan of input sequences? Let us go through the same reasoning as in the previous approach. Referring again to Fig. 13.5, it is seen that a fan of $2^{nH(X|Y)}$ most likely (and equiprobable) input sequences of the typical set of $X^n$ corresponds to any output sequence $y_j$ of the typical set of $Y^n$. Define $M$ as the number of *nonconfusable input fans*, i.e., the fans that are attached to distinguishable output sequences. The number of input sequences corresponding to the $M$ distinguishable outputs is, thus, $M2^{nH(X|Y)}$. Since there are $2^{nH(X)}$ elements in the typical set $X^n$, we have $M2^{nH(X|Y)} \leq 2^{nH(X)}$ and, hence,

$$\begin{aligned} M &\leq 2^{n[H(X)-nH(X|Y)]} \\ &= 2^{nH(Y;X)} = 2^{nH(X;Y)} \leq 2^{nC}, \end{aligned} \tag{13.35}$$

where we used the property of mutual information $H(Y;X) = H(X;Y)$. The result in Eq. (13.35) leads to the same condition expressed in Eq. (13.34), i.e., $R \leq C$, which defines the channel capacity as the maximum rate for accurate transmission.

### Approach 3

The question we ask here is: can one construct a code for which the transmission error can be made arbitrarily small? We show that there is at least one such code. Assume a source $X$ for which the channel capacity is achieved, i.e., $H(X;Y) = C$, and a channel rate $R \leq C$. We can define a code with a set of $2^{nR}$ codewords ($2^{nR} \leq 2^{nC}$), chosen successively at random from the typical set of $2^{nH(X)}$ sequences, with a uniform probability. Assume that the codeword $x_i$ is input to the channel and the output sequence $y$ is received. There is a possibility of transmission error if at least one other input sequence $x_{k \neq i}$ can result in the same output sequence $y$. The error probability $p_e \equiv p(x \neq x_i)$ of the event $x \neq x_i$ occurring satisfies:

$$
\begin{aligned}
p_e &= p(x_1 \neq x_i \text{ or } x_2 \neq x_i \text{ or } \ldots x_{2^{nR}} \neq x_i) \\
&\leq p(x_1 \neq x_i) + p(x_2 \neq x_i) + \cdots p(x_{2^{nR}} \neq x_i).
\end{aligned}
\tag{13.36}
$$

Specifying the right-hand side, we obtain:

$$
p_e \leq (2^{nR} - 1)\frac{2^{nH(X|Y)}}{2^{nH(X)}}. \tag{13.37}
$$

This result is justified as follows: (a) there exist $2^{nR} - 1$ sequences $x_{k \neq i}$ in the codeword set that are different from $x_i$, and (b) the fraction of input sequences, which output in $y$ is $2^{nH(X|Y)}/2^{nH(X)}$. It follows from Eq. (13.37) that

$$
\begin{aligned}
p &< 2^{nR}\frac{2^{nH(X|Y)}}{2^{nH(X)}} \\
&= 2^{nR}2^{n[H(X|Y)-H(X)]} = 2^{-n(C-R)},
\end{aligned}
\tag{13.38}
$$

which shows that since $R < C$ the probability of error can be made arbitrarily small with code sequences of suitably long lengths $n$, which is another statement of the CCT.

The formal demonstration from the original Shannon paper, completed with the three above approaches, establishes the existence of codes yielding error-free transmission, with arbitrary accuracy, under the sufficient condition $R \leq C$. What about the proof of the converse? Such a proof would establish that error-free transmission codes make $R \leq C$ a necessary condition. The converse proof can be derived in two ways.

The first and easier way, which is not complete, consists of assuming an absolute "zero error" code, and showing that this assumption leads to the necessary condition $R \leq C$. Recall that the originator source has $2^{nR}$ possible message codewords of $n$-bit length. With a zero-error code, any output bit $y_i$ of a received sequence must correspond exclusively to an input message codeword bit $x_i$. This means that the knowledge of $y_i$ is equivalent to that of $x_i$, leading to the conditional probability $p(x_i|y_i) = 1$, and to the conditional channel entropy $H(X|Y) = 0$. As we have seen, the entropy of the $n$-bit message codeword source, $X^n$, is $H(X^n) = nH(X)$. We can also assume that the message codewords are chosen at random, with a uniform probability distribution $p = 2^{-nR}$, which leads to $H(X^n) = nR$ and, thus, $H(X) = R$. We now substitute the above results into the definition of mutual information $H(X;Y)$ to obtain:

$$\begin{aligned} H(X;Y) &= H(X) - H(Y|X) \leftrightarrow \\ H(X) &= H(X;Y) + H(Y|X) \leftrightarrow \\ R &= H(X;Y) + 0 \leq \max H(X;Y) = C, \end{aligned} \tag{13.39}$$

which indeed proves the necessary condition $R \leq C$. However, this demonstration does not constitute the actual converse proof of the CCT, since the theorem concerns codes that are only asymptotically zero-error codes, as the block length $n$ is increased indefinitely.

The second, and formally complete way of demonstrating the converse proof of the CCT is trickier, as it requires one to establish and use two more properties. The whole point may sound, therefore, overly academic, and it may be skipped, taking for granted that the converse proof exists. But the more demanding or curious reader might want to go the extra mile. To this intent, the demonstration of this converse proof is found in Appendix L.

We shall next look at the practical interpretation of the CCT in the case of noisy, symmetric binary channels. The corresponding channel capacity as given by Eq. (13.5), is $C = 1 - f(\varepsilon) = 1 + \varepsilon \log \varepsilon + (1 - \varepsilon)\log(1 - \varepsilon)$, where $\varepsilon$ is the noise parameter or, equivalently, the bit error rate (BER). If we discard the cases of the ideal or noiseless channel ($\varepsilon = f(\varepsilon) = 0$) and the useless channel ($\varepsilon = 0.5$, $f(\varepsilon) = 1$), we have $0 < |f(\varepsilon)| < 1$ and, hence, the capacity $C < 1$, or strictly less than one bit per channel use. According to the CCT, the code rate for error-free transmission must satisfy $R < C$, here $R < C < 1$. Thus block codes $(n, n)$ of any length $n$, for which $R = n/n = 1$ are not eligible. Block codes $(n, k)$ of any length $n$, for which $R = k/n > C$ are not eligible either. The eligible block codes must satisfy $R = k/n < C$. To provide an example, assume for instance $\varepsilon = 0.01$, corresponding to $\text{BER} = 1 \times 10^{-3}$. We obtain $C = 0.9985$, hence the CCT condition is $R = k/n < 0.9985\ldots$ Consider now the following possibilities:

First, according to Chapter 11, the smallest linear block code, $(7, 4)$, has a length $n = 7$, and a rate $R = 4/7 = 0.57$. We note that although this code is consistent with the condition $R < C$, it is a poor choice, since we must use $n - k = 3$ parity bits for $k = 4$ payload bits, which represents a heavy price for a bit-safe communication. But let us consider this example for the sake of illustration. As a Hamming code of distance $d_{\min} = 3$, we have learnt that it has the capability of correcting no more than $(d_{\min} - 1)/2 = 1$ bit errors out of a seven-bit block. If a single error occurs in a block sequence of 1001 bits (143 blocks), corresponding to $\text{BER} \approx 1 \times 10^{-3}$, this error will be absolutely corrected, with a resulting BER of 0. If, in a block sequence of 2002 bits, exactly two errors occur within the same block, then only one error will be corrected, and the corrected BER will be reduced to $\text{BER} \approx 0.5 \times 10^{-3} = 5 \times 10^{-4}$. Since the condition $\text{BER} \approx 0$ is not achieved when there is more than one error, this code is, therefore, not optimal in view of the CCT, despite the fact that it satisfies the condition $R < C$. Note that the actual corrected error rate (which takes into account all error-event possibilities) is $\text{BER} < 4 \times 10^{-6}$, which is significantly smaller yet not identical to zero, and is left as an exercise to show.

Second, the CCT points to the existence of better codes $(n, nR)$ with $R < 0.9985\ldots$, of which (255, 231), with $R = 231/255 = 0.905 < C$, for instance, is an eligible one. From Chapter 11, it is recognized as the Reed–Solomon code $RS(255, 231) = (n = 2^m - 1, k = n - 2t)$ with $m = 8$, $t = 12$, and code rate $R = 231/255 = 0.905 < C$. As we have seen, this RS code is capable of correcting up to $t = 12$ bit errors per 255-bit block, yielding, in this case, absolute error correction. Yet the finite possibility of getting 13 errors or more within a single block mathematically excludes the possibility of achieving BER = 0. The probability $p(e > 12)$ of getting more than 12 bit errors in a single block is given by the formula

$$\begin{aligned} p(e > 12) &= \sum_{i-13}^{255} C_{255}^{i} \varepsilon^{i} (1 - \varepsilon)^{255-i} \\ &= p(13) + p(14) + \cdots + p(255), \end{aligned} \tag{13.40}$$

with (as assumed here) $\varepsilon = 10^{-3}$. Retaining only the first term $p(13)$ of the expansion (13 errors exactly), we obtain:

$$\begin{aligned} p(13) &= C_{255}^{13} \varepsilon^{13} (1 - \varepsilon)^{255-13} \\ &= \frac{255!}{13!\,242!} 10^{-13 \times 3} (1 - 10^{-3})^{242}, \end{aligned} \tag{13.41}$$

which, after straightforward computation, yields $p(13) \approx 1.8 \times 10^{-18}$. Because of the rapid decay of other higher-order probabilities, it is reasonable to assume, without any tedious proof, that $p(e > 12) < 1.0 \times 10^{-18}$, corresponding to an average single-bit error within one billion billion bits, which means that this RS code, with rate $R = 0.905 < 0.9985 = C$, is pretty much adequate for any realistic applications. Yet there surely exist better codes with rates close to the capacity limit, as we know with error-correction capabilities arbitrarily close to the $BER = 0$ absolute limit.

To conclude this chapter, the CCT is a powerful theorem according to which we know the existence of coding schemes enabling one to send messages through a noisy channel with arbitrarily small transmission errors. But it is important to note that the CCT does not provide any indication or clue as to how such codes may look like or should even be designed! The example code that we have used as a proof of the CCT consists in randomly choosing $2^{nR}$ codewords from the input typical set of size $2^{nH(X)}$. As we have seen, such a code is asymptotically optimal with increasing code lengths $n$, since it provides exponentially decreasing transmission errors. However, the code is impractical, since the corresponding look-up table $(y_j \leftrightarrow x_i)$, which must be used by the decoder, increases in size exponentially. The task of information theorists is, therefore, to find more practical coding schemes. Since the inception of the CCT, several families of codes capable of yielding suitably low transmission errors have been developed, but their asymptotic rates $(R = (\log M)/n)$ still do not approach the capacity limit $(C)$.

## 13.4 Exercises

**13.1** (M): Consider the binary erasure channel with input and output PDF,

$$p(X) = [p(x_1), p(x_2)],$$
$$P(Y) = [p(y_1), p(\phi), p(y_2)],$$

and the transition matrix,

$$P(Y|X) = \begin{pmatrix} 1-\varepsilon & 0 \\ \varepsilon & \varepsilon \\ 0 & 1-\varepsilon \end{pmatrix}.$$

Demonstrate the following three relations:

$$H(Y) = f(\varepsilon) + (1-\varepsilon)f(q),$$
$$H(Y|X) = f(\varepsilon),$$
$$H(X;Y) = H(Y) - H(Y|X) = (1-\varepsilon)f(q),$$

where $q = p(x_1)$ and the function $f(u)$ is defined as $(0 \leq u \leq 1)$:

$$f(u) = -u \log u - (1-u)\log(1-u).$$

**13.2** (M): Determine the bit capacity of a four-input, four-output quaternary channel with quaternary sources $X = Y = \{0, 1, 2, 3\}$, as defined by the following conditions on conditional probabilities:

$$p(y|x) = \begin{cases} 0.5 & \text{if } y = x \pm 1 \bmod 4 \\ 0 & \text{otherwise.} \end{cases}$$

**13.3** (T): Determine the capacity of the four-input, eight-output channel characterized by the following transition matrix:

$$P(Y|X) = \frac{1}{4}\begin{pmatrix} 1 & 0 & 0 & 1 \\ 1 & 0 & 0 & 1 \\ 1 & 1 & 0 & 0 \\ 1 & 1 & 0 & 0 \\ 0 & 1 & 1 & 0 \\ 0 & 1 & 1 & 0 \\ 0 & 0 & 1 & 1 \\ 0 & 0 & 1 & 1 \end{pmatrix}.$$

**13.4** (M): Assuming an uncorrected bit-error-rate of BER $\approx 1 \times 10^{-3}$, determine the corrected BER resulting from implementing the Hamming code (7, 4) in a binary channel. Is this code optimal in view of the channel coding theorem?

**13.5** (B): Evaluate an upper bound for the corrected bit-error-rate when using the Reed–Solomon code $RS(255, 231)$ over a binary channel with noise parameter $\varepsilon = 10^{-3}$ (*clue*: see text).

# 14 Gaussian channel and Shannon–Hartley theorem

This chapter considers the continuous-channel case represented by the *Gaussian channel*, namely, a continuous communication channel with Gaussian additive noise. This will lead to a fundamental application of Shannon's coding theorem, referred to as the *Shannon–Hartley theorem* (SHT), another famous result of information theory, which also credits the earlier 1920 contribution of *Ralph Hartley*, who derived what remained known as the *Hartley's law* of communication channels.[1] This theorem relates channel capacity to the signal and noise powers, in a most elegant and simple formula. As a recent and little-noticed development in this field, I will describe the *nonlinear channel*, where the noise is also a function of the transmitted signal power, owing to channel nonlinearities (an exclusive feature of certain physical transmission pipes, such as optical fibers). As we shall see, the modified SHT accounting for nonlinearity represents a major conceptual progress in information theory and its applications to optical communications, although its existence and consequences have, so far, been overlooked in textbooks. This chapter completes our description of *classical* information theory, as resting on Shannon's works and founding theorems. Upon completion, we will then be equipped to approach the field of *quantum information theory*, which represents the second part of this series of chapters.

## 14.1 Gaussian channel

Referring to Chapter 6, a *continuous communications channel* assumes a continuous originator source, $X$, whose symbol alphabet $x_1, \ldots, x_i$ can be viewed as representing time samples of a continuous, real variable $x$, which is associated with a continuous *probability distribution function* or PDF, $p(x)$. The variable $x$ can be conceived as representing the *amplitude* of some signal waveform (e.g., electromagnetic, electrical, optical, radio, acoustical). This leads one to introduce a new parameter, which is the *signal power*. This is the power associated with the physical waveform used to propagate the symbols through a transmission pipe. In any physical communication channel, the signal power, or more specifically the *average signal power* $P$, represents a practical

[1] Hartley established that, given a peak voltage $S$ and accuracy (noise) $\Delta S = N$, the number of distinguishable pulses is $m = 1 + S/N$. Taking the logarithm of this result provides a measure of maximum available information: this is Hartley's law. See: http://en.wikipedia.org/wiki/Shannon%E2%80%93Hartley_theorem.

constraint, which (as we shall see) must be taken into account to evaluate the channel capacity.

As a well-known property, the power of a signal waveform with amplitude $x$ is proportional to the square of the amplitude $x^2$. Overlooking the proportionality constant, the corresponding average power (as averaged over all possible symbols) is, thus, given by

$$P = \int_X x^2 p(x) \mathrm{d}x = \langle x^2 \rangle_X. \tag{14.1}$$

Chapter 6 described the concept and properties of *differential entropy*, which is the entropy $H(X)$ of a continuous source $X$. It is defined by an integral instead of a discrete sum as in the discrete channel, according to:

$$H(X) = -\int_X p(x) \log p(x) \mathrm{d}x \equiv \langle \log p(x) \rangle_X. \tag{14.2}$$

As a key result from this chapter, it was established that under the average-power constraint $P$, the Gaussian PDF is the one (and the only one) for which the source entropy $H(X)$ is maximal. No other continuous source with the same average power provides greater entropy, or average symbol uncertainty, or mean information contents. In particular, it was shown that this maximum source entropy, $H_{\max}(X)$, reduces to the simple closed-form expression:

$$H_{\max}(X) = \log \sqrt{2\pi \mathrm{e} \sigma^2}, \tag{14.3}$$

where $\sigma^2$ is the variance of the source PDF.

By sampling the signal waveform, the originating continuous source is actually transformed into a discrete source. Thus the definition of *discrete* entropy must be used, instead of the differential or continuous definition in Eq. (14.2), but the result turns out to be strictly identical.[2] In the following, we will, therefore, consider the Gaussian

[2] As can be shown through the following. Assume the Gaussian PDF for the source $X$

$$p(x) = \frac{1}{\sigma_{\text{in}}\sqrt{2\pi}} \exp\left(-\frac{x^2}{2\sigma_{\text{in}}^2}\right),$$

where $x$ is the symbol-waveform amplitude, and $\sigma_{\text{in}}^2 = \langle x^2 \rangle = P_{\text{S}}$ is the symbol variance, with $P_{\text{S}}$ being the average symbol or waveform power. By definition, the source entropy corresponding to discrete samples $x_i$ is

$$H(X) = -\sum_i p(x_i) \log_2 p(x_i),$$

which can be developed into

$$\begin{aligned} H(X) &= -\sum_i \frac{1}{\sigma_{\text{in}}\sqrt{2\pi}} \exp\left(-\frac{x_i^2}{2\sigma_{\text{in}}^2}\right)\left[\log_2\left(\frac{1}{\sqrt{\sigma_{\text{in}}^2 2\pi}}\right) - \frac{x_i^2}{2\sigma_{\text{in}}^2}\log_2(\mathrm{e})\right] \\ &\equiv \frac{1}{2}\log_2\left(2\pi\sigma_{\text{in}}^2\right)\sum_i p(x_i) + \frac{1}{2\sigma_{\text{in}}^2}\log_2(\mathrm{e})\sum_i x_i^2 p(x_i). \end{aligned}$$

The samples $x_i$ can be chosen sufficiently close for the following approximations to be valid: $\sum_i p(x_i) = 1$, and $\sum_i x_i^2 p(x_i) = \langle x^2 \rangle = \sigma_{\text{in}}^2$. Finally, we obtain

$$H(X) = \frac{1}{2}\log_2\left(2\pi\sigma_{\text{in}}^2\right) + \frac{1}{2\sigma_{\text{in}}^2}\log_2(\mathrm{e}) = \frac{1}{2}\log_2\left(2\pi \mathrm{e}\sigma_{\text{in}}^2\right) = \log\sqrt{2\pi \mathrm{e}\sigma_{\text{in}}^2}.$$

channel to be discrete, which allows for considerable simplification in the entropy computation. However, I shall develop in footnotes or as exercises the same computations while using continuous-channel or differential entropy definitions, and show that the results are strictly identical. The conclusion from this observation is that there is no need to assume that the Gaussian channel uses a discrete alphabet or samples of a continuous source.

The *Gaussian channel* is defined as a (discrete or continuous) communication channel that uses a Gaussian-alphabet source as the input, and has an intrinsic *additive noise*, also characterized by a noise source $Z$ with Gaussian PDF. The channel noise is said to be *additive*, because given the input symbol $x$ (or $x_i$, as sampled at time $i$), the output symbol $y$ (or $y_i$) is given by the sum

$$y = x + z,$$

or

$$y_i = x_i + z_i, \tag{14.4}$$

where $z$ (or $z_i$) is the amplitude of the channel noise.

Further, it is assumed that:

(a) The average noise amplitude is zero, i.e., $\langle z \rangle = 0$;
(b) The noise power is finite, with variance $\sigma_{\text{ch}}^2 = \langle z^2 \rangle = N$;
(c) There exists no correlation between the noise and the input signal.

These three conditions make it possible to calculate the average output power $\sigma_{\text{out}}^2$ of the Gaussian channel, as follows:

$$\begin{aligned} \sigma_{\text{out}}^2 &= \langle y^2 \rangle = \langle (x+z)^2 \rangle = \langle x^2 + 2xz + z^2 \rangle \\ &= \langle x^2 \rangle + 2\langle x \rangle \langle z \rangle + \langle z^2 \rangle \\ &= \sigma_{\text{in}}^2 + \sigma_{\text{ch}}^2 \equiv P + N, \end{aligned} \tag{14.5}$$

where $\sigma_{\text{in}}^2 = \langle x^2 \rangle = P$ is the (average) input signal power. Consistently with the definition in Eq. (14.3), the entropies of output source $Y$ and the noise source $Z$ are given by:

$$H(Y) = \log_2\left(\sqrt{2\pi e \sigma_{\text{out}}^2}\right) \tag{14.6}$$

and

$$H(Z) = \log_2\left(\sqrt{2\pi e \sigma_{\text{ch}}^2}\right), \tag{14.7}$$

respectively.

The *mutual information* of the Gaussian channel is defined as:

$$H(X;Y) = H(X) - H(X|Y) = H(Y) - H(Y|X), \tag{14.8}$$

where $H(Y|X)$ is the *equivocation or conditional entropy*. Given a Gaussian-distributed variable $y = x + z$, where $x, z$ are independent, the conditional probability of measuring

$y$ at the output given that $x$ is the input is $p(y|x) = p(z)$. Substituting this property in the definition of conditional entropy $H(Y|X)$, we obtain:

$$\begin{aligned} H(Y|X) &= -\sum_j \sum_i p(x_i, y_i) \log p(y_j|x_i) \\ &= -\sum_j \sum_i p(y_j) p(y_j|x_i) \log p(y_j|x_i) \\ &= -\sum_j \sum_i p(y_j) p(z_i) \log p(z_i) \\ &= -\sum_j p(y_j) \sum_i p(z_i) \log p(z_i) \\ &= H(Z) \\ &= \log \sqrt{2\pi e \sigma_{\text{ch}}^2}. \end{aligned} \tag{14.9}$$

As previously stated, the same result as in Eq. (14.9) can be obtained using the differential or continuous definition of conditional entropy, i.e., to recall from Chapter 6:

$$H(Y|X) = -\iint p(y|x) p(x) \log p(y|x) \mathrm{d}x \mathrm{d}y. \tag{14.10}$$

The computation of the double integral in Eq. (14.10) is elementary, albeit relatively tedious to carry out, so we shall leave it here as a "math" exercise.

We consider next the *capacity* of the Gaussian channel. By definition, the channel capacity $C$ is given by the maximum of the mutual information $H(X; Y) = H(Y) - H(Y|X)$. Using the previous results in Eqs. (14.6) and (14.7) for $H(Y)$ and $H(Y|X) = H(Z)$, we obtain:

$$\begin{aligned} H(X; Y) &= H(Y) - H(Y|X) = H(Y) - H(Z) \\ &= \frac{1}{2} \log \left(\pi e \sigma_{\text{out}}^2\right) - \frac{1}{2} \log \left(2\pi e \sigma_{\text{ch}}^2\right) \\ &= \frac{1}{2} \log_2 \left(\frac{\sigma_{\text{out}}^2}{\sigma_{\text{ch}}^2}\right) = \frac{1}{2} \log_2 \left(\frac{\sigma_{\text{in}}^2 + \sigma_{\text{ch}}^2}{\sigma_{\text{ch}}^2}\right), \end{aligned} \tag{14.11}$$

and using $\sigma_{\text{out}}^2 = \sigma_{\text{in}}^2 + \sigma_{\text{ch}}^2$, $\sigma_{\text{in}}^2 = P$, $\sigma_{\text{ch}}^2 = N$:

$$H(X; Y) = \frac{1}{2} \log_2 \left(1 + \frac{P}{N}\right). \tag{14.12}$$

The mutual information $H(X; Y)$ defined in Eq. (14.12) also corresponds to the channel capacity $C$. This is because (given the signal power constraint $P$) the Gaussian distribution maximizes the source entropies $H(X)$ or $H(Y)$, making $H(X; Y)$ an upper bound for mutual information. Thus, we can write, equivalently:

$$C = \frac{1}{2} \log_2 \left(1 + \frac{P}{N}\right). \tag{14.13}$$

In the above, the quantity $P/N$ is called the *signal-to-noise ratio*, or SNR. The fundamental result in Eq. (14.13), which establishes that the capacity of a "noisy channel" is proportional to $\log(1 + \text{SNR})$ is probably the most well known and encompassing

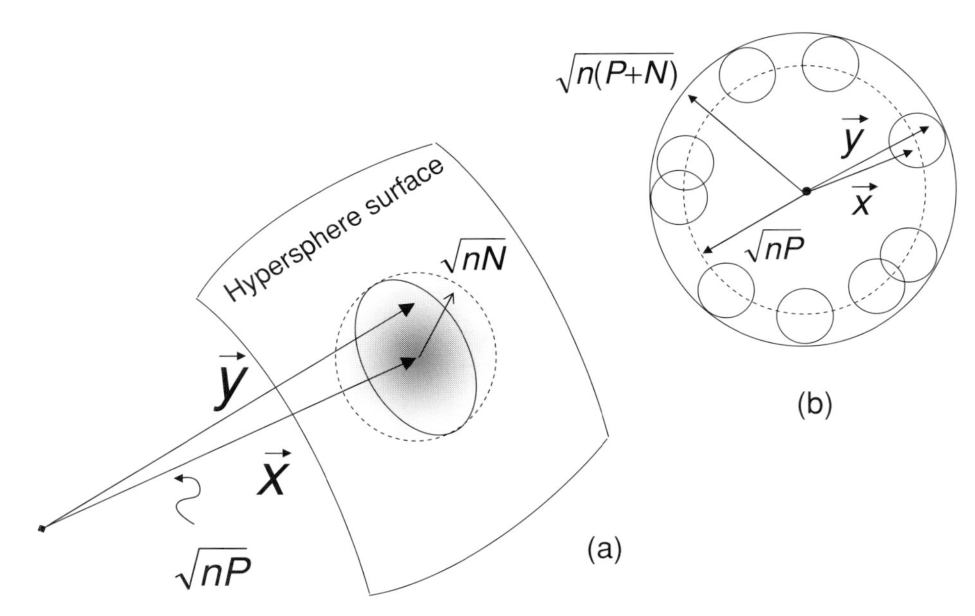


**Figure 14.1** (a) Location of input message $x$, defined by vector $\vec{x}$, and likely location of output message $y$, defined by vector $\vec{y}$, on the surface of a hypersphere of dimension $n$; (b) as (a), but for $n = 2$.

conclusion of all Shannon's information theory. Given the SNR characteristics of a communication channel with noise and signal power limitation, it provides an upper bound for the code rate at which information can be transmitted with *arbitrarily small errors*. Before expanding on the implications of this result for communication systems, it is worth providing a plausibility argument for the existence of the corresponding codes with low error probability. The argument immediately following can be found with different variants in most IT textbooks.

Consider an input message (or codeword) consisting of a sequence of $n$ discrete waveform samples with random amplitudes $x_i$ of zero mean and mean power $\langle x_i^2 \rangle = P$. This message defines a vector $\vec{x} = (x_1, x_2, \ldots, x_n)$ in an $n$-dimensional space, which is characterized by a length $|\vec{x}| = \sqrt{x_1^2 + x_2^2 + \cdots + x_n^2} = \sqrt{nP}$. Thus, we can view the input message as lying on any point $x = (x_1, x_2, \ldots, x_n)$, which is most probably located close to the surface of a hypersphere centered about the origin and having a radius of $\sqrt{nP}$. Likewise, the received or output message corresponds to an $n$-vector $\vec{y} = (y_1, y_2, \ldots, y_n)$ with $y_i = x_i + z_i$, where $\langle z_i \rangle = 0$ and $\langle z_i^2 \rangle = N$. Thus, the output vector $\vec{y}$ has the length $|\vec{y}| = \sqrt{n(P + N)}$ with deviation $\sqrt{nN}$. The output message is likely to fall on any point $y = (y_1, y_2, \ldots, y_n)$ inside a small hypersphere centered about $x = (x_1, x_2, \ldots, x_n)$ and having a radius of $\sqrt{nN}$, as illustrated in Fig. 14.1.

For any output message $y$ to represent nonconfusable input messages $x$ (i.e., most likely to correspond to a unique input message $x$), the small spheres should be nonintersecting or disjoint. Here comes the key question: how many such disjoint spheres can be fitted inside the hypersphere of radius $\sqrt{n(P + N)}$?

The answer comes as follows. The volume $V_n$ of a hypersphere with dimension $n$ and radius $r$ is given by the formula $V_n = A_n r^n$, with the coefficient $A_n$ defined by:

$$A_n = \frac{\pi^{n/2}}{\Gamma\left(1 + \frac{n}{2}\right)}, \tag{14.14}$$

where $\Gamma(m)$ is the gamma function.[3] The maximum number $M$ of disjoint spheres that can be fitted inside the volume $V_n$ is, thus,

$$M = \frac{A_n\left(\sqrt{n(P+N)}\right)^n}{A_n\left(\sqrt{nN}\right)^n} = \left(1 + \frac{P}{N}\right)^{n/2} \equiv 2^{\frac{n}{2}\log_2\left(1+\frac{P}{N}\right)}. \tag{14.15}$$

The result in Eq. (14.15) shows that there exist $M = 2^{nR}$ possible messages with nonconfusable inputs, corresponding to the *channel* or *code rate*

$$R = \frac{1}{2}\log_2\left(1 + \frac{P}{N}\right). \tag{14.16}$$

Since $R$ is the maximum rate for which the messages have nonconfusable inputs (meaning that receiving and decoding errors can be made arbitrarily small), the value found in Eq. (14.16) represents an upper bound, which corresponds to the *channel capacity*. Thus, the demonstration confirms the fundamental result found in Eq. (14.13).

The unit of channel capacity is *bits* or, more accurately, *bits per channel use*.[4] Then how does channel capacity relate to channel *bandwidth*? We shall now address this question as follows. Define $B$ as the signal bit rate. To recover bits from signal waveforms, a basic engineering principle states that the waveform must be sampled at the *Nyquist sampling rate* of $f = 2B$ per unit time (i.e., twice the bit rate). A continuous communication channel with a capacity $C_{\text{bit}}$, which is used at the rate $f = 2B$, gives us the *capacity per unit time* $C_{\text{bit/s}} = 2BC_{\text{bit}}$.

Substituting the above result into Eq. (14.13), we obtain

$$C_{\text{bit/s}} = B\log_2\left(1 + \frac{P}{N}\right). \tag{14.17}$$

By dividing both sides in Eq. (14.17) by $B$, we obtain another definition of channel capacity, which now has the dimensionless unit of *(bit/s)/Hz*:

$$C_{\text{(bit/s)/Hz}} = \log_2\left(1 + \frac{P}{N}\right). \tag{14.18}$$

The results in either Eq. (14.17) or Eq. (14.18) are known as the *Shannon–Hartley theorem (SHT)*, based on an earlier contribution from Hartley to Shannon's analysis (see previously). The SHT is also referred to as the *Shannon–Hartley law*, the *information*

[3] The gamma function has the property $\Gamma(m+1) = m!$, with $\Gamma(1/2) = \sqrt{\pi}$, $\Gamma(3/2) = \sqrt{\pi}/2$, and $\Gamma(m + 1/2) = \Gamma(1/2)\frac{1.3.5,\ldots,(2m-1)}{2^m}$. It can readily be checked that $V_n = A_n r^n$ with the definition in Eq. (14.14) yields $V_2 = \pi r^2$, $V_3 = (4/3)\pi r^3$, etc.

[4] Channel capacity is conceptually similar to the capacity of transportation means. Thus, if for instance an airplane or a truck can transport 300 passengers or 20 tons, respectively, this corresponds to capacities of 300 passengers or 20 tons per single use of the airplane or of the truck. Using these transportation means several times does not increase the capacity per use, as was also shown in Appendix L.

*capacity theorem*, or the *band-limited capacity theorem*. To recall, SHT or its different appellations relate to Shannon's "second theorem" or "[noisy] channel-capacity theorem." As expressed in Eq. (14.17), the SHT provides the *maximum bit rate achievable* ($C_{\text{bit/s}}$) *given the channel bandwidth, additive noise background, and signal power constraint, for which information can be transmitted with an arbitrarily small BER*. The SHT variant in Eq. (14.18) provides the maximum bit rate achievable *per unit of bandwidth* (Hz), under said conditions. The bit rate normalized to bandwidth is referred to as *information spectral density*, or ISD. The (bit/s)/Hz ISD, is often referred to as *spectral efficiency*. This would be a correct appellation if the ISD was bounded to a maximum of unity, corresponding to 100% efficiency, or 1 (bit/s)/Hz. However, there are a variety of modulation formats, which make it possible to encode more than 1 bit/s in a single Hz of frequency spectrum, as discussed further on. Regardless of the possible existence of such modulation formats, the SHT states that the ISD is actually *unbounded*. In the limit of high signal-to-noise ratios ($\text{SNR} = P/N$), we find $C \approx \log_2(\text{SNR})$. This result means that every twofold increase (or 3 dB, in the decibel scale[5]) of the SNR corresponds to an ISD increase of one additional (bit/s)/Hz. For instance, $\text{SNR} = 2^{10} = 1024 = +30.1\,\text{dB}$ corresponds to an ISD of $\log_2(1 + 2^{10}) \approx 10\,(\text{bit/s})/\text{Hz}$. The SHT, thus, states that given any SNR constraint, there exist some codes (and associated modulation formats) for which *error-free* transmission is possible within an ISD limit of $C_{(\text{bit/s})/\text{Hz}} = \log_2(1 + \text{SNR})$. As with the (discrete) channel coding theorem (Chapter 13), how to find or to design such codes is, however, not specified by the theory.

In the above SHT formulation, it is possible to introduce another important parameter overlooked so far, which is the *noise bandwidth*. Assume a communication channel with a capacity per unit time of $C_{\text{bit/s}}$ and a signal bandwidth $B$. The mean signal power, $P$, and the mean energy per bit, $E_{\text{S}}$, are related through $P = E_{\text{S}} C_{\text{bit/s}}$. The noise power, $N$, and the mean noise energy per bit, $E_{\text{N}}$, are related through $N = E_{\text{N}} B_{\text{Hz}}$. Thus, the SNR can also be expressed in the form

$$\text{SNR} = \frac{P}{N} = \frac{E_{\text{S}} C_{\text{bit/s}}}{E_{\text{N}} B}, \tag{14.19}$$

which explicitly relates the bit SNR to the bandwidth $B$. Replacing the above result in Eq. (14.17) yields

$$C_{\text{bit/s}} = B \log_2 \left(1 + \frac{E_{\text{S}} C_{\text{bit/s}}}{E_{\text{N}} B}\right), \tag{14.20}$$

then

$$C_{\text{bit/s}} = \log_2 \left(1 + \frac{E_{\text{S}}}{E_{\text{N}}} C_{(\text{bit/s})/\text{Hz}}\right). \tag{14.21}$$

[5] For any nonnegative real number $q_{\text{linear}}$, the decibel scale is defined as $q_{\text{dB}} = 10 \log_{10}(q_{\text{linear}})$. For instance, $20\,\text{dB} = 100_{\text{linear}}$, $-10\,\text{dB} = 0.1_{\text{linear}}$, and, in particular, $3.0102\,\text{dB} \approx 3\,\text{dB} = 2_{\text{linear}}$.

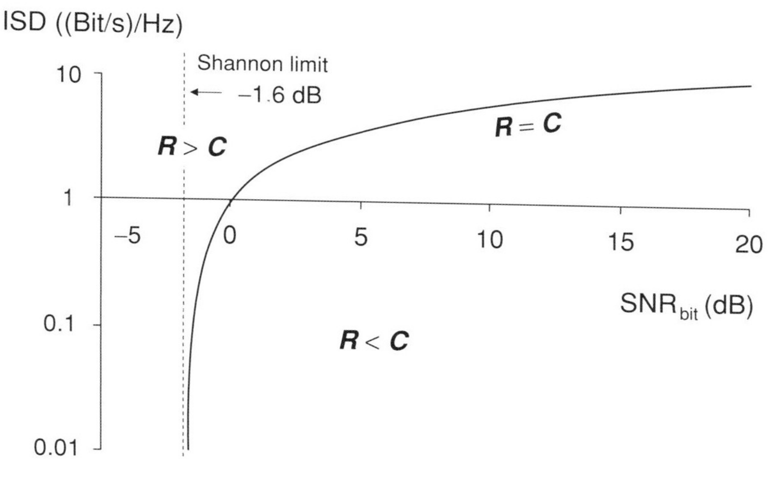


**Figure 14.2** Bandwidth-efficiency diagram plotting information spectral density (ISD) vs. signal-to-noise ratio (bit SNR), according to Eq. (14.22). The two regions $R < C$ and $R > C$, corresponding to bit rate $R$ and channel capacity $C_{\text{bit/s}}$, which are separated by the capacity-boundary curve ($R = C$) are also shown. The Shannon limit corresponds to SNR $= -1.6$ dB.

Base-two exponentiation of both sides of Eq. (14.21), and the introduction of $\text{SNR}_{\text{bit}} = E_{\text{S}}/E_{\text{N}}$ as defining the *bit SNR*,[6] we finally obtain:

$$\text{SNR}_{\text{bit}} = \frac{2^{C_{(\text{bit/s})/\text{Hz}}} - 1}{C_{(\text{bit/s})/\text{Hz}}}. \tag{14.22}$$

This formula now provides an explicit and universal relation between the bit SNR ($\text{SNR}_{\text{bit}}$) and the channel ISD ($C_{(\text{bit/s})/\text{Hz}}$). It is, thus, possible to plot the function $C_{(\text{bit/s})/\text{Hz}} = f(\text{SNR}_{\text{bit}})$, as shown in Fig. 14.2, which is referred to as the *bandwidth-efficiency diagram*.

As observed from Fig. 14.2, the boundary curve $C_{(\text{bit/s/Hz})} = f(\text{SNR}_{\text{bit}})$ defines two regions in the plane. If $R_{\text{bit/s}}$ is the bit rate, these two regions correspond to $R_{\text{bit/s}} < C_{\text{bit/s}}$ and $R_{\text{bit/s}} > C_{\text{bit/s}}$, respectively. The interpretation of the bandwidth-efficiency diagram is then:

- In the region $R_{\text{bit/s}} < C_{\text{bit/s}}$, there exist codes for which transmission can be achieved with arbitrary low error (BER);
- In the region $R_{\text{bit/s}} > C_{\text{bit/s}}$, there is no code making transmission possible at arbitrarily low error.

To avoid any misinterpretation, it is *always* possible to transmit information with some finite or nonzero BER in any location of the diagram. However, only in the lower-right region $R_{\text{bit/s}} < C_{\text{bit/s}}$ of the diagram can the BER be made arbitrarily small with a proper code choice code, leading to arbitrarily error-free transmission.

[6] The ratio $\text{SNR}_{\text{bit}} = E_{\text{S}}/E_{\text{N}}$ is often referred to in textbooks as $E_{\text{b}}/N_0$, where $E_{\text{b}} = E_{\text{S}}$ is the bit energy and $N_0 = E_{\text{N}}$ is the noise energy or noise spectral density.

We also observe from Fig. 14.2 that the boundary curve $C_{(\text{bit/s})/\text{Hz}} = f(\text{SNR}_{\text{bit}})$ has an asymptotic limit at $\text{SNR}_{\text{bit}} = -1.6\,\text{dB}$, which is referred to as the *Shannon limit*. As the figure indicates, when the SNR approaches this lower bound, the ISD rapidly vanishes to become asymptotically zero as $\text{SNR}_{\text{bit}} \to -1.6\,\text{dB}$. On the right-hand side of the boundary curve, there exist some codes for which error-free transmission is possible. But as observed from the figure, the closer one approaches the Shannon-limit SNR, the lower the ISD and the smaller the number of bits that can be effectively transmitted per unit time and unit bandwidth. The limiting condition $\text{SNR}_{\text{bit}} = -1.6\,\text{dB}$ corresponds to a *useless* channel with an ISD of 0 (bit/s)/Hz.

The bandwidth-efficiency diagram suggests two strategies for optimizing the transmission of information:

- Horizontally, or at fixed ISD, by increasing the bit-SNR to reduce the BER;
- Vertically, or at fixed SNR, to increase the ISD and, hence, to increase the bit rate.

At this point, we, thus, need to address two issues: (a) the relation between the waveform modulation format and the ISD, and (b) for a given modulation format, the relation between SNR and BER (or *symbol error rate*, SER, for multi-level formats). With such knowledge, we will then be able to fill out the diagram with different families of points, i.e., to plot *iso*-format or *iso*-SER/BER curves, and observe the corresponding tradeoffs.

An overview of the different types of modulation formats is beyond the scope of this book. Here, we shall use a limited selection of modulation formats and their ISD, SNR, or SER properties to provide illustrative cases of bandwidth-efficiency tradeoffs. The following assumes some familiarity with telecommunications signaling (i.e., intensity, frequency, phase, amplitude, and multi-level waveform modulation).[7]

## Nonreturn-to-zero (NRZ) format

The *nonreturn-to-zero* (NRZ) format, also called *on–off keying* (OOK), corresponds to the most basic and common modulation scheme. Bit encoding consists of turning the signal power on or off during each bit period, to represent 1 or 0, respectively. Its ISD is intrinsically limited to 1 (bit/s)/Hz[8] and its BER is given by the generic formula:

$$\text{BER} = \frac{1}{2}\text{erfc}\left(\frac{1}{2}\sqrt{\frac{\text{SNR}_{\text{bit}}}{2}}\right), \tag{14.23}$$

where $\text{erf}(x)$ is the *complementary error function*.[9]

[7] S. Haykin, *Digital Communications* (New York: J. Wiley & Sons, 1988), J. G. Proakis, *Digital Communications*, 4th edn. (New York: McGraw Hill, 2001).

[8] Actually, the 1 (bit/s)/Hz ISD for NRZ/OOK signals represents the asymptotic limit of channel capacity, which is rapidly reached for $\text{SNR}_{\text{bit}} \geq 5\,\text{dB}$. The demonstration of such a result is left as an exercise.

[9] By definition, $\text{erfc}(x) = 1 - \text{erf}(x)$ with $\text{erf}(x) = \frac{2}{\sqrt{\pi}}\int_0^x e^{-y^2}dy$ being the error function. The latter can be approximated through the expansion: $\text{erf}(x) = 1 - (a_1 t + a_2 t^2 + a_3 t^3 + a_4 t^4 + a_5 t^5)\exp(-x^2)$, with $t = 1/(1+px)$, $p = 0.3275$, $a_1 = 0.2548$, $a_2 = -0.2844$, $a_3 = 1.4214$, $a_4 = -1.4531$ and $a_5 = 1.0614$. Consistently with the exact definition, in the limit $x \to 0$ we have $\text{erf}(x) = 0$. It is customary to define $Q = \frac{1}{2}\sqrt{\text{SNR}_{\text{bit}}}$, hence from Eq. (14.23), $\text{BER} = \text{erfc}(Q/\sqrt{2})/2$ with $\text{erfc}(x) = 1 - \text{erf}(x)$. For $Q > 2$,

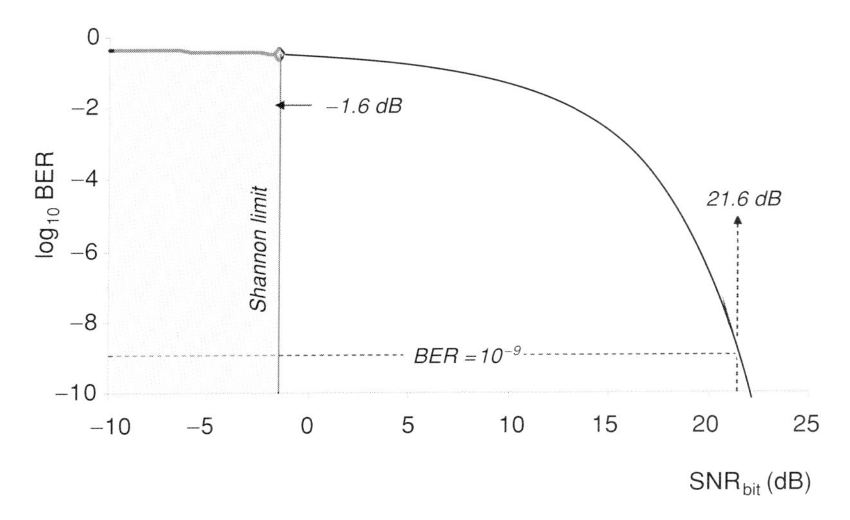


**Figure 14.3** Bit error rate (BER) as a function of bit signal-to-noise ratio ($SNR_{bit}$) for NRZ- or OOK-modulated signals.

Figure 14.3 shows the plot $BER = f(SNR_{bit})$, along with the Shannon limit. It is seen from the figure that the specific value $SNR_{bit} = +21.6$ dB yields a BER of $10^{-9}$, corresponding to one bit error (on average) for every one billion bits transmitted. For a long time in the modern history of telecommunications, such a BER has been considered to represent the standard for "error-free" transmission. In the limit of vanishing signals, or $SNR_{bit} \to 0$ or $SNR_{bit}(dB) \to -\infty$, the erf function vanishes and we have $BER = 1/2$. Since the BER is a probability, this result means that the bits 0 and 1 have equal probabilities of being detected, regardless of the input message sequence, which corresponds to a useless communication channel.

The shaded region in Fig. 14.3 defines the impossibility of finding *error-correction* codes (ECC) able to improve the BER at a given $SNR_{bit}$ value, while right-hand side region indicates that the possibility exists. A detailed introduction to the subject of ECC is provided in Chapter 11. Here, I shall provide a heuristic example illustrating the actual effect of error correction on the (otherwise uncorrected) BER performance.

Figure 14.4 shows an example of ECC implementation where the physically achievable signal-to-noise ratio is $SNR_{bit} = 16.6$ dB. At the receiver level, and before error-correction decoding, the bit error rate is about $BER = 10^{-3.5}$. After correction, the bit error rate is seen to improve $BER = 10^{-9}$. As the figure indicates, the horizontal translation between the two BER curves (uncorrected on top, and corrected at bottom) corresponds to a virtual, but effective SNR shift of +5 dB. As formally described in Chapter 11, this SNR shift is referred to as the *coding gain*.

the BER can be approximated by $BER = \exp(-Q^2/2)/(Q\sqrt{2\pi})$, which yields $BER = 10^{-9}$ for $Q = 6$ or $SNR_{bit} = 4Q^2 = 144 = +21.58$ dB. Note that different SNR definitions apply to optical telecommunication systems. A standard definition is $SNR \approx Q^2$, which gives $SNR \approx +15.5$ dB for $BER = 10^{-9}$.

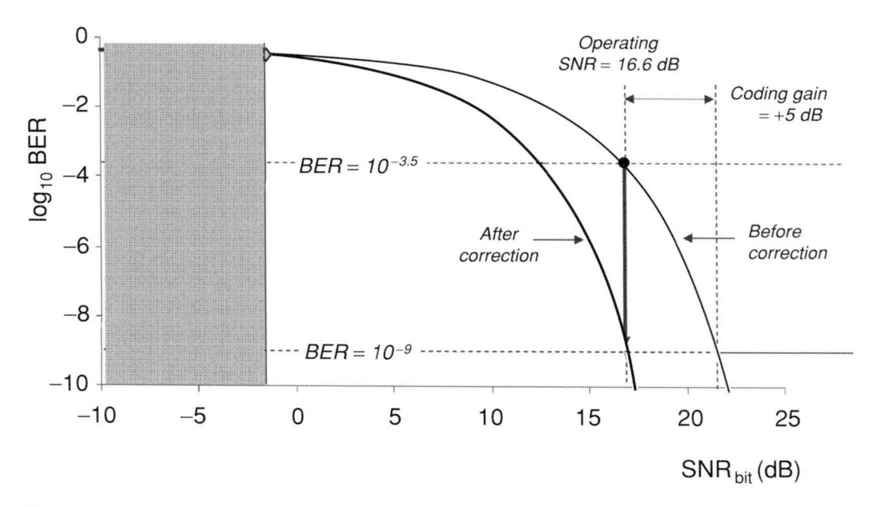


**Figure 14.4** Illustration of the effect of error-correction coding (ECC) on the BER performance (NRZ/OOK), showing a coding gain of +5 dB improving the BER from BER $= 10^{-3.5}$ to BER $= 10^{-9}$.

### *M*-ary frequency-shift-keying (*M*-FSK) format

The *M-ary frequency-shift-keying* (*M*-FSK) format is based on multi-level frequency modulation, where $M$ is a power of two, and coherent detection. The ISD associated with $M$-ary FSK modulation is given by $\text{ISD} = (2\log_2 M)/M$. Hence, for $M = 2, 4, 8, 16, 32, 64, 128$ we have $\text{ISD} = 1, 1, 0.75, 0.5, 0.31, 0.18, 0.1$ (bit/s)/Hz, respectively. The ISD, thus, *decreases* as the number of modulation levels $M$ increases, which is a drawback specific to this modulation format, which is not bandwidth efficient. The symbol error rate (SER) of $M$-FSK signals is given by the following approximation (valid for $\text{SER} \leq 10^{-3}$):

$$\text{SER} \approx \frac{M-1}{2}\text{erfc}\left(\sqrt{\frac{\text{SNR}_{\text{bit}}}{2}\log_2 M}\right). \tag{14.24}$$

### *M*-ary phase-shift-keying (*M*-PSK) format

The *M-ary phase-shift-keying* (*M*-PSK) format is based on multi-level phase modulation. Like $M$-FSK, it requires coherent detection. The ISD associated with $M$-ary PSK modulation is given by $\text{ISD} = \log_2 M$, which shows that it increases with $M$. Hence, for $M = 2, 4, 8, 16, 32, 64, 128$ we have $ISD = 1, 2, 3, 4, 5, 6, 7$ (bit/s)/Hz, respectively. The SER of $M$-PSK signals is given by:

$$\text{SER} = \text{erfc}\left[\sin\left(\frac{\pi}{M}\right)\sqrt{\text{SNR}_{\text{bit}} \times \log_2 M}\right]. \tag{14.25}$$

### *M*-ary quadrature amplitude modulation (*M*-QAM) format

The *M-ary quadrature amplitude modulation* ($M$-QAM) format is based on coherent multi-level amplitude modulation, using the combination of two quadratures.[10] As in the previous case, the ISD associated with $M$-ary QAM modulation is ISD $= \log_2 M$. The SER of $M$-QAM signals is given by:

$$\mathrm{SER} = 2\left(1 - \frac{1}{\sqrt{M}}\right)\mathrm{erfc}\left[\sqrt{\frac{3\langle \mathrm{SNR_{bit}}\rangle}{2(M-1)}\log_2 M}\right], \tag{14.26}$$

where $\langle \mathrm{SNR_{bit}}\rangle$ represents the average bit SNR. This averaging is required because the waveform energy depends on the particular symbol being transmitted, unlike the previous modulation formats.

Using Eqs. (14.23)–(14.26), and the numerical approximation for erfc($x$), one can compute the SER associated with each modulation format for different values of $M$. Since we are interested in *iso*-SER plots, we may choose, for instance, SER $= 10^{-3}$, $10^{-5}$, $10^{-11}$ for each of these plots and find the corresponding $\mathrm{SNR_{bit}}$ values numerically in each case.[11] The result is shown in Fig. 14.5.[12] One observes from the figure that there exist qualitative differences between the different formats. Comparing first binary formats ($M = 2$, SER $\equiv$ BER), it is seen that at equal BERs, NRZ requires substantially higher SNRs than any other format. This feature is explained by the fact that the other formats use *coherent receivers*, which have higher sensitivity: for equal SNRs, coherent receivers yield lower BERs than the direct-detection receivers used in NRZ. All binary formats have an ISD of 1 (bit/s)/Hz.[13] Therefore, the only two degrees of freedom are the format and the BER. Binary QAM is seen to have the lowest SNR requirement at given BER. The choice of a modulation format being primarily dictated by practical engineering considerations (complexity of transmitter or receiver, feasibility, reliability, cost), the BER is not necessarily a decisive factor. Furthermore, we know from the SHT that for any SNR satisfying $\mathrm{SNR_{bit}} > -1.6$ dB, there exist (error-correction) codes making it possible to reduce the BER to arbitrarily small values. How practical such codes are to implement, and how powerful (coding gain) remain two key questions for system design.

Consider next the $M$-ary format plots shown in Fig. 14.5. An additional degree of freedom is provided by the number of modulation levels, $M$, which makes it possible to modify the ISD. As previously seen, $M$-FSK is not bandwidth-efficient

[10] The modulated signal has the form $f(t) = a(t)\cos(\omega t) + b(t)\sin(\omega t)$, where $a(t)$, $b(t)$ are independently modulated with $M$ levels, forming a set of $M^2$ coding possibilities or *constellations*.

[11] This computation can be simply achieved by using a computer spreadsheet to tabulate the SER vs. SNR functions, and then by finding the SNRs corresponding to the three reference SERs.

[12] Figure originally published in E. Desurvire, *Survival Guides Series in Global Telecommunications, Signaling Principles, Network Protocols and Wireless Systems* (New York: J. Wiley & Sons, 2004). See also S. Haykin, *Digital Communications* (New York: J. Wiley & Sons, 1988), J. G. Proakis, *Digital Communications*, 4th edn. (New York: McGraw Hill, 2001) for other data points.

[13] A note at the end of this chapter shows that the capacity is asymptotically $C \to 1$ bit for increasing SNRs; convergence is rapid, thus the approximation $C \approx 1$ bit is valid for SNRs greater than 5 dB.

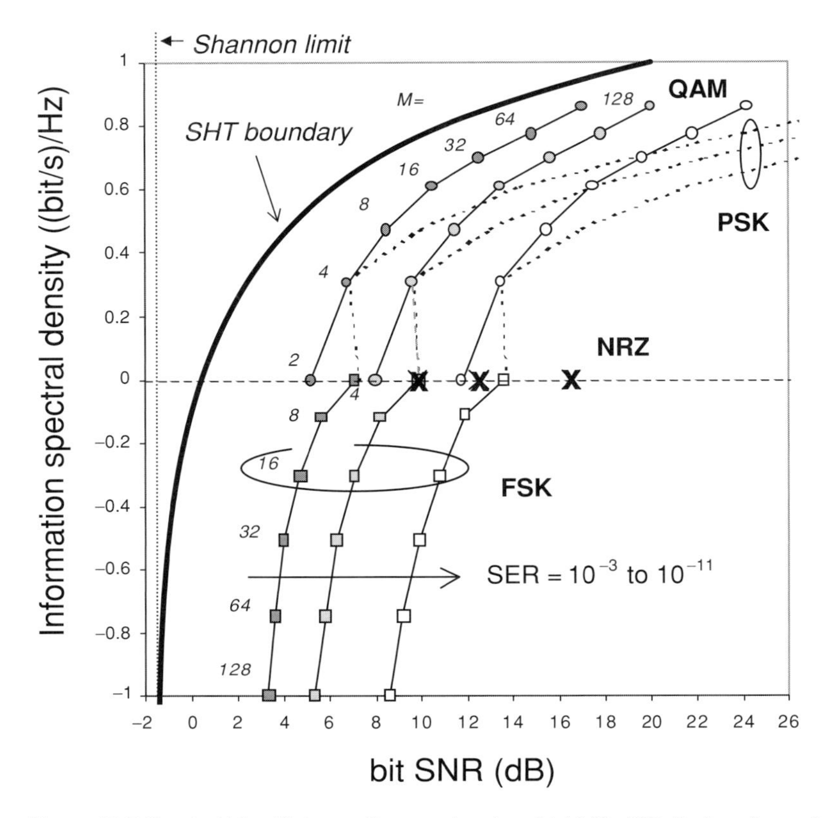


**Figure 14.5** Bandwidth-efficiency diagram showing (bit/s)/Hz ISD (in $\log_{10}$) as a function of bit SNR (in dB) for different coherent $M$-ary modulation formats, $M$-QAM, $M$-PSK, and $M$-FSK, and for error rates of SER $= 10^{-3}$ (dark symbols), SER $= 10^{-5}$ (shaded symbols), and SER $= 10^{-11}$ (open symbols). For clarity, the symbols are not shown for $M$-PSK (the ISD being identical to that of QAM for each $M$). The crosses ($\times$) on the horizontal axis correspond to NRZ/OOK modulation for BER $= 10^{-3}, 10^{-5}, 10^{-11}$.

and multi-level modulation decreases the ISD: a 128-FSK format has an ISD of 0.1 (bit/s)/Hz, which is ten times less than NRZ. Compared with $M$-QAM and $M$-PSK, however, $M$-FSK is seen to have a lower SNR requirement for a given BER, but this advantage does not compensate for the large bandwidth inefficiency. It is also noteworthy that the $M$-FSK *iso*-SER curves move away from the SHT boundary as $M$ increases.

Focusing now on $M$-QAM and $M$-PSK, we observe from the figure that increasing ISD (though increasing $M$) requires higher SNRs. The SNR requirement increases far more rapidly with $M$-PSK. For instance, at SER $= 10^{-3}$, the formats 64-PSK and 64-QAM approximately require SNR $= 26$ dB and SNR $= 15$ dB, respectively, corresponding to an SNR difference of more than 10 dB. From this analysis, $M$-QAM appears to be the champion format in combined terms of ISD and SNR requirements. It is noteworthy that as $M$ is increases, the *iso*-SER curves of $M$-QAM and $M$-PSK move closer to the SHT boundary, contrary to $M$-FSK. The closer any point in these curves get to the SHT boundary, the more sensitive the system SER is to SNR and signal power. Also, the more complex the error-correction codes to bring the BER to low values. Given constraints in power and noise, in BER requirement after code correction, in code-correction complexity, and, finally, in modulation-format technologies, the difficult task for communications engineers is to find the right tradeoffs between academic and practical engineering solutions.

## 14.2 Nonlinear channel

Shannon's information theory and all its subsequent developments have always been based on the assumption of *channel linearity*. The principle of channel linearity can be summarized by two basic conditions:

(a) Assuming signal input to the channel with power $P$, the signal output power is of the form $P' = \lambda P$, where $\lambda$ is a real constant that is power independent ($\lambda < 1$ for lossy channels, $\lambda = 1$ for transparent or lossless channels, and $\lambda > 1$ for channels with internal amplification or gain);
(b) The channel noise $N$ is added linearly to the signal power (additive noise), namely the total output power is of the form $P'' = \kappa P + \mu N$, where $\mu$ is a real constant that is power independent.

Condition (a) implies that the signal can be increased to an arbitrarily high power level, without any changes in the channel's properties, as provided by information theory. Condition (b) implies that the transmission pipe in the channel is free from any power-dependent signal degradation effects. In the physical world, such effects are referred to as *nonlinearities*. Such nonlinearities are usually not observed in vacuum or atmospheric space.[14] However, they are observed in transmission media such as optical fibers, for instance. The advent of high-speed optical telecommunications in the last 20 years has come with the discovery of various types of nonlinearity, namely, Raman and Brillouin scattering, self- and cross-phase modulation, and four-wave mixing. Such nonlinearities cause power limitations (hence SNR and BER limitations) in fiber-optic systems. In the following, I shall focus on fiber-optic systems as an illustration of the *nonlinear communications channel*.

An optical fiber is a highly transparent transmission medium. However, optical signals are attenuated on propagation therein, according to an exponential law $P' = \lambda P = P\exp(-\alpha L)$, where $L$ is the fiber length and $\alpha$ the attenuation coefficient. For lengths of 50 km and 100 km, the relative signal attenuation or power loss is typically 90% ($\lambda = 0.1$) or 99% ($\lambda = 0.01$) respectively. To compensate for fiber loss, fiber-optic systems use *in-line optical amplifiers*. Such amplifiers are periodically placed every 50–100 km (depending on applications) to produce a signal gain $g = 1/\lambda$, such that the

[14] At least with the power levels used in telecommunications. The atmosphere is made of atomic and molecular matter, which may exhibit nonlinear absorption or scattering at certain frequencies and powers, and a vacuum is nonlinear from the high-energy physics or quantum-electrodynamics viewpoint.

signal power at the amplifier output is $P' = g\lambda P \equiv P$. The amplified fiber-optic link is said, therefore, to be transparent. However, the amplification process produces additive noise, which accumulates linearly along the transparent link, each amplification stage contributing to a noise increment.

Overlooking any effect contributed by the optical fiber, an amplified fiber-optic link is intrinsically linear.[15] A simple justification comes as follows. At the end of the link, the optical signal is converted by the receiver into a *photocurrent*. It can be shown that the dominant noise component in the receiver photocurrent output is of the form:

$$\sigma^2 = 2PN, \tag{14.27}$$

where $P$ is the optical power input to the (transparent) link and, hence, to the end receiver, and $N$ is the optical amplifier noise accumulated along the link, which is power-independent.[16] The electrical signal power is of the form

$$P_e = P^2. \tag{14.28}$$

The electrical signal-to-noise ratio (SNR) is, thus,

$$\text{SNR} = \frac{P_e}{\sigma^2} = \frac{P^2}{2PN} \equiv \frac{P}{2N}, \tag{14.29}$$

whose noise denominator is now seen to be independent of the signal power. Thus, the SNR scales linearly with the input (optical) power, which is the linearity condition for the communication channel. But the transmission medium, the optical fiber, is intrinsically nonlinear. It is beyond the scope of this book to detail the different types and causes of fiber nonlinearities.

The introduction of *channel nonlinearity* in the field of information theory is relatively recent (2001).[17] As we shall see, this new hypothesis is rich in implications in the analysis of channel capacity, in particular for optical transmission systems. We shall, thus, assume that the nonlinear channel noise, $N^*$, has the power-dependent form

$$N^*(P) = N + sP, \tag{14.30}$$

where $N$ is the linear channel noise ($P = 0$) and $sP$ is the *power-dependent* contribution to the channel noise, which is attributed to optical fiber nonlinearities. The nonlinearity parameter $s$ can be written as

$$s = 1 - \exp\left[-\left(\frac{P}{P_{\text{th}}}\right)^2\right], \tag{14.31}$$

[15] As in any amplifiers, the amplification factor (gain) of optical amplifiers saturates with increasing signal power. However, optical amplifiers deployed in transmission systems are designed to yield constant gain regardless of signal power changes.

[16] See, for instance: E. Desurvire, *Erbium-Doped Fiber Amplifiers, Devices and System Developments* (New York: J. Wiley & Sons, 2002), Ch. 3.

[17] P. P. Mitra and J. B. Stark, Nonlinear limits to the information capacity of optical fiber communications. *Nature*, **411** (2001), 1027; for a tutorial introduction, see also E. Desurvire, *Erbium-Doped Fiber Amplifiers, Devices and System Developments* (New York: J. Wiley & Sons, 2002), Ch. 3.

where $P_{\text{th}}$ is the *power threshold* at which nonlinearity comes into effect. The exact definition of $P_{\text{th}}$ in relation to the various fiber-optics link parameters can be found in the aforementioned references. The above definition of the nonlinearity parameter $s$ shows that nonlinearity is negligible, or the optical channel is linear, for $P \ll P_{\text{th}}$ or $s \to 0$ (or $sP \ll N$), which, according to Eq. (14.30) gives the power-independent, additive channel noise $N^*(P) \approx N$. In the general case, we see from the relation $N^*(P) = N + sP$ that a fraction $sP$ of signal power is effectively converted into nonlinear noise. Because of energy conservation, the received signal power at the end of the link is not $P$, but rather $P^* = P(1-s)$. The actual SNR is now $\text{SNR}^* = P^*/N^*$, which reduces to $\text{SNR}^* = P/N$ in the linear limit $P \ll P_{\text{th}}$ ($s \to 0$). We can now evaluate the nonlinear channel capacity. According to the *Shannon–Hartley theorem* (SHT), Eq. (14.18), the channel capacity ((bit/s)/Hz) is given by:

$$\begin{aligned} C &= \log\left(1 + \frac{P^*}{N^*}\right) \\ &= \log\left(1 + \frac{P(1-s)}{N+sP}\right) \\ &= \log\left(1 + \frac{P}{N}\frac{1-s}{1+s\dfrac{P}{N}}\right). \end{aligned} \tag{14.32}$$

Replacing the definition of the nonlinearity parameter $s$ in the above yields:

$$\begin{aligned} C &= \log\left\{1 + \frac{P}{N}\frac{\exp\left[-\left(\dfrac{P}{P_{\text{th}}}\right)^2\right]}{1 + \dfrac{P}{P_{\text{th}}}\dfrac{P_{\text{th}}}{N}\left(1 - \exp\left[-\left(\dfrac{P}{P_{\text{th}}}\right)^2\right]\right)}\right\} \\ &= \log\left\{1 + \frac{\dfrac{P}{N}}{\exp\left(\dfrac{P}{P_{\text{th}}}\right)^2 + \dfrac{P}{P_{\text{th}}}\dfrac{P_{\text{th}}}{N}\left(\exp\left(\dfrac{P}{P_{\text{th}}}\right)^2 - 1\right)}\right\}. \end{aligned} \tag{14.33}$$

The result in Eq. (14.33) shows that in the *high-power limit* $P/P_{\text{th}} \to \infty$, the channel capacity reduces to $C = \log(1+\varepsilon)$ with $\varepsilon \to 0$, or $C \to 0$. *Nonlinearity, thus, asymptotically obliterates channel capacity*. Since in the linear regime channel, capacity increases with signal power, we then expect that a maximum can be reached at some power value. It can be shown that such a maximum capacity is reached for the optimal signal power $P_{\text{opt}}$ as analytically defined:

$$P_{\text{opt}} = \left(\frac{NP_{\text{th}}^2}{2}\right)^{\frac{1}{3}}. \tag{14.34}$$

Replacing this definition of $P_{opt}$ into Eq. (14.33) yields the maximum achievable channel capacity $C_{max}$:

$$C_{max} = \frac{2}{3}\log\left(\frac{2}{3\sqrt{3}}\frac{P_{th}}{N}\right). \tag{14.35}$$

A different model that I developed,[18] which takes into account the *quantum nature of amplifier noise* leads to an alternative definition $C'$ of the channel capacity:

$$\begin{aligned} C' &= \log\left(1 + \frac{P(1-s)}{1+2N+s\left(1+\dfrac{1}{1-s}\right)P}\right) \\ &= \log\left(1 + \frac{P}{N}\frac{1-s}{2+\dfrac{1}{N}+s\left(1+\dfrac{1}{1-s}\right)\dfrac{P}{N}}\right). \end{aligned} \tag{14.36}$$

The new definition of capacity in Eq. (14.36) yields different expressions for the optimal power and maximum capacity, namely $P'_{opt}$ and $C'_{max}$, as defined by:

$$P'_{opt} = \left[\frac{(1/2+N)P_{th}^2}{2}\right]^{\frac{1}{3}}, \tag{14.37}$$

$$C'_{max} = \frac{2}{3}\log\left[\frac{2\sqrt{2}P_{th}}{3\sqrt{3}(1+2N)}\right]. \tag{14.38}$$

We note from Eqs. (14.34)–(14.35) and Eqs. (14.37)–(14.38) that the optimal powers $P_{opt}$, $P'_{opt}$ and maximal capacities $C_{max}$, $C'_{max}$ actually have very similar values. With respect to the classical model, the quantum model of the nonlinear channel provides a small correction to the maximum SNR of the order of $2^{-1/3} \approx 0.8$, which, in base-2 logarithms, represents about 30%.

Figure 14.6 shows plots of thee nonlinear-channel capacities $C$, $C'$ (according to the two above models), as plotted vs. the linear SNR, i.e., $\text{SNR} = P/N$, as expressed in decibels,[19] with different parameter choices for $N$ and $P_{th}$. The SHT capacity corresponding to the linear-channel case is also shown. Figure 14.6(a) corresponds to the worst case of a channel with a nonlinear threshold relatively close to the linear noise ($N = 5$ and $P_{th} = 15$). In contrast, Figure 14.6(b) corresponds to the case of a comparatively low linear-noise channel having a relatively high nonlinear threshold ($N = 1$ and $P_{th} = 100$). It is seen from the figure that in both cases, and regardless of the nonlinear model (classical or quantum), the channel capacity is bounded to a power-dependent maximum, unlike in the linear information theory where from SHT the capacity increases as log(SNR).

[18] E. Desurvire, *Erbium-Doped Fiber Amplifiers, Devices and System Developments* (New York: J. Wiley & Sons, 2002), Ch. 3.

[19] With $\text{SNR(dB)} = 10\log_{10}(\text{SNR})$.

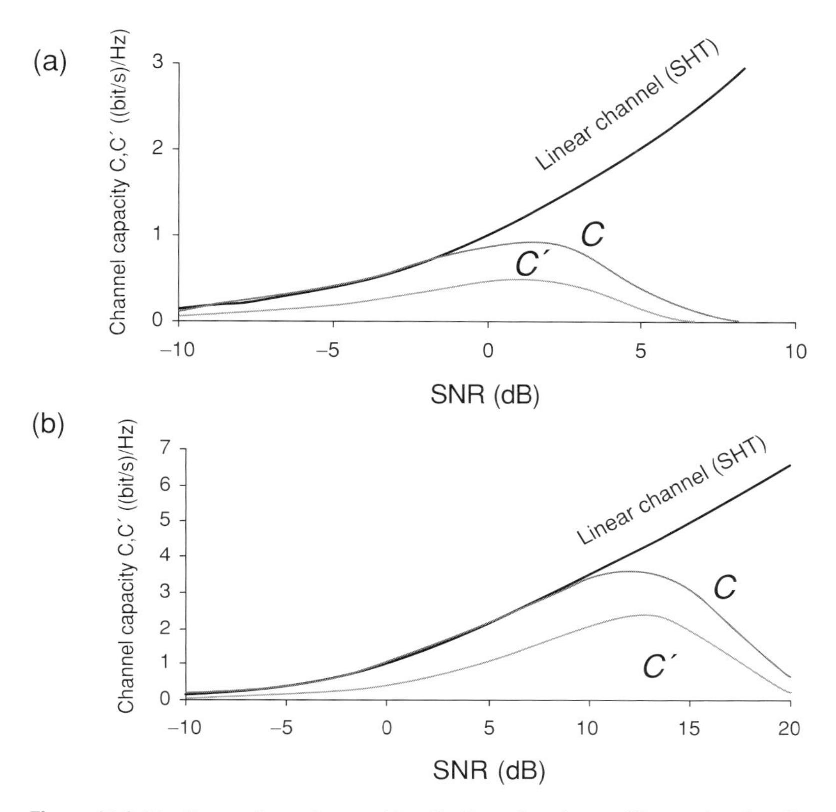


**Figure 14.6** Nonlinear channel capacities $C$, $C'$, as functions of linear signal-to-linear-noise ratio $P/N$ according to classical and quantum models with SHT linear-channel capacity shown for reference (thick line), with dimensionless parameters (a) $N = 5$ and $P_{\text{th}} = 15$, and (b) $N = 1$ and $P_{\text{th}} = 100$.

The consequences of these developments concerning *nonlinear channels* remain to be fully explored. It is important to note that the assumption of channel nonlinearity does not affect the key conclusions of Shannon's (linear) information theory, namely the channel coding theorem and the SHT. Indeed, error-correction codes are used in realistic (nonlinear) optical communication systems to enhance SNR and BER performance. The fact that in these nonlinear optical channels the noise is power-dependent does not affect the effectiveness of such codes. The key conclusion is that the capacity limits imposed by Shannon's information theory still apply, despite the new complexities introduced by channel nonlinearity, which introduces yet another upper bound. It should not be interpreted, however, that nonlinearity defines an "ultimate capacity limit" for optical communication channels, present or future. Indeed, optical fiber design makes it possible effectively to reduce fiber nonlinearity by increasing the fiber core size (referred to as the "effective area"), so as to decrease the nonlinearity threshold, $P_{\text{th}}$, and, hence, effectively achieve the operating conditions $P \ll P_{\text{th}}$ ($s \to 0$). Furthermore, there exist several types of "countermeasures" and power–SNR tradeoffs to alleviate optical nonlinearities, which it is beyond the purpose of this chapter to describe. Overall, the key conclusion is that the optical communication channel, albeit nonlinear, can be made to operate as a linear channel, which is, therefore, essentially compliant to the capacity limits set up by Shannon's coding theorem.

## 14.3 Exercises

**14.1** (M): The signal-to-noise ratio in a continuous channel is SNR = +12.5 dB. What is the channel capacity? Suggest a Reed–Solomon block code for which the capacity could be approached within approximately one bit.

**14.2** (T): Show that the differential conditional entropy

$$H(Y|X) = -\iint p(y|x)p(x)\log p(y|x)\mathrm{d}x\mathrm{d}y$$

reduces after analytical computation to

$$H(Y|X) = \log\sqrt{2\pi\mathrm{e}\sigma_{\mathrm{ch}}^2}.$$

**14.3** (T): Show that the capacity of a Gaussian channel with NRZ/OOK modulation rapidly converges with increasing signal power to the upper limit:

$$\mathrm{C} = 1\,\mathrm{(bit/s)Hz}.$$

*Clues*:

(a) Assume a binary input signal $x_i \in X = \{x_1, x_2\}$ taking the values $x_1 = a$ and $x_2 = -a$, and a continuous output $y \in Y$,

(b) Define the mutual information as

$$H(X;Y) = \sum_i \int_Y \mathrm{d}y p(x_i, y,)\log\left[\frac{p(x_i, y)}{p(x_i)p(y)}\right],$$

(c) Define the conditional probability $p(y|x_i)$ as:

$$p(y|x_i) = \frac{1}{\sigma\sqrt{2\pi}}\exp\left[-\frac{(y-x_i)^2}{2\sigma^2}\right],$$

where $\sigma^2$ is the channel noise variance.

(d) Show that the channel capacity is a function of the "bit SNR" parameter $S^2 \equiv a^2/(2\sigma^2)$ and numerically determine the limit for increasing $S^2$.

# 15 Reversible computation

This chapter makes us walk a few preliminary, but decisive, steps towards quantum information theory (QIT), which will be the focus of the rest of this book. Here, we shall remain in the classical world, yet getting a hint that it is possible to think of a different world where computations may be reversible, namely, without any loss of information. One key realization through this paradigm shift is that "information is physical." As we shall see, such a nonintuitive and striking conclusion actually results from the age-long paradox of Maxwell's demon in thermodynamics, which eventually found an elegant conclusion in Landauer's principle. This principle states that the erasure of a single bit of information requires one to provide an energy that is proportional to $\log 2$, which, as we know from Shannon's theory, is the measure of information and also the entropy of a two-level system with a uniformly distributed source. This consideration brings up the issue of irreversible computation. Logic gates, used at the heart of the CPU in modern computers, are based on such computation irreversibility. I shall then describe the computers' von Newman's architecture, the intimate workings of the ALU processing network, and the elementary logic gates on which the ALU is based. This will also provide some basics of Boolean logic, expanding on Chapter 1, which is the key to the following logic-gate concepts. As I shall describe, a novel concept for logic gates (and, hence, their associated circuits or networks), can be designed to provide truly reversible computation.

## 15.1 Maxwell's demon and Landauer's principle

The QIT story begins with a strange machine devised in 1871 for a thought experiment by the physicist J. C. Maxwell. Such a machine – if one were ever able to construct it – would violate the *second principle of thermodynamics*! As Maxwell describes in his theory of heat:[1]

One of the best established facts in thermodynamics is that it is impossible in a system enclosed in an envelope which permits neither change of volume nor passage of heat and in which both the temperature and the pressure are everywhere the same, to produce any inequality of temperature or of pressure without the expenditure of work. This is the second law of thermodynamics, and

[1] See http://en.wikipedia.org/wiki/Maxwell%27s_demon; http://users.ntsource.com/~neilsen/papers/demon/node3.html.

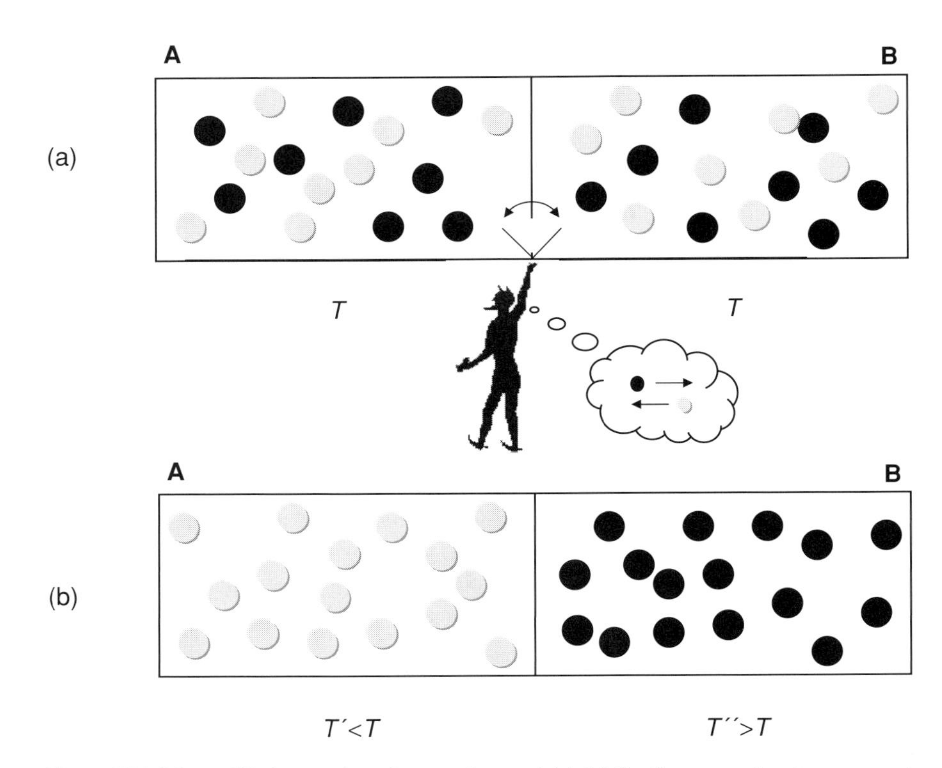


**Figure 15.1** Maxwell's demon thought experiment: (a) initially, the gas molecules are evenly spread between two parts A and B (dark = fast, clear = slow), with homogeneous temperature $T$, and (b) after the demon sorted molecules according to speed, showing lower temperature $T'$ in part A and higher temperature $T''$ in part B.

it is undoubtedly true as long as we can deal with bodies only in mass, and have no power of perceiving or handling the separate molecules of which they are made up. But if we conceive a being, whose faculties are so sharpened that he can follow every molecule in its course, such a being, whose attributes are still as essentially finite as our own, would be able to do what is at present impossible to us. For we have seen that molecules in a vessel full of air at uniform temperature are moving with velocities by no means uniform, though the mean velocity of any great number of them, arbitrarily selected, is almost exactly uniform. Now let us suppose that such a vessel is divided into two portions, A and B, by a division in which there is a small hole, and that a being, who can see the individual molecules, opens and closes this hole, so as to allow only the swifter molecules to pass from A to B, and only the slower molecules to pass from B to A. He will thus, without expenditure of work, raise the temperature of B and lower that of A, in contradiction to the second law of thermodynamics.

A basic illustration of Maxwell's machine is provided in Fig. 15.1. As Maxwell describes it, the machine consists of a gas container or "vessel," which is separated by a wall into two chambers, A and B. The gas absolute temperature $T$ is uniform,[2]

[2] To recall, the *absolute temperature* is expressed in units of kelvin (K). It has a one-to-one correspondence with the Celsius (C) temperature scale, with absolute zero being $0\,\mathrm{K} = -273\,^\circ\mathrm{C}$ and the melting-ice temperature being $0\,^\circ\mathrm{C} = +273\,\mathrm{K}$.

and is given by the formula $E = (3/2)k_B T$, where $E$ is the average kinetic energy of the gas molecules ($k_B$ = Boltzmann's constant). Both temperature and energy represent macroscopic measures, as individual molecules don't have a "temperature" and have different velocities of their own, some moving slower and some moving faster. As the figure shows, an imaginary, witty but not malicious creature (later dubbed "Maxwell's demon") guards a trapdoor, which allows individual molecules to pass from A to B or the reverse, without friction or inertia. The demon checks out the molecules' velocities and allows only the faster ones ($E' > E$) to pass from A to B, and only the slower ones ($E' < E$) to pass from B to A. After a while, most of the fast molecules are found in B, while most of the slow ones are in A. As a result, the gas temperature in A ($T'$) is lower than that in B ($T''$), with $T' < T < T''$.

The simultaneous heating of part B and cooling of part A, which is accomplished by the demon without deploying energy or work, is in contradiction to the second law of thermodynamic. This second law stipulates, in its simplest formulation, that: "Heat does not flow spontaneously from a cold to a hot body of matter." Since heat is a measure of the average kinetic energy of the gas, it has a natural tendency to diffuse from hot to cold bodies, and not the reverse. Another formulation of the second law is that: "In an isolated system, entropy as a measure of disorder or randomness can only increase." Let us have a closer look at this entropy notion. In Appendix A, we have seen that a system of particles that can be randomly distributed in $W$ arrangement possibilities into a set of distinct boxes, called "microstates," is characterized by an entropy $H$. This entropy represents the infinite limit, normalized to the number of particles, of the quantity $\log W$. Hence, the greater $W$ or $H$, the more randomly distributed the system, which justifies the notion of entropy as a physical measure of disorder. By separating and ordering the gas molecules into two distinct families (slow and fast), the demon decreases the global randomness of the system, since there is a smaller number of microstate arrangements $W$ for each of the initial molecules, the system's entropy $\log W$ is, thus, decreased, which is indeed in contradiction to the second law because the demon apparently did not perform any work.

At the time, the conclusion reached by Maxwell to solve this paradox was that the second principle only applies to large numbers of particles as a statistical law, i.e., to be used with certain confidence at a macroscopic, rather than at a microscopic level. In the first place, Maxwell's paradox was never intended to challenge the second law. Yet, it kept puzzling scientists for generations with many unsolved issues. In 1929, for instance, L. Szilárd observed that in order to measure molecular velocities, the demon should spend some kind of energy, such as that of photons from a flashlight, to spot the slow or fast molecules, and then decide from the observation whether or not to open the trapdoor. Each of such measurements and the generation of photons would then increase the system's entropy, which should balance out the other effect of entropy decrease. In 1951, L. Brillouin observed that in order to distinguish molecules by use of light photons, the photon energy would have to supersede that of the ambient electromagnetic radiation, thus bringing a quantum of heat $k_B T$ into the system. To move the trapdoor, the "demon" must also be material. Thus, some form of heat transfer should occur during the process, resulting in the heating of both the demon and the gas, canceling the cooling effect of molecule separation. In 1963, R. Feynmann observed that the demon would not be able to get rid of the heat gained in his molecule measurements, ending up in "shaking into Brownian motion" with the inability to perform the task!

Most of the above conjectures, along with the multiplication of improved "demon engines," contributed to turn Maxwell's initial paradox into an ever-deepening enigma. This is how the concept of *information entropy* came into the picture. Szilárd considered a demon engine processing just a single gas molecule. If the demon knew where the molecule was located, then he could move a frictionless piston to confine it where it belonged, without any work expenditure! Then the pressure effect of the molecule on the piston could be used to generate work, such as lifting a weight. The intuitive, yet not consolidated idea, was that the entropy bookkeeping could be balanced should some positive entropy be associated with the *information* concerning molecules or the demon's *memory* itself. Either the demon keeps track of any of its binary decisions (open trapdoor to molecule coming from A or B) or it erases them. In the first case, the demon's memory space expands, which goes together with an entropy increase (number of accessible memory states in the demon's mind). At some point, however, the memory space is exhausted, and information must be erased. In the second case, the memory is cleared one bit at a time prior to the next binary decision. Whether the memory is cleared at once or bit after bit after each demon operation, the action of erasure is *irreversible*.[3] According to thermodynamics, irreversible processes generate positive entropy. Therefore, *the erasure of information must increase the entropy of the surrounding environment*. Such a revolutionary concept was initially suggested by J. von Neumann[4] and then formalized in 1961 by R. Landauer. The key conclusion takes the form of *Landauer's principle* (or the *Landauer bound* or *limit*, or the *von Neumann–Landauer bound*):

The energy dissipated into the environment by the erasure of a single bit of information is at least equal to $k_B T \log 2$. The environment entropy increases by at least $k_B \log 2$.

Landauer's principle, thus, provides a *lower bound* on the energy dissipation and entropy increase associated with information erasure. It is not coincidental that this minimum is proportional to log 2 (natural logarithm). Without pretending to provide a formal demonstration of this result, consider the following. The memory to be erased can be viewed as a binary source $X = \{1, 0\}$ having a uniform distribution, i.e., $p(1) = p(0) = 1/2$ (the probability that the memory bit is either 1 or 0). As we have seen in Chapter 3, the *self-information* associated with any of the two events $x \in X$ is $I(x) = -\log_e[p(x)] \equiv -\log_e(1/2) = \log_e 2$, as expressed in *nats*. The source's *entropy* $H(X)$ corresponds to the average information

$$\begin{aligned}\langle I(x)\rangle &= -p(1)\log_e[p(1)] - p(0)\log_e p[(0)] \\ &= 0.5\log_e 2 + 0.5\log_e 2 = \log_e 2,\end{aligned}$$

[3] For a given memory address, the initial state is the bit value 0 or 1. Once cleared, the bit value is reset to 0 (for instance). From this final state, it is not possible to decide whether the initial state was the bit value 0 or 1, and so on for all previously recorded states, which illustrates the irreversibility of the erasure process.

[4] See http://en.wikipedia.org/wiki/Landauer%27s_principle.

with $H(X) = k_B\langle I(X)\rangle = k_B \log_e 2$ in the units of physics. According to Landauer's principle, clearing the memory, thus, corresponds to the conversion into energy of the memory's average information, which corresponds to the heat $Q = k_B T\langle I\rangle = k_B T \log_e 2$. The cleared memory can be viewed as a single-event source $X' = \{0\}$ with probability $p(0) = 1$, corresponding to zero information ($I(0) = -\log_e[p(0)] = -\log_e(1) = 0$) and zero entropy $H' = k_B\langle I(0)\rangle = 0$. The difference in memory entropy before and after erasure is, thus, $\Delta H = H' - H = -k_B \log_e 2$. This is the amount of entropy that is communicated to the environment. These values represent a lower bound because the physical erasure process must also generate some form of heat, for instance passing a current through a memory transistor gate. The key conclusion of Landauer's principle can be summarized into the following statement:

Information is physical.

This means that the creation, erasure, manipulation, or processing of information involves physical laws. This represents a new awareness, which contrasts with the classical background of Shannon's information theory, and calls for more questions, answers, and paradoxes. These are to be found in *quantum information theory* (QIT), which will be addressed shortly. At this stage, we must at least be convinced that there is more to information than a mere mathematical or ethereal definition, as we have been used and trained through education to believe intuitively.

Landauer's principle brings up another interesting question, which concerns *computing*. When one performs hand calculations on a blackboard, one is obliged at some point to erase it, while saving the last results and other useful parameters somewhere in the corner of the blackboard. Very large blackboards, with different folding or sliding panels, could fit a full math course. But the information must be cleared for the next course. This shows that processing information (computation) requires information erasure, because of the finite size of the memory. A computer that would keep and store the information regarding all of its intermediate calculation steps (namely the memory contents at each step, or the full record of the memory changes) would rapidly "choke on its own garbage." Such a necessary erasure yet obliterates the information of the *computing history*, which makes the computation *irreversible*. For computation, the term "irreversible" means that the computed information output cannot lead one to know the information input to the computation.[5]

The concept of computing irreversibility, however, seems to be in contradiction with the laws of physics. Indeed, if one fully knows the state of a system at a given time $t$, for instance, the distribution of molecules and their velocities inside a closed box,

[5] For instance, the addition of two integer numbers $a$ and $b$, given the fact that the input information $(a, b)$ is then erased, is irreversible: to any output $c$ correspond $c + 1$ possible input pairs. If $a$ and $b$ are real numbers, the number of input-pair possibilities is infinite. Another example of irreversibility is the operation of functional derivation: given any input function $f_{\text{in}} = f(x)$, the output is $f_{\text{out}} = f'(x)$. Reversing the algorithm yields $f^*(x) = \int_{x_0}^{x} f'(x)\text{d}x = f(x) + C$ where $C = -f(x_0)$ is an arbitrary constant. The knowledge of $f_{\text{out}} = f'(x)$, thus, makes it possible to know $f^* = f(x) + C = f_{\text{in}}(x) + C$. This may represent useful information, but $f^*$ and $f_{\text{in}}$ differ from each other by an infinity of possible real constants $C$, which illustrates the irreversibility of the computation algorithm.

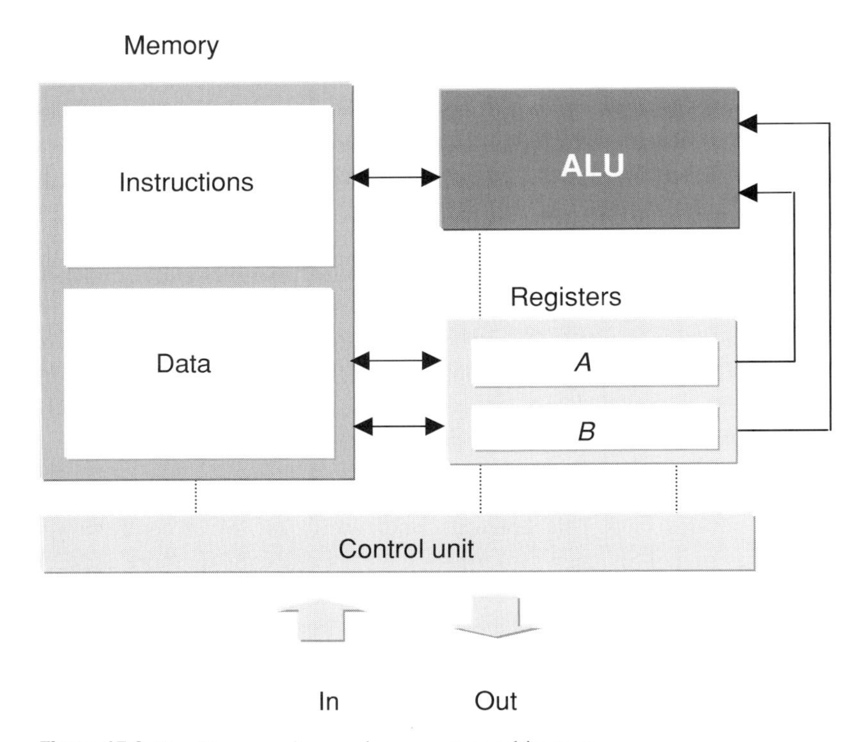


**Figure 15.2** Von Neumann's generic computer architecture.

the laws of mechanics make it possible to compute the system at any time prior to $t$. This is like playing a video of a billiard ball game in reverse. From this perspective, a physicist should suspect that it is possible to effect *reversible computations*, without any heat dissipation or entropy increase. This brings the issue of the existence of *reversible logic*. In the next two sections, I shall describe logic gates, which are used in the heart of computer microcircuits, and then show that it is possible to define reversible logic gates and circuits.

## 15.2 From computer architecture to logic gates

According to the 1945 design of computer architecture by von Neumann (VN),[6] the core of a standard computer is made of a *memory*, and of an *algorithmic and logical unit* (ALU), as illustrated in Fig. 15.2. It is seen that the VN architecture is completed with a control unit and the interfaces enabling the computer to communicate with the outside word and its peripherals. The memory is a device containing the instructions to initialize and run the computer with its interfaces and peripherals. It includes the program to be performed, in the form of a stack of instructions, and the data before, during, and after

[6] See, for instance, http://en.wikipedia.org/wiki/Von_Neumann_architecture.

computation. As the figure shows, the ALU works with two dedicated memory buffers, called *registers*, or sometimes "accumulators." The two registers ($A$ and $B$) are input from the memory with binary data words $a_1a_2, \ldots, a_k$ and $b_1b_2, \ldots, b_k$ of identical size, typically $k = 32$ bits, 64 bits, and, today, 128 bits. This size is the same as that of the cells in the memory, and also of the address used to index the memory cells (the memory, thus, has $2^k$ different addresses). Registers are literally used as scratchpads for the ALU to perform various logical operations between the two data words $\{a_i\}$ and $\{b_i\}$, such as addition, subtraction, multiplication, division, comparison, complementation, rotational permutation, overflow test, and zero-divide test, for the main ones. These operations are performed by an internal circuit of the ALU, which is formed of a complex network of logic gates, whose functions can be switched at will through a set of control bits. A basic example of a four-bit ALU circuit is provided later in this chapter (one may yet have a look at Fig. 15.6 for a flavor), which will come after I have described the principles of the elementary logic gates themselves.

All logical operations between the register data and the ALU are instructed from a "micromachine" program, which decodes the higher-level instructions fed by the computer memory into micro-instructions making the ALU capable to handle the register bits. The ALU is complete with an internal buffer register. Today, the integrated version of the ALU, internal buffer, and micromachine decoder represents what is referred to as a CPU, for *central processing unit*.[7] The CPU is, thus, a machine to "crunch bits" from the higher-level instructions of the central memory, hence the popular name "computer," which today means far more than just a calculator, although the appellation remains conceptually exact. Note that the VN architecture is not the only one possible. Indeed, several VN machines, or CPUs can be set to work in parallel, each being dedicated to an independent fraction of the computing algorithm, or more simply being assigned a share of the number-crunching groundwork. This is referred to as *parallel* or *multi-processor architecture*, to be viewed either as a magnified VN master machine ruling over many CPU slaves, or a "democratic" group of VN machines putting their effort towards some common-interest goals. The VN architecture yet remains crucial to the orchestration of the ensemble, whether the master is a ruling dictator or simply a discreet facilitator.

Before taking a closer look at the characteristics of binary logic gates forming the ALU circuitry, it is worth considering the *Turing machine* (TM), which was described in Chapter 7. To recall, the generic TM architecture, which is illustrated in Fig. 7.1, consists of a bidirectional tape of infinite length, and a head, which is able to read and write bits or symbols onto the tape. The TM is set into motion according to a stack of instructions, called an "action table." This action table is the program from which the tape input data are processed bit by bit (or symbol by symbol), resulting in the generation of output data, which are recorded on the tape. It includes a "halting instruction," which forces the machine to halt on completion of the task. The tape is a memory of infinite size, which puts no restriction as to the complexity of the task, the corresponding computing algorithm and the amount of data to be processed. Finally, a *universal Turing*

[7] See http://en.wikipedia.org/wiki/Central_processing_unit.

*machine* (UTM) belongs to a class of TMs, which, through a minimal set of action table instructions, are able to emulate any other TM, which is achieved by means of a tape program. The most important conclusion about TMs (and for that matter, UTMs) is that they can compute "anything that a real computer can compute." Namely, *any problem that can be solved through some form of computing process or algorithm devised by a mathematician can be solved by a TM.* This stunning statement is referred to as the *Church–Turing thesis*. Thus, TMs, and more so UTMs, represent the most elementary form of computing machine ever conceivable. In this view, why are modern computers based on VN architecture instead of that of a TM? As discussed in Chapter 7, the TM implementation is comparatively impractical, for two main reasons. First, as we have seen, the TM requires an infinite memory space, which is not physically possible. Second, the finite size of a realistic computer memory puts a constraint on the management of the memory space, which in turn impacts on the program structure and algorithms. The TM algorithms may be more general or universal, but their implementation in a computing machine of finite memory size may in some cases be intractable, if not physically impossible. We can now see some key differences and similarities between TM and VN computers. While the UTM represents the most universal and elementary computer architecture, the VN machine represents its conceptual approximation in the real world. Both have programs based on man-thought algorithms, and physical input–output interfaces. We may think of the ALU as a TM that uses parts of its memory as an input–output tape, and the other part as the place where the action table is located. But this does not resolve the problem of memory finiteness. Finally, another key difference is that the VN machine is not subject to the "halting problem," whereby a TM could run forever. As discussed in Chapter 7, the halting problem is generally "undecidable," meaning that given an action table and an input tape, there exists no formal proof that a TM will necessarily halt! In contrast, a VN machine does not have a halting problem. We may not attempt to prove the point here, but simply stated, the finite size of the memory must eventually cause the VN machine to "choke on its own garbage," unlike the TM, which may run indefinitely! But unlike the idealized TM, the realistic VN computer can be programmed to stop automatically if caught in an infinite loop, or trespassing a preset CPU time limit. As a matter of fact, modern VN computers are fully capable of emulating any universal TM, and for this reason they are said to be *Turing complete, Turing equivalent*, or *computationally universal*.[8] This does not mean that such a UTM emulation would be any quicker, more efficient, or easier; probably to the contrary in most general cases. The key difference remains the finiteness of the VN computer memory.

We shall now focus on the VN computer and look in more detail at the ALU and its elementary computing circuits. As we have seen, the ALU processes the codeword bits that are loaded in the two registers $A$ and $B$. The ALU is able to interpret high-level instructions from the memory into basic micromachine operations. In short, a high-level instruction may be stated in a form such as "$c = a + b$," which the ALU interprets as "load the operands $a, b$ pointed by the corresponding memory addresses

[8] See http://en.wikipedia.org/wiki/Turing-complete.

into the registers $A$ and $B$, perform the addition of the two, and return the result $c$ to the dedicated memory address. Such an operation calls for a microprogram, which must process the two binary operands $a, b$ ($a = a_1 a_2, \ldots, a_k$, $b = b_1 b_2, \ldots, b_k$) on a bit-by-bit ($a_i, b_i$) basis.[9] Another possible instruction may be "IF $a > b$, THEN." Here, the ALU is asked to verify indeed if $a > b$, so that in the event the answer is yes (or TRUE, or 1 in logic language) the high-level program may jump to another instruction. The ALU must then subtract $b$ from $a$, and return TRUE, or 1, if $z = a - b$ is identified as being neither negative nor zero. It is clear that such a test requires a microprogram, which sequentially compares every bit in the two registers $A, B$. The same is true for operations such as multiplication and division, for which the ALU must also return the extra information "overflow = TRUE/FALSE" and "zero divide = TRUE/FALSE." Here, we will not tarry on the micromachine programming concept and structure, but rather focus on the bit-to-bit operations ($a_i \leftrightarrow b_i$) in the registers $A, B$, which are performed by a preset circuitry of *logic gates*, acting as a controlled network. As mentioned earlier, a basic example of such a network is provided later in this chapter (Fig. 15.6). But to understand how the network operates under any micro-instructions, it is essential to review the different types of logic gate, as well as their possible basic circuit arrangements.

The three most elementary logic gates, which may have one or two input bits, but have a single output bit, are called NOT, AND, and OR, as illustrated in Fig. 15.3. Calling $a, b$ the bit operands, these gates correspond to the three *Boolean operators* noted, respectively and equivalently (see Chapter 1):

$$\begin{aligned} \neg a,\ \bar{a} \quad &\text{for} \quad \text{NOT } a, \\ a \vee b,\ a + b \quad &\text{for} \quad a \text{ OR } b, \\ a \wedge b,\ a \times b \quad &\text{for} \quad a \text{ AND } b. \end{aligned}$$

The "truth tables" shown in the figure list all possible outcomes of any logical operation. For instance, the output $c = a \wedge b$ (AND gate) is nonzero only in the case $a = b = 1$. The three other cases, $(a, b) = (00), (10), (01)$ yield $c = 0$. Thus, the result $c = 1$ has only one cause while the result $c = 0$ has three indistinguishable causes, illustrating that the AND operation is effectively nonreversible. It is observed from the truth tables in Fig. 15.3 that only the NOT gate performs a reversible computation (the output to input bit correspondence being unique).

The three gates NOT, OR, and AND make it possible to build a second group of elementary logic gates, shown in Fig. 15.4, called NAND (NOT AND) and XOR (eXclusive

[9] Let $A = a_0 a_1, \ldots, a_n$ and $B = b_0 b_1, \ldots, b_n$ be two binary numbers. Proceeding from the lowest-weight bit ($i = 0$) to the highest-weight bit ($i = n$), a first Boolean circuit computes both $x_i = a_i \oplus b_i$ and $c_i^{(1)} = a_i \wedge b_i$. The output $c_i^{(1)}$ provides a "carry" bit, which is 1 if the two bits ($a_i, b_i$) are equal to 1, and 0 otherwise. A second circuit stage then computes $y_i = x_i \oplus c_{i-1}$, $c_i^{(2)} = x_i \oplus c_{i-1}$ and $c_i = c_i^{(1)} + c_i^{(2)}$, where $c_{i-1}$ is the carry bit resulting from the previous addition stage (by convention, $c_{-1} = 0$). The result $y_i$ represents the $i$th bit of $A + B$ and the result $c_i$ is the carry to use for the next computation stage $i + 1$. It is left to the reader as an exercise to draw the Boolean circuits corresponding to these different computations stages, as well as the circuit for a full $k$-bit binary adder.

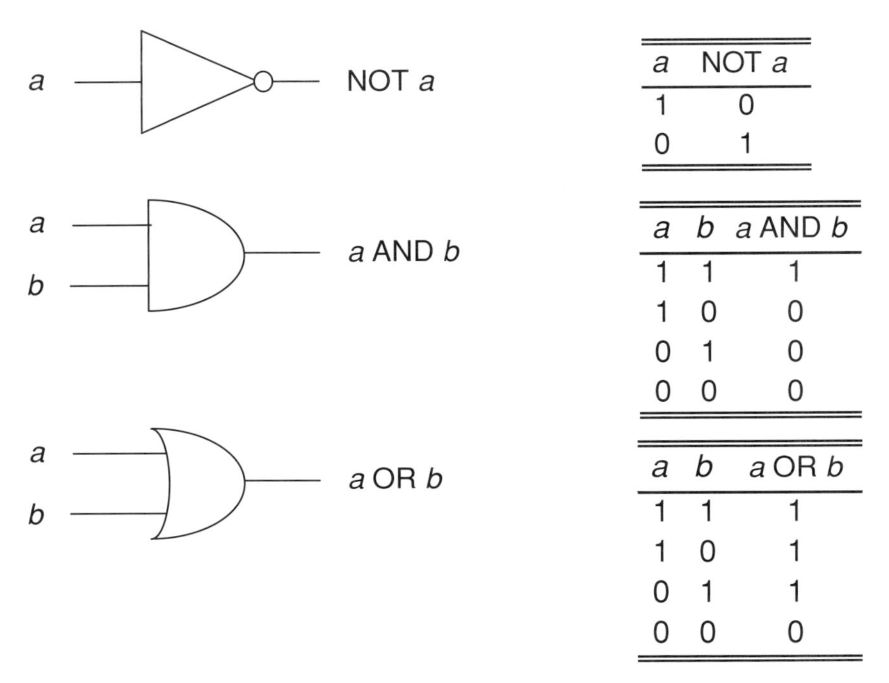


| $a$ | NOT $a$ |
|---|---|
| 1 | 0 |
| 0 | 1 |

| $a$ | $b$ | $a$ AND $b$ |
|---|---|---|
| 1 | 1 | 1 |
| 1 | 0 | 0 |
| 0 | 1 | 0 |
| 0 | 0 | 0 |

| $a$ | $b$ | $a$ OR $b$ |
|---|---|---|
| 1 | 1 | 1 |
| 1 | 0 | 1 |
| 0 | 1 | 1 |
| 0 | 0 | 0 |

**Figure 15.3** The three elementary logic gates: NOT, AND, and OR, along with their corresponding truth tables.

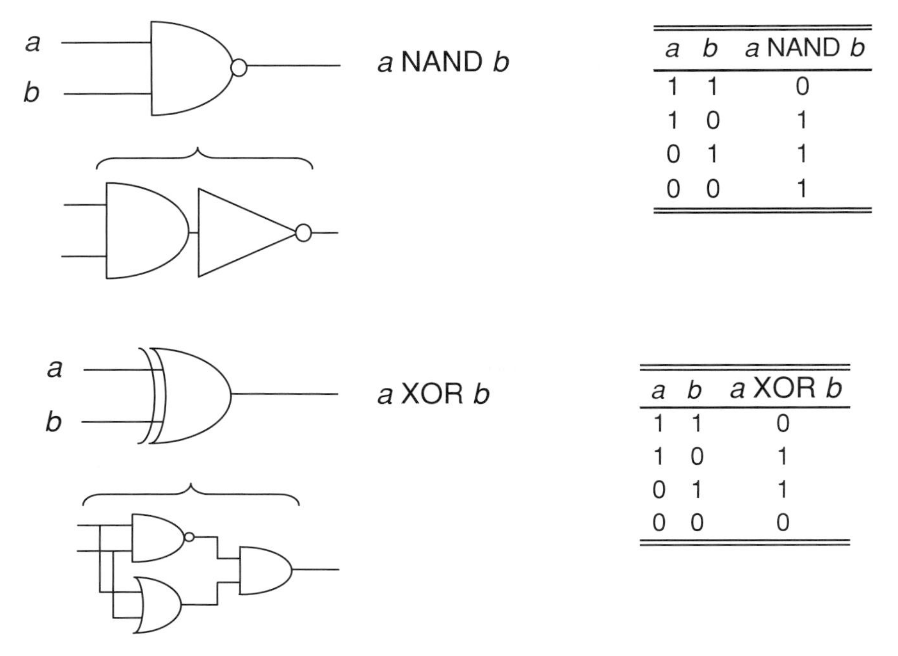


| $a$ | $b$ | $a$ NAND $b$ |
|---|---|---|
| 1 | 1 | 0 |
| 1 | 0 | 1 |
| 0 | 1 | 1 |
| 0 | 0 | 1 |

| $a$ | $b$ | $a$ XOR $b$ |
|---|---|---|
| 1 | 1 | 0 |
| 1 | 0 | 1 |
| 0 | 1 | 1 |
| 0 | 0 | 0 |

**Figure 15.4** Second group of elementary logic gates, NAND and XOR, along with their corresponding truth tables, as built from the first group shown in Fig. 15.3.

OR). In Boolean logic, the corresponding operators are:

$$\begin{aligned} x \,\bar{\wedge}\, y \quad &\text{for} \quad x \text{ NAND } y, \\ x \oplus y \quad &\text{for} \quad x \text{ XOR } y, \end{aligned}$$

respectively. Note the small open circle, or "bubble," which indicates the logical complementation at the gate's output (in some cases, the bubble can be placed at the input to complement some of the input data bits). The figure shows that the NAND gate is built through a succession of AND and NOT gates. The XOR gate is an essential function for performing modulo-2 addition of two binary numbers. This gate can be built in many different ways. The figure shows one possible Boolean circuit utilizing AND, NAND, and OR gates, according to the heuristic formula

$$a \oplus b = (a \,\bar{\wedge}\, b) \wedge (a \vee b). \tag{15.1}$$

Here are three other possible ways to effect a XOR computation, which the reader can check out using truth tables:

$$\begin{aligned} a \oplus b &= (a \vee b) \wedge (\bar{a} \vee \bar{b}) \\ &= (a \wedge \bar{b}) \vee (\bar{a} \wedge b) \\ &= (a \,\bar{\wedge}\, \bar{b}) \,\bar{\wedge}\, (\bar{a} \,\bar{\wedge}\, b). \end{aligned} \tag{15.2}$$

It is seen that the last solution in this equation makes exclusive use of NAND ($\bar{\wedge}$) and NOT ($\bar{x}$) operations. The NOT can also be realized with NAND, taking into account the property $\bar{a} = a \,\bar{\wedge}\, a$. Thus, we obtain from Eq. (15.2)

$$a \oplus b = [a \,\bar{\wedge}\, (b \,\bar{\wedge}\, b)] \,\bar{\wedge}\, [(a \,\bar{\wedge}\, a) \,\bar{\wedge}\, b], \tag{15.3}$$

which shows that an XOR gate can be built exclusively with a Boolean circuit of NAND gates. It is easily established that the NAND gate can be used as a universal building block in Boolean circuits to construct all elementary logic gates. This property is illustrated in Fig. 15.5.

We have, thus, identified the basic constitutive elements of computation: *logic gates* (NOT, AND, OR, XOR) with NAND representing a universal building block, and *wires* to realize any Boolean circuit, from elementary to complex. We can then consider other types of logic gates derived from these.

First, the XOR gate can also be used to perform yet another logical function called *controlled-NOT*, corresponding to the *CNOT* gate. Given $(a, x)$ as inputs, with $a$ being called the *control bit*, the output of CNOT is $a \oplus x$. For $a = 0$, we have $a \oplus x = x$; and for $a = 1$, we have $a \oplus x = \bar{x}$. Thus the value of $x$ is flipped if the control bit is set to $a = 1$ and is left unchanged otherwise, hence the name "controlled-NOT" gate.

Another Boolean-circuit component, which was only implicitly introduced, is the wire splitter that enables one to use the same bit to feed different gate inputs. This logical splitting function is referred to as *FANOUT*. Another logical function of interest is *CROSSOVER* (also called *SWAP*), which consists of switching two bits $(a, b)$ into $(b, a)$, depending on the value of a control bit $c$. As we shall see later, the function CROSSOVER is the key to building *reversible* logic gates. Here, we shall overlook more complex logical-gate arrangements, whereby the wires can form *circuit loops*,

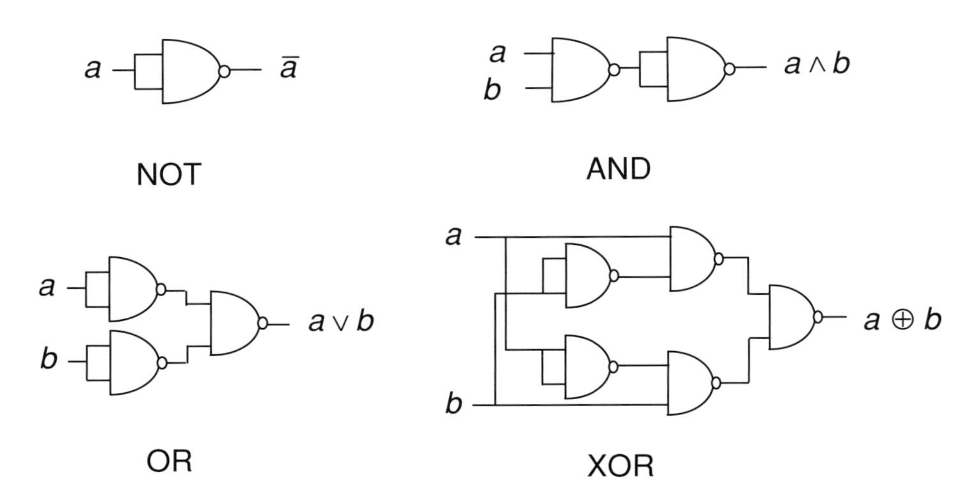


**Figure 15.5** Construction of logic gates NOT, AND, OR, and XOR from Boolean circuits of NAND gates.

thus enabling output bits to be fed back into the circuit that computes them. Such feedback loops are used to create "*bistable*" or *two-state* logical devices. The corresponding devices are called *FLIP-FLOP* gates, and are activated by a clock signal. A FLIP-FLOP gate can have different output possibilities, or *states*, depending on (a) the input bits, and (b) the previous state in which it is found at each clock cycle. There exist several types of elementary FLIP-FLOP gates,[10] which find many applications for data registers and counters, and computer memories. Finally, any single-port logic gate whose output is equal to the input performs the function *REPEAT*, as does a "repeater" in digital electronics. This is the case of the CNOT gate with the control bit set to $a = 0$.

Having, now, an understanding of the various logic gates and functions, it is well worthwhile to provide a basic example of ALU (algorithmic and logical unit) circuitry. Figure 15.6 shows the impressive logical circuit of a four-bit ALU, an integrated circuit from the TTL (transistor-transistor logic) family referred to as *74181*.[11]

The following does not represent any authoritative description of the '181 (as professionals call it), but simply a brief academic overview. As may be counted from the figure, this ALU network includes 67 logic gates (namely: 38 AND, 12 NOR, 4 XOR, 4 NAND, and 9 NOT). The four-bit inputs $\bar{A} = (\bar{A}_0\bar{A}_1\bar{A}_2\bar{A}_3)$, $\bar{B} = (\bar{B}_0\bar{B}_1\bar{B}_2\bar{B}_3)$ (top) and output $\bar{F} = (\bar{F}_0\bar{F}_1\bar{F}_2\bar{F}_3)$ (bottom) are shown in open circles (note that, here, the operands are conventionally input in complemented logic). On top left, the "mode-control" input $M$ selects the type of computation to perform, namely logic with $M = H$ (for high voltage

[10] See, for instance: http://en.wikipedia.org/wiki/Flip-flop_(electronics), http://computer.howstuffworks.com/boolean3.htm, www.eelab.usyd.edu.au/digital_tutorial/part3/fl-types.htm.

[11] See (in French): http://fr.wikipedia.org/wiki/Unit%C3%A9_arithm%C3%A9tique_et_logique. Also see datasheet in (for instance):
www.ac-nancy-metz.fr/enseign/ssi/ressourcesP/Documentation/GE/TTL/74hc181.pdf.

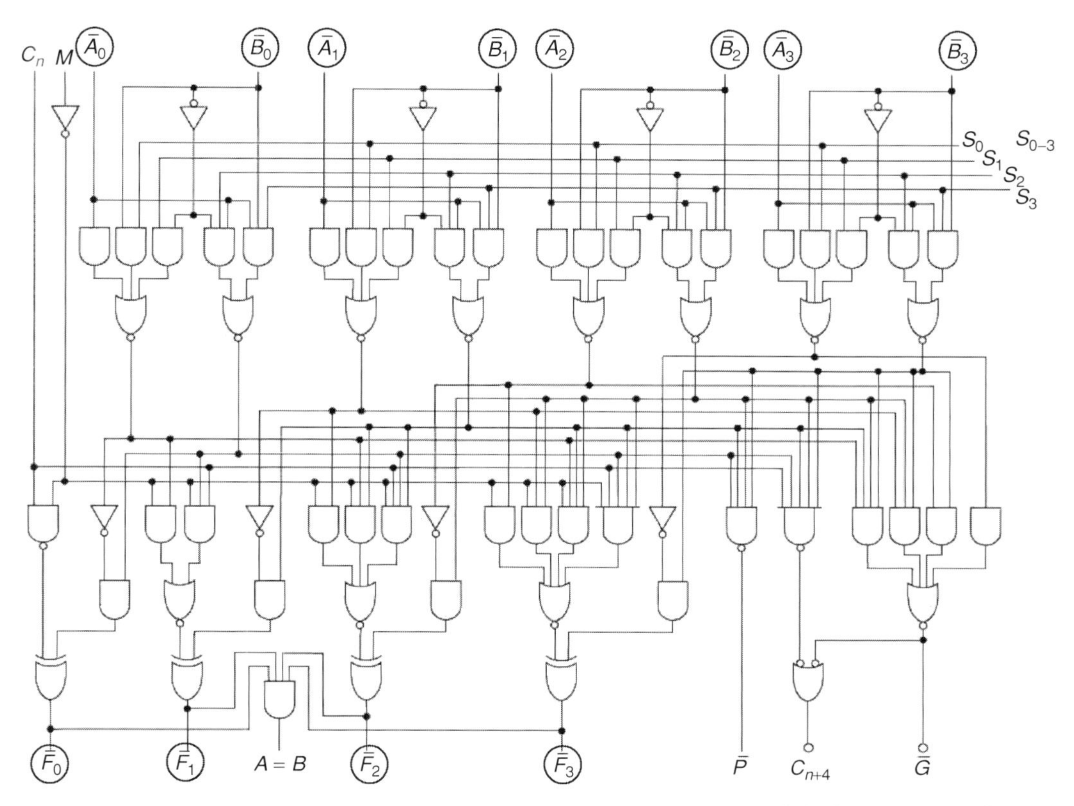


**Figure 15.6** Basic example of a four-bit ALU (inputs $\bar{A}_0\bar{A}_1\bar{A}_2\bar{A}_3$, $\bar{B}_0\bar{B}_1\bar{B}_2\bar{B}_3$, output $\bar{F}_0\bar{F}_1\bar{F}_2\bar{F}_3$).

level) and arithmetic with $M = L$ (for low voltage level). The logic mode ($M = H$) performs individual bit–bit logical operations between the two inputs $\bar{A}$, $\bar{B}$, with carry inhibition. The arithmetic mode ($M = L$) transfers carries at each individual bit–bit operation to the higher bit level. For instance, this is the difference between performing the operation $A \oplus B$, which consists in logically XORing the two operands on an individual bit-by-bit basis (without carry), and performing the operation $A + B$, which represents the arithmetic addition (with carry transfer) of the two binary numbers $A$, $B$. The circuit can perform either 16 logical operations or 16 arithmetic operations. Shown at the top right in Fig. 15.6 are four "mode select" bits $S = (S_0S_1S_2S_3)$. These bits control the $2^4 = 16$ possible functions, which may be either logical or arithmetic, according to the computational mode $M$. Also shown at the top left of the figure, an input $c_n$ and output $c_{n+4}$ (high or low values) are provisioned to enable the cascading of arithmetic computations over larger word sizes, for instance to form a 16-bit ALU network. Table 15.1 shows the 16 logical functions ($M = H$) performed by the ALU according to the 16 different selection modes $S$.

A detailed demonstration of the *74181-TTL* and other higher-level network arrangements (e.g., 16-bit ALU), which is animated through Java applets, is available on the

**Table 15.1** Logical functions $F = f(\bar{A}, \bar{B})$ performed from four-bit operands $\bar{A}$, $\bar{B}$ by the ALU network shown in Fig. 15.6, as defined by the 16 selection modes $S = (S_0\, S_1\, S_2\, S_3)$, with $S_i = 0$ for "low" and $S_i = 1$ for "high" voltage levels.

| $S_0$ | $S_1$ | $S_2$ | $S_3$ | $F = f(\bar{A}, \bar{B})$ | $S_0$ | $S_1$ | $S_2$ | $S_3$ | $F = f(\bar{A}, \bar{B})$ |
|---|---|---|---|---|---|---|---|---|---|
| 0 | 0 | 0 | 0 | $\bar{A}$ | 1 | 0 | 0 | 0 | $\bar{A} \wedge B$ |
| 0 | 0 | 0 | 1 | $\overline{A \wedge B}$ | 1 | 0 | 0 | 1 | $A \oplus B$ |
| 0 | 0 | 1 | 0 | $\bar{A} \vee B$ | 1 | 0 | 1 | 0 | $B$ |
| 0 | 0 | 1 | 1 | 1 | 1 | 0 | 1 | 1 | $A \vee B$ |
| 0 | 1 | 0 | 0 | $\overline{A \vee B}$ | 1 | 1 | 0 | 0 | 0 |
| 0 | 1 | 0 | 1 | $\bar{B}$ | 1 | 1 | 0 | 1 | $A \wedge \bar{B}$ |
| 0 | 1 | 1 | 0 | $\overline{\bar{A} \oplus \bar{B}}$ | 1 | 1 | 1 | 0 | $A \wedge B$ |
| 0 | 1 | 1 | 1 | $A \vee \bar{B}$ | 1 | 1 | 1 | 1 | $A$ |

Internet.[12] This basic example only provides a hint of the computing capabilities of logical circuits and more complex network arrangements. It just suffices to illustrate that the ALU can compute anything corresponding to a VN program, as decomposed into a series of micro-instructions and bit processing through such complex logic networks. We shall now move on to the core of the issue: the reversibility of computing and the need to find alternative ways to process information.

As we have learnt, the above description constitutes a basic inventory of classical logic gates and circuit networks, which (except for NOT gate) we know perform *nonreversible* operations. According to Landauer's principle, which was described earlier, the information erasure of each input bit corresponds to a minimum heat and entropy generation of $Q = k_B T \log 2$ and $H = k_B \log 2$, respectively. The power consumption of the logic gates and of the various electronic circuits (including digital amplifiers to compensate for signal propagation or splitting loss) also represents a source of heat and entropy, whose measures are orders of magnitude greater than Landauer's bound. The heat dissipation is estimated to be $Q' = 500 k_B T \log 2 \equiv 500Q$ for elementary logic gates, and $Q'' = 10^8 k_B T \approx 10^8 Q$ for the CPU in ordinary home computers.[13] Most of this heat, of resistive origin, could be alleviated if computers would be run at ultra-low temperatures ($T \to 0$), using "superconducting" electronics. If such a computer could ever be built, then one would reach the ultimate limits predicted by Landauer's bound. The key question, then, is whether or not it is possible to build computers with reversible logic. As previously hinted, physicists, who are attached to the concept of reversible processes, intuitively suspect that it should be possible to manipulate information in a *reversible* way, without heat dissipation or entropy increase. The corresponding computer should exclusively use *reversible gates*, which we shall analyze next.

[12] See: http://tams-www.informatik.uni-hamburg.de/applets/hades/webdemos/20-arithmetic/50-74181/demo-74181-ALU.html.

[13] See (respectively): M. A. Nielsen and I. L. Chuang, *Quantum Computation and Quantum Information* (Cambridge: Cambridge University Press, 2000) and V. Vedral and M. B. Plenio, Basics of quantum computation. *Prog. Quant. Electron.*, **22** (1998), 1–39.

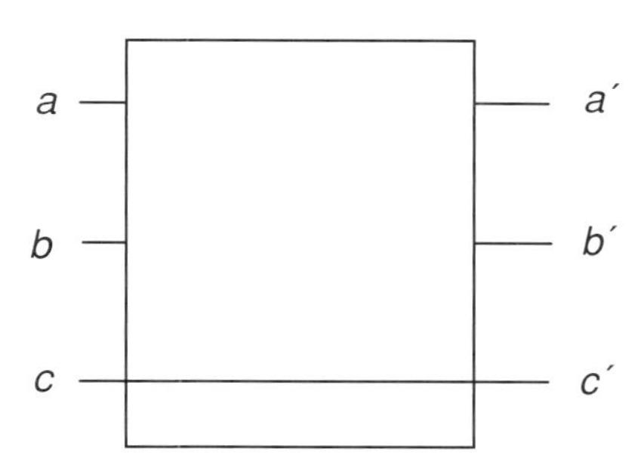


| a | b | c | a´ | b´ | c´ |
|---|---|---|---|---|---|
| 1 | 1 | 0 | 1 | 1 | 0 |
| 1 | 0 | 0 | 1 | 0 | 0 |
| 0 | 1 | 0 | 0 | 1 | 0 |
| 0 | 0 | 0 | 0 | 0 | 0 |
| 1 | 1 | 1 | 1 | 1 | 1 |
| 1 | 0 | 1 | 0 | 1 | 1 |
| 0 | 1 | 1 | 1 | 0 | 1 |
| 0 | 0 | 1 | 0 | 0 | 1 |

**Figure 15.7** Fredkin gate diagram, with corresponding truth table.

## 15.3 Reversible logic gates and computation

In this section, I shall describe a new class of logic gates, which make it possible to build a variety of *reversible-computing* circuits. The first two elementary gates having such a property are called the *Fredkin* and the *Toffoli* gates, which we shall consider now.

The *Fredkin gate* is illustrated in Fig. 15.7, along with its truth table. It is seen from the figure that the Fredkin gate has three input ports $(a, b, c)$ and three output ports $(a', b', c')$. Two input ports are used for the operand data $(a, b)$, the third port being used for a control bit $(c)$. The gate operates according to the following logic (see also the Figure's truth table):

- If $c = 0$, then $a, b$ remain unchanged $(a' = a, b' = b)$;
- If $c = 1$, then $a, b$ are swapped $(a' = b, b' = a)$;
- The control bit $c$ remains unchanged $(c' = c)$.

Figure 15.8(a) and (b) illustrate the first two basic operations of the Fredkin gate: REPEAT $(c = 0)$ and CROSSOVER or SWAP $(c = 1)$. But the gate can be configured to perform other functions of interest, as shown in Fig. 15.8(c) and (d): setting $a = 1, b = 0, c = x$ yields $a' = \bar{x}$ (NOT $x$) and $b' = c' = x$ (FANOUT $x$). Setting $a = 0, b = y, c = x$ yields $a' = x \wedge y$ ($x$ AND $y$) and $b' = \bar{x} \wedge y$ ($\bar{x}$ AND $y$). The input bits $a, b$, which are preset to constant values 0 or 1 in the Fredkin gate, as shown in Fig. 15.8(c) and (d), are referred to as *ancilla bits*. Such ancilla bits, which are the key to the operation of reversible gates, represent a novel concept in computational logic. Another novelty is that reversible gates produce extra or *garbage* bits, which are not useful to the rest of the computation (e.g., the bit $b' = \bar{x} \wedge y$ in Fig. 15.8(d). Last but not least, we notice from the Fredkin gate truth table (Fig. 15.7) that each output port configuration $(a', b', c')$ exclusively corresponds to a unique input port configuration $(a, b, c)$. Thus, there exists a mutual and unique correspondence between the input and the output information, hence, computation through the Fredkin gate is *reversible*.

Because the Fredkin gate can perform both NOT and AND logical functions, it can be used as a universal building block to construct any other logic gate. To prove this

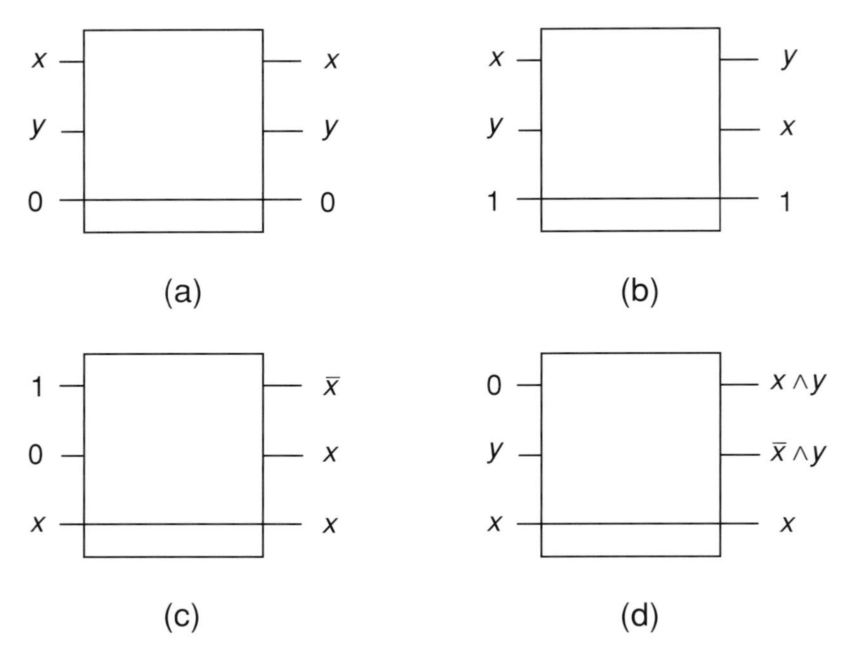


**Figure 15.8** Fredkin gate configurations (a) REPEAT, (b) CROSSOVER, (c) NOT [port $a'$] and FANOUT [ports $b'$ and $c'$], (d) AND [port $a'$] and NOT $x$ AND $y$ [port $b'$].

point, we need to introduce *De Morgan's law* (or theorem).[14] Such a theorem can be put in the form of the following equations:

$$\begin{aligned} \overline{a \vee b} &= \bar{a} \wedge \bar{b}, \\ \overline{a \wedge b} &= \bar{a} \vee \bar{b}. \end{aligned} \tag{15.4}$$

In ensemble theory, De Morgan's law equivalently translates into the relations $\overline{A \cup B} = \bar{A} \cap \bar{B}$ and $\overline{A \cap B} = \bar{A} \cup \bar{B}$, which can be readily verified from Venn diagrams (see Fig. 1.4). Using Eq. (15.4) we observe that the OR function can be generated through the identity

$$a \vee b = \overline{\overline{a \vee b}} = \overline{\bar{a} \wedge \bar{b}}, \tag{15.5}$$

which exclusively uses the NOT and AND gates. Using the above property and the relations in Eq. (15.1) (with $a \bar{\wedge} b = \overline{a \wedge b}$), we observe that the function XOR can be constructed either with NOT and AND gates or with NOT, AND, and OR gates.[15] Here, the key difference with the previous "classical" logic of the ALU in the VN computer is that all gate circuits constructed from Fredkin gates perform fully reversible computations, by virtue of the gate's reversible logic.

We consider next a second type of elementary reversible-logic gate, which is called the *Toffoli gate*. Its diagram and corresponding truth table are shown in Fig. 15.9. Like

[14] See http://en.wikipedia.org/wiki/De_Morgan%27s_laws.

[15] It is left as an exercise to construct different possibilities of XOR circuits exclusively based on Fredkin gates.

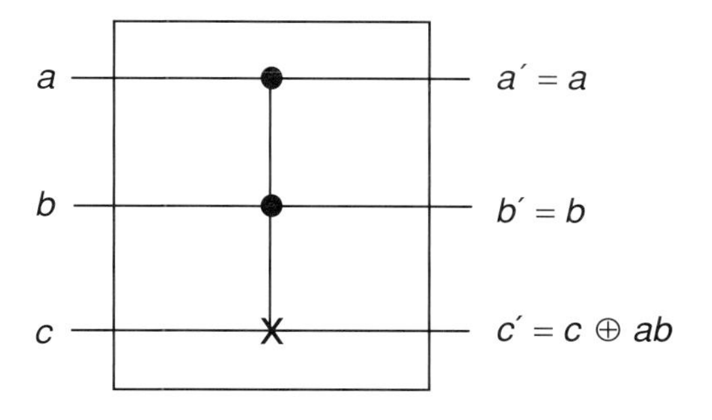


| a | b | c | a′ | b′ | c′ |
|---|---|---|---|---|---|
| 0 | 0 | 0 | 0 | 0 | 0 |
| 0 | 0 | 1 | 0 | 0 | 1 |
| 0 | 1 | 0 | 0 | 1 | 0 |
| 0 | 1 | 1 | 0 | 1 | 1 |
| 1 | 0 | 0 | 1 | 0 | 0 |
| 1 | 0 | 1 | 1 | 0 | 1 |
| **1** | **1** | **0** | **1** | **1** | **1** |
| **1** | **1** | **1** | **1** | **1** | **0** |

**Figure 15.9** Toffoli gate diagram, with corresponding truth table, showing bit-flipping of $c$ if the two control bits are set to $a = b = 1$.

the Fredkin gate, the Toffoli gate has three input ports $(a, b, c)$ and three output ports $(a', b', c')$. The difference is that $a, b$ are *control* bits ($a' = a, b' = b$) and $c$ is the *target* bit. By convention, control bits are marked with a dot (•) and the target bit with a cross (×). The Toffoli gate outputs $c' = c \oplus (a \wedge b)$, or $c' = c \oplus ab$ for short. The truth table in the figure shows that the value of $c$ is flipped (function NOT) when the two control bits are set to $a = b = 1$, and is left unchanged otherwise (function REPEAT). The Toffoli gate can, thus, be conceived as a *controlled-controlled-NOT* (CCNOT) gate. As we shall see next, the inputs $(a, b, c)$ can also play different roles of variable, control, or ancilla bits, with $c'$ remaining the gate's functional output bit, by definition.

Figure 15.10 illustrates that the logical functions NOT, AND, NAND, XOR, and FANOUT can be built from single Toffoli gates. The function OR can be built with two NOT and one NAND gates using the relation $a \vee b = \bar{a} \,\bar{\wedge}\, \bar{b}$. The figure also shows the CNOT gate, which can be viewed as a simplified representation of the XOR gate. The Toffoli gate, like the Fredkin gate, thus, represents a universal building block in the construction of reversible-logic circuits.

I shall now provide an example of a complex circuit exclusively constructed from CNOT and Toffoli gates, which is called the *plain adder*. A plain adder performs the addition of two binary numbers $A = a_0a_1, \ldots, a_n$ and $B = b_0b_1, \ldots, b_n$. This function entails not only the bit-by-bit addition operations ($a_i \oplus b_i$ with $i = 0 \ldots n$) but also the computation and addition of the *carry* bit $c_{i-1}$ at each stage $i$. To construct a plain adder, one, thus, needs two new gate circuits, which we call SUM and CARRY, and which are illustrated in Fig. 15.11. It is seen from the figure that SUM is built from a cascade of two CNOT gates.

The plain-adder circuit performs the double sum $x = a \oplus b$ (bit-by-bit addition of order $i$) and $y = x \oplus c$ (addition of the result with carry $c$ of order $i - 1$), with indices being omitted here for clarity. The circuit CARRY outputs the carry bit $c' = (ab) \oplus c(a \oplus b)$, where $c$ is the carry from order $i - 1$. It is easily checked that $c'$ is the carry of order $i$ of the sum $a \oplus b \oplus c$.[16] The plain-adder circuit, which performs the computation

[16] Indeed, (a) if $a = b = 0$ then $c' = 0 \oplus 0c = 0$; (b) if $a = b = 1$ then $c' = 1 \oplus 0c = 1$; (c) if $a \neq b$ then $c' = 0 \oplus 1c = c$.

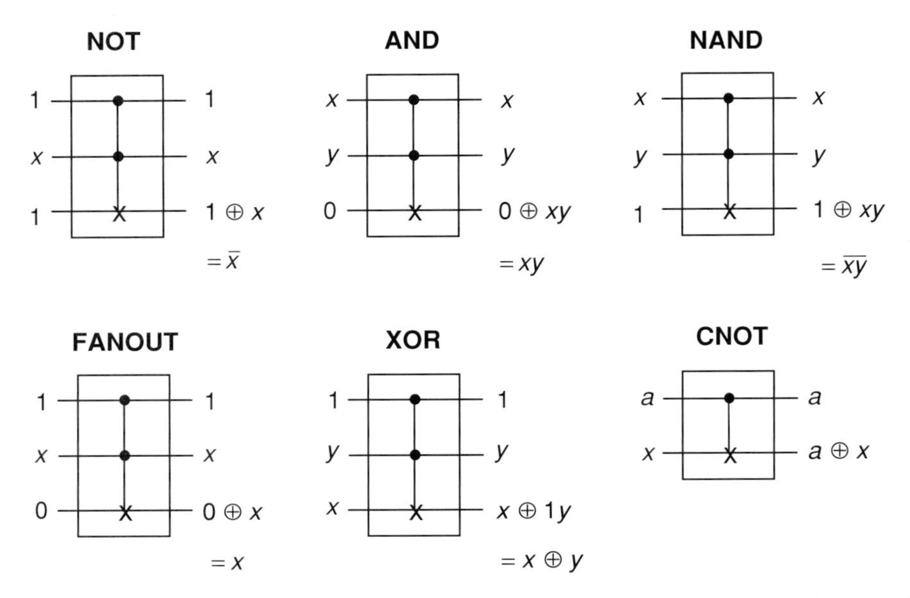


**Figure 15.10** Logic gates NOT, AND, NAND, FANOUT, and XOR, as built from single Toffoli gates. The gate CNOT is also shown.

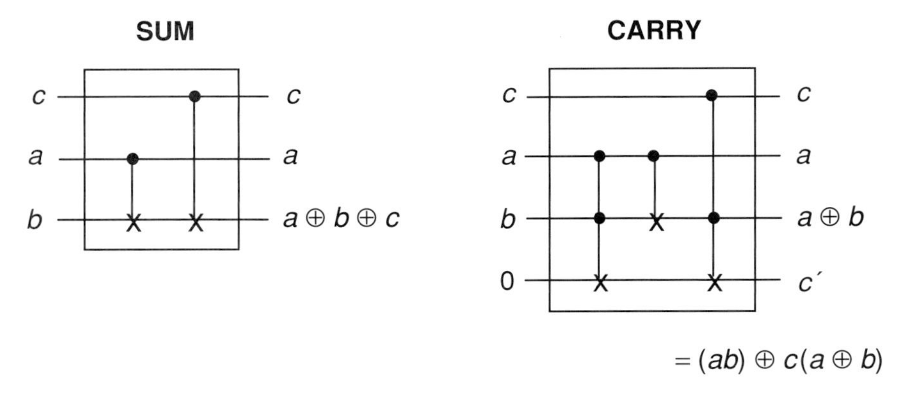


**Figure 15.11** Gate circuits SUM and CARRY, built from CNOT and Toffoli gates.

$D = A + B$ of two binary numbers $A = a_0a_1, \ldots, a_n$ and $B = b_0b_1, \ldots, b_n$, chosen here with $n = 2$ for simplicity,[17] is shown in Fig. 15.12.

The interpretation of the V-shaped circuit shown in the figure goes as follows. Consider first the CARRY gates. In the left part of the circuit (the descending branch of the V), we observe that the carry is computed through a cascade of three CARRY gates, up to the highest order ($i = 3$), which yields $c_3 = d_3$ (highest-weight bit of $D = A + B$). In the right part of the circuit (ascending branch of the V), the reverse operation is

[17] As adapted from V. Vedral and M. B. Plenio, Basics of quantum computation. *Prog. Quant. Electron.*, **22** (1998), 1–39.

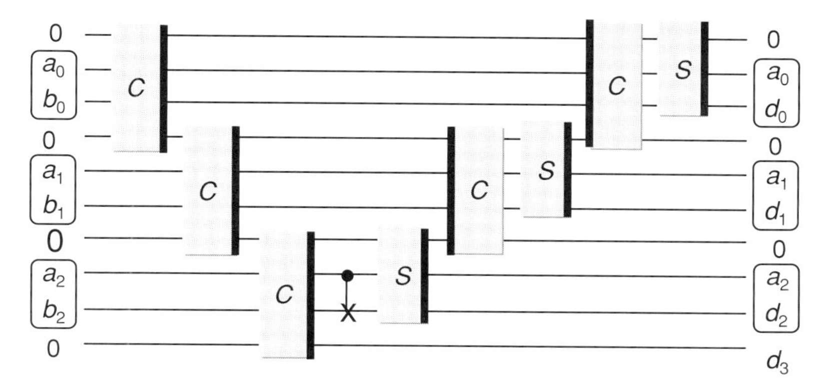


**Figure 15.12** Plain-adder gate circuit, performing the addition of two binary numbers $A = a_0a_1a_2$ and $B = b_0b_1b_2$, with output $A = a_0a_1a_2$ and $D = A + B = d_0d_1d_2d_3$. The gates $C$ and $S$ with a thick bar to the right correspond to the CARRY and SUM gates shown in Fig. 15.11. The gate $C$ with a thick bar to the left effects the reverse operation of CARRY.

effected,[18] which resets the bit values initially input to the previous CARRY gate in the sequence. Second, consider the three SUM gates in the right part of the circuit. These gates are seen to compute $d_i = a_i \oplus b_i \oplus c_{i-1}$ in descending order (from $i = 2$ to $i = 0$), with $c_0 = 0$. At the bottom of the V, one observes that the reverse CARRY gate is substituted with a CNOT gate. An easy verification shows that such an arrangement makes it possible to initiate the descending summation $d_i = a_i \oplus b_i \oplus c_{i-1}$ from the last CARRY gate outputs.

As expected from the property of its universal, reversible-logic gate components, the above plain-adder circuit is fully reversible, meaning that no information is lost (the simultaneous knowledge of $A$ and $D = A + B$ giving knowledge of $B$). The function of the circuit (also referred to as a "*quantum network*" in QIT jargon), can by symbolized by the notation

$$(A, B) \rightarrow (A, A + B). \tag{15.6}$$

In the same jargon, the operands $A, B$ are also called *registers*, by analogy with the classical or von Neumann (VN) computer, which was described in the previous section. A remarkable feature of the $(A, B) \rightarrow (A, A + B)$ network is that the many "garbage bits," which are generated by the intermediate computation steps, are eventually disposed of. Furthermore, the ancilla bits 0 used at the input (see Fig. 15.12) are all reset to their initial values, which makes them available for any future use in a larger quantum network. But we aren't finished yet with the plain-adder network: it also reserves some interesting surprises![19]

[18] The reverse operation of CARRY is effected by the same gate as shown in Fig. 15.12, but with the operations performed in the reverse order, as in a mirror version of the circuitry, or reading the diagram from right to left.

[19] M. A. Nielsen and I. L. Chuang, *Quantum Computation and Quantum Information* (Cambridge: Cambridge University Press, 2000); V. Vedral and M. B. Plenio, Basics of quantum computation. *Prog. Quant. Electron.*, **22** (1998), 1–39.

First, the plain-adder network can be used in the reverse order, i.e., feeding input-register data $A, B$ from right to left in Fig. 15.12, which yields the operations

$$(A, B) \to (A, A - B) \quad \text{for } A \geq B \tag{15.7}$$

and

$$(A, B) \to [A, 2^{n+1} - (B - A)] \quad \text{for } A < B \tag{15.8}$$

where $n + 1$ is the size of the second output register. In both cases ($A \geq B$ or $A < B$), the network performs the subtraction function. In the second case it can be shown that the bit of highest weight is always 1, which represents a "negative-sign bit" for the result of the difference $A - B$.

Second, the plain-adder network also makes it possible to perform *modular algebra* with the functions of *multiplication*, and *exponentiation*. Indeed, modular multiplication and exponentiation, i.e., $(A, B) \to (A, A \times B)$ and $(A, B) \to (A, A^B)$, respectively, can be performed using the properties

$$A \times B \bmod m = (\underbrace{A + A + \cdots + A}_{B\text{ times}}) \bmod m, \tag{15.9}$$

$$A^B \bmod m = (\underbrace{A \times A \times \cdots \times A}_{B\text{ times}}) \bmod m, \tag{15.10}$$

which apply given any modulus $m$. Modular algebra is the key to solving quantum-computation problems, such as the *factorization of integers into primes* (see Chapter 20).

In the forthcoming chapters, reversible-logic gates and their networks will not be used with "classical bits," but rather with "quantum bits," or qubits, which hold more interesting properties and surprises.

## 15.4 Exercises

**15.1** (B): Prove the following alternative definitions for XOR gates:

$$\begin{aligned} a \oplus b &= (a \vee b) \wedge (\bar{a} \vee \bar{b}) \\ &= (a \wedge \bar{b}) \vee (\bar{a} \wedge b) \\ &= (a \barwedge \bar{b}) \barwedge (\bar{a} \barwedge b). \end{aligned}$$

**15.2** (B): Prove De Morgan's law:

$$\begin{aligned} \overline{(a \wedge b)} &= \bar{a} \vee \bar{b}, \\ \overline{(a \vee b)} &= \bar{a} \wedge \bar{b}. \end{aligned}$$

**15.3** (T): Prove by Boolean algebra that the definition of XOR

$$a \oplus b = (a \barwedge b) \wedge (a \vee b)$$

is formally equivalent to any of the following three definitions:

$$\begin{aligned} a \oplus b &= (a \vee b) \wedge (\bar{a} \vee \bar{b}) \\ &= (a \wedge \bar{b}) \vee (\bar{a} \wedge b) \\ &= (a \bar{\wedge} \bar{b}) \bar{\wedge} (\bar{a} \bar{\wedge} b). \end{aligned}$$

(*Clue*: use the distributive property according to which $a \wedge (b \vee c) = (a \wedge b) \vee (a \wedge c)$).

**15.4** (B): What is the Boolean function of the gate circuit?

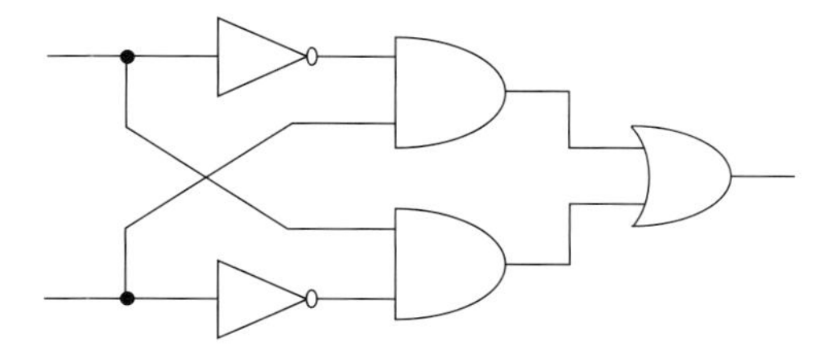

**15.5** (B): Through Boolean logic, show that the circuit arrangement of three successive CNOT gates, $A_{\text{CNOT}}$, $B_{\text{CNOT}}$, $A_{\text{CNOT}}$,

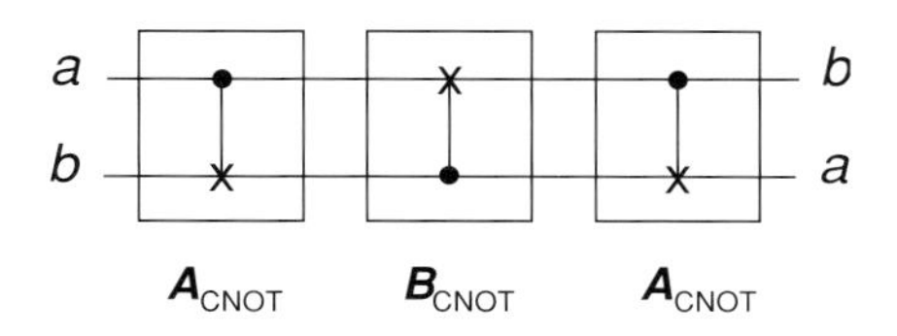


is equivalent to a SWAP gate.

# 16 Quantum bits and quantum gates

This chapter represents our first step into *quantum information theory* (QIT). The key to operating such a transition is to become familiar with the concept of the quantum bit, or *qubit*, which is a probabilistic superposition of the classical 0 and 1 bits. In the quantum world, the classical 0 and 1 bits become the *pure states* $|0\rangle$ and $|1\rangle$, respectively. It is as if a coin can be classically in either heads or tails states, but is now allowed to exist in a superposition of both! Then I show that qubits can be physically transformed by the action of *unitary matrices*, which are also called *operators*. I show that such qubit transformations, resulting from any qubit manipulation, can be described by rotations on a 2D surface, which is referred to as the *Bloch sphere*. The *Pauli matrices* are shown to generate all such unitary transformations. These transformations are *reversible*, because they are characterized by *unitary matrices*; this property always makes it possible to trace the input information carried by qubits. I will then describe different types of elementary quantum computations performed by elementary quantum gates, forming a veritable "zoo" of unitary operators, called $I$, $X$, $Y$, $Z$, $H$, CNOT, CCNOT, CROSSOVER or SWAP, controlled-$U$, and controlled-controlled-$U$. These gates can be used to form *quantum circuits*, involving any number of qubits, and of which several examples and tools for analysis are provided. Finally, the concept of *tensor product*, as progressively introduced through the above description, is eventually formalized. The chapter concludes with the intriguing *noncloning theorem*, according to which it is not possible to duplicate or "clone" a quantum state.

## 16.1 Quantum bits

In computer science, the binary digit, or *bit*, represents the elementary unit of information. It is a *scalar* with two possible values, 0 and 1. The bit can also be viewed as a *two-state* variable, which can be in either the state 0 or the state 1, with no possible other or intermediate states. In Shannon's theory, the bit is also the *unit of information*, but is a real number satisfying $0 \leq x \leq 1$, which expresses the average information available from a random binary source, which is called *entropy*. The results of Shannon's information theory, in fact, concern any information sources, including multi-level or multisymbolic ones, up to any integer number $M$. In these cases, the bit information available from the source is bounded according to $0 \leq x \leq \log_2 M$. When $M$ is a power of two ($M = 2^k$), it is possible to encode the different symbols into as many binary codewords, and the corresponding sources can be conceived as an extension of a binary source. Thus, each bit in a given codeword is allowed one out of two possible states, and this corresponds to the most fundamental representation of *classical* information.

In quantum information theory (QIT), and its derivative, quantum computation (QC), the elementary unit of information is the *quantum bit* or *qubit*. As we shall see, the striking property of the qubit is to escape any definition of being 0 or 1. It is correct, however, to say that it can be either 0 or 1. To clarify, somewhat, such a mystery, consider a closed box with a coin inside. We shake the box. The coin must then be resting in the heads or tails position (excluding here any other possibility, for simplicity). The question is, "What is the coin's position?" The intuitive and classical answer is, "It must be *either* heads *or* tails." According to the quantum definition, the coin is described by a qubit. The answer is that the coin position, or state, is "*neither* heads *nor* tails, but *a superposition of both*." As long as we do not open the box, we cannot know in which state the coin actually exists. By opening the box, we make a measurement of the coin state, and the outcome is a classical bit of information, namely heads or tails. This basic example provides an intuitive notion of the nature of the qubit, which I shall formalize through this chapter.

In the above, we have used different new terms, such as *state*, *superposition of states*, and *measurement*, which (to some readers) represents as many hints of the domain of *quantum mechanics*. In this introductory chapter, we shall approach the notion of qubit without approaching quantum mechanics any closer. Indeed, there is no need to present a more complicated picture if we can introduce the qubit and its properties and formalism by simpler means. We have realized from the preceding explanation that the qubit is a piece of information that combines in some ways the information of both 0 and 1. Therefore, we may simply view the qubit as a two-dimensional (2D) *vector*, with one dimension defining the 0 information component, and the other dimension the 1 information component.

We shall now formalize the 2D vector representation of the qubit. Given a 2D vector space with basis $\vec{u}$, $\vec{v}$, one can then define the qubit state, $\vec{q}$, under the linear combination

$$\vec{q} = \alpha\vec{u} + \beta\vec{v}, \tag{16.1}$$

where $\alpha, \beta$ are *complex numbers*,[1] which represent the vector's coordinates in the 2D space. Assuming that $(\vec{u}, \vec{v})$ is an orthonormal basis,[2] and using the column

[1] To recall essential basics, *complex numbers* $z$ are defined as $z = a + \mathrm{i}b$, where $(a, b)$ are real numbers and i the "pure imaginary" basis having the property $\mathrm{i}^2 = -1$. The real numbers $a$ and $b$ are called the *real* and the *imaginary* parts of $z$, respectively. The *length* or *modulus* of $z$ is defined as $|z| = \sqrt{a^2 + b^2}$. Complex numbers can be equivalently written as $z = |z|\mathrm{e}^{\mathrm{i}\theta}$ where $\theta = \tan^{-1}(b/a)$ is the *argument* of $z$, and where $\mathrm{e}^{\mathrm{i}\theta} \equiv \cos\theta + \mathrm{i}\sin\theta$ is the *imaginary-exponential* function. The *complex-conjugate* of $z$, indifferently called $z^*$ or $\bar{z}$, is defined as $z^* = \bar{z} = a - \mathrm{i}b = |z|\mathrm{e}^{-\mathrm{i}\theta}$. A key property is $z^*z = z\bar{z} = |z|^2 = a^2 + b^2$.

[2] To recall from vector-space algebra, an orthonormal basis is any set of $n$ vectors $\vec{u}_1, \vec{u}_2, \ldots, \vec{u}_n$, which satisfy for all $i, j = 1, \ldots, n$ the scalar-product condition $\vec{u}_i \cdot \vec{u}_j = \delta_{ij}$, where $\delta_{ij}$ is the *Kronecker symbol* ($\delta_{ij} = 1$ for $i = j$ and $\delta_{ij} = 0$ for $i \neq j$). For $i = j$, the condition gives $\vec{u}_i^2 = |\vec{u}_i|^2 = 1$, or $|\vec{u}_i| = 1$.

representation for vectors, we have

$$\vec{q} = \alpha\vec{u} + \beta\vec{v} = \alpha\begin{pmatrix}1\\0\end{pmatrix} + \beta\begin{pmatrix}0\\1\end{pmatrix}. \tag{16.2}$$

As detailed further on in this chapter, in QIT a vector or *qubit state* $\vec{q}$ is noted instead as $|q\rangle$, where the arrow $|\,\rangle$ is called a *ket* (as in the second half of the word *bracket*). Consistently, and by analogy with classical bits, the two elementary states $\vec{u}$ and $\vec{v}$ are noted $|0\rangle$ and $|1\rangle$, respectively. The qubit can, thus, be defined as the vector linear combination:

$$|q\rangle = \alpha|0\rangle + \beta|1\rangle. \tag{16.3}$$

This shows that a qubit can be conceived as a *superposition of the two elementary orthogonal states* $|0\rangle$ and $|1\rangle$. In QIT jargon, the complete set of states $|0\rangle$, $|1\rangle$ is referred to as the *computational basis*. The two complex coordinates $\alpha$, $\beta$, are also called the *qubit amplitudes*.

It is seen from the above that, contrary to the classical information *bits*, the *qubit* can be both 0 and 1, or, more accurately, a linear superposition of states 0 and 1. Unlike classical bits, which correspond to only two possible points on a line ($x = 0$ or $x = 1$), qubits can correspond to an infinity of points in a 2D space. We shall see later how such points can be represented visually.

As we have seen from Chapter 15, the classical or von Neumann computer retrieves information by reading the bit contents in its inner memory. Any such reading operation is, thus, analogous to a physical measurement of the memory contents. Likewise, the information carried by qubits can be measured. The key difference between the classical bit and the qubit is that, as we have seen, the latter is a superposition of the states $|0\rangle$ and $|1\rangle$ in a 2D space. It is *not* possible to physically measure the two complex amplitudes $\alpha$, $\beta$, which define the qubit state. This amounts to saying that *it is not possible to "measure" a qubit state*. Rather, any physical measurement of a qubit yields a classical bit, which must be either 0 or 1. Such a measurement is not deterministic, but associated with certain probabilities. The probability that the qubit measure yields the bit 0 is proportional to $|\alpha|^2$, and that of yielding the bit 1 is proportional to $|\beta|^2$. Since there exist only two measurement possibilities, we must have

$$|\alpha|^2 + |\beta|^2 = 1. \tag{16.4}$$

Since $|\alpha|^2 + |\beta|^2$ is the length or magnitude of the qubit vector, this result expresses the property that the qubit is a *unitary vector*. Thus a qubit with $\alpha = 0$ or $\beta = 0$ exists in the pure state $|0\rangle$ or $|1\rangle$. The corresponding measurements yield the exact result 0 with probability $|\alpha^2| = 1$, or 1, with probability $|\beta^2| = 1$, respectively. Another case of interest is $|\alpha|^2 = |\beta|^2 = 1/2$, where the measurement outcome has equal chances of being 0 or 1. In general, there is no way to tell the measurement outcome, even if the qubit state is known, because the measurement is probabilistic. Given the property in Eq. (16.4) concerning the qubit amplitudes, the most general definition of the qubit is, therefore,

$$|q\rangle = \frac{\alpha'}{\sqrt{|\alpha'|^2 + |\beta'|^2}}|0\rangle + \frac{\beta'}{\sqrt{|\alpha'|^2 + |\beta'|^2}}|1\rangle, \tag{16.5}$$

where $\alpha'$, $\beta'$ are any complex numbers.

To provide an illustration, let us come back to the analogy of the closed box containing a coin, and tell the story again with the qubit formalism. The box has been shaken to the point that it is impossible to tell on which side, heads or tails, the coin may be found. We can assume without loss of generality that the coin is "fair," and that it must rest on either side at the bottom of the box, so that the probabilities of the coin being on heads or on tails are strictly equal. This defines two equiprobable "coin states;" call them here $|H\rangle$ and $|T\rangle$. As long as we don't open the box, the coin remains in the *superposition* of states corresponding to the qubit:

$$|q\rangle = \frac{1}{\sqrt{2}}|H\rangle + \frac{1}{\sqrt{2}}|T\rangle \tag{16.6}$$

with the amplitudes satisfying $|\alpha|^2 = |\beta|^2 = (1/\sqrt{2})^2 \equiv 1/2$. Thus, from the quantum viewpoint, the coin inside the closed box does exist in *both* heads and tails states. The action of opening the box and checking the physical position of the coin yields the qubit projection $|H\rangle$ or $|T\rangle$, which corresponds to either of the classical bits 0 = heads or 1 = tails, respectively, with equal probabilities. This measurement is seen to *collapse* the coin state $|q\rangle$ into one of its elementary states $|H\rangle$ or $|T\rangle$, which yields the *observable* value referred to as heads or tails. The counterintuitive, or nonclassical, conclusion is that as long as no measurement (or state collapse) is effected, the coin dwells in the superposition of states defined by Eq. (16.6). Note that this superposition is often referred to as $|+\rangle$, or "cat state," as I shall explain.

It is possible to think of many other examples of superpositions of states. For instance, consider a toddler in her room on an early Sunday morning. Is she still sleeping (state $|S\rangle$) or is she awake and playing (state $|A\rangle$)? By 8 a.m. the parents know that there is a fair chance that she must be asleep, say 90%. By 9 a.m., they know that the odds are 50%. But if the parents don't get up and check, the quantum view is that by 8 a.m., the child must be in the state $|S\rangle/\sqrt{0.9} + |A\rangle/\sqrt{0.1}$ and by 9 a.m. in the state $(|S\rangle + |A\rangle)/\sqrt{2}$, which we have called the state $|+\rangle$. Only by opening the child's bedroom door can the actual measurement be made, which results in the state collapse into $|S\rangle$ or $|A\rangle$, giving the classical information 0 = sleeping or 1 = awake. As uncanny as this example may sound, it illustrates the quantum concept of "superposition of states," now with the notion of a time evolution from the initial pure state $|S\rangle$ to the final pure state $|A\rangle$. One may view our example as a gentler, yet conceptually equivalent, version of the *Schrödinger's cat* thought experiment.[3] The stories of the tossed coin,

[3] In a thought experiment, Schrödinger imagines a cat that is placed inside a sealed box. The box contains a can of poison gas, a radioactive substance, and a Geiger counter. The decay rate of the radioactive substance is such that after one hour, there is a 50% chance that one atom decays, which is recorded by the Geiger counter. In this event, the counter activates the opening of the poison-gas can, resulting in the cat's instant death. At any time, the (poor) cat's state can be viewed as a superposition of the pure states $|\text{dead}\rangle$ or $|\text{alive}\rangle$,

of the sleeping toddler, and of Schrödinger's cat illustrate the same paradox: does the state of a macroscopic object or system require an outside observer to be defined, or is it self-defined independent of outside observation? Our intuition tells us that such an object or system cannot exist in two states at the same time, and, therefore, it must be its own "observer." The quantum-mechanics viewpoint breaks with intuition and affirms the contrary: that objects or systems can, indeed, exist in multiple states, and that only the observer intervention defines what the actual state turns out to be. About elementary information, our classical mind training requires that a bit is absolutely defined as either being 0 or 1 (regardless of possible measurement mistakes). With the qubit, we must now retrain our mind to accept the fact that a quantum of information (the qubit) is in a $0/1$ superposition state, whose outcome in observed reality remains undefined until some measurement is performed. The next two chapters will clarify and further develop such a most intriguing notion!

To summarize the above description so far, a qubit must be conceived as a *two-dimensional bit*, whose coordinates in that space represent probability amplitudes. Since two coordinates define the qubit, it is possible to represent it as a unique point on the surface of a sphere of unity radius, called a *Bloch sphere*, which is described in Appendix M. In this appendix, it is shown that the most general definition of the qubit, within an unobservable phase factor, is

$$|q\rangle = \cos\frac{\theta}{2}|0\rangle + \sin\frac{\theta}{2}e^{i\varphi}|1\rangle. \tag{16.7}$$

The two angles $\theta$, $\varphi$, thus, uniquely define the position of a point on the surface of a sphere, just like latitude and longitude on the Earth, and as illustrated in Fig. 16.1. It is seen from the figure that the pure qubits $|0\rangle$ or $|1\rangle$ correspond to the cases $\theta = 0$ or $\theta = \pi$, respectively, which occupy the north and south poles of the Bloch sphere. The key conclusion is that the qubit information corresponds to an *infinite number of states*, which are continuously distributed onto the surface of the Bloch sphere.

The above description concerned *single qubits*, corresponding to *single classical bits*. It is possible to define higher-order qubits, which correspond to two classical bits or to even longer binary codewords. Since there exist four possible pairs of classical bits,

which begins from a certain $|\text{alive}\rangle$ and evolves over time towards a certain $|\text{dead}\rangle$. After the 1-hour delay, the probabilities of the two states are equal, and the cat dwells in the state superposition

$$|+\rangle = \frac{|\text{dead}\rangle + |\text{alive}\rangle}{\sqrt{2}},$$

hence, the name "cat state." At any time, one cannot be sure if the cat is dead or alive, and this information requires one to make a measurement by opening the box. Such a measurement results into the collapse of the cat's state into either of the pure states $|\text{dead}\rangle$ or $|\text{alive}\rangle$. The initial purpose of Schrödinger's though experiment was to illustrate that such a quantum view must be incomplete: the cat cannot be both dead *and* alive at the same time! And there is no need for someone to open the box to define in which state the cat actually exists. Yet such a view is consistent with the so-called "*Copenhagen interpretation*," according to which systems can exist in such a superposition of states until reaching state collapse through physical observation. A fine argument, which reconciles this interpretation with Schrödinger's cat paradox, is the fact that a cat is a macroscopic or classical system, and, therefore, the microscopic quantum interpretation may not apply. See discussion, and more puzzling arguments in (for instance): http://en.wikipedia.org/wiki/Schr%C3%B6dinger's_cat.

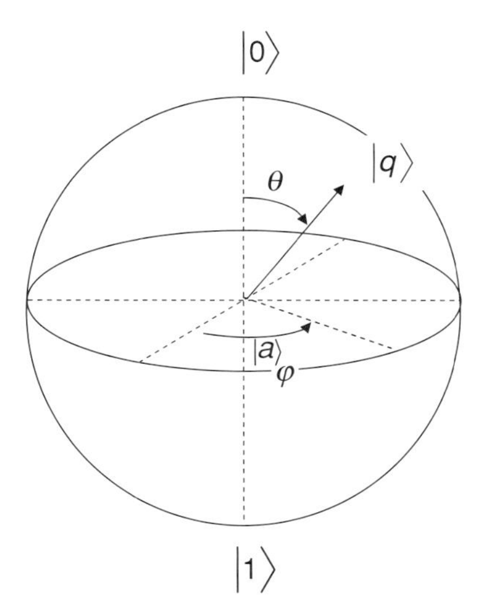


**Figure 16.1** Representation of qubit $|q\rangle$ by a point on the surface of the Bloch sphere, as defined by the "colatitude" angle $\theta$ and the "longitude" angle $\varphi$. The north and south poles correspond to the pure qubit states $|0\rangle$ and $|1\rangle$, respectively.

namely 00, 01, 10, and 11, we can call the corresponding elementary two-qubit (or 2-qubit) states $|00\rangle$, $|01\rangle$, $|10\rangle$, and $|11\rangle$, and, similarly to Eq. (16.3), we obtain the most general definition for any 2-qubit state:

$$|q\rangle = \alpha_{00}|00\rangle + \alpha_{01}|01\rangle + \alpha_{10}|10\rangle + \alpha_{11}|11\rangle. \tag{16.8}$$

In this definition, the complex amplitudes $\alpha_{ij}$ must satisfy the property

$$\sum_{ij} \left|\alpha_{ij}\right|^2 = 1. \tag{16.9}$$

From the previous description concerning 1-qubits, it is clear that each of the terms $|\alpha_{ij}|^2$ in the above summation represents the probability of finding the 2-qubit in the state $|ij\rangle$, or, equivalently, of measuring the classical bit pair $ij$. We shall note here that the notation $|ij\rangle$ for a pair of qubits $|i\rangle$, $|j\rangle$, or 2-qubit $|i\rangle|j\rangle$ actually anticipates the concept of *tensor state*, which is explained in the last section of this chapter.

We can even simplify the picture of a 2-qubit by looking at it as a "four-sided die," such as a tetrahedron, hidden in the box. Such a die has four *classically measurable* outcomes (00, 01, 10, and 11), but in the quantum world there exist an infinity of superpositions of 2-qubit pairs, as I shall now establish.

Indeed, the 2-qubit state defined in Eq. (16.8) can also be rewritten in the form

$$\begin{aligned} |q\rangle &= \sqrt{|\alpha_{00}|^2 + |\alpha_{01}|^2}\left(\frac{\alpha_{00}|00\rangle + \alpha_{01}|01\rangle}{\sqrt{|\alpha_{00}|^2 + |\alpha_{01}|^2}}\right) \\ &\quad + \sqrt{|\alpha_{10}|^2 + |\alpha_{11}|^2}\left(\frac{\alpha_{10}|10\rangle + \alpha_{110}|11\rangle}{\sqrt{|\alpha_{10}|^2 + |\alpha_{11}|^2}}\right) \\ &\equiv \beta_0|q_0\rangle + \beta_1|q_1\rangle, \end{aligned} \tag{16.10}$$

with $\beta_0 = \sqrt{|\alpha_{00}|^2 + |\alpha_{01}|^2}$ and $\beta_1 = \sqrt{|\alpha_{10}|^2 + |\alpha_{11}|^2}$, and

$$\begin{cases} |q_0\rangle = \dfrac{1}{\sqrt{|\alpha_{00}|^2 + |\alpha_{01}|^2}} (\alpha_{00}|00\rangle + \alpha_{01}|01\rangle) \equiv \gamma_{00}|00\rangle + \gamma_{01}|01\rangle \\ |q_1\rangle = \dfrac{1}{\sqrt{|\alpha_{10}|^2 + |\alpha_{11}|^2}} (\alpha_{00}|10\rangle + \alpha_{01}|11\rangle) \equiv \gamma_{10}|10\rangle + \gamma_{01}|11\rangle. \end{cases} \tag{16.11}$$

The development in Eq. (16.11) illustrates that any 2-qubit state can be conceived as the superposition of *two 2-qubits*, which we call here $|q_0\rangle$ (a 2-qubit with the first bit 0), and $|q_1\rangle$ (a 2-qubit with the first bit being 1), respectively. According to all expectation, should we have fully assimilated the earlier story of the 1-qubit measurement, if we attempt to measure *only the first bit* of any 2-qubit state, the latter must collapse into $|q_0\rangle$ or $|q_1\rangle$, according to the outcome of the measurement being the 0 or 1 classical bits, with associated probabilities $|\beta_0|^2$ and $|\beta_1|^2$, respectively. The resulting states $|q_0\rangle$ or $|q_1\rangle$, which represent superposition of 2-qubit states (with either 0 or 1 as the first bit), are referred to as *post-measurement states*.

The above has shown that any 2-qubit state can actually be decomposed into a variety of possible superposition of 2-qubit pairs, as selected from the family $\{|00\rangle, |01\rangle, |10\rangle, |11\rangle\}$. Of particular interest is the specific subgroup of 2-qubit pair superpositions $|ij\rangle \pm |kl\rangle$, where $i \neq k$ or $j \neq l$, conventionally defined according to:

$$\begin{cases} |\beta_{00}\rangle \equiv \dfrac{1}{\sqrt{2}}[|00\rangle + |11\rangle] \\ |\beta_{01}\rangle \equiv \dfrac{1}{\sqrt{2}}[|01\rangle + |10\rangle] \\ |\beta_{10}\rangle \equiv \dfrac{1}{\sqrt{2}}[|00\rangle - |11\rangle] \\ |\beta_{11}\rangle \equiv \dfrac{1}{\sqrt{2}}[|01\rangle - |10\rangle]. \end{cases} \tag{16.12}$$

The family of 2-qubit superpositions defined in the above inventory is referred to as *Bell states* or *EPR pairs*, the EPR acronym being short for the names Einstein, Podolsky, and Rosen. The Bell or EPR states exhibit quite intriguing and useful properties for quantum computing, as will be described in Chapter 18.

The key lesson we have learnt so far is that, unlike classical bits, *qubits are not physically observable*. But, as I shall describe next, qubits can be physically manipulated! Such a manipulation, which transforms the qubit state, is of consequence in the measurable or experimental domain. Hence, we may conceive of a *quantum computer*, a machine that processes information not from bits but from qubits. The fact that qubits are a probabilistic superposition of states introduces new dimensions and perspectives in computing power, which represents a key justification for QIT.

## 16.2 Basic computations with 1-qubit quantum gates

In Chapter 15, we have described the principles of *reversible-logic gates*, as based on classical bit inputs and controls. Here, we shall explore how such gates can perform elementary computations by exclusive use of qubits, instead of classical bits. As we will discover, *quantum computation* based on these two principles is quite different in nature and in power, when compared with classical computation obtained from mere Boolean-logic gates. We shall first consider operations with single qubit states, and then move to more complex ones involving two and three qubit states.

As we have seen earlier in this chapter, a single qubit (or 1-qubit), $|q\rangle$, is defined through its complex amplitudes $(\alpha, \beta)$ with respect to a given orthonormal base or pure states $(|0\rangle, |1\rangle)$, also referred to as *computational basis*. A quantum gate, call it $A$, has the effect of transforming an input qubit state $|q\rangle$ into an output qubit state $|q'\rangle$, using predefined control qubits. Thus, the relation between the input and output states can be expressed in the equivalent matrix-vector relation:

$$|q'\rangle = A|q\rangle. \tag{16.13}$$

In quantum vocabulary, the entity $A$ that transforms $|q\rangle$ into $|q'\rangle$ is referred to as an *operator*. Since the vector $|q'\rangle$ must remain on the Bloch sphere surface (so that its squared amplitudes correspond to actual probabilities), it must be a unitary vector. For this reason, the transformation from $|q\rangle$ to $|q'\rangle$ is called *unitary*, and $A$ must be a *unitary operator*. We shall now further develop the notion of unitary transformation and the relation between unitary operator and matrix.

For simplicity and to escape any further conceptual burden, so far we have assumed that qubits are equivalent to vectors in a 2D space. This view is simplistic, but accurate. Consistently, we may assume that the transformation operator $A$ may be defined through a $2 \times 2$ matrix, whose coefficients are defined by four complex numbers, $a_{ij}$. From the definitions Eq. (16.3) and Eq. (16.13), we, thus, obtain the matrix-vector equation in the $|0\rangle, |1\rangle$ base:

$$\begin{aligned} |q'\rangle &= A|q\rangle \\ &= \begin{pmatrix} a_{11} & a_{12} \\ a_{21} & a_{22} \end{pmatrix} \begin{pmatrix} \alpha \\ \beta \end{pmatrix} \\ &= \begin{pmatrix} a_{11}\alpha + a_{12}\beta \\ a_{21}\alpha + a_{22}\beta \end{pmatrix} \\ &\equiv (a_{11}\alpha + a_{12}\beta)\,|0\rangle + (a_{21}\alpha + a_{22}\beta)\,|1\rangle \\ &\equiv \alpha'|0\rangle + \beta'|1\rangle. \end{aligned} \tag{16.14}$$

Such a relation between the input $(\alpha, \beta)$ and output $(\alpha', \beta')$ amplitudes does not account for the control qubits in the quantum gate, but for the time being this will suffice for our description.

The result in Eq. (16.14) shows that the action of the quantum gate, or that of the quantum operator $A$, modifies the state amplitudes of the initial superposition, according to the transformations $\alpha \rightarrow \alpha' = a_{11}\alpha + \alpha_{12}\beta$ and $\beta \rightarrow \alpha_{21}\alpha + \alpha_{22}\beta$. Since the output state amplitudes $(\alpha', \beta')$ must correspond to probabilities, it is required that

$|\alpha'|^2 + |\beta'|^2 = 1$. As I have mentioned earlier, such a transformation and the associated operator/matrix, must be *unitary*.[4]

In the following, we shall consider five examples of $2 \times 2$ unitary operators or matrices and their effect on qubit transformation, namely: the *Pauli matrices* (called *I*, *X*, *Y*, and *Z*-gates) and the *Hadamard matrix* (also called a *H-gate*).

## Pauli matrices or I, X, Y, Z-gates

The *Pauli matrices* (named after the physicist W. Pauli) or *I, X, Y, Z-gates* are defined as:

$$\begin{aligned} I \equiv \sigma_0 &= \begin{pmatrix} 1 & 0 \\ 0 & 1 \end{pmatrix} \\ X \equiv \sigma_x = \sigma_1 &= \begin{pmatrix} 0 & 1 \\ 1 & 0 \end{pmatrix} \\ Y \equiv \sigma_y = \sigma_2 &= \begin{pmatrix} 0 & -i \\ i & 0 \end{pmatrix} \\ Z \equiv \sigma_z = \sigma_3 &= \begin{pmatrix} 1 & 0 \\ 0 & -1 \end{pmatrix}. \end{aligned} \tag{16.15}$$

The different notations $I = \sigma_0, X = \sigma_x = \sigma_1, Y = \sigma_y = \sigma_2$ and $Z = \sigma_z = \sigma_3$ are usually found in the related literature for historical reasons. It is easily established that the squares of all the above matrices are equal to unity or $X^2 = Y^2 = Z^2 = (\sigma_i)^2 = I^2 = I$.

The first matrix $I$ is the *identity matrix*, which corresponds to the classical REPEAT gate and leaves the input qubit $|q\rangle = \alpha|0\rangle + \beta|1\rangle$ invariant, according to

$$\begin{aligned} |q'\rangle &= I|q\rangle \\ &= \begin{pmatrix} 1 & 0 \\ 0 & 1 \end{pmatrix} \begin{pmatrix} \alpha \\ \beta \end{pmatrix} \\ &= \begin{pmatrix} \alpha \\ \beta \end{pmatrix} \equiv \alpha|0\rangle + \beta|1\rangle \equiv |q\rangle. \end{aligned} \tag{16.16}$$

The second matrix $X$ (or $X$-Pauli) corresponds to the classical *NOT gate*. Indeed, we obtain

$$\begin{aligned} |q'\rangle &= X|q\rangle \\ &= \begin{pmatrix} 0 & 1 \\ 1 & 0 \end{pmatrix} \begin{pmatrix} \alpha \\ \beta \end{pmatrix} \\ &= \begin{pmatrix} \beta \\ \alpha \end{pmatrix} \equiv \beta|0\rangle + \alpha|1\rangle, \end{aligned} \tag{16.17}$$

[4] A transformation and associated matrix $A$ is *unitary* if the following identity is satisfied $A^{-1} = (A^{\mathrm{T}})^* \equiv A^+$. In this definition, $A^{-1}$ is the *inverse matrix* of $A$ (such that $A^{-1}A = AA^{-1} = I$), $A^{\mathrm{T}}$ is the *transposed matrix* of $A$ (with coefficients $a_{ij} \to a_{ji}$), and $^*$ denotes complex conjugation ($a_{ij} \to a_{ij}^*$). The symbol $^+$ stands for Hermitian conjugation, which combines both transposition and complex conjugation ($a_{ij} \to a_{ji}^*$), i.e., $A^+$ is the *Hermitian conjugate* of the matrix $A$.

showing that the elementary qubits $|0\rangle$, $|1\rangle$ are switched into their counterparts $|1\rangle$, $|0\rangle$, which for a state superposition $|q\rangle = \alpha|0\rangle + \beta|1\rangle$ amounts to *swapping the amplitude probabilitie* according to $(\alpha, \beta) \to (\beta, \alpha)$. Thus, if the input $|q\rangle$ is in the pure state, for instance $|q\rangle = |0\rangle$ (or $\alpha = 1$, $\beta = 0$), the output state is $|q'\rangle = |1\rangle$, and the reverse with $|q\rangle = |1\rangle$, which gives $|q'\rangle = |0\rangle$. In these two limiting cases, this operation, indeed, corresponds to that of a classical NOT gate. But generally, it is also correct to call $X, \sigma_1, \sigma_x$ a *quantum* NOT gate.

The action of the $Y$-gate is both to swap the amplitudes and to introduce a $\pi$ phase shift between the two states of the initial superposition,[5] according to

$$\begin{aligned} |q'\rangle &= Y|q\rangle \\ &= \begin{pmatrix} 0 & -i \\ i & 0 \end{pmatrix} \begin{pmatrix} \alpha \\ \beta \end{pmatrix} \\ &= \begin{pmatrix} -i & \beta \\ i & \alpha \end{pmatrix} \\ &= -i(\beta|0\rangle - \alpha|1\rangle) \equiv \mathrm{e}^{\mathrm{i}\gamma}(\beta|0\rangle + \mathrm{e}^{\mathrm{i}\pi}\alpha|1\rangle), \end{aligned} \tag{16.18}$$

where $\gamma = -\pi/2$ is an immeasurable, or "unobservable" phase constant.

Finally, the action of the $Z$-gate is only to introduce a $\pi$ phase shift between the two states of the initial superposition, according to

$$\begin{aligned} |q'\rangle &= Z|q\rangle \\ &= \begin{pmatrix} 1 & 0 \\ 0 & -1 \end{pmatrix} \begin{pmatrix} \alpha \\ \beta \end{pmatrix} \\ &= \begin{pmatrix} \alpha \\ -\beta \end{pmatrix} \equiv \alpha|0\rangle - \beta|1\rangle \equiv \alpha|0\rangle + \mathrm{e}^{\mathrm{i}\pi}\beta|1\rangle. \end{aligned} \tag{16.19}$$

As can also be easily checked, the three Pauli matrices $X$, $Y$, $Z$ exhibit the following properties:

$$\begin{cases} \sigma_1\sigma_2 = i\sigma_3 \\ \sigma_2\sigma_3 = i\sigma_1 \\ \sigma_3\sigma_1 = i\sigma_2 \\ \sigma_i\sigma_j = -\sigma_j\sigma_i (i \neq j). \end{cases} \tag{16.20}$$

One defines the *commutator* of two operators or matrices $A$, $B$ as $[A, B] = AB - BA$. Two operators or matrices are said to *commute* if $[A, B] = 0$. Then we observe from the last equation in Eq. (16.20) that for $i \neq j$ the Pauli matrices do not commute, i.e., $\lfloor \sigma_i, \sigma_j \rfloor = \sigma_i\sigma_j - \sigma_j\sigma_i = 2\sigma_i\sigma_j$, and, in the general case

$$\lfloor \sigma_i, \sigma_j \rfloor = 2\delta_{ij}\sigma_i\sigma_j, \tag{16.21}$$

where $\delta_{ij}$ is the *Kronecker* symbol.[6] We, thus, have $[\sigma_1, \sigma_2] = 2\sigma_1\sigma_2 = 2\mathrm{i}\sigma_3$, $[\sigma_2, \sigma_3] = 2\sigma_2\sigma_3 = 2\mathrm{i}\sigma_1$ and $[\sigma_3, \sigma_1] = 2\sigma_3\sigma_1 = 2\mathrm{i}\sigma_2$. The commutator between any two Pauli

[5] With complex numbers, a change of sign corresponds to a phase shift of $\pm\pi$, or a multiplying factor of $\mathrm{e}^{\pm\mathrm{i}\pi} = -1$, since $\mathrm{e}^{\pm\mathrm{i}\pi} = \cos(\pm\pi) + \mathrm{i}\sin(\pm\pi) = -1$.

[6] Namely: $\delta_{ij} = 1$ for $i = j$, and $\delta_{ij} = 0$ for $i \neq j$.

matrices can be generalized in the formula

$$[\sigma_i, \sigma_j] = 2\mathrm{i}\varepsilon_{ijk}\sigma_k, \tag{16.22}$$

where $\varepsilon_{ijk}$ is the *Levi–Civita* symbol.[7] Likewise, using the property $(\sigma_i)^2 = I$, we can generalize Eq. (16.20) in the formula:

$$\sigma_i\sigma_j = \mathrm{i}\varepsilon_{ijk}\sigma_k + \delta_{ij}I. \tag{16.23}$$

These properties of the Pauli matrices or $I$, $X$, $Y$, $Z$-gates are used extensively in quantum computing, as will be illustrated. In Appendix N, I show that the Pauli matrices constitute a universal base, making it possible to generate any *unitary* $2 \times 2$ *matrices* or *unitary operators* $U$, which represent the universal building blocks for 2-qubit quantum gates and circuits.

## Hadamard matrix gate or H-gate

The *Hadamard matrix gate*, or *H-gate*, is defined as:

$$H = \frac{1}{\sqrt{2}}\begin{pmatrix} 1 & 1 \\ 1 & -1 \end{pmatrix} \equiv \frac{1}{\sqrt{2}}(X + Z). \tag{16.24}$$

Using results in Eqs. (16.17) and (16.19), we can interpret the action of the Hadamard gate as follows:

$$\begin{aligned} |q'\rangle &= H|q\rangle \\ &= \frac{1}{\sqrt{2}}(X + Z)|q\rangle \\ &= \frac{1}{\sqrt{2}}(X|q\rangle + Z|q\rangle) \\ &= \frac{1}{\sqrt{2}}[(\beta|0\rangle + \alpha|1\rangle) + (\alpha|0\rangle - \beta|1\rangle)] \\ &= \alpha\frac{|0\rangle + |1\rangle}{\sqrt{2}} + \beta\frac{|0\rangle - |1\rangle}{\sqrt{2}} \\ &\equiv \alpha|+\rangle + \beta|-\rangle. \end{aligned} \tag{16.25}$$

It is seen from the above that the Hadamard gate transforms any input qubit $|q\rangle = \alpha|0\rangle + \beta|1\rangle$ into the superposition $|q\rangle = \alpha|+\rangle + \beta|-\rangle$, where $|+\rangle, |-\rangle$ represent a new *pure states basis*. As we shall see later, the two pure-state bases $|0\rangle, |1\rangle$ and $|+\rangle, |-\rangle$ play a fundamental role in quantum computing.

The action of the above $2 \times 2$ quantum gates on input qubits is summarized in Table 16.1.

[7] With the following definition (see http://en.wikipedia.org/wiki/Levi-Civita_symbol)

$$\varepsilon_{ijk} = \begin{cases} +1 & \text{if } (i, j, k) \text{ is } (1, 2, 3), (2, 3, 1) \text{ or } (3, 1, 2), \\ -1 & \text{if } (i, j, k) \text{ is } (3, 2, 1), (1, 3, 2) \text{ or } (2, 1, 3), \\ 0 & \text{otherwise: } i = j \text{ or } j = k \text{ or } k = i. \end{cases}$$

**Table 16.1** Action of the elementary $2 \times 2$ quantum gates on input qubit $|q\rangle = \alpha|0\rangle + \beta|1\rangle$.

| Gate $U$ | Output $\|q'\rangle = U\|q\rangle$ | Action |
|---|---|---|
| Identity, $I, \sigma_0$ | $\alpha\|0\rangle + \beta\|1\rangle$ | Invariant |
| $X, \sigma_1, \sigma_x$ | $\beta\|0\rangle + \alpha\|1\rangle$ | Swaps amplitudes |
| $Y, \sigma_2, \sigma_y$ | $\beta\|0\rangle + e^{i\pi}\alpha\|1\rangle$ | Swaps amplitudes and $\pi$ shift between amplitudes |
| $Z, \sigma_3, \sigma_z$ | $\alpha\|0\rangle + e^{i\pi}\beta\|1\rangle$ | $\pi$ shift between amplitudes |
| Hadamard, $H$ | $\alpha\|+\rangle + \beta\|-\rangle$ | Switches to $\|+\rangle, \|-\rangle$ basis |

For illustration purposes, consider two examples showing the action of *cascades* of $2 \times 2$ quantum gates:

$$\begin{aligned} U &= XZX \rightarrow \\ U|q\rangle &= XZX(\alpha|0\rangle + \beta|1\rangle) \\ &= XZ(\beta|0\rangle + \alpha|1\rangle) \\ &= X(\beta|0\rangle + e^{i\pi}\alpha|1\rangle) \\ &= e^{i\pi}\alpha|0\rangle + \beta|1\rangle \\ &= e^{i\pi}(\alpha|0\rangle + e^{-i\pi}\beta|1\rangle) \\ &\equiv \alpha|0\rangle - \beta|1\rangle, \end{aligned} \tag{16.26}$$

$$\begin{aligned} U &= HX \rightarrow \\ U|q\rangle &= HX(\alpha|0\rangle + \beta|1\rangle) \\ &= H(\beta|0\rangle + \alpha|1\rangle) \\ &\equiv \beta|+\rangle + \alpha|-\rangle. \end{aligned} \tag{16.27}$$

Finally, we must note that the $2 \times 2$ Pauli and Hadamard matrix gates correspond to *reversible 1-qubit computations*. Indeed, any input qubit can be retrieved through the double operations $X^2 = Y^2 = Z^2 = H^2 = I^2 = I$. More generally, any gate corresponding to a unitary matrix or operator $U$ is reversible by the application of the Hermitian conjugate $U^+$, since, by definition, $U^+U = I$. This property also applies to gate cascades $UVW\ldots$ of unitary operators, which have for their Hermitian conjugate $(UVW\ldots)^+ = \cdots W^+V^+U^+$, thus,

$$(UVW\ldots)^+UVW = \cdots W^+V^+U^+UVW = \cdots W^+V^+VW = \cdots W^+W = \cdots I.$$

## 16.3 Quantum gates with multiple qubit inputs and outputs

With the background from the previous section on $2 \times 2$ quantum gates and that from reversible logic gates introduced in Chapter 15, we can now describe the matrix representation and operation of various higher-order quantum gates, which have two or more 1-qubit inputs and outputs. The elementary gates of this type are called CNOT, CROSSOVER, controlled-$U$, controlled-SWAP, Toffoli, and CCNOT, which I shall review in the following.

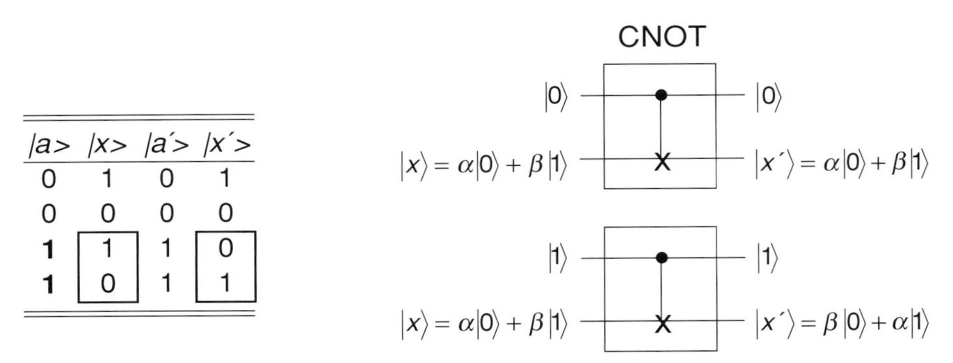

| \|a> | \|x> | \|a´> | \|x´> |
|---|---|---|---|
| 0 | 1 | 0 | 1 |
| 0 | 0 | 0 | 0 |
| **1** | 1 | 1 | 0 |
| **1** | 0 | 1 | 1 |



**Figure 16.2** CNOT gate with control qubit $|a\rangle$ and target qubit $|x\rangle$, with truth table and gate output $|a\rangle$, $|x'\rangle$ resulting from target qubit superposition $|x\rangle = \alpha|0\rangle + \beta|1\rangle$.

## CNOT gate

Consider first the CNOT gate, whose diagram is shown in Fig. 16.2. The gate has a control qubit $|a\rangle = |0\rangle$ or $|a\rangle = |1\rangle$, and a target qubit $|x\rangle = \alpha|0\rangle + \beta|1\rangle$. Here, we shall assume that the control qubits are in pure states, namely, $|0\rangle, |1\rangle$. The figure shows the action of the CNOT gate with the corresponding truth table. It is seen from the figure that the target qubit $|x\rangle$ is left unchanged when the control qubit is $|a\rangle = |0\rangle$ while in the case $|a\rangle = |1\rangle$ the amplitudes $\alpha, \beta$ of $|x\rangle$ are swapped which, as we have seen in the previous section, represents the quantum NOT version of a qubit.

But how can we define the CNOT gate *matrix*, since here we have a control qubit? To answer this, recall that in the case of 1-qubit gates, the matrix is defined in the computational base $|0\rangle, |1\rangle$. In the case of the CNOT gate, the matrix must be defined in the *extended computational base*, noted $\{|a\rangle|x\rangle\} = \{|0\rangle|0\rangle, |0\rangle|1\rangle, |1\rangle|0\rangle, |1\rangle|1\rangle\}$, which covers all possible gate inputs according to the truth table shown in Fig. 16.2, with $|a\rangle = |0\rangle$ or $|a\rangle = |1\rangle$ and $|x\rangle = |0\rangle$ or $|x\rangle = |1\rangle$ (the matrix applying to any superposition thereof). As will be described further on in this chapter, the state $|a\rangle|x\rangle$ is referred to as the *tensor product* of the *kets* $|a\rangle$ and $|x\rangle$. In the computational base $\{|0\rangle|1\rangle\}$, the input *tensor state* $|a\rangle|x\rangle$ is represented by a four-dimensional column-vector with coordinates $u_1, u_2, u_3, u_4$,

$$|a\rangle\,|x\rangle = \begin{bmatrix} u_1 \\ u_2 \\ u_3 \\ u_4 \end{bmatrix}, \tag{16.28}$$

which satisfies $|u_1|^2 + |u_2|^2 + |u_3|^2 + |u_4|^2 = 1$. The output state of the gate is then given by the matrix-vector relation $|a'\rangle|x'\rangle = A|a\rangle|x\rangle$, where $A$ is the gate's $4 \times 4$ matrix. In the case of the CNOT gate, and as expressed in the tensor base

**Table 16.2** Transformation of input qubit $|a\rangle|x\rangle$ of coordinates $(u_1, u_2, u_3, u_4)$ in computational basis $\{|0\rangle|0\rangle, |0\rangle|1\rangle, |1\rangle|0\rangle, |1\rangle|1\rangle\}$ into output qubit $|a'\rangle|x'\rangle = A_{\text{CNOT}}|a\rangle|x\rangle$.

| $u_1$ | $u_2$ | $u_3$ | $u_4$ | $\|a\rangle\|x\rangle$ | $\|a'\rangle\|x'\rangle$ | Observations |
|---|---|---|---|---|---|---|
| 1 | 0 | 0 | 0 | $\|0\rangle\|0\rangle$ | $\|0\rangle\|0\rangle$ | Invariant |
| 0 | 1 | 0 | 0 | $\|0\rangle\|1\rangle$ | $\|0\rangle\|1\rangle$ | Invariant |
| 0 | 0 | 1 | 0 | $\|1\rangle\|0\rangle$ | $\|0\rangle\|1\rangle$ | $x$ qubit flipped |
| 0 | 0 | 0 | 1 | $\|1\rangle\|1\rangle$ | $\|1\rangle\|0\rangle$ | $x$ qubit flipped |
| 1 | 1 | 0 | 0 | $\frac{1}{\sqrt{2}}(\|0\rangle\|0\rangle + \|0\rangle\|1\rangle) = \|0\rangle\|+\rangle$ | $\frac{1}{\sqrt{2}}(\|0\rangle\|0\rangle + \|0\rangle\|1\rangle) = \|0\rangle\|+\rangle$ | Invariant |
| 0 | 0 | 1 | 1 | $\frac{1}{\sqrt{2}}(\|1\rangle\|0\rangle + \|1\rangle\|1\rangle) = \|1\rangle\|+\rangle$ | $\frac{1}{\sqrt{2}}(\|1\rangle\|0\rangle + \|1\rangle\|1\rangle) = \|1\rangle\|+\rangle$ | Invariant |
| $\alpha$ | $\beta$ | 0 | 0 | $\alpha\|0\rangle\|0\rangle + \beta\|0\rangle\|1\rangle$ $= \|0\rangle(\alpha\|0\rangle + \beta\|1\rangle)$ | $\alpha\|0\rangle\|0\rangle + \beta\|0\rangle\|1\rangle$ $= \|0\rangle(\alpha\|0\rangle + \beta\|1\rangle)$ | Invariant |
| 0 | 0 | $\alpha$ | $\beta$ | $\alpha\|1\rangle\|0\rangle + \beta\|1\rangle\|1\rangle$ $= \|1\rangle(\alpha\|0\rangle + \beta\|1\rangle)$ | $\beta\|1\rangle\|0\rangle + \alpha\|1\rangle\|1\rangle$ $= \|1\rangle(\beta\|0\rangle + \alpha\|1\rangle)$ | $x$ qubit amplitudes swapped |

$\{|a\rangle|x\rangle\} = \{|0\rangle|0\rangle, |0\rangle|1\rangle, |1\rangle|0\rangle, |1\rangle|1\rangle\}$, the matrix takes the form:

$$A_{\text{CNOT}} = \begin{bmatrix} I & 0 \\ 0 & X \end{bmatrix} \equiv \begin{bmatrix} 1 & 0 & 0 & 0 \\ 0 & 1 & 0 & 0 \\ 0 & 0 & 0 & 1 \\ 0 & 0 & 1 & 0 \end{bmatrix}. \tag{16.29}$$

The $2 \times 2$ reduced form of the above gate matrix shows that states of the form $|0\rangle|x\rangle$ are invariant (sub-matrix $I$), while states of the form $|1\rangle|x\rangle$ have the target qubit $|x\rangle$ flipped (sub-matrix $X$). Although somewhat tedious, it is useful to verify now the above result by applying the gate matrix $A_{\text{CNOT}}$ to the input state $|a\rangle|x\rangle$. From Eqs. (16.28) and (16.29), we obtain:

$$\begin{aligned} |a'\rangle|x'\rangle = A_{\text{CNOT}}|a\rangle|x\rangle = \begin{bmatrix} u'_1 \\ u'_2 \\ u'_4 \\ u'_3 \end{bmatrix} = \begin{bmatrix} 1 & 0 & 0 & 0 \\ 0 & 1 & 0 & 0 \\ 0 & 0 & 0 & 1 \\ 0 & 0 & 1 & 0 \end{bmatrix} \begin{bmatrix} u_1 \\ u_2 \\ u_3 \\ u_4 \end{bmatrix} \\ = u_1|0\rangle|0\rangle + u_2|0\rangle|1\rangle + u_4|1\rangle|0\rangle + u_3|1\rangle|1\rangle. \end{aligned} \tag{16.30}$$

The right-hand side of Eq. (16.30) can now be developed according to different input possibilities for $|a\rangle|x\rangle$, i.e., concerning the control qubit $|a\rangle$ and the target qubit $|x\rangle$. Table 16.2 shows the result with the target qubit $|x\rangle$ as being in either a pure state (first four lines) or a superposition of states (last four lines). As expected, the table illustrates that the CNOT gate leaves the target qubit $|x\rangle = |0\rangle$ or $|1\rangle$ invariant when the control qubit is set to $|a\rangle = |0\rangle$. If the target qubit is a superposition $|x\rangle = \alpha|0\rangle + \beta|1\rangle$, the amplitudes $(\alpha, \beta)$ are either conserved ($|a\rangle = |0\rangle$) or swapped ($|a\rangle = |1\rangle$). In the specific case $\alpha = \beta = 1$, the target qubit remains invariant regardless of the control qubit $|a\rangle$, as expected. It is left as an exercise to analyze the action of the CNOT gate

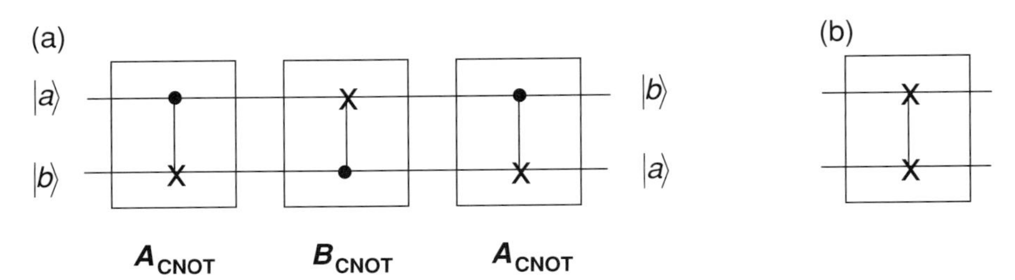


**Figure 16.3** (a) Quantum gate circuit based on the concatenation of three CNOT gates, with corresponding matrices $A_{\mathrm{CNOT}}$ (control qubit at top) and $B_{\mathrm{CNOT}}$ (control qubit at bottom). (b) Equivalent circuit representation (CROSSOVER or SWAP).

with the control qubit $|a\rangle$ being now in a superposition of states, and show that for certain combinations of input qubits $|a\rangle|x\rangle$, the CNOT gate can generate any of the four *EPR* or *Bell states*, as defined in Eq. (16.2).

It is easily verified that the matrix $A_{\mathrm{CNOT}}$ is unitary and that its inverse matrix is $A_{\mathrm{CNOT}}^{-1} = A_{\mathrm{CNOT}}$, or $A_{\mathrm{CNOT}}^{2} = I$ ($I = 4 \times 4$ identity matrix). This last result is expected, since the repeated action of CNOT (with same control qubit, by inherent circuit construction) must leave the target qubit invariant.

## CROSSOVER or SWAP gate

It is also possible to build circuits made of concatenated CNOT gates, where the control and target qubits exchange roles, i.e., an output target qubit from a first gate serving as an input control qubit for the next gate. Consider, for instance, the three-gate circuit shown in Fig. 16.3, which (as we shall see) corresponds to the CROSSOVER or SWAP operator. The circuit is seen to include in the middle a CNOT gate arranged upside down, thus, using the target output qubit of the first CNOT gate as a control for the second CNOT gate. For simplicity, we assume that the control qubits that are input to CNOT gates must be in a pure state. Under this assumption, this whole circuit must be input only with pure states $|a\rangle, |b\rangle = |0\rangle, |1\rangle$. The matrix representation of the CNOT gate makes it possible to construct the circuit shown in Fig. 16.3, with care to properly define the matrix $B_{\mathrm{CNOT}}$ corresponding to the reversed arrangement. As an exercise, one may show that in the computational base $\{|0\rangle|0\rangle, |0\rangle|1\rangle, |1\rangle|0\rangle, |1\rangle|1\rangle\}$ the matrix $B_{\mathrm{CNOT}}$ takes the form

$$B_{\mathrm{CNOT}} = \begin{bmatrix} 1 & 0 & 0 & 0 \\ 0 & 0 & 0 & 1 \\ 0 & 0 & 1 & 0 \\ 0 & 1 & 0 & 0 \end{bmatrix}. \tag{16.31}$$

A second exercise is to compute the circuit matrix $A_{\mathrm{SWAP}} = A_{\mathrm{CNOT}} B_{\mathrm{CNOT}} A_{\mathrm{NOT}}$ from the matrix definitions in Eqs. (16.29) and (16.30). The computation yields

$$A_{\mathrm{SWAP}} = \begin{bmatrix} 1 & 0 & 0 & 0 \\ 0 & 0 & 1 & 0 \\ 0 & 1 & 0 & 0 \\ 0 & 0 & 0 & 1 \end{bmatrix}. \tag{16.32}$$

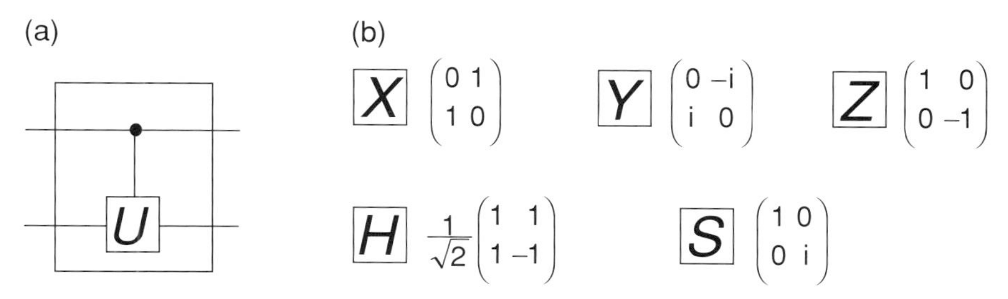


**Figure 16.4** (a) Controlled-$U$ gate, (b) corresponding matrix gate possibilities $U = X, Y, Z, H$, and $S$.

It is seen from the coefficients in the matrix $A_{\text{SWAP}}$ that the input tensor states $|0\rangle|0\rangle$ and $|1\rangle|1\rangle$ are left invariant (owing to the 1 coefficients in the matrix-diagonal), while the input tensor states $|0\rangle|1\rangle$ and $|1\rangle|0\rangle$ are swapped (owing to the 1 off-diagonal coefficients). The circuit gate thus corresponds to a CROSSOVER or SWAP function, as Fig. 16.3 indicates with the outputs $|b\rangle$, $|a\rangle$. The equivalent gate representation is also shown at right in the figure.

## Controlled-*U* gates

In the category of *controlled-U* gates, $U$ stands for any quantum gate with a $2 \times 2$ unitary (but not necessarily Hermitian) matrix. For instance, $U = X, Y, Z, H$ (Pauli and Hadamard gates) and $S$ ($\pi/2$ phase gate), as illustrated in Fig. 16.4 with their $2 \times 2$ matrices represented in the $|0\rangle, |1\rangle$ state basis. The gate *controlled-X* is the same as CNOT and its matrix (previously called $A_{\text{CNOT}}$) is shown in Eq. (16.29). We notice again from this equation that this $4 \times 4$ matrix can also be represented in the reduced form

$$\text{controlled-}X = A_{\text{CNOT}} = \begin{bmatrix} I & 0 \\ 0 & X \end{bmatrix}, \tag{16.33}$$

where the top-left side corresponds to the REPEAT or invariant function (control qubit set to $|0\rangle$, corresponding to $2 \times 2$ identity matrix $I$) and the bottom-right side corresponds to the NOT function (control qubit set to $|1\rangle$, corresponding to $2 \times 2$ matrix $X$). Thus, we can most generally define the matrix of any *controlled-U* gate (e.g., $U = X, Y, Z, H, S \ldots$) according to

$$\text{controlled-}U = \begin{bmatrix} I & 0 \\ 0 & U \end{bmatrix}. \tag{16.34}$$

Most generally, any controlled-$U$ gate corresponds to a target qubit rotation characterized by $U = R_{\vec{n}}(\theta)$, where $R_{\vec{n}}(\theta)$ is the *rotation operator* associated with the transformation $U$ on the Bloch sphere, as characterized by a rotation angle $\theta$ about the axis parallel to the unitary vector $\vec{n}$ (see Appendix N for related definition and properties).

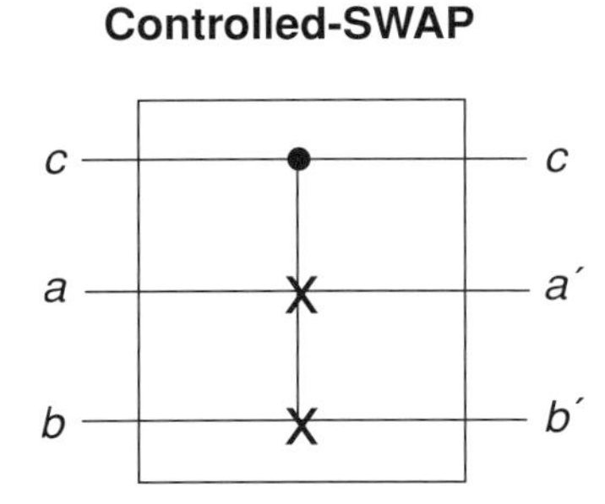


| $\|c\rangle$ | $\|a\rangle$ | $\|b\rangle$ | $\|c'\rangle$ | $\|a'\rangle$ | $\|b'\rangle$ |
|---|---|---|---|---|---|
| 0 | 0 | 0 | 0 | 0 | 0 |
| 0 | 0 | 1 | 0 | 0 | 1 |
| 0 | 1 | 0 | 0 | 1 | 0 |
| 0 | 1 | 1 | 0 | 1 | 1 |
| 1 | 0 | 0 | 1 | 0 | 0 |
| 1 | **0** | **1** | 1 | **1** | **0** |
| 1 | **1** | **0** | 1 | **0** | **1** |
| 1 | 1 | 1 | 1 | 1 | 1 |

**Figure 16.5** Controlled-SWAP gate with truth table.

## Controlled-SWAP gate

The schematic representation of a controlled-SWAP gate and its qubit truth table are shown in Fig. 16.5. This gate represents a particular case of the *Fredkin* gate previously described in Chapter 15, and represented in Fig. 15.7 (with the control qubit $c$ at the bottom). Since this is a $3 \times 3$ gate, the state basis has eight tensor elements $\{|c\rangle|a\rangle|b\rangle\} = \{|0\rangle|0\rangle|0\rangle, |0\rangle|0\rangle|1\rangle, \ldots, |1\rangle|1\rangle|1\rangle\}$, with the first state $|c\rangle$ corresponding to the control qubit. In this basis, and using the definition of $A_{\mathrm{SWAP}}$ in Eq. (16.32), the $8 \times 8$ matrix $A_{\text{C-SWAP}}$ takes the reduced and explicit forms:

$$A_{\text{C-SWA}P} = \begin{bmatrix} I & 0 \\ 0 & A_{\mathrm{SWAP}} \end{bmatrix} \equiv \begin{bmatrix} 1 & 0 & 0 & 0 & 0 & 0 & 0 & 0 \\ 0 & 1 & 0 & 0 & 0 & 0 & 0 & 0 \\ 0 & 0 & 1 & 0 & 0 & 0 & 0 & 0 \\ 0 & 0 & 0 & 1 & 0 & 0 & 0 & 0 \\ 0 & 0 & 0 & 0 & 1 & 0 & 0 & 0 \\ 0 & 0 & 0 & 0 & 0 & 0 & 1 & 0 \\ 0 & 0 & 0 & 0 & 0 & 1 & 0 & 0 \\ 0 & 0 & 0 & 0 & 0 & 0 & 0 & 1 \end{bmatrix}. \tag{16.35}$$

## Toffoli or CCNOT gate

In Chapter 15, we have seen that the *Toffoli* gate corresponds to the logical function controlled-controlled-NOT, or CCNOT. To recall, in the classical version of the Toffoli gate, $a, b$ are two control bits, and $c$ is the "target" bit to process. The gate then outputs $a, b, c'$ with $c' = c \oplus ab$, meaning that, classically, the Toffoli gate is an XOR gate with $ab$ and $c$ as inputs, together with the conservation of the control bits $a$, $b$, to ensure computational reversibility. Here, consider that the input–output information is not about "bits" but "qubits." Defining the two control qubits as $|c_1\rangle, |c_2\rangle$, and the target bit as $|a\rangle$, it is easily established that in the basis $\{|c_1\rangle|c_2\rangle|a\rangle\}$ the corresponding $8 \times 8$ matrix

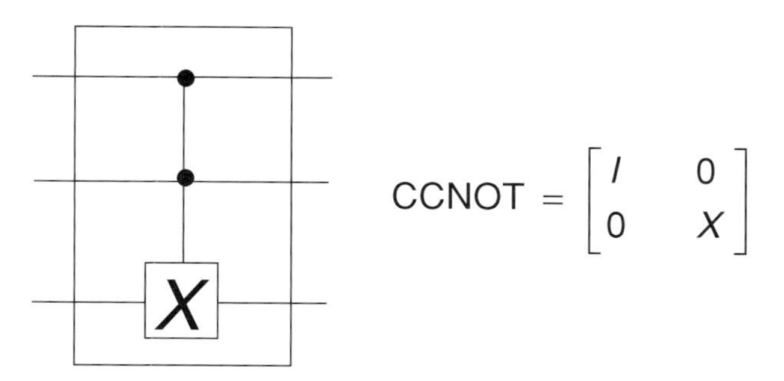


**Figure 16.6** Controlled-controlled-$X$ or CCNOT or Toffoli gate.

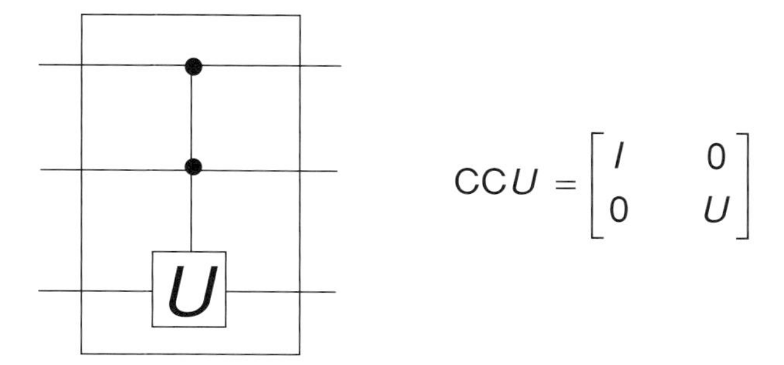


**Figure 16.7** Controlled-controlled-$U$, or CC$U$, gate with corresponding $8 \times 8$ matrix in reduced form.

takes the form

$$A_{\text{Toffoli}} = \begin{bmatrix} I & 0 \\ 0 & X \end{bmatrix} \equiv \begin{bmatrix} 1 & 0 & 0 & 0 & 0 & 0 & 0 & 0 \\ 0 & 1 & 0 & 0 & 0 & 0 & 0 & 0 \\ 0 & 0 & 1 & 0 & 0 & 0 & 0 & 0 \\ 0 & 0 & 0 & 1 & 0 & 0 & 0 & 0 \\ 0 & 0 & 0 & 0 & 1 & 0 & 0 & 0 \\ 0 & 0 & 0 & 0 & 0 & 1 & 0 & 0 \\ 0 & 0 & 0 & 0 & 0 & 0 & 0 & 1 \\ 0 & 0 & 0 & 0 & 0 & 0 & 1 & 0 \end{bmatrix}, \tag{16.36}$$

where $I$ is the $6 \times 6$ identity matrix and $X$ is the $2 \times 2$ quantum NOT matrix. The quantum circuit representation of the CCNOT gate and its reduced matrix are provided in Fig. 16.6. It is clear that the CCNOT gate transforms the target qubit $|a\rangle$ into $X|a\rangle$, if and only if $|c_1\rangle = |c_2\rangle = |1\rangle$ and leaves $|a\rangle$ invariant when $|c_1\rangle = |c_2\rangle = |0\rangle$.

### Controlled-controlled-$U$ or CC$U$ gate

This is a 3-qubit gate similar to the CCNOT or Toffoli gate, except that one uses any unitary transform $U$ for the $2 \times 2$ operator, see Fig. 16.7, along with the corresponding reduced matrix. It is clear that the CC$U$ gate transforms the target qubit $|a\rangle$ into $U|a\rangle$, if and only if $|c_1\rangle = |c_2\rangle = |1\rangle$ and leaves $|a\rangle$ invariant when $|c_1\rangle = |c_2\rangle = |0\rangle$.

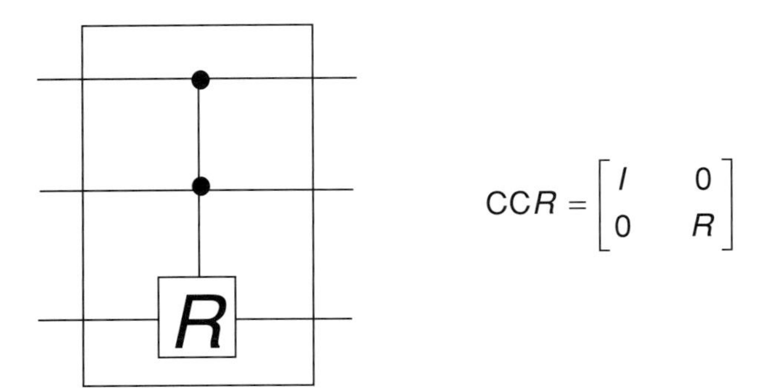


**Figure 16.8** Deutsch gate or CC$R$ gate, with corresponding $8 \times 8$ matrix in reduced form, where $R$ is a unitary rotation operator.

### Deutsch or CC*R* gate

A particular type of CC$U$ gate, called the *Deutsch gate*, is obtained when one chooses $U \equiv R_{\vec{n}}(\theta)$, as shown in Fig. 16.8. To make the Deutsch gate, a restriction applies to the rotation angle $\theta$. Indeed, in the Deutsch gate the angle $\theta$ should be *incommensurate with* $\pi$, which means that $\theta/\pi$ is *not* a rational fraction.[8] With such a property, any qubit $|v\rangle$ on the Bloch sphere that lies at an angle $\pm x\theta$ from the gate's target qubit $|u\rangle$ can be reached with arbitrary precision by applying the CC$R$ gate a finite number of times $k$, i.e., $|w\rangle = R^k(\pm\theta)|u\rangle = R(\pm k\theta)|u\rangle$ can be made arbitrarily close to $|v\rangle$ if $k/x \approx 1$. In particular, the rotation angle $k\theta$ can be made arbitrarily close to $\pi/2$, which makes the Deutsch gate closely similar to a Toffoli gate ($R(\pi/2) = X$). Consider, finally, that single $2 \times 2$ rotations defined as $R \equiv R_{\vec{n}}(\theta)$ make it possible to transform any input qubit $|u\rangle$ into any output qubit $|v\rangle$ on the Bloch sphere. A controlled-$R$ gate has the same complete transformation capability on the 2-qubit space, and a CC$R$ or Deutsch gate on the 3-qubit space. It is beyond the scope of this chapter to establish formally that, actually, quantum circuits based only on 3-qubit CC$R$ or Deutsch gates and CCNOT gates are capable of achieving any $n$-qubit transformations in the $n$-qubit space.

## 16.4 Quantum circuits

The matrix representation of 2- and 3-qubit gates may look somewhat impractical to handle, except in generic cases when they can be put in some reduced form. Therefore, it would seem that quantum-gate circuits with multiple gates and control qubits are not easy to model and analyze. In reality, however, gate circuits are far simpler to handle! I shall illustrate this through a few examples. Consider first the 2-qubit quantum circuit involving single-qubit gates ($J$, $K$) and controlled-$U$ gates as shown in Fig. 16.9, with, for instance, $J = X$. The circuit is seen to involve two different controlled-$U$ gates ($U$, $U'$) and two single-qubit gates ($X$, $K$), where $U$, $U'$, $K$ are any unitary gates. We do not need to calculate the corresponding matrix. Instead, consider the evolution of the

[8] A *rational fraction* or *rational number* can be expressed as the ratio $a/b$ of two integers $a$, $b$.

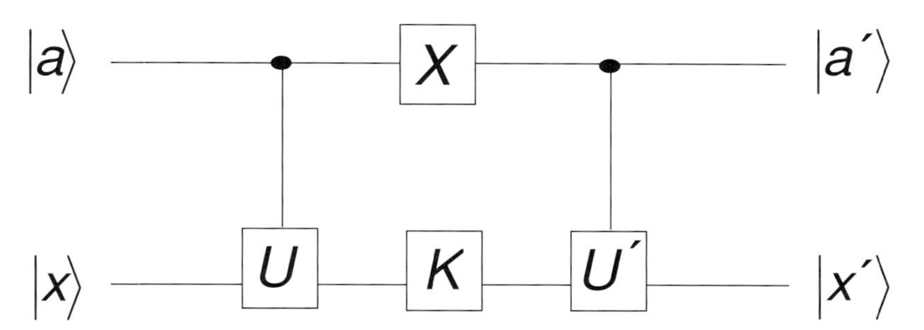


**Figure 16.9** Basic example of 2-qubit quantum circuit.

input tensor state $|a\rangle|x\rangle$ through the circuit, assuming separately $|a\rangle = |0\rangle$ and $|a\rangle = |1\rangle$, with each of the arrows ($\rightarrow$) representing the crossing of a gate:

$$\begin{aligned} &|0\rangle|x\rangle \rightarrow |0\rangle|x\rangle \rightarrow X|0\rangle K|x\rangle = |1\rangle U'K|x\rangle \\ &|1\rangle|x\rangle \rightarrow |1\rangle U|x\rangle \rightarrow X|1\rangle KU|x\rangle = |0\rangle KU|x\rangle. \end{aligned} \tag{16.37}$$

If, for instance, we let $U = Y$, $K = S$ and $U' = Z$ (according to the standard definitions of $Y$, $Z$, $S$ in Fig. 16.4), we obtain:

$$\begin{aligned} |0\rangle|x\rangle &\rightarrow |1\rangle ZS|x\rangle = |1\rangle ZS(\alpha|0\rangle + \beta|1\rangle) \\ &= |1\rangle Z(\alpha|0\rangle + \mathrm{i}\beta|1\rangle) \\ &= |1\rangle(\alpha|0\rangle - \mathrm{i}\beta|1\rangle) \\ &= \alpha|1\rangle|0\rangle - \mathrm{i}\beta|1\rangle|1\rangle \\ |1\rangle|x\rangle &\rightarrow |0\rangle SY|x\rangle = |0\rangle SY(\alpha|0\rangle + \beta|1\rangle) \\ &= |0\rangle S(\beta|0\rangle - \alpha|1\rangle) \\ &= |0\rangle(\beta|0\rangle - \mathrm{i}\alpha|1\rangle) \\ &= \beta|0\rangle|0\rangle - \mathrm{i}\alpha|0\rangle|1\rangle. \end{aligned} \tag{16.38}$$

Letting $\beta = 0$ or $\alpha = 1$ in Eq. (16.38), we obtain the transformation:

$$\begin{aligned} u_1|0\rangle|0\rangle &+ u_2|0\rangle|1\rangle + u_3|1\rangle|0\rangle + u_4|1\rangle|1\rangle \\ &\rightarrow u_1|1\rangle|0\rangle - \mathrm{i}u_2|1\rangle|1\rangle - \mathrm{i}u_3|0\rangle|1\rangle + u_4|0\rangle|0\rangle \\ &= u_4|0\rangle|0\rangle - \mathrm{i}u_3|0\rangle|1\rangle + u_1|1\rangle|0\rangle - \mathrm{i}u_2|1\rangle|1\rangle, \end{aligned} \tag{16.39}$$

which corresponds to the circuit matrix and its reduced form:

$$A = \begin{bmatrix} 0 & 0 & 0 & 1 \\ 0 & 0 & -\mathrm{i} & 0 \\ 1 & 0 & 0 & 0 \\ 0 & -\mathrm{i} & 0 & 0 \end{bmatrix} \equiv \begin{bmatrix} 0 & -\mathrm{i}SX \\ S^+ & 0 \end{bmatrix}. \tag{16.40}$$

As a second illustrative example, consider next the 2-qubit quantum circuit shown in Fig. 16.10, which involves two CNOT gates, three gates $A$, $B$, $C$, and a $\delta$-phase gate $S$.[9] It is further assumed that $ABC = I$. As before, the evolution of the input tensor state

[9] A $\delta$-phase gate has for matrix $S = \begin{pmatrix} 1 & 0 \\ 0 & \mathrm{e}^{\mathrm{i}\delta} \end{pmatrix}$.

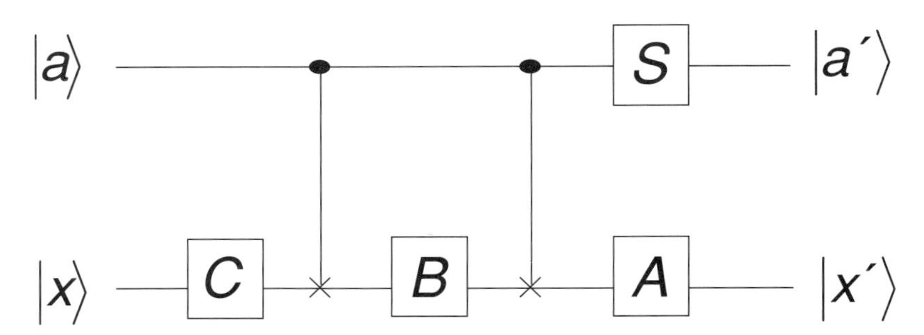


**Figure 16.10** Two-qubit quantum circuit with $ABC = I$ and $\delta$-phase gate $S$.

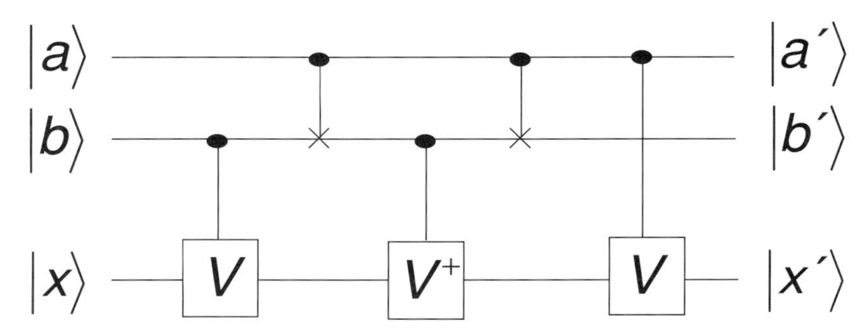


**Figure 16.11** Three-qubit quantum circuit with $V^2 = U$.

$|a\rangle|x\rangle$ through the circuit is calculated through the following:

$$\begin{aligned}&|0\rangle|x\rangle \to |0\rangle C|x\rangle \to |0\rangle C|x\rangle \to |0\rangle BC|x\rangle \to |0\rangle BC|x\rangle \to S|0\rangle ABC|x\rangle = |0\rangle|x\rangle \\ &|1\rangle|x\rangle \to |1\rangle C|x\rangle \to |1\rangle XC|x\rangle \to |1\rangle BXC|x\rangle \\ &\quad \to |1\rangle XBXC|x\rangle \to \mathrm{e}^{\mathrm{i}\delta}|1\rangle AXBXC|x\rangle.\end{aligned} \tag{16.41}$$

Letting $U = \mathrm{e}^{\mathrm{i}\delta} AXBXC$, the above transformation reduces to

$$\begin{aligned}&|0\rangle|x\rangle \to |0\rangle|x\rangle \\ &|1\rangle|x\rangle \to \mathrm{e}^{\mathrm{i}\delta}|1\rangle \mathrm{e}^{-\mathrm{i}\delta} U|x\rangle = |1\rangle U|x\rangle.\end{aligned} \tag{16.42}$$

The result shows that when $|a\rangle = |0\rangle$ the circuit leaves the target qubit $|x\rangle$ invariant, and when $|a\rangle = |1\rangle$ the target qubit is transformed into $U|x\rangle$. This is the definition of the *controlled-$U$* gate. Actually, the equivalence of the circuit shown in Fig. 16.10 and the controlled-$U$ gate stems from *Euler's theorem*, which is demonstrated in Appendix N. The theorem states that for any $2 \times 2$ unitary transformation $U$, there exist three matrices $A, B, C$ satisfying $ABC = I$ and for which $U = \mathrm{e}^{\mathrm{i}\delta} AXBXC$, where $\delta$ is an arbitrary phase.

As a third illustrative example, consider next the 3-qubit quantum circuit shown in Fig. 16.11, which involves two CNOT gates, and three controlled-$U$ gates based on a unitary operator $V$ and its Hermitian conjugate $V^+$. The property $V^2 = U$ is also assumed. Let us walk the input state $|a\rangle|b\rangle|x\rangle$ through the quantum circuit, considering the four possibilities for the control qubits $|a\rangle|b\rangle = |0\rangle|0\rangle, |0\rangle|1\rangle, |1\rangle|0\rangle, |1\rangle|1\rangle$. This

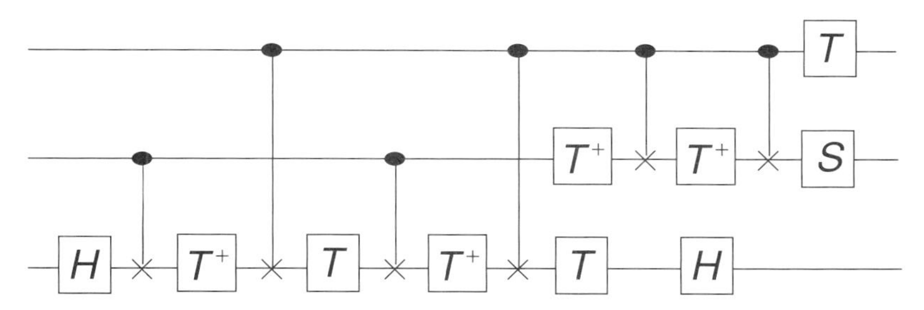


**Figure 16.12** Three-qubit quantum circuit with Hadamard gates ($H$), $\pm\pi/4$-phase gates ($T$, $T^+$), and a single $\pi/2$-phase gate ($S$), which is equivalent to a CCNOT or Toffoli gate.

gives:

$$\begin{aligned}
&|0\rangle|0\rangle|x\rangle \to |0\rangle|0\rangle|x\rangle \to |0\rangle|0\rangle|x\rangle \to |0\rangle|0\rangle|x\rangle \to |0\rangle|0\rangle|x\rangle \to |0\rangle|0\rangle|x\rangle \\
&|0\rangle|1\rangle|x\rangle \to |0\rangle|1\rangle V|x\rangle \to |0\rangle|1\rangle V|x\rangle \to |0\rangle|1\rangle V^+V|x\rangle \\
&\quad = |0\rangle|1\rangle|x\rangle \to |0\rangle|1\rangle|x\rangle \to |0\rangle|1\rangle|x\rangle \\
&|1\rangle|0\rangle|x\rangle \to |1\rangle|0\rangle|x\rangle \to |1\rangle|1\rangle|x\rangle \to |1\rangle|1\rangle V^+|x\rangle \to |1\rangle|0\rangle V^+|x\rangle \to |1\rangle|0\rangle VV^+|x\rangle \\
&\quad = |1\rangle|0\rangle|x\rangle \\
&|1\rangle|1\rangle|x\rangle \to |1\rangle|1\rangle V|x\rangle \to |1\rangle|0\rangle V|x\rangle \to |1\rangle|0\rangle V|x\rangle \to |1\rangle|1\rangle V|x\rangle \to |1\rangle|1\rangle VV|x\rangle \\
&\quad \equiv |1\rangle|1\rangle U|x\rangle.
\end{aligned} \tag{16.43}$$

We observe from the above result that when $|a\rangle|b\rangle \neq |1\rangle|1\rangle$ the circuit leaves the target qubit $|x\rangle$ invariant, and when $|a\rangle|b\rangle = |1\rangle|1\rangle$ the target qubit is transformed into $U|x\rangle$. This is the definition of a *controlled-controlled-$U$*, or CC$U$ gate, whose representation and matrix are shown in Fig. 16.7. It is straightforward to verify that in the case

$$V = \frac{1-\mathrm{i}}{2}(I + \mathrm{i}X) = \frac{1-\mathrm{i}}{\sqrt{2}}\mathrm{e}^{\mathrm{i}\frac{\pi}{4}X}, \tag{16.44}$$

we have $U = V^2 = X$, and the CC$U$ gate reduces to the previously described CCNOT or *Toffoli* gate (Fig. 16.6).[10]

As another example, consider the elaborate quantum circuit shown in Fig. 16.12, which includes Hadamard gates ($H$), $\pm\pi/4$-phase gates ($T$, $T^+$), and a single $\pi/2$-phase gate ($S$). It is left as an exercise to establish that this quantum circuit is actually a possible equivalent realization of a *Toffoli* or CCNOT gate.

As a final example of complex quantum circuits, consider the *plain-adder* circuit described in Chapter 15 for reversible computation with classical bits. We can now conceive of it as a quantum circuit capable of performing plain addition with qubits. The corresponding building blocks are the quantum gate circuits SUM and CARRY, which are based on CNOT and CCNOT or Toffoli gates, as shown in Fig. 16.13. Given two qubit operands $|x\rangle$, $|y\rangle$ and a carry qubit $|c\rangle$, the outputs $|x \oplus y\rangle$, $|x \oplus y \oplus c\rangle$ represent the quantum equivalents of XOR or modulo-2 addition. Note the qubit output $|c'\rangle$ in the CARRY gate circuit; this represents the carry result of the plain addition between

[10] For the exponential-operator representation in the right-hand side, see Appendix N.

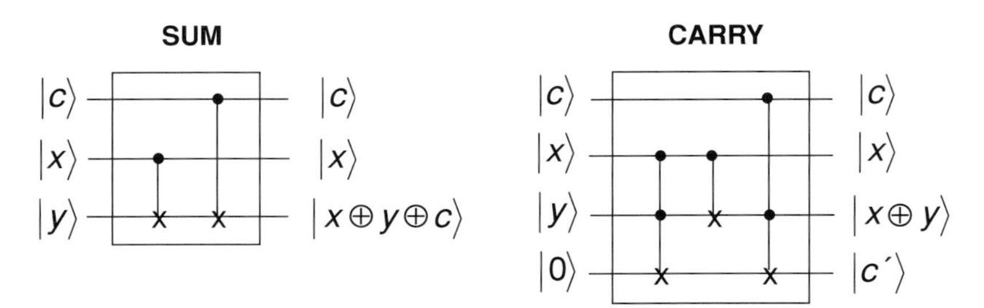


**Figure 16.13** Quantum gate circuits SUM and CARRY, built from CNOT and CCNOT or Toffoli gates.

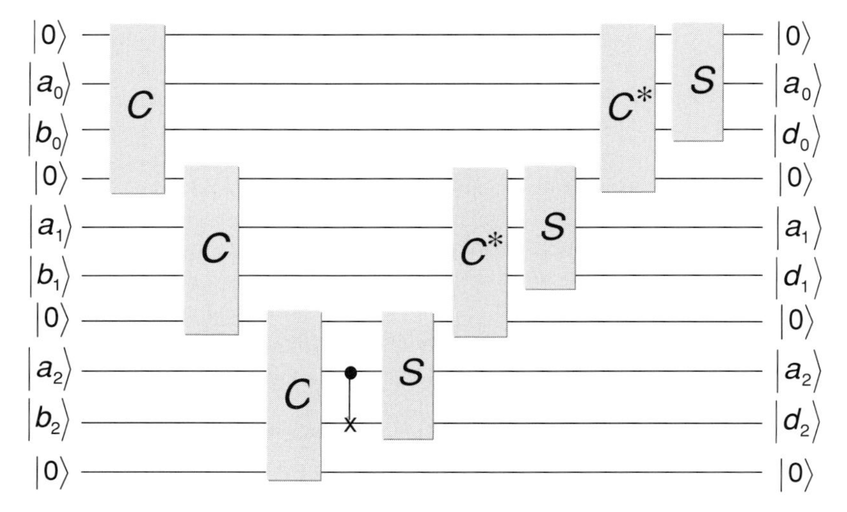


**Figure 16.14** Quantum plain-adder for two qubit registers $|a_0\rangle|a_1\rangle|a_2\rangle$, $|b_0\rangle|b_1\rangle|b_2\rangle$, as based on gate circuits SUM ($S$) and CARRY ($C$) shown in Fig. 16.13. The gate $C^*$ corresponds to the reverse operation of the CARRY gate.

$|x\rangle$, $|y\rangle$ and $|c\rangle$. Also note the *ancilla* bit $|0\rangle$ at the CARRY gate input. A 10-qubit plain-adder circuit, which is homologous to that described in Chapter 15 for classical computing, is shown in Fig. 16.14. This circuit performs the quantum addition of two qubit registers containing 3-qubits each, namely $|a_0\rangle|a_1\rangle|a_2\rangle$, $|b_0\rangle|b_1\rangle|b_2\rangle$, and outputs the result as $|d_0\rangle|d_1\rangle|d_2\rangle$. As discussed in Chapter 15, this circuit architecture can be extended to perform plain addition with registers of any size.[11] As also discussed in the previous chapter, the circuit can be reversed (i.e., traversed from right to left) to perform subtraction, including the generation of a "negative sign" qubit. Furthermore, the concatenation of the circuit makes it possible to perform operations such as multiplication and exponentiation in modular algebra. One must not conclude, however, that quantum

[11] As adapted from V. Vedral and M. B. Plenio, Basics of quantum computation. *Prog. Quant. Electron.*, **22** (1998), 1–39.

algorithms simply emulate classical ones, with qubits instead of bits! (Chapters 19 and 20 will prove far differently).

## 16.5 Tensor products

In previous sections, we made use of the notion of *tensor product* for qubits, to describe quantum gates with more than one qubit input. This section will make it possible to clarify such a notion, which is fundamental to the understanding of quantum computing with multiple states. In the description, we shall proceed gradually, from elementary to advanced.

Given two qubits $|a\rangle, |b\rangle$ we called the *tensor product* the 2-qubit $|a\rangle|b\rangle$. From now on, we shall write the result with the new notation $|a\rangle|b\rangle \equiv |a\rangle \otimes |b\rangle$, with the sign $\otimes$ standing for the tensor-product operation. Given the computational base $V = \{|0\rangle, |1\rangle\}$, which defined a 2D qubit space, we can form the 4D extended base defining a 4D 2-qubit space $V \otimes V$ according to:

$$\begin{aligned} V \otimes V &= \{|0\rangle, |1\rangle\} \otimes \{|0\rangle, |1\rangle\} \\ &= \{|0\rangle|0\rangle, |0\rangle|1\rangle, |1\rangle|0\rangle, |1\rangle|1\rangle\} \\ &\equiv \{|0\rangle \otimes |0\rangle, |0\rangle \otimes |1\rangle, |1\rangle \otimes |0\rangle, |1\rangle \otimes |1\rangle\}. \end{aligned} \tag{16.45}$$

In the extended base, $|a\rangle \otimes |b\rangle$ is represented by four complex coordinates $(u_1, u_2, u_3, u_4)$ to be determined. Assuming that $|a\rangle = a_0|0\rangle + a_1|1\rangle$ and $|b\rangle = b_0|0\rangle + b_1|1\rangle$, we obtain

$$\begin{aligned} |a\rangle \otimes |b\rangle &= (a_0|0\rangle + a_1|1\rangle) \otimes (b_0|0\rangle + b_1|1\rangle) \\ &= a_0b_0|0\rangle \otimes |0\rangle + a_0b_1|0\rangle \otimes |1\rangle + a_1b_0|1\rangle \otimes |0\rangle + a_1b_1|1\rangle \otimes |1\rangle, \end{aligned} \tag{16.46}$$

hence $(u_1, u_2, u_3, u_4) \equiv (a_0b_0, a_0b_1, a_1b_0, a_1b_1)$. In the above, we have implicitly made use of the linearity and distribution-over-addition properties of the tensor product:

$$\begin{cases} \lambda(|x\rangle \otimes |y\rangle) = \lambda|x\rangle \otimes |y\rangle = |x\rangle \otimes \lambda|y\rangle \\ |x\rangle \otimes (|y\rangle + |z\rangle) = |x\rangle \otimes |y\rangle + |x\rangle \otimes |z\rangle. \end{cases} \tag{16.47}$$

The notion of tensor product, along with the same properties, applies to any two $nD$ and $mD$ spaces defined by computational basis $V = \{|\alpha_1\rangle, |\alpha_2\rangle, \ldots, |\alpha_n\rangle\}$ and $W = \{|\beta_1\rangle, |\beta_2\rangle, \ldots, |\beta_m\rangle\}$. Hence, given

$$|a\rangle = \sum_{i=1}^{n} a_i|\alpha_i\rangle, \quad |b\rangle = \sum_{j=1}^{m} b_j|\beta_j\rangle, \tag{16.48}$$

we have

$$|a\rangle \otimes |b\rangle = \sum_{i=1}^{n}\sum_{j=1}^{m} a_i b_i |\alpha_i\rangle \otimes |\beta_i\rangle. \tag{16.49}$$

A qubit $|a\rangle$ that is tensored with itself $n$ times will be noted $|a\rangle^{\otimes n}$, namely,

$$|a\rangle^{\otimes n} = |a\rangle \otimes |a\rangle\otimes, \ldots, \otimes|a\rangle_{n\,\text{times}}. \tag{16.50}$$

The tensor state $|a\rangle \otimes |b\rangle$, often noted $|a, b\rangle$ for simplification, is also referred to as a *joint state*. Such a joint state corresponds to the description of two quantum systems, themselves being in the states $|a\rangle$ and $|b\rangle$, respectively. Thus, $|a\rangle \otimes |a\rangle$ corresponds to the description of two systems being in the same state $|a\rangle$. However, such a possibility should not lead one to the conclusion that given a quantum system in state $|a\rangle$, it is possible to duplicate this state into a second quantum system, so as to obtain the joint state $|a\rangle \otimes |a\rangle$. It simply cannot be achieved. This is a result of the noncloning theorem, which is described in Section 16.6.

Next, I shall introduce the notion of tensor product for *linear operators* (which so far have been referred to as matrices operating on the qubit vector space). Assume an operator $A$ defined on the $|a\rangle$ qubit space and an operator $B$ defined on the $|b\rangle$ qubit space. The *operator tensor product* $A \otimes B$ applies to the tensor states $|a\rangle \otimes |b\rangle$ and is defined according to the following distribution rule:

$$A \otimes B(|a\rangle \otimes |b\rangle) = A|a\rangle \otimes B|b\rangle. \tag{16.51}$$

An operator $A$ that is tensored with itself $n$ times will be noted $A^{\otimes n}$, namely,

$$A^{\otimes n} = A \otimes A\otimes, \ldots, \otimes A_{n \text{ times}}. \tag{16.52}$$

The operator tensor product $A \otimes B$ also satisfies the useful properties according to which *conjugation*, *transposition*, and *Hermitian conjugation*[12] are *distributive*. Namely:

$$\begin{cases} (A \otimes B)^* = A^* \otimes B^* \\ (A \otimes B)^{\mathrm{T}} = A^{\mathrm{T}} \otimes B^{\mathrm{T}} \\ (A \otimes B)^+ = A^+ \otimes B^+. \end{cases} \tag{16.53}$$

These three properties stem from the definition of the operator tensor product in Eq. (16.51), in which the operations of transposition and complex or Hermitian conjugation are clearly distributive.

How can we derive the *matrix* of the tensor operator $A \otimes B$? The rule, which is referred to as the *Kronecker product*, is quite simple. Assume that $A$ is represented by a $n \times m$ matrix ($n$ lines, $m$ columns), with coefficients $A_{ij}$ ($i = 1, \ldots, n$, $j = 1, \ldots, m$). We have, by definition

$$A \otimes B = \begin{pmatrix} A_{11}B & A_{12}B & \cdots & A_{1m}B \\ A_{21}B & A_{22}B & \cdots & A_{2m}B \\ \vdots & \vdots & \cdots & \\ A_{n1}B & A_{n2}B & \cdots & A_{nm}B \end{pmatrix}. \tag{16.54}$$

Thus, in the above reduced form, $A \otimes B$ is an $n \times m$ matrix with coefficients $A_{ij}B$. If $B$ is represented by a $p \times q$ matrix ($p$ lines, $q$ columns), then $A \otimes B$ is clearly an $np \times mq$ matrix. The Kronecker-product rule also applies to single-column matrices, or vectors,

[12] To recall, for any operator $A$ of matrix coefficients $A_{ij}$, the conjugate operator $A^*$ has for coefficients $A^*_{ij}$ (complex conjugate of $A_{ij}$), the transposed operator $A^{\mathrm{T}}$ has for coefficients $A_{ji}$, and the Hermitian-conjugate (or adjoint) operator $A^+$ has for coefficients $A^*_{ji}$. Also, the transposed and Hermitian conjugate of the product $AB$ are $(AB)^{\mathrm{T}} = B^{\mathrm{T}}A^{\mathrm{T}}$ and $(AB)^+ = B^+A^+$, respectively.

which enables one to calculate the tensor product $|a\rangle \otimes |b\rangle$. Let us next examine some illustrative examples.

First, consider the case of the qubit tensor product $|a\rangle \otimes |b\rangle$ in bases $V = \{|\alpha_1\rangle, |\alpha_2\rangle\}$ and $W = \{|\beta_1\rangle, |\beta_2\rangle, |\beta_3\rangle\}$. The corresponding matrices are single column, with $A_{i1} = a_i$ and $B_{i1} = b_i$. We obtain:

$$|a\rangle \otimes |b\rangle = \begin{pmatrix} a_1|b\rangle \\ a_2|b\rangle \end{pmatrix} = \begin{bmatrix} a_1 \begin{pmatrix} b_1 \\ b_2 \\ b_3 \end{pmatrix} \\ a_2 \begin{pmatrix} b_1 \\ b_2 \\ b_3 \end{pmatrix} \end{bmatrix} = \begin{pmatrix} a_1 b_1 \\ a_1 b_2 \\ a_1 b_3 \\ a_2 b_1 \\ a_2 b_2 \\ a_2 b_3 \end{pmatrix}, \tag{16.55}$$

which is the expected result.

Second, consider the tensor product $X \otimes Y$ of the two Pauli matrices $X, Y$. We obtain:

$$\begin{aligned} X \otimes Y &= \begin{pmatrix} 0 \times Y & 1 \times Y \\ 1 \times Y & 0 \times Y \end{pmatrix} \\ &= \begin{pmatrix} 0 & Y \\ Y & 0 \end{pmatrix} = \begin{pmatrix} 0 & 0 & 0 & -\mathrm{i} \\ 0 & 0 & \mathrm{i} & 0 \\ 0 & -\mathrm{i} & 0 & 0 \\ \mathrm{i} & 0 & 0 & 0 \end{pmatrix}. \end{aligned} \tag{16.56}$$

It is left as an easy exercise to verify the property $X \otimes Y(|a\rangle \otimes |b\rangle) = X|a\rangle \otimes Y|b\rangle$, which, as we have seen, applies to any pairs of operators $A$, $B$ and qubits $|a\rangle$, $|b\rangle$.

An interesting case of $n$-tensored operator is provided by $H^{\otimes n}$, where $H$ is the Hadamard gate. Assume the extended computational base $V^n = \{|0\rangle, |1\rangle\}^n = \{|a\rangle\}$, with $|a\rangle$ symbolizing any of the $n$-qubits base element generated by the $n$-tensor product $|a\rangle = |v_1\rangle \otimes |v_2\rangle \otimes \cdots \otimes |v_n\rangle$ where $v_i = 0$ or $v_i = 1$ $(i = 1, \ldots, n)$. As a general property, it can be shown that

$$H^{\otimes n}|a\rangle = \frac{1}{\sqrt{2^n}} \sum_{V^n} (-1)^{a*b} |b\rangle, \tag{16.57}$$

where $|b\rangle = |w_1\rangle \otimes |w_2\rangle \otimes \cdots |w_n\rangle$ is any base element of $V^n$, and $a^*b$ is a scalar defined as:

$$a^*b = v_1 w_1 + v_2 w_2 + \cdots + v_n w_n = \sum_{i=1}^{n} v_i w_i. \tag{16.58}$$

With the tensor-product tools developed in this section, it is proposed as a closing exercise to establish the property in Eq. (16.57) for the case $n = 2$, then by induction for the general case.

## 16.6 Noncloning theorem

Given a quantum system $A$, in any state $|\psi\rangle$, and a second quantum system $B$, in any pure state $|s\rangle$, is it possible to duplicate the first state into the second? Such a "cloning" operation would correspond to the transformation:

$$|\psi\rangle \otimes |s\rangle \rightarrow U(|\psi\rangle \otimes |s\rangle) = |\psi\rangle \otimes |\psi\rangle \equiv |\psi, \psi\rangle, \tag{16.59}$$

where $U$ is a unitary tensor operator. Assume that such an operator $U$ exists and applies to any state $|\psi\rangle$ of $A$. Let $|\phi\rangle$ be another state of $A$ such that $|\phi\rangle \neq |\psi\rangle$. We must also be able to duplicate it into $B$ according to:

$$U(|\phi\rangle \otimes |s\rangle) = |\phi\rangle \otimes |\phi\rangle \equiv |\phi, \phi\rangle. \tag{16.60}$$

By linearity of the transformation, we must also have for any state mixture $|\chi\rangle = \lambda|\psi\rangle + \mu|\phi\rangle$ of $A$, where $\lambda$, $\mu$ are two complex numbers:

$$U(|\chi\rangle \otimes |s\rangle) = |\chi\rangle \otimes |\chi\rangle \equiv |\chi, \chi\rangle. \tag{16.61}$$

If we develop the left-hand side of Eq. (16.61) we obtain:

$$\begin{aligned} U(|\chi\rangle \otimes |s\rangle) &= U(\lambda|\psi\rangle + \mu|\phi\rangle) \otimes |x\rangle \\ &= \lambda|\psi\rangle \otimes |\psi\rangle + \mu|\phi\rangle \otimes |\phi\rangle \\ &\equiv \lambda|\psi, \psi\rangle + \mu|\phi, \phi\rangle, \end{aligned} \tag{16.62}$$

while the right-hand side in Eq. (16.61) yields:

$$\begin{aligned} |\chi\rangle \otimes |\chi\rangle &= (\lambda|\psi\rangle + \mu|\phi\rangle) \otimes (\lambda|\psi\rangle + \mu|\phi\rangle) \\ &= \lambda^2|\psi\rangle \otimes |\psi\rangle + \mu\lambda(|\psi\rangle \otimes |\phi\rangle + |\phi\rangle \otimes |\psi\rangle) + \mu^2|\phi\rangle \otimes |\phi\rangle \\ &\equiv \lambda^2|\psi, \psi\rangle + \mu\lambda|\psi, \phi\rangle + \mu\lambda|\phi, \psi\rangle + \mu^2|\phi, \phi\rangle. \end{aligned} \tag{16.63}$$

Equating Eqs. (16.62) and (16.63) yields

$$\lambda(\lambda - 1)|\psi, \psi\rangle + \mu\lambda|\psi, \phi\rangle + \mu\lambda|\phi, \psi\rangle + \mu(\mu - 1)|\phi, \phi\rangle \equiv 0. \tag{16.64}$$

Assuming that $|\psi\rangle$, $|\phi\rangle$ are pure states, the above equation implies that $\mu\lambda = 0$ and, thus, $|\chi\rangle = |\psi\rangle$ or $|\chi\rangle = |\phi\rangle$. This result means that if there exists an operator $U$ that can clone two pure states $|\psi\rangle$, $|\phi\rangle$, this operator cannot clone any of their mixtures $|\chi\rangle = \lambda|\psi\rangle + \mu|\phi\rangle$, which is a quite restrictive conclusion.

We are then left with the open question: does any cloning operator $U$ exist in the first place? The answer is straightforward, but it requires one to use the *inner product* of states, which is introduced in Chapter 17. Suffice it here to provide the result: *there always exist a unitary operator $U$ capable of cloning a pure state $|\psi\rangle$, or any pair of pure states $|\psi\rangle$* and $|\bar{\psi}\rangle$ (see Chapter 17 for proof). As we have previously seen, however, such an operator cannot clone the mixture $|\chi\rangle = \lambda|\psi\rangle + \mu|\bar{\psi}\rangle$. The key conclusion is that, except for the limiting case of pure-state pairs, *it is not possible to clone quantum states in the general case*. This fundamental result is known as the *noncloning theorem*.

In the specific case of qubits, the two possible bases of pure states are $\{|s_1\rangle, |s_2\rangle\} \equiv \{|0\rangle, |1\rangle\}, \{|+\rangle, |-\rangle\}$. With our knowledge of quantum gates, it is trivial that we can find operators capable of "cloning" pure states into each other. Thus the exception about pure states does not weaken in any sense the generality of the *noncloning theorem*. In particular, there is no quantum gate capable of executing the equivalent of the classical FANOUT gate (Chapter 15), for any state other than a pure state.

## 16.7 Exercises

**16.1** (B): Show by two different methods that the Hadamard gate $H$ corresponds to a unitary transformation.

**16.2** (T): Prove the property according to which the three operators $R_k(\gamma)$ ($k = x, y, z$) rotate any qubit $|q\rangle$ on the Bloch sphere by an angle $\gamma$ about the axis $k$, in the counterclockwise direction.

**16.3** (B): Prove the following properties of the rotation operators:

$$\begin{aligned} R_i(2\theta)R_i(-2\theta) &= I \\ [R_i(2\theta), R_j(2\theta')] &= -2\mathrm{i}\varepsilon_{ijk} \sin\theta \sin\theta' + \sin(\theta - \theta') \\ [R_i(2\theta), \sigma_j] &= 2\varepsilon_{ijk}\sigma_k \sin\theta. \end{aligned}$$

**16.4** (B): Determine the parameters $\vec{n}, \theta$ associated with any unitary transformation $U$ according to the definition:

$$U = \exp[\mathrm{i}(\vec{n} \cdot \vec{\sigma})\theta].$$

*Clue*: use the generic definition of unitary matrices for $U$:

$$U = \mathrm{e}^{\mathrm{i}\delta} \begin{pmatrix} a & b \\ -\bar{b} & \bar{a} \end{pmatrix}.$$

**16.5** (M): Prove that any $2 \times 2$ *unitary* matrix $U$ can be expressed from the two rotation operators $R_y$, $R_z$ according to the product:

$$U = \mathrm{e}^{\mathrm{i}\delta} R_z(\alpha) R_y(\beta) R_z(\gamma),$$

where $\alpha, \beta, \gamma, \delta$ are real numbers (Euler's theorem).

**16.6** (M): Considering the quantum-gate circuit

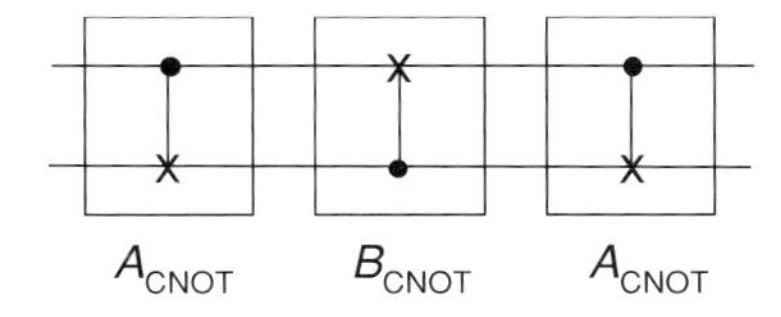


show first that the matrix of the intermediate $B_{\text{CNOT}}$ gate takes the form

$$B_{\text{CNOT}} = \begin{bmatrix} 1 & 0 & 0 & 0 \\ 0 & 0 & 0 & 1 \\ 0 & 0 & 1 & 0 \\ 0 & 1 & 0 & 0 \end{bmatrix},$$

assuming the computational base $\{|0\rangle|0\rangle, |0\rangle|1\rangle, |1\rangle|0\rangle, |1\rangle|1\rangle\}$. Then prove by matrix multiplication that the circuit is equivalent to a CROSSOVER or SWAP gate.

**16.7** (M): Analyze the action of a CNOT gate with control qubit $|a\rangle$ in a superposition of states (assume $|a\rangle = \gamma|0\rangle + \delta|1\rangle$ and $|x\rangle = \alpha|0\rangle + \beta|1\rangle$). Consider then the two cases $\gamma = \delta = 1/\sqrt{2}$ and $\gamma = -\delta = 1/\sqrt{2}$ for the control qubit. What are the possible output states? Show that the same result is obtained with all inputs in pure states and with a Hadamard gate placed on the control input path.

**16.8** (T): Prove that the quantum circuit

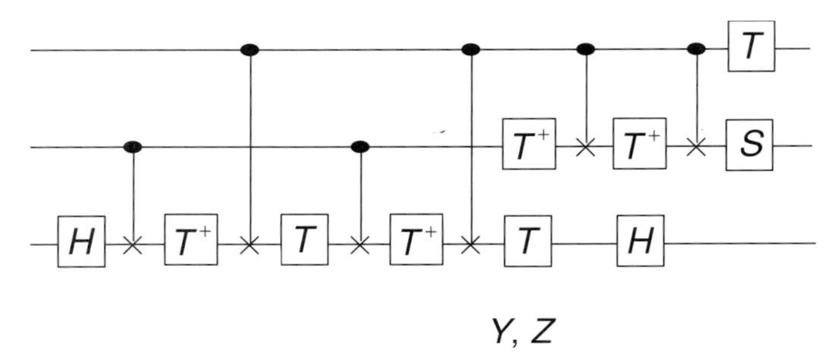


which includes Hadamard gates ($H$), $\pm\pi/4$-phase gates ($T$, $T^+$), and a single $\pi/2$-phase gate ($S$), is an equivalent realization of a CCNOT or Toffoli gate.

**16.9** (B): Given the Pauli matrices $Y$, $Z$, calculate the tensor product $Y \otimes Z$. Given two qubits $|a\rangle$, $|b\rangle$ and their tensor product $|a\rangle \otimes |b\rangle$, show that

$$Y \otimes Z(|a\rangle \otimes |b\rangle) = Y|a\rangle \otimes Z|b\rangle.$$

**16.10** (T): Given the computational base $V = \{|0\rangle, |1\rangle\}$ and the Hadamard operator $H$, show that the two-fold tensor product $H^{\otimes 2} = H \otimes H$ satisfies

$$H^{\otimes 2}|a\rangle = \frac{1}{2}\sum_{V^n}(-1)^{a*b}|b\rangle,$$

where $|a\rangle = |v_1\rangle \otimes |v_2\rangle$ and $|b\rangle = |w_1\rangle \otimes |w_2\rangle$ are any elements of $V^2$ (with $v_i, w_j = 0, 1$) and

$$a^*b = v_1 w_1 + v_2 w_2.$$

Then prove by induction that, in the general case,

$$H^{\otimes n}|a\rangle = \frac{1}{\sqrt{2^n}}\sum_{V^n}(-1)^{a^*b}|b\rangle,$$

with

$$a^*b = v_1 w_1 + v_2 w_2 + \cdots + v_n w_n = \sum_{i=1}^{n} v_i w_i.$$

# 17 Quantum measurements

This chapter is concerned with the *measure of quantum states*. This requires one to introduce the subtle notion of *quantum measurement*, an operation that has no counterpart in the classical domain. To this effect, we first need to develop some new tools, starting with *Dirac notation*, a formalism that is not only very elegant but is relatively simple to handle. The introduction of Dirac notation makes it possible to become familiar with the *inner product* for quantum states, as well as different properties for operators and states concerning *projection*, *change of basis*, *unitary transformations*, *matrix elements*, *similarity transformations*, *eigenvalues* and *eigenstates*, *spectral decomposition* and *diagonal representation*, *matrix trace* and *density operator or matrix*. The concept of density matrix makes it possible to provide a very first and brief hint of the analog of Shannon's entropy in the quantum world, referred to as *von Neumann's entropy*, to be further developed in Chapter 21. Once we have all the required tools, we can focus on *quantum measurement* and analyze three different types referred to as *basis-state measurements*, *projection* or *von Neumann measurements,* and *POVM* measurements. In particular, POVM measurements are shown to possess a remarkable property of *unambiguous quantum state discrimination* (UQSD), after which it is possible to derive "absolutely certain" information from unknown system states. The more complex case of *quantum measurements in composite systems* described by joint or tensor states is then considered. Although of a more abstract and formal character than any previous ones, this chapter is crucial to the understanding of the rest of these chapters, which are concerned with the *manipulation of qubits*. This is not only because of the importance of being comfortable with the Dirac formalism, but also of the need to have conceptually assimilated the basics of quantum measurement.

## 17.1 Dirac notation

In this section, I shall introduce so-called *Dirac notation*, which are used in quantum mechanics. When compared with the basic math formalism used in engineering, Dirac notation looks quite esoteric, if not highly involved and complex. But as we shall see, such notation is straightforward to assimilate, and quite easy to handle after familiarization. Dirac notation is used to recapitulate the basic properties of linear operators and their action on quantum states (here, qubits), which leads to a simple formalization of the concept of *quantum measurement*.

In Chapter 16, we defined the *qubit* as a 2D vector noted $|q\rangle$. In the orthonormal basis of pure states $V = \{|0\rangle, |1\rangle\}$, the qubit has complex coordinates $\alpha, \beta$, so that $|q\rangle = \alpha|0\rangle + \beta|1\rangle$. As we have seen, higher-dimension qubits can be defined from the extended basis $V^n = \{|x\rangle\}$, where $|x\rangle$ is the $n$th tensor product of the pure state $|\delta_i\rangle = |0\rangle$ or $|1\rangle$, namely:

$$\begin{aligned} |x\rangle &= |\delta_1\rangle \otimes |\delta_2\rangle \otimes \cdots \otimes |\delta_n\rangle \\ &\equiv |\delta_1, \delta_2, \ldots, \delta_n\rangle \equiv |\delta_1 \delta_2 \cdots \delta_n\rangle, \end{aligned} \tag{17.1}$$

where the two equivalent notations in the right-hand side can be used for lightening purposes. In this extended space, the qubit $|q\rangle$ can, thus, be expanded into

$$|q\rangle = \sum_{x \in V^n} x|x\rangle, \tag{17.2}$$

where $x$ represents the complex coordinate of $|q\rangle$ with respect to the $n$-qubit basis element $|x\rangle$.

Regardless of any basis dimension $n$, the qubit or vector $|q\rangle$ is also referred to as a "*ket*." Actually, the ket represents the *Dirac notation* for a *quantum state* in the space defined by $V = \{|0\rangle, |1\rangle\}$, which is referred to as *Hilbert space*. Here, we do not need to elaborate further on the notions of quantum state (as related to the physical world) and of Hilbert space (as defining the continuum of such quantum states). As it turns out, the concept of qubit as a *vector*, rather than as a quantum state, even as representing an oversimplification, is accurate and wholly sufficient to grasp Dirac formalism.

## Inner product

In a 2D vector space, the *scalar product* of two vectors $x = (x_1, x_2)$ and $y = (y_1, y_2)$ is defined as the real number $x \cdot y = x_1 y_1 + x_2 y_2$. In particular, the self-product $x \cdot x = x_1^2 + x_2^2 \equiv |x|^2$ corresponds to the vector's length or *modulus*. In the qubit space $V$ (meaning "defined by any orthogonal basis such as $V$"), we can introduce the concept of the *inner product of two qubits*, $|q\rangle = \alpha|0\rangle + \beta|1\rangle$ and $|q'\rangle = \alpha'|0\rangle + \beta'|1\rangle$ according to:

$$|q\rangle \cdot |q'\rangle = \bar{\alpha}\alpha' + \bar{\beta}\beta', \tag{17.3}$$

where $\bar{\alpha}, \bar{\beta}$ are the complex conjugates of the coordinates $\alpha, \beta$, respectively. In particular, we have $|q\rangle \cdot |q\rangle = |\alpha|^2 + |\beta|^2$, which as a real number, represents the modulus of the qubit $|q\rangle$. In vector notation, Eq. (17.3) can be written as a line-vector–column-vector product:

$$|q\rangle \cdot |q'\rangle = \overline{|q\rangle}^{\mathrm{T}} |q'\rangle = (\bar{\alpha}, \bar{\beta}) \begin{pmatrix} \alpha \\ \beta \end{pmatrix} = \bar{\alpha}\alpha' + \bar{\beta}\beta', \tag{17.4}$$

where $\overline{|q\rangle}^{\mathrm{T}}$ is the conjugate-transposed of $|q\rangle$. Here is a first opportunity to show how Dirac notation comes in handy. Indeed, define the "bra" $\langle q|$ as representing the conjugate-transposed vector $\overline{|q\rangle}^{\mathrm{T}}$. With this notation, Eq. (17.4) becomes the "bra-ket:"

$$\langle q|q'\rangle = \bar{\alpha}\alpha' + \bar{\beta}\beta' \tag{17.5}$$

And, hence, for the qubit modulus, $\langle q|q\rangle = |\alpha|^2 + |\beta|^2$. As another property, we note from Eq. (17.5) that $\langle q|q'\rangle = \overline{\langle q'|q\rangle}$, namely that the inner product of qubit states is *noncommutative*, unlike the scalar product of 2D vectors with real coordinates.

As we shall see next, there is more in the "bra" notation than just a convenient way to express the inner product between two qubit states. Indeed, consider $n$-qubits $|q\rangle$, $|q'\rangle$ in the space $V^n$, with, for $|q'\rangle$, the expansion in Eq. (17.2) and, for the bra, $\langle q|$ (consistently with definition):

$$\langle q| = \sum_{x\in V^n} \bar{x}\langle x|. \tag{17.6}$$

We can now develop the inner product $\langle q|q'\rangle$ according to

$$\langle q|q'\rangle = \left(\sum_{x\in V^n} \bar{x}\langle x|\right)\left(\sum_{x'\in V^n} x'|x'\rangle\right) = \sum_{x\in V^n}\sum_{x'\in V^n} \bar{x}x'\langle x|x'\rangle. \tag{17.7}$$

Since $|x\rangle$, $|x'\rangle$ are pure states from the orthonormal basis $V^n$, we have $\langle x|x'\rangle = \delta_{ss'}$ where $\delta_{ss'}$ is the *Kronecker symbol* ($\delta_{ss'} = 1$ for $x = x'$ and $\delta_{ss'} = 0$ otherwise). Hence the explicit definition of inner product when the two qubits are expressed in the same basis $V^n$:

$$\langle q|q'\rangle = \sum_{x\in V^n}\sum_{x'\in V^n} \bar{x}x'\delta_{xx'}. \tag{17.7}$$

Which, for $|q'\rangle = |q\rangle$, reduces to the modulus:

$$\langle q|q\rangle = \sum_{x\in V^n} |x|^2. \tag{17.8}$$

In particular, effecting a left product by $\langle x|$ to Eq. (17.2) we have

$$\langle x|q\rangle = \sum_{x'\in V^n} x\langle x|x'\rangle = \sum_{x'\in V^n} x\delta_{xx'} = x, \tag{17.9}$$

which shows that, expectedly, $\langle x|q\rangle = x$ is the projection of $|q\rangle$ over the pure state $|x\rangle$. Substituting this result into the definition in Eq. (17.2), we obtain

$$|q\rangle = \sum_{x\in V^n} x|x\rangle = \sum_{x\in V^n} \langle x|q\rangle|x\rangle \equiv \sum_{x\in V^n} |x\rangle\langle x|q\rangle, \tag{17.10}$$

which represents another way of expanding $|q\rangle$ over the basis $V^n$.

In Chapter 16, we have shown that it is generally not possible to "clone" quantum states, a property referred to as the *noncloning theorem*. As was mentioned without proof, the only exceptions to this theorem concern any pure state, or any pair of pure states. With the inner product introduced above, it is quite easy to show this.[1]

## Projection operators

In view of the expansion of $|q\rangle$ in Eq. (17.10), we can introduce $U_x = |x\rangle\langle x|$ as *the operator projecting any qubit onto the state* $|x\rangle$. To show this, we first observe that

$$U_x|x\rangle = (|x\rangle\langle x|)|x\rangle = |x\rangle\langle x|x\rangle = |x\rangle\delta_{xx} \equiv |x\rangle \tag{17.11}$$

and

$$\begin{aligned} U_x|q\rangle &= (|x\rangle\langle x|)\sum_{x'\in V^n} x'|x'\rangle \\ &= \sum_{x'\in V^n} x'|x\rangle\langle x|x'\rangle = |x\rangle\sum_{x'\in V^n} x'\delta_{xx'} \equiv x|x\rangle, \end{aligned} \tag{17.12}$$

which shows that $|x\rangle$ is invariant by $U_x$ and also that $|q\rangle$ is projected on its basis component $x|x\rangle$. With such a definition of projector operator, we also have from Eq. (17.10):

$$|q\rangle = \left(\sum_{x\in V^n} |x\rangle\langle x|\right)|q\rangle, \tag{17.13}$$

from which we obtain

$$\sum_{x\in V^n} |x\rangle\langle x| = I. \tag{17.14}$$

The above property, which is referred to as the *completeness* (or *closure*) *relation*, expresses the fact that the complete sum of all projection operators over the pure states of $V^n$ is the identity operator. More generally, a unitary operator $U$ is a *projector* if it satisfies the property $U^2 = U$ (projecting twice is the same as projecting once). In the case of $U_x = |x\rangle\langle x|$, we have, indeed,

$$\begin{aligned} U_x^2 &= U_x U_x = (|x\rangle\langle x|)(|x\rangle\langle x|) \\ &= |x\rangle\langle x|x\rangle\langle x| = |x\rangle\delta_{xx}\langle x| = |x\rangle\langle x| \equiv U_x \end{aligned} \tag{17.15}$$

[1] Referring to Section 16.6, given three pure states $|s\rangle, |\psi\rangle, |\phi\rangle$, we assume a unitary operator $U$ capable to achieve the following cloning transformations:

$$\begin{aligned} U(|\psi\rangle \otimes |x\rangle) &= |\psi\rangle \otimes |\psi\rangle, \\ U(|\phi\rangle \otimes |s\rangle) &= |\phi\rangle \otimes |\phi\rangle. \end{aligned}$$

Taking the inner product of the two right-hand sides and using the properties $U^+U = I$ and $\langle s|s\rangle = 1$, yields:

$$\begin{aligned} (\langle\phi| \otimes \langle s|)U^+U(|\psi\rangle \otimes |x\rangle) &= (\langle\phi| \otimes \langle\phi|)(|\psi\rangle \otimes |\psi\rangle) \leftrightarrow \langle\phi|\psi\rangle\langle s|s\rangle \\ &= \langle\phi|\psi\rangle\langle\phi|\psi\rangle \leftrightarrow \langle\phi|\psi\rangle = \langle\phi|\psi\rangle^2, \end{aligned}$$

which shows that either $\langle\phi|\psi\rangle = 1$ or $\langle\phi|\psi\rangle = 0$, corresponding to $|\phi\rangle = |\psi\rangle$ or $|\phi\rangle, |\psi\rangle$ being orthogonal states.

and

$$U_x|x'\rangle = (|x\rangle\langle x|)|x'\rangle = |x\rangle\langle x|x'\rangle = \delta_{xx'}|x\rangle. \tag{17.16}$$

These two results confirm that $U_x = |x\rangle\langle x|$ is the unique projector on the pure state $|x\rangle$. It is clear that for any state $|q\rangle$ of unity length (such as a qubit), the operator $|q\rangle\langle q|$ is the projector over the state $|q\rangle$, which is left as an easy exercise to demonstrate.

### Change of basis

Let $W^n = \{|y\rangle\}$ be a new orthonormal basis. Consistently, we can expand $|q\rangle$ over $W^n$, using a new system of coordinates $y$ defined over the states $|y\rangle$. How can we relate the new coordinates $y$ to the old coordinates $x$? Here again, Dirac notations turn out to be quite handy. First, let us insert the completeness relation of $W^n$ into the $V^n$ expansion according:

$$\begin{aligned} |q\rangle &= \sum_{x\in w^n} x|x\rangle = \sum_{x\in V^n} x\left(\sum_{y\in V^n} |y\rangle\langle y|\right)|x\rangle \\ &= \sum_{y\in V^n}\left(\sum_{x\in V^n} x\langle y|x\rangle\right)|y\rangle \equiv \sum_{y\in V^n} y|y\rangle. \end{aligned} \tag{17.17}$$

The result in Eq. (17.17), thus, yields the definition of the new $y$ coordinates with respect to the old $x$ coordinates:

$$y = \sum_{x\in V^n} \langle y|x\rangle x. \tag{17.18}$$

### Unitary transformations

As seen in Chapter 16, a unitary transformation is characterized by an operator $U$ satisfying the property $U^+U = I$, where $U^+$ is the *Hermitian conjugate* of $U$. Define $|q'\rangle = U|q\rangle$ as the output state resulting from the transformation of input state $|q\rangle$. Consistently with previous definition of the "bra," we can equivalently write $\langle q'| = \langle q|U^+$ as the transformation of $\langle q|$ into $\langle q'|$. Developing the inner product $\langle q'|q'\rangle$ and using the unitary condition $U^+U = I$ yields

$$\begin{aligned} \langle q'|q'\rangle &= ((\langle q'|U^+)(U|q\rangle) \\ &= \langle q|U^+U|q\rangle = \langle q|I|q\rangle \equiv \langle q|q\rangle, \end{aligned} \tag{17.19}$$

which shows that *unitary transformations conserve the state modulus*. Furthermore, *unitary transformations also conserve the inner product*. This is readily verified given $|a'\rangle = U|a\rangle$ and $|b'\rangle = U|b\rangle$, which yields $\langle a'|b'\rangle = \langle a|U^+U|b\rangle \equiv \langle a|b\rangle$.

### Operator matrix elements

Given the basis $V^n = \{|x\rangle\}$, the action of operator $U$ on the qubit or state $|q\rangle$, which yields $|q'\rangle = U|q\rangle$, can be characterized by a change of coordinates from $x$ to $x'$, as follows:

$$\begin{aligned}|q'\rangle = U|q\rangle = U\left(\sum_{x\in V^n} x|x\rangle\right) &= \sum_{x\in V^n} xU|x\rangle \\ = \sum_{x\in V^n} x\left(\sum_{x'\in V^n}|x'\rangle\langle x'|\right)U|x\rangle &= \sum_{x\in V^n} x\sum_{x'\in V^n}|x'\rangle\langle x'|U|x\rangle \\ = \sum_{x'\in V^n}\left(\sum_{x\in V^n}\langle x'|U|x\rangle x\right)|x'\rangle &\equiv \sum_{x'\in V^n} x'|x'\rangle\end{aligned} \tag{17.20}$$

with

$$x' = \sum_{x\in V^n}\langle x'|U|x\rangle x. \tag{17.21}$$

In Eq. (17.21), the complex coefficients $\langle x'|U|x\rangle$ represents the *matrix elements* of the operator $U$, as expressed in the basis $V^n$.

Assume next a *different* orthonormal basis $W^n = \{|y\rangle\}$. What are the matrix elements of $U$ in this new basis? To answer this question, we effect the following development:

$$\begin{aligned}\langle y'|U|y\rangle &= \langle y'|\left(\sum_{x'\in V^n}|x'\rangle\langle x'|\right)U\left(\sum_{x\in V^n}|x\rangle\langle x|\right)|y\rangle \\ &= \sum_{x'\in V^n}\sum_{x\in V^n}\langle y'|x'\rangle\langle x'|U|x\rangle\langle x|y\rangle \equiv \sum_{x\in V^n}\sum_{x'\in V^n} T_{y'x'}U_{x'x}T_{xy}\end{aligned} \tag{17.22}$$

with

$$T_{yx} = \overline{T_{xy}} = T^+_{yx} = \langle y|x\rangle, \tag{17.23}$$

where $T$ is called the *transition operator*. The result in Eq. (17.22) shows that the matrix $U$ in basis $W^n$ (call it $\tilde{U}$) can be calculated from the matrix $U$ in basis $V^n$ according to the transformation:

$$\tilde{U} = TUT^+, \tag{17.24}$$

which is referred to as *similarity transformation*. It is easily established that $T^+T = TT^+ = I$ in any basis (noticing that $T_{yx} = T^+_{yx}$ does not imply $T = T^+$, which is left as an exercise to show).

## Eigenvalues and eigenstates

Given an operator $A$, any state $|v\rangle$ such that $A|v\rangle = \lambda|v\rangle$, with $\lambda$ being a complex number, is called an *eigenstate* of $A$. The corresponding number $\lambda$ is called the *eigenvalue* of $|v\rangle$. The eigenvalues are the solutions of the *characteristic equation* $|A - \lambda I| = 0$, where $|Q|$ (also sometimes noted $\det|Q|$) is the *determinant* of the matrix $Q$.[2] It is customary to label the eigenstates after their eigenvalues, thus, the eigenstate $|\lambda_i\rangle$ implicitly satisfies $A|\lambda_i\rangle = \lambda_i|\lambda_i\rangle$. Two or more eigenstates may have

[2] For a $2\times 2$ matrix $Q = \begin{pmatrix} a & c \\ b & d\end{pmatrix}$, the determinant is $|Q| = ad - bc$. For a general definition see, for instance, http://en.wikipedia.org/wiki/Determinant.

the same eigenvalue, in which case the eigenvalue is said to be *degenerate* (in this case the corresponding eigenstates may be labeled as $|\lambda_i^{(1)}\rangle, |\lambda_i^{(2)}\rangle \cdots$). We also have $\langle\lambda_i|A^+ = \bar{\lambda}_i\langle\lambda_i|$, thus, $\langle\lambda_i|A^+A|\lambda_i\rangle = |\lambda_i|^2\langle\lambda_i|\lambda_i\rangle$. If $A$ is Hermitian ($A = A^+$), we have $\langle\lambda_i|A^+A|\lambda_i\rangle = \langle\lambda_i|AA|\lambda_i\rangle = \lambda_i^2\langle\lambda_i|\lambda_i\rangle$, which shows that $\lambda_i^2 = |\lambda_i|^2$, or that *the eigenvalues of a Hermitian operator are real*. It can be shown that *any linear operator has at least one eigenvalue and eigenvector* (the characteristic equation having at least one root). The complete set of eigenstates defines the operator's *eigenspace*. A key property of the eigenstates $|\lambda_i\rangle$ of any operator $A$ is that they form an *orthonormal basis*, i.e., $\langle\lambda_i|\lambda_j\rangle = \delta_{ij}$. Therefore, the completeness relation applies to the set of eigenstates:

$$\sum_i |\lambda_i\rangle\langle\lambda_i| = I. \tag{17.25}$$

Using the completeness relation, we can write the matrix elements of $A$ in the form:

$$\begin{aligned}\langle x'|A|x\rangle &= \langle x'|A\left(\sum_i |\lambda_i\rangle\langle\lambda_i|\right)|x\rangle = \sum_i \langle x'|A|\lambda_i\rangle\langle\lambda_i|x\rangle \\ &= \sum_i \langle x'|\lambda_i|\lambda_i\rangle\langle\lambda_i|x\rangle \equiv \langle x'|\left(\sum_i \lambda_i|\lambda_i\rangle\langle\lambda_i|\right)|x\rangle,\end{aligned} \tag{17.26}$$

which gives

$$A = \sum_i \lambda_i|\lambda_i\rangle\langle\lambda_i|. \tag{17.27}$$

The definition in Eq. (17.27), which decomposes $A$ into a sum of eigenstate projectors $|\lambda_i\rangle\langle\lambda_i|$, is called the *diagonal representation*, or *spectral decomposition* of $A$. For instance, the diagonal-operator representation of the Pauli matrix $Z$ is:

$$Z = \begin{pmatrix}1 & 0\\ 0 & -1\end{pmatrix} = |0\rangle\langle 0| - |1\rangle\langle 1|. \tag{17.28}$$

It is left as an easy exercise to determine the eigenstates and diagonal representations of the Pauli matrix $Y$ and the Hadamard matrix $H$. More generally, in the diagonal representation of $A$, the matrix elements $\langle\lambda_i|A|\lambda_j\rangle$ can be developed as follows:

$$\begin{aligned}\langle\lambda_i|A|\lambda_j\rangle &= \langle\lambda_i|\left(\sum_k \lambda_k|\lambda_k\rangle\langle\lambda_k|\right)|\lambda_j\rangle = \sum_k \lambda_k\langle\lambda_i|\lambda_k\rangle\langle\lambda_k|\lambda_j\rangle \\ &= \sum_k \lambda_k\delta_{ik}\delta_{kj} \equiv \lambda_i\delta_{ij},\end{aligned} \tag{17.29}$$

which shows that all nondiagonal elements of the matrix are identically zero, while the diagonal elements are equal to the eigenvalues $\lambda_i$, i.e.,

$$A = \begin{pmatrix}\lambda_1 & 0 & \cdots & 0\\ 0 & \lambda_2 & \cdots & \vdots\\ \vdots & \vdots & \ddots & 0\\ 0 & \cdots & 0 & \lambda_n\end{pmatrix}. \tag{17.30}$$

An operator matrix that can be transformed into the diagonal form shown in Eq. (17.30) is said to de *diagonalizable*. For this, its $n \times n$ matrix must have $n$ eigenstates. Clearly, if the matrix has fewer than $n$ eigenstates, it cannot be diagonalized, as there remain some nonzero off-diagonal elements in the matrix. An operator is said to be *normal* if it satisfies $AA^+ = A^+A$. A remarkable property: *any normal operator is diagonalizable, and any diagonalizable operator is normal.*[3] Consequently, since they are normal, *Hermitian operators are diagonalizable.*

## Matrix trace

The *trace* of a matrix $A$, noted $\text{tr}(A)$ is the sum of all its diagonal elements $A_{ii}$, according to

$$\text{tr}(A) = \sum_{i=1}^{n} A_{ii}. \tag{17.31}$$

It is easily established that for any two operators $A$, $B$ and complex number $\alpha$, we have the properties:

$$\begin{cases} \text{tr}(\alpha A) = \alpha\ \text{tr}(A) \\ \text{tr}(A + B) = \text{tr}(A) + \text{tr}(B) \\ \text{tr}(AB) = \text{tr}(BA). \end{cases} \tag{17.32}$$

In the diagonal representation (Eq. (17.30)), it is clear that *the trace of a matrix is the sum of its eigenvalues*. But the trace of a matrix does not depend on the choice of the

[3] This can be shown through the following. Since $A$ is a linear operator, it must have at least one eigenvalue, say $\lambda$. Define $P = |\lambda\rangle\langle\lambda|$ as the projector on the corresponding eigenspace $|\lambda\rangle$ (where $AP = \lambda P$), and $Q = I - P$ the projector on the orthogonal subspace (hence $P + Q = I$ and $PQ = QP = 0$). We also have $P^2 = P$, $Q^2 = Q$, and $P$, $Q$ are both Hermitian ($P^+ = P$, hence $Q^+ = Q$). We develop $A$ according to the following:

$$\begin{aligned} A &= (P + Q)A = (P + Q)A(P + Q) = PAP + QAP + PAQ + QAQ \\ &= P\lambda P + \lambda QP + PAQ + QAQ = \lambda P + PAQ + QAQ. \end{aligned}$$

Next we can show that $PAQ = 0$ if $A$ is normal. Indeed, we have $AA^+|\lambda\rangle = A^+A|\lambda\rangle = \lambda A^+|\lambda\rangle$, which shows that $|\lambda'\rangle = A^+|\lambda\rangle$ is also an eigenstate of $A$ and belongs to its eigenspace. Therefore, $QA^+ = 0$, hence, $QA^+P = (PAQ)^+ = 0 = PAQ$. We, thus, obtain $A = \lambda P + QAQ$. Finally, we can show that $QAQ$ is normal. We have

$$\begin{aligned} (QAQ)(QA^+Q) &= QAQQA^+Q \\ &= QAQA^+Q = QA(I - P)A^+Q \\ &= QA(I - P)A^+Q = QAA^+Q - QAPA^+Q \\ &= QA^+AQ - \lambda QPA^+Q = QA^+AQ \\ &= QA^+(P + Q)AQ = Q(A^+P)AQ + QA^+QAQ \\ &= QA^+QQAQ = (QA^+Q)(QAQ), \end{aligned}$$

which proves that $QAQ$ is normal. We have found that $A$ is decomposed as the sum $A = \lambda P + A'$, with $P$ being diagonal in the eigenspace $|\lambda\rangle$, and $A' = QAQ$ being normal. Since $A'$ is a linear operator, it must have at least one eigenvalue, say $\lambda'$, thus, it can be decomposed into $A' = \lambda' P' + A''$, where $P'$ is diagonal in the eigenspace $|\lambda'\rangle$, and $A''$ is normal. By induction, we conclude that there exists an orthonormal basis $\{|\lambda\rangle, |\lambda'\rangle \ldots\}$ of the same dimension as $A$ for which the matrix $A$ is diagonal.

basis representation. Indeed, using the basis transformation in Eq. (17.24), and the third property in Eq. (17.32), we obtain:

$$\mathrm{tr}(\tilde{U}) = \mathrm{tr}(TUT^{+}) = \mathrm{tr}(T^{+}TU) \equiv \mathrm{tr}(U). \tag{17.33}$$

## Density operator or matrix

Let $V^n = \{|x_i\rangle\} = \{|x_1\rangle, |x_2\rangle, \ldots, |x_n\rangle\}$ an orthonormal basis for the space of quantum states $|\psi\rangle$, i.e., satisfying $\langle x_i|x_j\rangle = \delta_{ij}$. In this basis, the states have a unique decomposition, which takes the form

$$|\psi\rangle = x_1|x_1\rangle + x_2|x_2\rangle + \cdots + x_n|x_n\rangle = \sum_{i=1}^{n} x_i|x_i\rangle, \tag{17.34}$$

where $x_i$ $(i = 1, \ldots, n)$ represents the complex coordinates. If the modulus or length of $|\psi\rangle$ is unity, we have

$$\langle\psi|\psi\rangle = \sum_{i=1}^{n}\sum_{j=1}^{n} \bar{x}_i x_j \langle x_i|x_j\rangle = \sum_{i=1}^{n} |x_i|^2 = 1. \tag{17.35}$$

As we have seen in Chapter 16, for qubits, the number $p_i = |x_i|^2$ represents the *probability of finding (or measuring) the state $|\psi\rangle$ in the basis state $|x_i\rangle$*. Hence, the coordinates $x_i$ represent complex *amplitude probabilities*. We can now define the *density operator or matrix* associated with the state $|\psi\rangle$ as:

$$\begin{aligned}\rho &= |x_1|^2|x_1\rangle\langle x_1| + |x_2|^2|x_2\rangle\langle x_2| + \cdots + |x_n|^2|x_n\rangle\langle x_n| \\ &= \sum_{i=1}^{n} |x_i|^2|x_i\rangle\langle x_i|.\end{aligned} \tag{17.36}$$

The density matrix is *diagonal*, since its elements $\rho_{ij}$ are given by

$$\begin{aligned}\rho_{ij} &= \langle x_i|\rho|x_j\rangle \\ &= \langle x_i|\left(\sum_{k=1}^{n} |x_k|^2|x_k\rangle\langle x_k|\right)|x_j\rangle \\ &= \sum_{k=1}^{n} |x_k|^2\langle x_i|x_k\rangle\langle x_k|x_j\rangle \\ &= \sum_{k=1}^{n} |x_k|^2\delta_{ik}\delta_{kj} \equiv |x_i|^2\delta_{ij}.\end{aligned} \tag{17.37}$$

Hence, the density matrix operator takes the diagonal matrix representation:

$$\rho = \begin{pmatrix} |x_1|^2 & 0 & \cdots & 0 \\ 0 & |x_2|^2 & \cdots & \vdots \\ \vdots & \vdots & \ddots & 0 \\ 0 & \cdots & 0 & |x_n|^2 \end{pmatrix} = \begin{pmatrix} p_1 & 0 & \cdots & 0 \\ 0 & p_2 & \cdots & \vdots \\ \vdots & \vdots & \ddots & 0 \\ 0 & \cdots & 0 & p_n \end{pmatrix}. \tag{17.38}$$

It is immediately noted that the trace of the density matrix is unity, since

$$\mathrm{tr}(\rho) = \sum_{i=1}^{n} |x_i|^2 = \sum_{i=1}^{n} p_i = 1. \tag{17.39}$$

We now have the tools to make a short hint at *quantum information theory*. This may also constitute a nice reward for having gone through the lengthy description of Dirac notation!

Let us introduce a new operator, called $U \log U$, where $U$ is assumed to be diagonal with nonnegative coefficients. To calculate the matrix coefficients of $U \log U$, we must first define $\log U$. Assume, then, a linear operator $V$, which satisfies $U = \exp(V)$, which defines $V = \log U$. Formally, the exponential operator is determined by the infinite series:

$$U = \exp(V) = \sum_{n}^{\infty} \frac{V^n}{n!}. \tag{17.40}$$

Since $U$ is diagonal, any of the powers $V^n$ must be diagonal. The diagonal coefficients of $U$ are, thus, given by $U_{ii} = \exp(V_{ii})$ or $V_{ii} = \log(U_{ii})$. The matrix $W = UV = U \log U$ is also diagonal. It is clear that its coefficients are given by $W_{ii} = U_{ii}V_{ii} = U_{ii}\log(U_{ii})$. This result shows that the matrix $W = U \log U$ is analytically defined for any diagonal matrix $U$ with coefficients $U_{ii} \geq 0$.[4] We conclude that the density matrix $U = \rho$, for which the coefficients are nonnegative, is an eligible candidate for the operator $U \log U$. We, thus, have $(\rho \log \rho)_{ii} = \rho_{ii} \log \rho_{ii}$ and the matrix definition

$$\begin{aligned} \rho \log \rho &= \begin{pmatrix} |x_1|^2 \log |x_1|^2 & 0 & \cdots & 0 \\ 0 & |x_2|^2 \log |x_2|^2 & \cdots & \vdots \\ \vdots & \vdots & \ddots & 0 \\ 0 & \cdots & 0 & |x_n|^2 \log |x_n|^2 \end{pmatrix} \\ &= \begin{pmatrix} p_1 \log p_1 & 0 & \cdots & 0 \\ 0 & p_2 \log p_2 & \cdots & \vdots \\ \vdots & \vdots & \ddots & 0 \\ 0 & \cdots & 0 & p_n \log p_n \end{pmatrix}. \end{aligned} \tag{17.41}$$

Finally, we find that the trace of $\rho \log \rho$ is given by the expression:

$$\mathrm{tr}(\rho \log \rho) = \sum_{i=1}^{n} |x_i|^2 \log |x_i|^2 = \sum_{i=1}^{n} p_i \log p_i. \tag{17.42}$$

Based on our background of *Shannon's information theory* (Chapter 4), we can heuristically define an "*entropy*" $H$ for the quantum state described by the density matrix $\rho$ in the form

$$H = -\sum_{i=1}^{n} p_i \log p_i, \tag{17.43}$$

[4] By application of the property $\lim_{x\to 0}(x \log x) = 0$, which defines the function $x \log x$ analytically for any real $x \geq 0$.

which, from Eq. (17.42), yields:

$$H = -\mathrm{tr}(\rho \log \rho). \tag{17.44}$$

The above result quite elegantly connects the concept of Shannon's entropy to the corresponding "notion" in quantum information theory, which is based on the density-matrix operator. The definition $H = -\mathrm{tr}(\rho \log \rho)$ is referred to as *von Neumann's entropy*, as I shall describe in Chapter 21. Anticipating a key result, *the entropy corresponding to a qubit state* $|q\rangle = \alpha|0\rangle + \beta|1\rangle$, with $p = |\alpha|^2 = 1 - |\beta|^2$, is given by

$$\begin{aligned} H &= -|\alpha|^2 \log |\alpha|^2 - |\beta|^2 \log |\beta|^2 \\ &= -p \log p - (1-p)\log(1-p) \equiv f(p). \end{aligned} \tag{17.45}$$

In the result in Eq. (17.45), we recognize the Shannon entropy of a two-event source $X^2 = \{0, 1\}$, corresponding to the two possible states of a classical information bit. As we have seen, however, the qubit is a superposition of *both* information states, which we referred to as $|0\rangle$, $|1\rangle$, with corresponding probabilities $|\alpha|^2$, $|\beta|^2$. We now have the required conceptual tools to analyze the notion of *quantum measurement*.

## 17.2 Quantum measurements and types

In this section, I introduce and analyze the concepts associated with different types of *quantum measurement*. A general definition for *quantum measurement operators* will first be introduced. This definition will then be applied to *measurements in the orthonormal basis*, to the *projective (or von Neumann) measurements* and to the so-called *POVM measurements*.

Through Dirac notation we have made a formal description of the *quantum states* and their various transformation properties through the action of *linear operators*. Such a description did not require any quantum-mechanics background, because in Chapters 15 and 16 we obtained a solid view of the world of *qubits*, as described by 2D complex vectors, and their *operator transformations*, as described by unitary rotations on the Bloch sphere. The simpler world of qubits, thus, offers a convenient introduction to the greater view of quantum states $|\psi\rangle$ and their linear operator transformations $|\psi'\rangle = A|\psi\rangle$. In the same spirit of conceptual simplification, we can view a *quantum system* as being a *physical* system with which one can associate a quantum state $|\psi\rangle$, as expressed onto some pure-state basis $V^n = \{|x\rangle\}$, and a set of linear operators $\{A\}$. We are now interested in learning about what can be *physically measured* in such a system, and how it may possibly be measured.

The following will show that there are different approaches for performing physical measurements in quantum systems. Assume, first, that for a quantum system with $n$ pure states, there exist a certain number $n$ of possible measurements, which we index by $m$ ($m = 1, \ldots, n$). We then introduce the most general definition of a *measurement operator*, $M_m$, the collection of which forms the finite operator set $\{M_m\}$. Calling $p(m)$

the probability of measuring $m$, we have by postulate:

$$p(m) = \langle \psi | M_m^+ M_m | \psi \rangle. \tag{17.46}$$

Since the probabilities must sum to one, we also have

$$\begin{aligned} \sum_{m=1}^{n} p(m) &= \sum_{m=1}^{n} \langle \psi | M_m^+ M_m | \psi \rangle \\ &= \langle \psi | \left( \sum_{m=1}^{n} M_m^+ M_n \right) | \psi \rangle = 1, \end{aligned} \tag{17.47}$$

which implies the "completeness" relation:

$$\sum_m M_m^+ M_m = I. \tag{17.48}$$

As a second postulate, the measurement $M_m$ causes the system state $|\psi\rangle$ to be transformed into the *post-measurement* state $|\psi'\rangle = \gamma M_m |\psi\rangle$, where $\gamma$ is a complex number. The probability $q$ that the system is in the post-measurement state $|\psi'\rangle$ is unity. We, thus, obtain

$$\begin{aligned} q &= \langle \psi' | \psi' \rangle = \langle \psi | \bar{\gamma} M_m^+ M_m \gamma | \psi \rangle \\ &= |\gamma|^2 \langle \psi | M_m^+ M_m | \psi \rangle \equiv |\gamma|^2 p(m) = 1, \end{aligned} \tag{17.49}$$

which shows that $|\gamma| = 1/\sqrt{p(m)}$, hence $\gamma = e^{i\delta}|\gamma| = e^{i\delta}/\sqrt{p(m)}$. Within an "unobservable" phase term $e^{i\delta}$, which we shall overlook here, we can set $\gamma = 1/\sqrt{p(m)}$, and the post-measurement state is now completely defined as:

$$|\psi'\rangle = \frac{1}{\sqrt{p(m)}} M_m |\psi\rangle = \frac{1}{\sqrt{\langle \psi | M_m^+ M_m | \psi \rangle}} M_m |\psi\rangle. \tag{17.50}$$

Using the expansion in Eq. (17.34) of the input state $|\psi\rangle$, we obtain the expansion of the post-measurement state $|\psi'\rangle$ according to

$$\begin{aligned} |\psi'\rangle &= \frac{1}{\sqrt{\langle \psi | M_m^+ M_m | \psi \rangle}} M_m |\psi\rangle \\ &= \frac{1}{\sqrt{\langle \psi | M_m^+ M_m | \psi \rangle}} M_m \sum_{i=1}^{n} x_i |x_i\rangle \\ &= \sum_{i=1}^{n} \frac{x_i}{\sqrt{\langle \psi | M_m^+ M_m | \psi \rangle}} M_m |x_i\rangle. \end{aligned} \tag{17.51}$$

We have, thus, obtained a general expression for the post-measurement state $|\psi'\rangle$, given the quantum measurement operator $M_m$. What could be the result of two successive measurements from different operators $M_m$ and $L_l$? The answer is that the two successive measurements (in this order) are equivalent to a *single measurement* of operator definition $K_{lm} = L_l M_m$, which is left as an exercise to demonstrate.

An important consequence of the above general definition for the quantum-measurement operator $M_m$ is that two states $|\psi\rangle$ and $e^{i\delta}|\psi\rangle$, which differ by the phase factor $e^{i\delta}$, have the same measurement properties and outcome. Hence such a phase factor is said to be "unobservable." It should not be concluded, however, that in quantum mechanics, *phase* is definitely unobservable, with no corresponding measurement operator. Indeed, more recent (1989) and less known work has shown that *Hermitian phase-measurement operators and phase eigenstates can truly be defined.*[5] Here, we shall overlook this academic note, and abide by the above general definition of quantum-measurement operators, for which phase (and multiplying factors $e^{i\delta}$ in quantum states) is indeed "unobservable."

We shall consider next three possibilities for the measurement operator $M_m$.

### Quantum measurements in the orthonormal basis

We look for a quantum measurement operator $M_m$ that projects the input state $|\psi\rangle$ into the pure state $|x_m\rangle$ from the orthonormal basis $V = \{|x\rangle\} = \{|x_1\rangle, |x_2\rangle, \ldots, |x_n\rangle\}$, namely, having the action $|\psi'\rangle = M_m|\psi\rangle = \mu|x_n\rangle$, where $\mu$ is a complex number. As we know, the projector over $|x_m\rangle$ is defined as $P_m = |x_m\rangle\langle x_m|$, with $P_m^+ P_m = P_m$. The set of Hermitian operators $\{P_m = |x_m\rangle\langle x_m|\}$ satisfies the completeness relation, hence $M_m = |x_m\rangle\langle x_m|$ is a valid measurement operator. We first derive the bracket $\langle\psi|M_m^+ M_m|\psi\rangle$ according to

$$\begin{aligned}\langle\psi|M_m^+ M_m|\psi\rangle &= \langle\psi|P_m|\psi\rangle \\ &= \left(\sum_{i=1}^{n} \bar{x}_i\langle x_i|\right)|x_m\rangle\langle x_m|\left(\sum_{j=1}^{n}|x_j\rangle x_j\right) \\ &= \sum_{i=1}^{n}\sum_{j=1}^{n}\bar{x}_i x_j \delta_{im}\delta_{jm} \equiv |x_m|^2,\end{aligned} \tag{17.52}$$

which shows that *the probability of measuring m* (or equivalently, *the probability that $|\psi\rangle$ is in the state $|x_m\rangle$*) is $p(m) = |x_m|^2$. Then, with substitution of $M_m = |x_m\rangle\langle x_m|$ into

[5] See, for background reference: E. Desurvire, *Erbium-Doped Fiber Amplifiers, Device and System Developments* (New York: J. Wiley & Sons, 2002), pp. 97–111. As an advanced topic for quantum-mechanics specialists, attention is drawn to *phase states* $|\delta_m\rangle$, which have eigenvalues $\delta_m$, and which are defined for electromagnetic fields by

$$|\delta_m\rangle = \frac{1}{\sqrt{s+1}}\sum_{k=0}^{s} e^{ik\delta_m}|k\rangle,$$

where $|k\rangle$ is a photon-number state, and the series applies in the limit $s \to \infty$. For full description, see: S. M. Barnett and D. T. Pegg, On the Hermitian phase operator. *J. Modern Optics*, **36** (1989), 7–19 and D. T. Pegg and S. M. Barnett, Phase properties of the quantized in single-mode electromagnetic field. *Phys. Rev. A*, **39** (1989), 1665.

Eqs. (17.51) and (17.52), we obtain

$$\begin{aligned}
|\psi'\rangle &= \frac{1}{\sqrt{\langle\psi|M_m^+ M_m|\psi\rangle}} M_m|\psi\rangle \\
&= \frac{1}{\sqrt{|x_m|^2}} |x_m\rangle\langle x_m| \sum_{i=1}^{n} x_i |x_i\rangle \\
&= \sum_{i=1}^{n} \frac{x_i}{|x_m|} |x_m\rangle\langle x_m|x_i\rangle \\
&= \sum_{i=1}^{n} \frac{x_i}{|x_m|} |x_m\rangle \delta_{mi} \equiv \frac{x_m}{|x_m|} |x_m\rangle.
\end{aligned} \tag{17.53}$$

The result in Eq. (17.53) indicates that the post-measurement state is indeed of the form $\mu|x_m\rangle$, but with $\mu = x_m/|x_m|$ being a complex number of unity modulus, or $\mu = \mathrm{e}^{\mathrm{i}\delta_m}$ being an "unobservable" phase term that can be overlooked.

We will use the above results in Chapter 18, which concerns *qubit measurements in the orthonormal basis.*

## Projective or von-Neumann measurements

Another type of quantum measurement whose operator satisfies the above two postulates is referred to as *projective* or *von-Neumann measurement.* As we shall see, it is not formally different from a measurement in an orthonormal basis, except that we shall now take for this basis the one formed by the eigenstates $\{|\lambda_m\rangle\}$ of a given Hermitian operator $A$. Define the corresponding eigenvalues as $\{\lambda_m\}$. Since $A$ is Hermitian, we know from the previous section that the eigenvalues are *real numbers*. For this reason, the operator $A$ could be associated with "real" physical quantities to characterize the system, hence the operator $A$ is referred to as an *observable*.

Given the set of projectors $P_m = |\lambda_m\rangle\langle\lambda_m|$, the spectral decomposition of $A$ is given in Eq. (17.27), which I reproduce for convenience:

$$A = \sum_{m=1}^{n} \lambda_m P_m = \sum_{m=1}^{n} \lambda_m |\lambda_m\rangle\langle\lambda_m|. \tag{17.54}$$

As we have previously established, the projector $P_m = |\lambda_m\rangle\langle\lambda_m|$ is a valid measurement operator and, hence, the probability of measuring the eigenvalue $\lambda_m$ from a system in the state $|\psi\rangle$, as decomposed in the eigenstates basis $\{|\lambda_m\rangle\}$ is given by

$$p(\lambda_m) = \langle\psi|P_m|\psi\rangle, \tag{17.55}$$

and we have for the post-measurement state:

$$\begin{aligned}
|\psi'\rangle &= \frac{1}{\sqrt{p(\lambda_m)}} P_m|\psi\rangle \\
&= \frac{1}{\sqrt{\langle\psi|P_m|\psi\rangle}} P_m|\psi\rangle.
\end{aligned} \tag{17.56}$$

Substituting the expansion of $|\psi\rangle$, with complex coordinates $x_i$, over the eigenstate basis $\{|\lambda_i\rangle\}$, we obtain for the measurement probability $p(\lambda_m)$ and corresponding post-measurement state $|\psi'\rangle$:

$$\begin{aligned} p(\lambda_m) &= \langle\psi|P_m|\psi\rangle \\ &= \left(\sum_{i=1}^{n}\langle\lambda_i|\bar{x}_i\right)|\lambda_m\rangle\langle\lambda_m|\left(\sum_{j=1}^{n}x_j|\lambda_j\rangle\right) \\ &= \sum_{i=1}^{n}\sum_{j=1}^{n}\bar{x}_i x_j \delta_{im}\delta_{jm} = |x_m|^2, \end{aligned} \tag{17.57}$$

$$\begin{aligned} |\psi'\rangle &= \frac{1}{\sqrt{p(\lambda_m)}}P_m|\psi\rangle \\ &= \frac{1}{x_m}|\lambda_m\rangle\langle\lambda_m|\sum_{i=1}^{n}x_i|\lambda_i\rangle \\ &= \frac{1}{x_m}|\lambda_m\rangle\sum_{i=1}^{n}x_i\delta_{im} \equiv |\lambda_m\rangle. \end{aligned} \tag{17.58}$$

We will use the above two results in Chapter 18, which concerns *projective qubit measurements*.

Let us now analyze some interesting properties of the projective or von-Neumann measurement.

Since each eigenvalue $\lambda_m$ has an associated measurement probability, $p(\lambda_m)$, we can define the *average value* or *mean of the observable measurement* as follows:

$$\begin{aligned} \langle A\rangle \equiv \langle\lambda\rangle &= \sum_{m=1}^{n}\lambda_m p(\lambda_m) = \sum_{m=1}^{n}\lambda_m\langle\psi|P_m|\psi\rangle \\ &= \langle\psi|\sum_{m=1}^{n}\lambda_m P_m|\psi\rangle \equiv \langle\psi|A|\psi\rangle. \end{aligned} \tag{17.59}$$

The nice conclusion from the above result is that the average value of an observable $A$ is defined as the bracket $\langle\psi|A|\psi\rangle$. Likewise, we can define the *mean-square* and *variance of the observable* as follows:

$$\langle A^2\rangle \equiv \langle\lambda^2\rangle = \sum_{m=1}^{n}\lambda_m^2 p(\lambda_m) \equiv \langle\psi|A^2|\psi\rangle, \tag{17.60}$$

$$\Delta A^2 \equiv \langle\lambda^2\rangle - \langle\lambda\rangle^2 = \langle A^2\rangle - \langle A\rangle^2 \tag{17.61}$$

(the notation $\Delta A$ corresponds to a real number, not an operator). We have, thus, obtained the definitions of the two main statistical parameters characterizing any observable measurement, namely the mean and the variance of the observed physical quantity. We call the *uncertainty* of the observable $A$ the measured standard deviation $\sqrt{\Delta A^2} = \Delta A$. Similar measurements can be performed with any other observable quantity $B$, with uncertainty $\Delta B$. A fundamental result in quantum mechanics, known as the *Heisenberg uncertainty principle*, states that given two Hermitian operators $A$, $B$ the product of their uncertainties measured in the system state $|\psi\rangle$ must satisfy:

$$\Delta A \Delta B \geq \frac{1}{2}|\langle\psi|[A, B]|\psi\rangle|, \tag{17.62}$$

where $[A, B] = AB - BA$ is the commutator of $A$, $B$ (see demonstration in Appendix O). In the particular case (called *conjugate observables*) where $[A, B] = \pm i$, and where $|\psi\rangle$ has a modulus of unity, we have

$$\Delta A \Delta B \geq \frac{1}{2}, \tag{17.63}$$

which represents the more well-known form of the uncertainty principle. The key *conclusion* of this principle could be stated as follows: *two independent measurements of noncommuting observables cannot both reach arbitrary accuracy*. The product of the corresponding uncertainties has a *nonzero lower bound* given by the right-hand side of Eq. (17.62), which we may call $\varepsilon$. Thus, a relatively accurate measurement of $A$, e.g., $\Delta A = 1/N$ where $N$ may be any large number, implies that the accuracy in the measurement of $B$ cannot be better than $N\varepsilon$. Thus, if $N \to \infty$, we have $\Delta A \to 0$, corresponding to absolute accuracy in the observable $A$, while we have $\Delta B \to \infty$, corresponding to an absolute "indetermination" of the observable $B$. It would be incorrect, however, to conclude that in any case, a first measurement of $A$ actually influences that of $B$, and the reverse. The two measurements are assumed to be independent, both in terms of time and order sequence. Only in the case where the two observables commute can the two measurements reach arbitrary accuracy.

## POVM measurements

A third type of measurement is referred to *as positive-operator-valued-measure*, or *POVM*. Define a finite set of Hermitian operators $\{E_m\}$ with any $E_m$ being a *positive* operator (meaning $\langle\psi|E_m|\psi\rangle \geq 0$ for any normalized state $|\psi\rangle$), and altogether satisfying the completeness relation

$$\sum_m E_m = I. \tag{17.64}$$

The operators $E_m$ that satisfy the above conditions are then said to be *POVM operators*. As we have seen in the previous section, any Hermitian operator $A$ can be represented by a diagonal matrix with nonnegative coefficients, hence, there exists a diagonal operator $B = \sqrt{A}$, such that $B^2 = A$, and which is also Hermitian ($B_{ii}^2$ is real, so $B_{ii}$ is). Therefore, for each $E_m$ we can associate a Hermitian operator $M_m = \sqrt{E_m}$ that satisfies $M_m^+ M_m = E_m$, and consequently the two above POVM conditions. According to the general definition, the set $\{M_m\}$ is, thus, a truly eligible set of *measurement operators*, even if their number, with respect to the number $n$ of pure states, is *not* specified. The projection or von-Neumann measurements, with the set $\{M_m = |\lambda_m\rangle\langle\lambda_m|\}$ being defined by the $n$ eigenstates of an observable, represent a specific case of POVM set.

I shall now illustrate an important application of POVM, which no other measurement type can provide. Consider a simple example in the basis $V = \{|0\rangle, |1\rangle\}$. Assume the

**Table 17.1** Probabilities associated with different POVM measurement possibilities when the input state is either $|\psi\rangle = |0\rangle$ (left) or $|\psi\rangle = |+\rangle$ (right).

| Input state $\|\psi\rangle = \|0\rangle$ | Input state $\|\psi\rangle = \|+\rangle$ |
|---|---|
| $p(1) = \langle 0\|E_1\|0\rangle = 0$ | $p(1) = \langle +\|E_1\|+\rangle = \frac{u}{2}$ |
| $p(2) = \langle 0\|E_2\|0\rangle = \frac{u}{2}$ | $p(2) = \langle +\|E_2\|+\rangle = 0$ |
| $p(3) = \langle 0\|(I - E_1 - E_2)\|0\rangle = 1 - \frac{u}{2}$ | $p(3) = \langle -\|(I - E_1 - E_2)\|-\rangle = 1 - \frac{u}{2}$ |

three POVM operators:

$$\begin{aligned} E_1 &= u|1\rangle\langle 1| \\ E_2 &= u\frac{|0\rangle - |1\rangle}{\sqrt{2}}\frac{\langle 0| - \langle 1|}{\sqrt{2}} \equiv u|-\rangle\langle -| \\ E_3 &= I - E_1 - E_2, \end{aligned} \tag{17.65}$$

with $u = \sqrt{2}/(1 + \sqrt{2})$. By summation of the above three equations, we immediately observe that the completeness relation is satisfied. It is also easily verified that the three operators $E_1, E_2, E_3$ are positive (it is left as an exercise to prove that in the general case, the condition $u \leq 2/3$ must be satisfied). We have, thus, defined *a valid POVM set* of three measurement operators from the two-element basis $V$.

The following will show how the above POVM measurement can be applied. Assume that the system state $|\psi\rangle$ has been prepared in such a way that it is *certain to be either* $|\psi\rangle = |0\rangle$ *or* $|\psi\rangle = |+\rangle$, to the exclusion of any other possibility. It is also assumed that we have this key information about the two system-state possibilities, prior to any measurement, but we don't know, a priori, which ones the system chooses to be in. The probabilities associated with our three possible measurements associated to each case are summarized in Table 17.1.

First, the table indicates that our measurement *fails to convey any information* in two cases:

- If we use $E_1$ when the input is $|\psi\rangle = |0\rangle$;
- If we use $E_2$ when the input is $|\psi\rangle = |+\rangle$.

This is because, in each case, the measurement operators project the input state on its orthogonal counterpart, namely $|1\rangle$ or $|-\rangle$, respectively, resulting in the measurement probabilities for $E_1$ of $p(E_1) = \langle 0|E_1|0\rangle = 0$ or $p(E_2) = \langle +|E_2|+\rangle = 0$, and showing that these measurement have no possible outcomes.

Second, the table indicates that there is a finite probability $u/2$ that we obtain a measurement

- Of $E_1$, if we use $E_1$ when the input is $|\psi\rangle = |+\rangle$;
- Of $E_2$, if we use $E_2$ when the input is $|\psi\rangle = |0\rangle$.

Then comes a nice subtlety in the interpretation of these two measurement possibilities. Indeed, if we happen to measure $E_1$ the system cannot be in state $|\psi\rangle = |0\rangle$, since $E_1$ is a projector on $|1\rangle$. In this case, we reach the *absolute conclusion* that the system must have been in state $|\psi\rangle = |+\rangle$. Likewise, if we happen to measure $E_2$, the system cannot be in state $|\psi\rangle = |+\rangle$, since $E_2$ is a projector on $|-\rangle$. In this case, we reach the *absolute conclusion* that the system must have been in state $|\psi\rangle = |0\rangle$. However, we obtain this absolute information only with probability $u/2$, meaning that there is a finite probability $1 - u/2$ that the two measurements fail to convey any information.

If we were to use the measurement $E_3$, Table 17.1 shows that the measurement works in all cases ($|\psi\rangle = |0\rangle$ or $|+\rangle$), but with a probability $1 - u/2$. By definition, $E_3$ is a projector on all states that are neither $|1\rangle$ or $|-\rangle$. Thus, the positive outcome of any $E_3$ measurement only tells us that the system state is neither $|1\rangle$ or $|-\rangle$, but we already know this for a fact, which represents no information! Therefore, there is no point in using $E_3$ as a means to measure the system state.

Let me, then, clarify what is meant by "positive outcome" and "failure" of any of the above measurements, using a figurative analogy with a physical measurement. Compare the system state $|\psi\rangle$ to a light source that randomly emits in two possible color tones, either $A$ or $B$ (standing for $|0\rangle$ or $|+\rangle$, respectively), these tones being invisible to the naked eye. Our measurement consists of determining which color tone is emitted by the source by observing it through a set of "magic filters," called 1, 2, 3, (for $E_1$, $E_2$, $E_3$). Such filters have the following strange properties: filter 1 does not react to tone $A$, while it makes tone $B$ visible to the eye with probability $p$ (or $u/2$); filter 2 does not react to tone $B$, while it makes tone $A$ visible to the eye with same probability $p$; and filter 3 does not react to any tone other than $A$, $B$ (but this is not useful here) but makes tone $A$, $B$ visible with probability $1 - p$ (or $1 - u/2$). Basically, if we choose either filter 1 or filter 2, we have a chance $p$ of seeing something, and $1 - p$ of seeing nothing!

Figure 17.1 shows what we can see through the magic filters, according to the twelve possible cases, namely, determined by the two source-tone possibilities $A$, $B$ and our three possible magic-filter choices 1, 2, 3. A bright spot indicates that we observe something, corresponding to a "positive" measurement. The absence of a spot, or dark image indicates a "negative" or "failed" measurement. The figure shows that the combinations $(A, 1)$ and $(B, 2)$ are certain to fail, while the other combinations $(A, 2)(B, 1)$ or $(A, 3)(B, 3)$ have a finite chance, $p$ or $1 - p$, respectively, of succeeding. The successful measurements in the two cases $(A, 2)$ or $(B, 1)$, as marked with a cross ($\times$), correspond to absolute certainty that the source tone is $A$ or $B$, respectively. The other two successful measurements $(A, 3)$ $(B, 3)$, like all failed measurements, do not convey any information, as mentioned earlier. The success of our measurement and the absolute conclusion therein, is, thus, dependent on our filter choice (1 or 2), which is essentially a matter of guesswork.

From any academic standpoint, it remains debatable whether the above fictitious "physical measurement" through a set of "magic filters" may, in some way, help clarify the essence of a true quantum POVM measurement. If it has any merit, however, it helps to stress the point that *quantum measurements are based on the observer's choice of a measurement operator* (the magic filter). The measurement or observation of an outcome may either succeed or fail (seeing or not seeing a spot). In the case of success,

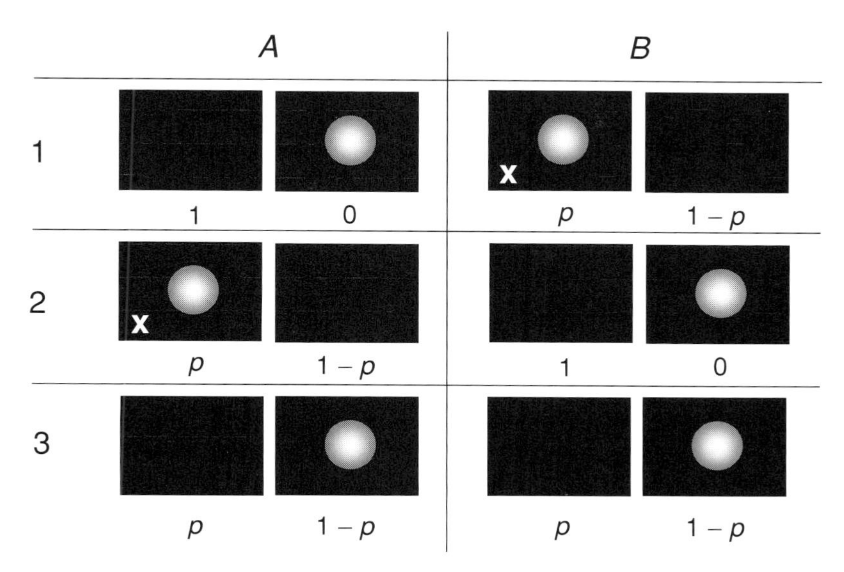


**Figure 17.1** Fictitious analogy of POVM measurement of tones $A$, $B$ and a magic-filter set 1, 2, 3 to characterize a source. Given the tones $A$, $B$ and choices 1, 2, 3, a positive observation of the tone is characterized by the bright spot on the images shown, with various probabilities $0, 1, p, 1-p$. The two images marked with a cross ($\times$) correspond to absolutely certain measurements that the source tone is $A$ or $B$.

the observer may or may not be able to derive an absolute conclusion as to what state was observed (the source tone), depending on his or her choice.

Leaving fiction behind, we can conclude that POVM measurements make it possible to get *absolutely certain* information about the input state $|\psi\rangle$, despite the fact that it is, a priori, unknown, just as the right measurement operators are unknown. We have noted that while *the corresponding measurements may sometimes fail, the conclusion is never wrong when they succeed*. This remarkable property is referred to as *unambiguous quantum state discrimination*, of UQSD. Remarkably, while UQSD has no classical explanation, or no counterpart in classical physics, it can be implemented in actual physical measurements, for instance, in *quantum cryptography* (see Chapter 25).

## 17.3 Quantum measurements on joint states

In Chapter 16, we introduced the concept of *tensor states*, which correspond to systems being described by more than one quantum state, e.g., $|\psi_1\rangle$ and $|\psi_2\rangle$. Here, we shall call such systems *composite systems*. As we have seen, the corresponding state can be written in the tensor form $|\psi\rangle = |\psi_1\rangle \otimes |\psi_2\rangle$, or $|\psi\rangle = |\psi_1\rangle|\psi_2\rangle$, or, for short $|\psi\rangle = |\psi_1, \psi_2\rangle$, and even, sometimes, $|\psi\rangle = |\psi_1\psi_2\rangle$, e.g., the notations $|\psi\rangle = |i\rangle \otimes |j\rangle \equiv |i\rangle|j\rangle \equiv |i, j\rangle \equiv |ij\rangle$ are all equivalent. These tensor states are also referred to as *joint states*. Regardless of the notation conciseness for joint states, however, it is important to keep in mind the underlying tensor product concept, which is also associated with that of *tensor operators*. Thus, the tensor operator $U = A \otimes B$ acts on the joint state $|\psi\rangle = |\psi_1\rangle \otimes |\psi_2\rangle$ according to the rules $U|\psi\rangle = A|\psi_1\rangle \otimes B|\psi_2\rangle$ and $\langle\psi|U^+ = \langle\psi_1|A^+ \otimes \langle\psi_2|B^+$, hence, the imperative need of keeping the single quantum state notation in the right order sequence inside the ket $|\cdot\rangle$, and choose the appropriate notation for the tensor products when operators are being used.

We are now interested in defining a generic measurement operator for joint states. The following may look complicated at first but, as we shall see, the result makes complete sense in view of what we have learnt from the previous section. Assume first a single-state space $V$, as defined by the orthonormal basis $V^n = \{|x_i\rangle\}$ (for simplicity we identify here the "space" and one corresponding "basis" with the same letter $V$). Assume next that $\{M_m\}$ represents a valid set of measurement operators for $V$ (meaning that $M_m^+ M_m$ are positive and verify the completeness relation). For the set $\{M_m\}$ there is a space $M$ of *orthonormal* states $\{|m\rangle\}$, corresponding to each of the different possibilities of *post-measurement states* resulting from the action of $M_m$. I shall refer to $M$ as an *ancilla space*. Let $|\tilde{m}\rangle$ be a fixed state of $M$, to be used as an arbitrary reference, which we might index $|\tilde{m}\rangle = |0\rangle$. We can now work with the joint states $|\psi\rangle|0\rangle = |\psi, 0\rangle$ from the extended space $V \otimes M$ of joint states $|\psi\rangle|m\rangle = |\psi, m\rangle$. Next, define the tensor operator $U$ whose action on $|\psi, 0\rangle$ results in:

$$U|\psi, 0\rangle = \sum_m M_m|\psi, m\rangle. \tag{17.66}$$

We then have for any two states $|\psi\rangle, |\psi'\rangle$:

$$\begin{aligned}
\langle\psi', 0|U^+U|\psi, 0\rangle &= \sum_k \sum_m \langle\psi', k|M_k^+ M_m|\psi, m\rangle \\
&= \sum_k \sum_m \langle\psi'|M_k^+ M_m|\psi\rangle \delta_{km} \\
&= \sum_m \langle\psi'|M_m^+ M_m|\psi\rangle \\
&= \langle\psi'|\sum_m M_m^+ M_m|\psi\rangle \equiv \langle\psi'|\psi\rangle.
\end{aligned} \tag{17.67}$$

The above result shows that the transformation $U$ preserves the inner product of all states $|\psi, 0\rangle$ as a unitary operator for the subspace $V \otimes \{|0\rangle\}$. Actually, we can make $U$ a unitary operator for the entire space $V \otimes M$, which applies to any state $|\psi, m\rangle$,[6]

[6] Indeed, define $U'$ as the sum of two projectors on the two orthogonal spaces $V \otimes \{|0\rangle\}$ and $V \otimes \{|m \neq 0\rangle\}$ for any state $|\psi\rangle \in V$, as follows: $U' = U|\psi, 0\rangle\langle\psi, 0| + \sum_{m\neq 0} |\psi, m\rangle\langle\psi, m|$. It is clear that since $U$ is unitary, $U'$ is unitary. We also have $U'|\psi, 0\rangle = U'|\psi, 0\rangle$ and, for $k \neq 0$, $U'|\psi, k\rangle = U|\psi, 0\rangle\langle\psi, k| + \sum_{m\neq 0} |\psi, m\rangle\langle\psi, m|\psi, k\rangle = U\delta_{0k} + \sum_{m\neq 0} |\psi, m\rangle\delta_{mk} \equiv |\psi, k\rangle$.

according to the alternative definition:

$$
\begin{aligned}
U_k|\psi, k\rangle &= \sum_m (M_m \otimes |k\rangle\langle m|)|\psi, k\rangle \\
&= \sum_m M_m|\psi\rangle \otimes (|k\rangle\langle m|k\rangle) \\
&= \sum_m M_m|\psi\rangle \otimes (|k\rangle\delta_{mk}) \\
&= \sum_m M_m|\psi, m\rangle \otimes (|k\rangle\delta_{mk}).
\end{aligned}
\tag{17.68}
$$

As an exercise, it is possible to check that $\langle\psi', k'|U_{k'}^{+}U_k|\psi, k\rangle \equiv \langle\psi'|\psi\rangle\delta_{kk'}$ which shows that the new definition of $U$ (i.e., $U_k$) in Eq. (17.68) is indeed unitary over $V \otimes M$. Having found the above unitary operator $U$ that conserves the inner product of all joint states $|\psi, k\rangle$ in $V \otimes M$, we can now introduce the *joint-state measurement* (JSM) *operator* $E_m$ as

$$
E_m = U^{+}P_m U,
\tag{17.69}
$$

with

$$
P_m = I_V \otimes |m\rangle\langle m|,
\tag{17.70}
$$

where $I_V$ is the identity operator on $V$ and $|m\rangle\langle m|$ the projector over $|m\rangle \in M$.[7] Next, we shall calculate the bracket function $p(m) = \langle\psi, k|E_m|\psi, k\rangle$, which, using Eq. (17.66) with $|0\rangle = |k\rangle$, is:

$$
\begin{aligned}
p(m) &= \langle\psi, k|U^{+}P_m U|\psi, k\rangle \\
&= \left(\sum_p \langle\psi, p|M_p^{+}\right)(I_V \otimes |m\rangle\langle m|)\left(\sum_q M_q|\psi, q\rangle\right) \\
&= \sum_p\sum_q \langle\psi|M_p^{+}M_q|\psi\rangle\langle p|m\rangle\langle m|q\rangle \\
&= \sum_p\sum_q \langle\psi|M_p^{+}M_q|\psi\rangle\delta_{pm}\delta_{mq} \\
&\equiv \langle\psi|M_m^{+}M_m|\psi\rangle.
\end{aligned}
\tag{17.71}
$$

We immediately recognize from the result that $p(m)$ is the *measurement probability associated with the operator $M_m$ on the state $|\psi\rangle \in V$*, as defined in Eq. (17.46). After Eq. (17.50), and writing the JSM operator $E_m = (U^{+}P_m^{+})P_m U$, we obtain the *joint post-measurement state* as the result of the transformation:

$$
\begin{aligned}
|\psi', k'\rangle &= \frac{1}{\sqrt{p(m)}}P_m U|\psi, k\rangle \\
&= \frac{1}{\sqrt{\langle\psi|M_m^{+}M_m|\psi\rangle}}M_m|\psi, m\rangle.
\end{aligned}
\tag{17.72}
$$

[7] The full definition of the JSM operator actually being $E_m = (P_m U)^{+}P_m U = U^{+}P_m^{+}P_m U \equiv U^{+}P_m U$, where we used the projector and Hermitian properties of $P_m$.

This final result shows that after the measurement, the state $|\psi\rangle$ of space $V$ is transformed into

$$|\psi'\rangle = \frac{1}{\sqrt{\langle\psi|M_m^+ M_m|\psi\rangle}} M_m|\psi\rangle, \tag{17.73}$$

while the *ancilla* state $|k\rangle$ is transformed into $|m\rangle$. In the rest of this book, we will not have to be concerned about the *ancilla space* $M$. Such a space is just used here as a mathematical tool to allow the construction of a quantum measurement operator for joint states. Bearing this in mind, we can now expand the conclusion to measurements of joint states of any dimension from $V^{\otimes n} \otimes M^{\otimes n}$, by means of the $n$-ancilla space, $M^{\otimes n}$, i.e.,

$$|\psi_1, \psi_2, \ldots, \psi_n, k_1, k_2, \ldots, k_n\rangle = |\psi_1\rangle \otimes |\psi_2\rangle \otimes \cdots \otimes |\psi_n\rangle \otimes |k_1\rangle \otimes k_2 \otimes \cdots \otimes |k_n\rangle. \tag{17.74}$$

We, thus, know that it is possible to define a measurement operator $E_{im}$ that can act on the subspace pair $|\psi_i\rangle$, $|k_i\rangle$ and collapse it into the pair $M_m|\psi_i\rangle$, $|m\rangle$, such that

$$\begin{aligned} &E_{im}|\psi_1, \psi_2, \ldots, \psi_i, \ldots, \psi_n, k_1, k_2, \ldots, k_i, \ldots, k_n\rangle \\ &= \left\{ \begin{array}{l} |\psi_1\rangle \otimes |\psi_2\rangle \otimes \cdots \otimes \dfrac{M_m}{\sqrt{\langle\psi_i|M_m^+ M_m|\psi_i\rangle}}|\psi_i\rangle \otimes \cdots |\psi_n\rangle \\ \otimes|k_1\rangle \otimes k_2 \otimes \cdots \otimes |k_{i-1}\rangle \otimes |m\rangle \otimes |k_{i+1}\rangle \otimes \cdots |k_n\rangle \end{array} \right\}. \end{aligned} \tag{17.75}$$

We may just forget now about the ancilla bits and consider simply that

$$E_{im}|\psi_1, \psi_2, \ldots, \psi_i, \ldots, \psi_n\rangle \equiv |\psi_1\rangle \otimes |\psi_2\rangle \otimes \cdots \otimes \frac{M_m}{\sqrt{\langle\psi_i|M_m^+ M_m|\psi_i\rangle}}|\psi_i\rangle \otimes \cdots \otimes |\psi_n.\rangle \tag{17.76}$$

To provide an illustration, assume that $M_m = |m\rangle\langle m|$. Letting $|\psi_i\rangle = \sum_n x_n^{(i)}|n\rangle$, we obtain

$$\begin{aligned} E_{im}|\psi_1, \ldots, \psi_n\rangle &= |\psi_1\rangle \otimes |\psi_2\rangle \otimes \cdots \otimes \frac{x_m^{(i)}}{|x_m^{(i)}|}|m\rangle \otimes \cdots \otimes |\psi_n\rangle \\ &= \frac{x_m^{(i)}}{|x_m^{(i)}|}|\psi_1, \ldots, \psi_{i-1}, m, \psi_{i+1}, \ldots, \psi_n\rangle \\ &\equiv |\psi_1, \ldots, \psi_{i-1}, m, \psi_{i+1}, \ldots, \psi_n\rangle, \end{aligned} \tag{17.77}$$

as obtained within an unobservable phase factor $e^{i\delta} = x_m^{(i)}/|x_m^{(i)}|$.

Finally, consider for simplicity two joint states in 2-state space $V \otimes V$, namely $|\phi\rangle = |\phi_1, \phi_2\rangle$ and $|\psi\rangle = |\psi_1, \psi_2\rangle$, and their superposition $|\chi\rangle$ with complex coefficients $\alpha$, $\beta$ (satisfying $|\alpha|^2 + |\beta|^2 = 1$):

$$|\chi\rangle = \alpha|\phi\rangle + \beta|\psi\rangle = \alpha|\phi_1, \phi_2\rangle + \beta|\psi_1, \psi_2\rangle. \tag{17.78}$$

Owing to the linearity and distributiveness of the measurement operators, we have

$$\begin{aligned} E_1|\chi\rangle &= E_1(\alpha|\phi\rangle + \beta|\psi\rangle) = \alpha E_1|\phi_1, \phi_2\rangle + \beta E_1|\psi_1, \psi_2\rangle = \alpha|1, \phi_2\rangle + \beta|1, \psi_2\rangle \\ E_2|\chi\rangle &= E_2(\alpha|\phi\rangle + \beta|\psi\rangle) = \alpha E_2|\phi_1, \phi_2\rangle + \beta E_2|\psi_1, \psi_2\rangle = \alpha|\phi_1, 2\rangle + \beta|\psi_1, 2\rangle \end{aligned} \tag{17.79}$$

and

$$E_1 E_2 |\chi\rangle = E_2 E_1 |\chi\rangle = \alpha |1, 2\rangle + \beta |1, 2\rangle \equiv (\alpha + \beta) |1, 2\rangle, \quad (17.80)$$

which actually illustrates the great simplicity of joint-state measurements and their properties. These can now be applied to the case of single and multiple *qubit systems*, which are described in Chapter 18.

## 17.4 Exercises

**17.1** (B): Calculate the moduli and scalar products of the 2-qubits

$$|q\rangle = \mathrm{i}|00\rangle + |01\rangle - |11\rangle,$$
$$|q'\rangle = |00\rangle + |01\rangle - \mathrm{i}|10\rangle + |11\rangle.$$

**17.2** (B): Show that $|q\rangle\langle q|$ is the projector operator over the qubit $|q\rangle$.

**17.3** (B): Determine the eigenstates and diagonal representations of the Pauli matrix $Y$ and Hadamard matrix $H$.

**17.4** (T): Show that the transition operator $T$ defined by its matrix element

$$T_{xy} = \langle x|y\rangle$$

in different orthogonal bases $V^n = \{|x\rangle\}$ and $W^n = \{|y\rangle\}$ satisfies the property

$$T^+ T = T T^+ = I.$$

Then show that $T \neq T^+$ in the general case.

**17.5** (M): Show that two successive measurements from different operators $M_m$ and $L_l$ (in this order) are equivalent to a single measurement of operator definition $K_{lm} = L_l M_m$.

**17.6** (T): Show that the three operators $\{E_1, E_2, E_3\}$ defined by

$$E_1 = u|1\rangle\langle 1|,$$
$$E_2 = v(|0\rangle - |1\rangle)(\langle 0| - \langle 1|),$$
$$E_3 = I - E_1 - E_2,$$

form a complete POVM set over the space $V = \{|0\rangle, |1\rangle\}$ if the constants $u, v$ satisfy the two conditions

$$0 < u \leq \frac{2}{3}, \quad v = \frac{u}{2}.$$

# 18 Qubit measurements, superdense coding, and quantum teleportation

This chapter is concerned with the measure of *information contained in qubits*. This can be done only through *quantum measurement*, an operation that has no counterpart in the classical domain. I shall first describe in detail the case of *single qubit* measurements, which shows under which measurement conditions "classical" bits can be retrieved. Next, we consider the measurements of higher-order or $n$-qubits. Particular attention is given to the *Einstein–Podolsky–Rosen* (EPR) or *Bell states*, which, unlike other joint tensor states, are shown being *entangled*. The various single-qubit measurement outcomes from the EPR–Bell states illustrate an effect of causality in the information concerning the other qubit. We then focus on the technique of *Bell measurement*, which makes it possible to know which Bell state is being measured, yielding two classical bits as the outcome. The property of EPR–Bell state entanglement is exploited in the principle of *quantum superdense coding*, which makes it possible to transmit classical bits at twice the classical rate, namely through the generation and measurement of a single qubit. Another key application concerns *quantum teleportation*. It consists of the transmission of quantum states over arbitrary distances, by means of a common EPR–Bell state resource shared by the two channel ends. While quantum teleportation of a qubit is instantaneous, owing to the effect of quantum-state collapse, it is shown that its completion does require the communication of *two classical bits*, which is itself limited by the speed of light. We then briefly consider the possibility of denser teleportation schemes, taking, for example, the case of *two qubits*. Quantum entanglement is also shown to be applicable to the *teleportation quantum gates*. This opens the perspective of *distributed quantum computing*: the possibility to manipulate qubits from a distance, or, in a futuristic view, to share the resources of remote quantum networks.

## 18.1 Measuring single qubits

This section describes the effect of quantum measurements on *qubits*. We shall first recall some key results obtained in previous chapters, and apply the principles to the case of single qubits and then *joint qubit* states, including the intriguing case of *entangled qubits*.

In Chapter 17, we learnt that given a space of quantum states $|\psi\rangle$ defined by an orthonormal base $V^n = \{|x_i\rangle\}$, there exist different possibilities of *quantum-measurement operator* sets, called $\{M_m\}$, corresponding to a number $n$ of possible measurements indexed by $m$ ($m = 1 \ldots n$). To recall, the two requirements for any such set are (a) that all operator products $M_m^+ M_m$ should be positive (which result in the *unitarity* of $M_m^+ M_m$), and (b) that they satisfy the *completeness relation* $\sum M_m^+ M_m = I$. The two key results are, first, that the *probability of measuring m* is given by the bracket:

$$p(m) = \langle \psi | M_m^+ M_m | \psi \rangle, \tag{18.1}$$

and, second that the measurement projects the input state $|\psi\rangle$ into the *post-measurement state* $|\psi'\rangle$ defined as

$$|\psi'\rangle = \frac{1}{\sqrt{p(m)}} M_m |\psi\rangle = \frac{1}{\sqrt{\langle \psi | M_m^+ M_m | \psi \rangle}} M_m |\psi\rangle. \tag{18.2}$$

Out of many possibilities for any quantum measurement, we can use the *measurement in the state basis*, where $M_m = |x_m\rangle\langle x_m|$ is the projector on the pure state $|x_m\rangle$. Given the state $|\psi\rangle = \sum x_i |x_i\rangle$, it is clear from the definition in Eq. (18.1) with $M_m^+ M_m = |x_m\rangle\langle x_m|$ that the probability that the measurement finds $|\psi\rangle$ in the state $|x_m\rangle$ is $p(m) = |x_m|^2$. As we have seen in Chapter 17, the post-measurement state is

$$|\psi'\rangle = \frac{x_m}{|x_m|} |x_m\rangle. \tag{18.3}$$

Let the input state be a qubit $|q\rangle = \alpha|0\rangle + \beta|1\rangle$, as defined from the orthonormal base $V = \{|0\rangle, |1\rangle\}$, and where (to recall) $\alpha, \beta$ are complex numbers satisfying $|\alpha|^2 + |\beta|^2 = 1$. It is clear that $\alpha, \beta$ represent the *complex probability amplitudes* of the qubit state, and that $|\alpha|^2$ or $|\beta|^2$ represent the probabilities of the qubit being found in either pure state $|0\rangle$ or $|1\rangle$ of the base $V$. According to Eq. (18.3), the corresponding post-measurement states are

$$\begin{cases} \dfrac{\alpha}{|\alpha|}|0\rangle = e^{i\delta_\alpha}|0\rangle \\ \dfrac{\beta}{|\beta|}|1\rangle = e^{i\delta_\beta}|1\rangle, \end{cases} \tag{18.4}$$

where $e^{i\delta_\alpha}$, $e^{i\delta_\beta}$ are unobservable phase factors, which can be omitted.

Since the qubit is a superposition of states $|q\rangle = \alpha|0\rangle + \beta|1\rangle$, the action of any measurement in the state-base $V$, as defined by the operators $M_0 = |0\rangle\langle 0|$ or $M_1 = |1\rangle\langle 1|$, is, thus, to project, or *collapse* it into one of the two states $|0\rangle$ or $|1\rangle$. Using $M_0$, the probability that $|q\rangle$ collapses into $|0\rangle$ is $p(0) = |\alpha|^2$. Using $M_1$, the probability that $|q\rangle$ collapses into $|1\rangle$ is $p(1) = |\beta|^2 = 1 - |\alpha|^2$. We can interpret the result of such a measurement as *reducing the qubit information into the classical bit values 0 or 1*, respectively. But unlike the classical case, there is no absolute certainty that the resulting information is, indeed, 0 or 1, when using measurement operators $M_0, M_1$, since either outcome is associated with some finite probabilities. For illustration, consider the following possibilities:

(a) $\alpha = 0, \beta = 1$, yielding $p(0) = 0,\ p(1) = 1$;
(b) $\alpha = \beta = 1/\sqrt{2}$, yielding $p(0) = p(1) = 1/2$;
(c) $\alpha = 1/\sqrt{3},\ \beta = 2/\sqrt{3}$, yielding $p(0) = 1/3,\ p(1) = 2/3$;
(d) $\alpha = 0.9, \beta = 0.1$, yielding $p(0) = 0.99,\ p(1) = 0.01$.

It is clear that in case (a), there is absolute certainty that the qubit measurement will result in the bit information 1. In case (b), the outcome has equal chances of being 0 or 1. In case (c) the odds on measuring 1 are twice those of measuring 0. In case (d), the odds are 99% for 0 and 1% for 1. It must be emphasized here that the above figures concern the measurement outcomes obtained by using the corresponding operators $M_0$, $M_1$. Here comes a subtlety: in case (c), for instance, we know that we have a $1/3$ chance of measuring 0 when using $M_0$. This means that there is a $2/3$ chance that this measurement *fails*. Likewise, there is a $1/3$ chance that the measurement of 1 fails when using $M_1$. But what does "measurement failure" mean? Basically, that whenever we choose to use a given measurement projector $M_m$, the input state is not absolutely certain of collapsing into the corresponding state $|x_m\rangle$. It will only happen "successfully" with probability $|x_m|^2$. If this collapse does not happen (with probability $1 - |x_m|^2$), then nothing is measured, and the measurement is, thus, *failed*. Also, the quantum state collapses into *nothingness*, which here means qubit annihilation and irreversible loss of its information. We can mathematically define such an annihilated, informationless state as $|q'\rangle = \emptyset$.

To clarify the picture, or to alleviate some possible unease with the notions of "successful" and "failed" qubit measurements, consider the following (fictitious) physical analogy, which we have already used in Chapter 17 for POVM measurements.

Compare the qubit $|q\rangle$ with a source that has two possible tones, $A$ or $B$ (standing for the states $|0\rangle$ or $|1\rangle$, respectively), and which are not visible to the eye. Our measurement consists of determining the color tones by observing this source through a set of two "magic filters," called $0, 1$ (for $M_0, M_1$). As we have seen earlier, such filters have strange properties. Indeed, filter 0 does not react to tone $B$, while it makes tone $A$ visible to the eye, but only with probability $p$ (or $|\alpha|^2$). Filter 1 does not react to tone $A$, while it makes tone $B$ visible to the eye, but only with probability $q = 1 - p$ (or $|\beta|^2$). Even stranger here (than the description in Chapter 17) is that the two filters $A, B$ are of complementary nature, since their probabilities of positive reaction are now complementary ($q = 1 - p$). Figure 18.1 shows the eye images obtained in the eight possible cases, according to the two *pure-tone* possibilities $A$ or $B$ (but not both) and their *mixture* $A, B$, and the two filter options $0, 1$. The observation of a bright spot indicates a positive, or successful, measurement (seeing something). The absence of a spot indicates a negative, or failed, measurement (not seeing anything).

Considering first the case of pure tones ($A$ or $B$), the figure indicates that there are actually two "successful" measurement outcomes, as characterized by the presence of the bright spot with *nonzero* probability. These successes correspond to the two input or filter cases $(A, 0)$, $(B, 1)$, which positively identify the tones to be $A$ or $B$ with *absolute certainty* ($p = 1$ for $A$, or $q = 1 - p = 1$ for $B$). In the two other cases, $(A, 1)$, $(B, 0)$, we never see any spot, or the measurement is always "failed." But here comes a subtlety: if we know prior to the measurement that the color tones are pure, the failed measurement is also a positive indication. Indeed, a failure to measure $A$ means an absolute certainty that the input is $B$, and the reverse. The certainty is the same regardless of our filter choice (0 or 1). In this case, the classical bit measurement from the qubit is 100% accurate. The

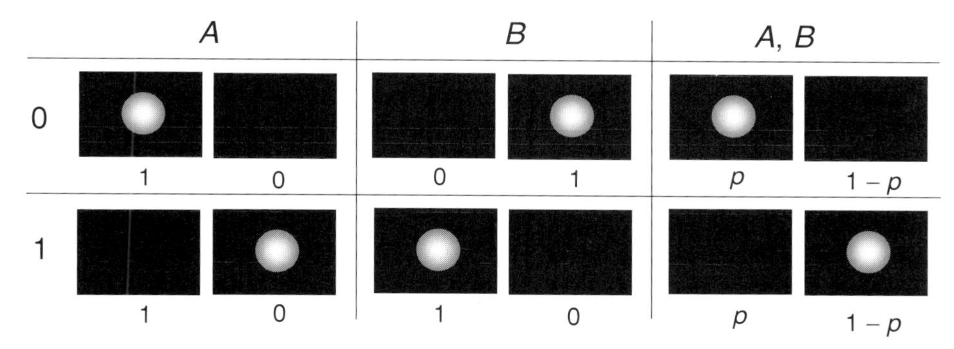


**Figure 18.1** Fictitious analogy of qubit measurement of tones $A$, $B$, or their mixture, and a magic-filter set 0, 1 to characterize a source. Given the tones $A$, $B$ and choices 1, 2, a positive observation of the tone is characterized by the bright spot on the images shown, with various probabilities 1, 0, $p$, $1-p$.

information gained is a 0 or 1 classical bit, in 100% correspondence with the qubit pure state. But to achieve this performance without any concern for the measurement option ($M_0$, $M_1$), it costs the tax of *one bit of advance information* (i.e., "pure input state?" = YES, NO). In the case of the tone mixture (Fig. 18.1, right) we observe only two measurement "successes," with probability $p$ when using the filter 0, and probability $1-p$ when using the filter 1. The information obtained in these two cases is a classical bit of information 0 or 1, as we have projected the qubit onto the corresponding pure states. But, as the figure indicates, there is a finite probability $1-p$ or $p$ that using either filter 0 or 1 results in a measurement failure, and in these cases we do not get any classical bit information. There is one exception to this, however: if we know in advance that the qubit is in the superposition $\alpha = \beta = 1/\sqrt{2}$ (case (b)), there are equal chances of success or failure, regardless of the measurement operator (or tone filter). There is no point in making any measurement, since we know for a fact that a successful outcome will give either a 0 or a 1 classical bit, with equal likelihood. We may just pick a bit value at random. The tax to pay for this result is *one bit of advance information* (i.e., "case (b)?" = YES, NO).

The above description, despite its questionable usefulness from a purist or academic standpoint, made it possible to "visualize" somewhat and get a better feel of the nature of quantum measurements, when applied to single qubits. It is noteworthy, however, that the fictitious "magic-filter" experiment can actually be implemented in the physical world, as described in Chapter 25 in the part concerning *quantum cryptography*. The lesson learnt is that *any "successful" qubit base-state measurement results in its collapse into one of the pure states, and the reduction of its intrinsic information into a classical 0 or 1 bit.*

In Chapter 17, we have also described another type of measurement referred to as *projective* or *von Neumann measurement*. In this approach, we can define the Hermitian *observable* $A$ according to its spectral decomposition over the eigenstates $|\lambda_0\rangle$, $|\lambda_1\rangle$

forming an orthonormal base $V = \{|\lambda_0\rangle, |\lambda_1\rangle\}$:

$$\begin{aligned} A &= \sum_{m=1}^{2} \lambda_m P_m = \sum_{m=1}^{2} \lambda_m |\lambda_m\rangle\langle\lambda_m| \\ &\equiv \lambda_0|\lambda_0\rangle\langle\lambda_0| + \lambda_1|\lambda_1\rangle\langle\lambda_1|, \end{aligned} \tag{18.5}$$

where $\lambda_0, \lambda_1$ are the corresponding eigenvalues, and $P_m = |m\rangle\langle m|$ the measurement operators. Next, we recall the two key results obtained in Chapter 17, while applying them to the case of qubits. First, given a qubit $|q\rangle = x_0|\lambda_0\rangle + x_1|\lambda_1\rangle$ as expressed in the eigenstate base $V$, the probability of measuring the eigenvalue $\lambda_m$ is given by $p(\lambda_m) = |x_m|^2$. With the usual qubit notations $|q\rangle = \alpha|\lambda_0\rangle + \beta|\lambda_1\rangle$, we, thus, obtain:

$$p(\lambda_0) = |\alpha|^2, \quad p(\lambda_1) = |\beta|^2. \tag{18.6}$$

Second, the post-measurement state $|q'\rangle$ is

$$|q'\rangle = |\lambda_m\rangle. \tag{18.7}$$

We may conclude from the two above results that the projective measurement of qubits is essentially similar to the previously described base measurement in the pure states, except that the measurements project the qubit onto the eigenstates of $A$. Such a conclusion is true, but it fails to convey an important property. Indeed, *if we know prior to the measurement that the qubit must be in one of the eigenstates*, then we are in the same situation as in the previous pure-state case. No matter whether we use $P_0 = |\lambda_0\rangle\langle\lambda_0|$ or $P_1 = |\lambda_1\rangle\langle\lambda_1|$ for the qubit measurement, we always retrieve a classical 0 or 1 bit. To recall, if we use $P_0$ when the qubit is in state $|\lambda_0\rangle$ or $P_1$ when the qubit is in state $|\lambda_1\rangle$, the measurements are certain to "succeed," yielding the classical 0 or 1 bits, and the collapsed qubit state remains invariant. But if we use the wrong operator (e.g., $P_0$ when the qubit is $|\lambda_1\rangle$), the measurement "fails," and the qubit state is annihilated. However, such a failure indicates that the qubit must have been in the other orthogonal state, and, therefore, we can conclude with certainty the value of the corresponding classical bit. The only penalty in this event is that we have lost the qubit, and it cannot be reused for another quantum computation, for instance as an input to a subsequent quantum gate within a quantum circuit. The key conclusion is that measurements can be destructive (even when informative) and, therefore, it is safer to proceed to quantum measurements only at the end of a purposeful computational chain. If intermediate measurements are required, one must ensure that these are always nondestructive, so that the collapsed qubit may be reused in the rest of the circuit computation.

In quantum-gate circuits (Chapter 16), any quantum measurement is conventionally represented by an "ammeter" symbol, as illustrated in Fig. 18.2. The two output lines shown at the right correspond to the collapsed or post-measurement state ($|\lambda_m\rangle$) and to the measurement value $\lambda_m$. In the case where the measurement fails, we may write $|q'\rangle = \emptyset$ for the input state annihilation.

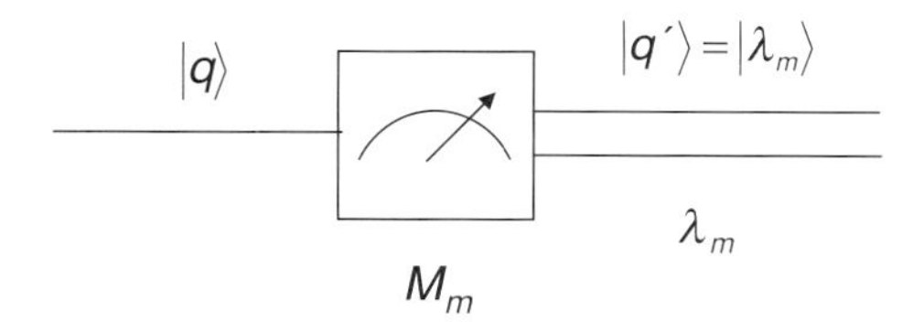


**Figure 18.2** Quantum-circuit representation of measurement apparatus to measure qubits with operator $M_m$, yielding measurement value $\lambda_m$.

## 18.2 Measuring *n*-qubits

In this section, I describe quantum measurements of higher-order or $n$-qubits, which have fundamental implications in both QIT and quantum computing (QC). This is a turning point, where we shall begin to grasp the strange beauty (and the oddity!) of the quantum properties of qubits.

As we have seen in Chapter 17, any base-state measurement $E_{im}$ on an $n$-state $|\psi\rangle$ from $V^{\otimes n}$ results in the collapsing effect

$$\begin{aligned} E_{im}|\psi\rangle &= E_{im}|\psi_1, \ldots, \psi_n\rangle \\ &= |\psi_1, \ldots, \psi_{i-1}, m, \psi_i, \ldots, \psi_n\rangle. \end{aligned} \tag{18.8}$$

In the case of an $n$-qubit $|q\rangle = |x_1, x_2, \ldots, x_n\rangle$, with $x_i = 0, 1$, we have:

$$\begin{aligned} E_{im}|q\rangle &= E_{im}|x_1, x_2, \ldots, x_n\rangle \\ &= |x_1, x_2, \ldots, x_{i-1}, m, x_{i+1}, \ldots, x_n\rangle. \end{aligned} \tag{18.9}$$

With this measurement tool in hand, we can now perform measurements on any qubit superpositions or mixtures in the subspace $V^{\otimes 2}$ (for instance), such as:

$$|q\rangle = u|x_1, x_2\rangle + v|x_1, x_2\rangle, \tag{18.10}$$

with $u, v$ being complex amplitudes satisfying $|u|^2 + |v|^2 = 1$, so that $|r\rangle$ has a modulus of one. For each qubit $i$ in the subspace $V$, there are two possible measurement operators, namely $E_{i1}$ and $E_{i2}$. To illustrate the action of these operators, consider the following example (using the compact notation $|x_i, x_j\rangle \equiv |x_i x_j\rangle$):

$$|q\rangle = \frac{1}{\sqrt{3}}(|00\rangle + |01\rangle + |10\rangle). \tag{18.11}$$

If we want to measure a classical bit 0 in the first qubit position ($i = 1$), we use $E_{10}$ and obtain

$$\begin{aligned} E_{10}|q\rangle &= \frac{1}{\sqrt{3}}(E_{10}|00\rangle + E_{10}|01\rangle + E_{10}|10\rangle) \\ &= \frac{1}{\sqrt{3}}(|00\rangle + |01\rangle + \emptyset) \\ &= \frac{1}{\sqrt{3}}(|00\rangle + |01\rangle). \end{aligned} \tag{18.12}$$

Thus, the probability of measuring 0 in the first qubit is:

$$p_{i=1}(0) = \langle q|E_{10}|q\rangle = \frac{1}{3}(\langle 00| + \langle 01|)(|00\rangle + |01\rangle) \equiv \frac{2}{3}. \tag{18.13}$$

It is now a straightforward exercise to evaluate mentally the other probabilities:

$$p_{i=1}(1) = \frac{1}{3}, \quad p_{i=2}(0) = \frac{2}{3}, \quad p_{i=2}(1) = \frac{1}{3}. \tag{18.14}$$

Next, we shall look at a peculiar family of some 2-qubits, which are of utmost interest for qubit measurements and for the "quantum manipulation" of information. In Chapter 15, we identified four state superpositions, referred to as *Einstein–Podolsky–Rosen (EPR)* or *Bell states (or pairs)*, and defined:

$$\begin{cases} |\beta_{00}\rangle \equiv \frac{1}{\sqrt{2}}(|00\rangle + |11\rangle) \\ |\beta_{01}\rangle \equiv \frac{1}{\sqrt{2}}(|01\rangle + |10\rangle) \\ |\beta_{10}\rangle \equiv \frac{1}{\sqrt{2}}(|00\rangle - |11\rangle) \\ |\beta_{11}\rangle \equiv \frac{1}{\sqrt{2}}(|01\rangle - |10\rangle). \end{cases} \tag{18.15}$$

These EPR–Bell states may seem innocuous at first glance, but they actually reserve a certain number of interesting properties, and as we shall see further down, really "stunning" applications.

As a first property, *there exists no tensor product of pure states or of qubits that is capable of generating any of the EPR–Bell states*. This is left as a straightforward exercise to check. Because the EPR–Bell states cannot be decomposed as a qubit tensor product, they are called *entangled states*. A faster way to realize the nature of entanglement is to observe that none of the states in Eq. (18.15) can be factorized into a pure-state product of the form $|i\rangle|j\rangle = |0\rangle|0\rangle, |0\rangle|1\rangle, |1\rangle|0\rangle, |1\rangle|1\rangle$. Thus, it is not possible to tell what the states of the first and the second qubits are.

While the EPR–Bell states are not tensor products of qubit states, they can, however, be generated by means of quantum gates. The circuit shown in Fig. 18.3, which comprises a *Hadamard* gate and a *CNOT* gate (Chapter 16) represents one possibility for the generation of EPR–Bell states. As we saw in that earlier chapter, the Hadamard gate transforms the input qubit $|0\rangle$ into the mixed state $|+\rangle = (|0\rangle + |1\rangle)/\sqrt{2}$, and the input qubit $|1\rangle$ into the mixed state $|-\rangle = (|0\rangle - |1\rangle)/\sqrt{2}$. Thus, for the quantum gate in Fig. 18.3, the tensor input $|0\rangle \otimes |y\rangle$ results in

$$\begin{aligned} (H|x\rangle) \otimes |x \oplus y\rangle &= (|x=0\rangle + |x=1\rangle) \otimes |x \oplus y\rangle/\sqrt{2} \\ &= (|0\rangle \otimes |0 \oplus y\rangle + |1\rangle \otimes |1 \oplus y\rangle)/\sqrt{2} \\ &\equiv (|0, y\rangle + |1, 1 \oplus y\rangle)/\sqrt{2}. \end{aligned} \tag{18.16}$$

Hence, $|0\rangle \otimes |0\rangle \to (|0,0\rangle + |1,1\rangle)/\sqrt{2} \equiv |\beta_{00}\rangle$ and $|0\rangle \otimes |1\rangle \to (|0,1\rangle + |1,0\rangle)/\sqrt{2} \equiv |\beta_{01}\rangle$. The two other results for the inputs $|1\rangle \otimes |y\rangle$ are also readily verified.

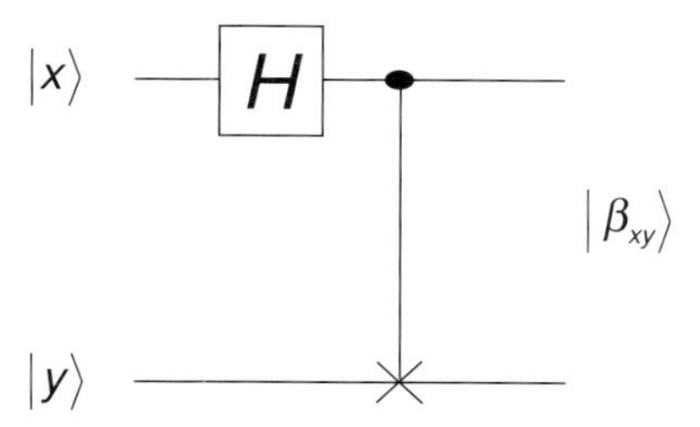


**Figure 18.3** Quantum circuit for the generation of EPR–Bell states $|\beta_{xy}\rangle$ as defined in Eq. (18.15).

As a second property, *the EPR–Bell states form an orthonormal basis for any 2-qubit* $|kl\rangle$ *and any superposition thereof.* Indeed, it is straightforward to see from Eq. (18.15) that

$$\begin{cases} |00\rangle = \dfrac{1}{\sqrt{2}}(|\beta_{00}\rangle + |\beta_{10}\rangle) \\ |01\rangle = \dfrac{1}{\sqrt{2}}(|\beta_{01}\rangle + |\beta_{11}\rangle) \\ |10\rangle = \dfrac{1}{\sqrt{2}}(|\beta_{01}\rangle - |\beta_{11}\rangle) \\ |11\rangle = \dfrac{1}{\sqrt{2}}(|\beta_{00}\rangle - |\beta_{10}\rangle), \end{cases} \tag{18.17}$$

hence, any 2-qubit superposition of the above pure $|ij\rangle$ states accepts a unique decomposition over the EPR–Bell states.

Third, we notice from the definitions in Eq. (18.5) that the EPR–Bell states represent a mix of 2-qubits $|ij\rangle$ and $|kl\rangle$ satisfying $i = 0, k = 1$, and $j, l = 0, 1$. For the measurement probabilities, this leads to the following property (as is easily mentally checked):

$$p_{i=1}(0) = p_{i=1}(1) = p_{i=2}(0) = p_{i=2}(1) = \frac{1}{2}. \tag{18.18}$$

This result shows that any measurement $E_{im}$ of the EPR–Bell states has a 50% chance of successfully projecting into any of the pure states. The 16 possible measurements resulting in the corresponding state collapses (phase factors $e^{-i\pi/2}$ being omitted) are summarized in Table 18.1. As expected, we observe from the table that the EPR–Bell states collapse into any of the pure states $|00\rangle, |01\rangle, |10\rangle, |11\rangle$. We see that using a positive measurement $E_{1m}$ necessarily collapses the EPR–Bell pair into the pure state $|m\rangle \otimes |x\rangle$, and a positive measurement $E_{2m}$ necessarily collapses the EPR–Bell pair into the pure state $|x\rangle \otimes |m\rangle$, where $x = 0, 1$. This has the consequence *that regardless of the four possible EPR–Bell pairs, a single positive measurement yields absolute information on one of the two qubits in the collapsed state.*

Because the first measurement gives only one qubit information, a second measurement would be necessary to get the missing information concerning the other qubit. But we do not need to make a second measurement if we know in advance the initial EPR–Bell pair. Indeed, Table 18.1 illustrates that the advance knowledge of $|\beta_{00}\rangle, |\beta_{01}\rangle, |\beta_{10}\rangle, |\beta_{11}\rangle$

**Table 18.1** State collapses resulting from single-qubit measurements $E_{im}$ on the EPR–Bell states $|\beta_{00}\rangle$, $|\beta_{01}\rangle$, $|\beta_{10}\rangle$, $|\beta_{11}\rangle$.

| | $\|\beta_{00}\rangle$ | $\|\beta_{01}\rangle$ | $\|\beta_{10}\rangle$ | $\|\beta_{11}\rangle$ |
|---|---|---|---|---|
| $E_{10}$ | $\|00\rangle$ | $\|01\rangle$ | $\|00\rangle$ | $\|01\rangle$ |
| $E_{11}$ | $\|11\rangle$ | $\|10\rangle$ | $\|11\rangle$ | $\|10\rangle$ |
| $E_{20}$ | $\|00\rangle$ | $\|10\rangle$ | $\|00\rangle$ | $\|10\rangle$ |
| $E_{21}$ | $\|11\rangle$ | $\|01\rangle$ | $\|11\rangle$ | $\|01\rangle$ |

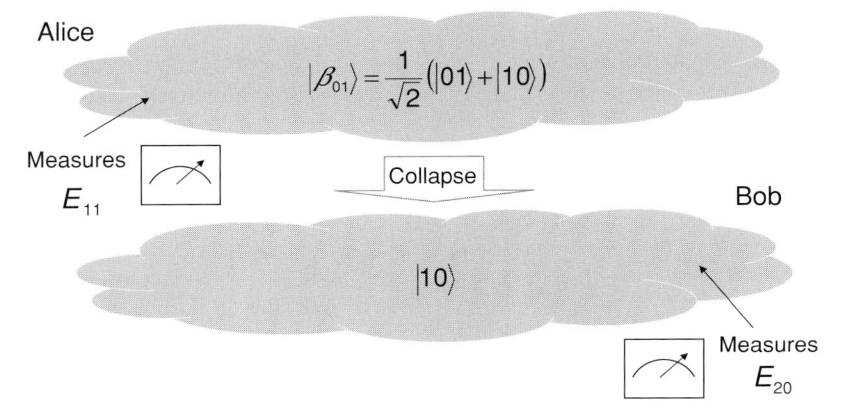


**Figure 18.4** Two successive measurements by Alice then Bob in an entangled quantum state, here $|\beta_{01}\rangle$.

and the choice of a single-measurement operator $E_{10}$, $E_{11}$, $E_{20}$, $E_{21}$ absolutely defines the collapsed state and, hence, the value of the two entangled qubits. For instance, if we use $|\beta_{11}\rangle$, the single measurement $E_{21}$ yields the 2-qubit $|01\rangle = |0\rangle \otimes |1\rangle$ and the absolute information therein. Thus, given the knowledge of the system EPR–Bell pair, *any positive measurement of one of the two qubits conditions the information knowledge of the other qubit.* Because the information of one qubit measurement conditions that of the second qubit measurement, the EPR–Bell states are said to be *entangled.* In QIT jargon, EPR–Bell states are also referred to as *ebits*, for "entanglement bits" (one ebit = one single entangled pair).

To provide a visual illustration of this, imagine two people A and B, who are traditionally named *Alice* and *Bob*, and who live far apart. However, both have access to a quantum system, which is in the entangled EPR–Bell state $|\beta_{01}\rangle$, as represented by a cloud in Fig. 18.4. The state remains in $|\beta_{01}\rangle$ until Alice unilaterally takes the initiative to measure the *first* qubit, through the operator $E_{11}$, for instance. This measurement immediately results in the EPR–Bell pair collapsing into the pure state $|10\rangle$. Bob's measurement of the second qubit can only yield $|0\rangle$ as a positive outcome. Bob and Alice can exchange roles. For instance, Bob may first measure the second qubit through $E_{20}$, resulting in the system collapsing into the state $|01\rangle$, and leaving $|0\rangle$ to Alice as the only positive measurement for the first qubit. In each case, the measurement of the first qubit conditions the measurement of the second, because of the effect of state collapse that occurs in between.

**Table 18.2** Outcome of Bell measurement with EPR–Bell states as input (normalization factor being overlooked).

| Input state $\lvert xy\rangle$ | Pre-measurement state | Cbit output | $m, m'$ |
|---|---|---|---|
| $\lvert\beta_{00}\rangle = \lvert 00\rangle + \lvert 11\rangle$ | $\lvert 00\rangle$ | 0 | 0 |
| $\lvert\beta_{01}\rangle = \lvert 00\rangle + \lvert 11\rangle$ | $\lvert 01\rangle$ | 0 | 1 |
| $\lvert\beta_{10}\rangle = \lvert 00\rangle - \lvert 11\rangle$ | $\lvert 10\rangle$ | 1 | 0 |
| $\lvert\beta_{11}\rangle = \lvert 01\rangle - \lvert 10\rangle$ | $\lvert 11\rangle$ | 1 | 1 |

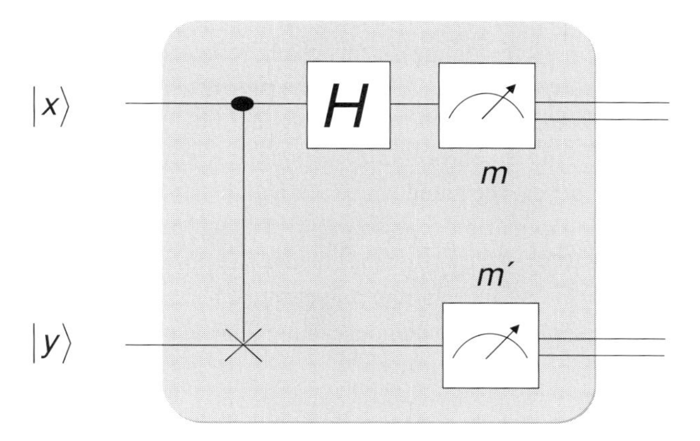


**Figure 18.5** Principle of Bell measurement or measurement in Bell states.

From any physical perception of our "classical world," the effect of state entanglement seems rather nonintuitive. We may illustrate it through a figurative experiment. Assume that we place two coins in two separate boxes. In each box, the position (or states) of the coins can be either heads or tails, but we have no awareness of which one. In the classical world, opening a box tells us only that the coin is heads or tails, but we cannot know of the other coin until we open the other box. In the quantum world, the coin states may be entangled. This means that opening only one box conditions the state of the coin in the other box. The point here is not about the advance knowledge of the entanglement possibilities, as represented by the four possible EPR–Bell states. Rather, it is the matter that *a single measurement in one physical system determines "for absolute certain" the state of the other physical system*. As described later in this chapter, the properties of EPR–Bell state entanglement are the key to applications called *quantum superdense coding* and *quantum teleportation*, both of which having no counterpart in the classical world.

## 18.3 Bell state measurement

In this section, I describe a 2-qubit measurement scheme, referred to as *measurement in the Bell basis*, or *Bell state measurement*. The principle is illustrated in Fig. 18.5. The circuit includes a Hadamard gate and a CNOT gate, followed by two qubit measurements.

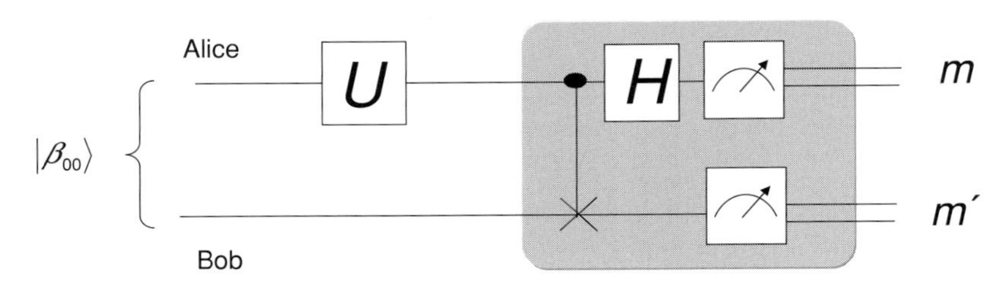


**Figure 18.6** Circuit used for superdense coding.

The measurements, which are made in the base $\{|00\rangle, |01\rangle, |10\rangle, |11\rangle\}$, output the classical bits $m, m'$. What happens if we input to this gate circuit any of the EPR–Bell states $|\beta_{mm'}\rangle$? It is left as a simple exercise to check that the pre-measurement states are $|mm'\rangle$ and the resulting measurement cbits (short for "classical bits") are $m, m'$, as shown in Table 18.2. This means that the apparatus is able to positively single out or identify any of the four EPR–Bell states, hence, the name "measurement in the Bell basis" or "Bell state measurement." Such a denomination may sound confusing, since the measurement is actually performed in the $\{|00\rangle, |01\rangle, |10\rangle, |11\rangle\}$ basis. The point is that while the circuit represents a true EPR–Bell state measurement apparatus, it can be used for other types of application, as will be shown next when describing *superdense coding* and *quantum teleportation*.

## 18.4 Superdense coding

In this section, I describe a first application of quantum entanglement, which enables *quantum superdense coding*. The key concept behind this appellation is the possibility for Alice to send to Bob *two* classical bits $m, m'$, which she has in her possession, by means of communicating only a single qubit. The generic circuit for superdense coding is represented in Fig. 18.6. Alice and Bob must share an EPR–Bell state, for instance $|\beta_{00}\rangle$. Alice controls a gate $U$, which she can change at will into $I$, $Z$, $X$, and $iY$. Bob has the control of a Bell-state measurement apparatus, as shown by the shaded subsystem in the figure. It can be checked as an easy exercise that the choice of $U$ transforms $|\beta_{00}\rangle$ into any EPR–Bell state $|\beta_{00}\rangle, |\beta_{10}\rangle, |\beta_{01}\rangle, |\beta_{11}\rangle$, according to

$$\begin{aligned} I|\beta_{00}\rangle &= |\beta_{00}\rangle;\ X|\beta_{00}\rangle = |\beta_{01}\rangle \\ Z|\beta_{00}\rangle &= |\beta_{10}\rangle;\ iY|\beta_{00}\rangle = |\beta_{11}\rangle. \end{aligned} \tag{18.19}$$

Thus, by switching the $U$ gate into $I$, $Z$, $X$, and $iY$, Alice can send Bob the classical bit pairs 00, 01, 10, and 11, respectively, at an information rate of *two bits per qubit*, which represents twice the classical information rate, hence the name superdense coding. Note that from any realistic telecom viewpoint, twofold improvements in information-rate capacities do not deserve such superlatives as "super" or "dense," or, worst of all, "superdense," notwithstanding the conceptual importance of the potential capacity expansion of the "old-fashioned" classical bit.

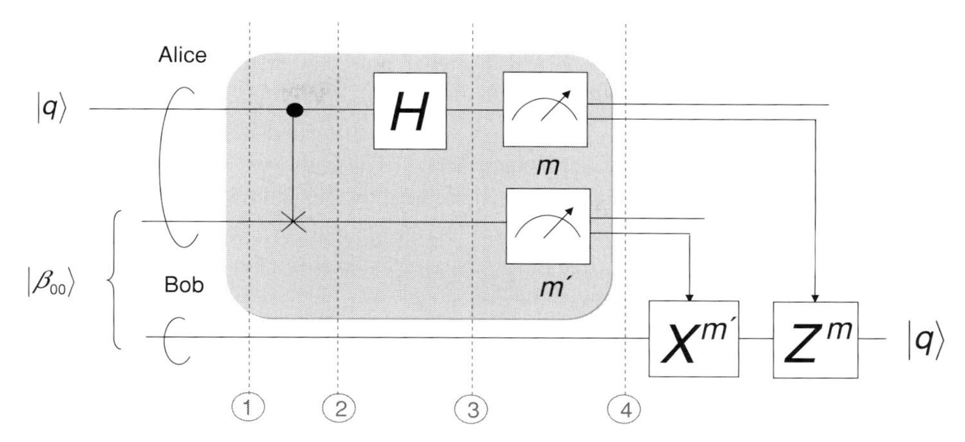


**Figure 18.7** Circuit used by Alice and Bob for quantum teleportation.

## 18.5 Quantum teleportation

The strange technique referred to as *quantum teleportation* (QT) makes it possible to *transmit an unknown qubit instantaneously over any distance.* Here, "distance" means any range spanning from next door's office to the opposite end of the Solar System, or even beyond in the Universe, despite the limitations imposed by Einstein's theory of relativity, the physical finiteness of the speed of light, and its invariance regardless of any reference frame.

To a wider audience, the word "teleportation" may raise false expectations in terms of the capability of moving physical objects, which we may call here "macrosystems," such as a coin, a cat or a human being (see Chapter 15)! The expectations are high indeed if public audiences "buy in" the quantum state superposition concept as applicable to the macroscopic world. At a microscopic scale, however, the notions of QT and "nonlocality" implications, are absolutely correct. Suffice it to recall here that in the quantum world, any elementary object, such as an atom or a particle, is, to all intents and purposes, a quantum system, and there are no conceptual limitations involved therein, both in terms of locality and physical distance. As we shall see, however, there is no contradiction between QT and relativity laws, when it comes to measuring actual information.

In this book, QT is, thus, our first quantum algorithm to "transport" quantum states, whose most elementary description is the *qubit*. The key to the QT algorithm is the use of an EPR–Bell state (or *ebit*), as a resource commonly shared between Alice and Bob, and making it possible for Alice to send any qubit to Bob. As we shall see, teleportation also requires the *communication of two classical bits* (cbits), using any classical communication channel, such as the telephone, or the Internet, for instance. This preliminary observation is completely sufficient to justify that indeed, teleportation does not violate relativity laws.

The principle of the generic QT circuit is illustrated in Fig. 18.7. As the figure shows, this is a 3-qubit circuit, which includes a *Bell-state measurement* apparatus (as shaded), which outputs the two cbits $m, m'$, and two gates $(X, Z)$, which are used according to the powers of $m, m'$. The two top wires are connected to Alice's side, while Bob uses the bottom wire for final measurements. The qubit that Alice wants to transmit to Bob is $|q\rangle = \alpha|0\rangle + \beta|1\rangle$, as shown at top left. Alice and Bob also share the resource of an EPR–Bell state, here, for instance, $|\beta_{00}\rangle$. The end result of the QT is that Bob is able to retrieve the qubit $|q\rangle = \alpha|0\rangle + \beta|1\rangle$, as shown at bottom right. At first glance, this QT circuit looks somewhat involved but, as we shall see, its operation turns out relatively easy to grasp, should one proceed step by step using the corresponding numbers shown in Fig. 18.7.

The initial state of the system, marked ①, is $|q\rangle \otimes |\beta_{00}\rangle$, namely, using the simplified 3-qubit notation:

$$\begin{aligned} |q\rangle \otimes |\beta_{00}\rangle &= (\alpha|0\rangle + \beta|1\rangle) \otimes \frac{1}{\sqrt{2}}(|00\rangle + |11\rangle) \\ &\equiv \frac{1}{\sqrt{2}}(\alpha|000\rangle + \alpha|011\rangle + \beta|100\rangle + \beta|111\rangle). \end{aligned} \tag{18.20}$$

Next, we evaluate the system state at ②. As seen from the figure, the CNOT gate controls the second qubit from the first qubit, with the equivalent of the logical XOR, i.e., $|x, y, z\rangle \mapsto |x, x \oplus y, z\rangle$. Hence, we have at ②:

$$\begin{aligned} &\frac{1}{\sqrt{2}}(\alpha|000\rangle + \alpha|011\rangle + \beta|100\rangle + \beta|111\rangle) \\ &\quad\mapsto \frac{1}{\sqrt{2}}(\alpha|000\rangle + \alpha|011\rangle + \beta|110\rangle + \beta|101\rangle). \end{aligned} \tag{18.21}$$

Next, we evaluate the system state at ③. The only change is the passing of the first qubit through the Hadamard gate $H$. To recall, $H|0\rangle = (|0\rangle + |1\rangle)/\sqrt{2}$ and $H|1\rangle = (|0\rangle - |1\rangle)/\sqrt{2}$. This transforms the 3-qubits in Eq. (18.20) according to the following (the underline to stress the qubit being acted on):

$$\begin{aligned} |\underline{0}00\rangle &\mapsto \frac{1}{\sqrt{2}}(|\underline{0}00\rangle + |\underline{1}00\rangle) \\ |\underline{0}11\rangle &\mapsto \frac{1}{\sqrt{2}}(|\underline{0}11\rangle + |\underline{1}11\rangle) \\ |\underline{1}10\rangle &\mapsto \frac{1}{\sqrt{2}}(|\underline{0}10\rangle - |\underline{1}10\rangle) \\ |\underline{1}01\rangle &\mapsto \frac{1}{\sqrt{2}}(|\underline{0}01\rangle - |\underline{1}01\rangle). \end{aligned} \tag{18.22}$$

Hence we have at ③:

$$\begin{aligned} &\frac{1}{\sqrt{2}}(\alpha|000\rangle + \alpha|011\rangle + \beta|110\rangle + \beta|101\rangle) \\ &\mapsto \frac{1}{2}\{\alpha|000\rangle + \alpha|100\rangle + \alpha|011\rangle + \alpha|111\rangle + \beta|010\rangle - \beta|110\rangle + \beta|001\rangle - \beta|101\rangle\} \\ &\equiv \left\{|00\rangle\frac{\alpha|0\rangle + \beta|1\rangle}{2} + |01\rangle\frac{\alpha|1\rangle + \beta|0\rangle}{2} + |10\rangle\frac{\alpha|0\rangle - \beta|1\rangle}{2} + |11\rangle\frac{\alpha|1\rangle - \beta|0\rangle}{2}\right\}. \end{aligned} \tag{18.23}$$

**Table 18.3** State collapse resulting from Alice's 2-qubit measurements $(m, m')$, and corresponding gates $X^{m'}Z^m$ to be used by Bob in order to retrieve the original qubit $|q\rangle = \alpha|0\rangle + \beta|1\rangle$.

| Alice measures | | | | |
|---|---|---|---|---|
| $m$ | $m'$ | Alice measures | Bob's 1-qubit | $X^{m'}Z^m$ |
| 0 | 0 | $\|00\rangle$ | $\alpha\|0\rangle + \beta\|1\rangle$ | $I$ |
| 0 | 1 | $\|01\rangle$ | $\alpha\|1\rangle + \beta\|0\rangle$ | $Z$ |
| 1 | 0 | $\|10\rangle$ | $\alpha\|0\rangle - \beta\|1\rangle$ | $X$ |
| 1 | 1 | $\|11\rangle$ | $\alpha\|1\rangle - \beta\|0\rangle$ | $XZ$ |

In the last expression in the right-hand side, the first two qubits, which are controlled by Alice, and the last qubit, which is Bob's, have been regrouped for clarity. Then come Alice's measurements of the first two qubits, with the corresponding classical-bit results $m, m'$ (respectively) obtained at ④. The rule has it that Alice must communicate the two classical bits, or *cbits*, $m, m'$ to Bob. It is clear that if Alice measures $|00\rangle$, the system state collapses into the qubit $|00\rangle(\alpha|0\rangle + \beta|1\rangle)$, as shown in Eq. (18.23), and so on for each of the four possible measurements. The outcomes of Alice's measurements, and the resulting state of Alice and Bob's qubits, are summarized in Table 18.3. The two cbits $m, m'$ communicated to Bob make it possible to define the gates $X^{m'}Z^m$ on Bob's wire (Fig. 18.7), as also shown in Table 18.3. It is seen from the table that in the first case (Alice measures $|00\rangle$) Bob's qubit has collapsed into the state $|q\rangle = \alpha|0\rangle + \beta|1\rangle$. Thus, Alice's original qubit $|q\rangle$ has been successfully "teleported" to Bob. From the two cbits $m, m' = 0, 0$, Bob is, thus, instructed to use the gates $X^0Z^0 = I$, namely, to leave his qubit unchanged. In the second case (Alice measures $|01\rangle$), Bob's qubit is $\alpha|1\rangle + \beta|0\rangle$, and the application of $X^0Z^1 = Z$ swaps the amplitudes $\alpha$, $\beta$ to yield $|q\rangle = \alpha|0\rangle + \beta|1\rangle$. The last two cases are also straightforward to analyze.

The principle of QT can, thus, be summarized as follows: (a) Because Alice and Bob share an ERP–Bell state, Alice's measurements cause the system to collapse and condition Bob's qubit; (b) Alice communicating the two classical bits describing her measurement makes it possible for Bob to retrieve Alice's qubit $|q\rangle$. Remarkably, Alice has no knowledge of the teleported $|q\rangle$. This point is quite important. Indeed, if Alice had this knowledge, she could communicate to Bob the full information required (amplitudes, base) for him to re-create the same qubit locally, and, therefore, they both would not need this QT apparatus. However, such a communication is complicated and quite resource consuming, should Bob need lots of qubits for his computations. And, most importantly, Alice would be able to communicate only qubits *known to her*, which is utterly restrictive in view of the QT potential. As a second observation, we note that QT does not violate the *noncloning theorem*. Indeed, the initial qubit $|q\rangle$ is collapsed by Alice's first measurement into the pure state $|m\rangle$, as seen from the top wire in Fig. 18.7.

The benefits of QT being now understood, we may then have a few legitimate questions. First, why use *classical bits* to determine which gates Bob should use? Indeed, Alice's measurements result in the collapse of her two qubits into the pure tensor state

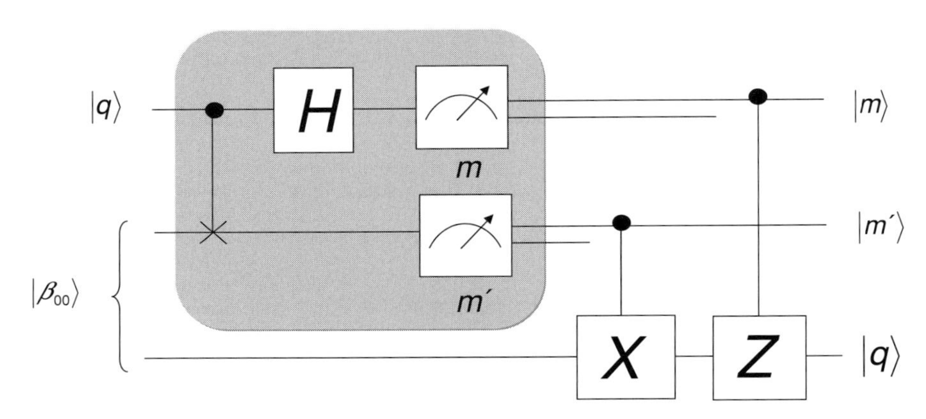


**Figure 18.8** Principle of quantum teleportation with an all-quantum gate circuit.

$|m\rangle|m'\rangle$, and each of the $|m\rangle$, $|m'\rangle$ qubits could be used as "automatic remote controls" for Bob's gates, i.e., controlled-$X$ and controlled-$Z$. This would give the all-quantum-gate QT circuit illustrated in Fig. 18.8. While such a circuit is logically equivalent to that of the QT shown in Fig. 18.7, it has several drawbacks. First, controlled-$U$ gates are more complex than simple $2 \times 2$ $U$ gates. But this is not as important as the fact that if Alice and Bob were located far apart, the control qubits $|m\rangle$, $|m'\rangle$, even as known to Alice, must be transported over this distance! It is, thus, far more sensible and straightforward for Alice to send the two bits $m, m'$ to Bob through any classical channel, such as the telephone or the Internet. On the other hand, if Bob is located nearby, or directly on Alice's premises, this whole qubit manipulation has no point, and the circuit shown in Fig. 18.8 reduces to an elegant, but needless *CROSSOVER* circuit (a circuit that just accurately transmits quantum information between Alice and Bob, who are in immediate reach of each other!). Regardless of these different observations, the QT circuits shown in the literature either use representations of Fig. 18.7 or of Fig. 18.8, or of both (the two outputs of any of Alice's measurement connecting to Bob's gates).

Two other legitimate questions are whether it is possible to use EPR–Bell states different from $|\beta_{00}\rangle$, and how Alice and Bob can set up an EPR–Bell state to share. Concerning the first question, the answer is yes, which is left as an exercise to show. In that case, however, the classical bits to be transmitted to Bob are different from $m$, $m'$ and Alice must use a logic look-up table to transform $m$, $m'$ into the right combinations, which depends on which EPR–Bell state has been used. Apart from this minor complication, the QT principle using the three EPR–Bell states $|\beta_{01}\rangle$, $|\beta_{10}\rangle$, $|\beta_{11}\rangle$ remains strictly identical. The answer to the second question is provided in Fig. 18.9. Both Alice and Bob must supply a qubit $|0\rangle$ of their own, for Alice to use as an *ancilla* qubit, and for Bob to use as a *target* qubit. The Hadamard–CNOT circuit shaded in the figure corresponds to that of Fig. 18.3, which, as we have seen, generates the entangled state $|\beta_{00}\rangle$, when input with the two qubits $|0\rangle$, $|0\rangle$.

As a summary and conclusion, we have learnt that it is possible to achieve the teleportation of a single unknown qubit through a quantum channel. For the two ends of the channel (Alice and Bob) the requirement is to share one EPR pair and two classical

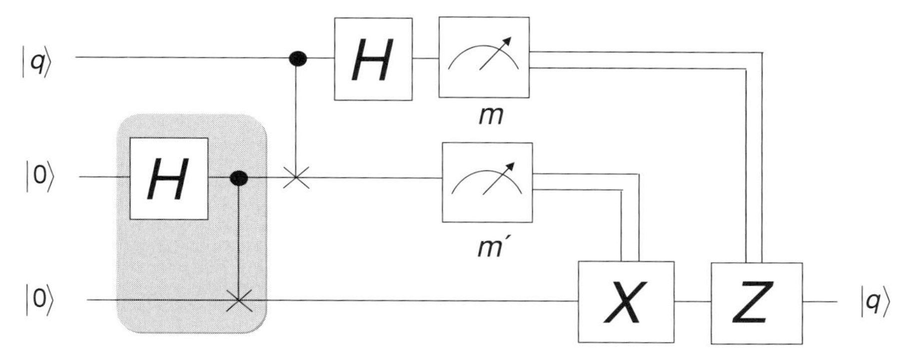


**Figure 18.9** Full implementation of quantum teleportation circuit with ancilla and target qubits $|0\rangle$, $|0\rangle$ at input.

bits of information. The fact that Alice must communicate the two classical bits to Bob shows that *QT cannot operate faster than the speed of light*, which is in agreement with a fundamental principle of the *theory of relativity*. However, we have seen that Alice's measurements *instantaneously* collapse the system into any of the four 3-qubit states given by Eq. (18.23), namely,

$$\begin{aligned} &|00\rangle(\alpha|0\rangle + \beta|1\rangle); \\ &|01\rangle(\alpha|1\rangle + \beta|0\rangle); \\ &|10\rangle(\alpha|0\rangle - \beta|1\rangle); \\ &|11\rangle(\alpha|1\rangle - \beta|0\rangle); \end{aligned} \tag{18.24}$$

which can be summarized in the general form: $|mm'\rangle|\psi\rangle$ with $|\psi\rangle = Z^{m'}X^m|q\rangle$ being the qubit accessible to Bob. It is clear that without any knowledge of the two bits $m, m'$, Bob is not able to make any sense of his qubit $|\psi\rangle$, other than performing meaningless random-measurement guesses. Bob cannot have any certainty either of Alice's measurement, through any prior agreement between them. No such agreement is possible, since by definition any of Alice's measurement of any of the four pure states $|mm'\rangle$ has $^1/_4$ chance of being successful. Alice and Bob could agree that only the measurement $|00\rangle$ will be systematically performed, which already represents one bit of advance information. Alice must then tell Bob if the measurement succeeded, and this is another bit of information, resulting in $m, m' = 0, 0$. Only then can Bob be certain that $|\psi\rangle = |q\rangle$. The key conclusion is that for successful qubit teleportation, Bob needs the two classical bits $m, m'$ from Alice. Should the QT apparatus use *qubits* instead of cbits to control Bob's gates, the conclusion remains the same: Alice's qubits $|m\rangle$, $|m'\rangle$ must be transmitted over the distance separating her from Bob. Any physical channel supporting qubits can be used for this transmission (for instance an optical fiber or a radio link), but the communication cannot be faster than the speed of light.[1]

[1] In any physical medium, the speed of an electromagnetic (EM) wave is given by $c = c_0/n$, where $n \geq 1$ is the medium's refractive index and $c_0$ is the speed of light in absolute vacuum. Thus, the condition $c \leq c_0$ is always satisfied for any EM-wave transmission.

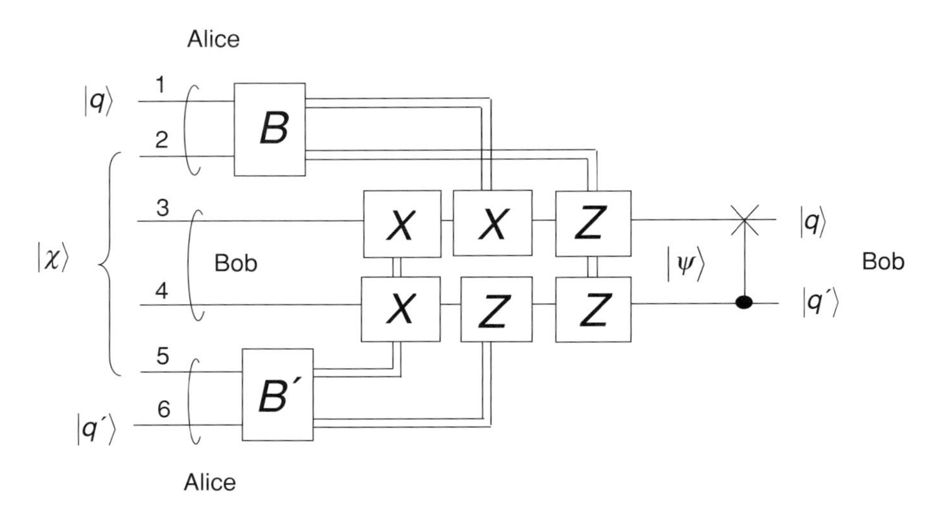


**Figure 18.10** Circuit implementation for simultaneous quantum teleportation of *two qubits*.

What about the possibility of teleporting an arbitrary number $N$ of different qubits simultaneously or, even better, teleporting at once an $N$-qubit, also called *quNit* or, equivalently, *qudit*, for $d$-qubit? The second case would achieve the "dream" of teleporting to another location any complex quantum system, like an atom, a molecule, a crystal, or even a fragment of DNA, as may be already envisioned by some futurists. It is beyond the scope of this chapter to address this issue and analyze, even superficially, what type of protocol and resources would be required to achieve *quNit teleportation*, in terms of shared entangled states and exchange of classical bits. We may, however, venture one step further in the description by considering the *teleportation of two qubits*. This description should suffice to illustrate the complexity of teleporting more than one qubit, and also to satisfy a legitimate curiosity about the potentials of teleporting an ensemble of quantum states.

A proposed circuit for the simultaneous teleportation of *two qubits* $|q\rangle, |q'\rangle$ from Alice is shown in Fig. 18.10, as an original, symmetrical variant of that shown in Gottesman and Chuang (1999).[2] As seen from the figure, Alice and Bob share a 4-qubit entangled resource, $|\chi\rangle$, which I shall specify later. The two boxes labeled $B$, with two cbit outputs each, correspond to Bell-state measurements. The rest of the circuit includes three controlled-$X$ and controlled-$Z$ gates with Alice's four cbits as control signals, to be communicated to Bob via a classical channel. The circuit output, as received remotely by Bob, is defined as $|\psi\rangle = C_{43}|q\rangle_3|q'\rangle_4$ with $|q'\rangle_4$ controlling $|q\rangle_3$. To retrieve $|q\rangle, |q'\rangle$, Bob simply needs to pass the two qubits through the same CNOT gate, as shown at the very right of Fig. 18.10. A possible circuit implementation to generate the entangled state $|\chi\rangle$ is illustrated in Fig. 18.11. The circuit is seen to use two identical EPR–Bell states $|\beta_{00}\rangle$ and one CNOT gate. Labeling the four quantum wires in Fig. 18.11 one to

[2] From D. Gottesman and I. L. Chuang, Quantum teleportation is a universal computational primitive. *Nature*, **402** (1999), 390–3, http://arxiv.org/PS_cache/quant-ph/pdf/9908/9908010v1.pdf.

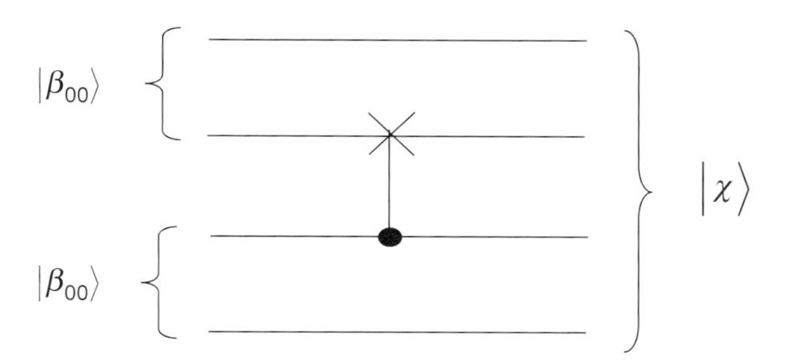


**Figure 18.11** Possible circuit implementation for generating the entangled state $|\chi\rangle$ in Fig. 18.10.

four, from top to bottom, we easily obtain the definition of $|\chi\rangle$:

$$\begin{aligned}\frac{1}{2}|\beta_{00}\rangle_{12}\otimes|\beta_{00}\rangle_{34} &= \frac{1}{2}\left(\frac{|00\rangle_{12}+|11\rangle_{12}}{\sqrt{2}}\right)\otimes\left(\frac{|00\rangle_{34}+|11\rangle_{34}}{\sqrt{2}}\right)\\ &= \frac{1}{4}(|00\rangle_{12}\otimes|00\rangle_{34}+|00\rangle_{12}\otimes|11\rangle_{34}\\ &\quad+|11\rangle_{12}\otimes|00\rangle_{34}+|11\rangle_{12}\otimes|11\rangle_{34})\\ &\equiv \frac{1}{4}(|0000\rangle+|0011\rangle+|1100\rangle+|1111\rangle)_{1234}.\end{aligned} \tag{18.25}$$

Note the factor ½, which was introduced to ensure proper normalization of the resulting tensor state. The subsequent application of the CNOT gate $C_{32}$ with qubit 3 as control, and qubit 2 as target finally yields:

$$\begin{aligned}|\chi\rangle &= \frac{1}{4}C_{32}(|0000\rangle+|0011\rangle+|1100\rangle+|1111\rangle)_{1234}\\ &\equiv \frac{1}{4}(|0000\rangle+|0111\rangle+|1100\rangle+|1011\rangle)_{1234}.\end{aligned} \tag{18.26}$$

It is seen from the above definition that $|\chi\rangle$ is a 4-qubit entangled state, which cannot be expressed as any tensor product of the form $|a\rangle_i\otimes|bcd\rangle_{jkl}$ or $|ab\rangle_{ij}\otimes|cd\rangle_{kl}$, with any permutations of the indices $i, j, k, l = 1, 2, 3, 4$.

It is a rather involved and tedious task to check that the circuit in Fig. 18.10, together with the communication to Bob of Alice's measured cbits $n, m, n', m'$, actually results in the *simultaneous teleportation of the two qubits* $|q'\rangle, |q\rangle$. However, this represents an excellent test case to apply all that we have learnt in this chapter, and the previous ones as well, in terms of Pauli gates, tensor states, entangled states, measurements in the Bell-state basis and quantum teleportation. For this specific reason, the full details of the demonstration have been outlined in Appendix P. Going through this appendix, it is observed that the 6-qubit pre-measurement state, $|\Phi''\rangle$, involves no less than 64 terms! But the twin measurement executed thereupon from Alice results in the instant collapse of $|\Phi''\rangle$ into a single state of the form $|nmn'm'\rangle_{1256}\otimes|\theta\rangle_{34}$. After hearing from Alice what the four cbits $n, m, n', m'$ are, it is just a routine for Bob to set up his gate circuit according to the sequence $X_3^{n'}X_4^{n'}X_3^{n}Z_4^{m'}Z_3^{m}Z_4^{m}$, as indicated in Fig. 18.10. If the whole exercise shows that teleporting a number of qubits greater than one is possible, it certainly illustrates that the corresponding quantum circuits become increasingly complex as this number increases.

*Superdense quantum teleportation* is about maximizing the information contents that can be communicated through a quantum channel, along with the necessary tax of classical information of cbits. As evoked earlier, there exist many sophisticated quantum circuits for the teleportation of single *quNits* of dimension $n$, as defined over an orthonormal base $V = \{|i\rangle\}_{i=0\ldots n-1}$ by

$$|q\rangle = \gamma_0|0\rangle + \gamma_1|1\rangle + \gamma_2|2\rangle + \cdots + \gamma_{n-1}|n-1\rangle, \tag{18.27}$$

where $\gamma_i (i = 0 \ldots n-1)$ are complex numbers. The first candidate immediately above the qubit is the *qutrit* $|q\rangle = \gamma_1|1\rangle + \gamma_2|2\rangle + \gamma_3|3\rangle$, which is a quantum superposition of the classical *trit* or three-level information with values $0, 1, 2$. Teleporting a single qutrit requires Alice and Bob to share *entangled 3-qutrit states*, for instance the one defined as $|\Phi^+\rangle = (|000\rangle + |111\rangle + |222\rangle)/\sqrt{3}$, which is known as a *three-level GHZ (Greenberger–Horne–Zeilinger)* state (the two-level being the first of the four Bell states, $|\beta_{00}\rangle$). The extra dimension opens up a large variety of approaches for teleportation algorithms, in particular where a third party, called *Charles*, comes into the picture! The mediation from Charles simplifies Alice's task in terms of Bell-basis measurements, this time in a 2-qutrit base. These considerations, which are beyond the scope of this book, show at least that, mathematically speaking, teleportation in $N$ dimensions knows no actual limits.

## 18.6 Distributed quantum computing

The principle of quantum teleportation by means of EPR–Bell states can also be extended to *quantum gates*. The basic concept is to act on remote qubits from a distance, which is referred to as *distributed quantum computation*. One way to achieve this would be for Alice and Bob to have their own independent quantum circuits, and use qubit teleportation back and forth to perform computations on each other's qubits. For this, they would use a classical channel to exchange cbits, not only to teleport their qubit data successfully, but also to instruct each other of the computations they want to see the other party performing. For instance, if Alice wants to remotely execute a CNOT operation onto Bob's qubit, Bob needs first to send Alice his qubit, and Alice to send the qubit resulting from her CNOT gate back to Bob. The single operation of this "remote" CNOT computation, thus, consumes two EPR–Bell states (or ebits), and four classical bits (or cbits).

Would it be possible to avoid such a complicated procedure, and instead conceive of quantum gates being capable of acting directly on *remote qubits*? The answer is yes, and the corresponding technique is called *quantum gate teleportation*. Within the scope of this chapter, we may not venture into the complex details of quantum gate teleportation, but I provide a generic example to show that it is possible, and also to show how this

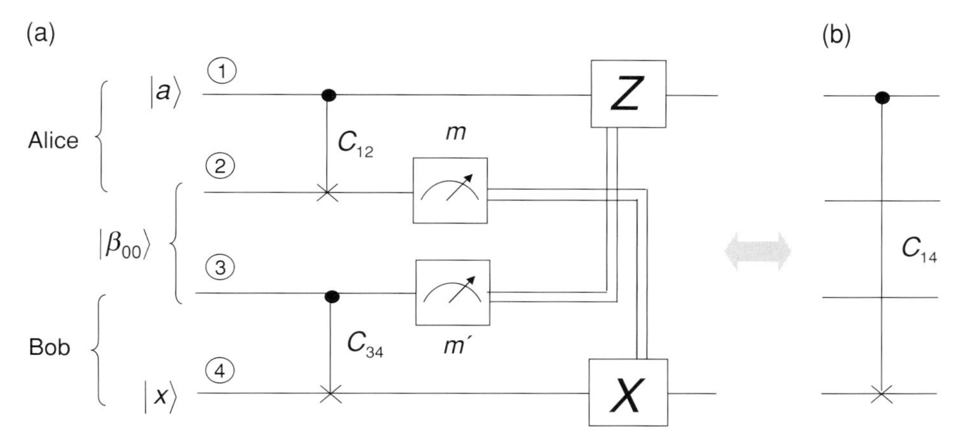


**Figure 18.12** (a) 4-qubit circuit used for quantum teleportation of CNOT gate; (b) equivalent distributed CNOT gate.

can work. Here, I shall describe the quantum teleportation of a CNOT gate,[3] which as we have learnt from Chapter 16, is one of the universal building blocks for quantum circuits. The principle of CNOT-gate teleportation is illustrated in Fig. 18.12. Locally, Alice and Bob have the control of two qubits each, which are labelled 1 and 2 for Alice (the first being called $|a\rangle$), and 3 and 4 for Bob (the last being called $|x\rangle$). As the figure indicates, the qubits 2 and 3 are entangled through the EPR–Bell state $|\beta_{00}\rangle$, thus, qubits 3 and 4 represent a *shared ebit*. Both Alice and Bob have local CNOT gates, called $C_{12}$ and $C_{34}$, respectively. Furthermore,

- Alice *first* measures qubit 3 in the base $\{|0\rangle, |1\rangle\}$, yielding a cbit $m$ (namely, $m = 0$ if the qubit is $|0\rangle$ and $m' = 1$ if the qubit is $|1\rangle$);
- Bob *then* measures qubit 3 in the different base $\{|+\rangle, |-\rangle\}$, yielding a cbit $m'$ (namely, $m' = 0$ if the qubit is $|+\rangle$ and $m' = 1$ if the qubit is $|-\rangle$).

Alice and Bob then use a classical channel to communicate to each other their respective cbit measurement. To ensure the proper sequence order, Bob won't proceed with his own measurement until he has received cbit $m$ from Alice. The cbit data $m, m'$ are then used locally by Alice and Bob to control their $Z$ or $X$ gates placed on the 1 or 4 qubit paths, respectively, as indicated in Fig. 18.12.

Now I shall explain why the whole circuit shown in Fig. 18.12 is equivalent to a distributed CNOT or $C_{14}$ gate, which uses Alice's qubit 1, or $|a\rangle$, as the control, and Bob's qubit 4, or $|x\rangle$, as the target. Call $|\psi_{ij}\rangle$ the initial qubit pairs, labeled by $i, j$. Hence, $\psi_{14} = |a\rangle \otimes |x\rangle \equiv |ax\rangle$ (tensor state) and $\psi_{23} \equiv |\beta_{00}\rangle$ (entangled state). Call $|0+\rangle_{23}, |0-\rangle_{23}, |1+\rangle_{23}, |1-\rangle_{23}$, or $|0+\rangle, |0-\rangle, |1+\rangle, |1-\rangle$, for simplicity, the post-measurement states generated by Alice and Bob's measurements on the qubit pair 2 and 3. The demonstration of the circuit equivalence into a $C_{14}$ gate holds into the ad-hoc

[3] See: Y.-F. Huang, X.-F. Ren, Y.-S. Zhang, L.-M. Duan, and G.-C. Guo, Experimental teleportation of a quantum controlled-NOT gate. *Phys. Rev. Letts.*, **93** (2004), id. 240501; see: http://arxiv.org/abs/quant-ph/0408007.

formula:

$$C_{34}C_{12}(|\psi_{14}\rangle \otimes |\beta_{00}\rangle) = \left\{ \begin{array}{l} |0+\rangle \otimes Z^0X^0C_{14}|\psi_{14}\rangle \\ +|0-\rangle \otimes Z^1X^0C_{14}|\psi_{14}\rangle \\ +|1+\rangle \otimes Z^0X^1C_{14}|\psi_{14}\rangle \\ -|1-\rangle \otimes Z^1X^1C_{14}|\psi_{14}\rangle \end{array} \right\}. \quad (18.28)$$

The proof of this equation is left as a tedious, yet relatively easy exercise. From the background of quantum teleportation we have gained in the previous section, and taking the result for granted, the interpretation is, however, straightforward. Indeed, assume for instance that Alice and Bob's measurements result in the 2–3 state collapse into $|0+\rangle$, yielding the cbits $0, 0$. The result indicates that the 4-qubit output is $|0+\rangle \otimes C_{14}|\psi_{14}\rangle$ and, in particular, that the qubit pair 1 and 4 is $C_{14}|\psi_{14}\rangle \equiv CNOT|ax\rangle \equiv |a, a \oplus x\rangle$. Alice does not know Bob's target qubit $|x\rangle$, and Bob does not know Alice's control qubit $|a\rangle$, but both know for certain that the CNOT operation has been duly executed. Likewise, in the case of a $|0-\rangle$ collapse, it is found from the formal proof that the output is $X_4C_{14}|\psi_{14}\rangle$, with $X_4$ being carried upon the qubit 4, hence the need to apply $X$ again on this qubit (recalling that $X^2 = I$), to transform the result into $C_{14}|\psi_{14}\rangle$. The same reasoning applies to the other two collapse cases $|1+\rangle$ and $|1-\rangle$.

In summary, the above description exemplifies that it is possible to achieve the quantum teleportation of a CNOT gate, with the only resource of one shared ebit and two cbits. Various teleportation schemes exist for other gate types, for instance controlled-$Z$ (also called *CSIGN* for "controlled sign"), which involve somewhat more gate-intensive quantum circuits and ebit resources.[4]

## 18.7 Exercises

**18.1** (B): Show that there exists no tensor product of single qubits able to generate any of the Bell–EPR states:

$$|\beta_{00}\rangle \equiv \frac{1}{\sqrt{2}}(|00\rangle + |11\rangle),$$

$$|\beta_{01}\rangle \equiv \frac{1}{\sqrt{2}}(|01\rangle + |10\rangle),$$

$$|\beta_{10}\rangle \equiv \frac{1}{\sqrt{2}}(|00\rangle - |11\rangle),$$

$$|\beta_{11}\rangle \equiv \frac{1}{\sqrt{2}}(|01\rangle - |10\rangle).$$

**18.2** (M): Show that quantum teleportation is possible with EPR–Bell states $|\beta_{01}\rangle, |\beta_{10}\rangle, |\beta_{11}\rangle$.

[4] See, for instance: www.iqc.ca/activities/projects/msilva01.php.

**18.3** (B): Show that a Bell measurement of any of the Bell states $|\beta_{mm'}\rangle$ outputs the two classical bits $m, m'$.

**18.4** (B): Given the EPR–Bell states $|\beta_{00}\rangle, |\beta_{01}\rangle, |\beta_{10}\rangle, |\beta_{11}\rangle$ and the Pauli operators $X, Y, Z$ acting on the first qubit, show the following identities used for superdense coding:

$$\begin{aligned} X|\beta_{00}\rangle &= |\beta_{01}\rangle \\ iY|\beta_{00}\rangle &= |\beta_{11}\rangle \\ Z|\beta_{00}\rangle &= |\beta_{10}\rangle. \end{aligned}$$

**18.5** (T): Given the 4-qubit circuit

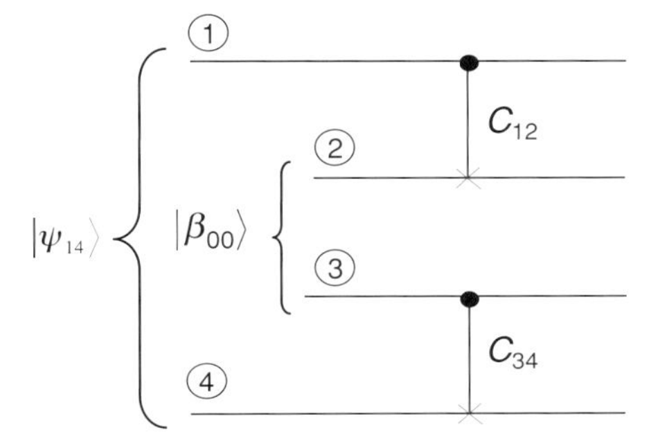


with $|\psi_{ij}\rangle$ as the input qubit pairs labeled with $|\psi_{23}\rangle \equiv |\beta_{00}\rangle$ (EPR–Bell state), show that the following equality holds:

$$C_{34}C_{12}(|\psi_{14}\rangle \otimes |\beta_{00}\rangle) = \frac{1}{2}\left\{\begin{array}{l} |0+\rangle \otimes Z^0 X^0 C_{14}|\psi_{14}\rangle \\ +|0-\rangle \otimes Z^1 X^0 C_{14}|\psi_{14}\rangle \\ +|1+\rangle \otimes Z^0 X^1 C_{14}|\psi_{14}\rangle \\ -|1-\rangle \otimes Z^1 X^1 C_{14}|\psi_{14}\rangle \end{array}\right\},$$

where

(i) $|0+\rangle, |0-\rangle, 1+\rangle, |1-\rangle$ are tensor products of the $|0\rangle, |1\rangle$ states with the $|+\rangle, |-\rangle$ states $(|\pm\rangle = (|0\rangle \pm |1\rangle)/\sqrt{2})$;

(ii) $X, Z$ are the usual Pauli matrices, used here to the powers 0 or 1, and applying on qubit 1 for $Z$ and qubit 4 for $X$;

(iii) $C_{ij}$ is the CNOT gate for the 2-qubit $|\psi_{ij}\rangle$, with $i$ as the control qubit and $j$ as the target qubit.

# 19 Deutsch–Jozsa, quantum Fourier transform, and Grover quantum database search algorithms

This mathematically intensive chapter takes us through our first steps in the domain of *quantum computation (QC) algorithms*. The simplest of them is the *Deutsch algorithm*, which makes it possible to determine whether or not a Boolean function is constant for any input. The key result is that this QC algorithm provides the answer at once, whereas in the classical case it would take two independent calculations. I describe next the generalization of the former algorithm to $n$ qubits, referred to as the *Deutsch–Jozsa algorithm*. Although they have no specific or useful applications in quantum computing, both algorithms represent a most elegant means of introducing the concept of *quantum computation parallelism*. I then describe two most important QC algorithms, which nicely exploit quantum parallelism. The first is the *quantum Fourier transform* (QFT), for which a detailed analysis of QFT circuits and quantum-gate requirements is also provided. As will be shown in the next chapter, a key application of QFT concerns the famous *Shor's algorithm*, which makes it possible to factor numbers into primes in terms of polynomials. The second algorithm, no less famous than Shor's, is referred to as the *Grover quantum database search*, whose application is the identification of database items with a quadratic gain in speed.

## 19.1 Deutsch algorithm

Our exploration of *quantum algorithms* shall begin with the solution of a very basic problem: finding whether or not a Boolean function $f(x)$ is a constant. The solution is given by the *Deutsch algorithm*, which illustrates a fundamental property of quantum computing, namely *parallelism*. As we shall indeed see, quantum computation makes it possible to evaluate $f(x)$ simultaneously for all values of the variable $x$, in contrast to classical computing (i.e., *von Neumann* architecture, Chapter 15) where such an evaluation must be performed one variable at a time, or through as many computers working in parallel.

Assume then a Boolean function $f(x)$ where $x$ is a binary variable. We can conceive of a 2-qubit quantum circuit based on a unitary operator $U_f$ that achieves the transformation

$$U_f|x, y\rangle = |x, y \oplus f(x)\rangle, \tag{19.1}$$

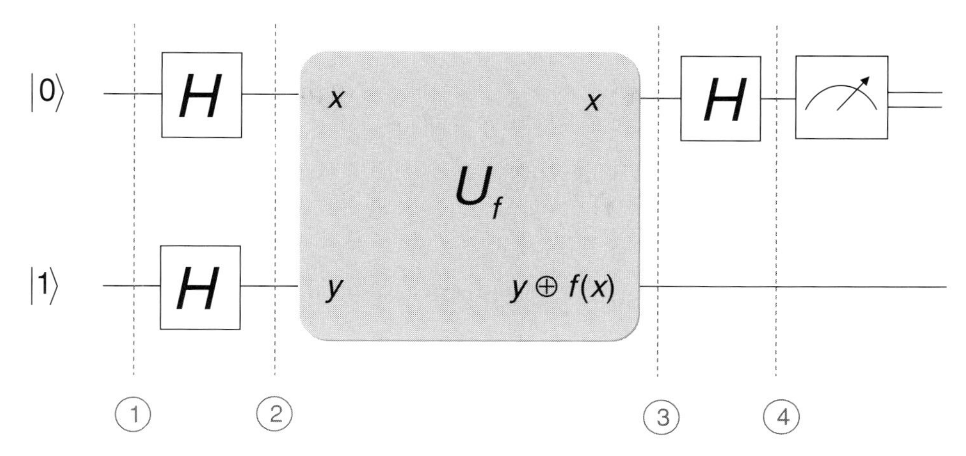


**Figure 19.1** Quantum circuit for implementing the Deutsch algorithm.

where the simplified notation $|u, v\rangle = |u\rangle \otimes |v\rangle$ stands for tensor product. Consider next the quantum circuit arrangement shown in Fig. 19.1, where $H$ are Hadamard gates and $|0\rangle, |1\rangle$ are two ancilla qubits. We call $x, y$ the inputs resulting from the Hadamard transforms on the two ancilla qubits, and note that the action of the "black box" gate $U_f$ on $x, y$ results in the outputs $x, y \oplus f(x)$. Let us now analyze the evolution of the tensor state, from left to right, through the locations marked ①②③④ in the figure, starting from $|\psi_1\rangle = |01\rangle = |0\rangle \otimes |1\rangle$ at ①:

$$
② : \quad \begin{aligned} |\psi_2\rangle &= H_1 H_2 |\psi_1\rangle \\ &= H_1|0\rangle \otimes H_2|1\rangle \\ &= |+\rangle \otimes |-\rangle \\ &= \frac{1}{2}\frac{|0\rangle + |1\rangle}{\sqrt{2}} \otimes \frac{|0\rangle - |1\rangle}{\sqrt{2}} \\ &= \frac{1}{4}(|00\rangle - |01\rangle + |10\rangle - |11\rangle) \end{aligned} \tag{19.2}
$$

(note the factor ½ introduced in the right-hand side, to ensure normalization of the tensor product of mixed states). Then, using the definition in Eq. (19.1):

$$
③ : \quad \begin{aligned} U_f |\psi_2\rangle &= \frac{1}{4} U_f (|00\rangle - |01\rangle + |10\rangle - |11\rangle) \\ &\equiv \frac{1}{4}(|0, 0 \oplus f(0)\rangle - |0, 1 \oplus f(0)\rangle + |1, 0 \oplus f(1)\rangle - |1, 1 \oplus f(1)\rangle). \end{aligned} \tag{19.3}
$$

Then, ④:

$$
\begin{aligned} |\psi_3\rangle &= H_1 |\psi_2\rangle \\ &= \frac{1}{4} H_1 (|0, 0 \oplus f(0)\rangle - |0, 1 \oplus f(0)\rangle + |1, 0 \oplus f(1)\rangle - |1, 1 \oplus f(1)\rangle) \\ &= \frac{1}{4}(|+, 0 \oplus f(0)\rangle - |+, 1 \oplus f(0)\rangle + |-, 0 \oplus f(1)\rangle - |-, 1 \oplus f(1)\rangle) \end{aligned}
$$

$$= \frac{1}{4\sqrt{2}}(|0, 0 \oplus f(0)\rangle + |1, 0 \oplus f(0)\rangle - |0, 1 \oplus f(0)\rangle - |1, 1 \oplus f(0)\rangle$$

$$+ |0, 0 \oplus f(1)\rangle - |1, 0 \oplus f(1)\rangle - |0, 1 \oplus f(1)\rangle + |1, 1 \oplus f(1)\rangle)$$

$$= \frac{1}{4\sqrt{2}}\{|0\rangle(|0 \oplus f(0)\rangle + |0 \oplus f(1)\rangle - |1 \oplus f(0)\rangle - |1 \oplus f(1)\rangle) \tag{19.4}$$

$$+ |1\rangle(|0 \oplus f(0)\rangle - |0 \oplus f(1)\rangle - |1 \oplus f(0)\rangle + |1 \oplus f(1)\rangle)\}$$

$$\equiv \frac{1}{2\sqrt{2}}\left\{|0\rangle\left(\frac{|f(0)\rangle + |f(1)\rangle}{2} - \frac{|1 \oplus f(0)\rangle + |1 \oplus f(1)\rangle}{2}\right)\right.$$

$$\left. + |1\rangle\left(\frac{|f(0)\rangle - |f(1)\rangle}{2} - \frac{|1 \oplus f(0)\rangle - |1 \oplus f(1)\rangle}{2}\right)\right\}.$$

Next consider the two cases (a) $a = f(0) = f(1)$, and (b) $a = f(0) \neq f(1) = \bar{a}$. From Eq. (19.4), we obtain:

$$|\psi_3\rangle_{f(0)=f(1)}$$

$$= \frac{1}{2\sqrt{2}}\left\{|0\rangle\left(\frac{|a\rangle + |a\rangle}{2} - \frac{|1 \oplus a\rangle + |1 \oplus a\rangle}{2}\right) + |1\rangle\left(\frac{|a\rangle - |a\rangle}{2} - \frac{|\bar{a}\rangle - |\bar{a}\rangle}{2}\right)\right\}$$

$$= \frac{|0\rangle(|a\rangle - |\bar{a}\rangle)}{2\sqrt{2}}$$

$$|\psi_3\rangle_{f(0)\neq f(1)}$$

$$= \frac{1}{2\sqrt{2}}\left\{|0\rangle\left(\frac{|a\rangle + |\bar{a}\rangle}{2} - \frac{|\bar{a}\rangle + |a\rangle}{2}\right) + |1\rangle\left(\frac{|a\rangle - |\bar{a}\rangle}{2} - \frac{|\bar{a}\rangle - |a\rangle}{2}\right)\right\}$$

$$= \frac{|1\rangle(|a\rangle - |\bar{a}\rangle)}{2\sqrt{2}}. \tag{19.5}$$

Finally, we note that $f(0) \oplus f(1) = 0$ if $f(0) = f(1)$ and $f(0) \oplus f(1) = 1$ if $f(0) \neq f(1)$. Using this property, and the definition of a Boolean variable $|a\rangle - |\bar{a}\rangle = \pm(|0\rangle - |1\rangle)$, we obtain from Eq. (19.5) within an unobservable phase $e^{i\pi} = -1$:

$$|\psi_3\rangle \equiv |f(0) \oplus f(1)\rangle \frac{|0\rangle - |1\rangle}{\sqrt{2}}. \tag{19.6}$$

The result shows that measuring the first qubit, namely, $|f(0) \oplus f(1)\rangle$, provides the answer "at once" to Deutsch's problem. Indeed, if we measure $|f(0) \oplus f(1)\rangle = |0\rangle$, we conclude that $f(0) = f(1)$ and, therefore, that $f(x)$ is a constant, and if we measure $|f(0) \oplus f(1)\rangle = |1\rangle$, we conclude the opposite. Thus, Deutsch's algorithm makes it possible to solve the problem without having to compute $f(0)$, $f(1)$ separately, unlike in the classical case. This simple problem magnifies the property of *quantum parallelism*, which is the key to quantum computing.

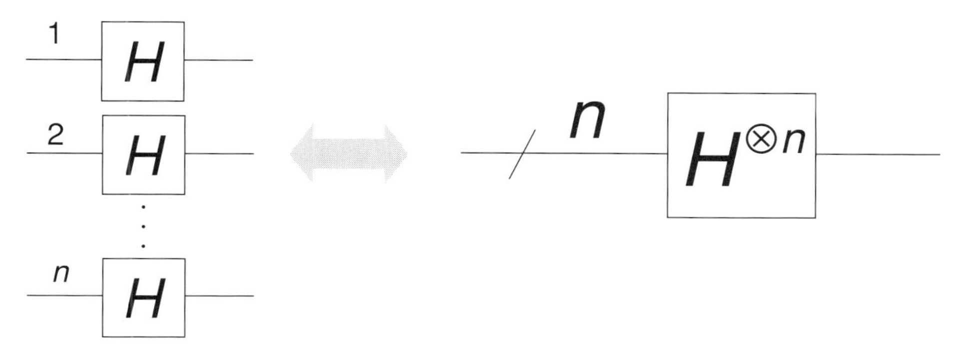


**Figure 19.2** Equivalent representation for $n$-qubit wires, with $H$ gates deployed in parallel.

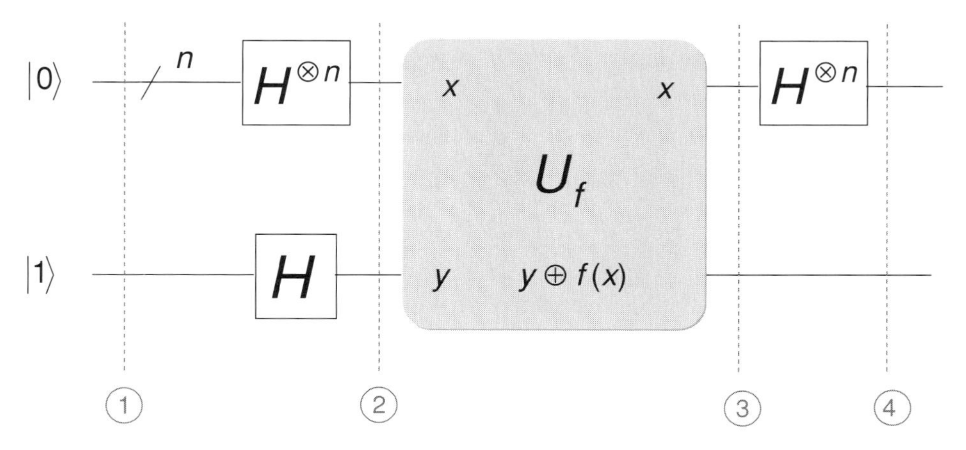


**Figure 19.3** Quantum circuit for implementing the Deutsch–Jozsa algorithm.

## 19.2 Deutsch–Jozsa algorithm

In this section, we consider the generalization of Deutsch's algorithm, which is referred to as the Deutsch–Jozsa algorithm. Now $f(x)$ is a function of any integer $x \in \{0, 1, 2 \dots 2^{n-1}\}$, which outputs a Boolean $f(x) \in \{0, 1\}$. What is known as *Deutsch's problem* is to find a way to tell, for all possible values of $x$, whether $f(x)$ is constant, or $f(x)$ is *balanced*, meaning in this last case that the output is 0 or 1 for exactly one half of the possible inputs. The answer to Deutsch's problem takes the form of a quantum algorithm, whose circuit is similar to the previous one. The difference is that $x$ is not a single Boolean but a binary number, which we shall distribute over as many qubits. As a new convention for quantum circuits, we shall represent $n$-qubit wires with $H$ gates deployed in parallel, according to Fig. 19.2. The Deutsch–Jozsa circuit is represented in Fig. 19.3 (recalling from Chapter 17 the notation $A^{\otimes n}$ for $n$-tensored operators $A$). The circuit is seen to have $n$ ancilla qubits $|0\rangle$ and one ancilla qubit $|1\rangle$ as inputs, which pass together through $n + 1$ parallel Hadamard gates. The input tensor state at ① is $|\psi_1\rangle = (|0\rangle \otimes |0\rangle \dots \otimes |0\rangle)^{n\ times} \otimes |1\rangle \equiv |0\rangle^{\otimes n} \otimes |1\rangle$. At ② in Fig. 19.3, we obtain[1]

$$\begin{aligned}|\psi_2\rangle &= H^{\otimes n} H_2 |\psi_1\rangle \\ &= H^{\otimes n}|0\rangle^{\otimes n} \otimes H|1\rangle \\ &= |+\rangle^{\otimes n}\,|-\rangle \\ &= \frac{1}{\sqrt{2^n}}\left(|0\rangle + |1\rangle\right)^{\otimes n}\,\frac{|0\rangle - |1\rangle}{\sqrt{2}} \\ &= \frac{1}{\sqrt{2^n}}\left[\sum_{x_i\in\{0,1\}} |x_1 x_2 \ldots x_n\rangle\right]\frac{|0\rangle - |1\rangle}{\sqrt{2}} \\ &= \frac{1}{\sqrt{2^n}}\left[\sum_{x=0}^{2^n-1} |x\rangle\right]\frac{|0\rangle - |1\rangle}{\sqrt{2}}.\end{aligned} \tag{19.7}$$

We call $|x\rangle$ the *query register*, similarly to the "register" in the classical von Neumann architecture (Chapter 15) the difference being that it is made of *qubits*. At ③, we obtain[2]

$$\begin{aligned}|\psi_3\rangle &= U_f\,|\psi_2\rangle \\ &= \sum_x \frac{|x\rangle}{\sqrt{2^n}}\,\frac{|0\oplus f(x)\rangle - |1\oplus f(x)\rangle}{\sqrt{2}} \\ &\equiv \sum_x \frac{(-1)^{f(x)}|x\rangle}{\sqrt{2^n}}\,|-\rangle\,.\end{aligned} \tag{19.8}$$

And at ④, after passing the top $n$-qubit through the parallel gate $H^{\otimes n}$, we obtain:

$$\begin{aligned}|\psi_4\rangle &= H^{\otimes n}\,|\psi_3\rangle \\ &= \frac{1}{\sqrt{2^n}}\left[\sum_x (-1)^{f(x)} H^{\otimes n}|x\rangle\right]|-\rangle\,.\end{aligned} \tag{19.9}$$

To develop the right-hand side in Eq. (19.9), we must calculate $H^{\otimes n}|x\rangle = H^{\otimes n}\,|x_1 x_2 \ldots x_n\rangle$. It is an easy exercise to establish that:

$$H^{\otimes n}|x\rangle = \sum_z (-1)^{x\cdot z}\,|z\rangle, \tag{19.10}$$

where $x\cdot z = x_1 z_1 + x_2 z_2 + \cdots + x_n z_n$ is a scalar product modulo 2. Combining Eqs. (19.9) and (19.10), we then obtain:

$$\begin{aligned}|\psi_4\rangle &= \frac{1}{2^n}\left[\sum_z \sum_x (-1)^{f(x)+x\cdot z}\,|z\rangle\right]|-\rangle \\ &\equiv |\Psi\rangle\,|-\rangle\,,\end{aligned} \tag{19.11}$$

[1] In the last two equations, I have introduced the equivalent $n$-qubit notations $\sum_x |x\rangle = \sum_{x_i} |x_1 x_2 \ldots x_n\rangle = (|0\rangle + |1\rangle)^{\otimes n}$ with $x_i = 0, 1$ and $x = 0, 1, 2 \ldots 2^{n-1}$. The correspondence is easily checked with $n = 2$ and then by induction.

[2] The last equality in Eq. (19.8) is straightforward: given $x$, if $f(x) = 0$ then $|0\oplus f(x)\rangle - |1\oplus f(x)\rangle = |0\rangle - |1\rangle \equiv (-1)^0\,(|0\rangle - |1\rangle)$; if $f(x) = 1$ then $|0\oplus f(x)\rangle - |1\oplus f(x)\rangle = 1 - |0\rangle \equiv (-1)^1\,(|0\rangle - |1\rangle)$; hence, $|0\oplus f(x)\rangle - |1\oplus f(x)\rangle \equiv (-1)^{f(x)}\,(|0\rangle - |1\rangle) \equiv (-1)^{f(x)}\,|-\rangle$ in the general case.

with

$$|\Psi\rangle = \frac{1}{2^n}\sum_z u_{xz}\,|z\rangle, \tag{19.12}$$

$$u_{xz} = \sum_x (-1)^{f(x)+x\cdot z}. \tag{19.13}$$

The $n$-qubit $|\Psi\rangle$ defined in Eqs. (19.11) and (19.12) represents the output query register. As we shall see, the corresponding amplitudes $u_{xz}$ contain all the information needed to answer *Deutsch's problem*. To show this, consider the two cases of interest: $f(x)$ is constant, or $f(x)$ is balanced. If $f(x) = a$ ($a = \pm 1$ being a constant), all the amplitudes $u_{xz}$ of $|z\rangle = |0\rangle^{\otimes n}$ are equal to $u_{x0} = (-1)^{f(x)} = (-1)^a = \pm 1$. Since $|\Psi\rangle$ has a length of unity, this means that all other amplitudes $u_{x,z\neq 0}$ must be zero. The query registers thus displays $|\Psi\rangle = |0\rangle^{\otimes n}$, i.e., all output qubits are equal to $|0\rangle$. A single measurement by projecting the register over $|0\rangle^{\otimes n}$, yielding $\langle\Psi|\Psi\rangle^{\otimes n} = 1$, suffices to prove the point. Consider next the second case: if $f(x)$ is balanced, this means that the amplitude $u_{x0}$ of $|z\rangle = |0\rangle^{\otimes n}$ is zero, as the terms $(-1)^{f(x)+x\cdot 0} = (-1)^{f(x)}$ in the sum in Eq. (19.3) cancel each other (one half yielding $+1$, the other half yielding $-1$). For any $x$ we thus have the property $\sum_x (-1)^{f(x)} = 0$, indicating that $f(x)$ is, indeed, balanced. In this case, the query register cannot output $|\Psi\rangle = |0\rangle^{\otimes n}$, meaning that at least one of the register qubits must be $|1\rangle$. A single projection test yielding $\langle\Psi \mid 0\rangle^{\otimes n} = 0$ suffices to prove the point.

In summary, the Deutsch–Jozsa algorithm demonstrates the capability of *quantum parallelism* over an arbitrary number of qubits $n$. It takes a *single calculation* and a *single measurement* of the output register $|\Psi\rangle$ to obtain the answer to Deutsch's problem. In contrast, a classical computation would require at least "$2^n/2$ plus one" measurements with random inputs to obtain the same answer within reasonable confidence, and $2^n$ measurements for absolute confidence.

While this discussion makes a point about the power of *quantum parallelism*, it should not be concluded that evaluating the properties of the function $f(x)$ is of any particular interest to quantum computing algorithms. Rather, the Deutsch and Deutsch–Jozsa algorithms must be regarded as representing a most elegant introduction to the concept.

## 19.3 Quantum Fourier transform algorithm

In this section, I shall describe the *quantum Fourier transform (QFT)*, which lies at the root of several important quantum-computing algorithms, for some revolutionary applications to be developed in Chapter 20. To avoid any misconception, QFT was not developed with the purpose of boosting the speed of Fourier transforms, as they can be implemented with classical bits in a von Neumann computer through the so-called *fast Fourier transform* (FFT) algorithm. Rather, QFT opens up some new perspectives in quantum computation, and we may realize this only through the next chapter. Here,

I shall outline the formal concept and show how QFT can be practically implemented through relatively simple circuits based on $2 \times 2$ and controlled $2 \times 2$ quantum-gates.

Let us begin by recalling the definition of the Fourier transform, as it applies to a discrete function or $N$-vector. In elementary calculus, and electrical and telecom engineering, the *discrete Fourier transform* (DFT) of an $N$-vector $x = (x_0, x_1 \ldots x_{N-1})$ with complex coefficients $x_k$ is a well known operation. It results in the generation of $N$ *Fourier components*, $y_n$ ($k = 0, 1 \ldots N-1$), as defined by the linear expansion:[3]

$$y_n = \frac{1}{\sqrt{N}} \sum_{k=0}^{N-1} x_k e^{ik\frac{2n\pi}{N}}. \tag{19.14}$$

It is seen that for each index $k$ the terms contributing to the series expansion in Eq. (19.14) involve powers of the discrete frequency (or tone) $f_n = 2n\pi/N$, which range from $f_0 = 0$ to $f_{N-1} = 2\pi(N-1)/N$. Thus, $y_n$ is the complex amplitude of the Fourier component at frequency $f_n$, the ensemble of which forms the frequency spectrum of $x$. Similarly, the *inverse discrete Fourier transform* (IDFT) is defined as:

$$x_k = \frac{1}{\sqrt{N}} \sum_{k=0}^{N-1} y_n e^{-in\frac{2k\pi}{N}}. \tag{19.15}$$

From this background, we can now introduce the *quantum Fourier transform* (QFT) of an orthonormal basis $\{|k\rangle\}_{k=0\ldots N-1} \equiv \{|0\rangle, |1\rangle \ldots |N-1\rangle\}$. Such a transform is enabled by a *unitary operator*, which we call QFT, and which acts on the $N$ basis states $|k\rangle$ in a way exactly similar to that in Eq. (19.14):

$$\text{QFT}\,|n\rangle = \frac{1}{\sqrt{N}} \sum_{k=0}^{N-1} e^{ik\frac{2n\pi}{N}} |k\rangle. \tag{19.16}$$

It is left as a nontrivial, but elementary exercise to show that the QFT transformation is, indeed, unitary. Next, let $|\psi\rangle$ be a qubit of dimension $N$ with coordinates $x_k$ in the basis $\{|k\rangle\}$, i.e.,

$$|\psi\rangle = \sum_{k=0}^{N-1} x_k |k\rangle. \tag{19.17}$$

The *quantum Fourier transform of the qubit* $|\psi\rangle$, which we note $|\tilde{\psi}\rangle$, is defined as:

$$\begin{aligned} |\tilde{\psi}\rangle &= \text{QFT}\,|\psi\rangle \\ &\equiv \frac{1}{\sqrt{N}} \sum_{n=0}^{N-1} \sum_{k=0}^{N-1} x_k\, e^{ik\frac{2n\pi}{N}} |k\rangle. \end{aligned} \tag{19.18}$$

This result establishes the formal correspondence between $|\tilde{\psi}\rangle$ and $|\psi\rangle$. The *inverse* QFT, noted $\text{QFT}^+$, is similarly defined through Eq. (19.15), which gives:

$$\begin{aligned} |\psi\rangle &= \text{QFT}^+ |\tilde{\psi}\rangle \\ &\equiv \frac{1}{\sqrt{N}} \sum_{k=0}^{N-1} \sum_{n=0}^{N-1} y_n e^{-in\frac{2k\pi}{N}} |n\rangle. \end{aligned} \tag{19.19}$$

[3] See, for instance, http://en.wikipedia.org/wiki/Discrete_Fourier_transform.

As a matter of fact, the QFT is far simpler than the above formulas make it appear. Let us show this by considering the elementary cases $N = 2$ and $N = 3$. Applying the QFT definition for the basis transformation, we obtain in the first case:

$$\begin{aligned} |y_n\rangle &= \frac{1}{\sqrt{2}} \sum_{k=0}^{1} \mathrm{e}^{\mathrm{i}k\frac{2n\pi}{2}} |x_k\rangle \\ &= \frac{1}{\sqrt{2}} \left(|0\rangle + \mathrm{e}^{\mathrm{i}k\pi} |1\rangle\right), \end{aligned} \tag{19.20}$$

hence:

$$\begin{cases} |y_0\rangle = \dfrac{1}{\sqrt{2}} (|0\rangle + |1\rangle) \equiv |+\rangle \\ |y_1\rangle = \dfrac{1}{\sqrt{2}} (|0\rangle + \mathrm{e}^{\mathrm{i}\pi} |1\rangle) = \dfrac{1}{\sqrt{2}} (|0\rangle - |1\rangle) \equiv |-\rangle . \end{cases} \tag{19.21}$$

This result indicates that in the case $N = 2$, the QFT reduces to the *Hadamard* transformation with quantum matrix gate $H$. In the case $N = 3$, we obtain:

$$\begin{aligned} |y_n\rangle &= \frac{1}{\sqrt{3}} \sum_{k=0}^{2} \mathrm{e}^{\mathrm{i}k\frac{2n\pi}{3}} |x_k\rangle \\ &= \frac{1}{\sqrt{3}} \left(|0\rangle + \mathrm{e}^{\mathrm{i}\frac{2n\pi}{3}} |1\rangle + \mathrm{e}^{\mathrm{i}\frac{4n\pi}{3}} |2\rangle\right), \end{aligned} \tag{19.22}$$

hence:

$$\begin{cases} |y_0\rangle = \dfrac{1}{\sqrt{3}} (|0\rangle + |1\rangle + |2\rangle) \\ |y_1\rangle = \dfrac{1}{\sqrt{3}} \left(|0\rangle + \mathrm{e}^{\mathrm{i}\frac{2\pi}{3}} |1\rangle + \mathrm{e}^{\mathrm{i}\frac{4\pi}{3}} |2\rangle\right) \\ |y_2\rangle = \dfrac{1}{\sqrt{3}} \left(|0\rangle + \mathrm{e}^{\mathrm{i}\frac{4\pi}{3}} |1\rangle + \mathrm{e}^{\mathrm{i}\frac{8\pi}{3}} |2\rangle\right). \end{cases} \tag{19.23}$$

We may represent the above results through a *symmetric* matrix $M$ with coefficients $M_{nk} = M_{kn} = \exp(\mathrm{i}2nk\pi/N)/\sqrt{N} \equiv \omega^{nk}/\sqrt{N}$, where $\omega = \exp(2\mathrm{i}\pi/N)$ and $n, k = 0 \ldots N-1$. The $N = 3$ matrix, thus, takes the form:

$$M = \frac{1}{\sqrt{3}} \begin{pmatrix} 1 & 1 & 1 \\ 1 & \omega & \omega^2 \\ 1 & \omega^2 & \omega^4 \end{pmatrix}. \tag{19.24}$$

It is clear that, in general, the matrix $M$ takes the following form, referred to as the *Vandermonde matrix*:

$$M = \frac{1}{\sqrt{N}} \begin{pmatrix} 1 & 1 & 1 & \cdots & 1 \\ 1 & \omega & \omega^2 & \cdots & \omega^{N-1} \\ 1 & \omega^2 & \omega^4 & \cdots & \omega^{2(N-1)} \\ 1 & \omega^3 & \omega^6 & \cdots & \omega^{3(N-1)} \\ \vdots & \vdots & \vdots & \ddots & \vdots \\ 1 & \omega^{N-1} & \omega^{2(N-1)} & \cdots & \omega^{(N-1)(N-1)} \end{pmatrix}. \tag{19.25}$$

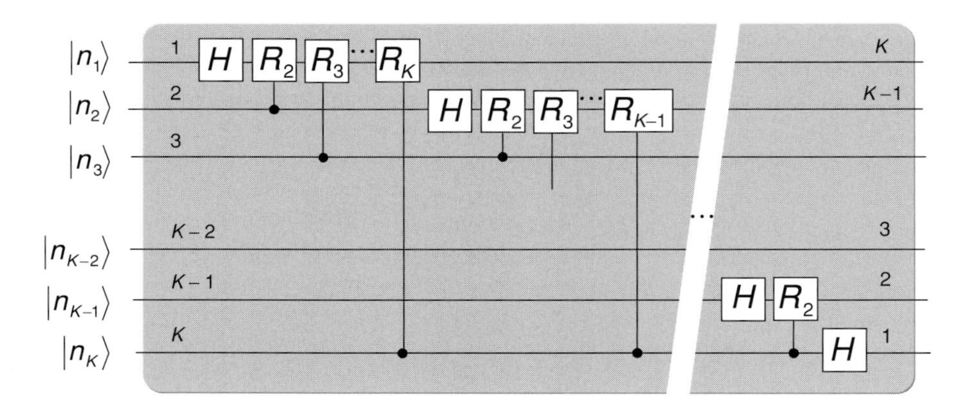


**Figure 19.4** Quantum circuit for QFT implementation.

Substituting the Vandermonde matrix into the definitions in Eqs. (19.18) and (19.19) we also obtain the general definition of the QFT, $\mathrm{QFT}^+$ qubit transformations:

$$\begin{cases} |\tilde{\psi}\rangle = \mathrm{QFT}\,|\psi\rangle = \displaystyle\sum_{k=0}^{N-1}\sum_{n=0}^{N-1} x_k M_{nk}|k\rangle \\ |\psi\rangle = \mathrm{QFT}^+|\tilde{\psi}\rangle = \displaystyle\sum_{n=0}^{N-1}\sum_{k=0}^{N-1} y_n \bar{M}_{kn}|n\rangle, \end{cases} \tag{19.26}$$

or, equivalently, for the relations between Fourier and inverse-Fourier components $y_n, x_k$:

$$\begin{cases} y_n = \langle n|\tilde{\psi}\rangle = \displaystyle\sum_{k=0}^{N-1} M_{nk}x_k \\ x_k = \langle k|\psi\rangle = \displaystyle\sum_{n=0}^{N-1} \bar{M}_{kn}y_n. \end{cases} \tag{19.27}$$

We have, thus, obtained the most general and very simple definition of the quantum Fourier operator through its corresponding matrix and its inverse, as applying to a single qubit of dimension $N$. We must now find an appropriate quantum circuit, taking $|\psi\rangle = (x_0, x_1, \ldots x_{N-1})$ as the input, and yielding $|\tilde{\psi}\rangle = (y_0, y_1, \ldots y_{N-1})$ as the output, which would also represent an $N$-qubit QFT gate. To this purpose, we want to avoid the complicated implementation of an $N \times N$ Vandermonde-matrix gate, and rather seek for a greatly simplified circuit based on $2 \times 2$ gates. As this search is a bit tedious and somewhat tricky in formal derivation, here I shall directly give the conclusion, while the complete derivation is detailed in Appendix Q. However, the result turns out to be relatively simple. The QFT circuit and its constituent $2 \times 2$ gates are represented in Fig. 19.4. As a first observation, there are only $K = \log_2 N$ qubit inputs, and as many qubit outputs, labeled in the reverse order. To explain the reduced circuit size, consider, for simplicity, the case $K = 3$ ($N = 8$). The relation between the $K = 3$ inputs $\{|n_1\rangle, |n_2\rangle, |n_3\rangle\}$ and the $N = 8$ basis states $\{|0\rangle, |1\rangle, |2\rangle, |3\rangle, |4\rangle, |5\rangle, |6\rangle, |7\rangle\}$ is given by the binary–tensor representation of the latter, namely, $\{|000\rangle, |001\rangle, |010\rangle, |011\rangle, |100\rangle, |101\rangle, |110\rangle, |111\rangle\}$. Thus, the QFT of the basis state $|6\rangle = |110\rangle$ is given by the circuit input $\{|0\rangle, |1\rangle, |1\rangle\}$, for instance (starting from the bit of lower weight). As it turns out, the order of the output qubits is reversed, and the full implementation of the QFT requires an additional series of SWAP gates. Such a SWAP operation can be performed by a cascade of three CNOT gates (see Fig. 16.3).

The second observation is that the circuit in Fig. 19.4 is exclusively made of unitary gates, namely *Hadamard* ($H$) and *controlled-phase* ($R_n$) gates. This establishes the fact that QFT is both a *unitary* and *reversible* transformation (using the circuit from left to right corresponding to the *inverse QFT*). As shown in Appendix Q, the controlled-phase gates are defined as follows:

$$R_n = \begin{pmatrix} 1 & 0 \\ 0 & e^{2i\frac{\pi}{2^n}} \end{pmatrix}, \tag{19.28}$$

noting that this $2 \times 2$ representation is a reduced one (from Chapter 16, we know that a controlled-$U$ gate has a $4 \times 4$ matrix). As detailed in Appendix Q, the output of the QFT circuit corresponds to the overall tensor product

$$\text{QFT}|n\rangle = \frac{1}{2^{K/2}}(|0\rangle_1 + e^{2i\pi\Omega_1}|1\rangle_1) \otimes (|0\rangle_2 + e^{2i\pi\Omega_2}|1\rangle_2) \otimes \ldots \otimes (|0\rangle_K + e^{2i\pi\Omega_K}|1\rangle_K), \tag{19.29}$$

where each output $|m\rangle = (|0\rangle_m + e^{2i\pi\Omega_m}|1\rangle_m)$ is characterized by the phase factor:

$$\Omega_m = \sum_{l=1}^{m} \frac{n_{K-m+l}}{2^l}. \tag{19.30}$$

I shall now provide two examples, to illustrate how the QFT circuit operates. Consider first the case $K = 2$ ($N = 4$), corresponding to a 2-qubit quantum circuit. Figure 19.5 shows the corresponding layout, consistent with that shown in Fig. 19.4. The SWAP gate to restore the order of qubits is also included for QFT circuit completion. According to Eq. (19.30), and after the swapping operation, the circuit output is:

$$\begin{aligned} \text{QFT}\,|n_1 n_2\rangle &= \frac{1}{2}(|0\rangle_2 + e^{2i\pi\Omega_2}|1\rangle_2) \otimes (|0\rangle_1 + e^{2i\pi\Omega_1}|1\rangle_1) \\ &= \frac{1}{2}(|00\rangle + e^{2i\pi\Omega_1}|01\rangle + e^{2i\pi\Omega_2}|10\rangle + e^{2i\pi(\Omega_1+\Omega_2)}|11\rangle)_{21} \\ &= \frac{1}{2}(|00\rangle + e^{2i\pi\Omega_1}\,|10\rangle + e^{2i\pi\Omega_2}\,|01\rangle + e^{2i\pi(\Omega_1+\Omega_2)}\,|11\rangle)_{12} \\ &= \frac{1}{2}\left(|00\rangle + e^{2i\pi\frac{n_2}{2}}\,|10\rangle + e^{2i\pi\left(\frac{n_1}{2}+\frac{n_2}{4}\right)}\,|01\rangle + e^{2i\pi\left(\frac{n_2}{2}+\frac{n_1}{2}+\frac{n_2}{4}\right)}\,|11\rangle\right)_{12} \\ &\equiv \frac{1}{2}\left(|00\rangle + \omega^{2n_2}\,|10\rangle + \omega^{2n_1+n_2}\,|01\rangle + \omega^{2n_1+3n_2}\,|11\rangle\right)_{12}, \end{aligned} \tag{19.31}$$

where I have introduced $\omega = \exp(2i\pi/2^K) \equiv \exp(i\pi/2)$. In the orthonormal basis $\{|n_1 n_2\rangle\} = \{|00\rangle, |01\rangle, |10\rangle, |11\rangle\}$ (taking note here of the binary ordering), we obtain

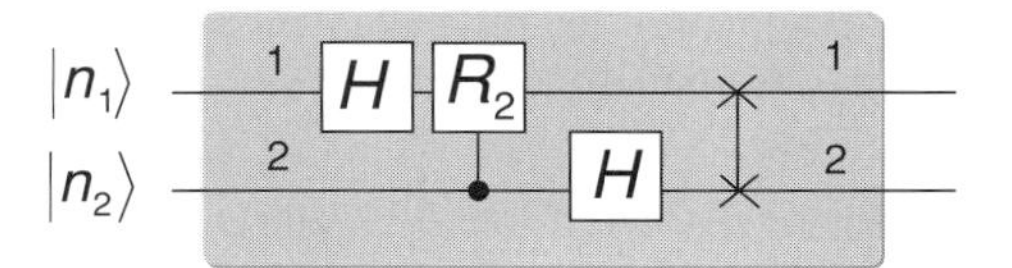


**Figure 19.5** QFT circuit for $K = 2$ ($N = 4$).

from Eq. (19.31) the following matrix representation of the QFT circuit:

$$
\begin{aligned}
M &= \frac{1}{2}\begin{pmatrix} 1 & 1 & 1 & 1 \\ 1 & i & -1 & -i \\ 1 & -1 & 1 & -1 \\ 1 & -i & -1 & i \end{pmatrix} \\
&\equiv \frac{1}{\sqrt{2^2}}\begin{pmatrix} 1 & 1 & 1 & 1 \\ 1 & \omega & \omega^2 & \omega^3 \\ 1 & \omega^2 & \omega^4 & \omega^6 \\ 1 & \omega^3 & \omega^6 & \omega^9 \end{pmatrix}.
\end{aligned} \tag{19.32}
$$

The right-hand side of Eq. (19.32) is recognized as the $4 \times 4$ *Vandermonde matrix*, which was defined in Eq. (19.25) in the general case, i.e., for all $K$, with $\omega = \exp(2i\pi/2^K)$. This result is, therefore, consistent with all QFT definitions previously derived in this section. An interesting application example is the quantum Fourier transform of the *EPR–Bell states*, $|\beta_{ij}\rangle$, which is left as a straightforward exercise to calculate, using the above matrix (there is no big point about the result, except for the conclusion that QFT maintains a state of entanglement, under a more complex state mixture).

The QFT circuit shown in Fig. 19.6 corresponds to the case $K = 3$ ($N = 8$), including the SWAP ordering operation. There is no point in expanding Eq. (19.29) again to define the 3-qubit circuit output explicitly. We simply need to substitute $\omega = \exp(2i\pi/2^3) \equiv \exp(i\pi/4)$ into the $8 \times 8$ Vandermonde matrix in Eq. (19.25), which, taking into account the $\omega$ periodicity yields:

$$
\begin{aligned}
M &= \frac{1}{\sqrt{2^3}}\begin{pmatrix}
1 & 1 & 1 & 1 & 1 & 1 & 1 & 1 \\
1 & \omega & \omega^2 & \omega^3 & \omega^4 & \omega^5 & \omega^6 & \omega^7 \\
1 & \omega^2 & \omega^4 & \omega^6 & \omega^8 & \omega^{10} & \omega^{12} & \omega^{14} \\
1 & \omega^3 & \omega^6 & \omega^9 & \omega^{12} & \omega^{15} & \omega^{18} & \omega^{21} \\
1 & \omega^4 & \omega^8 & \omega^{12} & \omega^{16} & \omega^{20} & \omega^{24} & \omega^{28} \\
1 & \omega^5 & \omega^{10} & \omega^{15} & \omega^{20} & \omega^{25} & \omega^{30} & \omega^{35} \\
1 & \omega^6 & \omega^{12} & \omega^{18} & \omega^{24} & \omega^{30} & \omega^{36} & \omega^{42} \\
1 & \omega^7 & \omega^{14} & \omega^{21} & \omega^{28} & \omega^{35} & \omega^{42} & \omega^{49}
\end{pmatrix} \\
&\equiv \frac{1}{2\sqrt{2}}\begin{pmatrix}
1 & 1 & 1 & 1 & 1 & 1 & 1 & 1 \\
1 & \omega & \omega^2 & \omega^3 & \omega^4 & \omega^5 & \omega^6 & \omega^7 \\
1 & \omega^2 & \omega^4 & \omega^6 & 1 & \omega^2 & \omega^4 & \omega^6 \\
1 & \omega^3 & \omega^6 & \omega & \omega^4 & \omega^7 & \omega^2 & \omega^5 \\
1 & \omega^4 & 1 & \omega^4 & 1 & \omega^4 & 1 & \omega^4 \\
1 & \omega^5 & \omega^2 & \omega^7 & \omega^4 & \omega & \omega^6 & \omega^3 \\
1 & \omega^6 & \omega^4 & \omega^2 & 1 & \omega^6 & \omega^4 & \omega^2 \\
1 & \omega^7 & \omega^6 & \omega^5 & \omega^4 & \omega^3 & \omega^2 & \omega
\end{pmatrix}.
\end{aligned} \tag{19.33}
$$

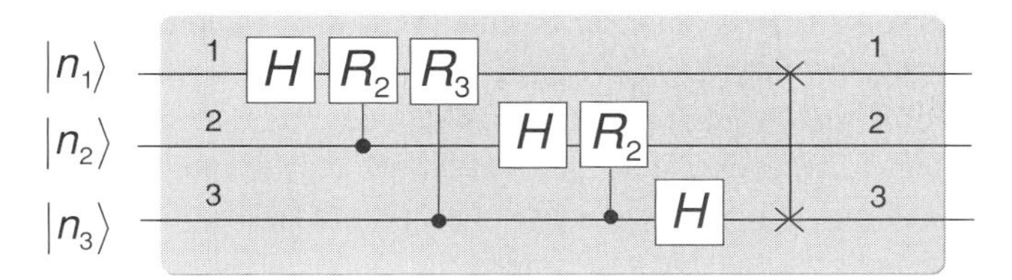


**Figure 19.6** QFT circuit for $K = 3$ ($N = 8$).

Is the QFT algorithm any faster than its classical counterpart, which is known as *fast Fourier transform* (*FFT*)? To answer this question, let us analyze the number of gates traversed by the signals for each circuit type. Referring back to Fig. 19.4, we observe that a $K$ circuit requires crossing a number of gates of $K + (K - 1) + (K - 2) + \ldots 2 + 1 = \sum_{k=1}^{K} k = K(K + 1)/2$, plus $K/2$ or $(K - 1)/2$ ($K$ even or odd) twin gates to perform the SWAP operation, each amounting to three CNOT gates. We can conclude that the computation time for a QFT circuit of $K$ inputs to generate the output is *of the order of* $K^2$, noted $O(K^2)$. In computer language, it is said that this algorithm (or *decision problem*) belongs to the *complexity class P*, as the implementation in a sequential machine can be performed *in polynomial time*.[4]

In contrast, the number of gates required in a classical FFT circuit for the same computation (data size $N = \log_2 K$) is $K \log K = 2^N \log_2 2^N = N2^N$. Thus, the FFT computation time is of the order $O(N2^N)$, which is exponential in circuit size, as opposed to quadratic in the QFT case. Such a comparison remains, however, questionable, because the quantum circuit does not operate on data bits, but *quantum states*. Further processing circuits would be required to measure the "classical bit" or cbit counterparts, but even this observation fails to convey any equitable comparison, as such measurements are, by nature, indeterministic. Simply put, the QFT is exponentially faster than FFT, but the signal amplitudes cannot be measured, unlike in the classical case. The picture becomes even darker if we consider that *there is currently no known technique to prepare the input states* $|n_k\rangle$ *of the QFT circuit*, except in the very limiting cases $K = 1$ or $K = 2$.

The quantum Fourier transform may, thus, appear to be fully impractical, or just an interesting mathematical curiosity. As we shall see in Chapter 20, however, the QFT algorithm is the key to solving one of the major computation problems, namely the *factoring of numbers into primes*, famously known as the *Shor algorithm*.

## 19.4 Grover quantum database search algorithm

In this section, I describe another famous QC algorithm, known after its conceiver as the *Grover quantum database search* (GQDS). Like the previously described QFT, this algorithm nicely exploits the property of quantum parallelism.

[4] This notion is not to be confused with the time required for solution *verification*, which is referred to as an NP (indeterministic polynomial time) problem. See more on this topic, for instance in: http://en.wikipedia.org/wiki/Complexity_classes_P_and_NP.

The problem to solve here is quite basic: how to identify an item out of an unsorted database of $N$ distinct elements? On average, such a task would require $N/2$ searches, with a worst case of $N$ searches and a luckiest case of one single search. The computation time to find any item is, therefore, of the order of $N$, or $O(N)$. As we shall see, the beauty in the GQDS algorithm is that the complexity of the search task becomes $O(\sqrt{N})$, which means *a quadratic increase in computing speed.*

Here, I shall first describe the GQDS algorithm and then analyze its elementary quantum-circuit implementation and requirements.[5]

There is no loss in generality in assuming that the size of the database is $N = 2^n$, where $n$ is some nonzero integer. We require a quantum space $V$ of dimension $N \geq 2$, and a known observable $\Omega$ on this space. This observable defines an orthonormal basis of $N$ eigenstates $|x\rangle$ labeled $|0\rangle, |1\rangle, \ldots, |N-1\rangle$, and known eigenvalues labeled $\lambda_0, \lambda_1, \ldots, \lambda_{N-1}$. Then we shall associate a unique eigenstate or eigenvalue with each item in the database. The database search problem is now a matter of measuring an eigenvalue of interest, call it $\lambda_\omega$, associated with the state $|\omega\rangle$, which represents some specific item $\omega$ in the original database, and which we want to find through the search algorithm.

As I will show further on, it is possible to construct the eigenstate superposition

$$|s\rangle = \frac{1}{\sqrt{N}} \sum_{x=0}^{N-1} |x\rangle, \tag{19.34}$$

noting that $\langle s|s\rangle = 1$ since the eigenstate basis $\{|x\rangle\}_{x=0\ldots N-1}$ is orthonormal. Any eigenstate $|\omega\rangle$ has the same projection $\langle\omega|s\rangle = 1/\sqrt{N}$, and given our item $\omega$, the probability of measuring $\lambda_\omega$ (or finding $|s\rangle$ in the state $|\omega\rangle$) is $|\langle\omega|s\rangle|^2 = 1/N$, consistently with a classical database search algorithm. Let us then introduce a unitary operator $U_\omega$, referred to as the "oracle," and defined by

$$U_\omega = I - 2|\omega\rangle\langle\omega|. \tag{19.35}$$

It is immediately verified that the oracle has the following properties: $U_\omega|x\rangle = -|\omega\rangle$ if $x = \omega$ and $U_\omega|x\rangle = |x\rangle$ if $x \neq \omega$. Thus, the action of the oracle on the superposition $|s\rangle$ is to change the sign of the amplitude of the component $|\omega\rangle$ in $|s\rangle$, namely:

$$\begin{aligned}
|\psi\rangle &= U_\omega|s\rangle \\
&= U_\omega\left(\frac{1}{\sqrt{N}}\sum_{x=0}^{N-1}|x\rangle\right) \\
&= \frac{1}{\sqrt{N}}U_\omega\left(\sum_{\substack{x=0 \\ x\neq\omega}}^{N-1}|x\rangle + |\omega\rangle\right)
\end{aligned}$$

[5] See, for instance, http://en.wikipedia.org/wiki/Grover's_algorithm with notations consistent with those used in this section. The description here is inspired from the tutorial paper: C. Lavor, L. R. U. Manssur, and R. Portugal, Grover's algorithm: quantum database search, (2003), which can be downloaded from http://arxiv.org/abs/quant-ph/0301079.

$$= \frac{1}{\sqrt{N}} \left( \sum_{\substack{x=0 \\ x \neq \omega}}^{N-1} U_\omega |x\rangle + U_\omega |\omega\rangle \right)$$

$$\equiv \frac{1}{\sqrt{N}} \sum_{\substack{x=0 \\ x \neq \omega}}^{N-1} |x\rangle - \frac{1}{\sqrt{N}} |\omega\rangle. \tag{19.36}$$

We can also rewrite the result in Eq. (19.36) in the form:

$$|\psi\rangle = U_\omega |s\rangle = |s\rangle - \frac{2}{\sqrt{N}} |\omega\rangle. \tag{19.37}$$

We can then interpret the action of the oracle $U_\omega$ of the state $|s\rangle$ as a small rotation through an angle $\theta$. Indeed, we have

$$\begin{aligned} \cos\theta &= \langle s|\psi\rangle \\ &= \langle s|s\rangle - \frac{2}{\sqrt{N}} \langle s|\omega\rangle \\ &= 1 - \frac{2}{\sqrt{N}} \frac{1}{\sqrt{N}} \\ &\equiv 1 - \frac{2}{N} \end{aligned} \tag{19.38}$$

(the angle $\theta$ being small, if one assumes that the database size, $N$, is large). Next we introduce a second operator $U_s$ according to

$$U_s = 2\,|s\rangle\langle s| - I, \tag{19.39}$$

and define the Grover operator as $G = U_s U_\omega$. Let us see next the action of this operator on $|s\rangle$, which we shall refer to as a *Grover iteration*. From Eq. (19.37), we have:

$$\begin{aligned} |\psi_1\rangle &= G|s\rangle = U_s U_\omega |s\rangle = U_s |\psi\rangle = U_s \left( |s\rangle - \frac{2}{\sqrt{N}} |\omega\rangle \right) \\ &= U_s |s\rangle - \frac{2}{\sqrt{N}} U_s |\omega\rangle \\ &= (2|s\rangle\langle s| - I)\,|s\rangle - \frac{2}{\sqrt{N}} (2|s\rangle\langle s| - I)\,|\omega\rangle \\ &= 2|s\rangle\langle s|s\rangle - |s\rangle - \frac{4}{\sqrt{N}} |s\rangle\langle s|\omega\rangle + \frac{2}{\sqrt{N}} |\omega\rangle \\ &= |s\rangle - \frac{4}{\sqrt{N}} \frac{1}{\sqrt{N}} |s\rangle + \frac{2}{\sqrt{N}} |\omega\rangle \\ &\equiv \left(1 - \frac{4}{N}\right) |s\rangle + \frac{2}{\sqrt{N}} |\omega\rangle. \end{aligned} \tag{19.40}$$

We observe from the result in Eq. (19.40) that the action of $G$ on $|s\rangle$ is to increase the amplitude of the eigenstate component $|\omega\rangle$, again by rotation. The corresponding rotation angle $\theta'$ is given by

$$\begin{aligned}\cos\theta' &= \langle s|\psi_1\rangle = \langle s|\left[\left(1-\frac{4}{N}\right)|s\rangle + \frac{2}{\sqrt{N}}|\omega\rangle\right]\\ &= \left(1-\frac{4}{N}\right)\langle s|s\rangle + \frac{2}{\sqrt{N}}\langle s|\omega\rangle\\ &= 1-\frac{4}{N}+\frac{2}{\sqrt{N}}\frac{1}{\sqrt{N}} \equiv 1-\frac{2}{N} \equiv \cos\theta,\end{aligned} \tag{19.41}$$

and according to Eq. (19.38), which shows that the rotation angles $\theta, \theta'$ due to $U_\omega$ and $G$ are equal in absolute value. It is easy to show that $|\psi\rangle = U_\omega|s\rangle$ and $|\psi_1\rangle = G|s\rangle$ are, in fact, rotating by that same absolute angle $\theta = \theta'$ in opposite directions away from $|s\rangle$, forming an angle $\theta'' = 2\theta$. Indeed, from Eqs. (19.37) and (19.40) we obtain:

$$\begin{aligned}\cos\theta'' &= \langle\psi|\psi_1\rangle = \left[\langle s| - \frac{2}{\sqrt{N}}\langle\omega|\right]\left[\left(1-\frac{4}{N}\right)|s\rangle + \frac{2}{\sqrt{N}}|\omega\rangle\right]\\ &= \left(1-\frac{4}{N}\right)\langle s|s\rangle - \frac{2}{\sqrt{N}}\left(1-\frac{4}{N}\right)\langle\omega|s\rangle + \frac{2}{\sqrt{N}}\langle s|\omega\rangle - \frac{2}{\sqrt{N}}\frac{2}{\sqrt{N}}\langle\omega|\omega\rangle\\ &= 1-\frac{4}{N}-\frac{2}{\sqrt{N}}\left(1-\frac{4}{N}\right)\frac{1}{\sqrt{N}}+\frac{2}{\sqrt{N}}\frac{1}{\sqrt{N}}-\frac{4}{N}\\ &\equiv 1-\frac{8}{N}+\frac{8}{N^2} \equiv 2\left(1-\frac{2}{N}\right)^2 - 1 \equiv 2\cos^2\theta - 1 \equiv \cos 2\theta.\end{aligned} \tag{19.42}$$

Finally, it is straightforward from the definitions in Eq. (19.40) that the cosine projection of $|\psi_1\rangle$ over the target state $|\omega\rangle$ takes the value $\langle\omega|\psi_1\rangle = (3-4/N)/\sqrt{N}$, to compare with the initial projection $\langle\omega|s\rangle = 1/\sqrt{N}$. The angle relations of the different states $|s\rangle$, $|\psi\rangle$, $|\psi_1\rangle$, and $|\omega\rangle$ and relevant projections $\langle\omega|s\rangle$, $\langle\omega|\psi_1\rangle$ are illustrated in Fig. 19.7. We observe from the figure that $G$ has rotated $|s\rangle$ into a state $|\psi_1\rangle$, which is closer to the target state $|\omega\rangle$, hence a projection being increased by a factor $3-4/N$ (which is greater than unity if $N > 2$). Should a measurement in the state $|\psi_1\rangle$ be made through the observable $\Omega$, the probability of measuring $\lambda_\omega$ would be

$$|\langle\omega|\psi_1\rangle|^2 = \frac{1}{N}\left(3-\frac{4}{N}\right)^2 = \frac{1}{N}\left(9-\frac{24}{N}+\frac{16}{N^2}\right) \approx \frac{9}{N}, \tag{19.43}$$

the upper limit on the right-hand side being obtained for large values of $N$. This result shows that in *a single Grover iteration, the probability of finding the searched item through a single-shot measurement has been increased by about tenfold with respect to the classical case*.

It is now tempting to iterate the Grover operation as many times as required to rotate the system state closer to the target state $|\omega\rangle$, thus, increasing the probability of successfully measuring $\lambda_\omega$ to the required accuracy. For this, we must establish an iterative formula that defines the state $|\psi_k\rangle$ after $k$ Grover iterations, and the corresponding success probability.

For the purpose of deriving the iteration formula, let us redefine the original eigenstate superposition in Eq. (19.34) by sorting out the $|\omega\rangle$ state from the series. Define first the

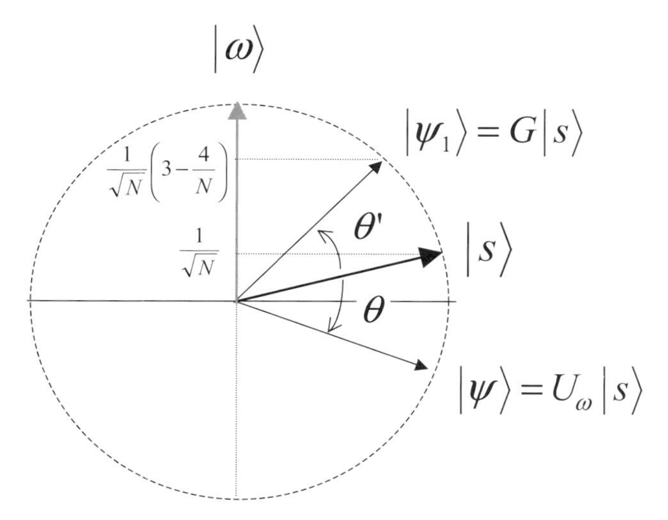


**Figure 19.7** Effect of the unitary operators $U_\omega$ and $G = U_s U_\omega$ on initial state $|s\rangle$, showing rotation by angles $\theta$ and $\theta' = \theta$ (respectively) corresponding to new states $|\psi\rangle = U_\omega|s\rangle$ and $|\psi\rangle = G|s\rangle$ (respectively). The cosine-projections of states $|s\rangle$ and $|\psi_1\rangle$ over the reference state $|\omega\rangle$ of the search algorithm are also indicated.

state $|u\rangle$ spanning the hyperspace orthogonal to $|\omega\rangle$ according to

$$\begin{aligned}|u\rangle &\equiv \frac{1}{\sqrt{N-1}}\sum_{\substack{x=0\\x\neq\omega}}^{N-1}|x\rangle \\ &= \sqrt{\frac{N}{N-1}}|s\rangle - \frac{1}{\sqrt{N-1}}|\omega\rangle.\end{aligned} \tag{19.44}$$

We can then write $|s\rangle$ according to the superposition of its two orthogonal component states $|u\rangle$, $|\omega\rangle$:

$$\begin{aligned}|s\rangle &= \frac{1}{\sqrt{N}}\sum_{x=0}^{N-1}|x\rangle \equiv \sqrt{1-\frac{1}{N}}|u\rangle + \frac{1}{\sqrt{N}}|\omega\rangle \\ &\equiv \cos\frac{\theta}{2}|u\rangle + \sin\frac{\theta}{2}|\omega\rangle,\end{aligned} \tag{19.45}$$

where $\theta$ is the rotation angle previously defined through $\cos\theta = 1 - 2/N$ (it is easily checked that $\cos\theta/2 = \sqrt{1-1/N}$). It is then an elementary (albeit nontrivial) exercise to show that $k$ Grover iterations result in the state

$$\begin{aligned}G^k|s\rangle &= (GG\ldots G)_{k\ \text{times}}|s\rangle \\ &= \cos\left[(2k+1)\frac{\theta}{2}\right]|u\rangle + \sin\left[(2k+1)\frac{\theta}{2}\right]|\omega\rangle.\end{aligned} \tag{19.46}$$

This result shows that $k$ applications of the Grover operator results in the rotation of the state $|s\rangle$ by an angle $k\theta$, as illustrated in Fig. 19.8. The probability $p(\omega)$ of finding the state $G^k|s\rangle$ in $|\omega\rangle$, and, hence, of measuring the eigenvalue $\lambda_\omega$ with the observable $\Omega$, is

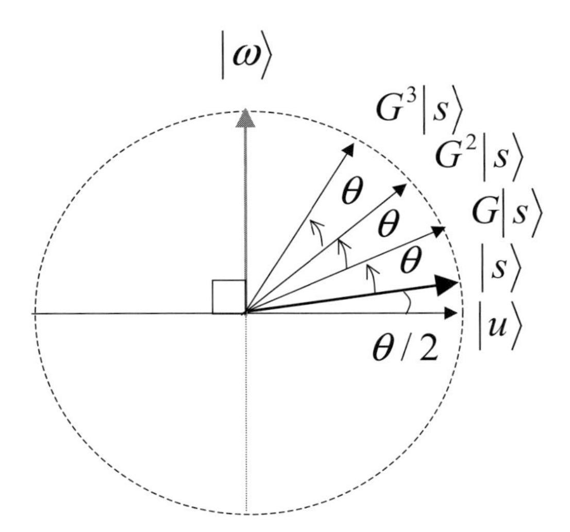


**Figure 19.8** Effect of successive applications (here from $k = 1$ to $k = 3$) of the Grover operator $G$ onto the initial state $|s\rangle$, showing incremental rotations by angle $\theta$ in the direction of state $|\omega\rangle$.

given by

$$p(\omega) = \left|\langle\omega|G^k|s\rangle\right|^2 = \sin^2\left[(2k+1)\frac{\theta}{2}\right]. \tag{19.47}$$

The probability $p(\omega)$ reaches a maximum for

$$k_{\max}\theta + \frac{\theta}{2} = \frac{\pi}{2}, \tag{19.48}$$

or

$$k_{\max} = \left\lfloor \frac{\pi - \theta}{2\theta} \right\rfloor. \tag{19.49}$$

From the definition in Eq. (19.45), $\sin\theta/2 = 1/\sqrt{N}$, which for large $N$ yields the approximation $\theta \approx 2/\sqrt{N}$ and, hence, the maximum number of iterations $k_{\max}$:

$$k_{\max} \approx \left\lfloor \frac{\pi}{4}\sqrt{N} - \frac{1}{2} \right\rfloor \approx \left\lfloor \frac{\pi}{4}\sqrt{N} \right\rfloor. \tag{19.50}$$

Clearly, the Grover search algorithm must be stopped after reaching the number of iterations $k = k_{\max}$, corresponding to the probability $p(\omega) \approx 1$. Figure 19.9 shows plots of the probability $p(\omega)$, a function of the number $k$ of Grover iterations, from quantum database sizes $N = 2^6 = 64$ to $N = 2^{14} = 16\,384$. Thanks to the $\sqrt{N}$ dependence of $k_{\max}$, it is seen that quantum databases as large as $N \approx 16\,000$ can successfully be searched with only $k_{\max} \approx 100$ Grover iterations, which represents a computation complexity in $O(\sqrt{N})$, to compare with $O(N)$ in the classical case.

Now that we are convinced of the benefits of the Grover algorithm, the next step is to analyze its quantum-circuit implementation. Consider first how the quantum database can be generated. In Section 19.2, concerning the Deutsch–Jozsa algorithm, we have

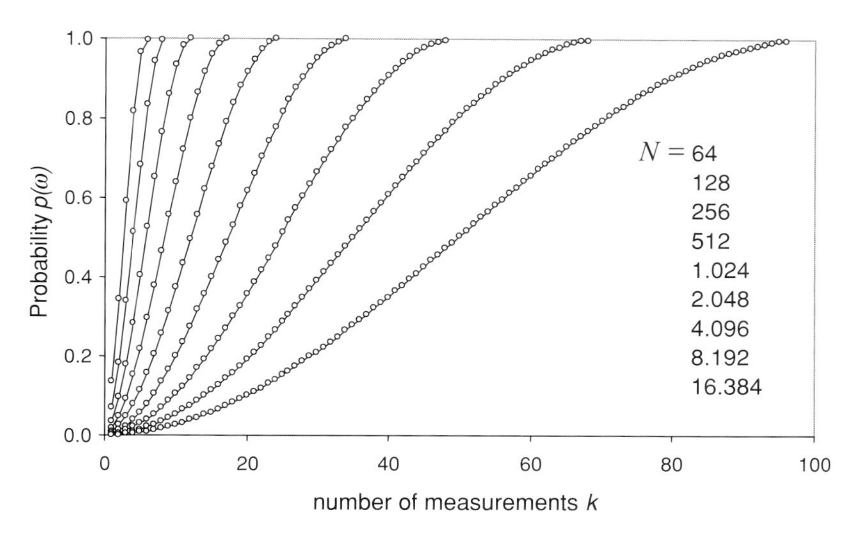


**Figure 19.9** Probability $p(\omega)$ of finding the state in $|\omega\rangle$ (or of measuring the eigenvalue $\lambda_\omega$) as a function of the number $k$ of Grover iterations, from quantum database sizes $N = 2^6 = 64$ (leftmost curve) to $N = 2^{14} = 16\,384$ (rightmost curve).

established and used in Eq. (19.7) the property according to which

$$\begin{aligned} |s\rangle &= \frac{1}{\sqrt{N}} \sum_{x=0}^{N-1} |x\rangle \\ &= H^{\otimes n}|0\rangle = \left( \frac{|0\rangle + |1\rangle}{\sqrt{2}} \right)^{\otimes n} \\ &= |+\rangle \otimes |+\rangle \otimes \ldots \otimes |+\rangle_{n\,\text{times}}, \end{aligned} \tag{19.51}$$

where $H^{\otimes n}$ is the $n$-tensor application of the Hadamard gate $H$. For instance, we have, for $N = 2^3 = 8$:

$$\begin{aligned} |s\rangle &= \frac{1}{\sqrt{2^3}} \sum_{x=0}^{7} |x\rangle \\ &= H^{\otimes 3}|0\rangle \\ &= \left( \frac{|0\rangle + |1\rangle}{\sqrt{2}} \right) \otimes \left( \frac{|0\rangle + |1\rangle}{\sqrt{2}} \right) \otimes \left( \frac{|0\rangle + |1\rangle}{\sqrt{2}} \right) \\ &\equiv \frac{1}{\sqrt{2^3}} (|000\rangle + |001\rangle + |010\rangle + |011\rangle + |100\rangle + |101\rangle + |110\rangle + |111\rangle). \end{aligned} \tag{19.52}$$

Thus, the state superposition $|s\rangle$, which is to become the quantum database, can be generated by using $n$ ancilla qubits $|0\rangle$ and $2 \times 2$ Hadamard gates $H$.

The second task is to define a quantum circuit for the "oracle" operator $U_\omega$. To recall, the oracle has the following properties: $U_\omega|x\rangle = -|\omega\rangle$ if $x = \omega$ and $U_\omega|x\rangle = |x\rangle$ if $x \neq \omega$. Let $f(x)$ be a function defined over $x = \{0, 1, \ldots, N-1\}$ with binary output $f(x) = 1$ if $x = \omega$ and $f(x) = 0$ if $x \neq \omega$. Assume that it is possible to build a quantum circuit (or quantum operator $U_\omega$) acting on the tensor state $|x\rangle|i\rangle$ (with $|i\rangle = |0\rangle$ or $|1\rangle$

being an ancilla qubit) yielding the following output:

$$U_\omega|x\rangle|i\rangle = |x\rangle|i \oplus f(\omega)\rangle. \tag{19.53}$$

It is clear from the above definitions that $U_\omega|x\rangle\,|i\rangle = |x\rangle\,|i\rangle$ if $x \neq \omega$ and $U_\omega|x\rangle\,|i\rangle = |x\rangle|i \oplus 1\rangle$ if $x = \omega$. More generally, if we let $|i\rangle = |-\rangle = (|0\rangle - |1\rangle)/\sqrt{2}$, we obtain

$$\begin{aligned} U_\omega|x\rangle|-\rangle &= U_\omega|x\rangle\frac{|0\rangle - |1\rangle}{\sqrt{2}} \\ &= |x\rangle\frac{|0 \oplus f(\omega)\rangle - |1 \oplus f(\omega)\rangle}{\sqrt{2}} \\ &= |x\rangle\frac{|f(\omega)\rangle - |1 \oplus f(\omega)\rangle}{\sqrt{2}} \\ &= \left\{ \begin{array}{l} |x\rangle\dfrac{|0\rangle - |1\rangle}{\sqrt{2}}, \text{ if } x \neq \omega \\ |x\rangle\dfrac{|1\rangle - |0\rangle}{\sqrt{2}}, \text{ if } x = \omega \end{array} \right\} \\ &\equiv (-1)^{f(\omega)}\,|x\rangle|-\rangle. \end{aligned} \tag{19.54}$$

Call the tensor state $|R\rangle = |s\rangle|-\rangle$ the "register" in the GQSD algorithm. If we apply the oracle $U_\omega$ to the register $|R\rangle$, and use the definitions in Eqs. (19.34) and (19.54), we obtain:

$$\begin{aligned} U_\omega\,|R\rangle &= U_\omega|s\rangle|-\rangle \\ &= U_\omega\left(\frac{1}{\sqrt{N}}\sum_{x=0}^{N-1}|x\rangle|-\rangle\right) \\ &= \frac{1}{\sqrt{N}}U_\omega\left(\sum_{x=0}^{N-1}|x\rangle|-\rangle\right) \\ &= \frac{1}{\sqrt{N}}\left(\sum_{\substack{x=0\\x\neq\omega}}^{N-1}U_\omega|x\rangle|-\rangle + U_\omega|\omega\rangle|-\rangle\right) \\ &= \frac{1}{\sqrt{N}}\left(\sum_{\substack{x=0\\x\neq\omega}}^{N-1}(-1)^0|x\rangle|-\rangle + (-1)^1|\omega\rangle|-\rangle\right) \\ &\equiv \frac{1}{\sqrt{N}}\left(\sum_{\substack{x=0\\x\neq\omega}}^{N-1}|x\rangle|-\rangle - |\omega\rangle\right)|-\rangle, \end{aligned} \tag{19.55}$$

which is precisely the oracle operation defined in Eq. (19.36). Since we have assumed that the oracle operator $U_s = I - 2|\omega\rangle\langle\omega|$ can be thus realized, we can also assume that similarly, one can realize the unitary transformation $U_s = 2|s\rangle\langle s| - I$, in similar ways (this being beyond the scope of this book).

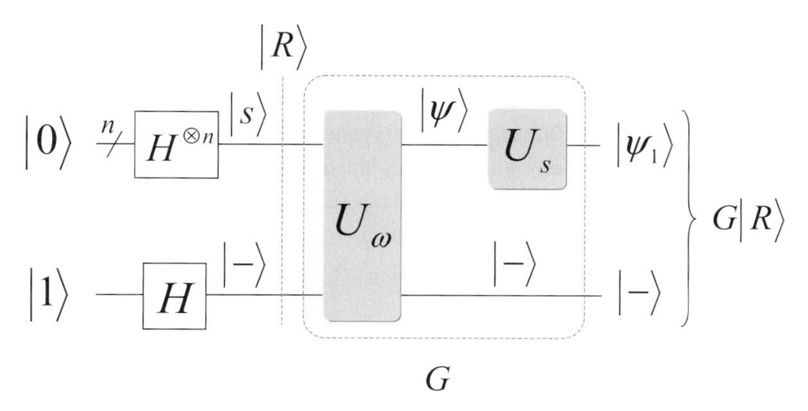


**Figure 19.10** Quantum-gate circuit showing the initialization of the register state $|R\rangle = |s\rangle|-\rangle$ from $n|0\rangle$ ancilla qubit and one $|1\rangle$ ancilla qubit, and $n+1$ Hadamard gates $H$. The circuit shown in the dashed-line box is the iterative Grover operator $G = U_s U_\omega$, with $U_s = 2|s\rangle\langle s| - I$.

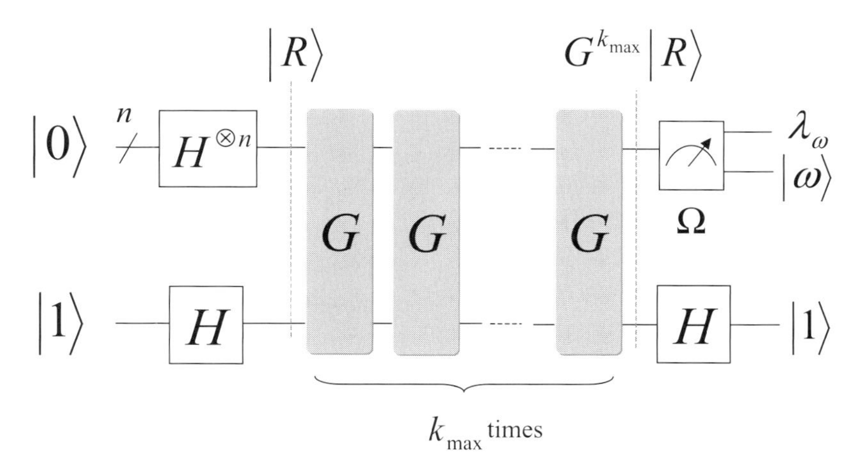


**Figure 19.11** Quantum-gate circuit architecture for the full Grover quantum database search algorithm implementation up to iteration $k = k_{\max}$ showing eventual measurement of eigenvalue $\lambda_\omega$ through the observable $\Omega$ (with success probability $p(\omega) \approx 1$).

The basic GQDS quantum-gate circuit architecture is shown in Fig. 19.10. From the figure, we observe the initialization of the register state $|R\rangle = |s\rangle|-\rangle$ from $n$ $|0\rangle$ ancilla qubits, one $|1\rangle$ ancilla qubit, and $n+1$ Hadamard gates $H$. The circuit shown in the dashed-line box is the iterative Grover operator $G = U_s U_\omega$, with $U_s = 2|s\rangle\langle s| - I$, which now acts on the register state $|R\rangle = |s\rangle$. It is only necessary to cascade the Grover circuit $G$ shown in Fig. 19.10 $k_{\max}$ times to obtain the full GQDS circuit shown in Fig. 19.11. As the figure shows, the algorithm implementation is complete with a measurement (eigenvalue $\lambda_\omega$) of the observable $\Omega$ with success probability $p(\omega) \approx 1$.

This description of the GQDS algorithm hopefully conveys the elegance of parallel computing through quantum circuits. Yet there are many questions remaining on the possible or even optimal circuit architectures for the "oracle" and Grover operators, let aside the practical physical implementation of the circuit at very large scales ($N \approx 10^3 - 10^9$), where the GQDS algorithm brings superior value. This reservation being stated, in a rising twenty-first century dominated by Internet communications, with its billions of host addresses, web pages, references, and databases, the physical possibility of implementing GQDS algorithms through practical quantum computers would, beyond any doubt, represent a most significant breakthrough. It would mean the possibility of elaborate search algorithms over the Internet with relatively fast access times and, in some applications, near-instantaneous access to practical information, however complex or remote!

## 19.5 Exercises

**19.1** (M): Show that for any $n$-qubit $|x\rangle = |x_1 x_2 \dots x_n\rangle$, $x_i \in \{0, 1\}$, and Hadamard gate $H$, we have:

$$H^{\otimes n}|x\rangle = \frac{1}{\sqrt{2^n}} \sum_z (-1)^{z \cdot x} |z\rangle,$$

where $|z\rangle = |z_1 z_2 \dots z_n\rangle$ and $x \cdot z = x_1 z_1 + x_2 z_2 + \dots + x_n z_n$ is a scalar product modulo 2.

**19.2** (M): Given the orthonormal basis $\{|x_k\rangle\}_{k=0\dots N-1}$, show that the transformation QFT, as defined by

$$|y_n\rangle = \frac{1}{\sqrt{N}} \sum_{k=0}^{N-1} e^{ik\frac{2n\pi}{N}} |x_k\rangle,$$

is unitary.

**19.3** (B): Calculate the quantum Fourier transform of the qubit:

$$|\psi\rangle = \frac{1}{\sqrt{14}}|0\rangle + \frac{2i}{\sqrt{14}}|1\rangle + \frac{3}{\sqrt{14}}|2\rangle.$$

**19.4** (B): Calculate the quantum Fourier transform of four EPR–Bell states $|\beta_{00}\rangle$, $|\beta_{01}\rangle$, $|\beta_{10}\rangle$, $|\beta_{11}\rangle$, as expressed in the ordered basis $V = \{|00\rangle, |01\rangle, |10\rangle, |11\rangle\}$.

**19.5** (M): Given the definitions of $G$, $|u\rangle$, $|\omega\rangle$, $|s\rangle$, and $|\psi_1\rangle = G|s\rangle$ in the text, demonstrate by induction the result in Eq. (19.45) concerning the $k$ action of the Grover operator $G$ on the state $|s\rangle$:

$$G^k|s\rangle = \cos\left[(2k+1)\frac{\theta}{2}\right]|u\rangle + \sin\left[(2k+1)\frac{\theta}{2}\right]|\omega\rangle.$$

# 20 Shor's factorization algorithm

This chapter describes what is generally considered to be one of the most important and historical contributions to the field of quantum computing, namely *Shor's factorization algorithm*. As its name indicates, this algorithm makes it possible to factorize numbers, which consists in their decomposition into a unique product of prime numbers. Other classical factorization algorithms previously developed have a complexity or computing time that increases exponentially with the number size, making the task intractable if not hopeless for large numbers. In contrast, Shor's algorithm is able to factor a number of any size in *polynomial time*, making the factorization problem tractable should a *quantum computer* ever be realized in the future. Since Shor's algorithm is based on several nonintuitive properties and other mathematical subtleties, this chapter presents a certain level of difficulty. With the previous chapters and tools readily assimilated, and some patience in going through the different preliminary steps required, such a difficulty is, however, quite surmountable. I have sought to make this description of Shor's algorithm as mathematically complete as possible and crack-free, while avoiding some academic considerations that may not be deemed necessary from any engineering perspective. Eventually, Shor's algorithm is described in only a few basic instructions. What is conceptually challenging is to grasp why it works so well, and also to feel comfortable with the fact that its implementation actually takes a fair amount of trial and error. The two preliminaries of Shor's algorithm are the *phase estimation* and the related *order-finding* algorithms. Both represent the purely quantum part of the approach: it cannot be implemented classically. Basically, phase estimation allows one to find the *periodicity* $r$ of a modular function by means of a multi-qubit quantum-gate circuit (Hadamard, controlled-$U$ gates, inverse-Fourier transform), followed by a probabilistic, quantum-mechanical measurement of the resulting qubit state, which yields a phase estimate $\varphi$. Order finding, from which the period $r$ is determined with high probability and (the measurement being successful) without ambiguity, represents a particular case of quantum phase estimation. Such a determination eventually rests on the implementation of *continued fraction expansion*, a classical algorithm that is straightforward to run with a computer. The requirement for $r$ to be the period is that the phase estimation $\varphi = s/r$, with $s$ an integer, is such that $s, r$ are co-prime. Since this does not happen systematically, there is a finite chance that the endeavor may fail. Any such event is not a failure of Shor's algorithm, but rather a call for another try in this specific implementation step. Such conditions of trial and error leading to factorizing success may sound strange to engineers – but they are really embedded in the algorithm game! The key feature to grasp is that the probability of such intermediate failures remains comparatively small, or innocuous for computing logistics, and that the chances of success, after only a few trials in the worst of all cases, are relatively high. As we shall see, all of the above steps are of *polynomial-time complexity*. A comparison is made with nonpolynomial algorithms, such as the *general number field sieve* (GNFS) approach, which is shown to require decades of CPU computing time to factorize numbers of 100-bits long! We then establish the connection between order finding and factorization of *composite* (or nonprime) numbers by using two basic theorems. The first theorem yields the two factors $N'$, $N''$ of any given composite $N$ such that $N = N' \times N''$, given the knowledge of the period. The second theorem establishes that the probability of the period meeting certain eligibility criteria is at least 75% for any composite. These two theorems combined validate and conclude Shor's factorization algorithm. The factorization of the composite $N = 15 = 3 \times 5$ is found in textbooks as the only illustrative example of Shor's algorithm. Here, we shall investigate the whole space of nontrivial composites $N \leq 100$, as an emulation of the quantum computer. It is possible to do this based on the fact that for such relatively small numbers, we can compute (with a basic home computer) all the possibilities associated with each step of Shor's algorithm, namely what the period-finding quantum circuit should yield, and the associated probabilities of success or failure. The result of this investigation is an original plot showing the probability $\Pi$ of successfully concluding the factorization of nontrivial composites $N \leq 100$ in a *single run*. The exercise helps one to grasp how Shor's algorithm would work when taking greater composite numbers. The last section, which briefly describes *public key cryptography* (PKC), is not completely out of place in this chapter. The purpose of this addition is to show how the PKC algorithm works, as based on the product $N = pq$ of two prime numbers $p, q$, whose factorization is indeed considered intractable by classical means. Should a quantum computer of corresponding computing power be implemented someday, the whole field of PKC-based cryptography, and Internet security for that matter, would be compromised overnight! Fortunately, this remains a distant perspective, while from this chapter, we know that the theory works mathematically.

## 20.1 Phase estimation

This section considers an algorithm referred to as *phase estimation*. The word "estimation" is correct, because the outcome of the algorithm is only what can be called a good estimate of the phase in a quantum system. This phase estimation will enable us to move another step further in our progression towards *Shor's factorization algorithm*, the second step being the *order-finding algorithm*, to be considered in the next section. These two steps are critical in the final understanding of Shor's algorithm, and this is why we ought to pay extra attention to the following. While based on lessons from previous chapters with no new conceptual difficulty, the quantum phase estimation and its application to order finding involve a few nontrivial properties and results, which must be fully assimilated at each step of the description.

Assume a unitary operator $U$, with eigenstate $|u\rangle$ of dimension $L$, and of unknown complex eigenvalue $\lambda_\phi = e^{2i\pi\varphi}$, where $\varphi$ is a real number such that $0 \le \varphi \le 1$, is to be determined through the aforementioned "phase estimation" algorithm. Also assume that we are capable of building a family of *controlled-$U^p$* operators, where $p = 2^0, 2^1, \ldots, 2^{K-1}$ is any power of two up to $K - 1$. The phase-estimation (PE) circuit comes in two stages, which we shall call here "front-end" and "back-end" modules, respectively. The PE front-end module is shown in Fig. 20.1. As we observe, its embodiment includes an input *register* of $K$ *ancilla* qubits $|0\rangle$, and the eigenstate $|u\rangle$, which is an $L$-qubit. Each of the qubits in the input register is submitted to a Hadamard transform $(H)$, the output of which drives an individual *controlled-$U^p$* gate, according to the array sequence shown at the bottom of the figure. It is easily established that the output register consists of $K$ qubits of the form $|0\rangle + e^{2i\pi p\phi}|1\rangle$ with $p = 2^0, 2^1, \ldots, 2^{K-1}$ and leaves invariant the $L$-qubit $|u\rangle$, as indicated in Fig. 20.1.[1] The tensor-product output can also be rewritten as a sum of $K + L$ qubits of the form $e^{2i\pi k\varphi}|k\rangle \otimes |u\rangle$:[2]

$$\begin{aligned} &\frac{1}{2^{K/2}}\left(|0\rangle + e^{2i\pi 2^{K-1}\varphi}|1\rangle\right)\left(|0\rangle + e^{2i\pi 2^{K-2}\varphi}|1\rangle\right)\ldots\left(|0\rangle + e^{2i\pi 2^{1}\varphi}|1\rangle\right) \\ &\quad\times\left(|0\rangle + e^{2i\pi 2^{0}\varphi}|1\rangle\right) \otimes |u\rangle = \frac{1}{\sqrt{N}}\sum_{k=0}^{N-1} e^{2i\pi k\varphi}|k\rangle \otimes |u\rangle, \end{aligned} \tag{20.1}$$

with $N = 2^K$. This result, in the summation form, will be used later. We shall now interpret the product form. Recall that $\varphi$ is a real number, such that $0 \le \varphi \le 1$. Any such number can be represented in binary form, which is noted $\varphi = 0.\varphi_1\varphi_2\varphi_3\ldots\varphi_K\ldots$, according to the definition:

$$\begin{aligned} \varphi &= 0.\varphi_1\varphi_2\varphi_3\ldots \\ &= \frac{\varphi_1}{2} + \frac{\varphi_2}{4} + \frac{\varphi_3}{8} + \cdots + \frac{\varphi_K}{2^K} + \cdots, \end{aligned} \tag{20.2}$$

[1] To show this, consider the effect of a controlled-$U^p$ gate, noted $CU^p$, at any stage in the circuit in Fig. 20.1. In each gate, the control qubit is $H|0\rangle = |0\rangle + |1\rangle$, within the factor $1/\sqrt{2}$. The tensor-product output is $CU^p[(|0\rangle + |1\rangle) \otimes |u\rangle]$ or, identically, $|0\rangle \otimes |u\rangle + |1\rangle \otimes U^p|u\rangle = |0\rangle \otimes |u\rangle + |1\rangle \otimes e^{2i\pi p\varphi}|u\rangle \equiv (|0\rangle \otimes e^{2i\pi p\varphi}|1\rangle) \otimes |u\rangle$. Clearly, the first and second terms in this tensor product belong to the first and second registers, respectively.

[2] This is established by developing the tensor product in the left-hand side, starting from the last two terms (or considering $K = 2$):

$$\begin{aligned} \left(|0\rangle + e^{2i\pi 2^1\varphi}|1\rangle\right)\left(|0\rangle + e^{2i\pi 2^0\varphi}|1\rangle\right) &= |0\rangle|0\rangle + |0\rangle e^{2i\pi 2^0\varphi}|1\rangle + e^{2i\pi 2^1\varphi}|1\rangle|0\rangle + e^{2i\pi 2^1\varphi}|1\rangle e^{2i\pi 2^0\varphi}|1\rangle \\ &= |00\rangle + e^{2i\pi 2^0\varphi}\,|01\rangle + e^{2i\pi 2^1\varphi}\,|10\rangle + e^{2i\pi(2^1+2^0)\varphi}\,|11\rangle \\ &\equiv e^{2i\pi 0\varphi}|0\rangle + e^{2i\pi 1\varphi}|1\rangle + e^{2i\pi 2\varphi}|2\rangle + e^{2i\pi 3\varphi}|3\rangle \\ &= \sum_{k=0}^{3} e^{2i\pi k\varphi}|k\rangle, \end{aligned}$$

noting that the tensor products $|ij\rangle$, with $i, j = 0, 1$, are noted $|i \times 2^0 + j \times 2^1\rangle$. It is clear that the full product yields the summation shown in the right-hand side in Eq. (20.1).

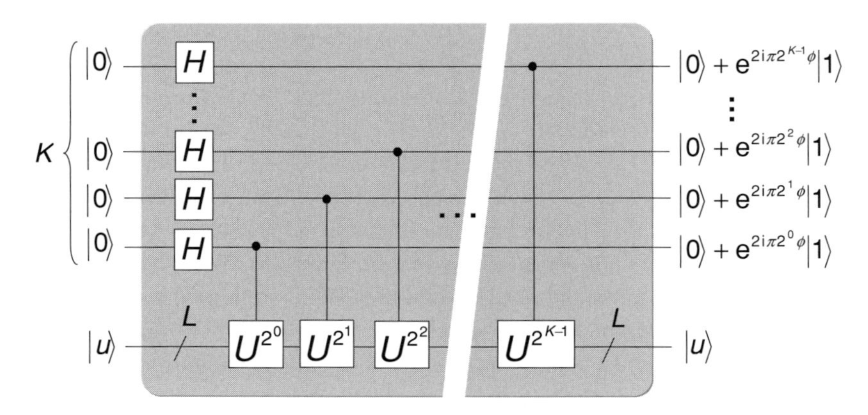


**Figure 20.1** Quantum circuit for phase estimation (front-end module).

where $\varphi_i = 0, 1$ are bits. In particular, we have

$$\begin{aligned} 2^{K-1}\varphi &= 2^{K-1}\left(\frac{\varphi_1}{2} + \frac{\varphi_2}{4} + \frac{\varphi_3}{8} + \cdots + \frac{\varphi_K}{2^K} + \cdots\right) \\ &= \{\varphi_1 2^{K-2} + \varphi_2 2^{K-3} + \cdots + \varphi_{K-1}2^0\} + \frac{\varphi_K}{2} + \frac{\varphi_{K+1}}{4} + \cdots \end{aligned} \tag{20.3}$$

and

$$\begin{aligned} 2^{K-2}\varphi &= 2^{K-2}\left(\frac{\varphi_1}{2} + \frac{\varphi_2}{4} + \frac{\varphi_3}{8} + \cdots + \frac{\varphi_K}{2^K} + \cdots\right) \\ &= \{\varphi_1 2^{K-3} + \varphi_2 2^{K-4} + \cdots + \varphi_{K-2}2^0\} + \frac{\varphi_{K-1}}{2} + \frac{\varphi_K}{4} + \frac{\varphi_{K+1}}{8} + \cdots \end{aligned} \tag{20.4}$$

Noting that the terms within braces { } in the right-hand side in Eqs. (20.3) and (20.4) are integers, we have for the imaginary exponentials:

$$\begin{cases} e^{2i\pi 2^{K-1}\varphi} = e^{2i\pi\left(\frac{\varphi_K}{2} + \frac{\varphi_{K+1}}{4} + \cdots\right)} \\ e^{2i\pi 2^{K-2}\varphi} = e^{2i\pi\left(\frac{\varphi_{K-1}}{2} + \frac{\varphi_K}{4} + \frac{\varphi_{K+1}}{8} + \cdots\right)} \\ \vdots \\ e^{2i\pi 2^0\varphi} = e^{2i\pi\left(\frac{\varphi_1}{2} + \frac{\varphi_2}{4} + \cdots \frac{\varphi_K}{2^K} + \frac{\varphi_{K+1}}{2^{K+1}} + \cdots\right)}. \end{cases} \tag{20.5}$$

Introduce the definition

$$\Omega_m = \sum_{l=1}^{m} \frac{\varphi_{K-m+l}}{2^l}, \tag{20.6}$$

and substitute it into Eq. (20.5) to obtain

$$\begin{cases} e^{2i\pi 2^{K-1}\varphi} = e^{2i\pi\Omega_1} e^{2i\pi\left(\frac{\varphi_{K+1}}{4} + \cdots\right)} \\ e^{2i\pi 2^{K-2}\varphi} = e^{2i\pi\Omega_2} e^{2i\pi\left(\frac{\varphi_{K+1}}{8} + \cdots\right)} \\ \vdots \\ e^{2i\pi 2^0\varphi} = e^{2i\pi\Omega_K} e^{2i\pi\left(\frac{\varphi_{K+1}}{2^{K+1}} + \cdots\right)}. \end{cases} \tag{20.7}$$

We then consider the specific case where $\varphi$ is exactly defined by $K$ bits, meaning that all bits $\varphi_{K+1}, \varphi_{K+2}, \ldots$ are identically zero. Substituting the results in Eq. (20.7) into

Eq. (20.1) under this assumption, and overlooking the output qubit $|u\rangle$, yields the first register output:

$$\frac{1}{2^{K/2}}(|0\rangle + e^{2i\pi\Omega_1}|1\rangle) \otimes (|0\rangle + e^{2i\pi\Omega_2}|1\rangle) \otimes \cdots \otimes (|0\rangle + e^{2i\pi\Omega_K}|1\rangle). \quad (20.8)$$

As based on Eqs. (19.29) and (19.30), the result in Eq. (20.8), together with the definition in Eq. (20.6), is immediately identified as being the *quantum Fourier transform* of the "phase" qubit $|\varphi\rangle = |\varphi_1\varphi_2 \ldots \varphi_K\rangle$. We can, thus, recover $|\varphi\rangle$, by performing the *inverse* Fourier transform of the output register, followed by a measurement in the computational basis, which yields the $K$ classical bits $\varphi_1\varphi_2 \ldots \varphi_K$ defining the phase. The "back-end" module of the PE circuit, thus, consists first of an *inverse Fourier transform circuit*, noted $FT^+$. This circuit is shown in Fig. 19.4, while conceived as being traversed from right to left. The second component of the back-end module is a $K$-qubit measurement gate, which restitutes the $\varphi_1\varphi_2 \ldots \varphi_K$ bits. The full PE circuit including front and back ends, is represented schematically in Fig. 20.2, assuming that the phase $\varphi$ is exactly defined by $K$ bits, or $2^K\varphi$ *is an integer*.

Consider next the more general case where $2^K\varphi$ is *not* an integer. Let us calculate the state of the output register after the circuit shown in Fig. 20.2. Starting from the input $|0\rangle^{\otimes K}|u\rangle$, and based on the result in Eq. (20.1) followed by the inverse Fourier transform, we have the following state evolution:

$$\begin{aligned}
|0\rangle^{\otimes K}|u\rangle &\to \frac{1}{\sqrt{N}}\sum_{k=0}^{N-1} e^{2i\pi k\varphi}|k\rangle|u\rangle \\
&\to \frac{1}{\sqrt{N}}\sum_{k=0}^{N-1} e^{2i\pi k\varphi}\left(\frac{1}{2^{K/2}}\sum_{n=0}^{N-1} e^{-\frac{2i\pi kn}{N}}|n\rangle\right)|u\rangle \\
&= \frac{1}{N}\sum_{n=0}^{N-1}\sum_{k=0}^{N-1} e^{-\frac{2i\pi kn}{N}} e^{2i\pi k\varphi}|n\rangle|u\rangle \\
&= \frac{1}{N}\sum_{n=0}^{N-1}\left\{\sum_{k=0}^{N-1}\left[e^{2i\pi(\varphi-\frac{n}{N})}\right]^k\right\}|n\rangle|u\rangle \\
&= \frac{1}{N}\sum_{n=0}^{N-1}\left\{\frac{1-e^{2i\pi(\varphi-\frac{n}{N})N}}{1-e^{2i\pi(\varphi-\frac{n}{N})}}\right\}|n\rangle|u\rangle.
\end{aligned} \quad (20.9)$$

The probability of measuring $n$ from the output register (or the probability of the $N$-qubit register being in state $|n\rangle$) is given by the square modulus of the corresponding amplitude, namely, from Eq. (20.9):

$$\begin{aligned}
p(n) &= \frac{1}{N^2}\left|\frac{1-e^{2i\pi(\varphi-\frac{n}{N})N}}{1-e^{2i\pi(\varphi-\frac{n}{N})}}\right|^2 \\
&\equiv \frac{1}{N^2}\frac{\sin^2\left[\pi\left(\varphi-\dfrac{n}{N}\right)N\right]}{\sin^2\left[\pi\left(\varphi-\dfrac{n}{N}\right)\right]}.
\end{aligned} \quad (20.10)$$

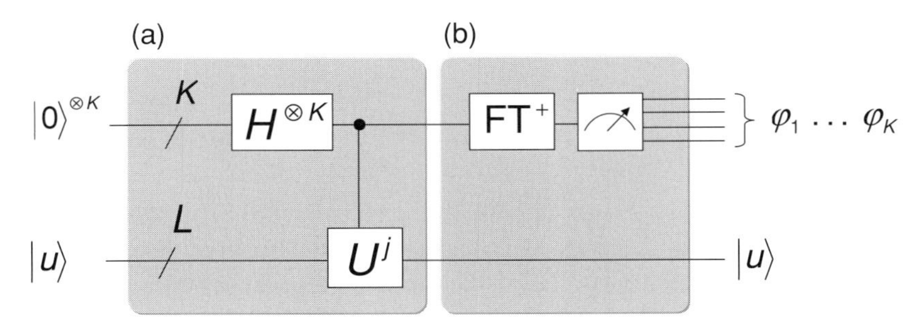


**Figure 20.2** Full quantum circuit for phase estimation, with (a) front-end and (b) back-end modules.

The measurement of $n$, with associated probability $p(n)$, corresponds to the phase estimation $\tilde{\varphi} = n/N$. The probability is maximal when $\delta = \varphi - \tilde{\varphi}$ is minimal. The probability defined in Eq. (20.10) rapidly vanishes as the error $\delta$ departs from this minimum. The conclusion is that *the measurement has the highest probability of yielding the closest approximation of the phase* $\varphi$. Based on the results in Eqs. (20.9) and (20.10), it is also established *that the circuit output is of the form* $|\tilde{\varphi}\rangle|u\rangle$, *where* $|\tilde{\varphi}\rangle$ *is a state superposition which, when measured, yields a fair approximation of the phase* $\varphi$.

It is an academic issue, which I shall not address here, to determine the size of $K$ required for obtaining a phase accuracy of $2^{-l}$ (or $l$ bits) with an arbitrary high probability of success. Suffice it to provide the result of the analysis: for this probability to be *at least* $1 - \varepsilon$, the rule is that the control register size must be $K = l + \lceil \log_2[2 + 1/(2\varepsilon)] \rceil$. It is readily checked that the probability's lower bound increases from 50% ($\varepsilon = 0.5$) to 99% ($\varepsilon = 0.01$) and 99.9% ($\varepsilon = 0.001$) with $K = l + 2$, $K = l + 6$, and $K = l + 9$, respectively, which illustrates the very rapid probability convergence with register size $K$, given the desired number of bits accuracy, $l$.

The above phase-estimation algorithm requires one to prepare the eigenstate $|u\rangle$ of the operator $U$ as the target register. What about the case where we have no prior knowledge of this eigenstate? The remaining possibility is to input to the circuit some other $N$-qubit $|\psi\rangle$, which we know how to prepare. This state can be uniquely decomposed over the orthonormal eigenstate, basis $\{|u\rangle\}$, according to $|\psi\rangle = \sum_u c_u|u\rangle$. By the principle of linearity of quantum gates, we then obtain for the circuit output $|\psi'\rangle = \sum_u c_u\,|\tilde{\varphi}_u\rangle \otimes |u\rangle$. On measuring the first register, we obtain a fair estimation of the phase $\varphi_u$, but this time associated with one possible eigenstate $|u\rangle$, with measurement probability $|c_u|^2$. Such a way to proceed, thus, yields a fair estimation of $\varphi_u$ from an unknown eigenstate $|u\rangle$ selected at random in the eigenstate basis. Stated without demonstration, but as intuitively expected, the probability of success for a phase-estimation accuracy of $2^{-l}$ (or $l$ bits) is at least $|c_u|^2\,(1 - \varepsilon)$ when the size register $K$ is chosen according to this rule.

The phase-estimation circuit, and, in particular, the "successful phase-approximation" algorithm, including the case of unknown eigenstates, are the key to solving the so-called *order-finding problem*, as we shall see in the next section.

**Table 20.1** Successive powers of $x = 4$ modulo $N = 13$, showing order $r = 6$.

| $p$ | $4^p$ | $4^p$ mod 13 | $p$ | $4^p$ | $4^p$ mod 13 | $p$ | $4^p$ | $4^p$ mod 13 |
|---|---|---|---|---|---|---|---|---|
| 0 | 1 | 1 | **6** | 4 096 | **1** | 12 | 16 777 216 | 1 |
| 1 | 4 | 4 | 7 | 16 384 | 4 | 13 | 67 108 864 | 4 |
| 2 | 16 | 3 | 8 | 65 536 | 3 | 14 | 268 435 456 | 3 |
| 3 | 64 | 12 | 9 | 262 144 | 12 | 15 | 1 073 741 824 | 12 |
| 4 | 256 | 9 | 10 | 1 048 576 | 9 | 16 | 4 294 967 296 | 9 |

## 20.2 Order finding

In this section, I shall describe the *order-finding algorithm*, which is based on the concept of number *order* in modular algebra. It will then be shown that there exists a unitary operator $U$ that allows one to determine this order through the previously described phase-estimation circuit. Let us introduce the "order" concept first. In the foregoing, we shall call $M = 2^K$ the number of Fourier components, and from now on use $N$ as the integer number that Shor's algorithm will attempt later to factorize into prime numbers.

Assume, then, two positive integer numbers $x$, $N$, such that $x < N$, with the two numbers having no common divisor other than unity. It is said that their *greatest common divisor* (GCD) is unity, or equivalently, that the two numbers are *co-prime*. By definition, *the order of $x$ modulo $N$* is the smallest nonzero integer $r$ satisfying

$$x^r = 1 \bmod N. \tag{20.11}$$

The order of $x$ can also be conceived as the *period* of the powers $x^0, x^1, x^2 \ldots$, modulo $M$. To give an example, let $x = 4$ and $N = 13$. The successive powers of 4 modulo 13 are listed in Table 20.1. It is seen that the period of the power series is $r = 6$, which also corresponds to the smallest nonzero integer for which $4^r = 1 \bmod 13$, according to the order definition in Eq. (20.11).

Next, I shall describe *how the quantum phase-estimation circuit makes it possible to determine $r$*, the order of $x$ modulo $N$, with a high probability of success and accuracy. For this, we first need to introduce the appropriate unitary operator $U$ and its corresponding eigenstates and eigenvalues. We assume that *given two integers $x$, $N$*, satisfying $x < N$ and $x$ being co-prime to $M$, there exists some operator $U_{x,N}$, which acts on the qubit $|y\rangle = |0\rangle, |1\rangle$ as:

$$U_{x,N}|y\rangle = |xy \bmod N\rangle. \tag{20.12}$$

In the following, I shall just note $U_{x,N} = U$ for simplicity.

Second, let $\{|u_s\rangle\}_{s=0,1\ldots r-1}$ be the set of $r$ eigenstates of $U$, with associated eigenvalues $\exp(2i\pi s/r)$, namely, satisfying

$$U\,|u_s\rangle = e^{\frac{2i\pi s}{r}}\,|u_s\rangle \tag{20.13}$$

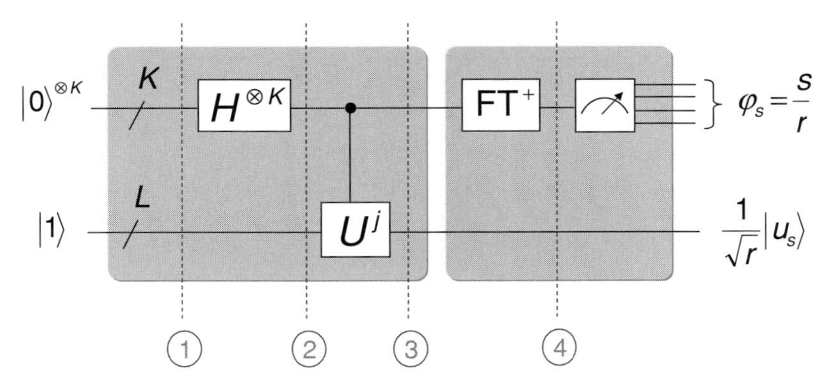


**Figure 20.3** Quantum circuit for order finding.

and with the phase $\varphi_s = s/r$ satisfying $0 \leq \varphi_s \leq 1$. Such eigenstates are defined according to

$$|u_s\rangle = \frac{1}{\sqrt{r}} \sum_{k=0}^{r-1} \mathrm{e}^{-\frac{2i\pi ks}{r}} |x^k \bmod N\rangle, \tag{20.14}$$

which is left as an easy exercise to prove. Finally, we shall observe that the above-defined eigenstates $\{|u_s\rangle\}$ satisfy the properties

$$\frac{1}{\sqrt{r}} \sum_{s=0}^{r-1} |u_s\rangle = |1\rangle \tag{20.15}$$

and

$$\frac{1}{\sqrt{r}} \sum_{s=0}^{r-1} \mathrm{e}^{\frac{2i\pi ks}{r}} |u_s\rangle = |x^k \bmod N\rangle, \tag{20.16}$$

which are also left as easy exercises to prove. The property in Eq. (20.15) is rather convenient, since we do not know how to define the state $|u_s\rangle$, which is a function of the unknown parameter $r$ to be estimated! Recall, indeed, from the previous section, that if we do not know the eigenstate to input to the phase-estimation circuit, we can still use an eigenstate superposition, here, namely, a known state $|1\rangle$ of some arbitrary qubit dimension $L$. The nice additional feature is that the coefficients in the superposition correspond to a *uniform* probability distribution $p(s) = |c_s|^2 = 1/r$, for the register measurement and phase estimation.

The next step will show that using $|1\rangle$ as the input to the second register of the phase-estimation circuit yields, as the output of the front-end module, a $K$ superposition of the states $|k\rangle \otimes |x^k \bmod N\rangle$. The property in Eq. (20.16) then makes it possible by inverse Fourier transform (back-end module) to obtain a fair estimation $\tilde{\varphi}_s$ of the actual phase $\varphi_s = s/r$. Let us see now in detail how the state evolves step by step through the phase-estimation circuit, which is represented in Fig. 20.3. We assume that the first, control, and second, target, register sizes are $K$ and $L = \lceil \log_2 N \rceil$, respectively, with the relation between the two to be specified later on. The controlled-$U^j$ gates perform the application of operator $U$, as defined in Eq. (20.12), to the power $j = 2^0, 2^1 \dots 2^{K-1} = 1 \dots K$. Consistently, we have for the target qubit $|1\rangle$

$$\begin{cases} CU^j(|0\rangle \otimes |1\rangle) = |1\rangle \\ CU^j(|1\rangle \otimes |1\rangle) = |x^j \bmod N\rangle. \end{cases} \tag{20.17}$$

As seen from Fig. 20.3, the combined register input (①) is

$$|\psi_1\rangle = |0\rangle^{\otimes K} \otimes |1\rangle. \tag{20.18}$$

After passing through the Hadamard gate (②) the state has evolved into[3]

$$\begin{aligned} |\psi_2\rangle &= \frac{1}{\sqrt{M}}(|0\rangle + |1\rangle)^{\otimes K} \otimes |1\rangle \\ &= \frac{1}{\sqrt{M}}\sum_{j=0}^{M-1}|j\rangle \otimes |1\rangle, \end{aligned} \tag{20.19}$$

with $M = 2^K$, as a recall. After application of the controlled-$U^j$ gate (③) we obtain the state

$$\begin{aligned} |\psi_3\rangle &= \frac{1}{\sqrt{M}}\sum_{j=0}^{M-1} CU^j\, |j\rangle \otimes |1\rangle \\ &= \frac{1}{\sqrt{M}}\sum_{j=0}^{M-1} |j\rangle \otimes |x^j \bmod N\rangle. \end{aligned} \tag{20.20}$$

Substituting then the property in Eq. (20.16), effecting the index change $j \to k$ for reading convenience, and introducing $\varphi_s = s/r$, we obtain, equivalently:

$$\begin{aligned} |\psi_3\rangle &= \frac{1}{\sqrt{M}}\sum_{k=0}^{M-1} |k\rangle \otimes \frac{1}{\sqrt{r}}\sum_{s=0}^{r-1} e^{2i\pi k\frac{s}{r}}\, |u_s\rangle \\ &= \sum_{s=0}^{r-1}\left(\frac{1}{\sqrt{M}}\sum_{k=0}^{M-1} e^{2i\pi k\varphi_s}\, |j\rangle\right) \otimes \frac{1}{\sqrt{r}}|u_s\rangle. \end{aligned} \tag{20.21}$$

Next, in ④ we apply the inverse Fourier transform to the first register, whose contents are expressed in parenthesis in Eq. (20.21). In this expression, we recognize the same state superposition as in Eq. (20.9). As we have previously seen, the inverse Fourier transform of this superposition leads to a fair approximation $|\tilde{\varphi}_s\rangle$ of the state $|\varphi_s\rangle$. Therefore, the

[3] For the tensor product to the summation conversion, see previous note, while setting $\varphi = 0$. In Eq. (20.20), we also used the property according to which

$$\begin{aligned} CU^j|j\rangle \otimes |1\rangle &= \left|x^{j_1 2^{K-1}} \bmod N\right\rangle \left|x^{j_2 2^{K-2}} \bmod N\right\rangle \cdots \left|x^{j_K 2^0} \bmod N\right\rangle \\ &= \left|x^{j_1 2^{K-1}} \times x^{j_2 2^{K-2}} \times \cdots \times x^{j_K 2^0} \bmod N\right\rangle \\ &= \left|x^{j_1 2^{K-1} + j_2 2^{K-2} + \cdots + j_K 2^0} \bmod N\right\rangle \\ &\equiv |x^j \bmod N\rangle. \end{aligned}$$

circuit output can be expressed in the form:

$$|\psi_4\rangle = \frac{1}{\sqrt{r}} \sum_{s=0}^{r-1} |\tilde{\varphi}_s\rangle \otimes |u_s\rangle . \tag{20.22}$$

A measurement of the first register projects the superposition in Eq. (20.22) into one of the $r$ states $|\varphi_s\rangle$, with uniform probability $p(s) = 1/r$, which then yields the ratio $s/r$, corresponding to the phase estimation $\tilde{\varphi}_s \approx \varphi_s = s/r$ associated with the eigenstate $|u_s\rangle$.

As discussed in the previous section, if the control register size, $K$, is set to $K = l + \lceil \log_2[2 + 1/(2\varepsilon)] \rceil$, the measurement $\tilde{\varphi}_s$ is an approximation of $\varphi_s = s/r$ that is accurate up to $2^{-l}$, with a probability of success of at least $1 - \varepsilon$. Assume here that we require an accuracy of $l = 2L + 1$ bits, a specific but most useful condition that will be justified later. The control register size is, thus, set to $K = 2L + 1 + \lceil \log_2[2 + 1/(2\varepsilon)] \rceil$.

The next step consists in the evaluation of the number $r$, given the knowledge of the $\tilde{\varphi}_s$ measurement, with $s \in \{0, 1, 2, \ldots, r - 1\}$ being random, and the fact that $s/r < 1$ is a *rational number*, i.e., the ratio of two bounded integers, $p, q$. The determination of $r$ requires a classical computation, which is based on the *continued fraction expansion* algorithm described in the next section.

## 20.3 Continued fraction expansion

In this section, I introduce the *continued fraction expansion* (or *continued fraction algorithm*), and show how it can be applied to determine an integer $r$ from a given rational number $s/r < 1$.

The principle of the expansion is to express a rational number $a = p/q$, with $p, q$ having no common factor, into a unique and finite suite, or expansion of positive integers $[a_0, a_1, \ldots, a_n]$ and satisfying

$$a = a_0 + \cfrac{1}{a_1 + \cfrac{1}{a_2 + \cfrac{1}{\cdots + \cfrac{1}{a_n}}}}. \tag{20.23}$$

To provide an example, take $a = 57/21$. We obtain

$$\begin{aligned} \frac{57}{21} &= 2 + \frac{15}{21} = 2 + \cfrac{1}{\cfrac{21}{15}} = 2 + \cfrac{1}{1 + \cfrac{6}{15}} \\ &= 2 + \cfrac{1}{1 + \cfrac{1}{\cfrac{15}{6}}} = 2 + \cfrac{1}{1 + \cfrac{1}{2 + \cfrac{3}{6}}} \equiv 2 + \cfrac{1}{1 + \cfrac{1}{2 + \cfrac{1}{2}}}, \end{aligned} \tag{20.24}$$

which yields the expansion $[2, 1, 2, 2]$. We note that the iteration of this "split and invert" expansion always stops at some point, since the numerator in the last fraction at the bottom at each step decreases, eventually reducing to one. The lesson learnt is that any rational number $a = p/q$ lends itself to a continued fraction expansion having a unique signature suite $[a_0, a_1, \ldots, a_n]$.

Given the measurement $\tilde{\varphi} = x$, with $x$ being some approximation of $\varphi = s/r$, our task is now to identify a rational number $s/r$, called *convergent* of $x$, which may closely approach $x$. To do this, we must use some key properties of the continued fraction expansion, which are described in Appendix R. Here, we shall only need the final property. This property states that given a rational number $x$, and two co-prime integer numbers $s, r$ satisfying

$$\left| \frac{s}{r} - x \right| \leq \frac{1}{2r^2}, \tag{20.25}$$

the ratio $s/r$ is convergent on $x$. This is where the choice of register size and phase accuracy we made in the previous section comes into the picture and becomes justified. With such a choice, indeed, we know that the phase estimation $x$ is accurate up to $2^{-l} = 1/2^{2L+1}$. Since we inherently have $r \leq N$, and also $N \leq 2^L$ by definition, we have $r \leq N \leq 2^L$, thus, $1/(2r^2) \leq 1/2^{2L+1}$ and, therefore, the condition in Eq. (20.25) applies. The continued fraction expansion of $\tilde{\varphi} = x$, can be computed classically through the algorithm described in Appendix R, as also illustrated with a practical numerical example. The algorithm yields a finite series of rational numbers $s'/r'$, which from the above condition, are known to be convergent on $s/r$. The convergent $s'/r'$ that most closely approaches the upper bound defined in Eq. (20.25) is the one for which $r' = r$, concluding the search.

The following discussion, which can be skipped to keep the focus on this chapter, addresses some fine points about the success probability and the implementation cost of the order-finding algorithm. Suffice it to state that the algorithm efficiently yields the order of $N$ and that the complexity or cost is $O(L^3)$, namely, that the answer can be obtained in *polynomial time*.

## Discussion

(i) Since the above determination of $s/r$ (and hence of $r$) is only probabilistic, there exists a finite chance that it may fail. As we have seen, the probability of failure is, at most, $\varepsilon$, which can be made arbitrarily small through an adequate choice of the control register size, $K$. Independently of such probability considerations, checking whether or not the determination is successful is immediate: it is only necessary to calculate $x^{r'} \bmod M$ and verify that the result is unity. In case of failure, the algorithm may be repeated, with the probability of failing again being, at most, $\varepsilon^2$. Another possibility of failure ($x^{r'} \bmod M \neq 1$) is that $s$ and $r$ turn out to have a common factor, or be not co-prime. In this case the continued fraction algorithm yields a multiple of $r$. Recall that the phase estimation produces an estimate of $s/r$ with $s \in \{0, 1, 2, \ldots, r-1\}$ being a uniformly distributed random number. There is, indeed, a finite chance that $s, r$ is not co-prime. It has been established from theory that the number of primes under a given integer $r$ is at least $r/\ln r$.[4] This means that any positive number less than $r$, as selected at random, has a probability of at least $1/\ln r$ of being prime. Therefore, the probability that $s, r$ $(0 < s < r, r < N)$ are co-prime is at least $1/\ln r > 1/\ln N$. Hence, it takes one to implement the phase measurement up to $\ln N$ times to obtain with reasonably high probability a co-prime pair $s', r'$ for which $r' = r$.

(ii) What is the cost of implementing the order-finding algorithm? Looking at Fig. 20.3, we observe that the corresponding quantum circuit includes (a) $K$ Hadamard gates ($H^{\otimes K}$), (b) an $L$-qubit modular-exponentiation circuit ($U^j$), and (c) a $K$-qubit inverse Fourier transform circuit ($\text{FT}^+$). Since $K = 2L + 1 + \lceil \log_2[2 + 1/(2\varepsilon)] \rceil$, given $\varepsilon$ the number of required gates in case (a) is of the order $O(L)$, and in case (c) it is of the order $O(K^2) = O(L^2)$, as established in Chapter 19. The modular exponentiation in case (c) requires a number of gates of the order of $O(L^3)$. This is explained by the fact that computing $|x^j \bmod N\rangle$, with $j = 0, 1 \ldots 2^K$, requires up to $K - 1$ modulo $N$ squaring operations ($x^{2^1} = (x^{2^0})^2$, $x^{2^2} = (x^{2^1})^2, \ldots, x^{2^{K-1}} = (x^{2^{K-2}})^2$, with each squaring operation having a cost of $O(K^2)$ gates,[5] resulting in an overall cost of $O(K^3) \equiv O(L^3)$ gates. Finally, the continued fraction expansion, which is to be implemented with a classical computer, has a cost of $O(L^3)$ elementary operations. This is because the algorithm takes $L$ split-and-invert steps, since $s, r$ are $L$-bit integers, each of these steps requiring a number of basic arithmetic operations of the order $O(L^2)$ (see Appendix R, namely, the operations $u_n = 1/(u_{n-1} - a_{n-1})$ and the validation test $1/2q_n^2 - \Delta$ in Table R1). The key conclusion is that the order-finding algorithm has a complexity of $O(L^3)$ and, therefore, can be run in polynomial time.

## 20.4 From order finding to factorization

In this section, we describe how the quantum order-finding algorithm makes it possible to factorize composite numbers into primes, a problem also known as *prime decomposition*. Before going into the mathematical detail of the description, it is useful to outline the problem of factorizing composite numbers and its difficulty, as viewed from a classical computation perspective.

It is a fundamental property that all integer numbers can be uniquely generated by products of prime numbers and their powers. This is illustrated in Table 20.2 for numbers up to $N = 30$.[6] By definition, a composite number is a product of at least two prime numbers. For numbers up to a few hundreds, the task of factorizing can be performed

[4] The number of primes less than $x$ corresponds to the heuristic function called $\pi(x)$. The "prime number theorem" states that the number of primes not exceeding $x$ is asymptotic to $x/\ln x$. Actually, a finer approximation to $\pi(x)$ is $x/(\ln x - 1)$. For reference, see, for instance, http://primes.utm.edu/howmany.shtml.

[5] Indeed, it is easily checked that multiplying two numbers with $N$ and $M$ decimals (or bits) requires $N \times M$ individual multiplications and $N \times M - 1$ additions of the resulting terms. The number of gates is thus $O(N \times M)$, or $O(N^2)$ for a squaring operation ($N = M$).

[6] A complete list of prime factors up to $N = 1000$ is available at http://en.wikipedia.org/wiki/Table_of_prime_factors.

**Table 20.2** Factorization of integers into primes up to $N = 30$.

| $N$ | | $N$ | | $N$ | | $N$ | | $N$ | | $N$ | |
|---|---|---|---|---|---|---|---|---|---|---|---|
| 1 | – | 6 | $2 \times 3$ | 11 | 11 | 16 | $2^4$ | 21 | $3 \times 7$ | 26 | $2 \times 13$ |
| 2 | 2 | 7 | 7 | 12 | $2^2 \times 3$ | 17 | 7 | 22 | $2 \times 11$ | 27 | $3^3$ |
| 3 | 3 | 8 | $2^3$ | 13 | 13 | 18 | $2 \times 3^2$ | 23 | 23 | 28 | $2^2 \times 7$ |
| 4 | $2^2$ | 9 | $3^2$ | 14 | $2 \times 7$ | 19 | 19 | 24 | $2^3 \times 3$ | 29 | 29 |
| 5 | 5 | 10 | $2 \times 5$ | 15 | $3 \times 5$ | 20 | $2^2 \times 5$ | 25 | $5^2$ | 30 | $2 \times 3 \times 5$ |

mentally. This is because we have our multiplication tables memorized, which leaves out a list of the first prime numbers: 2, 3, 5, 7, 11, 13, 17, 19, 23, 29, 31, 37, 41, 43, 47, . . . It only takes a few divisions to hit one or two numbers from such a short list. For greater numbers that can be entered into a pocket calculator or a computer spreadsheet, the problem of factorizing is also elementary. The basic approach consists of the routine of successively attempting to divide $N$ by all known prime numbers lower than $\sqrt{N}$, and repeating such a trial division until complete prime factoring is obtained. Extensive lists of the known prime numbers, up to 15 million, are available on the Internet,[7] making it possible to implement such a trial division algorithm. More practical and efficient algorithms exist, however, named after Pollard, William, Lenstra, Fermat, Dixon, and Shank, and also under different variants of the so-called "number field sieve."[8] The most efficient one is the general number field sieve (GNFS) algorithm, which applies to numbers having more than 100 binary digits. It is not the point here to describe the GNFS algorithm and its variants, but only to mention that its complexity (the minimum computing program length, see Chapter 7) is defined by:[9]

$$C = \exp\left\{[\text{const} + \mathrm{O}(1)] \times (\ln N)^{\frac{1}{3}} (\ln \ln N)^{\frac{2}{3}}\right\}. \tag{20.26}$$

It can be shown that the best GNFS algorithm variant has a computational order of

$$\mathrm{O}\left\{\exp\left[\frac{64}{9}(\ln N)^{\frac{1}{3}} \times (\ln \ln N)^{\frac{2}{3}}\right]\right\} = \mathrm{O}\left\{\exp\left[\frac{64}{9}k^{\frac{1}{3}} \times (\ln k)^{\frac{2}{3}}\right]\right\}, \tag{20.27}$$

where $k \approx \log N$ is the binary size of the number $N$ to be factorized. Thus, classical implementation of prime decomposition is *subexponential*, but *superpolynomial* in number (bit) size. To show that the above conclusion is not innocuous, assume that the size is $k = 100$. We obtain from Eq. (20.27) the result $\mathrm{O}[f(N)] \approx 4.7 \times 10^{39}$. For $k = 1000$, we obtain $\mathrm{O}[f(N)] \approx 1.0 \times 10^{112}$. For a classical computer having the computing power of *one Giga* ($10^9$) *operations per second*, or $3 \times 10^{16}$ operations per year, just factoring a $k = 100$ number according to GNFS would require some 23 years, or, say, implementing about 100 such computers running in parallel for a duration of three months. We should now be convinced, if needed, that the factoring problem is nontrivial when it comes to relatively large numbers. This point is illustrated by the *RSA challenge*.[10] The company

[7] See, for instance: http://en.wikipedia.org/wiki/List_of_prime_numbers; www.math.utah.edu/~pa/math/primelist.html; http://primes.utm.edu/lists/small/1000.txt; http://primes.utm.edu/lists/small/millions/.

[8] See http://en.wikipedia.org/wiki/Integer_factorization.

[9] See http://en.wikipedia.org/wiki/General_number_field_sieve.

[10] See www.rsa.com/press_release.aspx?id=3520.

RSA offers substantial prizes for teams who may succeed in factoring composite numbers of various sizes, i.e., given $N$, finding primes $p, q$ such that $N = pq$. The latest challenge to be solved, called RSA_640, was reported on November 2005.[11] The number to be factorized had 193 digits, corresponding to 640 bits:

$$\begin{aligned} N = {} & 3\,107\,418\,240\,490\,043\,721\,350\,750\,035\,888\,567\,930\,037\,346\,022\,842\,727\,545 \\ & 720\,161\,948\,823\,206\,440\,518\,081\,504\,556\,346\,829\,671\,723\,286\,782\,437\,916\,272 \\ & 838\,033\,415\,471\,073\,108\,501\,919\,548\,529\,007\,337\,724\,822\,783\,525\,742\,386\,454 \\ & 014\,691\,736\,602\,477\,652\,346\,609, \end{aligned}$$

which decomposes into the two primes:

$$\begin{aligned} p = {} & 1\,634\,733\,645\,809\,253\,848\,443\,133\,883\,865\,090\,859\,841\,783\,670\,033\,092\,312\,181 \\ & 110\,852\,389\,333\,100\,104\,508\,151\,212\,118\,167\,511\,579, \\ q = {} & 1\,900\,871\,281\,664\,822\,113\,126\,851\,573\,935\,413\,975\,471\,896\,789\,968\,515\,493\,666 \\ & 638\,539\,088\,027\,103\,802\,104\,498\,957\,191\,261\,465\,571. \end{aligned}$$

The computation of $p, q$ took the equivalent of 30 CPU years, using 80 processors at 2.2 GHz clock cycle, and spread over 5 months of calendar time. The highest and yet unsolved challenge, RSA-2048, is a 2048-bit or 617-digit number. According to Eq. (20.27), its factorization is of the order $\mathrm{O}[f(N)] \approx 8.5 \times 10^{151}$. It is clear that solving this last challenge will require even more substantial computing power and resources, along with new and significant progress in factorization algorithms, which may take a few decades. It is tempting to speculate that, by that time, factorization of such big numbers might be routinely performed by quantum computers through Shor's algorithm! Only history will tell.

We now return to the core subject of this section, which is the connection between order finding and factoring. This connection requires two key theorems, which I am going successively to describe, comment, and illustrate with examples. It is assumed that $N$ is a composite number of size $L$ bits, and that $x$ is an integer number such that $1 \le x \le N - 1$. The notation GCD$(n, m)$ corresponds to the greatest common divider between two integers $n, m$.

THEOREM 20.1 *If $x$ is a nontrivial solution of $x^2 = 1$ mod $N$, then either* GCD$(x - 1, N)$ *or* GCD$(x + 1, N)$ *is a factor of $N$; such a factor is computable in* $\mathrm{O}(L^2)$ *operations.*

*In the above, the two trivial solutions of $x^2 = 1$ mod $N$ to be discarded are $x = \pm 1$ mod $N$ corresponding to $x = 1$ and $x = N - 1$. By assumption, therefore, the solution $x$ is in the range $1 < x < N - 1$. We have $x^2 - 1 = 0$ mod $N$, which shows that $N$ divides by $x^2 - 1 = (x + 1)(x - 1)$, or, equivalently, that $N$ has a common factor with either $x - 1$ or $x + 1$. We note that such a common factor cannot be $N$ itself, since $x - 1 < x + 1 < N$. Thus the factor is found by computing both* GCD$(x - 1, N)$ *and* GCD$(x + 1, N)$.

I shall now illustrate Theorem 1 through two basic examples.

[11] See http://mathworld.wolfram.com/RSANumber.html.

**Table 20.3** Values of $x^2 = z \bmod N$ for $N = 35$ and $1 < x < N - 1$.

| $x$ | $z$ | $x$ | $z$ | $x$ | $z$ | $x$ | $z$ | $x$ | $z$ |
|---|---|---|---|---|---|---|---|---|---|
| 1 | | 8 | 29 | 15 | 15 | 22 | 29 | 29 | 1 |
| 2 | 4 | 9 | 11 | 16 | 11 | 23 | 4 | 30 | 25 |
| 3 | 9 | 10 | 30 | 17 | 9 | 24 | 16 | 31 | 16 |
| 4 | 16 | 11 | 16 | 18 | 9 | 25 | 30 | 32 | 9 |
| 5 | 25 | 12 | 4 | 19 | 11 | 26 | 11 | 33 | 4 |
| **6** | **1** | 13 | 29 | 20 | 15 | 27 | 29 | 34 | |
| 7 | 14 | 14 | 21 | 21 | 21 | 28 | 14 | 35 | |

As a first example, assume the composite number $N = 35$. The values of $x^2 = z \bmod 35$ are listed in Table 20.3. We observe from the table that $x^2 = 1 \bmod 35$ has two nontrivial solutions $x_1 = 6$ and $x_2 = 29 \equiv -6 \bmod 35$, or, equivalently, $x = \pm 6 \bmod 35$. With the solution $x_1 = 6$, we obtain $\mathrm{GCD}(x_1 - 1, N) = \mathrm{GCD}(5, 35) \equiv 5$ and $\mathrm{GCD}(x_1 + 1, N) = \mathrm{GCD}(7, 35) \equiv 7$. With the other equivalent solution $x_2 = 29$, we obtain $\mathrm{GCD}(x_2 - 1, N) = \mathrm{GCD}(28, 35) = 7$ and $\mathrm{GCD}(x_2 + 1, N) = \mathrm{GCD}(30, 35) \equiv 5$. Thus, any of the solutions yield two GCDs that are both factors of $N = 35$.

As a second example, assume the composite number $N = 561$. With a tabulating spreadsheet, we find the solutions $x^2 = 1 \bmod 561$ to be $x_1 = 67$, $x_2 = 188$, $x_3 = 254$, $x_4 = 307$, $x_5 = 373$, and $x_6 = 494$, or, equivalently, $x = \pm 67, \pm 188, \pm 254$. We can find $\mathrm{GCD}(x_i \pm 1, N)$ by means of the extended Euclidian algorithm.[12] Given two numbers $a, b$ such that $a > b$, the algorithm can be summarized through the iterated operation:

$$(a, b) \to (a', b') = (b, a \bmod b), \tag{20.28}$$

which at the stage where $b' = 0$ yields $a' = GCD(a,b)$. Let us illustrate the algorithm through a basic example. Considering $a = 561$ and $b = 189$ for instance, we obtain the following iteration:

$$\begin{aligned}
(561, 189) &\to (a', b') = (189, 183) \\
(189, 183) &\to (a', b') = (183, 6) \\
(183, 6) &\to (a', b') = (6, 3) \\
(6, 3) &\to (a', b') = (3, 0).
\end{aligned}$$

The result obtained in the final iteration (where $b' = 0$) shows that $\mathrm{GCD}(561, 189) = 3$. Using the algorithm and a computing spreadsheet, we easily obtain for the solutions $x_1, x_2, x_3$:

$$\begin{aligned}
\mathrm{GCD}(x_1 - 1, N) &= \mathrm{GCD}(66, 561) \equiv 33 \\
\mathrm{GCD}(x_1 + 1, N) &= \mathrm{GCD}(68, 561) \equiv 17 \\
\mathrm{GCD}(x_2 - 1, N) &= \mathrm{GCD}(187, 561) \equiv 187 \\
\mathrm{GCD}(x_2 + 1, N) &= \mathrm{GCD}(189, 561) \equiv 3 \\
\mathrm{GCD}(x_3 - 1, N) &= \mathrm{GCD}(253, 561) \equiv 11 \\
\mathrm{GCD}(x_3 + 1, N) &= \mathrm{GCD}(255, 561) \equiv 51.
\end{aligned}$$

[12] See, for instance: http://en.wikipedia.org/wiki/Euclidean_algorithm.

It is readily checked that $561 = 33 \times 17 = 3 \times 187 = 11 \times 51 \equiv 3 \times 11 \times 17$, which shows that $\mathrm{GCD}(x_i \pm 1, N)$ is always a factor of $N$. We also observe that given a single solution $x_i$, one of the two corresponding GCD, i.e., $\mathrm{GCD}(x_i \pm 1, N)$, is a prime factor of $N$ (namely, 3, 11, or 17). The calculation of the six GCDs corresponding to the three solutions $x_1, x_2, x_3$, namely, $\mathrm{GCD}(x_1 \pm 1, N)$, $\mathrm{GCD}(x_2 \pm 1, N)$, and $\mathrm{GCD}(x_3 \pm 1, N)$, thus, makes it possible to factorize $N$ completely.

The lesson learnt from Theorem 1 and these two illustrative examples is that the knowledge of any single (nontrivial) solution of $x^2 = 1 \bmod N$ yields two factors of $N$. The complete factorization of $N$ can, thus, be achieved by repeating the process with the remaining factors (call any of these $N'$), provided that for each we can find a solution of $x^2 = 1 \bmod N'$. What is the computation cost of the algorithm used to determine the GCDs? There exist several possible answers to this question, depending on the algorithm choice and its implementation. It can be shown that for $L$ digit numbers, the extended Euclidian algorithm complexity is of the order of $\mathrm{O}(L^2)$ or better and, furthermore, that there exist other algorithms for which finding the GCD is reduced to $\mathrm{O}[L(\ln L)^2 \ln\ln L]$. The continued fraction expansion algorithm, which was described in Section 20.3, can be also implemented for this purpose but, as we have seen, its complexity is $\mathrm{O}(L^3)$. Here, we may only retain that finding the GCD through the extended Euclidian algorithm is, at most, of the order of $\mathrm{O}(L^2)$.

While Theorem 1 enables one to factorize a given composite number $N$ rapidly, it exclusively relies on prior knowledge of at least one nontrivial solution of $x^2 = 1 \bmod N$. But how can we get this knowledge? This is where the order-finding algorithm, which was described in Section 20.2, nicely comes to the rescue. As we have seen, given any integer $x$, such that $1 \leq x \leq N - 1$, the order of $x$ modulo $N$ is the smallest number $r$ for which $x^r = 1 \bmod N$. In the case where $r$ is *even*, let $y = x^{r/2}$. We, thus, have $y^2 = 1 \bmod N$, which shows that $y$ is a solution of the Theorem 1 equation. If $y \neq \pm 1 \bmod N$, then $y$ is a nontrivial solution of the equation and, according to the theorem, $\mathrm{GCD}(y \pm 1, N) = \mathrm{GCD}(x^{r/2} \pm 1, N)$ is a factor of $N$. In this case, the order-finding algorithm successfully yields a determination of a factor of $N$. If $y$ is a trivial solution, the algorithm fails, and the order-finding algorithm must be implemented again, using a different trial value for $x$. The same conclusion applies when $r$ turns out to be *odd*. The fact that the algorithm may fail should not be perceived as an embarrassing weakness of the factoring endeavor. It just takes a finite number of order-finding trials (each of $\mathrm{O}(L^3)$) to lead to a successful and conclusive step, and as many such trials to eventually achieve the full factorization. Actually, a second theorem shows that the convergence of the algorithm is strong. This theorem states:

THEOREM 20.2 *Given $N$, an odd composite integer with the factorization $N = p_1^{\alpha_1} p_2^{\alpha_2} \ldots p_m^{\alpha_k}$, where $p_i$ are prime numbers ($\alpha_i$ integers), and given an integer $x$ chosen at random in the interval $[1, N-1]$ and co-prime to $N$, the probability that $r$ is even and $y = x^{r/2}$ is a nontrivial solution of $y^2 = 1 \bmod N$ satisfies*

$$p \geq 1 - \frac{1}{2^m}. \tag{20.29}$$

This second theorem shows that the probability of successfully obtaining a factor of $N$ rapidly increases with the number $m$ of prime factors. In the most basic case $m = 2$ (as in the above-described RSA challenge), we have $p \geq 1 - 1/2^2 = 0.75$. This corresponds to a 25% probability of failure, which reduces to less than 0.5% in four successive trials! Since any composite is at least the factor of two primes, this failure probability actually represents an upper bound for all possible composites. Theorem 2 is also a very strong one: to implement the order-finding algorithm and find the factors of $N$ if successful, we are allowed to choose at random any integer $x$, provided it be co-prime to $N$. The test is only a matter of calculating GCD$(x, N)$, at a mere $O(L^2)$ cost. It is an academic issue to prove Theorem 2, which I shall not address here. The key lesson learnt is that the order-finding algorithm leads to efficient and rapid factorization of composite numbers. Basically, this whole description constitutes the essence and ingredients of Shor's factorization algorithm, to be summarized and illustrated in the next section.

## 20.5 Shor's factorization algorithm

The rewards of going through the preceding sections and tedious developments are now at hand. In fact, this section does not bring any new concept in this respect. It only consists in the formalization of *Shor's factorization algorithm*. There are many possible ways to formalize an algorithm, from a list of practical steps with programming flow charts, to more mathematically abstract and academic definitions. Here, I shall follow the first approach and also provide some illustrations. It is brought to the reader's attention that to date, *there exists no classical computer algorithm making it possible to implement order-finding in polynomial time*. The order-finding algorithm, the key constituent in Shor's factorization, is to be implemented in a hypothetical quantum-computing or quantum-phase-estimation circuit, as illustrated in Fig. 20.3. As we have seen, the rest is only a matter of classical computation with an overall $O(L^3)$ or $O[(\log N)^3]$ complexity.

We may first make a few assumptions to simplify the algorithm description. First, the composite number $N$ must be *odd*. The test is immediate, based on the value of the last or lowest-weight binary digit. Second, it must not be a trivial product of small prime numbers and their powers. For instance $N = 15, 27, 39, 144$ are trivial composites for the human brain. For a supercomputer, which has in its memory the list of the first 1000 prime numbers (for instance), and look-up tables of the type shown in Table 20.2 giving the basic factorizations, thereof, there exist millions of "trivial" composites that can be factorized in milliseconds or faster. In such cases, there is no point whatsoever in considering Shor's algorithm! Another trivial case (as viewed from a computer perspective) is when there exist two integers $a \geq 1$ and $b \geq 2$, such that $N = a^b$. It is an academic issue, not addressed here, to establish that the solution $a, b$ can be classically obtained in $O(L^3)$ time. The problem is, thus, reduced to the factoring of $a$. It can be assumed that given $N$, all of the above tests, forming a "preamble" prove negative, and this is where Shor's algorithm must be implemented, as described in the following.

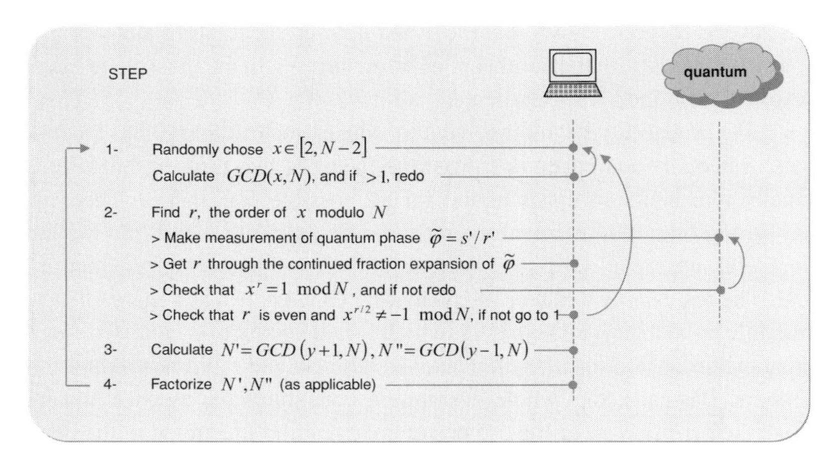


**Figure 20.4** Step-by-step implementation of Shor's algorithm for nontrivial composite number $N$ of bit size $L$, returning all factors of $N'$.

## Shor's algorithm

Here is a step-by-step description, which is also schematically illustrated in Fig. 20.4.

Step 1 Choose at random $x \in [2, N-2]$ and test if $x$, $N$ are co-prime. If the test fails, then another $x$ must be selected.

Step 2 Find $r$, the order of $x$ modulo $N$. This is done through the quantum phase measurement or estimation, $\tilde{\varphi} = s'/r'$, and the continued fraction expansion algorithm, which yields $r' = r$, as described in Section 20.3. We then check that $x^{r'} = 1 \bmod N$. As we have seen, there exists a small yet finite probability that this operation may fail, in which case this test is negative, and we may just try another phase measurement to obtain a different phase estimate, $\tilde{\varphi}$. Then we must ensure that $r$ is even, and also that $y = x^{r/2} \neq -1 \bmod N$ is a nontrivial solution.[13] If the test is negative, the whole process may be restarted at Step 1 (Theorem 2 ensures, however, that the success probability is high).

Step 3 Calculate GCD($y \pm 1$, $N$), which yields two factors of $N$, call them $N'$, $N''$. Then $N'$, $N''$ can be submitted to the aforementioned "preamble" tests, to identify whether further factorizing through Shor's algorithm may be warranted.

Step 4 As applicable, $N'$, $N''$ may be factorized in turn, starting again from step 1.

According to the above rendition of Shor's algorithm, the end result is a prime decomposition of two factors of $N$, according to $N = p_1^{\alpha_1} p_2^{\alpha_2} M$. In the case where $M > 1$, the "preamble" tests may be implemented for the possibility of trivial factoriza-

[13] As a matter of fact, given $r$ being even, the test $x^{r/2} \neq 1 \bmod N$ does not need to be performed, as this condition is implicitly verified should $x$ be selected from the interval $[2, N-1]$, or $x \neq 1$. This is because $r$ is, by definition, the smallest integer verifying $x^r = 1 \bmod N$. Since $r$ is even, then $r/2$ is an integer. Then if we had $x^{r/2} = 1 \bmod N$ with $x \neq 1$, the period of $x$ would be $r/2$ and not $r$, which proves the point.

tion. As applicable, Shor's algorithm may otherwise be called upon again to factorize $M$, and so on until we obtain the full prime-factor decomposition $N = p_1^{\alpha_1} p_2^{\alpha_2} \ldots p_m^{\alpha_k}$.

## 20.6 Factorizing $N = 15$ and other nontrivial composites

In this section, I shall consider illustrative examples of Shor's factorization algorithm, using *nontrivial* composite cases (the notion of trivial composites will be specified further down). This will also show how one can simplify the inverse Fourier transform operation ($\mathrm{FT}^+$) in the quantum phase-estimation circuit, and estimate the probability distribution function of the phase measurement.

Assume, for instance, the composite number $N = 15$. We must set $L = \lceil \log_2 N \rceil = 4$ for the size of the second or target register, and for an error probability of at most $\varepsilon = 0.25$, we must also set $K = 2L + 1 + \lceil \log_2[2 + 1/(2\varepsilon)] \rceil = 11$ for the size of the first or control register. Thus, $M = 2^K = 2^{11} = 2048$. We then select "at random" from the interval $[2, N-2]$ the value $x = 8$, which meets the requirement of being co-prime to $N$. The qubit tensor input to the phase-estimation circuit in Fig. 20.3 is $|\psi_1\rangle = |0\rangle^{\otimes K} \otimes |1\rangle$, Eq. (20.18). After passing through the Hadamard gate, it is transformed into (Eq. (20.19)):

$$\begin{aligned} |\psi_2\rangle &= \frac{1}{\sqrt{M}} \sum_{j=0}^{M-1} |j\rangle \otimes |1\rangle \\ &= \frac{1}{\sqrt{2^{13}}} \left( |0\rangle + |1\rangle + |2\rangle + \cdots + |M-1\rangle \right). \end{aligned} \tag{20.30}$$

After application of the controlled-$U^j$ gate, we obtain the state (Eq. (20.20)):

$$\begin{aligned} |\psi_3\rangle &= \frac{1}{\sqrt{M}} \sum_{j=0}^{M-1} |j\rangle \otimes |x^j \bmod N\rangle \\ &= \frac{1}{\sqrt{M}} \sum_{j=0}^{M-1} |j\rangle \otimes |8^j \bmod 15\rangle \\ &= \frac{1}{\sqrt{M}} \begin{pmatrix} |0\rangle|1\rangle + |1\rangle|8\rangle + |2\rangle|4\rangle + |3\rangle|2\rangle \\ + |4\rangle|1\rangle + |5\rangle|8\rangle + |6\rangle|4\rangle + |7\rangle|2\rangle \\ + |8\rangle|1\rangle + |9\rangle|8\rangle + |10\rangle|4\rangle + |11\rangle|2\rangle + \cdots \end{pmatrix} \\ &= \frac{1}{\sqrt{M}} \begin{bmatrix} (|0\rangle + |4\rangle + |8\rangle + \cdots)\,|1\rangle \\ +(|1\rangle + |5\rangle + |9\rangle + \cdots)\,|8\rangle \\ +(|2\rangle + |6\rangle + |10\rangle + \cdots)\,|4\rangle \\ +(|3\rangle + |7\rangle + |11\rangle + \cdots)\,|2\rangle \end{bmatrix}. \end{aligned} \tag{20.31}$$

The state $|\psi_3\rangle$ in Eq. (20.31) can also be put in the form:

$$|\psi_3\rangle = |u_1\rangle \otimes |1\rangle + |u_2\rangle \otimes |8\rangle + |u_3\rangle \otimes |4\rangle + |u_4\rangle \otimes |2\rangle, \tag{20.32}$$

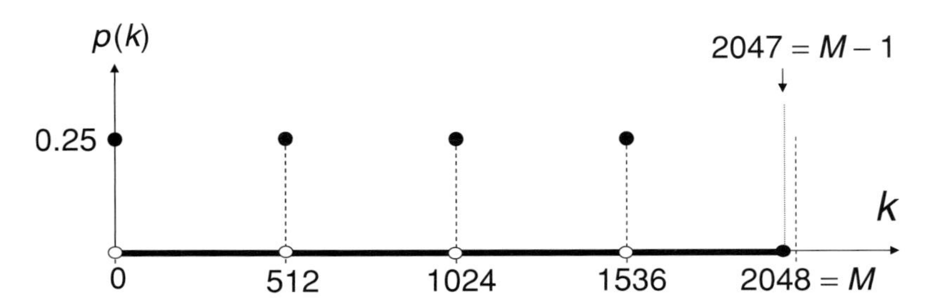


**Figure 20.5** Plot of the distribution $p(k) = |\alpha_k|^2$ in the definition interval $k \in [0, M-1]$.

with

$$\begin{cases} |u_1\rangle = \dfrac{1}{\sqrt{M}}(|0\rangle + |4\rangle + |8\rangle + \cdots) \\ |u_2\rangle = \dfrac{1}{\sqrt{M}}(|1\rangle + |5\rangle + |9\rangle + \cdots) \\ |u_3\rangle = \dfrac{1}{\sqrt{M}}(|2\rangle + |6\rangle + |10\rangle + \cdots) \\ |u_4\rangle = \dfrac{1}{\sqrt{M}}(|3\rangle + |7\rangle + |11\rangle + \cdots). \end{cases} \tag{20.33}$$

If at this stage (③ in Fig. 20.3) we perform a measurement of the second register, which is no longer used, Eq. (20.32) shows that we may obtain at random any of the values $z = 1, 8, 4$, or 2, with a uniform probability of 1/4. Such a measurement causes the first register to collapse into the corresponding state $|u_1\rangle$, $|u_2\rangle$, $|u_3\rangle$, or $|u_4\rangle$ with a new normalization factor of $1/\sqrt{4}$. We could also make the same measurement at ④, namely, after the first register has been submitted to the inverse Fourier transform, $\mathrm{FT}^+$, and even further down in the circuit, i.e., after the first register has been measured. Intuitively, the *measurement statistics* of the second register are not affected by the transformation and measurement of the first register, and the reverse. This principle can be stated as "any quantum wires whose qubits are not measured at the end of a quantum circuit may yet be assumed to be also measured, without affecting any other measurement statistics." It remains a fine point to analyze this principle academically, but here we shall take its validity for granted. Assume, then, that our measurement of the second register would be $z = 8$ (any other value giving identical results as the conclusion). Thus, in the first register, the input to the inverse Fourier transform circuit would be $|u_2\rangle$. After transformation, the first register output is

$$\mathrm{FT}^+ |u_2\rangle = \sum_{k=0}^{M-1} \alpha_k |k\rangle, \tag{20.34}$$

where $p(k) = |\alpha_k|^2$ represents the probability distribution of the final measurement of the first register. The detailed computation of the distribution $p(k)$ is given in Appendix S. The resulting distribution is plotted in Fig. 20.5. As we see from the figure, the distribution exhibits four equiprobable peaks ($p_{\max} = 0.25$) at the register locations $k = 0$, 512, 1024, 1536, with any other measurement probabilities being zero. This probability distribution represents the ideal measurement conditions with zero failure. As we have learnt, the actual measurement is successful only within some error probability $\varepsilon$, which is a function of the register size $K$, and which has been set in this example to $\varepsilon = 0.25$. To recall, this failure probability can be made as small as desired, hence making the measurement reality closer to the ideal probability distribution shown in Fig. 20.5. The existence of four equiprobable probability peaks betrays the periodicity of the function $x^n$ mod $N$, which is of the order $r = 4$. But the experimentalist has no preconceived idea of this, and is allowed to make only a few measurement attempts. Then it is equally likely for each of the measurements to hit the values $k_i = 0, 512, 1024$, or 1536. To recall, these measurements correspond to the phase $\tilde{\varphi} \approx \varphi = s/r$, with $0 \leq \varphi \leq 1$ being defined in $K$ bits. The measurement, thus, corresponds to the rational numbers $k_i/M = k_i/2^K = k_i/2^{13} = k_i/2048$, corresponding to the four possible determinations $\tilde{\varphi}$:

$$\left.\frac{0}{2048}\right|_{k_i=0}, \left.\frac{512}{2048}\right|_{k_i=512}, \left.\frac{1024}{2048}\right|_{k_i=1024}, \left.\frac{1536}{2048}\right|_{k_i=1536}.$$

The first determination, $k_i = 0$, does not give any clue as to the order $r$. Measuring it is, therefore, a failed attempt. Consider the next two other determinations, i.e., $k_i = 512, 1024$. We have $k_1 = 512/2048 = 1/4$ and $k_2 = 1024/2048 = 1/2$. These two rational numbers do not meet the condition in Eq. (20.25) since $(1, 4)$ and $(1, 2)$ are *not* co-prime. Therefore, they cannot be convergent on $\varphi$. The last possible measurement, $k_3 = 1536$ lends itself to the continued fraction expansion:

$$k_3 = \frac{1536}{2048} = \frac{1}{1 + \frac{1}{3}},$$

which, according to the algorithm (see Appendix R) gives the series of fractions $p_0/q_0 = 0/1$, $p_1/q_1 = 1/1$, and $p_2/q_2 = 3/4$. Since $(3, 4)$ are co-prime, the fraction 3/4 is a convergent of $\varphi$ and thus $r = q_2 = 4$ is of the order of $x$. Luckily, $r = 4$ is even, and, furthermore, we have $x^{r/2}$ mod $N = 8^2$ mod $15 = 4$ mod $15 \neq \pm 1$ mod 15, which shows that the solution $x^{r/2} = 8^2 = 64$ is valid (it can also be checked that $x^r = 8^4 = 1$ mod 15, but this test does not need to be made). Therefore, two factors $N'$, $N''$ of $N = 15$ are

$$N' = \text{GCD}(x^{r/2} - 1, N) = \text{GCD}(63, 15) \equiv 3,$$
$$N'' = \text{GCD}(x^{r/2} + 1, N) = \text{GCD}(65, 15) \equiv 5,$$

which concludes Shor's factorization algorithm. Since both factors $N'$, $N''$ are prime, the factorization of $N$ is also complete! As we have seen, however, the algorithm succeeded in this example because (a) one of the four possible measurements yielded a valid convergent of $\varphi$, (b) the order $r$ turned out to be even, and (c) the condition $x^{r/2} \neq -1$ mod 15 was satisfied. In the case where any of these prerequisites is not met, the algorithm would have failed. Provided we obtained a convergent of $\varphi$, it turned out that the random selection $x = 8$ was a "lucky" one. Then what are the other alternatives

**Table 20.4** Possibilities of selecting $x$ towards the factorization of $N = 15$ through Shor's algorithm, with corresponding order $r$, test $x^{r/2} \neq \pm 1 \bmod N$, and number $m$ of convergents.

| $x$ | $r$ | $x^{r/2} \bmod N$ | $m$ |
|---|---|---|---|
| 2 | 4 | 4 | 1 |
| 4 | 2 | 4 | 0 |
| 7 | 4 | 4 | 1 |
| 8 | 4 | 4 | 1 |
| 11 | 2 | 11 | 0 |
| 13 | 4 | 4 | 1 |

and chances of failure under any other random selection for $x$? The answer is provided in the following discussion.

## Discussion

To factorize $N = 15$, we must randomly select $x$ in the interval $x \in [2, N - 2]$, with the condition that $(x, N)$ be co-prime. This leaves $x = 2, 4, 7, 8, 11, 13$ as the only six possibilities of fully implementing the algorithm. We then use a computer spreadsheet to obtain the order $r$ "classically" (this being not part of Shor's algorithm, but just a computing means for the purpose of the analysis). The results are listed in Table 20.4, along with the test $x^{r/2} \bmod N$ and the number $m$ of convergents. It is seen from the table that the order of $x$ is either $r = 2$ $(x = 4, 11)$ or $r = 4$ $(x = 2, 7, 8, 13)$, and that both pass the test $x^{r/2} \neq \pm 1 \bmod N$. Consider the case $r = 4$, and assume the register size $K = 2^{11} = 2048$ ($L = 4$ bits). This case is wholly identical to the previously described example where $x = 8$ was assumed. Then if we implement Shor's algorithm (namely without any prior knowledge of $r$), we have equal probabilities of measuring $k = 0, 512, 1024, 1536$, corresponding to the determinations $\tilde{\varphi} = 0/2048, 512/2048, 1024/2048, 1536/2048$. As we have seen, only the fourth one, $\tilde{\varphi} = 1536/2048 = 3/4$, is a convergent of $\varphi$. Thus, only one measurement in four possibilities leads to a successful answer. If, on the other hand, our random selection is $x = 4, 11$, the probability distribution has only two peaks corresponding to the determinations $\tilde{\varphi} = 0/2048$ and $\tilde{\varphi} = 1024/2048 = 1/2$. Since none is a convergent of $\varphi$, the algorithm fails, and we must try it again with a different value of $x$. To summarize, the chances of randomly selecting the lucky values $x = 2, 7, 8, 13$ out of $N - 3 = 12$ possibilities in the interval $x = [2, N - 2]$ are $p_1^{(1)} = 4/12 = 1/3$, and the chances that the output measurement leads to a convergent are $p_2^{(1)} = 1/4$. On the other hand, the chances of randomly selecting the "unlucky" values $x = 4, 11$ are $p_1^{(1)} = 2/12 = 1/6$, and the chances that the output measurement leads to a convergent are $p_2^{(1)} = 0$. Overall, the *chances of the algorithm concluding successfully in a single run, based on any random selection of $x$ are given by the probability indicator*

$$\Pi = \sum_{i=1}^{2} p_1^{(i)} p_2^{(i)} = p_1^{(1)} p_2^{(1)} + p_1^{(2)} p_2^{(2)}, \tag{20.35}$$

**Table 20.5** First group of composite numbers $N = pq$ ($p$, $q = 3, 5, 7, \ldots, 37$) eligible for factorization, up to $N = 1147$, eliminating $N = a^b$ (as shaded).

| | 3 | 5 | 7 | 11 | 13 | 17 | 19 | 23 | 29 | 31 | 37 |
|---|---|---|---|---|---|---|---|---|---|---|---|
| 3 | 9 | 15 | 21 | 33 | 39 | 51 | 57 | 69 | 87 | 93 | 111 |
| 5 | | 25 | 35 | 55 | 65 | 85 | 95 | 115 | 145 | 155 | 185 |
| 7 | | | 49 | 77 | 91 | 119 | 133 | 161 | 203 | 217 | 259 |
| 11 | | | | 121 | 143 | 187 | 209 | 253 | 319 | 341 | 407 |
| 13 | | | | | 169 | 221 | 247 | 299 | 377 | 403 | 481 |
| 17 | | | | | | 289 | 323 | 391 | 493 | 527 | 629 |
| 19 | | | | | | | 361 | 437 | 551 | 589 | 703 |
| 23 | | | | | | | | 529 | 667 | 713 | 851 |
| 29 | | | | | | | | | 841 | 899 | 1073 |
| 31 | | | | | | | | | | 961 | 1147 |
| 37 | | | | | | | | | | | 1369 |

which gives here $\Pi = (1/3) \times (1/4) + 0 \equiv 1/12 = 8.3\%$. But the discussion does not end here. Indeed, this result actually assumes that the initial phase estimation $\tilde{\varphi}$ is infinitely accurate ($\varepsilon \ll 1$). As we chose $K = 2L + 1 + \lceil \log_2[2 + 1/(2\varepsilon)] \rceil = 11$, corresponding to an estimation success of at least $p = 1 - \varepsilon = 3/4$, the overall success chances in these global conditions are finally $p\Pi = (3/4) \times (1/12) = 1/16 = 6.2\%$. As the register size $K$ is increased, the chances of success asymptotically reach the limit $\Pi$. The key conclusion of this discussion is that, regardless of register size, *the implementation of Shor's algorithm is a trial and error process: some values of x do not work out, and in the best case, some register measurements fail to yield any convergent*, as we have seen with the $N = 15$ factorization example.

This discussion leads to another question: are some composites easier to factorize than others under Shor's algorithm, corresponding to higher values of the $\Pi$ indicator? In an attempt to address this question, we may analyze the periods of the first composite numbers and see which ones may present the least factoring difficulty or the highest $\Pi$. Consider, then, the composites of the type $N = pq$ (with $p, q$ primes), which satisfy the two conditions: (a) $N$ is not even, and (b) there are no integers $a, b$ ($a, b \geq 2$), such that $N = a^b$. With a list of the first primes limited to $p, q = \{3, 5, 7, 11, 13, 17, 19, 23, 29, 31, 37\}$, excluding $p, q = 2$, the set of eligible composites comes to 55 numbers ranging from $N = 15$ to $N = 1147$, as illustrated in Table 20.5. The selection listed in the table, thus, justifies that $N = 15$ is the first composite candidate to implement Shor's algorithm. Considering the highest number of the selection, $N = 1147$, determining the period would require one to compute the successive powers of $1145^p$, which scale as $10^{p \times \ln(1145)/\ln(10)} \approx 10^{3p}$ and, thus, rapidly leads to an overflow. If, on the other hand, we limit the scope to $N \leq 100$, i.e., $\max(N) = 91$, the computation remains tractable ($89^p \approx 10^{1.9p}$), at least with a personal computer. A period-finding computer program can easily be implemented,[14] yielding a tabulation of $r$ for each value $x \in [2, N - 2]$, along with the validation tests $\{x^{r/2}$ even

[14] A free executable for period finding (FINDPRIM.EXE) can also be downloaded from: http://users.pandora.be/nicvroom/progrm19.htm.

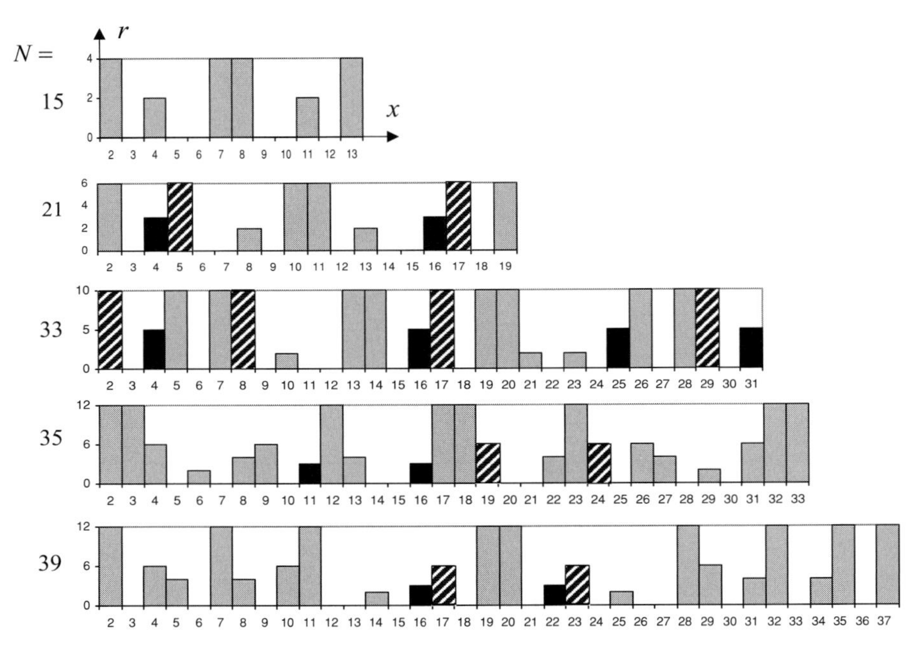


**Figure 20.6** Values $r$ of the order of $x$ modulo $N$ ($x$, $N$ co-prime) for composites $N = 15, 21, 33, 35, 39$. The hashed bars correspond to odd values of $r$, the shaded bars correspond to $r$ values for which $x^{r/2} = -1 \bmod N$.

and $x^{r/2} \neq \pm 1 \bmod N\}$. The results for $N = 15, 21, 33, 35, 39$ are shown in Fig. 20.6. The darkened data correspond to odd values of $r$, while the shaded data correspond to $r$ values for which $x^{r/2} = -1 \bmod N$. The graph, thus, makes it possible to calculate the probabilities $p_1^{(i)}$, $p_2^{(i)}$ for each subset of $x$ satisfying the algorithm requirements, and the indicator $\Pi = \sum_i p_1^{(i)} p_2^{(i)}$. In the case $N = 21$, for instance, we find $x = \{2, 10, 11, 19\}$ for $r = 6$ and $x = \{8, 13\}$ for $r = 2$, corresponding to $p_1^{(1)} = 4/18 = 2/9$ and $p_2^{(1)} = 2/18 = 1/9$, respectively. The associated values of $r$ give $p_2^{(1)} = 1/3$ and $p_2^{(2)} = 0$, respectively, yielding $\Pi = (2/9) \times (1/3) + 0 \equiv 2/27 = 7.4\%$.

Considering next the case $N = 33$, we find $x = \{5, 7, 13, 14, 19, 20, 26, 28\}$ for $r = 10$, $x = \{10, 21, 23\}$ for $r = 2$, corresponding to $p_1^{(1)} = 8/30 = 4/15$ and $p_1^{(2)} = 3/30 = 1/10$, respectively. For $r = 10$, the probability distribution exhibits 10 peaks at the $k$ locations:

$$\frac{0}{10}, \frac{1}{10}, \frac{2}{10}, \frac{3}{10}, \frac{4}{10}, \frac{5}{10}, \frac{6}{10}, \frac{7}{10}, \frac{8}{10}, \frac{9}{10},$$

corresponding to the irreducible fractions $p/q$

$$\frac{0}{10}, \frac{1}{10}, \frac{1}{5}, \frac{3}{10}, \frac{2}{5}, \frac{1}{2}, \frac{3}{5}, \frac{7}{10}, \frac{4}{5}, \frac{9}{10},$$

in which only six have $p, q$ as co-prime numbers. We, thus, have $p_2^{(1)} = 6/10 = 3/5$ and also $p_2^{(2)} = 0$, yielding $\Pi = (4/15) \times (3/5) + 0 \equiv 4/25 = 16.0\%$.

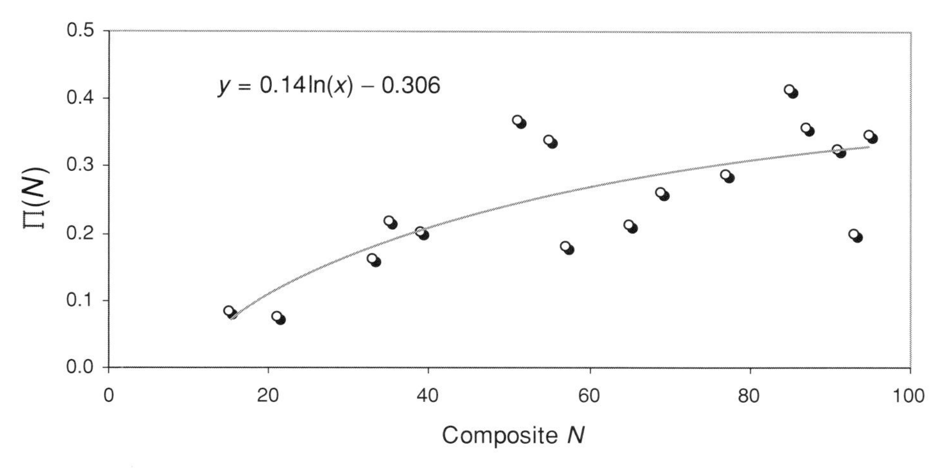


**Figure 20.7** Indicator function $\Pi(N)$, corresponding to the probability of successfully factoring $N$ in a single run of Shor's algorithm.

Similar analysis of the other $N$ values leads to the indicator function $\Pi(N)$. To recall, such a function represents the probability of successfully factorizing $N$ based on a single random selection of $x \in [2, N-2]$. The function $\Pi(N)$ is plotted in Fig. 20.7 for $N \leq 100$. We observe from the plot that $\Pi(N)$ increases with $N$ according to a logarithmic fit, along with some scattering. Such scattering is explained by either "lucky" or "unlucky" composites. For instance, $N = 15$ and $N = 51$ are lucky in the sense that all periods $r$ satisfy the two conditions of being even and such that $x^{r/2} \neq 1 \bmod N$ for any $x \in [2, N-1]$. This property maximizes the probability $p_1^{(i)}$. In contrast, for $N = 93$ (next to last point at right in Fig. 20.7) these conditions are not met for 1/3 of all the possible $r$ values, which explains the drop from the global trend. Given an eligible $r$, and independently of $N$, the second "luck" factor comes from the number of associated convergents, as measured by the probability $p_2^{(i)}$. For instance, $r = 2$ has no associated convergent, $r = 4$ has only one convergent with probability $p_2^{(i)} = 1/4 = 25\%$, $r = 10$ has 6 convergents with probability $p_2^{(i)} = 6/10 = 60\%$, and $r = 28$ has 24 convergents with probability$p_2^{(i)} = 24/28 = 85.7\%$. The product $p_1^{(i)} p_2^{(i)}$, thus, determines the weight taken by a given period $r$ out of the possibilities indexed by $i$ in the sum in Eq. (20.35) yielding $\Pi(N)$. For $N \leq 100$, Fig. 20.7 shows that the maximum is found at $N = 85$ with $\Pi(85) = 41.4\%$. The logarithmic fit shown in the figure ($y = 0.14 \ln \times - 0.306$) suggests that a probability of $\Pi = 50\%$ would be reached for $x = 316$, in the vicinity of the composites $N = 319 = 11 \times 29$ and $N = 323 = 17 \times 19$. As I am not aware that any plot similar to that shown in Fig. 20.7 has ever been published. It remains a reasonable conjecture that for composites larger than $N = 100$, $\Pi(N)$ asymptotically reaches 100%, with some scatter due to the aforementioned effects, with the contribution of $p_2^{(i)}$ becoming negligible as $N$ increases, as $s, r$ in the phase estimate $\varphi = s/r$ become more likely to be co-prime.

## 20.7 Public-key cryptography

This concluding section is about *public-key cryptography* (PKC). Such a topic is not completely out of place in this chapter on Shor's factorization algorithm because, as I shall describe, the secrecy involved in PKC heavily relies on the fact that factorizing large numbers is a task that is essentially intractable, which is true according to the best computing means and algorithms currently available. Since PKC is widely used for different applications in the Internet, and since its principle is relatively simple to understand, it is worthwhile to describe it briefly here in the context of this chapter.

Public-key cryptography was invented and developed by R. Rivest, A. Shamir and L. Adleman, a trio of people who gave the name to the *RSA standard*. The RSA principle feeds on the current fact that it is extremely difficult, if not completely intractable, to factorize large composite numbers, or decompose them into a product of primes. Given $N$, factorization (or prime decomposition) consists of finding the unique set of prime numbers $p_1, p_2, \ldots, p_m$ and powers $\alpha_1, \alpha_2, \ldots, \alpha_m$ such that $p_1^{\alpha_1} p_2^{\alpha_2} \ldots p_m^{\alpha_m} = N$. For instance, $N = 1000$ accepts the unique factorization or decomposition $2^3 \times 5^3 = 1000$. For the purpose of PKC, we may just use two sufficiently large (and assumedly different!) prime numbers $p, q$ to generate a "big" composite number $N = pq$, resting on the confidence that if anyone in the public domain knows $N$, that a person or entity would have a hard time or find it impossible to figure out the two primes $p, q$.

For instance, consider the number $N = 62\,615\,533$. To find its factorization, we need to divide it by prime numbers, trying them out one after another. In this example, the answer is $p = 7919$ and $q = 7907$, which are the highest two primes to be found at the top of a list of the first thousands.[15] This unfortunate choice made the factorization easy, since anyone can figure it out from such a list as representing the biggest product. What about $n = 15\,773\,077$? The answer ($p = 2383$, $q = 6619$) is less immediate, since it takes 200 division tests to find $q$ starting from the top of this first-thousand-prime list. Assume next that $p, q$ are selected from a huge list of known primes, for instance up to $10^9$, yielding composites $N = pq$ up to the order of $10^{18}$. Even at a rate of $10^9$ division tests per second (a state-class computing power that only a few may afford), this would leave about $10^9$ seconds or *31.7 years* to find the prime factors!

I shall now focus on the RSA standard and algorithm specifics, after recalling the basic underlying principle of *cryptography*. The principle of cryptography is to encode (or encrypt) a given "open-text" message, called *plaintext*, into a "secret" message, called *ciphertext*. Such words come from *cryptology*, the science, and history of cryptography,[16]

[15] For various lists of prime numbers, see, for instance: http://primes.utm.edu/lists/small/1000.txt, http://en.wikipedia.org/wiki/List_of_prime_numbers; http://primes.utm.edu/; www.prime-numbers.org/; www.rsok.com/~jrm/printprimes.html.

[16] For riveting and historical accounts of cryptology, see S. Singh, *The Code Book* (New York: Anchor Books, 1999), S. Levy, *Crypto* (Harmondsworth: Penguin Books, 2001), and D. Kahn, *The Codebreakers, the Story of Secret Writing* (New York: Scribner, 1967). For an easy but detailed overview of crypto-algorithms

whereby *text messages* were encrypted for the purposes of military, state, security, trade, or simply private communications. In the computer age, however, any "text" to be so encrypted may be anything from email to computer files, sensitive data, up to the payload of Internet frames (as used in the so-called IPv6 standard or other secure-communication protocols). Note that the word "cipher" indifferently designates the algorithm specifically used for encryption, or the ciphertext material itself.

The destinee of the ciphertext (and only this person or entity), should be able to make sense of it and translate it back into readable plaintext, an operation which is called *decryption*. For this, he or she must use a *secret key*. By definition, the secret key must be difficult to find or figure out, just like the combination of a bank safe, or a "good" computer or network password.[17] Traditional cryptography, thus, requires that the sender does communicate, by some indirect means, his or her key to the destinee, so that the latter may be able to complete the decryption. In contrast, PKC came as a revolution to this age-long principle by offering the possibility that the sender *does not know or use any secret key*. Indeed, PKC is based on the "asymmetric" principle of using two different cipher keys, one for encryption (public sender, call him Bob) and one for decryption (individual destinee, call her Alice). I shall now describe how this puzzling approach can make sense.

Define the two following keys:

- For *encryption*: a number $e$, such that it is relatively prime with $\phi(N) = (p-1)(q-1)$;
- For *decryption*: the number $d$, which satisfies $ed = 1 \bmod \phi$, or $d = e^{-1} \bmod \phi$.

We shall call $e$ the *public key* and $d$ the *private key*. As its name indicates, the public key should be available to anyone (Bob) who wants to send an encrypted message to Alice. On the other side, Alice keeps her "private" key in absolute secrecy. The number $N = pq$ (also called the *RSA modulus*) is known or accessible to the general public, Bob, or anyone else.

The operation of *encryption* with the public key $e$ consists first of decomposing the plaintext message into as many numerical blocks smaller than $N$. In the binary system, a 64-bit block represents a maximum number of $1.8 \times 10^{19}$, therefore, it is an eligible block size if $N \geq 2 \times 10^{19}$. For messages of arbitrary or random length, it is always possible to "pad" the blocks with zeros on the left (called "nulls" in crypto jargon), so that they fit a convenient standard size. Encryption of each of these blocks with number value $m$ is performed by calculating the (cipher) number

$$c = m^e \bmod N. \tag{20.36}$$

used from history to now, see E. Desurvire, *Global Telecommunications, Broadband Access, Optical Components and Networks, and Cryptography* (New York: J. Wiley and Sons, 2004), Ch. 3, pp. 345–478, from which this section on PKC is partially inspired. For an advanced description, see, for instance, B. Schneier, *Applied Cryptography* (New York: J. Wiley and Sons, 2006).

[17] If the key is easy or likely to be figured out by any third party, it is called a weak key.

For Alice, who is the only one to own and know the private key $d$, the operation of block-by-block decryption consists of computing

$$m' = c^d \bmod N. \tag{20.37}$$

Let's now look at the modular-arithmetic value of the $m'$ that Alice gets. It takes a few substitutions to figure out that the result comes to:

$$\begin{aligned} m' &= c^d \bmod N \\ &= (m^e)^d \bmod N \\ &= m^{ed} \bmod N \\ &= m^{k(p-1)(q-1)+1} \bmod N \\ &= m \times m^{k(p-1)(q-1)} \bmod N \\ &= m \times [m^{(p-1)(q-1)} \bmod N]^k \bmod N \\ &= m \times 1^k \bmod N \\ &= m. \end{aligned} \tag{20.38}$$

The result in Eq. (20.38) shows that Alice has fully and unambiguously retrieved the original plaintext block $m$. Note that for the above demonstration, we used different properties of modular arithmetic, and in particular *Euler's theorem*, as described in Appendix T.

We also used the pre-set property of Alice's key, $ed = 1 \bmod \phi$, from which we have $ed = k(p-1)(q-1) + 1$. Alice is, thus, able to decrypt any message sent to her from various "Bob" correspondents who may use her public key for this very purpose.

For illustration, consider a full example of RSA encryption and decryption.[18] First, here is Bob's private and intimate declaration to Alice:

| **Plaintext** | I | l | o | v | e | y | o | u |
|---|---|---|---|---|---|---|---|---|
| **PT-ASCII** | 1001001 | 1101100 | 1101111 | 1110110 | 1100101 | 1111001 | 1101111 | 1110101 |

Consistently with the RSA approach, Bob may encrypt his plaintext by *blocks*, which represent for instance two letters each. With the above ASCII message, a two-letter block is a number of $2 \times 7 = 14$ bits, corresponding to a maximum size of $2^{14} - 1 = 16\,383$. Let us assume that this is the standard. On her part, Alice also chose two primes $p, q$ whose product is greater or equal to 16 383, for instance $p = 73$ and $q = 227$, which gives $N = pq = 16\,571$, and $\phi = (p-1)(q-1) = 72 \times 226 = 16\,272$. To be effective, Alice made the number $N = 16\,571$ known to everyone, just like a phone number in the directory. Of course, the numbers $p, q, \phi$ remain only known to her.

[18] This example was previously published in E. Desurvire, *Global Telecommunications, Broadband Access, Optical Components and Networks, and Cryptography* (New York: J. Wiley and Sons, 2004).

Following the standard, Bob must convert his message into 2-letter blocks and the resulting blocks into decimal numbers, as follows:

| **Plaintext** | I | l | o | v | e | y | o | u |
|---|---|---|---|---|---|---|---|---|
| **PT-ASCII** | 1001001 | 1101100 | 1101111 | 1110110 | 1100101 | 1111001 | 1101111 | 1110101 |
| **Block** | 10 01 00 11 10 11 00 | | 11 01 11 11 11 01 10 | | 11 00 10 11 11 10 01 | | 11 01 11 11 11 01 01 | |
| **Decimal** | 9452 | | 14 326 | | 13 049 | | 14 325 | |

Alice has also chosen her public key, $e$. This public key must be co-prime with $\phi = 16\,272$. To be able to perform the encryption and decryption computations with a pocket calculator (just for this example's sake!), assume that Alice picked for her public key $e = 5$, a number that represents the smallest eligible value (3 and 4 divide into 16 272). Going to the public directory or Alice's website, Bob, like any other visitor, may read, "Alice's encryption instructions: please kindly use two-ASCII block encryption modulo 16 571 as the standard; my public key is number 5." No more instructions are needed for Bob to proceed to encryption.

Bob then proceeds to encrypt his love declaration according to the formula $c_i = m_i^e \bmod N$, where $m_i$ is the decimal block number $i$ in Bob's message sequence $m_1m_2m_3m_4$. This gives

$$\begin{cases} c_1 = m_1^5 \bmod 16\,571 = 9452^5 \bmod 16\,571 = 3704 \\ c_2 = m_2^5 \bmod 16\,571 = 14\,326^5 \bmod 16\,571 = 766 \\ c_3 = m_3^5 \bmod 16\,571 = 13\,049^5 \bmod 16\,571 = 475 \\ c_4 = m_4^5 \bmod 16\,571 = 14\,325^5 \bmod 16\,571 = 372. \end{cases} \quad (20.39)$$

In the above computations, it is absolutely essential that Bob make no truncation errors.[19] Having finished these computations, Bob emails his cipher message

$$c_1c_2c_3c_4 = 3704 - 766 - 475 - 372$$

to Alice; he may possibly include a few more words in order to identify himself (e.g., "Bob your classmate."), but this is not the point of the endeavor.

Alice is then the only person who is able to decrypt Bob's message. Her private key, $d$, is defined by the formula $ed = 1 \bmod \phi$, or $d = e^{-1} \bmod \phi$, namely $d = 5^{-1} \bmod 16\,272$. Thus, Alice needs to find the inverse of 5 modulo 16 272. Since 16 272 is not the product of two primes ($16\,272 = 16 \times 9 \times 113$), she can't use Euler's theorem. Instead, she

[19] For instance, with a pocket calculator the function $9452^5$ yields $7.544293311 \times 10^{19}$, but the last ten digits are missing! To avoid truncations from the pocket calculator, the trick is to use the modulus-arithmetic formula $|m^5|_N = |m^2|_N \times |m^2|_N \times |m|_N|_N$, in which the different terms and their successive products (after reducing each one to their modulo $N$ residue) fit in the calculator display size.

uses the *extended Euclidian algorithm*, which is described in Appendix T. It is not a coincidence that, in this appendix, I use the above numbers to illustrate this algorithm principle. The result is $d = 6509$. Such a big, 13-bit private key is the price for Alice to pay for having chosen a public key as small as $e = 5$, but this does not affect the generality of the demonstration. Alice then uses her private key to compute the following blocks:[20]

$$\begin{cases} m_1 = c_1^{6509} \bmod 16\,571 = 3704^{6509} \bmod 16\,571 = 9452 \\ m_2 = c_2^{6509} \bmod 16\,571 = 766^{6509} \bmod 16\,571 = 14\,326 \\ m_3 = c_3^{6509} \bmod 16\,571 = 475^{6509} \bmod 16\,571 = 13\,049 \\ m_4 = c_4^{6509} \bmod 16\,571 = 372^{6509} \bmod 16\,571 = 14\,325. \end{cases} \quad (20.40)$$

Then Alice converts the decimal numbers $m_1m_2m_3m_4$ into four 14-bit binary words, splits each word into two 7-bit groups, and finally translates the resulting blocks into ASCII, which yields and finally displays on Alice's screen:

| **Decimal** | 9452 | | 14 326 | | 13 049 | | 14 325 | |
|---|---|---|---|---|---|---|---|---|
| **Binary** | 10 01 00 11 10 11 00 | | 11 01 11 11 11 01 10 | | 11 00 10 11 11 10 01 | | 11 01 11 11 11 01 01 | |
| **ASCII** | 1001001 | 1101100 | 1101111 | 1110110 | 1100101 | 1111001 | 1101111 | 1110101 |
| **Plaintext** | I | l | o | v | e | y | o | u |

Bob and Alice both having computers to perform RSA and a connection to the Internet, the entire operation of encryption, transmission, and decryption of Bob's message (possibly completed with a few more statements!) took, in fact, milliseconds, as slowed down by the IP routing protocol and other switching bottlenecks. This is the same amount of time that after composing her reply, Alice will take to send Bob a PKC message using Bob's public key and Bob to retrieve into plaintext through his private key.

[20] One may wonder whether it is physically possible to compute numbers as monstrous as $100^{6509}$ or $1000^{6509}$ *exactly* with a personal computer or even worse, a pocket calculator. Indeed, if we make the conversion through the formula $a^x = 10^{x \log a / \log 10}$, we get for the first block $3704^{6509} = 10^{23\,228.4794} \approx 3.01 \times 10^{23\,228}$, namely a number with a whopping 23 229 digits! But surprisingly enough, such a computation can be performed with a cheap pocket calculator! Here is how to proceed. Consider, for instance, the last cipher block, $372^{6509}$. It takes a few minutes to calculate the following power series (modulo 16571): $372^2 = 5816$, $372^4 = 4445$, $372^8 = 5393$, etc., up to $372^{4096} = 8695$. Thus, $372^{6509}$ is the product $372^{4096} \times 372^{1024} \times 372^{1024} \times 372^{256} \times 372^{64} \times 372^{32} \times 372^8 \times 372^4 \times 372^1$. The result of each successive product must be reduced to its residue so that no truncation occurs. The same procedure must be followed with the other blocks $3704^{6509}$, $766^{6509}$ and $475^{6509}$. This point being made, there is no reason not to develop a simple computer program that can perform all these successive reduction tasks in a wink. With the present example, we have chosen a block format ($2^{14} - 1 = 16\,383$) whose decimal size allows computing any power of two, i.e., $(2^{14} - 1)^2 = 268\,402\,689$, and its successive multiples, with a pocket calculator. But with a personal computer, larger coding formats to encrypt four to eight ASCII or EBCDIC characters (i.e., $2^{32} - 1 = 4\,294\,967\,295$ or $2^{64} - 1 \approx 2 \times 10^{19}$) are possible, provided the program is designed to handle numbers up to $(2^{32} - 1)^2$ or $(2^{64} - 1)^2$ without truncation errors.

As we have learnt, the key feature of the PKC algorithm is that the composite $N = pq$ is definitely too "big" to be readily factorized by any third party maliciously attempting to intercept the communications between Alice and Bob. This is the reason why, so long as no large-scale quantum computer able to factorize $N$ is demonstrated, or no state-grade computer is assigned to the task, Alice and Bob can exchange their thoughts, safe from any third-party scrutiny.

## 20.8 Exercises

**20.1** (M): Show that for any two integers $x, N$, satisfying $x < N$ and any integer $s$, the state

$$|u_s\rangle = \frac{1}{\sqrt{r}} \sum_{k=0}^{r-1} \mathrm{e}^{-\frac{2\mathrm{i}\pi ks}{r}} |x^k \bmod N\rangle$$

is an eigenstate of the operator defined by

$$U_{x,N}|y\rangle = |xy \bmod N\rangle$$

with eigenvalue $\exp(2\mathrm{i}\pi s/r)$.

**20.2** (M): Given the eigenstates $|u_s\rangle$ defined as

$$|u_s\rangle = \frac{1}{\sqrt{r}} \sum_{k=0}^{r-1} \mathrm{e}^{-\frac{2\mathrm{i}\pi ks}{r}} |x^k \bmod N\rangle,$$

show that

$$\frac{1}{\sqrt{r}} \sum_{s=0}^{r-1} |u_s\rangle = |1\rangle.$$

**20.3** (M): Given the eigenstates $|u_s\rangle$ defined as

$$|u_s\rangle = \frac{1}{\sqrt{r}} \sum_{k=0}^{r-1} \mathrm{e}^{-\frac{2\mathrm{i}\pi ks}{r}} |x^k \bmod N\rangle,$$

show that

$$\frac{1}{\sqrt{r}} \sum_{s=0}^{r-1} \mathrm{e}^{\frac{2\mathrm{i}\pi ks}{r}} |u_s\rangle = |x^k \bmod N\rangle.$$

**20.4** (B): Show by induction that the two reals defined by

$$p_n = a_n p_{n-1} + p_{n-2}$$
$$q_n = a_n q_{n-1} + q_{n-2}$$

for $n \geq 2$, and

$$p_0 = a_0, q_0 = 1, \; p_1 = 1 + a_0 a_1, \; q_1 = a_1$$

for $n = 0, 1$, satisfy the relation

$$q_n p_{n-1} - p_n q_{n-1} = (-1)^n.$$

# 21 Quantum information theory

This chapter sets the basis of *quantum information theory (QIT)*. The central purpose of QIT is to qualify the transmission of either classical or quantum information over *quantum channels*. The starting point of the QIT description is *von Neumann entropy*, $S(\rho)$, which represents the quantum counterpart of Shannon's classical entropy, $H(X)$. Such a definition rests on that of the *density operator* (or *density matrix*) of a quantum system, $\rho$, which plays a role similar to that of the random-events source $X$ in Shannon's theory. As we shall see, there also exists an elegant and one-to-one correspondence between the quantum and classical definitions of the entropy variants *relative entropy*, *joint entropy*, *conditional entropy*, and *mutual information*. But such a similarity is only apparent. Indeed, one becomes rapidly convinced from a systematic analysis of the entropy's additivity rules that fundamental differences separate the two worlds. The classical notion of *information correlation* between two event sources for quantum states shall be referred to as *quantum entanglement*. We then define a quantum communication channel, which encodes and decodes classical information into or from quantum states. The analysis shows that the mutual information $H(X;Y)$ between originator and recipient in this communication channel cannot exceed a quantity $\chi$, called the *Holevo bound*, which itself satisfies $\chi \leq H(X)$, where $H(X)$ is the entropy of the originator's classical information source. It is shown that the Holevo bound is maximal ($\chi = H(X)$) when the set of density operators used for coding the message have *orthogonal supports* or *eigenspaces*.

## 21.1 Von Neumann entropy

The concept of *source entropy*, as introduced in Shannon's classical information theory, has been extensively described and developed in Chapters 4 to 6. To recall, a source $X$ of $n$ random events labeled $i$ ($I = 1 \ldots n$) and having associated probabilities $p_i$ (corresponding to the discrete probability distribution $p$), is characterized by an entropy $H(X)$, defined by

$$H(X) = -\sum_{i-1}^{n} p_i \log p_i = \left\langle \log \frac{1}{p} \right\rangle, \tag{21.1}$$

with the continuity property $x \log x = 0$ in the limit $x \to 0$ and the logarithm conventionally being in base two. Such entropy represents the average measurement of the source's *information contents* $\log(1/p_i)$ associated with each event. As we have seen in Chapter 4, the entropy is maximum when all the source events are equiprobable, namely when $p_i = 1/n$, which yields $H_{\max} \equiv \log n$.

Consider now a quantum system. This system may exist in a *quantum state* $|\psi\rangle$, which we shall assume here represents a statistical mixture of *pure states* $|x_i\rangle$.[1] Such pure states, which cannot be defined by any other mixtures of pure states, are orthogonal to each other and have unit length, such that $\langle x_i|x_j\rangle = \delta_{ij}$. The set of pure states $\{|x_i\rangle\} = \{|x_1\rangle, |x_2\rangle, \ldots, |x_n\rangle\}$, thus, defines an orthonormal basis for the $n$-dimensional space $V^n$ of all possible quantum states $|\psi\rangle$ defining the system. Consistently, any state $|\psi\rangle$ of $V^n$ accepts a unique decomposition of the form

$$|\psi\rangle = x_1|x_1\rangle + x_2|x_2\rangle + \cdots + x_n|x_n\rangle = \sum_{i=1}^{n} x_i|x_i\rangle, \tag{21.2}$$

where $x_i$ $(i = 1 \ldots n)$ are complex coordinates. We may choose to represent the quantum system with a state $|\psi\rangle$ of unit length, i.e., $\langle\psi|\psi\rangle = \sum_i |x_i|^2 = 1$, in which case the real number $p_i = |x_i|^2$, represents the probability of finding the state $|\psi\rangle$ in the pure state $|x_i\rangle$. In the quantum world, the system in the quantum state $|\psi\rangle$, thus, plays the role of a "random event" source and, naturally, the concepts of "information" and "entropy" may be associated with such a system.

To establish such a connection, we need to use the concept of *density operator* (or *density matrix)*, which was introduced in Chapter 17. As we have learnt, the density operator or matrix $\rho$ is an alternative way to define a quantum system in a given state $|\psi\rangle$. Formally,

$$\rho = \sum_{i=1}^{n} p_i |x_i\rangle\langle x_i|, \tag{21.3}$$

where $|x_i\rangle\langle x_i|$ is the projector (or measurement) operator on the basis state $|x_i\rangle$. It is clear that $\rho|x_i\rangle = p_i|x_i\rangle$, which shows that $|x_i\rangle$ *is an eigenstate of* $\rho$ *with associated eigenvalue* $p_i$. In the case where $|\psi\rangle$ is a *pure state*, e.g., $|\psi\rangle = |x_k\rangle$ with $p_i = \delta_{ik}$, and only in such a case, the density operator is simply given by $\rho = |\psi\rangle\langle\psi|$. As a general definition, *a pure state is a state that has 100% probability of being observed in a quantum system, or which is exactly known*. A given basis state (for instance $|0\rangle$ or $|1\rangle$ in a 2D space) *may or may not be a pure state*, according to whether this condition is fulfilled or not (see more on this further on).

As we have also seen in Chapter 17, the matrix elements $\rho_{ij}$ of the density operator satisfy $\rho_{ij} = \langle x_i|\rho|x_j\rangle \equiv |x_i|^2\delta_{ij} = p_i\delta_{ij}$, showing that the matrix is *diagonal in the computational basis* $\{|x_i\rangle\}$, as expected from the fact that it is the basis of eigenstates.

[1] The assumption according to which a system may be accurately or completely defined through a *quantum state* $|\psi\rangle$, is equivalent to the assumption that the system is *closed*, meaning that it is neither coupled nor entangled to any other unknown external system.

The matrix representation of $\rho$ is, thus:

$$\rho = \begin{pmatrix} p_1 & 0 & \cdots & 0 \\ 0 & p_2 & \cdots & \vdots \\ \vdots & \vdots & \ddots & 0 \\ 0 & \cdots & 0 & p_n \end{pmatrix}. \tag{21.4}$$

A key property is that the sum of diagonal elements, or *the trace of the density matrix satisfies* $\mathrm{tr}(\rho) = 1$, as for the definition of the probability distribution $p = \{p_i\}$. As shown in Chapter 17, *the trace of any matrix or operator is independent of the choice of the base representation* $\{|x_i\rangle\}$, or is invariant by unitary base transformation (see Eq. (17.33)).

A useful property is that $\mathrm{tr}(\rho^2) \leq 1$ applies for any density operator $\rho$, with equality only for the case of *pure states* ($\rho = |\psi\rangle\langle\psi|$), which is left as an exercise to establish. We may note that such a property does not depend on the choice of basis. This is in spite of the fact that a pure state may appear to be a mixed state in another basis! For instance, given $|0\rangle$ assumed a pure state in basis $\{|0\rangle, |1\rangle\}$, the same state appears to be "mixed" in the basis $\{|+\rangle, |-\rangle\}$ according to the equality $|0\rangle = (|+\rangle + |-\rangle)/\sqrt{2}$. But if $|0\rangle$ is a pure state, the system has $\rho = |0\rangle\langle 0|$ for density operator. It is then an easy exercise to show that $\rho = |0\rangle\langle 0|$ is actually transformed into $\tilde{\rho} = |+\rangle\langle +|$ in this new basis, which illustrates that *pure states remain pure states, independent of basis representation* and conserve their properties in any basis transformation. It is important, therefore, not to confuse basis states (as a projection possibility for the system) and pure states (as the only system state possibility).

Assume, next, an operator $U$ corresponding to a physical *observable* (meaning that $U$ is Hermitian or has real, nonzero eigenvalues, see Chapter 17). Let us now take a close look at the operator $\rho U$, and its diagonal matrix elements in the basis $\{|x_i\rangle\}$, namely $(\rho U)_{ii} = \langle x_i|\rho U|x_i\rangle$. From Eq. (21.3), we obtain

$$\begin{aligned} (\rho U)_{ii} &= \langle x_i|\rho U|x_i\rangle \\ &= \langle x_i| \left( \sum_{k=1}^{n} p_k |x_k\rangle\langle x_k| \right) U|x_i\rangle \\ &= \sum_{k=1}^{n} p_k \langle x_i|x_k\rangle\langle x_k|U|x_i\rangle \\ &= \sum_{k=1}^{n} p_k \delta_{ik} \langle x_k|U|x_i\rangle \\ &= p_i \langle x_i|U|x_i\rangle \equiv p_i \langle U\rangle_i . \end{aligned} \tag{21.5}$$

It is seen from this result that $(\rho U)_{ii}$ represents the *expectation value* $\langle U_i\rangle = \langle x_i|U|x_i\rangle$ of the observable $U$, should the system be measured in the state $|x_i\rangle$ (see Eq. (17.57)), affected by the corresponding probability weight $p_i$ of such a measurement event. The summation over all event possibilities, which is the *trace* of $\rho U$, takes the form

$$\begin{aligned}\mathrm{tr}(\rho U) &= \sum_{i=1}^{n} (\rho U)_{ii} \\ &= \sum_{i=1}^{n} p_i \langle x_i | U | x_i \rangle \\ &\equiv \langle U \rangle_\psi .\end{aligned} \tag{21.6}$$

This result shows that given a physical system in state $|\psi\rangle$ with associated density matrix $\rho$, the expected value $\langle U \rangle_\psi$ of the observable $U$ is simply given by the trace $\mathrm{tr}(\rho U)$. Actually, this result enormously simplifies the perspective on quantum measurements gained from Chapter 17. It will also be used to analyze the tricky issue of *composite quantum systems*, as described later.

In Chapter 17, we have also established that for any operator $U$ having a diagonal matrix with nonnegative coefficients, it is possible to associate an operator $U \log U$ with diagonal coefficients $(U \log U)_{ii} = U_{ii} \log U_{ii}$. In the case $U = \rho$, we obtain the matrix:

$$\rho \log \rho = \begin{pmatrix} p_1 \log p_1 & 0 & \cdots & 0 \\ 0 & p_2 \log p_2 & \cdots & \vdots \\ \vdots & \vdots & \ddots & 0 \\ 0 & \cdots & 0 & p_n \log p_n \end{pmatrix}. \tag{21.7}$$

From the above definition, the trace of $\rho \log \rho$, with a minus sign, is given by:

$$\begin{aligned} S(\rho) &= -\mathrm{tr}(\rho \log \rho) \\ &= -\sum_{i=1}^{n} p_i \log p_i . \end{aligned} \tag{21.8}$$

A comparison between Eqs. (21.8) and (21.1) shows that $S(\rho) = -\mathrm{tr}(\rho \log \rho)$ is strictly analogous to Shannon's classical entropy, $H(X)$, with $X = \{i\}_{i=1\ldots n}$ being a random-event source characterized by the probability distribution $\{p_i\}$. The key difference between $S(\rho)$ and $H(X)$ is that, in the former, *the source is a quantum system, characterized not by a probability distribution*, like the latter, *but by a density operator $\rho$*. We shall refer to $S(\rho)$ as *von Neumann entropy*. The von Neumann (VN) entropy, thus, represents a *quantum measure of the information contents in the quantum system or state* under consideration, referred to as a *quantum source $\rho$*. Since the VN entropy measure is always equivalent to that of a classical source, $H(X)$, it is *nonnegative,* or $S(\rho) \geq 0$ always applies.

In Chapter 17, it was shown that the trace of any operator is independent of the basis representation. Thus, the density operator transformation under change of basis, i.e., $\rho \to \tilde{\rho} = T \rho T^+$ where $T$ is a unitary operator, is of no effect on the VN entropy. Formally:

$$S(\tilde{\rho}) = S(T \rho T^+) = S(\rho). \tag{21.9}$$

This result shows that quantum information, as defined by the VN entropy, represents an incompressible feature in quantum systems, as is also the case for classical information in random-event sources, as defined by Shannon entropy. What is the *quantum information* contained in a qubit? The answer is straightforward. Assume a qubit of general definition

$$|q\rangle = \alpha|0\rangle + \beta|1\rangle, \tag{21.10}$$

where $|0\rangle$, $|1\rangle$ are two pure states in the 2D quantum space $V^2$, and $p = |\alpha|^2 = 1 - |\beta|^2$. From the definitions in Eqs. (21.3) and (21.4), we have

$$\rho = |\alpha|^2|0\rangle\langle 0| + |\beta|^2|1\rangle\langle 1| = p|0\rangle\langle 0| + (1-p)|1\rangle\langle 1|, \tag{21.11}$$

$$\rho = \begin{pmatrix} |\alpha|^2 & 0 \\ 0 & |\beta|^2 \end{pmatrix} = \begin{pmatrix} p & 0 \\ 0 & 1-p \end{pmatrix}, \tag{21.12}$$

and

$$\begin{aligned} S(\rho) &= -(|\alpha|^2 \log|\alpha|^2 + |\beta|^2 \log|\beta|^2) \\ &= -p\log p - (1-p)\log(1-p) \equiv f(p). \end{aligned} \tag{21.13}$$

In the result in Eq. (21.13), we recognize the Shannon entropy of a two-event source $X = \{0, 1\}$, corresponding to the two possible "states" of a classical bit (see Eqs. (4.13) and (4.14)). The average qubit information is, thus, equivalent to that of two classical bits, the amount depending on the weights $p_0$, $p_1$ in the state mixture. As described in Chapter 4, the function $f(p)$ has a maximum of unity for $p = 1/2$, corresponding to $|\alpha| = |\beta| = 1/\sqrt{2}$ (phases being arbitrary) or a uniform distribution $p_0 = p_1 = 1/2$, and for the VN entropy, $S_{\max}(\rho) \equiv \log 2 = 1$. In this case, the quantum information amounts to exactly one classical bit. The minimum of $f(p)$ is zero, which is reached either when $p_0 = 1$, $p_1 = 0$ or when $p_0 = 0$, $p_1 = 1$, meaning that the qubit is in a pure state, i.e., $|q\rangle = |0\rangle$ or $|q\rangle = |1\rangle$, giving $S_{\max}(\rho) \equiv 0$. In this case, there is no quantum information in the system, as there is no information in a single, deterministic classical bit.

It is clear that the VN entropy of an $n$-qubit (or *quNit*) in the quantum space $V^n$ always has the maximum $S_{\max}(\rho) \equiv \log n$ when the system is in the most homogeneous state superposition with a uniform probability distribution $p_i = 1/n$, hence corresponding to maximum quantum information. In the general case, we have $0 \leq S(\rho) \leq \log n$.

It is a generally possible that the state $|\psi\rangle$ can be a *mixture of nonorthogonal states*. In this case, the corresponding density matrix $\rho$ is nondiagonal. How, then, can we determine the VN entropy of the system? The answer is that we know it is possible to express the density operator in the computational basis of *eigenstates*, where the matrix is diagonal. Given the eigenvalues $\lambda_i$ $(i = 1 \ldots n)$ corresponding to the eigenstates $\{|\lambda_i\rangle\}_{i=1\ldots n}$, the VN entropy, thus, takes the form

$$S(\rho) = -\sum_{i=1}^{n} \lambda_i \log \lambda_i. \tag{21.14}$$

As an illustration (see also the exercises), assume the following density operator:

$$\begin{aligned}\rho &= \frac{2}{3}|0\rangle\langle 0| + \frac{1}{3}|-\rangle\langle -| \\ &= \frac{2}{3}|0\rangle\langle 0| + \frac{1}{3}\left(\frac{|0\rangle - |1\rangle}{\sqrt{2}}\right)\left(\frac{\langle 0| - \langle 1|}{\sqrt{2}}\right).\end{aligned} \tag{21.15}$$

In the computational basis $\{|0\rangle, |1\rangle\}$, it is easily established that the corresponding matrix is

$$\rho = \begin{pmatrix} \langle 0|\rho|0\rangle & \langle 0|\rho|1\rangle \\ \langle 1|\rho|0\rangle & \langle 1|\rho|1\rangle \end{pmatrix} \equiv \frac{1}{6}\begin{pmatrix} 5 & -1 \\ -1 & 1 \end{pmatrix}. \tag{21.16}$$

As recalled in Chapter 17, the eigenvalues $\lambda_i$ are given by the solutions $\lambda$ of the *characteristic equation*, i.e., $\det(\rho - \lambda I) = 0$. By definition, each eigenvalue corresponds to an eigenvector $|\lambda_i\rangle$, for which $\rho|\lambda_i\rangle = \lambda_i|\lambda_i\rangle$, with $\{|\lambda_i\rangle\}$ constituting an orthonormal basis. In the example, the solutions of the characteristic equation are easily found to be $\lambda = (3 \pm \sqrt{5})/6$, which, in the eigenstate-basis representation $\{|\lambda_i\rangle\}$, yields the diagonal density matrix:[2]

$$\begin{aligned}\tilde{\rho} &= \begin{pmatrix} \langle \lambda_1|\rho|\lambda_1\rangle & 0 \\ 0 & \langle \lambda_2|\rho|\lambda_2\rangle \end{pmatrix} \\ &\equiv \frac{1}{6}\begin{pmatrix} 3+\sqrt{5} & 0 \\ 0 & 3-\sqrt{5} \end{pmatrix}.\end{aligned} \tag{21.17}$$

We can then determine the VN entropy of the associated quantum system, which is in the state now represented as $|\psi\rangle = \sqrt{\lambda_1}|\lambda_1\rangle + \sqrt{\lambda_2}|\lambda_2\rangle$, as follows:

$$\begin{aligned}S(\tilde{\rho}) &= -\mathrm{tr}(\tilde{\rho}\log\tilde{\rho}) \\ &= -\left(\frac{3+\sqrt{5}}{6}\log\frac{3+\sqrt{5}}{6} + \frac{3-\sqrt{5}}{6}\log\frac{3-\sqrt{5}}{6}\right) \\ &\equiv 0.55004.\end{aligned} \tag{21.18}$$

This result concludes that this system contains quantum information amounting to about 0.55 classical bits, as measured when using the eigenstate basis $\{|\lambda_i\rangle\}$.

The connection between classical and quantum information, as measured by Shannon and von Neumann entropies, appears to be elegant and complete, almost to the extent of hiding the fundamental differences between their respective worlds. As the following sections will show, however, it would be largely erroneous to conclude from this connection that QIT brings nothing new to information theory.

[2] It is easily checked that the same density operator $\tilde{\rho}$ is obtained by starting from the computational basis $\{|+\rangle, |-\rangle\}$ instead of $\{|0\rangle, |1\rangle\}$.

## 21.2 Relative, joint, and conditional entropy, and mutual information

The properties of VN entropy (henceforth to be referred to simply as "entropy") are listed in this section, along with those of related entropy variants that represent the quantum counterparts of similar definitions in classical theory.

To recall the basics, the previous section showed that the VN entropy is nonnegative ($S(\rho) \geq 0$), is zero for systems in pure states ($S(\rho) = S(|\psi\rangle\langle\psi|) = 0$), and has a maximum of $\log n$ for systems of dimension $n$ made of a uniform superposition of states ($S(\rho) = \log n$). Finally, the entropy is invariant under basis transformation, i.e., $S(T\rho T^{+}) = S(\rho)$, where $T$ is a unitary matrix, which illustrates that quantum information is an incompressible feature in quantum systems.

I shall now describe more properties of VN entropy.

### Relative entropy

Our starting point in the investigation is the definition of *relative entropy*. This notion is the quantum equivalent to the classical relative entropy, or Kullback–Leibler distance, described in Chapter 5. Assume two quantum systems with density operators $\rho$ and $\sigma$. The relative entropy between these two systems, noted $S(\rho\|\sigma)$ is given by:

$$S(\rho\|\sigma) = \mathrm{tr}(\rho \log \rho) - \mathrm{tr}(\rho \log \sigma). \tag{21.19}$$

For any operator, the *kernel* is the space described by the set of eigenvectors having zero for eigenvalues, and the *support* is the space described by the set of eigenvectors with nonzero eigenvalues (the union of kernel and support being the complete eigenvector space). In the definition in Eq. (21.17), the term $\mathrm{tr}(\rho \log \sigma)$ is finite only in the case where the support of $\rho$ does not intersect with the kernel of $\sigma$, or is fully included in the support of $\sigma$ (meaning $(\rho \log \sigma)_{ii} \geq 0$ for all diagonal matrix elements). In the contrary case, we have $-\mathrm{tr}(\rho \log \sigma) = +\infty$, and the relative entropy is infinite. As a key property of relative entropy, we have

$$S(\rho\|\sigma) \geq 0, \tag{21.20}$$

where the equality stands uniquely for the case $\rho = \sigma$. Such a property is referred to as *Klein's inequality*. The basic proof of Klein's inequality is provided in Appendix U. The consequence is that for two quantum systems with density operators $\rho$ and $\sigma$, we have $S(\rho) = \mathrm{tr}(\rho \log \rho) \geq \mathrm{tr}(\rho \log \sigma)$, a property that we shall use later when considering additivity rules for entropy.

### Composite system in pure state

Consider, next, a *composite* system made of two subsystems $A$, $B$ in states $|\psi_A\rangle$, $|\psi_B\rangle$, and assume that it is in a *pure joint state*, namely $|\psi\rangle = |\psi_A\rangle|\psi_B\rangle$. Call $\rho_A$, $\rho_B$ the density operators of the subsystems $A$, $B$. Then we have the property

$$S(\rho_A) = S(\rho_B). \tag{21.21}$$

The proof of this property is based on the *Schmidt decomposition*, as described in Appendix V. According to the Schmidt decomposition (which remains to be familiarized with from this appendix), any pure joint state of a composite system can be expressed in the form

$$|\psi\rangle = \sum_i x_i |i_A\rangle|i_B\rangle, \tag{21.22}$$

where $x_i$ $(i = 1 \ldots n)$ are nonnegative numbers (called *Schmidt coefficients*) satisfying the property $\sum_i x_i^2 = 1$, and where $\{|i_A\rangle\}$, $\{|i_B\rangle\}$ are some orthonormal basis for the subsystems $A$, $B$. Since $|\psi\rangle$ is a pure state, its density operator is $|\psi\rangle\langle\psi|$. From Eq. (21.20), we obtain

$$\begin{aligned} |\psi\rangle\langle\psi| &= \left(\sum_i x_i \langle i_A|\langle i_B|\right)\left(\sum_k x_k |k_A\rangle|k_B\rangle\right) \\ &= \sum_{ik} x_i x_k |k_A\rangle\langle i_A| \otimes |k_B\rangle\langle i_B|. \end{aligned} \tag{21.23}$$

We now need to make sense of this result by introducing the notion of *partial trace*, which applies to composite systems. Let $|i_A\rangle$, $|j_A\rangle$ be any two basis states in the system $A$, and likewise, let $|k_B\rangle$, $|l_B\rangle$ be any two basis states in the system $B$. The partial trace $\mathrm{tr}_A(U)$ of the tensor operator $U = |i_A\rangle\langle j_A| \otimes |k_B\rangle\langle l_B|$ is then defined as:

$$\begin{aligned} \mathrm{tr}_A U &= \mathrm{tr}_A(|i_A\rangle\langle j_A| \otimes |k_B\rangle\langle l_B|) \\ &\equiv \mathrm{tr}(|i_A\rangle\langle j_A|) \times |k_B\rangle\langle l_B|. \end{aligned} \tag{21.24}$$

Likewise, the partial trace $\mathrm{tr}_B(U)$ is defined as

$$\begin{aligned} \mathrm{tr}_B U &= \mathrm{tr}_B(|i_A\rangle\langle j_A| \otimes |k_B\rangle\langle l_B|) \\ &\equiv \mathrm{tr}(|i_B\rangle\langle j_B|) \times |k_A\rangle\langle l_A|. \end{aligned} \tag{21.25}$$

It should also be noted that for any pair of orthogonal states $|i\rangle$, $|j\rangle$, and by definition of the tensor operator $|i\rangle\langle j|$, we have $\mathrm{tr}(|i\rangle\langle j|) = \langle i|j\rangle = \delta_{ij}$. We now apply the above definitions and this last property to calculate $\mathrm{tr}_A|\psi\rangle\langle\psi|$ and $\mathrm{tr}_B|\psi\rangle\langle\psi|$ based on the result in Eq. (21.23). We obtain:

$$\begin{aligned} \mathrm{tr}_A|\psi\rangle\langle\psi| &= \mathrm{tr}_A\left(\sum_{ik} x_i x_k |k_A\rangle\langle i_A| \otimes |k_B\rangle\langle i_B|\right) \\ &= \sum_{ik} x_i x_k \mathrm{tr}_A(|k_A\rangle\langle i_A|) \times |k_B\rangle\langle i_B| \\ &= \sum_{ik} x_i x_k \langle i_A|k_A\rangle \times |k_B\rangle\langle i_B| \\ &= \sum_{ik} x_i x_k \delta_{ik} |k_B\rangle\langle i_B| \\ &= \sum_i x_i^2 |i_B\rangle\langle i_B| \\ &\equiv \rho_B, \end{aligned} \tag{21.26}$$

$$\begin{aligned}
\mathrm{tr}_B|\psi\rangle\langle\psi| &= \mathrm{tr}_B\left(\sum_{ik} x_i x_k |k_A\rangle\langle i_A| \otimes |k_B\rangle\langle i_B|\right) \\
&= \sum_{ik} x_i x_k |k_A\rangle\langle i_A| \times \mathrm{tr}_B(|k_B\rangle\langle i_B|) \\
&= \sum_{ik} x_i x_k |k_A\rangle\langle i_A| \times \langle k_B|i_B\rangle \\
&= \sum_{ik} x_i x_k \delta_{ik} |k_A\rangle\langle i_A| \\
&= \sum_{i} x_i^2 |i_A\rangle\langle i_A| \\
&\equiv \rho_A.
\end{aligned} \tag{21.27}$$

In these two equations, the identification in the right-hand side with the subsystem density operators $\rho_A, \rho_B$ comes from the aforementioned property of the Schmidt coefficients, i.e., $\sum_i x_i^2 = 1$, making, de facto, $\{x_i^2\}$ the corresponding probability distribution. We must now realize that this result is quite interesting. Indeed, the only assumption that a composite system is in a pure state, thus, implies that $\rho_A = \rho_B$, and hence the property $S(\rho_A) = S(\rho_B)$, as was claimed in Eq. (21.21). The examples described at the end of this section provide more opportunities to apply the technique of partial tracing to the determination of subsystem entropy.

### Subadditivity inequality and quantum joint entropy

Assume a composite system $AB$ made of two subsystems $A, B$. Let $\rho_{AB}, \rho_A, \rho_B$ be the density operators of the system and its subsystems. The corresponding entropies $S(\rho_{AB}), S(\rho_A), S(\rho_B)$ are then linked by the *subadditivity inequality*:

$$S(\rho_{AB}) \leq S(\rho_A) + S(\rho_B), \tag{21.28}$$

where the equality stands when $\rho_{AB} = \rho_A \otimes \rho_B$, which corresponds to two subsystems with uncorrelated information. The entropy of the composite system, $S(\rho_{AB})$, which is the *quantum joint entropy*, is, therefore, less than or equal to the sum of the subsystem entropies. There is no surprise here, since this property appears to be the quantum equivalent of that concerning the Shannon entropy $H(X, Y)$ of joint sources (see Chapter 5), i.e.

$$H(X, Y) \leq H(X) + H(Y). \tag{21.29}$$

The apparent equivalence between the classical and quantum definition of joint entropy, as intuitive and innocuous as it may appear, is in fact deceiving. We shall immediately realize this fact by considering the case where $AB$ is in a pure state. From the previous property, we must have $S(\rho_A) = S(\rho_B)$, but also $S(\rho_{AB}) = 0$, since $AB$ is in a pure state. Thus, regardless of the information contained in subsystems $A$ or $B$, which may be finite, the joint system $AB$ contains no information! Such a possibility is definitely nonclassical and counterintuitive. Indeed, referring to the Venn diagrams in Fig. 5.2, where $H(X)$, $H(Y)$ are represented by sets with $H(X, Y) = H(X) \cup H(Y)$ there is no possibility for $H(X, Y) = \emptyset$ (empty set), except in the limiting case $H(X) = H(Y) = \emptyset$.

I shall now prove the subadditivity inequality introduced in Eq. (21.28). Basically, it is a consequence of *Klein's inequality*, which was discussed previously. We have shown that for any pair of quantum systems of equal dimension and having density operators $\rho, \sigma$, Klein's inequality can be translated into

$$\begin{aligned} -S(\rho) &= \mathrm{tr}(\rho \log \rho) \geq \mathrm{tr}(\rho \log \sigma) \\ \leftrightarrow S(\rho) &\leq -\mathrm{tr}(\rho \log \sigma). \end{aligned} \tag{21.30}$$

Substituting $\rho = \rho_{AB}$ and $\sigma = \rho_A \otimes \rho_B$ into the last inequality yields:

$$S(\rho_{AB}) \leq -\mathrm{tr}[\rho_{AB} \log(\rho_A \otimes \rho_B)]. \tag{21.31}$$

It is left as a relatively easy exercise to show that:

$$\log(\rho_A \otimes \rho_B) = \log(\rho_A) \otimes I_B + I_A \otimes \log(\rho_B), \tag{21.32}$$

where $I_A$, $I_B$ are the identity matrices of the subsystems $A$, $B$. Using this property in Eq. (21.31), and also the additive and commutative properties of the trace (see Eq. (17.32)) we obtain:

$$\begin{aligned} S(\rho_{AB}) &\leq -\mathrm{tr}\{\rho_{AB}[\log(\rho_A) \otimes I_B + I_A \otimes \log(\rho_B)]\} \\ &= -\mathrm{tr}[\rho_{AB} \log(\rho_A) \otimes I_B] - \mathrm{tr}[I_A \otimes \log(\rho_B)\rho_{AB}]. \end{aligned} \tag{21.33}$$

We must now make sense of the traces involved in the right-hand side of this equation. Recall first that $\rho_{AB}$ is the density operator of the composite system in state $|\psi\rangle$. From the discussion in the previous section, and in particular the result in Eq. (21.6), we have learnt that the expectation value of any observable $U$, is given by $\langle U \rangle_{AB} = \mathrm{tr}(\rho_{AB} U)$. There is no reason to assume that the same outcome would be obtained by the same measurement applied just to subsystem $A$ separately, namely $\langle U \rangle_A = \mathrm{tr}(\rho_A U) \neq \mathrm{tr}(\rho_{AB} U) = \langle U \rangle_{AB}$. This is because in the joint system $AB$, the subsystem $B$ may also contribute to positive outcomes for the same observable. On the other hand, intuition dictates that there must exist an operator $U'$ such that the expected value of $U'$ in the joint system $AB$ and that of $U$ in the subsystem $A$ do coincide, i.e.,

$$\langle U' \rangle_{AB} = \langle U \rangle_A \leftrightarrow \mathrm{tr}(\rho_{AB} U') = \mathrm{tr}(\rho_A U). \tag{21.34}$$

It is a fine academic issue to show that $U' = U \otimes I_B$, which expands $U$ to the full system space, is, in fact, the only candidate to solve such a measurement reconciliation problem. Here, we may just accept it as an intuitive postulate. Hence,

$$\begin{cases} \mathrm{tr}[\rho_{AB} \log(\rho_A) \otimes I_B] = \mathrm{tr}(\rho_A \log \rho_A) \equiv -S(\rho_A) \\ \mathrm{tr}[I_A \otimes \log(\rho_B)\rho_{AB}] = \mathrm{tr}(\rho_B \log \rho_B) \equiv -S(\rho_B), \end{cases} \tag{21.35}$$

to be followed by substitution in Eq. (21.33) to yield the inequality $S(\rho_{AB}) \leq S(\rho_A) + S(\rho_B)$, as claimed in Eq. (21.28), which is now established.

As also claimed earlier, the equality in Eq. (21.28) stands when (and only when) $\rho_{AB} = \rho_A \otimes \rho_B$, meaning that the entropies of the subsystems are additive, according to:

$$S(\rho_{AB}) \equiv S(\rho_A) + S(\rho_B). \tag{21.36}$$

Proving this equality is straightforward. Indeed, since $\rho_{AB} = \rho_A \otimes \rho_B$, we can now write $S(\rho_{AB}) \equiv -\mathrm{tr}[\rho_{AB} \log(\rho_A \otimes \rho_B)]$, and as we have seen from the results in Eqs. (21.32) to (21.35), the right-hand side is identical to $S(\rho_A) + S(\rho_B)$.

## Quantum mutual information

As we have learnt from Chapter 5, the *mutual information* $H(X; Y)$ measures the amount of information correlation between the two event sources $X, Y$, as defined by

$$H(X; Y) = H(X) + H(Y) - H(X, Y). \tag{21.37}$$

If the sources are uncorrelated, the equality in Eq. (21.29) stands and $H(X; Y)$ in Eq. (21.37) is zero. We may define *quantum mutual information* $S(\rho_A; \rho_B)$ through

$$S(\rho_A; \rho_B) = S(\rho_A) + S(\rho_B) - S(\rho_{AB}). \tag{21.38}$$

As in the classical case, $S(\rho_A; \rho_B)$ represents *the measure of information correlation between the two quantum subsystems* $A, B$. In the absence of any correlation ($\rho_{AB} = \rho_A \otimes \rho_B$), the equality in Eq. (21.28) stands, and $S(\rho_A; \rho_B)$ is zero. From Eq. (21.28), we note that $S(\rho_A; \rho_B) \geq 0$ in all cases, meaning that *the quantum mutual information is always nonnegative*. In the case where the joint system $AB$ is in a pure state, we have $S(\rho_{AB}) = 0$ and $S(\rho_A) = S(\rho_B)$, which implies that $S(\rho_A; \rho_B) = 2S(\rho_A) = 2S(\rho_B)$. Such a result is definitely nonclassical. Indeed, referring to the Venn diagrams in Fig. 5.2, where $H(X), H(Y)$ are represented by sets with $H(X; Y) = H(X) \cap H(Y)$, there is no possibility for $H(X; Y) = 2H(X)$ or $H(X; Y) = 2H(Y)$, except in the limiting case $H(X) = H(Y) = \emptyset$.

If two quantum systems have correlated information, we shall say from now on that their quantum states are *entangled*. The end of this section will provide examples of entangled states.

## Conditional entropy

Given two subsystems $A, B$ and their composite system $AB$, the *conditional entropy* $S(\rho_A|\rho_B)$ is defined as

$$S(\rho_A|\rho_B) = S(\rho_{AB}) - S(\rho_B), \tag{21.39}$$

and, likewise, $S(\rho_B|\rho_A) = S(\rho_{AB}) - S(\rho_A)$. This definition is the quantum equivalent of the classical case concerning joint events from two sources $X, Y$, namely (see Chapter 5):

$$H(X|Y) = H(X, Y) - H(X), \tag{21.40}$$

and, likewise, $H(Y|X) = H(X, Y) - H(Y)$. Where the information in the two subsystems is uncorrelated ($S(\rho_{AB}) \equiv S(\rho_A) + S(\rho_B)$), we have, from Eq. (21.39):

$$\begin{cases} S(\rho_A|\rho_B) = S(\rho_A) \\ S(\rho_B|\rho_A) = S(\rho_B), \end{cases} \tag{21.41}$$

which translates that the information of one subsystem is not affected by the knowledge of the information in the other subsystem, as expected in the uncorrelated case. Again, while we observe a nice parallel between the classical and the quantum definitions for conditional entropy, we should know for a fact (from preceding analysis) that it is deceptive! Indeed, in the case where $AB$ is in a pure state, we have $S(\rho_{AB}) = 0$ and, from Eq. (21.39), $S(\rho_A|\rho_B) = -S(\rho_B) \leq 0$ (as per the nonnegativity of $S(\rho_B)$) and, likewise, $S(\rho_B|\rho_A) = -S(\rho_A) \leq 0$. Hence, the *quantum conditional entropy can be negative*, which is definitely a nonclassical feature. Since, in this case, $S(\rho_A) = S(\rho_B)$, we also have

$$\begin{cases} S(\rho_A|\rho_B) = -S(\rho_A) \\ S(\rho_B|\rho_A) = -S(\rho_B), \end{cases} \tag{21.42}$$

which contrast with the correlated case expressed in Eq. (21.41).

## Triangle inequality

The subadditivity inequality in Eq. (21.28) provided an upper bound to the joint entropy $S(\rho_{AB})$, namely $S(\rho_{AB}) \leq S(\rho_A) + S(\rho_B)$. The *triangle inequality*, also referred to as the *Araki–Lieb inequality*, provides a *lower bound* to $S(\rho_{AB})$, according to

$$|S(\rho_A) - S(\rho_B)| \leq S(\rho_{AB}). \tag{21.43}$$

To prove this property, assume the existence of a third system $R$, such as the composite system $ABR$ in a pure state. The system $R$ is referred to as a *purifying system* for $AB$ (see Appendix W). From the subadditivity inequality we find

$$S(\rho_{AR}) \leq S(\rho_A) + S(\rho_R). \tag{21.44}$$

Because $ABR$ is in a pure state, we also find that $S(\rho_{AB}) = S(\rho_R)$ and $S(\rho_{AR}) = S(\rho_B)$ (see earlier subsection on *composite system in pure state*). Substituting these last two equalities into Eq. (21.44) we obtain $S(\rho_{AB}) \geq S(\rho_B) - S(\rho_A)$. Since $A$ and $B$ play a symmetric role, we also have $S(\rho_{AB}) \geq S(\rho_A) - S(\rho_B)$, and, hence, $S(\rho_{AB}) \geq \pm|S(\rho_A) - S(\rho_B)|$, which proves the property in Eq. (21.43). It can be shown (but the proof is not to be considered here) that the equality in Eq. (21.43) is given by the condition $\rho_{AR} = \rho_A \otimes \rho_R$, which means that the $A$, $R$ information is uncorrelated and, therefore, that all possible correlation of $A$ information with the rest of the world is exclusively with $B$.

## Concavity of entropy and entropy of system in random states

Assume a quantum system that can be in any random mixed state $|\psi_i\rangle$, according to some known probability distribution $p_i$. Each of these random states is associated with a density operator $\rho_i$. The *concavity* of a function $f(x)$ corresponds to the property $\langle f(x)\rangle \leq f(\langle x\rangle)$. Here, I shall establish that *entropy is a concave function of the density-operator variable* $\rho_i$, namely,

$$\langle S(\rho)\rangle \leq S(\langle \rho\rangle), \tag{21.45}$$

or, formally:

$$\sum_i p_i S(\rho_i) \leq S\left(\sum_i p_i \rho_i\right). \tag{21.46}$$

Furthermore, if one assumes that the set of operators $\rho_i$ *has support on orthogonal subspaces* (meaning that the set of all possible eigenstates form an orthonormal basis), an exact relation also exists between $S(\langle\rho\rangle)$ and $\langle S(\rho)\rangle$, which nicely relates to the Shannon entropy of a classical source $H(X) = -\sum_i p_i \log p_i$, according to:

$$S(\langle \rho\rangle) = \langle S(\rho)\rangle + H(X), \tag{21.47}$$

or, formally:

$$S\left(\sum_i p_i \rho_i\right) = \sum_i p_i S(\rho_i) + H(X). \tag{21.48}$$

The property expressed in Eq. (21.47) or Eq. (21.48) can easily be interpreted as follows: *the quantum information of a system whose states are random equals the mean information, as averaged over the individual states, plus the information on the probability distribution, which is the Shannon entropy*. Note that since $H(X)$ is nonnegative, this property also establishes the concavity of the entropy, as expressed in Eq. (21.45) or Eq. (21.46). In the case where the probability distribution is uniform, the Shannon entropy $H(X)$ is maximal, meaning that there is maximal uncertainty as to which quantum state the system is in. Such a situation also maximizes the entropy $S(\langle\rho\rangle)$ of the system. The deterministic case where the system has only one possible quantum state corresponds to $H(X) = 0$ and a minimum entropy.

The demonstration of the property expressed in Eq. (21.48) is relatively simple. Indeed, let $|\lambda_i^k\rangle$ and $\lambda_i^k$ be the eigenstates and eigenvalues for each quantum state associated with the density operator $\rho_i$. Hence $\rho_i$ has diagonal matrix elements $(\rho_i)_{kk} = \lambda_i^k$. Next, given *any* $i, k$ we observe that $|\lambda_i^k\rangle$ and $p_i\lambda_i^k$ are eigenvectors and eigenvalues of

$\langle\rho\rangle = \sum_j p_j \rho_j$,[3] making $\{|\lambda_i^k\rangle\}$ an orthonormal basis for $\langle\rho\rangle$, regardless of the choice of $i$. Hence, $\langle\rho\rangle$ has diagonal matrix elements $\langle\rho\rangle_{kk} = p_i\lambda_i^k$ and $(\log\langle\rho\rangle)_{kk} = \log(p_i\lambda_i^k)$. Using these different properties, we obtain

$$\begin{aligned}
S(\langle\rho\rangle) &= -\mathrm{tr}(\langle\rho\rangle \log\langle\rho\rangle) \\
&= -\sum_k \left[\left(\sum_i p_i\rho_i\right)_{kk} (\log\langle\rho\rangle)_{kk}\right] \\
&= -\sum_{ik} \left[p_i(\rho_i)_{kk}\,(\log\langle\rho\rangle)_{kk}\right] \\
&= -\sum_{ik} \left[p_i\lambda_i^k \log\left(p_i\lambda_i^k\right)\right] \\
&= -\sum_{ik} \lambda_i^k p_i \log p_i - \sum_{ik} \left[p_i\lambda_i^k \log\lambda_i^k\right] \\
&= -\sum_i p_i \log p_i \sum_k \lambda_i^k + \sum_i p_i\left(-\sum_k \lambda_i^k \log\lambda_i^k\right) \\
&\equiv H(X) + \sum_i p_i S(\rho_i),
\end{aligned} \tag{21.49}$$

which proves Eqs. (21.46) and (21.47). As we have seen, the property thus established rests on the assumption that the set of operators $\rho_i$ *has support on orthogonal subspaces*. If such a condition is not met, the equality in Eq. (21.47) or Eq. (21.48) does not hold and the originator-source entropy $H(X)$ is only an *upper bound*, namely:

$$S(\langle\rho\rangle) - \langle S(\rho)\rangle \le H(X), \tag{21.50}$$

or, formally:

$$S\left(\sum_i p_i\rho_i\right) - \sum_i p_i S(\rho_i) \le H(X). \tag{21.51}$$

Here, for the sake of conciseness and to keep the momentum, we shall not go through the detailed demonstration of the inequalities in Eqs. (21.50) and (21.51). The key point is to remember that the equality in these relations is not guaranteed, except in the case where the subspaces defined by $\rho_i$ are orthogonal. In all cases, *the entropy $H(X)$ remains the upper bound*. This general property for the *concavity of entropy* shall be used in the next section, concerned with the so-called *Holevo bound*.

This tedious inventory of entropy properties, which has been limited here to the most elementary ones, now calls for a few illustrative examples. I shall consider three basic possibilities for a composite quantum system $AB$ made of two subsystems $A$, $B$, namely:

[3] Indeed, we have

$$\begin{aligned}
\sum_j p_j\rho_j|\lambda_i^k\rangle &= \left(\sum_{jl} p_j\lambda_j^l|\lambda_j^l\rangle\langle\lambda_j^l|\right)|\lambda_i^k\rangle \\
&= \sum_{jl} p_j\lambda_j^l|\lambda_j^l\rangle\langle\lambda_j^l|\lambda_i^k\rangle \\
&= \sum_{jl} p_j\lambda_j^l|\lambda_i^k\rangle\delta_{ij}\delta_{kl} \\
&\equiv p_i\lambda_i^k|\lambda_i^k\rangle.
\end{aligned}$$

(1) when the subsystem information is *uncorrelated*, (2) when it is *correlated*, and (3) when the composite system is in *a pure state*. The last two examples will also highlight the interest and implications of using *partial tracing*, and illustrate the concept of *state entanglement*.

---

**Example 21.1:** ***Subsystems with uncorrelated information***
Consider, for instance, two subsystems $A$, $B$ in respective qubit states:[4]

$$\begin{cases} |\psi_A\rangle = \dfrac{1}{\sqrt{2}}(|0\rangle + |1\rangle)_A \\ |\psi_B\rangle = \dfrac{1}{\sqrt{3}}(|0\rangle + |1\rangle + |2\rangle)_B. \end{cases} \tag{21.52}$$

The joint state of the composite system is assumed to be

$$\begin{aligned} |\psi_{AB}\rangle &= |\psi_A\rangle \otimes |\psi_B\rangle \\ &= \frac{1}{2}(|0\rangle + |1\rangle)_A \otimes (|0\rangle + |1\rangle + |2\rangle)_B \\ &= \frac{1}{\sqrt{6}}(|00\rangle + |01\rangle + |02\rangle + |10\rangle + |11\rangle + |12\rangle)_{AB}. \end{aligned} \tag{21.53}$$

It is clear from the above definitions that the three systems $A$, $B$, $AB$ are expressed in their respective eigenstate bases and, thus, have density matrices of $\rho_A = I_A/2$, $\rho_B = I_B/3$, and $\rho_{AB} = I_{AB}/6$, respectively, with the property $\rho_{AB} = \rho_A \otimes \rho_B$. We therefore have

$$\begin{cases} S(\rho_A) = -\mathrm{tr}[(I_A/2)\log(I_A/2)] = -2 \times \dfrac{1}{2}\log\dfrac{1}{2} \equiv 1 \\ S(\rho_B) = -\mathrm{tr}[(I_B/3)\log(I_B/3)] = -3 \times \dfrac{1}{3}\log\dfrac{1}{3} \equiv 1.584 \\ S(\rho_{AB}) = -\mathrm{tr}[(I_{AB}/6)\log(I_{AB}/3)] = -6 \times \dfrac{1}{6}\log\dfrac{1}{6} \equiv 2.584, \end{cases} \tag{21.54}$$

which shows that $S(\rho_{AB}) = S(\rho_A) + S(\rho_B)$ and illustrates that *entropy is additive for subsystems with uncorrelated information* (namely satisfying $\rho_{AB} = \rho_A \otimes \rho_B$ or $|\psi_{AB}\rangle = |\psi_A\rangle \otimes |\psi_B\rangle$), or, equivalently, whose quantum states are *not entangled*.

---

**Example 21.2:** ***Subsystems with correlated information***
Consider, for instance, a composite system $AB$ with the joint state

$$|\psi_{AB}\rangle = \frac{1}{\sqrt{6}}(|00\rangle + 2|10\rangle + |11\rangle)_{AB}, \tag{21.55}$$

[4] For each subsystem, it is possible to choose the eigenstate basis $\{|i\rangle\}$ in which each component $\langle\psi|i\rangle$ is in phase with $|i\rangle$, hence $\langle\psi|i\rangle \geq 0$, meaning that the example where all components are positive is actually not restrictive.

which corresponds to the density operator

$$\rho_{AB} = \frac{1}{6}(|00\rangle\langle 00| + 4|10\rangle\langle 10| + |11\rangle\langle 11|). \tag{21.56}$$

It is straightforward to check that $\mathrm{tr}(\rho_{AB}^2) = 1/2 < 1$, which illustrates that $|\psi_{AB}\rangle$ is not a pure state. It is also a basic exercise to show that $|\psi_{AB}\rangle$, like other states having similar decomposition with any other nonzero coefficients, *cannot* be decomposed into a tensor product of two subsystem states $|\psi_A\rangle, |\psi_B\rangle$. Because of this, we conclude that the information in the two subsystems is *correlated*, or that the subsystem states are *entangled*.

So far, we only know that the quantum information in system $AB$ is:

$$\begin{aligned} S(\rho_{AB}) &= -2 \times (1/6)\log(1/6) - 0 \times \log(0) - 1 \times (4/6)\log(4/6) \\ &\equiv 1.251. \end{aligned} \tag{21.57}$$

Given the above assumptions, can we get any knowledge of the quantum information in subsystems $A, B$? The answer is yes, if we use the tool of *partial tracing*. Indeed, we may rewrite Eq. (21.55) as a sum of tensor products:

$$\rho_{AB} = \frac{1}{6}(|0\rangle\langle 0|_A \otimes |0\rangle\langle 0|_B + 4|1\rangle\langle 1|_A \otimes |0\rangle\langle 0|_B + |1\rangle\langle 1|_A \otimes |1\rangle\langle 1|_B). \tag{21.58}$$

Applying partial tracing over $A$ from the definition in Eq. (21.24), while using the property $\mathrm{tr}(|i\rangle\langle i|) = \langle i \mid i\rangle \equiv 1$, we obtain:

$$\begin{aligned} \mathrm{tr}_A(\rho_{AB}) &= \frac{1}{6}\mathrm{tr}_A(|0\rangle\langle 0|_A \otimes |0\rangle\langle 0|_B + 4|1\rangle\langle 1|_A \otimes |0\rangle\langle 0|_B + |1\rangle\langle 1|_A \otimes |1\rangle\langle 1|_B) \\ &= \frac{1}{6}[\mathrm{tr}_A(|0\rangle\langle 0|_A \otimes |0\rangle\langle 0|_B) + 4\mathrm{tr}_A(|1\rangle\langle 1|_A \otimes |0\rangle\langle 0|_B) + \mathrm{tr}_A(|1\rangle\langle 1|_A \otimes |1\rangle\langle 1|_B)] \\ &= \frac{1}{6}[\mathrm{tr}_A(|0\rangle\langle 0|_A) \times |0\rangle\langle 0|_B + 4\mathrm{tr}_A(|1\rangle\langle 1|_A) \times |0\rangle\langle 0|_B + \mathrm{tr}_A(|1\rangle\langle 1|_A) \times |1\rangle\langle 1|_B] \\ &= \frac{1}{6}(|0\rangle\langle 0|_B + 4|0\rangle\langle 0|_B + |1\rangle\langle 1|_B) \\ &= \frac{1}{6}(5|0\rangle\langle 0|_B + |1\rangle\langle 1|_B) \equiv \hat{\rho}_B. \end{aligned} \tag{21.59}$$

The operator $\hat{\rho}_B = \mathrm{tr}_A(\rho_{AB})$ is referred to as the *reduced density operator* of subsystem $B$. Likewise, we obtain

$$\mathrm{tr}_B(\rho_{AB}) = \frac{1}{6}(|0\rangle\langle 0|_A + 5|1\rangle\langle 1|_A) \equiv \hat{\rho}_A. \tag{21.60}$$

For the sake of curiosity, let us now calculate the tensor product $\hat{\rho}_A \otimes \hat{\rho}_B$ from Eqs. (21.59) and (21.60). We obtain

$$\begin{aligned} \hat{\rho}_A \otimes \hat{\rho}_B &= \frac{1}{36}(|0\rangle\langle 0|_A + 5|1\rangle\langle 1|_A) \otimes (5|0\rangle\langle 0|_B + |1\rangle\langle 1|_B) \\ &= \frac{1}{36}(5|00\rangle\langle 00| + |01\rangle\langle 01| + 25|10\rangle\langle 10| + 5|11\rangle\langle 11|)_{AB}. \end{aligned} \tag{21.61}$$

It is clear by comparison with Eq. (21.56) that $\rho_{AB} \neq \hat{\rho}_A \otimes \hat{\rho}_B$, meaning that, in the present example, *the tensor product $\hat{\rho}_A \otimes \hat{\rho}_B$ of the reduced density operators $\hat{\rho}_A, \hat{\rho}_B$ does not correspond to the density operator of the composite system*. This is not a surprise, since we established earlier that the joint state $|\psi_{AB}\rangle$ cannot be reduced to a tensor product.

Pushing curiosity further, we may calculate the quantum information, $S(\hat{\rho}_A)$, $S(\hat{\rho}_B)$, $S(\hat{\rho}_A \otimes \hat{\rho}_B)$ associated with the reduced density operators $\hat{\rho}_A, \hat{\rho}_B$, and $\hat{\rho}_A \otimes \hat{\rho}_B$, to find:

$$\begin{aligned} S(\hat{\rho}_A) &= S(\hat{\rho}_B) \\ &= -(1/6)\log(1/6) - (5/6)\log(5/6) \\ &\equiv 0.650, \\ S(\hat{\rho}_A \otimes \hat{\rho}_B) &= -2 \times (5/36)\log(5/36) - (1/36)\log(1/36) - (25/36)\log(25/36) \\ &\equiv 1.300 \\ &\equiv S(\hat{\rho}_A) + S(\hat{\rho}_B). \end{aligned} \tag{21.62}$$

Based on the information knowledge of the composite system, $S(\rho_{AB}) = 1.251$, as established in Eq. (21.56), we find that:

$$\begin{aligned} S(\rho_{AB}) &= 1.251 \\ &< S(\hat{\rho}_A) + S(\hat{\rho}_B) \\ &= 1.300. \end{aligned} \tag{21.63}$$

This result is consistent with the *subadditivity inequality* established earlier and formalized in Eq. (21.28). It is also consistent with the triangle inequality defined in Eq. (21.43). We may, thus, conclude, as a postulate but not as a formal demonstration, that $S(\hat{\rho}_A)$, $S(\hat{\rho}_B)$ *indeed represent the quantum information contents of the two subsystems $A, B$*, despite the fact that the density operator of the composite system, $\rho_{AB}$, does not reduce to the tensor product $\hat{\rho}_A \otimes \hat{\rho}_B$.

Without realizing it, this example and postulate helped make great progress in the analysis of information *correlation*! Indeed, we can now apply the definition of *quantum mutual information* in Eq. (21.38) to obtain in this example:

$$\begin{aligned} S(\rho_A; \rho_B) &= S(\hat{\rho}_A) + S(\hat{\rho}_B) - S(\rho_{AB}) \\ &= 1.300 - 1.251 \\ &= 0.049. \end{aligned} \tag{21.64}$$

We can, thus, conclude that *quantum information in subsystems $A, B$ is correlated*, as expressed by the mutual information $S(\rho_A; \rho_B) = 0.049$, corresponding to about 0.05 classical bit. From Eq. (21.39), we find the conditional entropy:

$$\begin{aligned} S(\rho_A|\rho_B) &= S(\rho_{AB}) - S(\hat{\rho}_B) \\ &= 1.251 - 0.650 \\ &= 0.601 \end{aligned} \tag{21.65}$$

with $S(\rho_B|\rho_A) = S(\rho_A|\rho_B)$. The knowledge of the information in one of the subsystems, thus, decreases the information (or uncertainty) in the other system by the amount $S(\rho_A) - S(\rho_B|\rho_A) = 0.650 - 0.601 = 0.049 = S(\rho_A;\rho_B)$, which corresponds to the mutual information in the composite system.

---

**Example 21.3:** ***Composite system in pure state***

As a final example, consider the joint state

$$|\psi_{AB}\rangle = \frac{1}{\sqrt{2}}(|00\rangle + |11\rangle)_{AB}. \tag{21.66}$$

This state, which is also noted $|\beta_{00}\rangle$, is one of the four *Bell–EPR states* (see Eqs. (16.2) and (18.15)). As we have learnt, the Bell–EPR states *cannot* be generated by the tensor product of two qubits, and they also form an orthonormal basis in the 2-qubit space (referred to as $\{|\beta_{00}\rangle, |\beta_{01}\rangle, |\beta_{10}\rangle, |\beta_{11}\rangle\}$). In this basis, $|\psi_{AB}\rangle = |\beta_{00}\rangle$ is, thus, a *pure state* and its density operator is, therefore, equal to $\rho_{AB} = |\psi_{AB}\rangle\langle\psi_{AB}|$. Substitution of the definition in Eq. (21.66) and rearrangement of the different terms into tensor products yields:

$$\begin{aligned}\rho_{AB} &= |\psi_{AB}\rangle\langle\psi_{AB}| \\ &= \frac{1}{2}(|00\rangle + |11\rangle)_{AB} \otimes (\langle 00| + \langle 11|)_{AB} \\ &= \frac{1}{2}(|00\rangle\langle 00| + |00\rangle\langle 11| + |11\rangle\langle 00| + |11\rangle\langle 11|)_{AB} \\ &= \frac{1}{2}\begin{pmatrix} |0\rangle\langle 0|_A \otimes |0\rangle\langle 0|_B + |0\rangle\langle 1|_A \otimes |0\rangle\langle 1|_B \\ +|1\rangle\langle 0|_A \otimes |1\rangle\langle 0|_B + |1\rangle\langle 1|_A \otimes |1\rangle\langle 1|_B \end{pmatrix}.\end{aligned} \tag{21.67}$$

Next, we calculate the reduced density operators $\hat{\rho}_A = \mathrm{tr}_B(\rho_{AB})$ and $\hat{\rho}_B = \mathrm{tr}_A(\rho_{AB})$, while using the property with the property $\mathrm{tr}(|i\rangle\langle j|) = \langle i|j\rangle \equiv \delta_{ij}$, to obtain:

$$\begin{aligned}\hat{\rho}_A &= \mathrm{tr}_B(\rho_{AB}) \\ &= \frac{1}{2}\begin{bmatrix} |0\rangle\langle 0|_A \mathrm{tr}(|0\rangle\langle 0|_B) + |0\rangle\langle 1|_A \mathrm{tr}(|0\rangle\langle 1|_B) \\ +|1\rangle\langle 0|_A \mathrm{tr}(|1\rangle\langle 0|_B) + |1\rangle\langle 1|_A \mathrm{tr}(|1\rangle\langle 1|_B) \end{bmatrix} \\ &= \frac{1}{2}(|0\rangle\langle 0|_A + |1\rangle\langle 1|_A) \\ &\equiv \frac{I_A}{2}.\end{aligned} \tag{21.68}$$

$$\begin{aligned}\hat{\rho}_B &= \mathrm{tr}_A(\rho_{AB}) \\ &= \frac{1}{2}[\mathrm{tr}(|0\rangle\langle 0|_A) \times |0\rangle\langle 0|_B + \mathrm{tr}(|0\rangle\langle 1|_A) \times |0\rangle\langle 1|_B + \mathrm{tr}(|1\rangle\langle 0|_A) \\ &\quad \times |1\rangle\langle 0|_\mathrm{B} + \mathrm{tr}(|1\rangle\langle 1|_A) \times |1\rangle\langle 1|_B] \\ &= \frac{1}{2}(|0\rangle\langle 0|_B + |1\rangle\langle 1|_B) \\ &\equiv \frac{I_B}{2},\end{aligned} \tag{21.69}$$

where $I_A, I_B$ are the identity matrices of subsystems $A, B$. We observe that $\mathrm{tr}(\hat{\rho}_A^2) = \mathrm{tr}(\hat{\rho}_B^2) = 1/2 < 1$, which shows that $\hat{\rho}_A, \hat{\rho}_B$ correspond to *mixed states*. For the sake of curiosity, let us now calculate the tensor product $\hat{\rho}_A \otimes \hat{\rho}_B$ from Eqs. (21.68) and (21.69). We obtain

$$\begin{aligned}\hat{\rho}_A \otimes \hat{\rho}_B &= \frac{I_A}{2} \otimes \frac{I_B}{2}\\ &= \frac{I_{AB}}{4}\\ &\equiv \frac{1}{4}(|00\rangle\langle 00| + |01\rangle\langle 01| + |10\rangle\langle 10| + |11\rangle\langle 11|).\end{aligned} \tag{21.70}$$

Comparison between Eq. (21.70) and Eq. (21.67) shows that, as expected, $\rho_{AB} \neq \hat{\rho}_A \otimes \hat{\rho}_B$. This provides an indication that the information in the subsystems $A$, $B$ is correlated or, equivalently, that the subsystem states are *entangled*.

Next, we calculate the quantum information, $S(\hat{\rho}_A)$, $S(\hat{\rho}_B)$ associated with the reduced density operators $\hat{\rho}_A$, $\hat{\rho}_B$, to find:

$$S(\hat{\rho}_A) = S(\hat{\rho}_B) = -2 \times (1/2)\log(1/2) \equiv 1. \tag{21.71}$$

Thus, *each of the subsystems $A$, $B$ has a quantum information equivalent to exactly one classical bit, which corresponds to the qubit state $|q\rangle = (|0\rangle \pm |1\rangle)/\sqrt{2}$ of maximal uncertainty*. Such a conclusion is also consistent with our previous observation according to which the two subsystems $A$, $B$ must be in mixed states ($\mathrm{tr}(\hat{\rho}_A^2) = \mathrm{tr}(\hat{\rho}_B^2) = 1/2 < 1$). The situation of the two subsystems, $A$, $B$, contrasts with that of the composite system, $AB$, which is in the pure state $|\beta_{00}\rangle$. By definition, this state is known exactly and, therefore, it has zero quantum information, meaning that $S(\rho_{AB}) = 0$. This is a most intriguing feature, without any counterpart in the classical world: *there is no information in the composite system, while there is maximal information in the two subsystems*! Formally, we obtain for the mutual information and the conditional entropy:

$$\begin{aligned}S(\rho_A; \rho_B) &= S(\hat{\rho}_A) + S(\hat{\rho}_B) - S(\rho_{AB})\\ &= 1 + 1 - 0 = 2,\end{aligned} \tag{21.72}$$

$$\begin{aligned}S(\rho_A|\rho_B) &= S(\rho_{AB}) - S(\hat{\rho}_B) = 0 - 1 \equiv -1\\ S(\rho_B|\rho_A) &= S(\rho_{AB}) - S(\hat{\rho}_A) = 0 - 1 \equiv -1.\end{aligned} \tag{21.73}$$

These results show that the two subsystems have the equivalent of two classical bits (two cbits) for mutual information, which shows that the 1-cbit information they each possess in fact constitute shared property! On the other hand, knowledge of the information contents of one subsystem *removes one cbit* from the information in the other subsystem, as *the conditional entropy is negative*. Since the other subsystem has one cbit of information content, there is actually no information left to measure! This observation means that *knowledge of one subsystem exactly conditions that of the other*, which illustrates the principle of *quantum entanglement*. Another way to understand this conditioning property is to consider the effect of measuring the information in one of the two subsystems. Assume, for instance, that our measurement finds subsystem $A$ in the state $|0\rangle$. The post-measurement state, which now characterizes subsystem $B$, is the pure state $|\psi_B\rangle = |0\rangle_B$. If our measurement finds subsystem $A$ in the state $|1\rangle$, the post-measurement state is the pure state $|\psi_B\rangle = |1\rangle_B$. Either measurement in subsystem

$A$, thus, results in the collapse of subsystem $B$ into a pure state. The same conclusion strictly applies if subsystem $B$ is measured first. Thus, in all possible cases, a measurement of one subsystem collapses the other into a pure state, which, as we know, contains zero information.

As we have seen in Chapter 18, quantum entanglement, as a characteristic property of the Bell–EPR states, can be used for the purpose of *quantum teleportation*. See more illustrations of the effect of quantum entanglement in the exercises.

## 21.3 Quantum communication channel and Holevo bound

We may now conceive a quantum version of Shannon's *communication channel* (Chapter 11), with the purpose of transmitting information from an originator entity to a recipient entity. Here, the words "communication" and "transmission" should not be interpreted in the engineering meaning of sending information through a wire from one point to another over some distance. Instead, the strict meaning is that of a message being communicated in some coded form, from an originator to a recipient, regardless of the physical transmission means, the overall channel being a *quantum system*. Let me now clarify how such a channel may be operated.

As in the classical case, the information is encapsulated into a message $X$, which represents a succession of *symbols* (or *letters*, or *codewords*), $x$, to be encoded by the originator from a possible alphabet (or code) of size $n$. The message symbols are associated with a probability $p_x$. After passing through a physical "transmission pipe," the symbols are then decoded by the recipient, to restore the classical information therein. In the quantum case, however, the information is to be carried by *quantum states*, which will now be referred to by their density operators $\rho_x$. From the simplest perspective, the operation of encoding consists, for the originator, of the "preparation" of the quantum states, $\rho_x$, which is made according to the message's probability distribution $p_x$. Here, the word "preparation" means setting up the conditions for a given quantum system to be exactly in the state $\rho_x$. The operation of decoding is, for the recipient, the action of performing "measurements" on each of the received quantum states. According to Chapter 17, such measurements can be made through a collection of $n$ Hermitian operators $\{E_m\}_{m=1\ldots n}$ referred to as a *POVM set*. The outcome of any measurement is a real positive number $y$ belonging to some alphabet $Y$ of size $n$. We note that from this description, $X$, $Y$ are classical sources with $x$, $y$ as associated random events, while the communication channel is quantum.

I shall now focus on the *mutual information,* $H(X;Y)$ associated with the above-described quantum communication channel. A key property is that the mutual information is bounded by a maximum, $\chi$, referred to as a *Holevo bound*, and defined by

$$H(X;Y) \leq S(\rho) - \sum_x p_x S(\rho_x) = \chi, \tag{21.74}$$

where $\rho = \sum_x p_x \rho_x = \langle \rho \rangle$ is the mean density operator, as averaged over the coding possibilities. Based on this notation, and combining Eq. (21.74) with the general property of *concavity of entropy* in Eqs. (21.50) and (21.51), the Holevo bound condition also gives

$$H(X;Y) \le \chi \le H(X), \tag{21.75}$$

where $H(X)$ is the Shannon entropy of the originator message source. *In a classical communication channel, the mutual information cannot exceed the entropy of the originator source*, namely $H(X;Y) = H(X) - H(X|Y)$ with $H(X|Y) \ge 0$ (see Chapter 5), hence, $H(X;Y) \le H(X)$. Thus, Eq. (21.74) is granted. However, the intermediate Holevo bound $\chi$ in this equation is nontrivial, except in the case $\chi = H(X)$, corresponding to the specific situation where $\rho_x$ have support on orthogonal states. In the contrary case, we have $\chi < H(X)$ and, hence, the condition $H(X;Y) < H(X)$. In such a condition, *there exists no possibility for the recipient to completely recover the originator's information $H(X)$ through any set of measurements on $Y$*. This constitutes a nonintuitive situation that cannot be experienced from any "classical viewpoint."

The formal proof of the Holevo bound is presented in Appendix X. The proof is tractable but, unfortunately, only to a certain extent! As described in this appendix, indeed, the full proof requires the introduction of an additional entropy property referred to as *strong subadditivity*.[5] Such a property, which has not be described in the earlier subsection, rests on two additional theorems, whose demonstration is rather mathematically involved. For this reason, and for the purpose of these chapters, we shall take "strong subadditivity" as a granted postulate. Going through the derivations in Appendix X, however, should not be viewed as a vain exercise. Rather, in addition to coming very close to an intuitive (albeit incomplete) proof of the Holevo bound, it also represents an opportunity to familiarize oneself with the concept of *quantum operations*, of which *quantum measurements* with POVM operators represent a very representative illustration.

I shall now illustrate the consequences of the Holevo bound through a few basic examples.

---

**Example 21.4:** ***A "useless" quantum communication channel***

Assume that the originator is able to prepare a quantum system in two possible orthogonal states $|q_1\rangle = (|0\rangle + |1\rangle)/\sqrt{2}$ and $|q_2\rangle = (|0\rangle - |1\rangle)/\sqrt{2}$, but with some relative uncertainty, according to a probability law defined by $p_1 = p$ and $p_2 = 1 - p$. Not being 100% certain, these two states are mixed states, having the corresponding and

[5] The property of "strong subadditivity" states that for any three quantum systems $A, B, C$, the following inequalities hold:

$$S(A) + S(B) \le S(A, C) + S(B, C)$$
$$S(A, B, C) + S(B) \le S(A, B) + S(B, C).$$

equal density operators:

$$\rho_1 = \rho_2 = \frac{1}{2}(|0\rangle\langle 0| + |1\rangle\langle 1|) = \frac{1}{2}\begin{pmatrix} 1 & 0 \\ 0 & 1 \end{pmatrix}. \tag{21.76}$$

The mean density operator is $\rho = p\rho_1 + (1-p)\rho_2 = \rho_1 = \rho_2$. Clearly, the VN entropy is the same for the three states $\rho, \rho_1, \rho_2$, i.e., $S(\rho) = S(\rho_1) = S(\rho_2) = -2 \times (1/2)\log(1/2) = 1$, hence, the Holevo bound $\chi$ is zero, meaning $H(X;Y) = 0$. Such a quantum communication channel has no mutual information available whatsoever in reserve and, in the sense of information theory, is, therefore, *useless*. Formally, this uselessness characteristic is because the qubit states used for *the message quantum states, albeit orthogonal as "symbols,"* have *strictly identical supports* (or eigenstate spaces). Basically, the originator does not have any control on his or her source to prepare any "communicable" information. Based on the fact that each of the proposed qubits contain exactly one classical bit (cbit), and that the two are orthogonal, this conclusion of uselessness sounds like a disillusionment concerning the communication potential. But it is now clear that their mixed-state nature render the communication attempt useless, no matter what the distribution defined by the parameter $p$.

---

**Example 21.5:** ***A quantum communication channel reduced to classical***

The originator (based on the previous experience) is now able to prepare his or her quantum system with the same orthogonal qubits, i.e., $|q_1\rangle = (|0\rangle + |1\rangle)/\sqrt{2}$ and $|q_2\rangle = (|0\rangle - |1\rangle)/\sqrt{2}$, but this time as *pure states*, contrary to the previous example. This means that in each possible preparation, there is absolute certainty that the outcome is either $|q_1\rangle$ or $|q_2\rangle$, as required. Then the two preparations that can be chosen have corresponding density-operators:

$$\begin{aligned} \rho_1 &= \frac{1}{2}(|0\rangle + |1\rangle)(\langle 0| + \langle 1|) = \frac{1}{2}\begin{pmatrix} 1 & 1 \\ 1 & 1 \end{pmatrix} \to \begin{pmatrix} 1 & 0 \\ 0 & 0 \end{pmatrix} \\ \rho_2 &= \frac{1}{2}(|0\rangle - |1\rangle)(\langle 0| - \langle 1|) = \frac{1}{2}\begin{pmatrix} 1 & -1 \\ -1 & 1 \end{pmatrix} \to \begin{pmatrix} 0 & 0 \\ 0 & 1 \end{pmatrix}. \end{aligned} \tag{21.77}$$

In the above, the arrows mean in each case that the density matrix is transformed from the basis $\{|0\rangle, |1\rangle\}$ into the eigenstate basis $\{|+\rangle, |-\rangle\} = \{|q_1\rangle, |q_2\rangle\}$. The originator then selects the qubits according to the message's probability law, here defined by $p_1 = p$ and $p_2 = 1 - p$, where $p$ is a parameter. The mean density operator is, therefore,

$$\rho = p\rho_1 + (1-p)\rho_2 = \begin{pmatrix} p & 0 \\ 0 & 1-p \end{pmatrix}. \tag{21.78}$$

Clearly, we have for the VN entropy $S(\rho) = -[p\log p - (1-p)\log(1-p)] = f(p) \equiv H(X)$ and $S(\rho_1) = S(\rho_2) = 0$ (pure states). Thus, the Holevo bound is $\chi = f(p) = H(X)$ meaning $H(X;Y) \leq H(X)$ as in the case of a classical communication channel. We note that the strict equality for the Holevo bound, $\chi = H(X)$, stems from the fact that the message states $\rho_1, \rho_2$ have orthogonal support. This result shows that the quantum communication channel is able to convey information that can be up to $H(X)$, which is the classical entropy of the originator's message source. Interestingly, the quantum "symbols" carry no information by themselves, since $S(\rho_1) = S(\rho_2) = 0$ are pure states. As we know, $H(X)$ is maximized for the choice $p = 0.5$, corresponding to a coin-tossing probability law.

---

**Example 21.6:** ***General case***

This example represents an interesting generalization of the previous situation. The originator is able to prepare single qubits as pure states, and the message to be encoded is a long string of 1 or 0 cbits with a uniform probability distribution. If the cbit is 0, the qubit symbol is chosen, for instance, to be $|0\rangle$, and if the cbit is 1, the qubit symbol is chosen to be $\cos\theta|0\rangle + \sin\theta|1\rangle$, where $\theta \in [0, \pi]$ is an arbitrary parameter. The density operators of the message's qubit symbols are

$$\begin{cases} \rho_1 = |0\rangle\langle 0| = \begin{pmatrix} 1 & 0 \\ 0 & 0 \end{pmatrix} \\ \rho_2 = (\cos\theta|0\rangle + \sin\theta|1\rangle)(\cos\theta\langle 0| + \sin\theta\langle 1|) \\ \quad = \cos^2\theta|0\rangle\langle 0| + \cos\theta\sin\theta|0\rangle\langle 1| + \cos\theta\sin\theta|1\rangle\langle 0| + \sin^2\theta|0\rangle\langle 0| \\ \quad = \begin{pmatrix} \cos^2\theta & \cos\theta\sin\theta \\ \cos\theta\sin\theta & \sin^2\theta \end{pmatrix}. \end{cases} \tag{21.79}$$

Each qubit symbol having the same probability $p_1 = p_2 = 1/2$, we obtain for the mean density operator:

$$\begin{aligned} \rho = \sum_{i=1}^{2} p_i\rho_i &= \frac{1}{2}\begin{pmatrix} 1 & 0 \\ 0 & 0 \end{pmatrix} + \frac{1}{2}\begin{pmatrix} \cos^2\theta & \cos\theta\sin\theta \\ \cos\theta\sin\theta & \sin^2\theta \end{pmatrix} \\ &\equiv \frac{1}{2}\begin{pmatrix} 1+\cos^2\theta & \cos\theta\sin\theta \\ \cos\theta\sin\theta & \sin^2\theta \end{pmatrix}. \end{aligned} \tag{21.80}$$

It is easily found through the characteristic equation that the eigenvalues of $\rho$ are $\lambda_{1,2} = (1 \pm \cos\theta)/2$, or $\lambda_1 = \cos^2(\theta/2)$ and $\lambda_2 = \sin^2(\theta/2) = 1 - \lambda_1$. The VN entropy of the originator's quantum source $(\rho)$ is, therefore:

$$\begin{aligned} S(\rho) &= -\lambda_1 \log\lambda_1 - \lambda_2 \log\lambda_2 \\ &= -\lambda_1 \log\lambda_1 - (1-\lambda_1)\log(1-\lambda_1) \\ &\equiv f(\lambda_1) \\ &\equiv f[\cos^2(\theta/2)]. \end{aligned} \tag{21.81}$$

Since $S(\rho_1) = S(\rho_2) = 0$ (pure states) the Holevo bound is

$$\chi = S(\rho) = f\left(\cos^2\frac{\theta}{2}\right). \tag{21.82}$$

A plot of the above-defined Holevo bound is provided in Fig. 21.1. It is seen that a maximum is reached for $x = 0.5$ or $\theta = \pi/2$, which corresponds to a choice for the

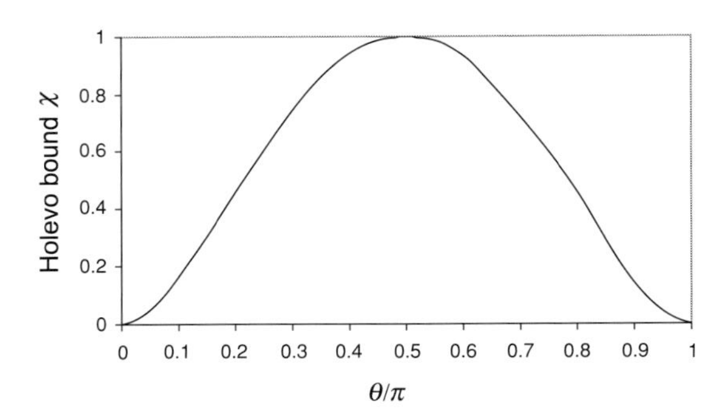


**Figure 21.1** Holevo bound $\chi$ as a function of the argument $\theta$, showing a maximum at $\theta = \pi/2$.

second qubit of $|q_2\rangle = |1\rangle$ or $\rho_2 = |1\rangle\langle 1|$. In this case, $H(X;Y) \leq \chi = H(X) = 1$, meaning that information can be transmitted through the quantum channel up to 1 bit, corresponding to the originator's source entropy. In any other case, or $\theta \neq \pi/2$, the Holevo bound is less than unity (the originator's source entropy) and there is no possibility for the original information to be fully retrieved by the recipient. Clearly, such a limitation is attributable to the fact that the set of density operators used to code the symbols does not have orthogonal support, meaning that they have a finite *overlap,* defined formally as $|\langle q_1|q_2\rangle|^2 = \cos^2\theta \neq 0$. The case $\theta = \pi/2$, where the two states $|q_1\rangle, |q_2\rangle$ are orthogonal, reduces the quantum channel to a classical one, as described in Example 2.

In summary, the key lesson to retain from the above description and examples is that it is possible to transmit (or "convey") classical information through a quantum communication channel. A prerequisite is that the alphabet of quantum symbols used to code information must be made of *pure states*, in which case the Holevo bound $\chi$ equals the VN entropy or average quantum information per symbol ($\chi = S(\rho)$). A second, optional, requirement is that the alphabet states be orthogonal, in which case the Holevo bound is maximized and is equal to the originator's source entropy ($\chi = H(X)$). In any case, the mutual information $H(X;Y)$, i.e., the amount of information that the originator and the recipient may be able to share, satisfies $H(X;Y) \leq \chi$.

The concept of the quantum communication channel will be revisited in Chapter 23, which considers the effect of channel corruption by *quantum noise*.

## 21.4 Exercises

**21.1** (B): Show that the quantum system with density matrix (basis representation $\{|0\rangle, |1\rangle\}$)

$$\rho = \frac{1}{2}\begin{pmatrix} 1 & 1 \\ 1 & 1 \end{pmatrix}$$

corresponds to the pure state

$$|+\rangle = \frac{1}{\sqrt{2}}(|0\rangle + |1\rangle).$$

**21.2** (B): Show that the density operator $\rho = |0\rangle\langle 0|$ of a pure state $|0\rangle$ can be transformed into the basis $\{|+\rangle, |-\rangle\}$ with the definition (using here the sign $\oplus$ for clarity in the summation):

$$\begin{aligned}\tilde{\rho} &= \frac{1}{2}(|+\rangle\langle +| \oplus |+\rangle\langle -| \oplus |-\rangle\langle +| \oplus |-\rangle\langle -|)\\ &= \frac{|+\rangle \oplus |-\rangle}{\sqrt{2}} \otimes \frac{\langle +| \oplus \langle -|}{\sqrt{2}}.\end{aligned}$$

**21.3** (B): Show that for any density operator $\rho$, the property

$$\mathrm{tr}(\rho^2) \leq 1$$

applies with equality only for the case of pure states.

**21.4** (M): Given a quantum system with the density matrix

$$\rho = \frac{1}{3}\begin{pmatrix} 2 & 1 \\ 1 & 1 \end{pmatrix},$$

determine the corresponding state $|\psi\rangle$ in the eigenstate basis and the von Neumann entropy.

**21.5** (M): Given the density matrices $\rho_A$, $\rho_B$ and identity matrices $I_A$, $I_B$ of two subsystems $A$, $B$ with dimensions $n \times n$ and $p \times p$, respectively, and the density matrix $\rho_{AB}$ of the joint system with dimension $n \times p$, show the following relations:
(1) $\rho_A \otimes \rho_B = (\rho_A \otimes I_B)(I_A \otimes \rho_B)$,
(2) $\log(\rho_A \otimes \rho_B) = \log(\rho_A) \otimes I_B + I_A \otimes \log(\rho_B)$,
and

(3) $$\begin{cases} \mathrm{tr}[\rho_{AB} \log(\rho_A \otimes I_B)] = \mathrm{tr}(\rho_A \log \rho_A) \\ \mathrm{tr}[\rho_{AB} \log(I_A \otimes \rho_B)] = \mathrm{tr}(\rho_B \log \rho_B), \end{cases}$$

using for (3) the property valid for any observable $U_X$ in a composite system $XY$

$$\mathrm{tr}[\rho_{XY}(U_X \otimes I_Y)] = \mathrm{tr}(\rho_X U_X)$$

with a similar relation applying to any observable $U_Y$.
(*Clue*: start from the operator tensor-product definition in Eq. (16.54), and assume, for simplicity and without loss of generality, that $n = 2$, $p = 3$)

**21.6** (B): Show that the 2-qubit state

$$|\psi\rangle = \frac{1}{\sqrt{3}}(|00\rangle + |10\rangle + |11\rangle)$$

(a) Is not a pure state;

(b) Cannot be decomposed as a tensor product $|\psi_A\rangle \otimes |\psi_B\rangle$ of some subsystem states $|\psi_A\rangle$, $|\psi_B\rangle$.

**21.7** (M): Determine the von Neumann entropy of the composite system having density operator

$$\rho = \frac{1}{4}(|00\rangle\langle 00| + |00\rangle\langle 11| + 2|10\rangle\langle 10| + |11\rangle\langle 00| + |11\rangle\langle 11|).$$

# 22 Quantum data compression

This chapter describes the principle of *compression* in quantum communication channels. The underlying concept is that it is possible to convey "faithfully" a quantum message with a large number of qubits, while transmitting a compressed version of this message with a reduced number of qubits through the channel. Beyond the mere notion of *fidelity*, which characterizes the quality of quantum message transmission, the description brings the new concept of *typicality* in the space defined by all possible "quantum codewords." The theorem of *Schumacher's quantum compression* states that for a qubit source with von Neumann entropy $S$, the message compression factor $R$ has $S + \varepsilon$ for the lower bound, where $\varepsilon$ is any nonnegative parameter that can be made arbitrarily small for sufficiently long messages (hence, $R \approx S$ is the best possible compression factor). An original graphical and numerical illustration of the effect of Schumacher's quantum compression and the evolution of the typical quantum-codeword subspace with increasing message length is provided.

## 22.1 Quantum data compression and fidelity

In this chapter, we have reached the stage where it is possible to start addressing the issues that are central to information theory, namely, "How efficiently can we code information in a quantum communication channel?" both in terms of economy of means – the concept of *data compression* – and accuracy of transmission – the concept of *message integrity* or *minimal data error*, referred to here as *fidelity*. In QIT, the concepts of compression and fidelity are, in fact, intimately and nicely related, as described in this section.

As described in the previous chapter, a quantum communication channel requires an alphabet of pure states, which are used to encode the classical information contained in message bits, letters, or codewords. In the classical case, a message like $m = 1001110101000110$ can be viewed either as a sequence of 16 individual bits that are transmitted one at a time, or as a single "block" of two-byte size (one byte is an eight-bit sequence). Similarly, we may conceive of a full quantum message not as a time sequence of $n$ qubit symbols, but as a "block" represented by the tensor state $|M\rangle = |x_1 x_2 \ldots x_n\rangle$, with each qubit $|x_i\rangle$ being randomly selected from a given symbol alphabet $\{|a_k\rangle\}_{k=1\ldots N}$ of size $N$. Since each symbol $|a_k\rangle$ is associated with an occurrence probability $p_k$, we can define a *symbol density operator* as:

$$\rho = \sum_{k=1}^{N} p_k |a_k\rangle\langle a_k|. \tag{22.1}$$

Consistently, the full symbol block is characterized by the *message density operator*:

$$\rho_M = \rho \otimes \rho \otimes \cdots \otimes \rho \equiv \rho^{\otimes n}. \tag{22.2}$$

I shall now provide a basic illustration of the above, which will also enable me to introduce the concept of *fidelity*. Assume a quantum communication channel with messages based on a two-symbol alphabet of pure states $\{|a\rangle, |b\rangle\}$, defined as:

$$\begin{cases} |a\rangle = |0\rangle = \begin{pmatrix} 1 \\ 0 \end{pmatrix} \\ |b\rangle = \frac{1}{\sqrt{2}}(|0\rangle + |1\rangle) = \frac{1}{\sqrt{2}}\begin{pmatrix} 1 \\ 1 \end{pmatrix}. \end{cases} \tag{22.3}$$

The classical message source, $X$, is assumed to be uniformly distributed, so the two quantum symbols have equal probabilities $p_a = p_b = 1/2$. Thus, the symbol density operator is given by

$$\begin{aligned} \rho &= p_a|a\rangle\langle a| + p_a|b\rangle\langle b| \\ &= \frac{1}{2}\begin{pmatrix} 1 & 0 \\ 0 & 0 \end{pmatrix} + \frac{1}{4}\begin{pmatrix} 1 & 1 \\ 1 & 1 \end{pmatrix} \\ &\equiv \frac{1}{4}\begin{pmatrix} 3 & 1 \\ 1 & 1 \end{pmatrix}. \end{aligned} \tag{22.4}$$

The eigenvalues of $\rho$ are easily found to be $\lambda_a = (1 + 1/\sqrt{2})/2 \equiv \cos^2(\pi/8)$ and $\lambda_a = (1 - 1/\sqrt{2})/2 \equiv \sin^2(\pi/8)$, respectively. The corresponding eigenvectors $|\lambda_a\rangle, |\lambda_b\rangle$ are

$$\begin{aligned} |\lambda_a\rangle &= \begin{pmatrix} \cos\frac{\pi}{8} \\ \sin\frac{\pi}{8} \end{pmatrix}, \\ |\lambda_b\rangle &= \begin{pmatrix} \sin\frac{\pi}{8} \\ -\cos\frac{\pi}{8} \end{pmatrix}. \end{aligned} \tag{22.5}$$

and in the eigenstate basis $\{|\lambda_a\rangle, |\lambda_b\rangle\}$, the density operator $\rho$ has the diagonal form:

$$\rho = \begin{pmatrix} \lambda_a & 0 \\ 0 & \lambda_b \end{pmatrix} = \begin{pmatrix} \cos^2\frac{\pi}{8} & 0 \\ 0 & \sin^2\frac{\pi}{8} \end{pmatrix}. \tag{22.6}$$

Finally, the VN entropy associated with any message symbol, or *per-qubit entropy* is

$$\begin{aligned} S(\rho) &= -\lambda_a \log \lambda_a - \lambda_b \log \lambda_b \\ &= -\lambda_a \log \lambda_a - (1 - \lambda_a)\log(1 - \lambda_a) \\ &\equiv f\left[\cos^2(\pi/8)\right] \\ &\equiv 0.6008. \end{aligned}$$

Let us now calculate the overlap between the eigenstates $|\lambda_a\rangle$, $|\lambda_b\rangle$ and the quantum symbols $|a\rangle$, $|b\rangle$. Using Eqs. (22.3) and (22.5), we obtain:

$$\begin{cases} |\langle\lambda_a|a\rangle|^2 = |\langle\lambda_a|b\rangle|^2 = \cos^2\dfrac{\pi}{8} = \lambda_a \approx 0.8535 \\ |\langle\lambda_b|a\rangle|^2 = |\langle\lambda_b|b\rangle|^2 = \sin^2\dfrac{\pi}{8} = \lambda_b \approx 0.1465. \end{cases} \tag{22.7}$$

The overlap between the subspaces defined or "spanned" by either $|\lambda_a\rangle$ or $|\lambda_b\rangle$ and the full space defined or spanned by $|a\rangle$, $|b\rangle$ is, thus, observed to be significantly greater in the first case (i.e., 85% vs. 15%, or $\lambda_a$ vs. $\lambda_b$).

We may now introduce the concept of *fidelity*, defined by

$$F = \langle\psi|\rho|\psi\rangle \equiv p_a|\langle\psi|a\rangle|^2 + p_b|\langle\psi|b\rangle|^2, \tag{22.8}$$

where $|\psi\rangle$ is an arbitrary measurement state. Here, since $p_a = p_b = 1/2$, the fidelity reduces to $F = (|\langle\psi|a\rangle|^2 + |\langle\psi|b\rangle|^2)/2$. From Eq. (22.7), it is clear that *the fidelity is a maximum for the measurement choice* $|\psi\rangle = |\lambda_a\rangle$, regardless of whether $|a\rangle$, $|b\rangle$ was sent by the originator, which gives $F = 0.853$. Hence, $\{|\psi\rangle\} = \{|\lambda_a\rangle\}$ corresponds to a one-dimensional "likely subspace" with which any message symbol is most strongly overlapping.

Let us now extend the above notion of "likely subspace" to the case of a message of three qubits long. The eight *equally-likely* message outcomes $|M\rangle$ are

$$|M\rangle = |aaa\rangle, |aab\rangle, |aba\rangle, |abb\rangle, |baa\rangle, |bab\rangle, |bba\rangle, |bbb\rangle. \tag{22.9}$$

Let $|\psi\rangle = |\lambda_i\lambda_j\lambda_k\rangle$ be an arbitrary measurement state with $\lambda_i, \lambda_j, \lambda_k = \lambda_a, \lambda_b$. Using the property $|\langle\psi|M\rangle|^2 = |\langle\lambda_i\lambda_j\lambda_k|xyz\rangle|^2 = |\langle\lambda_i|x\rangle|^2|\langle\lambda_j|y\rangle|^2|\langle\lambda_k|z\rangle|^2$, together with the result in Eq. (22.7), we obtain the space overlaps applicable to any message $|M\rangle$:

$$|\langle\lambda_a\lambda_a\lambda_a|M\rangle|^2 = \lambda_a^3 = \cos^6\frac{\pi}{8} = 0.6219, \tag{22.10}$$

$$\begin{aligned} |\langle\lambda_a\lambda_a\lambda_b|M\rangle|^2 &= |\langle\lambda_a\lambda_b\lambda_a|M\rangle|^2 \\ &= |\langle\lambda_b\lambda_a\lambda_a|M\rangle|^2 = \lambda_a^2\lambda_b = \cos^4\frac{\pi}{8}\sin^2\frac{\pi}{8} = 0.1067, \end{aligned} \tag{22.11}$$

$$\begin{aligned} |\langle\lambda_a\lambda_b\lambda_b|M\rangle|^2 &= |\langle\lambda_b\lambda_a\lambda_b|M\rangle|^2 \\ &= |\langle\lambda_b\lambda_b\lambda_a|M\rangle|^2 = \lambda_a\lambda_b^2 = \cos^2\frac{\pi}{8}\sin^4\frac{\pi}{8} = 0.0183, \end{aligned} \tag{22.12}$$

$$|\langle\lambda_b\lambda_b\lambda_b|M\rangle|^2 = \lambda_b^3 = \sin^6\frac{\pi}{8} = 0.0031. \tag{22.13}$$

It is seen that the *subspace* defined by $\Omega = \{|\lambda_a\lambda_a\lambda_a\rangle, |\lambda_a\lambda_a\lambda_b\rangle, |\lambda_a\lambda_b\lambda_a\rangle, |\lambda_b\lambda_a\lambda_a\rangle\}$ is the most likely, while its orthogonal subspace complement, $\Omega^\perp = \{|\lambda_a\lambda_b\lambda_b\rangle, |\lambda_b\lambda_a\lambda_b\rangle, |\lambda_b\lambda_b\lambda_a\rangle, |\lambda_b\lambda_b\lambda_b\rangle\}$, is the least likely. Such a statement means that making any measurement of the message using a state $|\psi\rangle$ from the eigenstate basis $\Lambda = \{|\lambda_a\rangle, |\lambda_b\rangle\}^{\otimes 3}$ is more likely to project the state in the subspace $\Omega$ than in the subspace $\Omega^\perp$, the corresponding probabilities being $p(\Omega) = \lambda_a^3 + 3\lambda_a^2\lambda_b = 0.6219 + 3 \times 0.1067 = 0.942 \equiv 1 - \delta$ and $p(\Omega^\perp) = 3\lambda_a\lambda_b^2 + \lambda_b^3 = 3 \times 0.0183 + 0.0031 = 0.058 \equiv \delta$, respectively. We may then define $E$ as the projector on the likely four-dimensional subspace

$\Omega$, according to

$$\begin{aligned} E &= \sum_{|\lambda_i\lambda_j\lambda_k\rangle\in\Omega} |\lambda_i\lambda_j\lambda_k\rangle\langle\lambda_i\lambda_j\lambda_k| \\ &= |\lambda_a\lambda_a\lambda_a\rangle\langle\lambda_a\lambda_a\lambda_a| + |\lambda_a\lambda_a\lambda_b\rangle\langle\lambda_a\lambda_a\lambda_b| \\ &\quad + |\lambda_a\lambda_b\lambda_a\rangle\langle\lambda_a\lambda_b\lambda_a| + |\lambda_b\lambda_a\lambda_a\rangle\langle\lambda_b\lambda_a\lambda_a|, \end{aligned} \tag{22.14}$$

or as expressed in the eigenstate basis $\Lambda$:

$$E = \begin{pmatrix} 1 & 0 & 0 & 0 & 0 & 0 & 0 & 0 \\ 0 & 1 & 0 & 0 & 0 & 0 & 0 & 0 \\ 0 & 0 & 1 & 0 & 0 & 0 & 0 & 0 \\ 0 & 0 & 0 & 1 & 0 & 0 & 0 & 0 \\ 0 & 0 & 0 & 0 & 0 & 0 & 0 & 0 \\ 0 & 0 & 0 & 0 & 0 & 0 & 0 & 0 \\ 0 & 0 & 0 & 0 & 0 & 0 & 0 & 0 \\ 0 & 0 & 0 & 0 & 0 & 0 & 0 & 0 \end{pmatrix}. \tag{22.15}$$

In the same eigenstate basis, the message's density operator $\rho_M = \rho^{\otimes n}$ takes the diagonal form:

$$\rho_M = \begin{pmatrix} \lambda_a^3 & 0 & 0 & 0 & 0 & 0 & 0 & 0 \\ 0 & \lambda_a^2\lambda_b & 0 & 0 & 0 & 0 & 0 & 0 \\ 0 & 0 & \lambda_a^2\lambda_b & 0 & 0 & 0 & 0 & 0 \\ 0 & 0 & 0 & \lambda_a^2\lambda_b & 0 & 0 & 0 & 0 \\ 0 & 0 & 0 & 0 & \lambda_a\lambda_b^2 & 0 & 0 & 0 \\ 0 & 0 & 0 & 0 & 0 & \lambda_a\lambda_b^2 & 0 & 0 \\ 0 & 0 & 0 & 0 & 0 & 0 & \lambda_a\lambda_b^2 & 0 \\ 0 & 0 & 0 & 0 & 0 & 0 & 0 & \lambda_b^3 \end{pmatrix}. \tag{22.16}$$

Based on the definitions in Eqs. (22.15) and (22.16), we clearly have

$$\operatorname{tr}(\rho_M E) = p(\Omega) = 1 - \delta. \tag{22.17}$$

We shall now refer to $\Omega$ as the *typical* subspace, which is reminiscent of the *typical set* of classical information theory described in Chapter 13. Likewise, any message of the type $|\psi_{\text{typ}}\rangle = |\lambda_a\lambda_a\lambda_a\rangle, |\lambda_a\lambda_a\lambda_b\rangle, |\lambda_a\lambda_b\lambda_a\rangle, |\lambda_b\lambda_a\lambda_a\rangle$ represents the quantum analog of the *typical sequences* described in that chapter. We shall now refer to them as *typical states*. At this stage, it is only an intuitive notion that as the message length $n$ is increased, the probability associated with the typical subspace $\Omega$ asymptotically increases and that associated with $\Omega^{\perp}$ asymptotically decreases, meaning that the uncertainty $\delta$ may become arbitrarily small, and from Eq. (22.17), we may be able to achieve the condition $\operatorname{tr}(\rho_M E) = p(\Omega) \approx 1$. These observations will be discussed again.

We are now going to exploit the possibility of partitioning the space into typical and atypical subspaces in order to achieve *quantum compression* in the message. Assume, for instance, that we want to reduce the 3-qubit message to a 2-qubit, compressed, message. The following describes how the originator must proceed. First, the originator should apply a unitary transformation $U$, such that the four *typical states* (indifferently called

$|\psi_{\text{typ}}\rangle$) are transformed into states of the form $|xy0\rangle = |xy\rangle \otimes |0\rangle$, and the other four (atypical) states are transformed into states of the form $|xy1\rangle = |xy\rangle \otimes |1\rangle$. This is just a specific way of re-encoding the original message $|\psi\rangle$. Call $|\psi'\rangle = U|\psi\rangle$ the result of this re-encoding. With the knowledge of the re-encoding transformation $U$, the recipient is then able to retrieve the original message state $|\psi\rangle$ by performing the reverse operation, namely $|\psi\rangle = U^{-1}|\psi'\rangle$.

This above notion of message re-encoding being understood, consider now that the originator has the option of implementing the following trick. Assume indeed that she or he performs a measurement on the third qubit of $|\psi'\rangle$:

- If the outcome of the fuzzy measurement is 0, then the message state (and its post-measurement) is of the form $|xy0\rangle$, a typical state;
- If the outcome of the fuzzy measurement is 1, then the message state (and its post-measurement) is of the form $|zk1\rangle$, an atypical state.

Based on these two fuzzy measurement outcomes, the originator then takes the corresponding actions:

- If the outcome is 0, the remaining two qubits $|xy\rangle$, which we call $|\psi_{\text{comp1}}\rangle$, are sent through the quantum channel;
- If the outcome is 1, a 2-qubit state $|\psi_{\text{comp2}}\rangle$ is sent through the quantum channel; such that $U^{-1}\left(|\psi_{\text{comp2}}\rangle \otimes |0\rangle\right) = |\lambda_a\lambda_a\lambda_a\rangle$, which is the most likely typical state.

From the recipient's end, it is understood as a rule that the qubit $|0\rangle$ must be appended to the incoming message, which, thus, becomes either $|\psi_{\text{comp1}}\rangle \otimes |0\rangle$ or $|\psi_{\text{comp2}}\rangle \otimes |0\rangle$. The rule also has it that the inverse transform $U^{-1}$ must be applied. From this "decompression" operation, the recipient, thus, obtains either one of the two states:

$$\begin{cases} |\psi''\rangle = U^{-1}(|\psi_{\text{comp1}}\rangle \otimes |0\rangle) = U^{-1}U|xy0\rangle = |\psi_{\text{typ}}\rangle \\ |\psi'''\rangle = U^{-1}(|\psi_{\text{comp2}}\rangle \otimes |0\rangle) = |\lambda_a\lambda_a\lambda_a\rangle. \end{cases} \tag{22.18}$$

On decompression, the recipient is, thus, able to retrieve the original message state $|\psi_{\text{typ}}\rangle$, while occasionally getting a "junk" or "best-guess" message $|\lambda_a\lambda_a\lambda_a\rangle$. This coding trick, along with the agreed rules, has, thus, made it possible for the originator to transmit the 3-qubit original message to the recipient in a compressed, 2-qubit message, and for the recipient to uncompress it into a 3-qubit message. But how *faithful* is this whole operation? To answer this question, we must express the density operator $\tilde{\rho}$ of the uncompressed message, after reconstruction by the recipient. The original message, $|\psi\rangle$, has density operator $|\psi\rangle\langle\psi|$. Its projection on the typical space $\Omega$ is $E|\psi\rangle\langle\psi|E^{+} = E|\psi\rangle\langle\psi|E$, which provides a first component of $\tilde{\rho}$, corresponding to the typical message states $|\psi_{\text{typ}}\rangle$. The second component of $\tilde{\rho}$, corresponding to the other atypical message states, can be heuristically written $\tilde{\rho}_{\text{junk}} = \langle\psi|(I - E)|\psi\rangle|\lambda_a\lambda_a\lambda_a\rangle\langle\lambda_a\lambda_a\lambda_a|$, where $I = I^{\otimes 3}$ is the identity matrix. Clearly, $I - E$ is the projector on the atypical space $\Omega^{\perp}$. Thus, $\langle\psi|(I - E)|\psi\rangle$ equals one if $|\psi\rangle$ is an atypical state, with corresponding projector $|\lambda_a\lambda_a\lambda_a\rangle\langle\lambda_a\lambda_a\lambda_a|$, and $\langle\psi|(I - E)|\psi\rangle$ is equal to zero otherwise. The transmitted and reconstructed message, thus, has the complete density operator

$$\begin{aligned}\tilde{\rho} &= E|\psi\rangle\langle\psi|E + \langle\psi|(I - E)|\psi\rangle|\lambda_a\lambda_a\lambda_a\rangle\langle\lambda_a\lambda_a\lambda_a| \\ &= E|\psi\rangle\langle\psi|E + \tilde{\rho}_{\text{junk}}.\end{aligned} \tag{22.19}$$

From the definition in Eq. (22.8), we obtain the fidelity

$$\begin{aligned}\tilde{F} &= \langle\psi|\tilde{\rho}|\psi\rangle \\ &= \langle\psi|E|\psi\rangle\langle\psi|E|\psi\rangle + \langle\psi|\tilde{\rho}_{\text{junk}}|\psi\rangle \\ &= |\langle\psi|E|\psi\rangle|^2 + \langle\psi|(I - E)|\psi\rangle|\langle\psi|\lambda_a\lambda_a\lambda_a\rangle|^2 \\ &= P(\Omega)^2 + P(\Omega^\perp) \times |\langle\psi|\lambda_a\lambda_a\lambda_a\rangle|^2 \\ &\equiv (1-\delta)^2 + \delta|\langle\psi|\lambda_a\lambda_a\lambda_a\rangle|^2.\end{aligned} \tag{22.20}$$

Substituting $\delta = 0.058$ and $|\langle\lambda_a\lambda_a\lambda_a|\psi\rangle|^2 = \lambda_a^3 = 0.6219$ from Eq. (22.10) into Eq. (22.20), we obtain

$$\tilde{F} = (1 - 0.058)^2 + 0.058 \times 0.629 \equiv 0.923. \tag{22.21}$$

The result $\tilde{F} = 0.923$ compares well with the fidelity value $F = 0.853$ obtained earlier. To recall, this is the fidelity obtained by "guessing" any missing qubit is likely to be $|\lambda_a\rangle$. Thus, the alternate approach of transmitting only the first two qubits as uncoded, and leaving the recipient the task of "guessing" the third missing one (likely to be $|\lambda_a\rangle$) is $F = 0.853$. Such a guess-based coding for compression does not have a fidelity as high as that of the above coding, for which $\tilde{F} = 0.923$.

The example has shown that it is possible to compress a quantum message of $n$-qubit size ($n = 3$) into a coded version of $m$-qubit size ($m = 2$), with relatively high fidelity. Here, the *compression factor* that was achieved is $\eta = m/n = 2/3 = 0.666\ldots = 66.66\%$. The next section will establish that for messages of asymptotically increasing length, the achievable compression factor is bounded according to

$$\eta \geq S(\rho), \tag{22.22}$$

where $S(\rho)$ is the VN entropy carried by each of the message symbols. This result is known as *Schumacher's (quantum coding) theorem*. In this example, the per-qubit entropy is $S(\rho) = 0.6008$, corresponding to a best compression factor of $\eta = 60.08\%$. Therefore, for 3-qubit messages based on the symbol density matrix defined in Eq. (22.6), there is no code capable of achieving compression down to a single qubit ($\eta' = 0.333\ldots$) with any high fidelity. Such a compression factor would require $S(\rho) \leq 0.333\ldots$ Based on Schumacher's theorem, the key conclusion is that *the lower the VN entropy per qubit, the higher the message compression potential*. We may, thus, view the VN entropy as representing *redundant information*, meaning susceptible to compression. In the limiting case $S(\rho) = 1$ (or $\rho = I/2$), no compression is possible. This density operator corresponds to the qubit of the type $|\pm\rangle = (|0\rangle \pm |1\rangle)/\sqrt{2}$, where the information randomness is evenly distributed between 0 and 1 cbits, as with a classical source having the maximal entropy $H(X) = 1$ bit. This observation is consistent with the fact that *there is no code that is capable of compressing a purely random bit sequence as generated by a classical source* $H(X^n) = n$. But as soon as randomness is no longer evenly distributed between 0 and 1 bits ($H(X^n) < n$), there is information redundancy and, hence, the possibility of reducing the sequence to a size $m < n$.

**Table 22.1** Probabilities associated with quantum messages of length $n = 4$, defining the typical space $\Omega = \{|\lambda_a\lambda_a\lambda_a\lambda_a\rangle, |\lambda_a\lambda_a\lambda_a\lambda_b\rangle, |\lambda_a\lambda_a\lambda_b\lambda_b\rangle, | \leftrightarrow\rangle\}$ with likelihood $1 - \delta = 0.9888$.

| Codeword type | Probability $p$ | Number $N$ | Probability $Np$ | $1-\delta$ | $\delta$ |
|---|---|---|---|---|---|
| $\lvert\lambda_a\lambda_a\lambda_a\lambda_a\rangle$ | 0.5308 | 1 | 0.5308 | | |
| $\lvert\lambda_a\lambda_a\lambda_a\lambda_b\rangle$ | 0.0911 | 4 | 0.3643 | | |
| $\lvert\lambda_a\lambda_a\lambda_b\lambda_b\rangle$ | 0.0156 | 6 | 0.0937 | 0.9888 | |
| $\lvert\lambda_a\lambda_b\lambda_b\lambda_b\rangle$ | 0.0027 | 4 | 0.0107 | | |
| $\lvert\lambda_b\lambda_b\lambda_b\lambda_b\rangle$ | 0.0005 | 1 | 0.0005 | | 0.0112 |
| | | 16 | 1.0000 | | |

I have stated earlier that as the message length $n$ is increased, the probability associated with the typical subspace $\Omega$ asymptotically increases, and that associated with $\Omega^{\perp}$ asymptotically decreases. This suggests that given a compression factor $\eta = S(\rho)$, the fidelity $F$ asymptotically increases with the message length $n$. Here, I shall heuristically verify this property by considering messages longer than three qubits, and evaluating the corresponding fidelity, while assuming $S(\rho) = 0.6008$ with the eigenvalues $\lambda_a = 0.8535, \lambda_b = 0.1465$.

The probabilities associated with messages of 4-qubit length ($n = 4$) are defined in Table 22.1. This table reads as follows. We call "codeword" any 4-qubit $|\lambda_i\lambda_j\lambda_k\lambda_l\rangle$ with $\lambda_i, \lambda_j, \lambda_k, \lambda_l = \lambda_a, \lambda_b$. We call "codeword type" any 4-qubit obtained by permutation of the same values $\lambda_i, \lambda_j, \lambda_k, \lambda_l$. Each type has a probability $p = \lambda_i\lambda_j\lambda_k\lambda_l$ and a number $N$ given by the associated combinatorics. For instance, the codeword type $|\lambda_a\lambda_a\lambda_b\lambda_b\rangle$ has $p = \lambda_a^2\lambda_b^2 = 0.0156$ and $N = C_4^2 = 6$, as shown in Table 22.1. The product $Np$, also shown in the table, corresponds to the likelihood of all messages based on a given codeword type. We define the typical space as $\Omega = \{|\lambda_a\lambda_a\lambda_a\lambda_a\rangle, |\lambda_a\lambda_a\lambda_a\lambda_b\rangle, |\lambda_a\lambda_a\lambda_b\lambda_b\rangle, | \leftrightarrow\rangle\}$, where $\leftrightarrow$ stands for all possible permutations of the index $i, j, k, l$, corresponding overall to $1 + 4 + 6 = 11$ "typical" codewords. We observe from the table that $\Omega$ has a total likelihood of $P(\Omega) = 1 - \delta = 0.5308 + 0.3643 + 0.0937 = 0.9888$, while for the complementary space $\Omega^{\perp}$, the likelihood is $p(\Omega^{\perp}) = \delta = 0.0112$. Similarly to the definition in Eq. (22.20), we obtain the fidelity:

$$\begin{aligned}\tilde{F}_{n=4} &= P(\Omega)^2 + P(\Omega^{\perp}) \times |\langle\psi \mid \lambda_a\lambda_a\lambda_a\lambda_a\rangle|^2 \\ &= (1-\delta)^2 + \delta\lambda_a^4 \\ &= (0.9888)^2 + 0.0112 \times (0.8535)^4 \\ &\equiv 0.983.\end{aligned} \tag{22.23}$$

This result compares favourably with the fidelity $\tilde{F}_{n=3} = 0.923$ obtained earlier for the compression of quantum messages of length $n = 3$, which illustrates the property that fidelity increases with message length. In the case $n = 4$, however, the compression factor is $\eta_{n=4} = 3/4 = 0.75$, which represents less compression than in the previous case, where $\eta_{n=3} = 2/3 = 0.66$. On the other hand, we cannot compress the 4-qubit message into a 2-qubit one, because $\lceil S(\rho) \times 4\rceil = \lceil 2.40\rceil = 3$ bits, meaning that the minimum length of the compressed message is three qubits.

Consider, next, the case $n = 5$, based on the same assumptions for $S(\rho)$, $\lambda_a$, $\lambda_b$. We have $\lceil S(\rho) \times 5 \rceil = \lceil 3.004 \rceil = 4$ bits, which means that the minimum length of the compressed message is four qubits. The associated compression factor is $\eta_{n=5} = 4/5 = 0.80$, to compare with $\eta_{n=4} = 3/4 = 0.75$ and $\eta_{n=3} = 2/3 = 0.66$ in the previous cases. Thus, as we expect the fidelity to be higher with messages of length $n = 5$, the compression performance is poorer than in the previous cases.

Consider, next, the case $n = 6$, based on the same assumptions for $S(\rho)$, $\lambda_a$, $\lambda_b$. We have $\lceil S(\rho) \times 6 \rceil = \lceil 3.608 \rceil = 4$ bits, which means that the minimum length of the compressed message is four qubits. Thus, a compression factor of $\eta_{n=6} = 4/6 = 0.66$ is achievable, which represents compressing the 6-qubit original message into a 4-qubit one. As the next section describes, the compression code requires the recipient to manipulate the transmitted states $|\psi_{\text{comp1}}\rangle \otimes |00\rangle$ or $|\psi_{\text{comp2}}\rangle \otimes |00\rangle$, with the first case corresponding to the typical or most likely subspace. I will also clarify how the dimension of the typical subspace can be defined formally.

## 22.2 Schumacher's quantum coding theorem

In this section, I shall formalize *Schumacher's quantum coding theorem*. The driving concept is that it is possible to encode a message with high fidelity when using quantum states from the typical subspace $\Omega$. The key property of $\Omega$ is that it asymptotically reaches a dimension close to $2^{nS(\rho)}$. It is useful to look back at Section 13.2 concerning typical sets. To recall, the typical set represents roughly $2^{nH(X)}$ bit strings of length $n$, referred to as *typical sequences*. Such typical sequences roughly contain $nq$ 1 bits and $n(1-q)$ 0 bits, with $H(X) = f(q)$ being the source entropy of the sequence bits, assumed to be generated independently, $(H(X^n) = nH(X))$. The fundamental property is that when $n$ becomes large, any typical sequence asymptotically has the probability $2^{-nH(X)}$of being observed. Thus, there is a one-to-one conceptual correspondence between the typical set of classical bit sequences, of size $2^{nH(X)}$, and the typical subspace $\Omega$ of quantum state messages, of dimension close to $2^{nS(\rho)}$.

Consider, now, the quantum message block $\rho_M = \rho^{\otimes n}$ of length $n$ defined in Eq. (22.2), where $\rho$ is the density operator associated with any of the individual message symbols, as defined in Eq. (22.1). The $2^n$ eigenvalues and eigenstates of $\rho_M$ are $\mu_1, \mu_2 \ldots \mu_n$ and $|\mu_1\rangle, |\mu_2\rangle \ldots |\mu_n\rangle$, respectively. To recall, the eigenvalues $\mu_i$ represent the probability that the message is in the state $|\mu_i\rangle$. Formally, the typical subspace $\Omega$ is defined by the set of eigenstates $\{|\mu_i\rangle\}$ for which the eigenvalues $\mu_i$ satisfy the double inequality:

$$2^{-n(S+\varepsilon)} \leq \mu_i \leq 2^{-n(S-\varepsilon)}, \tag{22.24}$$

where $S \equiv S(\rho)$ and $\varepsilon$ is a given positive real, which can be arbitrary small. We note that this double inequality is conceptually identical to that in Eq. (13.26), corresponding to the formal definition of typical sequences.

Let me immediately illustrate this definition of the typical subspace by means of the $n = 3$ example used in the previous section. For reading clarity, we recall here the

**Table 22.2** Lower bounds for the parameter $\varepsilon$, defining different possibilities for the typical subspace $\Omega$ and its complement $\Omega^{\perp}$.

| $\mu_i$ | $\varepsilon$ | | Typical subspace ($\Omega$) and complement ($\Omega^{\perp}$) | | |
|---|---|---|---|---|---|
| $\lambda_a^3$ | 0.6219 | 0.372 | $\Omega_1$ | | |
| $\lambda_a^2\lambda_b$ | 0.1067 | | | $\Omega_2$ | |
| $\lambda_a^2\lambda_b$ | 0.1067 | 0.475 | | | |
| $\lambda_a^2\lambda_b$ | 0.1067 | | | | $\Omega_3$ |
| $\lambda_a\lambda_b^2$ | 0.0183 | 1.323 | $\Omega_1^{\perp}$ | | |
| $\lambda_a\lambda_b^2$ | 0.0183 | | | $\Omega_2^{\perp}$ | |
| $\lambda_a\lambda_b^2$ | 0.0183 | | | | |
| $\lambda_b^3$ | 0.0031 | 2.171 | | | $\Omega_3^{\perp}$ |

density matrix of $\rho_M$:

$$\rho_M = \begin{pmatrix} \lambda_a^3 & 0 & 0 & 0 & 0 & 0 & 0 & 0 \\ 0 & \lambda_a^2\lambda_b & 0 & 0 & 0 & 0 & 0 & 0 \\ 0 & 0 & \lambda_a^2\lambda_b & 0 & 0 & 0 & 0 & 0 \\ 0 & 0 & 0 & \lambda_a^2\lambda_b & 0 & 0 & 0 & 0 \\ 0 & 0 & 0 & 0 & \lambda_a\lambda_b^2 & 0 & 0 & 0 \\ 0 & 0 & 0 & 0 & 0 & \lambda_a\lambda_b^2 & 0 & 0 \\ 0 & 0 & 0 & 0 & 0 & 0 & \lambda_a\lambda_b^2 & 0 \\ 0 & 0 & 0 & 0 & 0 & 0 & 0 & \lambda_b^3 \end{pmatrix}, \tag{22.25}$$

which shows that the $2^n$ eigenvalues $\mu_i$ of $\rho_M$ are of the form $\mu_{i=u,v} = \lambda_a^u\lambda_b^v$ with $u + v = n$. Taking the base-2 logarithm of Eq. (22.24), we have

$$\begin{aligned} S + \varepsilon \geq \frac{1}{n}\log\frac{1}{\mu_i} & \geq S - \varepsilon \\ \leftrightarrow \left|\frac{1}{n}\log\frac{1}{\mu_i} - S\right| & \leq \varepsilon. \end{aligned} \tag{22.26}$$

We note that the inequality is conceptually identical to that in Eq. (13.25), corresponding to the formal definition of typical sequences. Substituting $n = 3$, $S = 0.6008$, $\lambda_a = \cos^2\pi/8 = 0.853$, $\lambda_b = \sin^2\pi/8 = 0.146$, $\mu_1 = \lambda_a^3$, $\mu_2 = \mu_3 = \mu_4 = \lambda_a^2\lambda_b$, $\mu_5 = \mu_6 = \mu_7 = \lambda_a\lambda_b^2$, and $\mu_8 = \lambda_b^3$ in Eq. (22.26), we obtain the lower bounds for $\varepsilon$ listed in Table 22.2. We observe that $\varepsilon = 0.372$ (eigenvalue $\mu_1 = \lambda_a^3$) defines a first, 1D typical subspace $\Omega_1$, spanned by $|\lambda_a\lambda_a\lambda_a\rangle$. A second possible boundary, $\varepsilon = 0.475$ (eigenvalues $\mu_1$ and $\mu_2 = \mu_3 = \mu_4 = \lambda_a^2\lambda_b$), defines a 4D typical subspace $\Omega_2$ spanned by $\{|\lambda_a\lambda_a\lambda_a\rangle, |\lambda_a\lambda_a\lambda_b\rangle, |\lambda_a\lambda_b\lambda_a\rangle, |\lambda_b\lambda_a\lambda_a\rangle\}$. A third possible boundary, $\varepsilon = 0.475$ (eigenvalues $\mu_1$, $\mu_2$ and $\mu_5 = \mu_6 = \mu_7 = \lambda_a\lambda_b^2$) defines a 7D typical subspace $\Omega_3$ spanned by $\{|\lambda_a\lambda_a\lambda_a\rangle, |\lambda_a\lambda_a\lambda_b\rangle, |\lambda_a\lambda_b\lambda_a\rangle, |\lambda_b\lambda_a\lambda_a\rangle, |\lambda_a\lambda_b\lambda_b\rangle, |\lambda_b\lambda_b\lambda_a\rangle, |\lambda_b\lambda_a\lambda_b\rangle\}$. The remaining subspace, spanned by $|\lambda_b\lambda_b\lambda_b\rangle$, corresponds to the smallest atypical subspace, $\Omega^{\perp} = \Omega_3^{\perp}$.

**Table 22.3** Parameter $\delta$ and fidelity $\tilde{F}$ corresponding to the typical subspace choices for $\Omega$ and related dimension $\dim(\Omega)$.

| $\Omega$ | $\dim(\Omega)$ | $p(\Omega) = \sum_i \mu_i$ | $\delta$ | $\tilde{F}$ |
|---|---|---|---|---|
| $\Omega_1$ | 1 | 0.6219 | 0.378 | 0.622 |
| $\Omega_2$ | 4 | 0.9419 | 0.058 | 0.923 |
| $\Omega_3$ | 7 | 0.9969 | 0.003 | 0.996 |

In the previous section, we heuristically used $\Omega_2$ as "the" typical subspace, but it is now clear that the other subspaces $\Omega_1$ and $\Omega_3$ are also eligible under the intrinsic boundary definition in Eq. (22.24) or Eq. (22.26) for the corresponding eigenvalues, as defined by the parameter $\varepsilon$. Given the highest value of $\varepsilon$ in the typical subspaces $\Omega = \Omega_1, \Omega_2, \Omega_3$, we shall call any "codeword" $|\mu_i\rangle = |\lambda_j \lambda_k \lambda_l\rangle \in \Omega$ as "$\varepsilon$-typical."

To recall, the probability $p(\Omega)$ that any message belongs to the typical subspace $\Omega$ is given by the sum of the associated eigenvalues $\mu_i$. Formally, if $E$ is the projector onto $\Omega$, we have

$$p(\Omega) = \sum_{i=1}^{\dim(\Omega)} \mu_i = \mathrm{tr}(\rho_M E) \geq 1 - \delta, \tag{22.27}$$

where $\delta$ is a positive real and smaller than unity. Table 22.3 shows the values of the parameter $\delta$ corresponding to the typical subspaces shown in Table 22.2, and the corresponding *fidelity* as defined in the previous section by $\tilde{F} = (1 - \delta)^2 + \delta \lambda_a^3$, Eq. (22.20). It is seen from the table that the fidelity increases as the parameter $\delta$ decreases and as the dimension of the typical subspace $\dim(\Omega) = \mathrm{tr}(E)$, or the number of $\varepsilon$-typical codewords increases. We note that the case $\Omega = \Omega_1$, which has only one codeword, is a poor choice, since the fidelity is $\tilde{F} = 62.2\%$, corresponding to a "useless channel" almost half of the time. And what valuable information can be transmitted with only one codeword, independently of this consideration? The best choice is $\Omega = \Omega_3$, which has seven codewords (out of $2^3 = 8$ codeword possibilities), since the fidelity reaches the maximum; $\tilde{F} = 99.6\%$. But we have now reached a confusing situation where there apparently exist several possible choices for the typical subspace, which are all consistent with previous eligibility criteria. We, thus, need to develop the analysis further, bearing in mind that we want to show that the dimension of the typical subspace is close to$2^{nS(\rho)}$. In the above $n = 3$ example, we have $2^{nS(\rho)} = 2^{3\times 0.6008} = 3.488 \approx 4 = \dim(\Omega_2)$, which indicates that $\Omega_2$ is, in fact, the correct choice! But to reach such a conclusion, we must consider messages of asymptotically increasing length, and we will have to forget the previous $n = 3$ example, despite its usefulness in demonstrating the possibility of message compression.

As we have seen earlier, the two parameters $\varepsilon, \delta$, defined in Eqs. (22.26) and (22.27), determine a lower bound for the probability $p(\Omega) = \mathrm{tr}(\rho_M E)$ of a given message codeword to be $\varepsilon$-typical. More generally, we may state that *given the two parameters $\varepsilon, \delta$, there exists a message length n sufficiently long to provide the condition that $p(\Omega)$ is at least* $1 - \delta$. To keep the focus, we shall not worry here about formally demonstrating such a statement. Suffice it to take for granted as a postulate that, given $\varepsilon$, it is possible to make $p(\Omega)$ arbitrarily close to unity (with $\delta$ chosen sufficiently small), if the message length $n$ is sufficiently large. Consistently, the dimension of the typical subspace, $\dim(\Omega) = \mathrm{tr}(E)$, also increases. But can we show to what extent?

To answer the above question, I shall now introduce a key theorem concerning the dimension of the typical subspace. Given a message length $n$, the VN entropy $S(\rho)$ of the quantum-symbol source, the two parameters $\varepsilon$, $\delta$, then the dimension $\dim(\Omega)$ of the corresponding typical subspace $\Omega$ is bounded according to:

$$(1-\delta)2^{n(S-\varepsilon)} \leq \dim(\Omega) \leq 2^{n(S+\varepsilon)}. \tag{22.28}$$

The proof for this is actually straightforward. First, we sum each term in the double inequality in Eq. (22.24) from $i = 1$ to $i = \dim(\Omega)$ to obtain

$$\begin{gathered}\sum_{i=1}^{\dim(\Omega)} 2^{-n(S+\varepsilon)} \leq \sum_{i=1}^{\dim(\Omega)} \mu_i \leq \sum_{i=1}^{\dim(\Omega)} 2^{-n(S-\varepsilon)} \\ \leftrightarrow \\ \dim(\Omega)2^{-n(S+\varepsilon)} \leq p(\Omega) \leq \dim(\Omega)2^{-n(S-\varepsilon)}.\end{gathered} \tag{22.29}$$

Second, we substitute the two properties $1-\delta \leq p(\Omega)$ and $p(\Omega) \leq 1$ into Eq. (22.29) and then obtain

$$\begin{gathered}\begin{cases}\dim(\Omega)2^{-n(S+\varepsilon)} \leq p(\Omega) \leq 1 \\ 1-\delta \leq p(\Omega) \leq \dim(\Omega)2^{-n(S-\varepsilon)}\end{cases} \\ \leftrightarrow \\ \begin{cases}\dim(\Omega) \leq 2^{n(S+\varepsilon)} \\ \dim(\Omega) \geq (1-\delta)2^{n(S-\varepsilon)}\end{cases} \\ \leftrightarrow \\ (1-\delta)2^{n(S-\varepsilon)} \leq \dim(\Omega) \leq 2^{n(S+\varepsilon)},\end{gathered} \tag{22.30}$$

which is in Eq. (22.28) and, thus, proves the *typical subspace theorem*.

Our exploitation of the above theorem will be made purposefully simple, the goal being to convey only an intuitive proof of *Schumacher's quantum coding theorem* (usually called *Schumacher's noiseless channel coding theorem*) and, thus, avoid lengthier and more academically involved developments.

Define the parameter $R$ such that $R = S + \varepsilon$, with the condition $R \leq 1$ (or $0 < \varepsilon \leq 1 - S$). From the typical subspace theorem, the dimension of the typical subspace $\Omega$ satisfies $\dim(\Omega) \leq 2^{nR}$. Thus, we may use up to $2^{nR}$, $\varepsilon$-typical codewords, such that for any $\delta > 0$ the condition $p(\Omega) \geq 1 - \delta$ is satisfied for message lengths $n$ sufficiently large. In this case, we may say that the compression scheme is "reliable," i.e., corresponds to arbitrary high fidelity. Under these conditions, *Schumacher's quantum coding theorem* states that *there exists a reliable compression code, with arbitrarily high transmission fidelity, and compression factor* $\eta = R$. We note that $R$ *is strictly higher than the VN entropy* $S$, which sets an ultimate limit to the achievable compression factor ($R > S$). The second part of the theorem is that any compression code with $R < S$ is *not* reliable. Here, again to avoid lengthy mathematical developments, I shall only concentrate on the reliable compression code, and provide a simple description.

The quantum symbols forming the message are now assumed to be the eigenstates $|\lambda_i\rangle$ of the density operator $\rho$, and the possible messages are the eigenstates $|\mu_i\rangle$ of the operator $\rho_M = \rho^{\otimes n}$. The originator then proceeds to encode these into states of the form

$$|\mu_i'\rangle = U|\mu_i\rangle = |\mu_{\text{compl}}\rangle \otimes |0_{\text{rest}}\rangle, \tag{22.31}$$

where $|\mu_{\text{compl}}\rangle$ is a compressed state of $nR < n$ qubits, and $|0_{\text{rest}}\rangle = |00\ldots0\rangle = |0\rangle^{\otimes q}$ represents the remainder of the block, of length $q = n(1-R)$ qubits. The originator then makes "fuzzy" measurements on each of the qubits from the remainder state. If the outcome is $00\ldots0$, then $|\mu_i'\rangle$ has been projected onto the typical subspace $\Omega$, and $|\mu_{\text{compl}}\rangle$ is sent through the quantum channel. If the outcome is anything different, the originator sends an $nR$ qubit message $|\psi_{\text{comp2}}\rangle$, as a "junk" substitute, which is determined by the same rotation $U$ such that $U^{-1}\left(|\psi_{\text{comp2}}\rangle \otimes |0_{\text{rest}}\rangle\right) = |\mu_{\text{junk}}\rangle \in \Omega$. As in the previous ($n = 3$) example, the recipient just needs to append $|0_{\text{rest}}\rangle$ to the received message, followed by the inverse transformation $U^{-1}$ on this $n$-qubit state. If $E$ is the projector onto the typical subspace $\Omega$, and $|\mu_i\rangle$ the original message that has been sent, the corresponding density operator $\tilde{\rho}_i$ of the received/uncompressed message takes a form similar to that in Eq. (22.19):

$$\begin{aligned}\tilde{\rho}_i &= E|\mu_i\rangle\langle\mu_i|E + \langle\mu_i|(I-E)|\mu_i\rangle|\mu_{\text{junk}}\rangle\langle\mu_{\text{junk}}| \\ &= E|\mu_i\rangle\langle\mu_i|E + \tilde{\rho}_{\text{junk}}.\end{aligned} \tag{22.32}$$

The fidelity of the transmission must be averaged over all the message possibilities, each with probability $\mu_i$, as follows:

$$\tilde{F} = \sum_{i=1}^{2^n} \mu_i \langle\mu_i|\tilde{\rho}_i|\mu_i\rangle, \tag{22.33}$$

which, by substituting the result in Eq. (22.32), yields

$$\begin{aligned}\tilde{F} &= \sum_{i=1}^{2^n} \mu_i \langle\mu_i|(E|\mu_i\rangle\langle\mu_i|E)|\mu_i\rangle + \sum_{i=1}^{2^n} \mu_i \langle\mu_i|\tilde{\rho}_{\text{junk}}|\mu_i\rangle \\ &= \sum_{i=1}^{2^n} \mu_i (\langle\mu_i|E|\mu_i\rangle)^2 + tr(\tilde{\rho}_{\text{junk}}) \\ &\geq \sum_{i=1}^{2^n} \mu_i (\langle\mu_i|E|\mu_i\rangle)^2.\end{aligned} \tag{22.34}$$

The last inequality is caused because the junk contribution is nonnegative; i.e., $\text{tr}(\tilde{\rho}_{\text{junk}}) \geq 0$. Then we use the property that for any real $x$ we have $x^2 \geq 2x - 1$ because $(x-1)^2 \geq 0$. Thus, each term $x^2 = (\langle\mu_i|E|\mu_i\rangle)^2$ in the above right-hand side summation is greater than or equal to $2x - 1 = 2\langle\mu_i|E|\mu_i\rangle - 1$, which yields

$$\begin{aligned}\tilde{F} &\geq \sum_{i=1}^{2^n}\mu_i\left(2\langle\mu_i|E|\mu_i\rangle - 1\right)\\ &= 2\sum_{i=1}^{2^n}\langle\mu_i|\mu_i E|\mu_i\rangle - \sum_{i=1}^{2^n}\mu_i\\ &\equiv 2\mathrm{tr}(\rho_M E) - 1\\ &\equiv 2p(\Omega) - 1.\end{aligned} \tag{22.35}$$

From Schumacher's (quantum coding) theorem, for *any* $\delta > 0$ and sufficiently long messages we have $p(\Omega) \geq 1 - \delta$, therefore

$$\tilde{F} \geq 2p(\Omega) - 1 \geq 2(1-\delta) - 1 = 1 - 2\delta. \tag{22.36}$$

This final result indicates that *for sufficiently long messages, the fidelity $\tilde{F}$ can be made arbitrarily close to unity* (consistent with the limit $\delta \to 0$), which establishes that the compression code is "reliable." It is possible to show (but not described here for the sake of brevity) that any other coding choice, namely with $R < S$, or $R = S - \varepsilon$ (with $\varepsilon > 0$, $\delta > 0$), makes the fidelity of sufficiently long messages to be asymptotically bounded according to $\tilde{F} < \delta$, which is essentially zero as $\delta \to 0$.

## 22.3 A graphical and numerical illustration of Schumacher's quantum coding theorem

In this section, I shall provide an original illustration of Schumacher's quantum coding, both graphical and numerical. This will help visualize how the dimension of the typical subspace $\Omega$ asymptotically converges towards $2^{nS}$, as the message length $n$ increases and if the parameter $\varepsilon$ may be chosen to be arbitrarily small. The numerical application will also illustrate that there is still a large degree of freedom over the choice of $\varepsilon$ and the typical subspace possibilities to compress quantum messages.

Here, we shall assume the same parameters as in the example used in the two previous sections, i.e., $S(\rho) = 0.6008$ for the per-symbol VN entropy, and the two eigenvalues $\lambda_a = 0.8535$, $\lambda_b = 0.1465$. For a message length $n = 3$, we have $2^3 = 8$ possible messages defining an 8D eigenspace $\Lambda$. Any of these messages are based on the qubit permutations of four codeword types of the form $|\lambda_a\lambda_a\lambda_a\rangle$, $|\lambda_a\lambda_a\lambda_b\rangle$, $|\lambda_a\lambda_b\lambda_b\rangle$, $|\lambda_b\lambda_b\lambda_b\rangle$ with the respective measurement probabilities $\lambda_a^3$, $\lambda_a^2\lambda_b$, $\lambda_a\lambda_b^2$, $\lambda_b^3$, in descending order. Such probabilities also correspond to the eigenvalues: $\mu_1 = \lambda_a^3$, $\mu_2 = \mu_3 = \mu_4 = \lambda_a^2\lambda_b$, $\mu_5 = \mu_6 = \mu_7 = \lambda_a\lambda_b^2$, and $\mu_8 = \lambda_b^3$. According to Eq. (22.37), which I reproduce here for convenience, we can associate a parameter $\varepsilon$ with each of these four groups of eigenvalues:

$$\left|\frac{1}{n}\log\frac{1}{\mu_i} - S(\rho)\right| \leq \varepsilon. \tag{22.37}$$

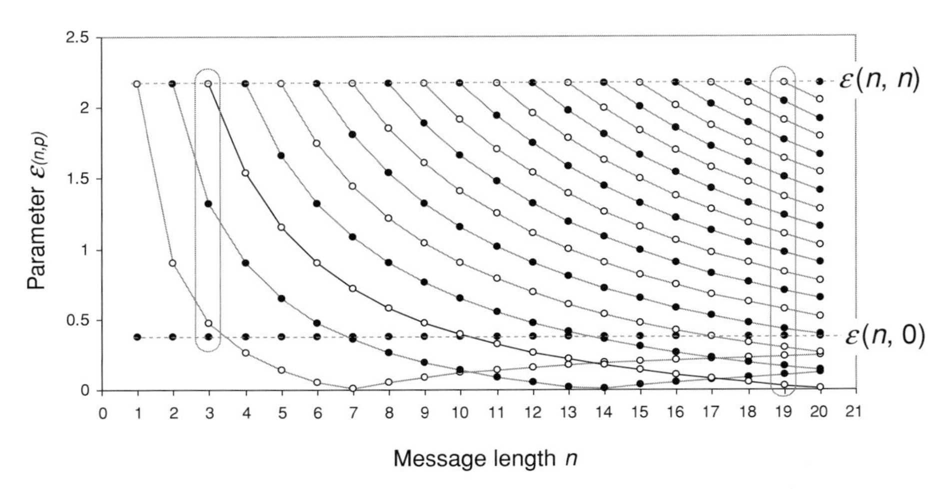


**Figure 22.1** Lower bound for the parameter $\varepsilon(n, p)$, calculated from Eq. (22.39), in the range $n = 1$ to $n = 20$.

Substituting the above values of $\mu_i$ with $n = 3$ and $S(\rho) = 0.6008$, we obtain four lower bounds for $\varepsilon$, namely, to call them by the same letter: $\varepsilon = 0.372, 0.475, 1.322, 2.170$. As we have seen in the previous section, this calculation does not tell us much as to how to define a typical subspace. Furthermore, only the first value $\varepsilon = 0.372$ gives a parameter $R < 1$ (namely $R = S + \varepsilon = 0.6008 + 0.3724 = 0.9732 < 1$), and there is nothing that we can conclude about any compression potential from such a result! But if we increase the message length $n$, and calculate all possible $\varepsilon$ values from Eq. (22.37), we obtain a much broader range of possibilities for the parameter $R$.

Given the message length $n$, it is straightforward to tabulate the suite of $n + 1$ eigenvalues according to the definition

$$\begin{cases} \lambda_a^n \\ \lambda_a^{n-1}\lambda_b \\ \lambda_a^{n-2}\lambda_b^2 \\ \vdots \\ \lambda_a^2\lambda_b^{n-2} \\ \lambda_b^n. \end{cases} \tag{22.38}$$

The parameter $\varepsilon(n, p)$ corresponds to each eigenvalue $\mu_i(n, p) = \lambda_a^{n-p}\lambda_b^p$, as defined by

$$\left| \frac{1}{n} \log \frac{1}{\mu_i(n, p)} - S(\rho) \right| \leq \varepsilon(n, p). \tag{22.39}$$

Figure 22.1 shows the values of the parameter $\varepsilon(n, p)$ in the range $n = 1$ to $n = 20$. The values are connected by full lines to guide the eye. The top and bottom horizontal lines correspond to $\varepsilon(n, n) = \varepsilon(\mu_i = \lambda_b^n)$ and $\varepsilon(n, 0) = \varepsilon(\mu_i = \lambda_a^n)$, respectively. The two sets circled at $n = 3$ and $n = 19$ help visualize the effect according to which an increasing

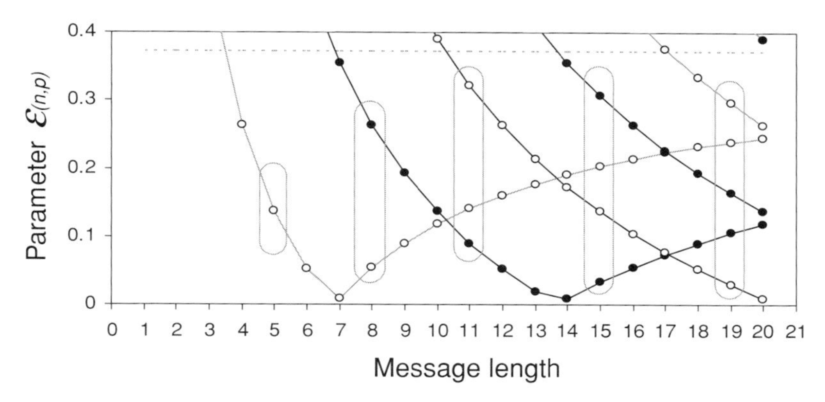


**Figure 22.2** Lower bound for the parameter $\varepsilon(n, p)$, calculated from Eq. (22.39), with zoom in the region $0 < \varepsilon(n, p) \leq 0.4$.

fraction of the $\varepsilon(n, p)$ values progressively form a cluster in the region between $\varepsilon = 0$ and $\varepsilon(n, 0) \leq 0.4$. This effect is also illustrated in Fig. 22.2, which shows a zoom in this region. It is clear that for a message length $n$ sufficiently large, any value $\varepsilon > 0$ can be reached by $\varepsilon(n, p)$. Furthermore, we observe that given a message length $n$ and any $\varepsilon > 0$ it is always possible to define a subset of codewords satisfying $\varepsilon(n, p) \leq \varepsilon$, as indicated in the figure by the circled clusters. In the case $n = 19$, for instance, the cluster consists of a subset of five codewords satisfying $\varepsilon(19, p) \leq 0.29681$, and this group of codewords is precisely the $\varepsilon$-typical subspace $\Omega$ we have been looking for! Let us take a closer look at the corresponding data. Table 22.4 shows, for each codeword type $\mu$ of length $n = 19$, from left to right: the individual codeword probability $p(\mu)$, the number of possible permutations $N$ of the codeword type $\mu$, the codeword-type probability $Np(\mu)$, the codeword parameter $\varepsilon$, and the associated value $R = S + \varepsilon$. We have defined our $\varepsilon$-typical subspace $\Omega$ as generated by the five codeword types for which $\varepsilon(19, p) \leq 0.29681$. As the table shows, these codeword types are of the form $|\lambda_a^{18}\lambda_b^1\rangle$, $|\lambda_a^{17}\lambda_b^2\rangle$, $|\lambda_a^{16}\lambda_b^3\rangle$, $|\lambda_a^{15}\lambda_b^4\rangle$, and $|\lambda_a^{14}\lambda_b^5\rangle$. Summing the $N$ data, it can be found that there exist 16 663 such codewords, out of 524 288 possibilities, representing only about 3% of the codeword space $\Lambda$. Summing the values $Np(\mu)$ in the table, one finds that the probability of any received codeword belonging to this typical subspace is $p(\Omega) = 0.90208$, which is relatively high. This result yields the lower bound $\delta = 1 - p(\Omega) = 0.09792$ and hence, using $\lambda_a = 0.8535$, the transmission fidelity $\tilde{F} = (1 - \delta)^2 + \delta\lambda_a^{19} = 0.81859 \approx 82\%$, which is fair. From the table, we also find that the maximum value of $R$, or the best compression factor achievable under this coding scheme, is given by the highest value of $\varepsilon$ in the subspace $\Omega$, namely $\varepsilon_{\max} = 0.29681$, or $R = 0.89769 \approx 89\%$.

Having defined the $\varepsilon$-typical subspace $\Omega$ (with the choice $\varepsilon \leq 0.29681$), the next step is to analyze the effect of increasing the message length $n$. It is only a matter of tabulating a spreadsheet that outputs the same data as shown in Table 22.4, along with the values $p(\Omega)$, $\delta = 1 - p(\Omega)$, and $\tilde{F} = (1 - \delta)^2 + \delta\lambda_a^n$, for each value of $n$. Care must be taken to effect the summation yielding $p(\Omega)$ only in the spreadsheet domain where $0 < \varepsilon \leq 0.29681$. The fidelity $\tilde{F}$ calculated for message lengths from $n = 19$ to $n = 100$

**Table 22.4** Defining an $\varepsilon$-typical subspace $\Omega$ with $\varepsilon \leq 0.29681$, in the case $n = 19$, corresponding to a maximum compression $R = 0.89769$.

| Codeword type $\mu$ | $p(\mu)$ | $N$ | $Np(\mu)$ | $\varepsilon$ | $R = S + \varepsilon$ |
|---|---|---|---|---|---|
| $\lambda_a^{19}\lambda_b^0$ | $4.936 \times 10^{-2}$ | 1 | – | 0.37243 | 0.97330 |
| $\lambda_a^{18}\lambda_b^1$ | $8.469 \times 10^{-3}$ | 19 | $1.609 \times 10^{-1}$ | 0.23858 | 0.83946 |
| $\lambda_a^{17}\lambda_b^2$ | $1.453 \times 10^{-3}$ | 171 | $2.485 \times 10^{-1}$ | 0.10473 | 0.70561 |
| $\lambda_a^{16}\lambda_b^3$ | $2.493 \times 10^{-4}$ | 969 | $2.416 \times 10^{-1}$ | 0.02911 | 0.62999 |
| $\lambda_a^{15}\lambda_b^4$ | $4.277 \times 10^{-5}$ | 3876 | $1.658 \times 10^{-1}$ | 0.16296 | 0.76384 |
| $\lambda_a^{14}\lambda_b^5$ | $7.339 \times 10^{-6}$ | 11628 | $8.534 \times 10^{-2}$ | 0.29681 | 0.89769 |
| $\lambda_a^{13}\lambda_b^6$ | $1.259 \times 10^{-6}$ | 27132 | – | 0.43066 | 1.03153 |
| $\lambda_a^{12}\lambda_b^7$ | $2.160 \times 10^{-7}$ | 50388 | – | 0.56451 | 1.16538 |
| $\lambda_a^{11}\lambda_b^8$ | $3.707 \times 10^{-8}$ | 75582 | – | 0.69835 | 1.29923 |
| $\lambda_a^{10}\lambda_b^9$ | $6.359 \times 10^{-9}$ | 92378 | – | 0.83220 | 1.43308 |
| $\lambda_a^{9}\lambda_b^{10}$ | $1.091 \times 10^{-9}$ | 92378 | – | 0.96605 | 1.56692 |
| $\lambda_a^{8}\lambda_b^{11}$ | $1.872 \times 10^{-10}$ | 75582 | – | 1.09990 | 1.70077 |
| $\lambda_a^{7}\lambda_b^{12}$ | $3.212 \times 10^{-11}$ | 50388 | – | 1.23375 | 1.83462 |
| $\lambda_a^{6}\lambda_b^{13}$ | $5.511 \times 10^{-12}$ | 27132 | – | 1.36759 | 1.96847 |
| $\lambda_a^{5}\lambda_b^{14}$ | $9.455 \times 10^{-13}$ | 11628 | – | 1.50144 | 2.10232 |
| $\lambda_a^{4}\lambda_b^{15}$ | $1.622 \times 10^{-13}$ | 3876 | – | 1.63529 | 2.23616 |
| $\lambda_a^{3}\lambda_b^{16}$ | $2.783 \times 10^{-14}$ | 969 | – | 1.76914 | 2.37001 |
| $\lambda_a^{2}\lambda_b^{17}$ | $4.775 \times 10^{-15}$ | 171 | – | 1.90298 | 2.50386 |
| $\lambda_a^{1}\lambda_b^{18}$ | $8.193 \times 10^{-16}$ | 19 | – | 2.03683 | 2.63771 |
| $\lambda_a^{0}\lambda_b^{19}$ | $1.406 \times 10^{-16}$ | 1 | – | 2.17068 | 2.77156 |

is shown in Fig. 22.3. It is seen that convergence towards 100% fidelity is relatively rapid (e.g., $\tilde{F} \approx 99\%$ or better, for $n \geq 70$). Interestingly, such a convergence goes through sawtooth-like oscillations. These are explained by the irregular variations of the $\varepsilon(n, p)$ cluster size and amplitude previously observed in Fig. 22.2.

The last open issue to discuss is the dimension of the typical subspace, $\dim(\Omega)$, corresponding to a given message length $n$. From Eq. (22.28), we know that $\dim(\Omega)$ is bounded according to $(1-\delta)2^{n(S-\varepsilon)} \leq \dim(\Omega) \leq 2^{n(S+\varepsilon)}$, which asymptotically becomes $2^{nS}$ as $\delta, \varepsilon \to 0$ and with sufficiently long messages. In the most general case ($\varepsilon > 0$, any message length $n$), we have

$$S - \varepsilon + \frac{\log_2(1-\delta)}{n} \leq \frac{\log_2[\dim(\Omega)]}{n} \leq S + \varepsilon. \tag{22.40}$$

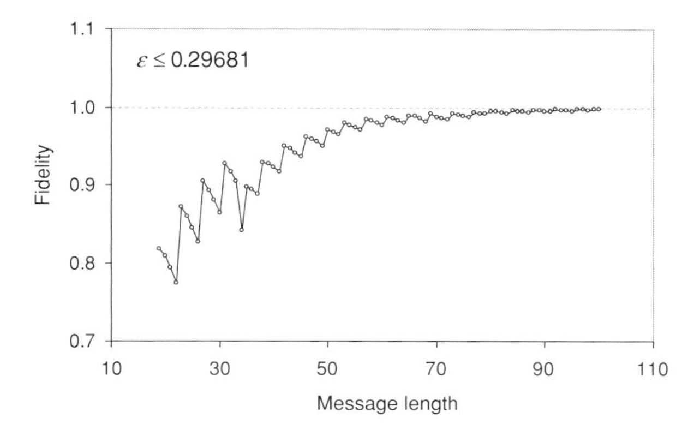


**Figure 22.3** Fidelity as a function of message length ($n = 19$ to $n = 100$), assuming the $\varepsilon$-typical subspace $\Omega$ with $\varepsilon \leq 0.29681$.

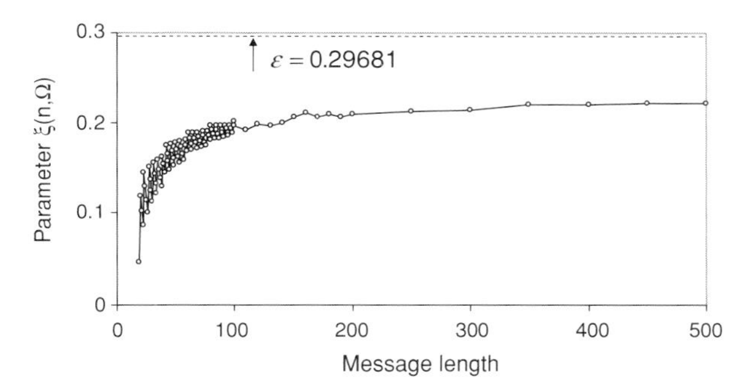


**Figure 22.4** Evolution of the parameter $\xi(n, \Omega)$ with increasing message length $n$.

Using the property $\log_2(1 - \delta)/n \approx \delta/n(\ln 2) \approx 0$, we can introduce the parameter $\xi$ for which the double inequality in Eq. (22.40) is very nearly equivalent to:

$$\xi(n, \Omega) = \left| \frac{\log_2[\dim(\Omega)]}{n} - S \right| \leq \varepsilon. \tag{22.41}$$

Figure 22.4 shows the evolution of the parameter $\xi(n, \Omega)$ with the message length $n$, as calculated from our previous example, up to $n = 500$. For each $n$, the value $\dim(\Omega)$ is estimated by summing the values of $N$ (number of $\varepsilon$-typical codewords for which $\varepsilon \leq 0.29681$), displayed in the third column in Table 22.4. For clarity, the plot shows successive data with an increment of one up to $n = 100$, then with an increment of 10 up to $n = 200$, then with an increment of 50. The figure shows that the parameter $\xi(n, \Omega)$ asymptotically converges towards the limit $\varepsilon = 0.29681$, which defines the typical subspace $\Omega$ for each $n$. We may, thus, write

$$\begin{aligned}
\lim_{n\to\infty} \xi(n, \Omega) &= \varepsilon \\
&\leftrightarrow \\
\lim_{n\to\infty} \frac{\log_2[\dim(\Omega)]}{n} &= S + \varepsilon = R \\
&\leftrightarrow \\
\lim_{n\to\infty} \dim(\Omega) &= \mathrm{e}^{n(S+\varepsilon)} = \mathrm{e}^{nR}
\end{aligned} \tag{22.42}$$

and since $\varepsilon$ can be chosen arbitrarily small, the limiting value for $\dim(\Omega)$ is $\mathrm{e}^{nS}$.

To summarize, the numerical example developed in this section has shown that given a message of length $n$, and any $\varepsilon > 0$ such that $\varepsilon = S - R$ (with $R < 1$), there exists a typical subspace $\Omega$, spanned by a subset of corresponding $\varepsilon$-typical codewords, where message compression is possible with high fidelity $\tilde{F}$. Consistently with the formal description, the example also showed that (a) for messages of increasing length, the fidelity $\tilde{F}$ can be made arbitrarily high, and (b) the dimension of the typical subspace $\dim(\Omega)$ asymptotically converges towards $e^{n(S+\varepsilon)}$. As the parameter $\varepsilon$ can be chosen to be arbitrarily small, the limiting value for the subspace dimension is $e^{nS}$, which corresponds to the best achievable compression factor $R \approx S$.

## 22.4 Exercises

**22.1** (T): Given two tensor states $|abc\rangle, |def\rangle$, where $|a\rangle, |b\rangle, |c\rangle, |d\rangle, |e\rangle, |f\rangle$ are qubits, provide a general formula for the space-overlap factor

$$|\langle abc \mid def\rangle|^2$$

and apply this formula to all the nine possible cases:

$$|a\rangle, |b\rangle, |c\rangle = \begin{pmatrix} 1 \\ 0 \end{pmatrix} \equiv |\alpha_1\rangle \text{ or } \frac{1}{\sqrt{2}}\begin{pmatrix} 1 \\ 1 \end{pmatrix} \equiv |\alpha_2\rangle$$

$$|d\rangle, |e\rangle, |f\rangle = \begin{pmatrix} \cos\alpha \\ \sin\alpha \end{pmatrix} \equiv |\lambda_1\rangle \text{ or } \begin{pmatrix} -\sin\alpha \\ \cos\alpha \end{pmatrix} \equiv |\lambda_2\rangle,$$

where $\alpha$ is a parameter.

(*Clue*: refer to the qubit tensor-product definition in Eq. (16.55) as applied to the present case.)

**22.2** (B): What is the best achievable compression factor for a quantum message of arbitrary long length, whose qubits are generated by the density operator

$$\rho = \frac{1}{3}\begin{pmatrix} 1 & 3 \\ 0 & 2 \end{pmatrix}?$$

**22.3** (M): A qubit source is characterized by the density operator

$$\rho = \frac{1}{3}\begin{pmatrix} 1 & 0 \\ 0 & 2 \end{pmatrix}.$$

Assuming a quantum message length of $n = 5$ qubits, determine the $\varepsilon$-typical subspace and corresponding quantum codewords for which $\varepsilon = 0.3$.

# 23 Quantum channel noise and channel capacity

This chapter introduces the notion of *noisy quantum channels*, and the different types of "quantum noise" that affect qubit messages passed through such channels. The main types of noisy channel reviewed here are the *depolarizing*, *bit-flip*, *phase-flip*, and *bit-phase-flip* channels. Then the *quantum channel capacity* $\chi$ is defined through the *Holevo–Schumacher–Westmoreland* (HSW) theorem. Such a theorem can conceptually be viewed as the elegant quantum counterpart of *Shannon's (noisy) channel coding theorem*, which was described in Chapter 13. Here, I shall not venture into the complex proof of the HSW theorem but only provide a background illustrating the similarity with its classical counterpart. The resemblance with the channel capacity $\chi$ and the *Holevo bound*, as described in Chapter 21, and with the classical *mutual information* $H(X;Y)$, as described in Chapter 5, are both discussed. For advanced reference, a hint is provided as to the meaning of the still not fully explored concept of *quantum coherent information*. Several examples of quantum channel capacity, derived from direct applications of the HSW theorem, along with the solution of the maximization problem, are provided.

## 23.1 Noisy quantum channels

The notion of "noisiness" in a classical communication channel was first introduced in Chapter 12, when describing *channel entropy*. Such a channel can be viewed schematically as a probabilistic relation between two random sources, $X$ for the originator, and $Y$ for the recipient. These sources are defined by symbol alphabets, $X = \{x_1, x_2\}$ and $Y = \{y_1, y_2\}$ for instance, and their associated probabilities $p(x_i)$, $p(y_j)$. An originator message is, thus, a string of $n$ symbols $x_i \in X$ randomly selected from $X$ according to the distribution $p(x_i)$, which is transformed at the recipient's end into a string of $n$ symbols $y_i \in Y$ randomly selected from $Y$ according to the distribution $p(y_i)$. An ideal or noiseless channel is such that the knowledge of any symbol $x_i$ input from the originator absolutely conditions the knowledge of the symbol $y_j$ obtained at the recipient's end. Thus, for any originator symbol $x_i$, there must be a recipient symbol $y_j$, such that $p(y_j|x_i) = 1$ and $p(y_j|x_{i' \neq i}) = 0$. The set of conditional probabilities $p(y_j|x_i)$ can be put in the form of a *transition matrix* $P(Y|X)$ and in the case of a noiseless channel with a two-symbol alphabet (binary channel) we must have

$$P(Y|X) = \begin{pmatrix} 1 & 0 \\ 0 & 1 \end{pmatrix} \text{ or } \begin{pmatrix} 0 & 1 \\ 1 & 0 \end{pmatrix}. \tag{23.1}$$

In the nonideal case where the channel is corrupted by noise, we have $p(y_j|x_i) = 1 - \varepsilon$, where $0 < \varepsilon < 1$ defines the *symbol error probability* (the probability of mistaking one symbol for another). The error probability can be the same for the two symbols (binary symmetric channel), which gives, for instance,

$$P(Y|X) = \begin{pmatrix} 1-\varepsilon & \varepsilon \\ \varepsilon & 1-\varepsilon \end{pmatrix}. \tag{23.2}$$

The above recall from the classical information theory will now help to conceive of the effect of noise in a *quantum* communication channel.

As we have seen in Chapter 22, a quantum message $M$ is a "block" represented by the tensor state $|M\rangle = |q_1 q_2 \ldots q_n\rangle$, with each qubit $|q_i\rangle$ being randomly selected from a given symbol alphabet $\{|x_k\rangle\}_{k=1\ldots N}$ of size $N$. Given the fact that each symbol $|x_k\rangle$ is associated with an occurrence probability $p_k$, it is possible to define a *symbol density operator* according to:

$$\rho = \sum_{k=1}^{N} p_k |a_k\rangle\langle a_k|. \tag{23.3}$$

Consistently, the full symbol block of length $n$ is characterized by the *density operator*:

$$\rho_M = \rho \otimes \rho \otimes \ldots \otimes \rho \equiv \rho^{\otimes n}, \tag{23.4}$$

which represents the originator's message $M$. As with the definition of a density operator, we have $\mathrm{tr}(\rho_M) = 1$. Consequently, some message $\sigma_M$ may be received at the recipient's end. The correspondence between the originator's and the recipient's messages may be defined through a transformation, or *quantum operation*, $\varepsilon$, such that

$$\sigma_M = \varepsilon(\rho_M). \tag{23.5}$$

Such a quantum operation $\varepsilon$ must be trace preserving, so that $\sigma_M$ is a density operator with $\mathrm{tr}(\sigma_M) = 1$. Since the message transformation from originator to recipient is fully defined by the operation $\varepsilon$, it is customary to refer to $\varepsilon$ as the *quantum channel* itself.

In the ideal or "noiseless" case, one would expect that the quantum channel $\varepsilon$ corresponds to some unitary (trace-preserving) transformation $U$, which gives

$$\sigma_M = \varepsilon(\rho_M) = U^+ \rho_M U. \tag{23.6}$$

Thus, according to the above operation the message is not invariant by transmission through the quantum channel ($\sigma_M \neq \rho_M$), but its integrity is fully conserved. Indeed, it only takes the recipient to apply the inverse operation $\varepsilon^{-1}$ defined by $U^{-1} = U^+$ to the received message $\sigma_M$, to obtain $\sigma'_M = \varepsilon^{-1}(\sigma_M) = U \sigma_M U^+ \equiv \rho_M$ and, hence, fully retrieve the originator's message. A trivial case of ideal or noiseless channel is given by $U = I^{\otimes n}$, where $I$ is the $2 \times 2$ identity matrix. A more general definition for any quantum channel $\varepsilon$ is

$$\varepsilon(\rho_M) = \sum_k U_k^+ \rho_M U_k. \tag{23.7}$$

The definition in Eq. (23.7) is referred to as the *operator-sum representation* of the channel, with $U_k$ being the *channel-operator elements*.

Consider next the case of *nonideal*, or *noisy quantum channels*. Clearly, such channels can be modeled by a quantum operation $\varepsilon(\rho_M)$, whose outcome $\sigma_M$ must be *random*. Here, I shall describe four basic types of noisy quantum channels, assuming that the message is made of a single qubit.

## Depolarizing channel

The first type of noisy quantum channel, called a *depolarizing* channel, is defined as follows:

$$\varepsilon(\rho) = p\frac{I}{2} + (1-p)\rho, \tag{23.8}$$

where $p$ is some probability distribution and $I$ is the identity matrix. According to this definition, there is a probability $1-p$ that the originator message $\rho$ is left invariant by the channel, and a probability $p$ that it is transformed into the message $\sigma = I/2$. Thus, the messages $\rho' = |0\rangle\langle 0|$ and $\rho'' = |1\rangle\langle 1|$, corresponding to the pure basis states $|0\rangle$ and $|1\rangle$, are transformed into

$$\begin{cases} \sigma' = \varepsilon(\rho'') = \dfrac{p}{2}\begin{pmatrix} 1 & 0 \\ 0 & 1 \end{pmatrix} + (1-p)\begin{pmatrix} 1 & 0 \\ 0 & 0 \end{pmatrix} = \begin{pmatrix} 1-\dfrac{p}{2} & 0 \\ 0 & \dfrac{p}{2} \end{pmatrix} \\ \sigma'' = \varepsilon(\rho'') = \dfrac{p}{2}\begin{pmatrix} 1 & 0 \\ 0 & 1 \end{pmatrix} + (1-p)\begin{pmatrix} 0 & 0 \\ 0 & 1 \end{pmatrix} = \begin{pmatrix} \dfrac{p}{2} & 0 \\ 0 & 1-\dfrac{p}{2} \end{pmatrix}. \end{cases} \tag{23.9}$$

The above-defined quantum operation $\varepsilon$ is referred to as the "depolarizing" channel, because *it transforms a pure basis state $\rho', \rho''$ into a uniformly mixed state $\sigma = I/2$ with a certain probability $p$*. From the above results, it would appear that the operation $\varepsilon(\rho)$ is in fact equivalent to the following pure-state transformation:

$$\begin{cases} |0\rangle \to \sqrt{1-\dfrac{p}{2}}|0\rangle \pm \sqrt{\dfrac{p}{2}}|1\rangle \\ |1\rangle \to \sqrt{\dfrac{p}{2}}|0\rangle \pm \sqrt{1-\dfrac{p}{2}}|1\rangle. \end{cases} \tag{23.10}$$

However, in view of the recipient's qubit definitions in Eq. (23.10), such an interpretation is ambiguous, since the relative phases ($\pm$) between the $|0\rangle$ and $|1\rangle$ qubits are undefined. Then which are the actual recipient qubits? The answer can be obtained by analyzing the quantum operation defined in Eq. (23.8). We observe that there is a probability $\varepsilon = 1-p$ that the communication channel accurately transmits the original message $\rho$. Thus, $\varepsilon$ represents the probability of the recipient obtaining *nonerrored states*

$|q\rangle_{\rm in} = |q\rangle_{\rm out} = |0\rangle, |1\rangle$, corresponding to original states. The probability that the recipient states are errored is $p$. In this event, the communication channel "ignores" $\rho$ and "chooses" instead to transmit $\sigma = I/2$, which is associated with the uniform mixed states $|\pm\rangle = (|0\rangle \pm |1\rangle)/\sqrt{2}$. In this event, the channel randomly outputs the errored states $|q\rangle_{\rm out} = |+\rangle, |-\rangle$, regardless of the input states $|q\rangle_{\rm in} = |0\rangle, |1\rangle$. The conclusion is that *there is a probability p that there is no mathematically defined relation between the channel input* ($|q\rangle_{\rm in}$) *and output* ($|q\rangle_{\rm out}$) *qubits*. Such an event corresponds to a *loss of coherence*, a unique quantum effect referred to as *decoherence*. It is beyond the scope of this book to analyze further the meaning and implication of quantum decoherence, as introduced by random channel noise. The only (but important!) lesson to be learnt here is twofold: (a) the quantum-channel transformation is accurately defined by the density matrix operation $\sigma = \varepsilon(\rho)$; (b) without supplemental information, the output states of the quantum channel cannot be accurately or predictably retrieved, because of the decoherence effect. However, it is possible to implement *quantum error correction*, and hence retrieve the input quantum message ($\rho$) and associated qubit symbols ($|q\rangle_{\rm in}$), as will be described in Chapter 24.

Let us now look at the VN entropy change through the depolarizing channel. For any pure-state message $\rho$, we have, by definition, $S(\rho) = 0$ (there is no uncertainty as to what message is being sent). From the recipient's end, and based on the result in Eq. (23.9), we have

$$\begin{aligned} S(\sigma) &= S(\sigma') = S(\sigma'') = -\,\mathrm{tr}\,(\sigma \log \sigma) \\ &= \frac{p}{2}\log\frac{p}{2} - \left(1-\frac{p}{2}\right)\log\left(1-\frac{p}{2}\right) \equiv f\left(\frac{p}{2}\right). \end{aligned} \tag{23.11}$$

Thus, the quantum channel has an uncertainty of the amount $S(\sigma) = f(p/2)$, which as we know from earlier chapters is maximal for the argument $p/2 = 1/2$ or $p = 1$, which gives $S(\sigma) = 1$ bit. Clearly, this limiting case corresponds to a *useless quantum channel*, where the received message is deterministically $\sigma = I/2$ and, no matter what the original message information is, the net uncertainty is one classical bit!

To conclude this example formally, it is left as a basic exercise to show that the depolarizing channel can be put in the operator-sum representation

$$\begin{aligned} \varepsilon(\rho) &= p\frac{I}{2} + (1-p)\rho \\ &\equiv \frac{q}{3}(X\rho X + Y\rho Y + Z\rho Z) + (1-q)\rho, \end{aligned} \tag{23.12}$$

where $q = 3p/4$ and $X, Y, Z$ are the *Pauli matrices* (see Chapter 16). This expression, which also holds in the general case where $\rho$ is nondiagonal, shows that the effect of noise is to transform with equal probability $q/3$ the originator's message $\rho$ into either $X\rho X$, or $Y\rho Y$ or $Z\rho Z$, and to leave it invariant with probability $1-q$.

### Bit-flip channel

As a second type of noisy quantum operation, I introduce next the *bit-flip* channel, as defined in the operator-sum representation:

$$\begin{aligned}\varepsilon(\rho) &= pX\rho X + (1-p)\rho \\ &\equiv U_0\rho U_0 + U_1\rho U_1,\end{aligned} \tag{23.13}$$

where, as usual, $p$ is some probability distribution, and where we have the following channel-operator elements:

$$\begin{cases} U_0 = \sqrt{p}X = \sqrt{p}\begin{pmatrix} 0 & 1 \\ 1 & 0 \end{pmatrix} \\ U_1 = \sqrt{1-p}I = \sqrt{1-p}\begin{pmatrix} 1 & 0 \\ 0 & 1 \end{pmatrix}. \end{cases} \tag{23.14}$$

It is straightforward to establish that the action of the first operation term $X\rho X$ in Eq. (23.13) is to transform the originator's message $\rho$ into the recipient message $\sigma = \varepsilon(\rho)$ with the flipped coefficients $\sigma_{11} = \rho_{22}$ and $\sigma_{22} = \rho_{11}$. Such a transformation has probability $p$ of occurring, while the originator's message is left unchanged ($\sigma = \rho$) with probability $1-p$. Effecting the calculation with the qubit message $|q\rangle = \alpha|0\rangle + \beta|1\rangle$ ($\alpha = \sqrt{\rho_{11}}, \beta = \sqrt{\rho_{22}}$), it is also straightforward to obtain the channel transformation:

$$|q\rangle \to |q'\rangle = \sqrt{(1-p)|\alpha|^2 + p|\beta|^2}|0\rangle + \sqrt{p|\alpha|^2 + (1-p)|\beta|^2}|1\rangle. \tag{23.15}$$

Thus, if the originator's message is the pure-state qubit $|0\rangle$ (i.e., $\alpha = 1, \beta = 0$), the recipient receives the message

$$|q'\rangle = \sqrt{1-p}|0\rangle + \sqrt{p}|1\rangle. \tag{23.16}$$

Clearly, the symbol-error (or cbit-error) probability is $\varepsilon = p$, meaning that the channel has the probability $p$ of outputting the errored message $|1\rangle$ when $|0\rangle$ was sent. The same conclusion applies in the opposite case, which illustrates the notion of *bit flipping* as a noise effect in quantum channels. It is left as an interesting exercise to analyze the effect of bit-flip transformation in the case where $\rho$ is nondiagonal. As expected, the outcome of the exercise is that an originator's qubit message $|q\rangle = \alpha|0\rangle + \beta|1\rangle$ is "flipped" through the noisy channel into a recipient's message $|q\rangle = \beta|0\rangle + \alpha|1\rangle$ under an outcome probability $p$. Finally, the entropy difference between a pure-state originator's message $\rho = |0\rangle\langle 0|$ and the recipient's message $\sigma = (1-p)|0\rangle\langle 0| + p|1\rangle\langle 1|$ is

$$\begin{aligned}\Delta S = S(\sigma) - S(\rho) &= [-(1-p)\log(1-p) - p\log p] - 0 \\ &\equiv f(1-p) = f(p),\end{aligned} \tag{23.17}$$

with the same result being obtained with $\rho = |1\rangle\langle 1|$. The maximum uncertainty ($\Delta S = 1$ or one cbit), corresponds to the case $p = 0.5$, meaning that, according to Eq. (23.13), there is a 50% chance that the qubit amplitudes are flipped when passing through the channel. Interestingly, there is no entropy change ($\Delta S = 0$) if the qubit amplitudes are deterministically flipped ($p = 1$). This is because there is no difference in information contents if the cbit polarity (the definition of 0 and 1, or $|0\rangle$ and $|1\rangle$) is *systematically* flipped.

### Phase-flip channel

As a third example of noisy quantum operation, I define next the *phase-flip* channel in the operator-sum representation:

$$\varepsilon(\rho) = pZ\rho Z + (1-p)\rho \equiv U_0\rho U_0 + U_1\rho U_1, \tag{23.18}$$

with the corresponding elements:

$$\begin{cases} U_0 = \sqrt{p}Z = \sqrt{p}\begin{pmatrix} 1 & 0 \\ 0 & -1 \end{pmatrix} \\ U_1 = \sqrt{1-p}I = \sqrt{1-p}\begin{pmatrix} 1 & 0 \\ 0 & 1 \end{pmatrix}. \end{cases} \tag{23.19}$$

It is again left as an exercise to show that the originator's message $|q\rangle = \alpha|0\rangle + \beta|1\rangle$ is transformed into $|q'\rangle = \alpha|0\rangle - \beta|1\rangle$ with probability $p$ and is left invariant otherwise. Thus, the action of the noisy channel is to flip (with probability $p$) the relative phase $e^{i\delta}$ randomly between the complex amplitudes $\alpha, \beta$ from $e^{i0}$ to $e^{i\pi}$. In the case of qubit states $|q\rangle = |0\rangle, |1\rangle$ it is found that the entropy change is zero ($\Delta S = 0$), clearly because no phase change can affect a single qubit. In the case $|q\rangle = |+\rangle$, or $\rho = |+\rangle\langle +|$, one finds the recipient's message:

$$\sigma = p\,|-\rangle\langle -| + (1-p)|+\rangle\langle +|, \tag{23.20}$$

which in the basis $\{|+\rangle, |-\rangle\}$ is diagonal, i.e.,

$$\sigma = \begin{pmatrix} p & 0 \\ 0 & 1-p \end{pmatrix}, \tag{23.21}$$

hence, the net entropy difference $\Delta S = S(\sigma) - S(\rho) = f(p) - 0 \equiv f(p)$, with the same result being obtained with $\rho = |-\rangle\langle -|$. The maximum uncertainty ($\Delta S = 1$ or one cbit), corresponds to the case $p = 0.5$, meaning that, according to Eq. (23.18), there is a 50% chance that the relative qubit phase is flipped by $e^{i\pi}$ when passing through the channel. Interestingly, there is no entropy difference ($\Delta S = 0$) if the phase is deterministically changed ($p = 1$). This is because there is no difference in information contents if the cbit polarity (the definition of 0 and 1 or $|+\rangle$ and $|-\rangle$) is systematically flipped.

### Bit-phase-flip channel

As a fourth example of noisy quantum operation, we define next the *bit-phase-flip* channel in the operator-sum representation:

$$\begin{aligned} \varepsilon(\rho) &= pY\rho Y + (1-p)\rho \\ &\equiv U_0\rho U_0 + U_1\rho U_1, \end{aligned} \tag{23.22}$$

with the corresponding elements:

$$\begin{cases} U_0 = \sqrt{p}Y = \sqrt{p}\begin{pmatrix} 0 & -i \\ i & 0 \end{pmatrix} \\ U_1 = \sqrt{1-p}I = \sqrt{1-p}\begin{pmatrix} 1 & 0 \\ 0 & 1 \end{pmatrix}. \end{cases} \tag{23.23}$$

It is again left as an exercise to show that the originator message $|q\rangle = \alpha|0\rangle + \beta|1\rangle$ is transformed into $|q'\rangle = \beta|0\rangle - \alpha|1\rangle$ with probability $p$ and is left invariant otherwise. The action of the noisy channel is to flip both the phase and the complex amplitudes. It is easily found that for any originator messages $\rho = |0\rangle\langle 0|$, $\rho = |1\rangle\langle 1|$, $\rho = |+\rangle\langle +|$ or $\rho = |-\rangle\langle -|$, the net entropy difference is $\Delta S = S(\sigma) - S(\rho) = f(p)$. It is clear that for $\rho = |0\rangle\langle 0|$, $\rho = |1\rangle\langle 1|$ this difference is only attributable to the bit-flipping effect, while for $\rho = |+\rangle\langle +|$ or $\rho = |-\rangle\langle -|$ it is attributable to the phase-flipping effect.[1]

Other noise processes, which are beyond the scope of these chapters but very important in the analysis of time evolution of qubits in quantum channels, concern *amplitude damping* and *phase damping*.

## 23.2 The Holevo–Schumacher–Westmoreland capacity theorem

In this section, I shall introduce the concept of *quantum channel capacity*, based on the *Holevo–Schumacher–Westmoreland* (HSW) theorem. The HSW theorem defines the channel capacity $\chi$ under which, at a rate $R < \chi$, it is possible to transmit quantum messages with arbitrary small probability of error or information loss. It can be viewed conceptually as the elegant quantum counterpart of *Shannon's channel coding theorem*, which was described in Chapter 13. Here, the term "capacity" refers, in fact, to the same concept as in *classical information*, while assuming that such information is encoded in quantum messages that are passed through noisy quantum channels.

The formal demonstration of the HSW theorem being particularity involved and tedious, it will not be considered here. For the general interest and purpose of this chapter, it will suffice to state the HSW theorem "as is:" given a quantum-symbol source $\{\rho_i\}$ with associated probability distribution $\{p_i = p(\rho_i)\}$, the capacity $\chi$ of a noisy quantum channel $\varepsilon$ providing the operation $\sigma_i = \varepsilon(\rho_i)$ is given by the maximum difference:

$$\begin{aligned} \chi &= \max[S\langle\sigma\rangle - \langle S(\sigma)\rangle] \\ &= \max_{\{p_i\}} \{S[\varepsilon(\langle\rho_i\rangle)] - \langle S[\varepsilon(\rho_i)]\rangle\}, \end{aligned} \tag{23.24}$$

[1] Flipping complex amplitudes leave $|+\rangle = (|0\rangle + |1\rangle)/\sqrt{2}$ invariant, while resulting in a global phase factor $e^{i\pi} = -1$, for $|-\rangle = (|0\rangle - |1\rangle)/\sqrt{2}$, which does not modify the definition of $|-\rangle$.

where the brackets $\langle\ \rangle$ indicate an averaging over the distribution $\{p_i\}$. This definition can be put into the more explicit or complete formulation:

$$\begin{aligned}\chi &= \max_{\{p_i\}}\left[S\left(\sum_i p_i\sigma_i\right) - \sum_i p_i S(\sigma_i)\right]\\ &= \max_{\{p_i\}}\left\{S\left[\varepsilon\left(\sum_i p_i\rho_i\right)\right] - \sum_i p_i S[\varepsilon(\rho_i)]\right\}.\end{aligned} \tag{23.25}$$

As with Shannon's channel coding theorem, *the capacity $\chi$ represents the maximum code rate for which the probability of transmission error can be made arbitrarily small*, assuming sufficiently long message lengths. The following provides a background and a basic interpretation for the HSW theorem, showing how it nicely parallels Shannon's channel coding theorem.

As we have seen in Chapter 22, a quantum message of length $n$ can be conceived as an $n$-qubit "block" represented by the tensor state $|M\rangle = |x_1x_2\ldots x_n\rangle$, with each qubit $|x_k\rangle$ being randomly selected with an occurrence probability $p_i$, from a given symbol alphabet $\{|a_i\rangle\}_{i=1\ldots N}$ of size $N$. We assume that the originator chooses among $2^{nR}$ message possibilities, where $R > 0$. This limited set of "quantum codewords" of length $n$ can be referred to as a "code," and $R$ as a *code rate*. A *symbol density operator* $\rho_{m_k}$ can be associated with any message qubit in position $k$, where $m_k$ is an index selected from the alphabet $i = 1, 2, \ldots, N$. Hence, the originator's message block or codeword is characterized by a density operator $\rho_M$ constructed from the $n$-tensor product:[2]

$$\rho_M = \rho_{m_1}\otimes\rho_{m_2}\otimes\cdots\otimes\rho_{m_n}. \tag{23.26}$$

We note that since the message block is defined as an $n$-tensor product, there is no entanglement between any of the qubits therein. The recipient message, or codeword is received as

$$\sigma_M = \sigma_{m_1}\otimes\sigma_{m_2}\otimes\cdots\otimes\sigma_{m_n}, \tag{23.27}$$

where

$$\sigma_{m_k} = \varepsilon(\rho_{m_k}). \tag{23.28}$$

Following the description in Chapter 22 (but not being concerned here about compression), it is clear that a specific codeword $\sigma_M$ corresponds to a set of $N' = N^n$ eigenvalues $\Lambda_{MK} = \lambda_{m_1k_1}\lambda_{m_1k_2}\ldots\lambda_{m_1k_{n1}}$ and eigenvectors $|\Lambda_{MK}\rangle = |\lambda_{m_1k_1}\lambda_{m_1k_2}\ldots\lambda_{m_1k_{n1}}\rangle$, such that $\sigma_M$ has the diagonal form

$$\begin{aligned}\sigma_M &= \sum_K \Lambda_{MK}|\Lambda_{MK}\rangle\langle\Lambda_{MK}|\\ &= \begin{pmatrix}\Lambda_{M1} & 0 & \cdots & 0\\ 0 & \Lambda_{M2} & \cdots & 0\\ \vdots & \vdots & \ddots & \vdots\\ 0 & 0 & \cdots & \Lambda_{MN'}\end{pmatrix},\end{aligned} \tag{23.29}$$

[2] For instance, $|M\rangle = |a_1a_2a_2a_3\rangle$ is represented by the density operator $\rho_M = \rho_1\otimes\rho_2\otimes\rho_2\otimes\rho_3$.

with $\sum_K \Lambda_{MK} = 1$. Next, define $\langle S\rangle = \langle S(\sigma)\rangle$ as the mean value of the per-symbol VN entropy of the received codeword, according to:

$$\langle S\rangle = \sum_i p_i S(\sigma_i). \tag{23.30}$$

Given any parameter $\varepsilon > 0$, and based on the concept described in Chapter 22, we can define the $\varepsilon$-typical subspace $\Omega$ for $\sigma_M$, based on the parameters $\Lambda_{MK}$, $\langle S\rangle$ and spanned by the eigenvectors $|\Lambda_{MK}\rangle$ according to the condition:

$$\left|\frac{1}{n}\log\frac{1}{\Lambda_{MK}} - \langle S\rangle\right| \leq \varepsilon. \tag{23.31}$$

Let $P_M$ be the projector onto this $\varepsilon$-typical subspace $\Omega$. Based on Eq. (22.30), we know that the dimension of this subspace, $\mathrm{tr}(P_M)$, is upper-bounded according to

$$\mathrm{tr}(P_M) \leq 2^{n(\langle S\rangle+\varepsilon)}. \tag{23.32}$$

We also know that for any $\sigma_M$ with sufficiently long length $n$ and for any $\delta$ such that $0 < \delta < 1$, the probability $p = p(\sigma_M \in \Omega)$ of projecting onto the typical subspace is lower-bounded according to:

$$p = \langle \mathrm{tr}(\sigma_M P_M)\rangle \geq 1 - \delta = p_{\min}, \tag{23.33}$$

where the brackets have the meaning of an *expectation value*, i.e., the projection probability as averaged over all possible message codewords $\sigma_M$. This result shows that the probability that any received codeword $\sigma_M$ belongs to the typical subspace has a lower bound $p_{\min}$, which is nonzero. I shall now introduce the notion of *error probability*, which is the probability that a POVM measurement of $\sigma_M$ by the recipient fails to identify $\sigma_M$ as one of the codewords from the typical set $\Omega$.

Consider, indeed, that given the typical set $\Omega = \{\sigma_M\}$ of size $2^{nR}$, there is a corresponding set $\{E_M\}$ of *POVM measurement operators* (see Chapter 17) of the same size. The POVM set is completed with the measurement operator $E_0$ for all other possible codewords, defined as

$$E_0 = I^{\otimes n} - \sum_{M\neq 0} E_M. \tag{23.34}$$

The probability $p_M$ of the recipient identifying a specific message $M$ in $\Omega$ while using a POVM measurement $E_M$ is given by

$$p_M = \mathrm{tr}(\sigma_M E_M) \tag{23.35}$$

and in the opposite case, where such identification is unsuccessful, the error probability $p_M^{\mathrm{error}}$ is:

$$p_M^{\mathrm{error}} = 1 - p_M = 1 - \mathrm{tr}(\sigma_M E_M). \tag{23.36}$$

If we assume that the originator *uniformly* selects the messages from the codeword set $\{\rho_M\}$, it is then possible to define an *average error probability* $\bar{p}_{\text{error}}$ through

$$\bar{p}_{\text{error}} = \frac{1}{2^{nR}} \sum_M p_M^{\text{error}}. \tag{23.37}$$

Define $\langle \bar{p}_{\text{error}} \rangle$ as the expectation value of $\bar{p}_{\text{error}}$ over all possible codewords. The formal demonstration of the HSW theorem then leads to the following result, which I shall here provide "as is:"

$$\langle \bar{p}_{\text{error}} \rangle \leq 4\delta + (2^{nR} - 1)2^{-n(\chi - 2\varepsilon)}, \tag{23.38}$$

with

$$\chi = S(\langle \sigma \rangle) - \langle S(\sigma) \rangle. \tag{23.39}$$

The result in Eq. (23.38) shows that provided the condition for the code rate $R$,

$$R < S(\langle \sigma \rangle) - \langle S(\sigma) \rangle = \chi, \tag{23.40}$$

and for long message lengths ($n \to \infty$) the expected error probability is upper-bounded according to $\langle \bar{p}_{\text{error}} \rangle \leq \delta' = 4\delta$ with $\delta$ being *any* real such that $0 < \delta < 1$, i.e., which can be chosen arbitrarily close to zero. The maximization of $\chi$, which yields the quantum channel *capacity*, is an issue at least as complex as in the case of classical communication channels. Suffice it here to infer that under the condition $R < \max(\chi)$ it is possible to find another class of quantum codes for which $\bar{p}_{\text{error}} < \delta''$, with $\delta''$ being arbitrarily close to zero. The key conclusion is, therefore:

Given a quantum channel $\varepsilon$, there exist quantum codes with long codeword size $n$ and code rate $R < \max(\chi)$, for which the error probability $\bar{p}_{\text{error}}$ (of the recipient failing to identify a codeword $\sigma_M = \varepsilon(\rho_M)$ given an originator codeword $\rho_M$) can be made arbitrarily small.

This conclusion defines the key condition under which "error-free" transmission of quantum messages through noisy quantum channels can be achieved. It is also possible to show the converse, namely: "for $R > \chi$ there exists no quantum code of any codeword size able to achieve error-free transmission," meaning that $\bar{p}_{\text{error}}$ is irremediably bounded away from zero.

As stated earlier, there exists a nice parallel between the HSW "capacity" theorem and Shannon's coding theorem for classical, noisy communication channels. To recall from Chapter 13, the classical channel capacity $C$ is defined as the mutual information maximum:

$$C = \max_{p(x)} H(X; Y). \tag{23.41}$$

The coding theorem states that:

Given a noisy transmission channel, there exist binary codes of sufficiently long codeword size, with $2^{nR}$ codewords of sufficient length $n$, and code rate $R < C$, for which an originator message can be transmitted to a recipient with arbitrary low error probability.

It is clear that Shannon', coding theorem and the HSW theorem are conceptually very similar. As in QIT, the typical set of codewords (or of typical sequences) is defined under the condition (Eq. (13.25)):

$$\left| \frac{1}{n} \log \frac{1}{p(x)} - H(X) \right| \leq \varepsilon, \tag{23.42}$$

where $p(x)$ is the *codeword* probability distribution associated with the *symbol* source $X$, and the entropy $H(X)$. This condition parallels that in Eq. (23.29), with $p(x) \to \Lambda_{MK}$ and $H(X) \to \langle S \rangle$. In the classical case, the dimension of the typical set is upper-bounded by $2^{n(H+\varepsilon)}$, while in the quantum case, the bound is $2^{n(\langle S \rangle + \varepsilon)}$, see Eq. (23.30). Because of this analogy, it is clear that the HSW theorem owes a great deal to Shannon's classical analysis, even if its background assumptions are hardly reducible to any classical conception. Then we may ask: "Is quantum channel capacity in any way reducible to Shannon's original concept?" The following discussion, although with no pretence at providing any academic proof, may guide towards some form of a satisfactory answer; we expected no!

We may now conclude this section with a discussion about the connection between the HSW theorem and the *Holevo bound* described in Chapter 21, and clarifying the difference between $\chi$ and the classical channel capacity $C$. Indeed, letting $\chi' = S(\langle \sigma \rangle) - \langle S(\sigma) \rangle$, we recognize from Chapter 21 that $\chi'$ is exactly the definition of the *Holevo bound* for the quantum source $\sigma$, and the HSW theorem seems simply to state that the channel capacity is given by $\chi = \max \chi'$. Then, apart from the maximization issue, we may wonder what is conceptually new and useful in this HSW theorem that has not already been captured into the Holevo bound concept. The answer to this apparent paradox is quite simple. In Chapter 21, we have considered the action of encoding and decoding classical information into or from a suite of quantum symbols $\rho_x$ from an alphabet $\{\rho_i\}$, to be conveyed through some ideal quantum channel. The operation of encoding is the "preparation" of the state $\rho_x$, which means that the originator sets up the quantum system in the state $\rho_x$. The operation of decoding is the "measurement" by the recipient of any received symbol $\rho_y$, using the adequate POVM set for the identification of each symbol possibility. The full encoding and decoding operation can be virtually reduced into a classical information channel that relates two random sources $X$ and $Y$. As we have learnt, the mutual information $H(X; Y)$ is bounded by $H(X)$, just as in the purely classical case, but the quantum channel introduces the extra (Holevo) bound $\chi$ such that

$$H(X; Y) \leq \chi \leq H(X). \tag{23.43}$$

As we have also seen, if the symbols $\rho_i$ have orthogonal support in $\{\rho_i\}$, meaning that different qubit symbols are necessarily orthogonal, then $\chi = H(X)$ and the Holevo bound is reduced to the originator source entropy, which obliterates any particularity of using a quantum channel as opposed to a purely classical one.

Consider next the case of the noisy quantum channel. The encoding and decoding operations are conceptually the same as previously, except that decoding (POVM measurement) is performed onto message *codewords*, defined by $\sigma_M = \sigma_{m_1} \otimes \sigma_{m_2} \otimes \cdots \otimes \sigma_{m_n}$, given the originator codeword $\rho_M = \rho_{m_1} \otimes \rho_{m_2} \otimes \cdots \otimes \rho_{m_n}$. In the received message $\sigma_M$, each symbol $\sigma_{m_k}$ is the result of a quantum operation $\varepsilon$ such that $\sigma_{m_k} = \varepsilon(\rho_{m_k})$. From statistical measurements, the recipient may then compute $\chi' = S(\langle\sigma\rangle) - \langle S(\sigma)\rangle = S[\langle\varepsilon(\rho)\rangle] - \langle S[\varepsilon(\rho)]\rangle$ for the source $\sigma$, but it is irrelevant to call this quantity a "Holevo bound," because of the quantum operation $\varepsilon$ introduced by the channel, which, for each recipient symbol $\sigma_{m_k}$, has corrupted the reference symbol $\rho_{m_k}$. If $\varepsilon$ is not a "noiseless" constant, therefore, the action of decoding is conceptually quite different from that assumed in the Holevo bound derivation. Furthermore, however tempting, we ought not to view $\chi'$ as representing some elaborated definition of "quantum mutual information" between the sources $\rho$ and $\sigma$ of two quantum systems $Q$ (originator) and $Q'$ (recipient). In Chapter 21, we have described the *quantum mutual information* $S(\hat{\rho}_A; \hat{\rho}_B)$ for a composite system $AB$ as

$$S(\hat{\rho}_A; \hat{\rho}_B) = S(\hat{\rho}_A) + S(\hat{\rho}_B) - S(\rho_{AB}), \tag{23.44}$$

where $\hat{\rho}_A = \mathrm{tr}_B(\rho_{AB})$ and $\hat{\rho}_B = \mathrm{tr}_A(\rho_{AB})$ are the partial-traced operators. Life in QIT would be so much simpler if $\max\lfloor S(\hat{\rho}_Q; \hat{\rho}_{Q'})\rfloor$ could represent the quantum channel capacity $\chi$, but this is definitely not the case! To cut through any other explanation, it is not possible to define the composite system $QQ'$ with density operator $\rho_{QQ'}$, simply because after the operation $\varepsilon$ the original system $Q$ does not exist anymore as it has been transformed into $Q'$! The physical parameter that can make sense is the difference $\Delta S = S(\rho_{Q'}) - S(\rho_Q) \equiv S(\sigma) - S(\rho)$, see further on.

The quantum counterpart of mutual information is, indeed, far more complex, and its description is beyond the scope of this chapter. Here, to quench our thirst, I may provide just a brief hint of a new concept referred to as *quantum coherent information*. Appendix W shows that given a quantum system $Q$ in a mixed state $|\psi_Q\rangle$ with density operator $\rho_Q \equiv \rho$, it is possible to define a reference quantum system $R$, such that the composite system $QR$ is in a pure state $|\psi_{QR}\rangle$, namely, for which $\rho_{QR} = |\psi_{QR}\rangle\langle\psi_{QR}|$. Such an operation is referred to as *state purification*. Upon the action of the quantum operation $\varepsilon$, the systems $Q$, $R$ are then transformed into the systems $Q'$, $R'$ with, in particular, $\rho_{Q'} \equiv \sigma = \varepsilon(\rho)$. One defines the *entropy exchange* of the action of $\varepsilon$ on the composite system $QR$ that transforms it into the composite system $Q'R'$ as

$$S_{\mathrm{E}}(\rho, \varepsilon) = S(\rho_{Q'R'}). \tag{23.45}$$

Then one defines the *quantum coherent information* as the difference

$$I(\rho, \varepsilon) = S(\sigma) - S_{\mathrm{E}}(\rho, \varepsilon). \tag{23.46}$$

In today's state of the art of QIT, the quantum coherent information $I(\rho, \varepsilon)$ is only "conjectured" as playing the same role as the mutual information

$$H(X; Y) = H(Y) - H(Y|X) \tag{23.47}$$

does in Shannon's classical theory. In particular, it can be shown that given the entropy difference between source and originator,

$$\Delta S = S(\sigma) - S(\rho), \tag{23.48}$$

the following property is always satisfied:

$$\Delta S + S_{\mathrm{E}}(\rho, \varepsilon) \geq 0. \tag{23.49}$$

This last result shows that the entropy change caused by the quantum operation $\varepsilon$ on both the quantum channel and the "outside world" is always nonnegative, which recalls the *second law of thermodynamics*, according to which the entropy of a closed system may only increase on any physical transformation or evolution from its initial state.

## 23.3 Capacity of some quantum channels

In this section, I apply the HSW theorem to evaluate the capacity of different types of quantum channel. We assume that the originator source has, for symbols, the pure states $\rho_i = |\psi_i\rangle\langle\psi_i|$ which are associated within a quantum message (or codeword) with some probability distribution $p_i$. For convenience, we introduce

$$\chi' = S\left[\varepsilon\left(\sum_i p_i \rho_i\right)\right] - \sum_i p_i S\left[\varepsilon\left(\rho_i\right)\right], \tag{23.50}$$

to which, after the HSW theorem, the channel capacity $\chi = \max(\chi')$ corresponds.

Before touching upon any examples, it is interesting to consider the case of ideal, constant or "noiseless" channels. This is the case where $\varepsilon\left(\rho_i\right) = \rho_i$ and, hence, from Eq. (23.50):

$$\chi'_{\text{ideal}} = \left(\sum_i p_i \rho_i\right) - \sum_i p_i S\left(\rho_i\right) \equiv S\left(\sum_i p_i \rho_i\right), \tag{23.51}$$

where we used the property that for any pure state $\rho_i$ the VN entropy $S(\rho_i)$ is zero. The result in Eq. (23.49) shows that in this ideal case, $\chi'$ actually represents the VN entropy of the originator's message source $S(\rho) = S(\langle\rho_i\rangle)$. The "ideal channel" capacity is given by $\chi_{\text{ideal}} = \max(\chi'_{\text{ideal}})$, or

$$\chi_{\text{ideal}} = \max(\chi'_{\text{ideal}}) = \max_{p_i}\left[S\left(\sum_i p_i \rho_i\right)\right] \equiv \max S(\rho). \tag{23.52}$$

This maximization problem, thus, addresses the question of how much classical information can be conveyed through any given message $\rho$ before transmission through any quantum channel. Intuitively, we expect that for nonideal channels we have the property

$$\chi \leq \chi_{\text{ideal}}, \tag{23.53}$$

meaning that, obviously, *noisy transmission does not improve the information of the message source*. This is just like in the classical case, with the mutual information satisfying $H(X;Y) \leq H(X)$, and with the capacity $C = \max[H(X;Y)] \leq \max[H(X)] = 1$ bit. The following examples will illustrate the fact that the quantum channel capacity $\chi$ cannot, indeed, exceed the highest-possible VN entropy of the originator source, i.e., $\chi_{\text{ideal}} = \max[S(\rho)]$.

**Example 23.1:** ***Depolarizing channel***

In Eq. (23.8) we have seen that this channel is defined by the operation

$$\varepsilon(\rho_i) = p\frac{I}{2} + (1-p)\rho_i. \tag{23.54}$$

Assuming the orthogonal symbols $\rho_1 = |0\rangle\langle 0|$ and $\rho_2 = |1\rangle\langle 1|$, we obtain

$$\begin{aligned}\varepsilon\left(\sum_i p_i\rho_i\right) &= \varepsilon\left[p_1\rho_1 + (1-p_1)\rho_2\right] \\ &= p\frac{I}{2} + (1-p)\left[p_1\rho_1 + (1-p_1)\rho_2\right] \\ &= \frac{p}{2}\begin{pmatrix}1 & 0\\ 0 & 1\end{pmatrix} + (1-p)p_1\begin{pmatrix}1 & 0\\ 0 & 0\end{pmatrix} + (1-p)(1-p_1)\begin{pmatrix}0 & 0\\ 0 & 1\end{pmatrix} \\ &= \frac{p}{2}\begin{pmatrix}1 & 0\\ 0 & 1\end{pmatrix} + (1-p)\begin{pmatrix}p_1 & 0\\ 0 & 1-p_1\end{pmatrix} \\ &\equiv \begin{pmatrix}\frac{p}{2} + (1-p)p_1 & 0\\ 0 & \frac{p}{2} + (1-p)(1-p_1)\end{pmatrix},\end{aligned} \tag{23.55}$$

and the corresponding VN entropy

$$S\left[\varepsilon\left(\sum_i p_i\rho_i\right)\right] = -\left[\begin{array}{l}\dfrac{p+2(1-p)p_1}{2}\log\dfrac{p+2(1-p)p_1}{2} \\ +\dfrac{p+2(1-p)(1-p_1)}{2}\log\dfrac{p+2(1-p)(1-p_1)}{2}\end{array}\right]. \tag{23.56}$$

As we have seen in the previous section, the VN entropy of the recipient symbols is $S[\varepsilon(\rho_1)] = S[\varepsilon(\rho_2)] = f(p/2)$, hence,

$$\sum_i p_i S\left[\varepsilon(\rho_i)\right] = p_1 f\left(\frac{p}{2}\right) + (1-p_1) f\left(\frac{p}{2}\right) \equiv f\left(\frac{p}{2}\right). \tag{23.57}$$

From Eq. (23.50), we finally obtain

$$\begin{aligned}\chi'(p, p_1) &= S\left[\varepsilon\left(\sum_i p_i\rho_i\right)\right] - \sum_i p_i S\left[\varepsilon(\rho_i)\right] \\ &= -\left\{\begin{array}{l}\left[\frac{p}{2} + (1-p)p_1\right]\log\left[\frac{p}{2} + (1-p)p_1\right] \\ +\left[\frac{p}{2} + (1-p)(1-p_1)\right]\log\left[\frac{p}{2} + (1-p)(1-p_1)\right]\end{array}\right\} - f\left(\frac{p}{2}\right).\end{aligned} \tag{23.58}$$

In the above equation, it is clear that the first term in braces, corresponding to the definition in Eq. (23.56), is maximized if we choose the two symbols to have equal probabilities $p_1 = 1 - p_2 = 1/2$, which gives

$$S\left[\varepsilon\left(\sum_i p_i\rho_i\right)\right] = -\left(\frac{1}{2}\log\frac{1}{2} + \frac{1}{2}\log\frac{1}{2}\right) = f\left(\frac{1}{2}\right) = 1. \tag{23.59}$$

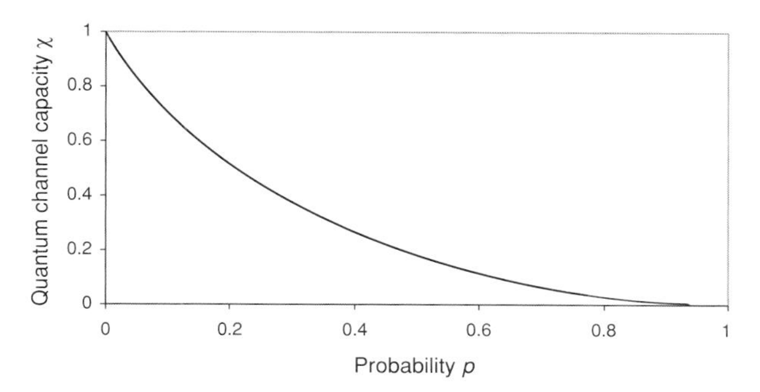


**Figure 23.1** Quantum capacity $\chi(p)$ of the depolarizing channel as a function of the depolarizing parameter $p$.

Thus, we obtain the channel capacity $\chi$:

$$\chi(p) = \max \chi'\left(p, \frac{1}{2}\right) \equiv 1 - f\left(\frac{p}{2}\right). \tag{23.60}$$

As expected from the analysis in the previous section, the capacity vanishes for $p = 1$, which corresponds to a "useless channel," projecting any random symbol $\rho_i$ onto $\sigma = \varepsilon(\rho_i) = I/2$ as shown from Eq. (23.54). The highest capacity $\chi(0) = 1$ bit is obtained for $p = 0$, in which case the quantum channel $\varepsilon$ is constant or "ideal," or "noiseless," i.e., $\sigma = \varepsilon(\rho_i) = \rho_i$, but this case is of no interest here. Figure 23.1 shows a plot of the depolarizing channel capacity as a function of the depolarizing probability parameter $p$. It is seen from the figure that, as expected, the channel capacity monotonously decays as the depolarizing parameter increases from $p = 0$ to $p = 1$. To provide a practical engineering interpretation of the above result, assume, for instance, that $p = 0.1$, which yields $f(p/2) \approx 0.25$ (see Fig. 4.7), and, thus, defines a channel capacity of $\chi = 0.75$ bit. The HSW theorem states that it is possible to transmit codewords of sufficient lengths with arbitrary error probability provided the code rate satisfies $R < \chi = 0.75$. Given a block code $(n, k)$, the code rate must satisfy $R = k/n < 0.75$, which means that the code must include 25% redundancy bits and 75% payload bits.

The exercises provide other application examples of the depolarizing channel capacity. Here, it is worth showing the results obtained when assuming the codeword symbols $|0\rangle, |+\rangle$ instead of $|0\rangle, |1\rangle$. It is a tedious but tractable exercise to show that the corresponding capacity takes the form:

$$\chi(p) = \max_{p_1} \left\{ \begin{array}{l} -\dfrac{1+(1-p)\sqrt{1-2p_1(1-p_1)}}{2} \log \dfrac{1+(1-p)\sqrt{1-2p_1(1-p_1)}}{2} \\ -\dfrac{1-(1-p)\sqrt{1-2p_1(1-p_1)}}{2} \log \dfrac{1-(1-p)\sqrt{1-2p_1(1-p_1)}}{2} \\ +(1-p_1)\dfrac{1+\sqrt{1-p(2-p)}}{2} \log \dfrac{1+\sqrt{1-p(2-p)}}{2} \\ +(1-p_1)\dfrac{1-\sqrt{1-p(2-p)}}{2} \log \dfrac{1-\sqrt{1-p(2-p)}}{2} - p_1 f\left(\dfrac{p}{2}\right) \end{array} \right\}. \tag{23.61}$$

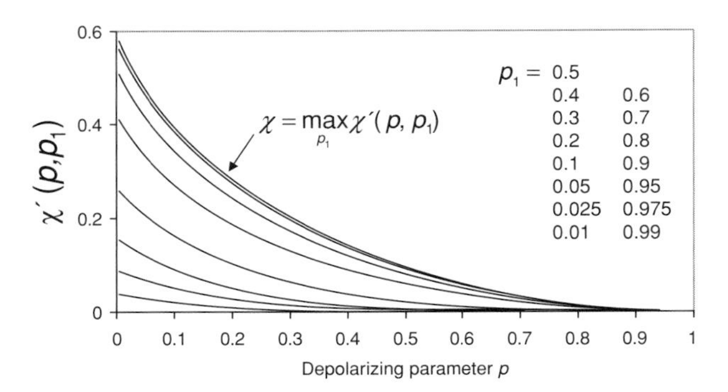


**Figure 23.2** Plots of $\chi'(p, p_1)$ as a function of the depolarizing parameter $p$ in depolarizing-channel, and corresponding channel capacity $\chi(p) = \max \chi'(p, p_1)$ (codewords formed by pure states $|0\rangle, |+\rangle$ with probabilities $p_1, 1 - p_1$, respectively).

The expression in brackets $\{\chi'(p, p_1)\}$ in Eq. (23.61) can be maximized by numerically solving the transcendental equation $\partial\chi'(p, p_1)/\partial p_1 = 0$. But we may as well directly infer the answer, as it will be shown further on. First, we may just plot the function $\chi'(p, p_1)$ for different values of the parameter $p_1$ (probability associated with $|0\rangle$), as illustrated in Fig. 23.2. From the figure, we first note the property $\chi'(p, p_1) = \chi'(p, 1 - p_1)$. Second, we observe that in all cases, the function $\chi'(p, p_1)$ is maximal at $p = 0$ (constant or noiseless channel) and zero at $p = 1$ (useless channel). Third, we see that the channel capacity $\chi(p) = \max \chi'(p, p_1)$ is achieved for $p_1 = 0.5$. This was expected, since this condition corresponds to *maximum uncertainty* in the occurrence of the $|0\rangle, |+\rangle$ qubits forming the codewords. From the property in Eq. (23.53), the constant or noiseless capacity, $\chi(0)$, must correspond to the (maximum possible) VN entropy of originator's source, $\max[S(\rho)] = \chi_{\text{ideal}}$, as discussed earlier and also formally established in the exercise. Substituting $p_1 = 0.5$ into Eq. (23.61), we obtain the following analytical form of the channel capacity:

$$\chi \equiv \chi'\left(p, \frac{1}{2}\right) = \left\{ \begin{aligned} & -\frac{1 + \frac{1-p}{\sqrt{2}}}{2} \log \frac{1 + \frac{1-p}{\sqrt{2}}}{2} \\ & -\frac{1 - \frac{1-p}{\sqrt{2}}}{2} \log \frac{1 - \frac{1-p}{\sqrt{2}}}{2} \\ & + \frac{1 + \sqrt{1 - p(2-p)}}{4} \log \frac{1 + \sqrt{1 - p(2-p)}}{2} \\ & + \frac{1 - \sqrt{1 - p(2-p)}}{4} \log \frac{1 - \sqrt{1 - p(2-p)}}{2} - \frac{1}{2} f\left(\frac{p}{2}\right) \end{aligned} \right\}. \tag{23.62}$$

Substituting $p = 0$ in the above, we get the upper bound for the channel capacity, which corresponds to the originator's maximum source entropy $\max[S(\rho)]$:

$$\begin{aligned}\chi_{\text{ideal}} &= \max S(\rho) \\ &= \chi'\left(0, \frac{1}{2}\right) \\ &\equiv -\left\{\frac{2+\sqrt{2}}{4}\log\frac{2+\sqrt{2}}{4} + \frac{2-\sqrt{2}}{4}\log\frac{2-\sqrt{2}}{4}\right\}.\end{aligned} \tag{23.63}$$

This example, thus, provides an illustration of the HSW maximization problem, leading, in this case, to a tractable analytical solution. It also illustrates the property $\chi(p) \le \chi_{\text{ideal}} = \max S(\rho)$, which can be used to guide the maximization problem solving.

---

**Example 23.2:** ***Bit-flip channel***

The quantum operation of this channel is defined in Eq. (23.13). Assuming the orthogonal symbols $\rho_1 = |0\rangle\langle 0|$ and $\rho_2 = |1\rangle\langle 1|$, we obtain

$$\begin{aligned}&\varepsilon\left(\sum_i p_i\rho_i\right) = \varepsilon\left[p_1\rho_1 + (1-p_1)\rho_2\right] \\ &= pX\left[p_1\rho_1 + (1-p_1)\rho_2\right]X + (1-p)\left[p_1\rho_1 + (1-p_1)\rho_2\right] \\ &= pp_1X\rho_1X + p(1-p_1)X\rho_2X + (1-p)p_1\rho_1 + (1-p)(1-p_1)\rho_2 \\ &= \left\{\begin{array}{l} pp_1\begin{pmatrix}0&1\\1&0\end{pmatrix}\begin{pmatrix}1&0\\0&0\end{pmatrix}\begin{pmatrix}0&1\\1&0\end{pmatrix} + p(1-p_1)\begin{pmatrix}0&1\\1&0\end{pmatrix}\begin{pmatrix}0&0\\0&1\end{pmatrix}\begin{pmatrix}0&1\\1&0\end{pmatrix} \\ + (1-p)p_1\begin{pmatrix}1&0\\0&0\end{pmatrix} + (1-p)(1-p_1)\begin{pmatrix}0&0\\0&1\end{pmatrix}\end{array}\right\} \\ &= pp_1\begin{pmatrix}0&0\\0&1\end{pmatrix} + p(1-p_1)\begin{pmatrix}1&0\\0&0\end{pmatrix} + \begin{pmatrix}(1-p)p_1&0\\0&0\end{pmatrix} + \begin{pmatrix}0&0\\0&(1-p)(1-p_1)\end{pmatrix} \\ &\equiv \begin{pmatrix}p+p_1-2pp_1&0\\0&1-[p+p_1-2pp_1]\end{pmatrix}.\end{aligned} \tag{23.64}$$

and the corresponding VN entropy

$$\begin{aligned}A &= S\left[\varepsilon\left(\sum_i p_i\rho_i\right)\right] \\ &= -\left\{\begin{array}{l}(p+p_1-2pp_1)\log(p+p_1-2pp_1) \\ +\left[1-(p+p_1-2pp_1)\right]\log\left[1-(p+p_1-2pp_1)\right]\end{array}\right\} \\ &\equiv f(p+p_1-2pp_1).\end{aligned} \tag{23.65}$$

Next, we calculate the quantum operation $\varepsilon$ onto $\rho_1$ and $\rho_2$ along with the corresponding entropies:

$$\begin{aligned}
\varepsilon(\rho_1) &= pX\rho_1 X + (1-p)\rho_1 \\
&\equiv p\begin{pmatrix} 0 & 0 \\ 0 & 1 \end{pmatrix} + (1-p)\begin{pmatrix} 1 & 0 \\ 0 & 0 \end{pmatrix} \\
&\equiv \begin{pmatrix} 1-p & 0 \\ 0 & p \end{pmatrix}, \\
&\rightarrow \\
S[\varepsilon(\rho_1)] &= -[(1-p)\log(1-p) + p\log p] \\
&\equiv f(1-p) \\
&\equiv f(p)
\end{aligned} \tag{23.66}$$

$$\begin{aligned}
\varepsilon(\rho_2) &= pX\rho_2 X + (1-p)\rho 2 \\
&\equiv p\begin{pmatrix} 1 & 0 \\ 0 & 0 \end{pmatrix} + (1-p)\begin{pmatrix} 0 & 0 \\ 0 & 1 \end{pmatrix} \\
&\equiv \begin{pmatrix} p & 0 \\ 0 & 1-p \end{pmatrix} \\
&\rightarrow \\
S[\varepsilon(\rho_2)] &\equiv f(p).
\end{aligned} \tag{23.67}$$

Hence, we obtain

$$\begin{aligned}
B &= \sum_i p_i S[\varepsilon(\rho_i)] \\
&= p_1 f(p) + (1-p_1) f(p) \\
&\equiv f(p).
\end{aligned} \tag{23.68}$$

Combining the results in Eqs. (23.65) and (23.68), we finally obtain the channel capacity

$$\begin{aligned}
\chi(p, p_1) &= \max_{p_1}(\chi') \\
&\equiv \max_{p_1}(A - B) \\
&\equiv \max_{p_1}[f(p + p_1 - 2pp_1) - f(p)].
\end{aligned} \tag{23.69}$$

We note that the argument satisfies the property $\chi'(p, p_1) = \chi'(p, 1-p_1)$. The maximization problem can be solved analytically by studying the partial derivatives $\partial\chi'(p, p_1)/\partial p_1$ and $\partial^2\chi'(p, p_1)/\partial p_1^2$, but we may observe, again, that the capacity cannot exceed the amount $\chi_{\text{ideal}} = \max S(\rho)$. Using the definition of the symbol density operator $\rho = p_1\rho_1 + (1-p_1)\rho_2$, we clearly have $S(\rho) = f(p_1)$, and $\chi_{\text{ideal}} = \max S(\rho)$, yielding $p_1 = 1/2$, as expected (maximum VN entropy is obtained for a uniform message-symbol distribution). Thus, we have for all $p$ the channel capacity:

$$\begin{aligned}
\chi(p) &= \chi'\left(p, \frac{1}{2}\right) \\
&= [f(p + 1/2 - 2p/2) - f(p)] \\
&\equiv 1 - f(p).
\end{aligned} \tag{23.70}$$

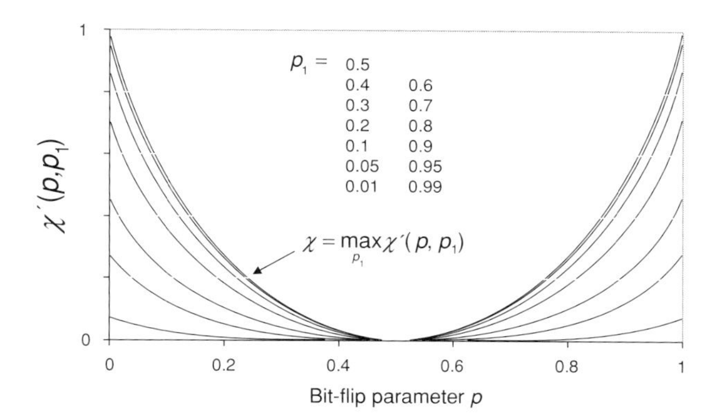


**Figure 23.3** Plots of $\chi'(p, p_1)$ as a function of the bit-flipping parameter $p$ in the bit-flip channel, and corresponding channel capacity $\chi(p) = \max \chi'(p, p_1)$ (codewords formed by pure states $|0\rangle, |+\rangle$ with probabilities $p_1, 1 - p_1$, respectively).

Figure 23.3 shows plots of $\chi'(p, p_1)$ for $p_1 = 0$ to $p_1 = 1$ through different increments (recalling that $\chi'(p, p_1) = \chi'(p, 1 - p_1)$), with the top curve $\chi(p) = \chi'(p, 1/2)$ representing the channel capacity. Interestingly, the channel capacity is seen to have two maxima, i.e., for $p = 0$ and $p = 1$. These two limiting cases correspond to the noiseless channel, and the "deterministic" bit-flip channel, respectively. A deterministic bit-flipping is simply a change of code polarity, meaning that the classical cbit codewords from originator to recipient are exactly inverted or complemented, which entails no information degradation or error. The other limiting situation is obtained for $p = 1/2$, meaning that the qubit has a 50% chance of being flipped and a 50% chance of being conserved in its integrity. The recipient's measurement amounts to a coin-flipping experiment, and all initial information is lost, which is the situation of the *useless channel*.

## 23.4 Exercises

**23.1** (B): Given the Pauli matrices $X, Y, Z$ (see definition in Chapter 16), show that for any $2 \times 2$ operator $A$ with unity trace, and for $q = 3p/4$, the following relation holds:

$$\frac{q}{3}(XAX + YAY + ZAZ) + (1 - q)A = p\frac{I}{2} + (1 - p)A.$$

**23.2** (M): Analyze the effect of the bit-flip quantum channel defined by

$$\sigma = \varepsilon(\rho) = pX\rho X + (1 - p)\rho,$$

with $p$ being a probability distribution and $X$ the Pauli matrix

$$X = \begin{pmatrix} 0 & 1 \\ 1 & 0 \end{pmatrix},$$

in the case where the density operator $\rho$ is nondiagonal. Determine the entropy change $\Delta S = S(\sigma) - S(\rho)$ in the case $\rho = |0\rangle\langle 0|$ and $\rho = |1\rangle\langle 1|$.

**23.3** (M): Analyze the effect of the phase-flip quantum channel as defined by

$$\sigma = \varepsilon(\rho) = pZ\rho Z + (1-p)\rho,$$

with $p$ being a probability distribution and $Z$ the Pauli matrix

$$Z = \begin{pmatrix} 1 & 0 \\ 0 & -1 \end{pmatrix},$$

in the case where the density operator $\rho$ is nondiagonal. Determine the entropy change $\Delta S = S(\sigma) - S(\rho)$ in the four cases $\rho = |0\rangle\langle 0|$, $\rho = |1\rangle\langle 1|$, $\rho = |+\rangle\langle +|$ and $\rho = |-\rangle\langle -|$.

**23.4** (M): Analyze the effect of the bit-phase-flip quantum channel as defined by

$$\sigma = \varepsilon(\rho) = pY\rho Y + (1-p)\rho,$$

with $p$ being a probability distribution and $Y$ the Pauli matrix

$$Y = \begin{pmatrix} 0 & -\mathrm{i} \\ \mathrm{i} & 0 \end{pmatrix},$$

in the case where the density operator $\rho$ is nondiagonal. Determine the entropy change $\Delta S = S(\sigma) - S(\rho)$ in the four cases $\rho = |0\rangle\langle 0|$, $\rho = |1\rangle\langle 1|$, $\rho = |+\rangle\langle +|$, and $\rho = |-\rangle\langle -|$.

**23.5** (M): Determine the quantum channel capacity corresponding to the depolarizing channel

$$\varepsilon(\rho_i) = p\frac{I}{3} + (1-p)\rho_i,$$

assuming the three-symbol alphabet

$$\rho_1 = |0\rangle\langle 0|,\ \rho_2 = |1\rangle\langle 1|,\ \rho_3 = |2\rangle\langle 2|$$

associated with probabilities $p_1, p_2, p_3$. Provide a plot of the capacity as a function of the depolarizing parameter $p$.

**23.6** (T): Determine the quantum channel capacity corresponding to the depolarizing channel

$$\varepsilon(\rho) = pZ\rho Z + (1-p)\rho,$$

assuming the two-symbol alphabet

$$\rho_1 = |0\rangle\langle 0|,\ \rho_2 = |+\rangle\,\langle +| = \frac{|0\rangle + |1\rangle}{\sqrt{2}}$$

associated with probabilities $p_1, p_2$.

*Hint*: plot

$$\chi'(p) = S(\langle\varepsilon\rangle) - \langle S(\varepsilon)\rangle$$

as a function of $p$, for different values of $p_1$ and justify that the capacity $\chi = \max(\chi')$ is achieved for $p_1 = p_2 = 1/2$; also show that $\chi(p) \leq \chi(0) = S(\rho)$, where $\chi(0)$ is the capacity of the constant or noiseless quantum channel and $S(\rho)$ is the VN entropy of the originator source.

# 24 Quantum error correction

This chapter deals with the subject of *quantum error correction* and the related *codes* (QECC), which can be applied to noisy quantum channels and quantum memories with the purpose of preserving or protecting the information integrity. I first describe the basics of *quantum repetition codes*, as applicable to *bit-flip* and *phase-flip* quantum channels. Then I consider the *9-qubit Shor code*, which has the capability of diagnosing and correcting any combination of bit-flip and phase-flip errors, up to one error of each type. Furthermore, it is shown that the Shor code is, in fact, capable of fully restoring qubit integrity under a continuum of bit or phase errors, a property that has no counterpart in the classical world of error-correction codes. But the exploration of QECC does not stop here! We shall discover the elegant *Calderbank–Shor–Steane* (CSS) codes, which have the capability of correcting any number of errors $t$, both bit-flip and phase-flip. As an application of the CSS code, I then describe the *7-qubit Hadamard–Steane code*, which can correct up to one error on single qubits. A corresponding quantum circuit, based on an original generator-matrix example, is presented.

## 24.1 Quantum repetition code

In Chapter 11, we saw that the simplest form of error-correction code (ECC) is the *repetition code*, based on the principle of *majority logic*. The background assumption is that in a given message sequence, or bit string, the probability of a bit error is sufficiently small for the majority of bits to be correctly transmitted through the channel. It then suffices to repeat each of the bits a certain number of times, at the cost of wasting the channel resource or "bandwidth." For instance, if one repeats each of the bits four times, the original bit sequence 0111 is encoded into (underscores introduced for reading clarity):

$$0000\_1111\_1111\_1111,$$

which may yield at the channel output

$$0010\_1111\_0111\_1011.$$

Clearly, a single bit error occurred in the first and last two blocks of the transmitted message, and the simple rule of *majority logic* suffices to detect such errors and revert the bits to the correct values. If $p$ is the probability of error for a single bit, the probability that there is more than one error in a single four-bit block is

$$\begin{aligned} p^{(4)} &= C_4^2 p^2(1-p)^2 + C_4^3 p^3(1-p) + C_4^4 p^4(1-p)^0 \\ &= 6p^2(1-p)^2 + 4p^3(1-p) + p^4 \\ &\equiv 4p^3 + 6p^2 - 9p^4. \end{aligned} \tag{24.1}$$

It is easily checked that $p^{(4)} < p$ for $p < 0.155$, and $p^{(4)} \le p/10$ for $p \le 0.0165$, hence, this repetition code is *reliable* (i.e., errors can be reliably corrected) provided $p$ is chosen to be sufficiently small.

Here, we shall apply the principle of repetition codes to *quantum channels*, which defines the new notion of *quantum error-correction coding* (QECC). We may conceptually expand our view of a "quantum channel" beyond that of a quantum communication system and conceive that it may also correspond to that of a *quantum memory*. Whether information is transmitted or stored, the principle remains the same: an originator must encode the information and input the result into a physical system, and a recipient must do the reverse operation. We may, thus, equivalently refer to either transmission or storage "fidelity" to qualify the quantum channel or system into or through which information is being passed. Such a notion will be developed later.

As an example of a noisy quantum system, consider the *bit-flip channel* described in Chapter 23. To recall, the bit-flip channel converts the qubits $|0\rangle$ and $|1\rangle$ into each other with probability $p$. Hence, the original qubit $|q\rangle = \alpha|0\rangle + \beta|1\rangle$ has probability $p$ of being transformed by the quantum channel into $|q\rangle = \alpha|1\rangle + \beta|0\rangle$. Based on the principle of the four-bit repetition code, it is then sensible to encode the input qubit $|q\rangle$ into

$$|\hat{q}\rangle = \alpha|0000\rangle + \beta|1111\rangle, \tag{24.2}$$

for instance (we could have chosen a 3-qubit repetition code as well). This is equivalent to encoding the basis states $|0\rangle, |1\rangle$ in the form

$$\begin{cases} |0\rangle \to |\hat{0}\rangle = |0000\rangle \\ |1\rangle \to |\hat{1}\rangle = |1111\rangle, \end{cases} \tag{24.3}$$

where $|\hat{0}\rangle, |\hat{1}\rangle$ have the meaning of the logical $|0\rangle, |1\rangle$ qubits. The quantum circuit for this repetition code, which is based on three CNOT gates (see Chapter 15), is shown in Fig. 24.1. After passing through the bit-flip channel, the 4-qubit basis states $|0000\rangle, |1111\rangle$ might have experienced zero, one, two, three, or four qubit flips or errors. We consider here the first two possibilities, whose detailed outcomes are summarized in Table 24.1. From the recipient's end, the detection and correction of errors is a matter of *syndrome diagnosis*. The notion of "syndrome" for linear block codes and cyclic codes was described in Chapter 11. Here, it is similarly applied to quantum error correction. The first step is to detect the existence of errors, namely of the occurrence of single qubit flips in the received codeword, which is achieved by means of an *error syndrome* measurement. Such a measurement consists of projecting the received codeword using

**Table 24.1** Outcome possibilities for the basis states $|0000\rangle$, $|1111\rangle$ on passing through a bit-flip channel, assuming up to one flip.

| Input | Output | Event |
|---|---|---|
| $\|0000\rangle, \|1111\rangle$ | $\|0000\rangle, \|1111\rangle$ | No bit flip |
| | $\|1000\rangle, \|0111\rangle$ | Bit flip on 1st qubit |
| | $\|0100\rangle, \|1011\rangle$ | Bit flip on 2nd qubit |
| | $\|0010\rangle, \|1101\rangle$ | Bit flip on 3rd qubit |
| | $\|0001\rangle, \|1110\rangle$ | Bit flip on 4th qubit |

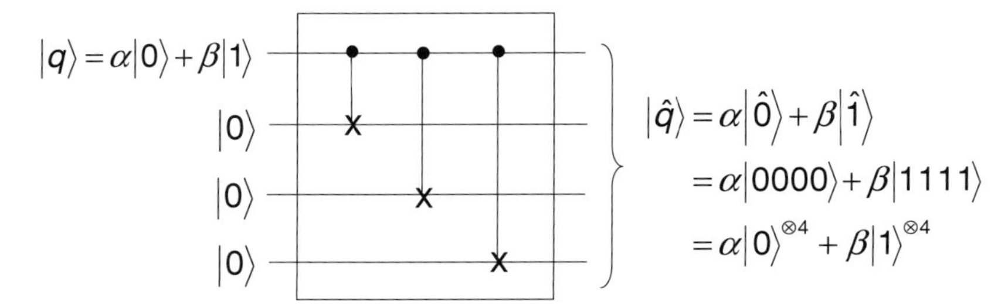


**Figure 24.1** Quantum circuit for encoding a qubit $|q\rangle$ into a 4-qubit repetition codeword $|\hat{q}\rangle$ before transmission to a bit-flip channel.

the following projector set:

$$\begin{cases} P_0 = |0000\rangle\langle 0000| + |1111\rangle\langle 1111| \\ P_1 = |1000\rangle\langle 1000| + |0111\rangle\langle 0111| \\ P_2 = |0100\rangle\langle 0100| + |1011\rangle\langle 1011| \\ P_3 = |0010\rangle\langle 0010| + |1101\rangle\langle 1101| \\ P_4 = |0001\rangle\langle 0001| + |1110\rangle\langle 1110|. \end{cases} \tag{24.4}$$

Assume that the received qubit is $|\hat{q}'\rangle = \alpha|0010\rangle + \beta|1101\rangle$. It is clear that a projection measurement through $P_3$ yields the certain outcome, or "expectation value" $\langle P_3\rangle = \langle\hat{q}'|P_3|\hat{q}'\rangle = 1$, with the *post-measurement* state being $|\hat{q}^*\rangle$. We note that the measurement does not yield any information about the amplitudes $\alpha, \beta$, thus, leaving the codeword unknown. Yet the measurement tells which type of error is to be corrected, and the original codeword can be restored to its full integrity by flipping again the errored qubit $i$ should any $P_{i=1,2,3,4}$ measurement yield $\langle\hat{q}'|P_i|\hat{q}'\rangle = 1$, while the measurement $\langle\hat{q}'|P_0|\hat{q}'\rangle = 1$ indicates with certainty that there are no errors.

The effectiveness of a quantum error-correction code can be qualified through *fidelity*, a notion which was introduced in Chapter 22. Assuming a noisy channel $\varepsilon$ with the encoded originator message $\rho = |\hat{q}\rangle\langle\hat{q}|$ and the recipient message $\varepsilon(\rho) = \rho'$, the fidelity can be defined as follows:

$$F = \sqrt{\langle\hat{q}|\rho'|\hat{q}\rangle}. \tag{24.5}$$

In the case of the bit-flip channel, we have (Eq. (23.11)):

$$\begin{aligned}\rho' &= \varepsilon(\rho) \\ &= pX\rho X + (1-p)\rho \\ &= pX|\hat{q}\rangle\langle\hat{q}|X + (1-p)|\hat{q}\rangle\langle\hat{q}|.\end{aligned} \tag{24.6}$$

From the definition in Eq. (24.5), the corresponding fidelity of the code is

$$\begin{aligned}F &= \sqrt{\langle\hat{q}|\rho'|\hat{q}\rangle} \\ &= \sqrt{\langle\hat{q}|pX|\hat{q}\rangle\langle\hat{q}|X|\hat{q}\rangle + (1-p)\langle\hat{q}|\hat{q}\rangle\,\langle\hat{q}|\hat{q}\rangle} \\ &\equiv \sqrt{p\,|\langle\hat{q}|X|\hat{q}\rangle|^2 + (1-p)}.\end{aligned} \tag{24.7}$$

Consider now some cases of interest. If we use no error correction, the two basic sets of pure-state symbols we can use are $|q\rangle = |0\rangle, |1\rangle$ and $|q\rangle = |+\rangle, |-\rangle$. As we know, the action of the Pauli matrix $X$ is to flip the states $|0\rangle, |1\rangle$ or $X|0\rangle = |1\rangle$ and $X|1\rangle = |0\rangle$, yielding $\langle q|X|q\rangle = 0$ and $F = \sqrt{1-p}$ in the first case. The fidelity, thus, ranges from $F = 1$ (noiseless channel, $p = 0$) to $F = 0$ ("certain" bit-flip channel, $p = 1$). In the second case, the action of $X$ leaves the states $|+\rangle, |-\rangle$ invariant, i.e., $X|+\rangle = (X|0\rangle + X|1\rangle)/\sqrt{2} = (|1\rangle + |0\rangle)/\sqrt{2} \equiv |+\rangle$ and $X|-\rangle = (X|0\rangle - X|1\rangle)/\sqrt{2} = (|1\rangle - |0\rangle)/\sqrt{2} = e^{i\pi}|-\rangle \equiv |-\rangle$, which yields $|\langle q|X|q\rangle| = 1$ and, hence, $F = 1$. The fidelity is, thus, maximal, even if the channel is "noisy" and no error-correction coding is implemented! Consider, next, the case with the error-correction code implementation according to Eq. (24.2), i.e., $|\hat{q}\rangle = \alpha|\hat{0}\rangle + \beta|\hat{1}\rangle$, with $\rho = |\hat{q}\rangle\langle\hat{q}|$. We now have a noisy quantum channel that has the capability of corrupting one, two, or more qubits, which calls for a definition of $\varepsilon(\rho)$ that is, in fact, different from Eq. (24.6). For each corruption pattern $x$, we need to introduce the corresponding quantum operation $\varepsilon_x(\rho)$. Here are the different corruption patterns with their operations:

$$\text{No error: } \varepsilon_0(\rho) = (1-p)^4 I^{\otimes 4}\rho I^{\otimes 4}, \tag{24.8}$$

$$\text{Exactly one error: } \varepsilon_a(\rho) = p(1-p)^3 \begin{pmatrix} X\otimes I^{\otimes 3}\rho X\otimes I^{\otimes 3} \\ +I\otimes X\otimes I^{\otimes 2}\rho I\otimes X\otimes I^{\otimes 2} \\ +I^{\otimes 2}\otimes X\otimes I\rho I^{\otimes 2}\otimes X\otimes I \\ +I^{\otimes 3}\otimes X\rho I^{\otimes 3}\otimes X \end{pmatrix}, \tag{24.9}$$

$$\text{Exactly two errors: } \varepsilon_b(\rho) = p^2(1-p)^2 \begin{pmatrix} X^{\otimes 2}\otimes I^{\otimes 2}\rho X^{\otimes 2}\otimes I^{\otimes 2} \\ +X\otimes I\otimes X\otimes I\rho X\otimes I\otimes X\otimes I \\ +X\otimes I^{\otimes 2}\otimes X\rho X\otimes I^{\otimes 2}\otimes X \\ +I\otimes X^{\otimes 2}\otimes I\rho I\otimes X^{\otimes 2}\otimes I \\ +I\otimes X\otimes I\otimes X\rho I\otimes X\otimes I\otimes X \\ +I^{\otimes 2}\otimes X^{\otimes 2}\rho I^{\otimes 2}\otimes X^{\otimes 2} \end{pmatrix}, \tag{24.10}$$

$$\text{Exactly three errors: } \varepsilon_c(\rho) = p^3(1-p) \begin{pmatrix} X^{\otimes 3}\otimes I\rho X^{\otimes 3}\otimes I \\ +X^{\otimes 2}\otimes I\otimes X\rho X^{\otimes 2}\otimes I\otimes X \\ +X\otimes I\otimes X^{\otimes 2}\rho X\otimes I\otimes X^{\otimes 2} \\ +I\otimes X^{\otimes 3}\rho I\otimes X^{\otimes 3} \end{pmatrix}, \tag{24.11}$$

$$\text{Exactly four errors: } \varepsilon_d(\rho) = p^4 X^{\otimes 4}\rho X^{\otimes 4}, \tag{24.12}$$

which makes up the overall channel definition

$$\rho' = \varepsilon(\rho) = \varepsilon_0(\rho) + \varepsilon_a(\rho) + \varepsilon_b(\rho) + \varepsilon_c(\rho) + \varepsilon_d(\rho). \tag{24.13}$$

Here, there is no point in developing the full expression of the recipient's qubit by substituting $\rho = |\hat{q}\rangle\langle\hat{q}| = (\alpha|\hat{0}\rangle + \beta|\hat{1}\rangle)(\bar{\alpha}\langle\hat{0}| + \bar{\beta}\langle\hat{1}|)$, which generates no less than $4 \times 16 = 64$ operator terms! Rather, it is sensible to conclude directly that *after* error-correction decoding, which would ideally detect and correct all possible error patterns, the transmitted qubit state takes the form:

$$\begin{aligned}(\rho')^* &= \varepsilon^*(\rho)\\ &= [(1-p)^4 + 4p(1-p)^3 + 6p^2(1-p)^2 + 4p^3(1-p) + p^4]|\hat{q}\rangle\langle\hat{q}|\\ &\equiv |\hat{q}\rangle\langle\hat{q}|.\end{aligned} \tag{24.14}$$

If the QECC only has the capability of correcting *up to a single error* (using the projectors $P_{i=1,2,3,4}$), we may write

$$\varepsilon^*(\rho) = [(1-p)^4 + 4p(1-p)^3 + \ldots]|\hat{q}\rangle\langle\hat{q}|, \tag{24.15}$$

with the missing terms corresponding to patterns of more than one error (see further in discussion about correcting higher-order error patterns). In this case, the fidelity is given by

$$\begin{aligned}F(p) &= \sqrt{\langle\hat{q}|\varepsilon^*(\rho)|\hat{q}\rangle}\\ &\equiv \sqrt{(1-p)^4 + 4p(1-p)^3 + \ldots} \geq \sqrt{(1-p)^4 + 4p(1-p)^3}\\ &= F_{\min}.\end{aligned} \tag{24.16}$$

In the above, $F_{\min}$ represents a lower bound for the fidelity corresponding to an error-correction capability of up to one qubit. We may compare the minimum fidelity $F_{\min}$ to that corresponding to the transmission of $|0\rangle$, $|1\rangle$ *without* error-correction coding, and for which $F = \sqrt{1-p}$, as established earlier. It is found numerically that $F_{\min} > F$ under the condition $p < p_{\max} = 0.23$, as illustrated in Fig. 24.2. This result means that for channels with $p \geq 0.23$, there is no point or merit whatsoever in implementing this quantum repetition code! In comparison, the 3-qubit repetition code yields a minimum fidelity of $F_{\min} = \sqrt{(1-p)^3 + 3p(1-p)^2}$ (which is straightforward to establish), to which the condition $p < 0.5 = p_{\max}$ corresponds. The fact that the channel bit-flip probability $p$ may not exceed some threshold value $p_{\max}$, and that this threshold decreases as we may increase the number of repeated qubits, nicely illustrates the intrinsic limitations of the quantum repetition code.

As we have seen from the above description, implementing the repetition code with longer codewords increases the chance of obtaining more than one (bit-flip) error. How about the possibility of correcting these higher-order error patterns? We may easily convince ourselves that, unfortunately, the repetition code does not offer such a possibility. Indeed, consider, for instance, the projection operator $P = |0011\rangle\langle 0011| + |1100\rangle\langle 1100|$. Using such a projection will tell with certainty that there are two errors. But there is no way of telling whether these errors occurred on the first two or the last two

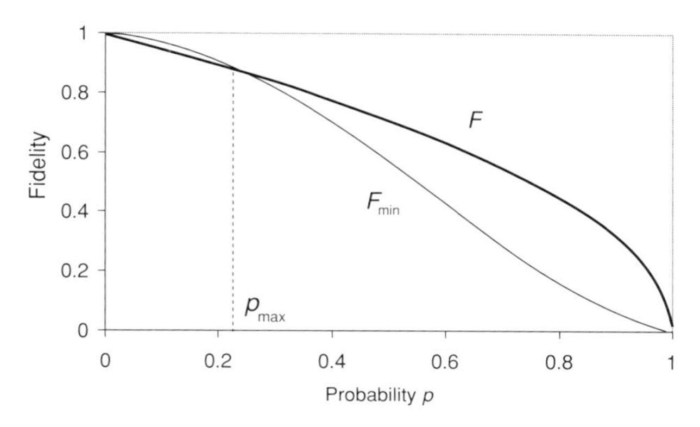


**Figure 24.2** Fidelity of 4-qubit repetition coding in bit-flip channel for no error correction ($F$), and up to one error correction ($F_{\min}$), showing improvement in the region $p < p_{\max} = 0.23$.

qubits, these two events being equiprobable. The same ambiguity prevails with the two other possible projectors, $P = |0101\rangle\langle 0101| + |1010\rangle\langle 1010|$ and $P = |0110\rangle\langle 0110| + |1001\rangle\langle 1001|$. The detection of three errors would take the same projectors as $P_{i=1,2,3,4}$ defined in Eq. (24.4), but this time with an ambiguity in favor of a single-error event. This shows that *the quantum repetition code is inadequate for higher-order error detection and correction.*

How about implementing the quantum repetition code in the *phase-flip channel*? In Chapter 11, we saw that the phase-flip channel transforms a qubit $|q\rangle = \alpha|1\rangle + \beta|0\rangle$ into $|q\rangle = \alpha|1\rangle - \beta|0\rangle$, or introduces a dephasing factor $e^{i\pi}$ between the two complex amplitudes $\alpha, \beta$. Clearly, the effect of this channel is to flip the states $|+\rangle$ and $|-\rangle$ into each other. Therefore, we can apply the same repetition-code principle as described previously, using, this time (and for instance), the 4-qubit encoding scheme:

$$\begin{cases} |+\rangle \to |\hat{+}\rangle = |++++\rangle \\ |-\rangle \to |\hat{-}\rangle = |----\rangle, \end{cases} \tag{24.17}$$

where $|\hat{+}\rangle, |\hat{-}\rangle$ have the meaning of the logical $|+\rangle, |-\rangle$ qubits. In the basis representation $\{|+\rangle, |-\rangle\}$, there is strictly no difference between this quantum repetition code and that applying to the bit-flip channel in the basis representation $\{|0\rangle, |1\rangle\}$. But we may use the code in Eq. (24.17) in the latter basis by means of *Hadamard gates*. As we have seen in Chapter 15, the Hadamard gate $H$ performs the following transformations:

$$\begin{cases} H|0\rangle = |+\rangle \\ H|1\rangle = |-\rangle \\ H|+\rangle = |0\rangle \\ H|-\rangle = |1\rangle. \end{cases} \tag{24.18}$$

The encoding circuit for this repetition code is similar to the one concerning the bit-flip channel (Fig. 24.1) with the inclusion of three Hadamard gates to effect this conversion, as shown in Fig. 24.3. The originator qubit $|q\rangle = \alpha|1\rangle + \beta|0\rangle$ is, thus, transformed into $|q\rangle = \alpha|\hat{+}\rangle + \beta|\hat{-}\rangle$.

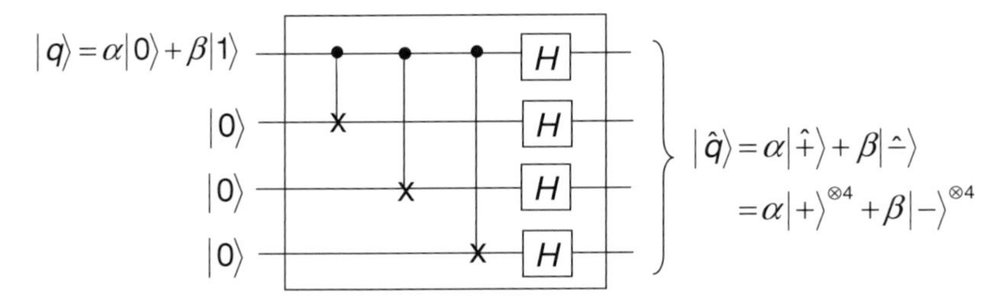


**Figure 24.3** Quantum circuit for encoding a qubit $|q\rangle$ into a 4-qubit repetition codeword $|\hat{q}\rangle$ before transmission to a phase-flip channel.

The error syndrome measurement can be performed with the same projector set $P_{i=1,2,3,4}$ as defined in Eq. (24.4), but with the 1, 0 symbols inside the bra–kets $\langle\cdot|$ / $|\cdot\rangle$ being switched into $+, -$, namely, $P_0 = |++++\rangle\langle++++| + |----\rangle\langle----|$, and so on. Another possibility is to pass each of the received codeword qubits through a Hadamard gate, corresponding to the operation $H^{\otimes 4}\varepsilon(|\hat{q}\rangle\langle\hat{q}|)$, which allows one to perform the error syndrome measurement through the same projector set $P_{i=1,2,3,4}$ as in the bit-flip channel. Clearly, the fidelity characteristics and limitations of the quantum repetition code for the phase-flip channel are strictly the same as in the case of the bit-flip channel.

I shall conclude this section by showing an important property, according to which the projector set $P_{i=1,2,3,4}$ is not the only tool to perform error syndrome measurements in bit- or phase-flip channels. This will be very useful in the next section, concerning the *Shor code*. Consider indeed that we have the possibility of performing partial measurements with the aim of comparing qubit *pairs*. For instance, we may compare the first and the second qubit, or the second and the third, or the first and the third, and so on. Should any comparison reveal a difference, then an error is detected. One must then perform as many such comparisons as necessary to obtain an unambiguous error syndrome, leading to certain error correction. Let us see how this approach can be implemented. With a 3-qubit bit-flip code, comparing qubit pairs consists of projecting the received state $\hat{\sigma} = \varepsilon(|\hat{q}\rangle\langle\hat{q}|)$ by using the following "observable" operators:

$$\begin{cases} Z_{12} = (|00\rangle\langle 00| + |11\rangle\langle 11|) \otimes I - (|01\rangle\langle 01| + |10\rangle\langle 10|) \otimes I \\ Z_{23} = I \otimes (|00\rangle\langle 00| + |11\rangle\langle 11|) - I \otimes (|01\rangle\langle 01| + |10\rangle\langle 10|) \\ Z_{13} = |0\rangle\langle 0| \otimes I \otimes |0\rangle\langle 0| + |1\rangle\langle 1| \otimes I \otimes |1\rangle\langle 1| \\ \qquad - (|0\rangle\langle 0| \otimes I \otimes |1\rangle\langle 1| + |1\rangle\langle 1| \otimes I \otimes |0\rangle\langle 0|)\,. \end{cases} \tag{24.19}$$

It is clear that the expectation value $\langle Z_{ij}\rangle = \langle\hat{q}|Z_{ij}|\hat{q}\rangle$ for any of the above three operators $Z_{ij}$ is equal to $+1$, should the qubits $i$ and $j$ match (no error), while it is equal to $-1$, should they not match (error). Let us leave here the consideration of being capable of physically measuring *observables*, such as $\pm 1$. The matter is simply that the partial measurement yields two possible logical answers: *error or no error* on the pair of qubits considered, corresponding to classical YES or NO information. Such information is the key to the elaboration of an error diagnostic, leading to corrective action. Since a single measurement (or YES or NO information) is not sufficient to tell which of the two qubits is actually errored (or flipped), we may perform as many successive measurements as required to reach an unambiguous error diagnostic. The nice feature of the approach is that these projective measurements do not alter the codeword, i.e., they leave it in the same post-measurement state! In the 3-qubit coding example, let us see that it merely suffices to perform two such measurements in a row. Indeed, assume we obtain the two measurements $\langle Z_{12}\rangle$ and $\langle Z_{13}\rangle$. If $\langle Z_{12}\rangle = \langle Z_{13}\rangle = 1$, there is no error and, therefore, no action is required. If $\langle Z_{12}\rangle = 1$ and $\langle Z_{13}\rangle = -1$, we have detected an error on the third qubit and, therefore, the corrective action is to flip it. We may reach the same conclusion for the measurement pair $\langle Z_{12}\rangle, \langle Z_{23}\rangle$ or $\langle Z_{13}\rangle, \langle Z_{23}\rangle$. Thus, any arbitrary choice of only two successive measurements of this type altogether yields $2^2 = 4$ possible syndrome diagnostics, covering all patterns from zero error to one single error concerning any of the three qubits. The result is the same as using the measurement set $P_{i=1,2,3,4}$, except that only two projection operators instead of four have been used. In any case, it is important to recall that the output codeword $|q'\rangle$ and the post-measurement states $P_i|q'\rangle$ or $Z_{ij}|q'\rangle$ are identical.

It is now only a straightforward technical matter to describe how the $Z_{ij}$ projectors defined in Eq. (24.19) can be realized. We recall from Chapter 16 the $Z$-gate, also referred to as the $\sigma_3$ or $\sigma_z$ *Pauli matrix*, whose action is to introduce a $\pi$ phase flip between the qubit amplitudes (see Table 16.2). It is left as an easy exercise to show that the measurement projector $Z_{ij}$, as defined in Eq. (24.19), in fact summarizes into the simultaneous application of $Z$ onto the qubits $i$ and $j$ in the codeword to be analyzed. Namely,

$$\begin{cases} Z_{12} = Z \otimes Z \otimes I \equiv Z^{\otimes 2} \otimes I \\ Z_{23} = I \otimes Z \otimes Z \equiv I \otimes Z^{\otimes 2} \\ Z_{13} = Z \otimes I \otimes Z. \end{cases} \tag{24.20}$$

We may write the above definitions in the compact form: $Z_{12} \equiv Z_1 Z_2 I_3 \equiv Z_1 Z_2$, $Z_{23} \equiv I_1 Z_2 Z_3 \equiv Z_2 Z_3$, and $Z_{13} \equiv Z_1 I_2 Z_3 \equiv Z_1 Z_3$, with the indices in the right-hand side referring to the action of the operator onto the corresponding qubits, and with the identity matrix applying to the remaining qubit being overlooked.

## 24.2 Shor code

In this section, I describe the *Shor code*, a QECC that concatenates the features of the bit-flip and phase-flip repetition codes, as implemented with *9-qubit* codewords. *The nice feature of the Shor code is that it can correct any single bit-flip or phase-flip error*, and as a matter of fact, *any error*, as will be established. The encoding principle proceeds as follows. First, consider the 3-qubit *phase-flip* repetition code. The quantum circuit is the same as shown in Fig. 24.3, with the last wire removed. As we have seen earlier, this code performs the following transformation:

$$\begin{cases} |0\rangle \to |\hat{0}\rangle = |+++\rangle \\ |1\rangle \to |\hat{1}\rangle = |---\rangle. \end{cases} \tag{24.21}$$

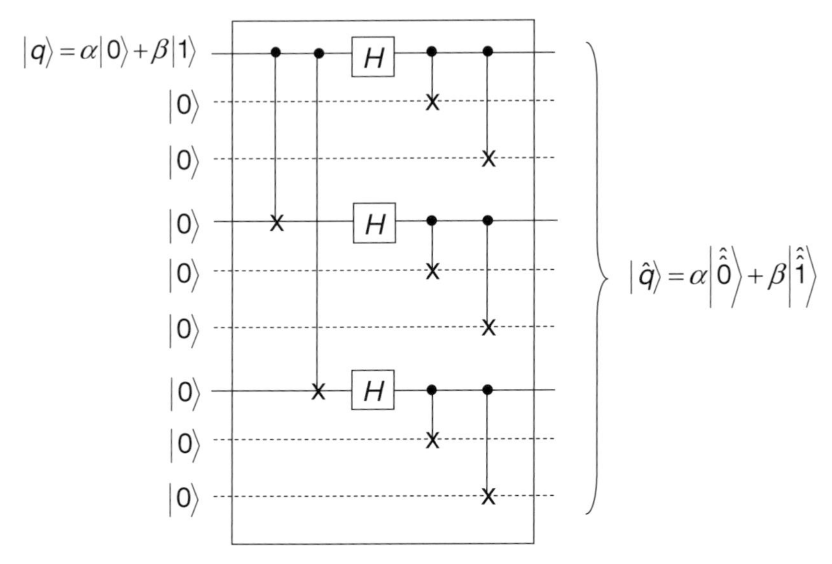


**Figure 24.4** Quantum circuit for encoding a qubit $|q\rangle$ into a 9-qubit repetition codeword $|\hat{\hat{q}}\rangle$ before transmission to a bit-phase-flip channel.

In a second stage, we consider the 3-qubit *bit-flip* repetition code. The quantum circuit is the same as shown in Fig. 24.1, with the last wire removed. Clearly, such a circuit encodes the qubits $|+\rangle = (|0\rangle + |1\rangle)/\sqrt{2}$ and $|+\rangle = (|0\rangle - |1\rangle)/\sqrt{2}$ according to

$$\begin{cases} |+\rangle \to |\hat{+}\rangle = \dfrac{|000\rangle + |111\rangle}{\sqrt{2}} \\ |-\rangle \to |\hat{-}\rangle = \dfrac{|000\rangle - |111\rangle}{\sqrt{2}}. \end{cases} \tag{24.22}$$

We notice from the above definition that the two qubits $|\hat{+}\rangle, |\hat{-}\rangle$ are *entangled*. They represent the 3-qubit extension of the *EPR* or *Bell states* $|\beta_{00}\rangle = (|00\rangle + |11\rangle)/\sqrt{2}$ and $|\beta_{10}\rangle = (|00\rangle - |11\rangle)/\sqrt{2}$, which were described in Chapter 18. Such a property will be used later, when considering error correction. The end result of applying successively, or *concatenating* the two encoding phases, as defined in Eqs. (24.19) and (24.20), is the following:

$$\begin{cases} |0\rangle \to \left|\hat{\hat{0}}\right\rangle = \dfrac{|000\rangle + |111\rangle}{\sqrt{2}} \dfrac{|000\rangle + |111\rangle}{\sqrt{2}} \dfrac{|000\rangle + |111\rangle}{\sqrt{2}} \\ |1\rangle \to \left|\hat{\hat{1}}\right\rangle = \dfrac{|000\rangle - |111\rangle}{\sqrt{2}} \dfrac{|000\rangle - |111\rangle}{\sqrt{2}} \dfrac{|000\rangle - |111\rangle}{\sqrt{2}}. \end{cases} \tag{24.23}$$

The corresponding encoding circuit is shown in Fig. 24.4. The first stage of the circuit, which includes two CNOT and three $H$ gates, performs the phase-flip repetition coding as defined in Eq. (24.19). The second stage, which includes six CNOT gates, performs the bit-flip repetition coding as defined in Eq. (24.20). For clarity, the ancillary qubits used for the second stage are shown with dashed lines.

**Table 24.2** Corrective actions in Shor code from phase-flip syndrome diagnostic based on measurements of $(X_{123456}, X_{456789})$ and the principle of majority logic.

| $\langle X_{123456}\rangle_{AB}$ | $\langle X_{456789}\rangle_{BC}$ | Corrective action |
|---|---|---|
| 1 | 1 | None |
| 1 | −1 | Flip phase in block $C$ |
| −1 | 1 | Flip phase in block $A$ |
| −1 | −1 | Flip phase in block $B$ |

In the Shor code, syndrome diagnosis and error correction works in a way conceptually similar to that in the bit-flip and phase-flip repetition codes described in the previous section, except that here we want to identify and correct *both types of error*. The task appears a bit tedious because the codewords are made of the *9-qubit* states, $|\hat{\hat{0}}\rangle$ and $|\hat{\hat{1}}\rangle$, defined in Eq. (24.23). I shall first consider the occurrences of phase-flip and bit-flip separately, and eventually show that both types of event can be corrected.

## Phase-flip error

For instance, calling the three 3-qubit blocks $A, B, C$, if a single phase-flip occurs on *any* qubit within block $A$, the channel outputs:

$$\begin{cases} |\hat{\hat{0}}\rangle \to |\hat{\hat{0}}'\rangle = \left.\dfrac{|000\rangle - |111\rangle}{\sqrt{2}}\right|_A \left.\dfrac{|000\rangle + |111\rangle}{\sqrt{2}}\right|_B \left.\dfrac{|000\rangle + |111\rangle}{\sqrt{2}}\right|_C \\ |\hat{\hat{1}}\rangle \to |\hat{\hat{1}}'\rangle = \left.\dfrac{|000\rangle + |111\rangle}{\sqrt{2}}\right|_A \left.\dfrac{|000\rangle - |111\rangle}{\sqrt{2}}\right|_B \left.\dfrac{|000\rangle - |111\rangle}{\sqrt{2}}\right|_C . \end{cases} \tag{24.24}$$

The first step in the syndrome diagnosis consists of comparing the relative phase of blocks $A, B$ then of blocks $B, C$ (or any other two pair selection), just like in the 1-qubit phase-flip channel case described earlier. Majority logic then dictates which block may have been corrupted and where the phase correction applies. It is left as an (easy) exercise to show that comparing the signs of $A, B$ blocks can be performed by measuring $X_{123456} = X_1X_2X_3X_4X_5X_6$, and likewise for the $B, C$ blocks with $X_{456789} = X_4X_5X_6X_7X_8X_9$, where $X_i$ is the Pauli matrix $X$ applied to the qubit $i$. The entire syndrome diagnostic, leading to the proper corrective action on the concerned block phase, thus, consists of performing the two measurements $(X_{123456}, X_{456789})$, while the other possibility is $(X_{123456}, X_{123789})$, both being applicable in any measurement order. Table 24.2 summarizes the corrective actions from the syndrome diagnostic based on $(X_{123456}, X_{456789})$.

To restore the phase in the corrupted block, say, for instance, $A$, the corrective action is to apply the operator $Z_1Z_2Z_3$. Indeed, because of the properties $Z|0\rangle = |0\rangle$ and $Z|1\rangle = -|1\rangle$, the phase of the 3-qubit block $A$ is flipped according to

$$\begin{aligned} Z_1Z_2Z_3\,(|000\rangle - |111\rangle) &= |000\rangle - (-1)^3|111\rangle \\ &\equiv |000\rangle + |111\rangle. \end{aligned} \tag{24.25}$$

## Bit-flip error

The syndrome diagnosis for bit-flip errors is basically the same as described in the earlier section. As in the phase-flip case, the diagnosis may proceed by blocks, considering successively the blocks $A, B, C$. As we have seen, the procedure consists of comparing qubit pairs $i, j$ and $j, k$ by means of $Z_{ij} = Z_i Z_j$ and $Z_{jk} = Z_j Z_k$ measurements. Considering any block $U = A, B, C$, assume for instance a bit-flip error happening on the second qubit of $|\hat{\hat{0}}\rangle_U$, such that the channel outputs:

$$\left|\hat{\hat{0}}'\right\rangle_U = \frac{|010\rangle + |101\rangle}{\sqrt{2}}. \tag{24.26}$$

We ought to notice that a single bit-flip error in $(|000\rangle + |111\rangle)/\sqrt{2}$ corrupts not only one but *both* qubit terms in the sum, because the state is *entangled*, as we have mentioned earlier. Thus, there is no possibility to have an error pattern of the type $(|010\rangle + |111\rangle)/\sqrt{2}$ where the second term would remain uncorrupted. This is a nice feature of the Shor code, which is going to be exploited in the syndrome diagnostic. To show this, we first apply $Z_{12}$ on block $U$ to obtain:

$$\begin{aligned}
&Z_{12}\left|\hat{\hat{0}}'\right\rangle_U \\
&= [(|00\rangle\langle 00| + |11\rangle\langle 11|) \otimes I - (|01\rangle\langle 01| + |10\rangle\langle 10|) \otimes I]\,\frac{|010\rangle + |101\rangle}{\sqrt{2}} \\
&= (|00\rangle\langle 00| + |11\rangle\langle 11|) \otimes I\,\frac{|010\rangle + |101\rangle}{\sqrt{2}} - (|01\rangle\langle 01| + |10\rangle\langle 10|) \otimes I\,\frac{|010\rangle + |101\rangle}{\sqrt{2}} \\
&= \begin{pmatrix} |00\rangle\langle 00| \otimes I\,\dfrac{|010\rangle + |101\rangle}{\sqrt{2}} + |11\rangle\langle 11| \otimes I\,\dfrac{|010\rangle + |101\rangle}{\sqrt{2}} \\ - |01\rangle\langle 01| \otimes I\,\dfrac{|010\rangle + |101\rangle}{\sqrt{2}} - |10\rangle\langle 10| \otimes I\,\dfrac{|010\rangle + |101\rangle}{\sqrt{2}} \end{pmatrix} \\
&= 0 + 0 - \frac{|010\rangle}{\sqrt{2}} - \frac{|101\rangle}{\sqrt{2}} \\
&\equiv -\frac{|010\rangle + |101\rangle}{\sqrt{2}}.
\end{aligned} \tag{24.27}$$

Hence, the measurement

$$\begin{aligned}
\langle Z_{12}\rangle_U &= \left\langle\hat{\hat{0}}'\right| Z_{12}\left|\hat{\hat{0}}'\right\rangle_U \\
&= \frac{\langle 010| + \langle 101|}{\sqrt{2}}\left(-\frac{|010\rangle + |101\rangle}{\sqrt{2}}\right) \\
&\equiv -1.
\end{aligned} \tag{24.28}$$

We notice here that the post-measurement state is the same as the measured state $|\hat{\hat{0}}'\rangle$, within an unobservable phase factor $e^{i\pi}$. Hence, the possibility of continuing the

**Table 24.3** Corrective actions in Shor code from bit-flip syndrome diagnostic based on measurements of $(Z_{12}, Z_{23})$ in entangled 3-qubit block $U$, and the principle of majority logic.

| $\langle Z_{12}\rangle_U$ | $\langle Z_{23}\rangle_U$ | Corrective action |
|---|---|---|
| 1 | 1 | None |
| 1 | −1 | Flip 3rd qubit |
| −1 | 1 | Flip 1st qubit |
| −1 | −1 | Flip 2nd qubit |

syndrome measurements! Next, we can apply $Z_{23}$ on block $U$ to obtain, in turn:

$$\begin{aligned}
Z_{23}\left|\hat{\hat{0}}'\right\rangle_U &= [I\otimes(|00\rangle\langle 00| + |11\rangle\langle 11|) - I\otimes(|01\rangle\langle 01| + |10\rangle\langle 10|)]\,\frac{|010\rangle + |101\rangle}{\sqrt{2}} \\
&= \begin{pmatrix} I\otimes|00\rangle\langle 00|\dfrac{|010\rangle + |101\rangle}{\sqrt{2}} + I\otimes|11\rangle\langle 11|\dfrac{|010\rangle + |101\rangle}{\sqrt{2}} \\ -\,I\otimes|01\rangle\langle 01|\dfrac{|010\rangle + |101\rangle}{\sqrt{2}} - I\otimes|10\rangle\langle 10|\dfrac{|010\rangle + |101\rangle}{\sqrt{2}} \end{pmatrix} \\
&= 0 + 0 - \frac{|101\rangle}{\sqrt{2}} - \frac{|010\rangle}{\sqrt{2}} \\
&\equiv -\frac{|101\rangle + |010\rangle}{\sqrt{2}} \\
&\equiv Z_{12}\left|\hat{\hat{0}}'\right\rangle_U .
\end{aligned} \tag{24.29}$$

Hence, the measurement

$$\langle Z_{23}\rangle_U = \langle Z_{12}\rangle_U \equiv -1. \tag{24.30}$$

The above two measurements, each yielding $-1$, are sufficient to diagnose that the second qubit in block $U$ is corrupted. If we are not convinced, we may perform a third measurement with $Z_{13}$, which, according to expectation, will yield $+1$ (this being left as an exercise). Such a result indicates that the first and third qubits are the same, which confirms the previous diagnostic. Table 24.3 summarizes the different diagnostic and corrective action possibilities in the general case. Clearly, the complete diagnostic and error-correction procedure requires one to perform the two comparative measurements $Z_{12}, Z_{23}$ on each of the three blocks $A, B, C$, representing altogether six measurements and $4^3 = 64$ possible bit-flip error diagnostics. It is also clear that a bit flip error occurring on any block $U$ of $|\hat{\hat{0}}\rangle$ replicates on the same block $U$ of $|\hat{\hat{1}}\rangle$ (observing from Eq. (24.23) that these two blocks only differ by their relative phase factor).

The above description concerned the detection and correction of either *phase-flip* or *bit-flip* errors occurring in the encoded 9-qubits $|\hat{\hat{0}}\rangle$ and $|\hat{\hat{1}}\rangle$. It is clear that since the syndrome measurements leave the qubit invariant, both error types can be corrected in any order. What about error correction on encoded 9-qubits of the form

$$\left|\hat{\hat{q}}\right\rangle = \alpha\left|\hat{\hat{0}}\right\rangle + \beta\left|\hat{\hat{1}}\right\rangle, \tag{24.31}$$

corresponding to an originator message $|q\rangle = \alpha|0\rangle + \beta|1\rangle$? To answer this question, we may first define the channel output qubit as

$$\left|\hat{\hat{q}}'\right\rangle = \alpha\left|\hat{\hat{0}}'\right\rangle + \beta\left|\hat{\hat{1}}'\right\rangle. \tag{24.32}$$

As we have seen earlier, the *phase-flip* syndrome diagnosis consists of making two successive measurements of the type (*ijklmn* $= 123456, 123789, 456789$):

$$\begin{aligned}
\left\langle X_{ijklmn}\right\rangle &= \left\langle \hat{\hat{q}}'\right| X_{ijklmn}\left|\hat{\hat{q}}'\right\rangle \\
&= \left(\bar{\alpha}\left\langle\hat{\hat{0}}'\right| + \bar{\beta}\left\langle\hat{\hat{1}}'\right|\right) X_{ijklmn}\left(\alpha\left|\hat{\hat{0}}'\right\rangle + \beta\left|\hat{\hat{1}}'\right\rangle\right) \\
&= \begin{pmatrix} |\alpha|^2\left\langle\hat{\hat{0}}'\right| X_{ijklmn}\left|\hat{\hat{0}}'\right\rangle + \bar{\alpha}\beta\left\langle\hat{\hat{0}}'\right| X_{ijklmn}\left|\hat{\hat{1}}'\right\rangle \\ +\alpha\bar{\beta}\left\langle\hat{\hat{1}}'\right| X_{ijklmn}\left|\hat{\hat{0}}'\right\rangle + |\beta|^2\left\langle\hat{\hat{1}}'\right| X_{ijklmn}\left|\hat{\hat{1}}'\right\rangle \end{pmatrix} \\
&\equiv |\alpha|^2\left\langle\hat{\hat{0}}'\right| X_{ijklmn}\left|\hat{\hat{0}}'\right\rangle + |\beta|^2\left\langle\hat{\hat{1}}'\right| X_{ijklmn}\left|\hat{\hat{1}}'\right\rangle \\
&\equiv \pm 1.
\end{aligned} \tag{24.33}$$

In the above result, we first used the property that the measurement $X_{ijklmn}$ leaves the state invariant, therefore, the nondiagonal elements in the right-hand side in Eq. (24.33) are zero. Since the same phase-flip error must be found in the two qubits $|\hat{\hat{0}}'\rangle$ and $|\hat{\hat{1}}'\rangle$, the two nonzero matrix elements must be equal, and considering the property $|\alpha|^2 + |\beta|^2 = 1$, we finally have $\langle X_{ijklmn}\rangle = \pm 1$, corresponding to no $(+1)$ or one $(-1)$ phase-flip error between the blocks $A, B$. Thus, syndrome diagnostic and error correction can be implemented on the encoded qubit $|\hat{\hat{q}}'\rangle = \alpha|\hat{\hat{0}}'\rangle + \beta|\hat{\hat{1}}'\rangle$ in exactly the same way as described earlier for the qubits $|\hat{\hat{0}}'\rangle$ and $|\hat{\hat{1}}'\rangle$. Clearly, the same conclusion applies to the case of *bit-flip* errors with two successive measurements $Z_{ij}$ $(ij = 12, 23, 13)$ on the individual blocks $U = A, B, C$, yielding measurement outcomes $\langle Z_{ij}\rangle_U = \pm 1$. The key conclusion is that *the Shor code can be implemented to correct both phase-flip and bit-flip errors on any qubit of the general form* $|q\rangle = \alpha|0\rangle + \beta|1\rangle$.

The powerful error-correction capability of the Shor code does not end here. Indeed, *we may conceive of this code as being able to correct a true continuum of error events*, while simply using the discrete set of the above-described syndrome operators!

To show, this recall from Chapter 23 that a noisy quantum channel can be described by a trace-preserving quantum operation $\varepsilon$ of the form

$$\varepsilon(\rho_M) = \sum_k U_k^+ \rho_M U_k, \tag{24.34}$$

which is referred to as its *operator-sum representation*. In this definition, $\rho_M = |\psi_M\rangle\langle\psi_M|$ is the originator message density matrix, which most generally is an $n$th tensor product of symbol states $\rho_i = |q_i\rangle\langle q_i|$ selected from a qubit symbol alphabet $\{|q_i\rangle\}$. The operators $U_k$, called *channel-operator elements*, are responsible for various sources of noise impairment. For single qubits $(\rho_i)$, the operator elements take the form: $U_1 = \sqrt{p_1}X$ for the bit-flip channel, $U_2 = \sqrt{p_2}Z$ for the phase-flip channel, $U_3 = \sqrt{p_3}Y$ for the bit-phase-flip channel, and $U_0 = \sqrt{p_0}I = \sqrt{1 - p_1 - p_2 - p_3}I$ for the noiseless channel, with $\{p_n\}$ representing the corresponding probability distribution. We may define $U_{ki}$ as the operator of noise-type $k$ acting on the qubit $i$ in the Shor codeword $|\hat{q}\rangle$. Thus, we have:

$$\begin{aligned} |\hat{q}'\rangle_i &= \sum_k U_{ki}|\hat{q}\rangle_i \\ &= (p_{0i}I_i + p_{1i}X_i + p_{2i}Z_i + p_{3i}Y_i)\,|\hat{q}\rangle_i, \end{aligned} \tag{24.35}$$

with $\{p_{ni}\}$ being the corresponding probability distribution. We observe that through the weighted action of the operator $I, X, Y, Z$, the above-defined quantum channel defines a *continuous* qubit transformation onto the Bloch sphere. As we have seen, the Shor code makes it possible to perform syndrome diagnosis and corrections onto any discrete, single-error type, as caused by the operator elements $U_{ki}$. Therefore, *any qubit passed through this channel, and corrupted by an "error continuum," can be effectively restored in its full original integrity*. This remarkable property has no counterpart in the classical world of error-correction codes.

## 24.3 Calderbank–Shor–Steine (CSS) codes

In this section, I describe a new class of quantum error-correction codes referred to as the *Calderbank–Shor–Steine* (CSS) codes, which have the capability of correcting up to $t$ bit-flip and phase-flip qubit errors. The construction of CSS codes is based on the use of two classical *linear block codes*, as described in Chapter 11. To recall, a linear block code $C$ is defined as a set $(n, k)$ of $|C| = 2^k$ codewords $x$ of bit length $n$. It is characterized by:[1]

(i) A *generator matrix* $G$, of dimension $n \times k$, which, for any $n$-vector $z$ to be encoded, yields the corresponding codeword $x = Gz$;
(ii) A *parity-check matrix* $H$, of dimension $n \times (n - k)$, which satisfies $Hx = 0$ for all codewords $x \in C$.

Let $(n, k_1)$ and $(n, k_2)$, with $k_1 > k_2$ define two linear block codes $C_1, C_2$ of sizes $|C_2| < |C_1|$. To construct a CSS code, we require the following two conditions:

(a) $C_2 \subset C_1$: *all codewords of* $C_2$ *belong to* $C_1$;
(b) $C_1$ *and* $C_2^{\perp}$ *have a bit-error correction capability of* $t$.

[1] In Chapter 11, we used the *left* vector-matrix multiplication $y^{\mathrm{T}}U$, where $y^{\mathrm{T}}$ is a line vector, instead of the *right* vector-matrix multiplication $Uy$, where $y$ is a column vector, which is only a matter of convention. Also we previously called $x$ the vectors to be encoded and $y$ the resulting codewords, while in this chapter we shall use $x$ and $y$ to designate codewords from two different linear block codes.

In condition (b), $C_2^{\perp}$ is called the *dual* of $C_2$. Given a block code with $G$, $H$ as generator and parity-check matrices, respectively, the dual of $C$, noted $C^{\perp}$, is a unique block code having $H^{\mathrm{T}}$, $G^{\mathrm{T}}$ for *generator* and *parity-check* matrices, respectively (the matrix $U^{\mathrm{T}}$ is the transposed matrix of $U$).

Given two linear block codes $C_1, C_2$ satisfying the above conditions, we can construct a CSS code as follows. Let $x, y$ be two $n$-bit codewords, such that $x \in C_1$ and $y \in C_2$. It is possible to define a quantum state $|x + y\rangle$, where $+$ indicates here bit-wise *addition modulo* 2 (or in Boolean logic, the exclusive OR, noted $\oplus$). For instance, if $x = 00101$ and $y = 10111$, we have $|x + y\rangle = |10010\rangle$. With such a definition at hand, given $x \in C_1$ we are able to construct all possible quantum states $|x + y\rangle$ with $y \in C_2$ as well as the normalized sum:

$$|x + C_2\rangle \equiv \frac{1}{\sqrt{|C_2|}} \sum_{y \in C_2} |x + y\rangle. \tag{24.36}$$

Given the number $2^{k_1} = |C_1|$ of codewords $x$ in $C_1$, how many orthogonal quantum states $|x + C_2\rangle$ can be, thus, generated? The answer to this question is $|C_1|/|C_2| = 2^{k_1 - k_2}$, which stems from *group theory* (GT) and the fine notion of *cosets*.[2] Here, I shall not venture into GT, but leave it as an interesting exercise to establish that the $C = (7, 4)$ *Hamming code*, (described in Chapter 11) forms a group $(C, \oplus)$ under the bit-wise addition $\oplus$ and to make the inventory of its various cosets.

A key property is that any element $x \in C_1$ must belong to *one and only one* coset of $C_2$. The same coset $x + C_2 = x' + C_2$ corresponds to two different elements $x, x' \in C_1$ satisfying the property $x - x' \in C_2$ and, hence $|x + C_2\rangle = |x' + C_2\rangle$.[3] It is clear that if two different elements $x, x' \in C_1$ do not belong to the same coset ($x + C_2 \neq x' +$

[2] I provide here a simplified description of *cosets*, which also explains the notation $x + C_2$. First, it is important to be familiar with, or to quickly revisit the basics of *groups* and *subgroups*, for instance through the links: http://en.wikipedia.org/wiki/Group_%28mathematics%29, http://en.wikipedia.org/wiki/Subgroup. Then assume two *commutative groups* $C_1$ and $C_2$ with additive operation $+$, such that $C_2 \subset C_1$ is a subgroup of $C_1$. The cosets of $C_2$ in $C_1$, noted $x_i + C_2$, are defined for all elements $x_i \in C_1$ as follows

$$x_i + C_2 \equiv \{x_i + y_j\}_{y_j \in C_2}.$$

Taking an illustrative example, assume $C_1 = (0, 1, 2, 3)$, with $+$ representing the addition modulo 4, and which has the only "nontrivial" subgroup $C_2 = (0, 2)$. The *cosets* of $C_2$ in $C_1$ are

$$\begin{aligned}
0 + C_2 &\equiv (0, 2) = C_2 \\
1 + C_2 &\equiv (1, 3) \\
2 + C_2 &\equiv (2, 0) = (0, 2) = C_2 \\
3 + C_2 &\equiv (3, 1) = (1, 3) = 1 + C_2.
\end{aligned}$$

It is seen that there are two distinct cosets of $C_2$ in $C_1$, including $C_2$ itself, which are $(0, 2)$ and $(1, 3)$. In the general case with groups and subgroups of finite sizes $|C_1|, |C_2|$, *Lagrange's theorem* states that the number of cosets is given by the ratio $|C_1|/|C_2|$. See also http://en.wikipedia.org/wiki/Coset, http://en.wikipedia.org/wiki/Lagrange%27s_theorem_%28group_theory%29, and Exercise (24.4), which studies the $(7, 4)$ Hamming code as a group under the bit-wise addition operation.

[3] Indeed, taking into account that $0 \in C_2$ ($C_2$ is a *subgroup* with the identity element under $\oplus$) and assuming $x - x' \in C_2$, we have

$$\begin{aligned}
x + C_2 &= \{x + 0, x + (x - x'), \ldots\} = \{x, x', \ldots\} \\
x' + C_2 &= \{x' + 0, x' + (x - x'), \ldots\} = \{x', x, \ldots\},
\end{aligned}$$

$C_2$) there is no $y, y' \in C_2$ such that $x + y = x' + y'$ and, therefore, $|x + C_2\rangle$ must be orthogonal to $|x' + C_2\rangle$. Based on GT, the number of distinct cosets is $|C_1|/|C_2|$, hence the quantum space spanned by the states $|x + C_2\rangle$, as generated from the codewords $x \in C_1$, has a dimension $|C_1|/|C_2| = 2^{k_1 - k_2}$. This quantum space corresponds to a new class of quantum code, noted CSS($C_1$, $C_2$), which is to be pronounced, "CSS *code of* $C_1$ *over* $C_2$."

Next, I shall describe how the CSS($C_1$, $C_2$) code can correct up to $t$ bit-flip ($X$) and phase-flip ($Z$) errors. Just as in the classical ECC case, the signature of errors, whether of the $X$- or $Z$-type, is an $n$-vector with 1 s indicating the position where any error occurred. Define $e_X$ and $e_Y$ as the corresponding vectors. If $x$ is an $n$-bit codeword and $|x\rangle$ the corresponding quantum state, the applications of the combined error pattern $e_X, e_Y$ shall transform the state $|x\rangle$ into the errored state $|x^*\rangle$ according to:

$$|x^*\rangle = (-1)^{x \cdot e_Y} |x + e_X\rangle, \qquad (24.37)$$

where $x \cdot e_Y$ is the dot product of $x$ and $e_Y$. For instance, $|x\rangle = |10011\rangle$, $e_X = (0, 1, 0, 1, 0)$ ($X$ errors on the second and fourth qubits) and $e_Y = (1, 0, 0, 1, 1)$ ($Z$ errors on the first and last two qubits) yield $x \cdot e_Y = 1.1 + 0.0 + 0.0 + 1.1 + 1.1 \equiv 3$ and $|x + e_X\rangle = |1 + 0\rangle|0 + 1\rangle|0 + 0\rangle|1 + 1\rangle|1 + 0\rangle \equiv |11001\rangle$, and the errored state $|x^*\rangle = (-1)^3|11001\rangle \equiv -|11001\rangle$. When the error pattern $e_X, e_Y$ is applied to the CSS codeword $|x + C_2\rangle$ defined in Eq. (24.36), the resulting errored state is

$$|(x + C_2)^*\rangle \equiv \frac{1}{\sqrt{|C_2|}} \sum_{y \in C_2} (-1)^{(x+y) \cdot e_Y} |x + y + e_X\rangle. \qquad (24.38)$$

The next step consists of the introduction of the parity-check matrix $\tilde{H}$ to detect the $t$ errors of the $X$ type (not to be confused here with the Hadamard gate $H$). This is achieved by appending an $n$-qubit ancilla $|0\rangle$ to the recipient's codeword $|x^*\rangle = |x + e_X\rangle$, and passing the result through a quantum circuit that achieves the transformation $|x^*\rangle|0\rangle \to |x^*\rangle|\tilde{H}x^*\rangle$. Recall from Chapter 11 that if $x$ is a codeword, then $\tilde{H}x = 0$, and hence $|\tilde{H}x^*\rangle = |\tilde{H}(x + e_X)\rangle \equiv |\tilde{H}e_X\rangle$. Measuring each of the qubits in the ancilla $|\tilde{H}e_X\rangle$ yields the syndrome vector $\tilde{H}e_X$ and, hence, the full error diagnosis $e_X$, which is common to each of the series terms in Eq. (24.38). The ancilla can then be discarded and each of the errored qubits can be corrected (or back flipped) by applying $X$ gates in the corresponding circuit wires, to obtain the $X$-corrected state

$$\left|(x + C_2)^*\right\rangle_X \equiv \frac{1}{\sqrt{|C_2|}} \sum_{y \in C_2} (-1)^{(x+y) \cdot e_Y} |x + y\rangle. \qquad (24.39)$$

A quantum circuit performing the operation $|z\rangle|0\rangle \to |z\rangle|Uz\rangle$ for any $n \times n$ matrix $U$ (here with $U \equiv \tilde{H}$ and $|z\rangle \equiv |(x + C_2)^*\rangle$) can be realized by exclusively using CCNOT (controlled-CNOT) gates, as illustrated in Fig. 24.5. The design concept of such a circuit can be grasped by analyzing the basic functionality $|z\rangle \to |Uz\rangle$ with $U$

where we used the (equivalent) bit-wise addition properties $x + x = 0$ and $x' = -x'$. The same result is obtained with the assumption $x + x' \in C_2$. It shows that the cosets $x + C_2$ and $x' + C_2$ have two elements in common, which necessarily implies that $x + C_2 = x' + C_2$ and, hence, $|x + C_2\rangle = |x' + C_2\rangle$.

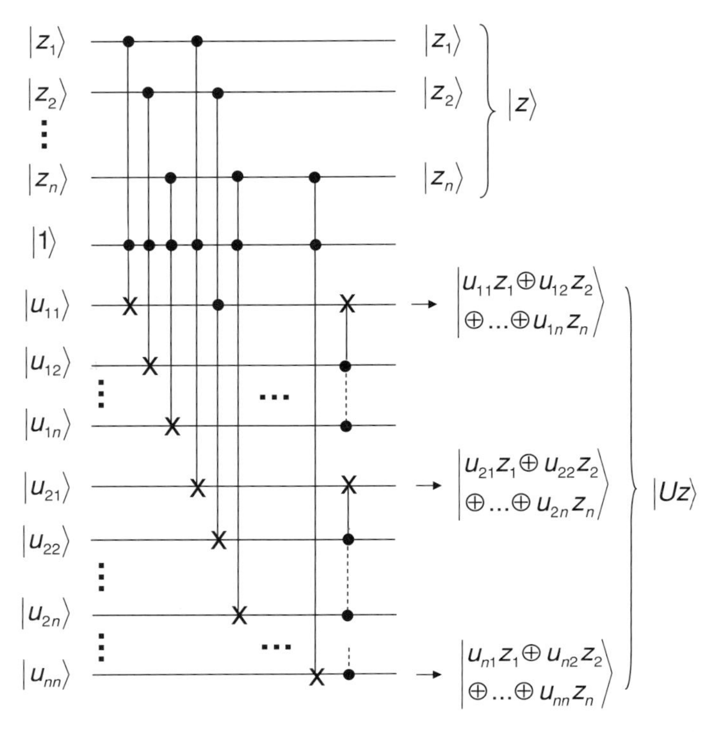


**Figure 24.5** Quantum circuit to achieve the transformation $|z\rangle\,|0\rangle \to |z\rangle\,|Uz\rangle$, given an $n$-qubit $|z\rangle$ and an $n \times n$ matrix $U$ with binary coefficients.

being a $2 \times 2$ matrix with binary coefficients, which is left as an interesting exercise. The figure actually shows that the binary coefficients $u_{ij} = 0, 1$ of $U$ can be directly used as corresponding ancilla qubits $|u_{ij}\rangle = |0\rangle, |1\rangle$, with the input state of the circuit being $|z\rangle\,|1\rangle|u_{11}\ldots u_{1n}\rangle|u_{21}\ldots u_{2n}\rangle\ldots|u_{n1}\ldots u_{nn}\rangle$. After discarding the useless (or "garbage") computation qubits, the circuit output is indeed $|z\rangle\,|Uz\rangle$. To summarize, we have shown that the CSS($C_1$, $C_2$) code can correct all $t$ errors of the $X$ type, as allowed by the $C_1$ linear block code, and the syndromes $|\tilde{H}e_X\rangle$ and $\tilde{H}e_X$, can be generated through a basic quantum circuit of CCNOT gates (Fig. 24.5), followed by the corresponding qubit measurements and bit-flip corrections.

The correction of phase-flip, or $Z$-type errors, is achieved by first passing each qubit of the $X$-corrected state $|(x + C_2)^*\rangle_X$ through a *Hadamard* gate, i.e., to obtain the state $H^{\otimes n}|(x + C_2)^*\rangle_X$. In Chapter 19, when describing the *Deutsch–Joszsa algorithm*, we have established (as supported through Exercise 19.1) that the action of $H^{\otimes n}$ on any $n$-qubit $|x\rangle$ yields the transformation

$$H^{\otimes n}|x\rangle = \frac{1}{\sqrt{2^n}}\sum_z (-1)^{z \cdot x}\,|z\rangle\,, \tag{24.40}$$

where $z$ is any possible $n$-bit combination. Applying this property to the state $|(x + C_2)^*\rangle_X$ in Eq. (24.39), namely, for each of the terms $|x + y\rangle$ in the sum, we obtain:

$$H^{\otimes n}|(x+C_2)^*\rangle_X \equiv \frac{1}{\sqrt{|C_2|\,2^n}}\sum_z \sum_{y\in C_2}(-1)^{(x+y)\cdot(z+e_Y)}|z\rangle. \tag{24.41}$$

By introducing $z' = z + e_Y$, we may rewrite the above in the form

$$H^{\otimes n}|(x+C_2)^*\rangle_X \equiv \frac{1}{\sqrt{|C_2|\,2^n}}\sum_{z'} \sum_{y\in C_2}(-1)^{(x+y)\cdot z'}|z'+e_Y\rangle. \tag{24.42}$$

To reduce the right-hand side in Eq. (24.42) to a single summation, we first isolate the term involving the dot-product $y\cdot z'$ as follows

$$H^{\otimes n}|(x+C_2)^*\rangle_X \equiv \frac{1}{\sqrt{|C_2|\,2^n}}\sum_{z'}\left[\sum_{y\in C_2}(-1)^{y\cdot z'}\right](-1)^{x\cdot z'}|z'+e_Y\rangle. \tag{24.43}$$

It can be shown through a (not so trivial but tractable) exercise that the sum in brackets $[\cdot]$ is equal to $|C_2|$ for all $z'\in C_2^\perp$ ($C_2^\perp$ being the dual of the $C_2$ code), and zero otherwise ($z'\notin C_2^\perp$). Hence, the simplification of the definition in Eq. (24.43):

$$H^{\otimes n}|(x+C_2)^*\rangle_X \equiv \sqrt{\frac{|C_2|}{2^n}}\sum_{z'\in C_2^\perp}(-1)^{x\cdot z'}|z'+e_Y\rangle. \tag{24.44}$$

Except for the normalization factor and the phase terms, the above expression is similar to that in Eq. (24.38) concerning the bit-flip errored state, but here with $e_Y$ as the error pattern and $C_2^\perp$ as the code (as opposed to $e_X$ and $C_2$, respectively). A similar error-correction procedure can, thus, be implemented. First, we append an $n$-qubit ancilla $|0\rangle$ to this state, transforming each term in the sum in Eq. (24.44) into $|z'+e_Y\rangle|0\rangle$, then pass the result through a quantum circuit to achieve the transformation

$$\begin{aligned}|z'+e_Y\rangle|0\rangle \to |z'+e_Y\rangle|\tilde{H}^\perp(z'+e_Y)\rangle &= |z'+e_Y\rangle|\tilde{H}^\perp z'+\tilde{H}^\perp e_Y\rangle\rangle \\ &\equiv |z'+e_Y\rangle|\tilde{H}^\perp e_Y\rangle\rangle,\end{aligned} \tag{24.45}$$

where $\tilde{H}^\perp$ is the parity-check matrix of the code $C_2^\perp$, for which $\tilde{H}^\perp z'=0$. Measuring each qubit in the ancilla $|\tilde{H}^\perp e_Y\rangle$ yields the syndrome vector $\tilde{H}^\perp e_Y$ and, hence, the full error diagnosis $e_Y$, which is common to each of the series terms in Eq. (24.44). The ancilla can then be discarded and each of the errored qubits can be corrected (or back flipped) by applying $Y$ gates in the corresponding circuit wires, to obtain the $Y$-corrected state

$$H^{\otimes n}|(x+C_2)^*\rangle_{XY} \equiv \sqrt{\frac{|C_2|}{2^n}}\sum_{z'\in C_2^\perp}(-1)^{x\cdot z'}|z'\rangle. \tag{24.46}$$

The final step consists of passing each qubit in the above state through a Hadamard gate, i.e., to obtain the state $H^{\otimes n}H^{\otimes n}|(x+C_2)^*\rangle_{XY}$. Here, there is no point in going through the detailed calculation of such an operation, because as a useful property, $H$ and $H^{\otimes n}$ are self-inverse operators ($HH=I$, $H^{\otimes n}H^{\otimes n}=I^{\otimes n}$). If we let $e_Y=0$ in Eq. (24.44), we obtain

$$\begin{aligned} H^{\otimes n}|(x+C_2)^*\rangle_X &\equiv \sqrt{\frac{|C_2|}{2^n}} \sum_{z' \in C_2^\perp} (-1)^{x \cdot z'} |z'\rangle \\ &\equiv H^{\otimes n}|x + C_2\rangle, \end{aligned} \tag{24.47}$$

which is precisely the same state as $H^{\otimes n}|(x+C_2)^*\rangle_{XY}$, and also the $H^{\otimes n}$ transform of the error-free state $|x + C_2\rangle$! Thus, a second application of $H^{\otimes n}$ on the states defined in either Eq. (24.46) or Eq. (24.47) yields

$$\begin{aligned} H^{\otimes n} H^{\otimes n}|(x+C_2)^*\rangle_{XY} &= H^{\otimes n} H^{\otimes n}|(x+C_2)^*\rangle_X \\ &= |x + C_2\rangle \\ &\equiv \frac{1}{\sqrt{|C_2|}} \sum_{y \in C_2} |x + y\rangle, \end{aligned} \tag{24.48}$$

which yields our initial CSS-encoded state $|x + C_2\rangle$. Thus, the second round of correction concerning $Z$-errors has successfully restored the CSS codeword in its full integrity.

The next section concerning the *Steane code* provides an applied illustration of the CSS$(C_1, C_2)$ codes.

## 24.4 Hadamard–Steane code

The *Hadamard–Steane code*, also sometimes called the *Steane* code, belongs to the CSS$(C_1, C_2)$ family. It has the same bit-flip and phase-flip error correction capability as the earlier-described *Shor code*, namely, up to one error in either or both cases, but it uses *7-qubit codewords* as opposed to nine qubits in the second case. It is based on the *Hamming code* $C_1 = C = (7, 4)$, which was described in Chapter 11, and its dual $C_2 = C^\perp$. To recall, a possible parity-check matrix $\tilde{H}$ for the $(7, 4)$ Hamming code, which we used in that chapter,[4] is defined as:

$$\tilde{H} = \begin{pmatrix} 1 & 1 & 0 & 1 & 1 & 0 & 0 \\ 1 & 0 & 1 & 1 & 0 & 1 & 0 \\ 0 & 1 & 1 & 1 & 0 & 0 & 1 \end{pmatrix}. \tag{24.49}$$

[4] Some other possible parity-check matrices $\tilde{H}$ for the $(7, 4)$ Hamming code commonly used in the literature or the Internet are

$$\tilde{H} = \begin{pmatrix} 1 & 0 & 0 & 1 & 0 & 1 & 1 \\ 0 & 1 & 0 & 1 & 1 & 0 & 1 \\ 0 & 0 & 1 & 0 & 1 & 1 & 1 \end{pmatrix},$$

$$\tilde{H} = \begin{pmatrix} 1 & 1 & 0 & 1 & 0 & 0 & 1 \\ 0 & 1 & 0 & 1 & 1 & 1 & 0 \\ 1 & 1 & 1 & 0 & 0 & 0 & 0 \end{pmatrix},$$

$$\tilde{H} = \begin{pmatrix} 0 & 0 & 0 & 1 & 1 & 1 & 1 \\ 0 & 1 & 1 & 0 & 0 & 1 & 1 \\ 1 & 0 & 1 & 0 & 1 & 0 & 1 \end{pmatrix}.$$

**Table 24.4** Block codewords $Y$ of the Hamming code $C_1 = (7, 4)$, as produced from the generator matrix $\tilde{G} = \tilde{H}^{\mathrm{T}}$, according to $Y = X\tilde{G}$. The parity bits are shown in bold.

| Message word $X$ | Block code $Y$ |
|---|---|
| 0000 | 0000 **000** |
| 0001 | 0001 **111** |
| 0010 | 0010 **011** |
| 0011 | 0011 **100** |
| 0100 | 0100 **101** |
| 0101 | 0101 **010** |
| 0110 | 0110 **110** |
| 0111 | 0111 **001** |
| 1000 | 1000 **110** |
| 1001 | 1001 **001** |
| 1010 | 1010 **101** |
| 1011 | 1011 **010** |
| 1100 | 1100 **011** |
| 1101 | 1101 **100** |
| 1110 | 1110 **000** |
| 1111 | 1111 **111** |

In the convention of *left* matrix-vector multiplication, the corresponding generator matrix $\tilde{G}$ is:[5]

$$\tilde{G} = \begin{pmatrix} 1 & 0 & 0 & 0 & 1 & 1 & 0 \\ 0 & 1 & 0 & 0 & 1 & 0 & 1 \\ 0 & 0 & 1 & 0 & 0 & 1 & 1 \\ 0 & 0 & 0 & 1 & 1 & 1 & 1 \end{pmatrix}. \tag{24.50}$$

Note that the above definitions correspond to a code expressed in a *systematic form*, as shown by the fact that the right $3 \times 3$ sub-matrix of $\tilde{H}$ and the left $4 \times 4$ sub-matrix of $\tilde{G}$ are identity matrices. The $|C_1| = 2^k = 2^4 = 16$ block codewords $Y = X\tilde{G}$ of $C_1 = (7, 4)$, which were already listed in Chapter 11, are reproduced here for convenience in Table 24.4.

Consider next the *dual code* $C_2 = C^{\perp}$. By definition, its parity-check matrix is $\hat{H} = \tilde{G}^{\mathrm{T}}$, with corresponding generator matrix $\hat{G} = \tilde{H}^{\mathrm{T}}$. From the definitions in Eqs. (24.49)

[5] In Chapter 11 we used the convention of *left* vector-matrix multiplication. Thus, the codewords are generated according to the product $Y = X\tilde{G}$, see Eq. (11.2) in Chapter 11. Under this convention, for an $(n, k)$ code with $m = n - k$, the systematic form of the generator and parity-check matrices are $\tilde{G} = [I_k | P_{k\times m}]$ and $\tilde{H} = [(P^T)_{m\times k} | I_m]$, respectively. Thus, for the $(7, 4)$ Hamming code $(m = 3)$, $\tilde{G}$ is a $4 \times 7$ matrix and $\tilde{H}$ is a $3 \times 7$ matrix.

**Table 24.5** Block codewords $Y$ of the dual code $C_2 = C_1^{\perp} = (7, 3)$ of the Hamming code $C_1 = (7, 4)$, as produced from the generator matrix $\hat{G} = \tilde{H}^{\mathrm{T}}$, where $\tilde{H}$ is the parity-check matrix of $C_1$, according to $Y = \hat{G}X$. The parity bits are shown in bold.

| Message word $X$ | Block code $Y$ |
|---|---|
| 000 | **0000** 000 |
| 001 | **0111** 001 |
| 010 | **1011** 010 |
| 011 | **1100** 011 |
| 100 | **1101** 100 |
| 101 | **1010** 101 |
| 110 | **0110** 110 |
| 111 | **0001** 111 |

and (24.50), we obtain:

$$\hat{H} = \tilde{G}^T = \begin{pmatrix} 1 & 0 & 0 & 0 \\ 0 & 1 & 0 & 0 \\ 0 & 0 & 1 & 0 \\ 0 & 0 & 0 & 1 \\ 1 & 1 & 0 & 1 \\ 1 & 0 & 1 & 1 \\ 0 & 1 & 1 & 1 \end{pmatrix}, \tag{24.51}$$

$$\hat{G} = \tilde{H}^T = \begin{pmatrix} 1 & 1 & 0 \\ 1 & 0 & 1 \\ 0 & 1 & 1 \\ 1 & 1 & 1 \\ 1 & 0 & 0 \\ 0 & 1 & 0 \\ 0 & 0 & 1 \end{pmatrix}. \tag{24.52}$$

As defined in Eq. (24.52), the $7 \times 3$ generator matrix $\hat{G}$ of the dual code $C_2$ generates seven-bit codewords from three-bit message words, this time using the *right* vector-matrix multiplication, i.e., $Y = \hat{G}X$. Since $\hat{G}$ has the systematic form $\hat{G} = [P_{4\times 3}|I_3]$, and since $\hat{G}\hat{H}^{\mathrm{T}} = \tilde{H}^{\mathrm{T}}\tilde{G} = 0 = \tilde{G}\tilde{H}^{\mathrm{T}} = \hat{H}^{\mathrm{T}}\hat{G}$, the dual code is a valid $(n, k) = (7, 3)$ code (note that it is *not* a Hamming code since $n \neq 2^{n-k} - 1$). The $|C_2| = 2^k = 2^3 = 8$ block codewords $Y = \hat{G}X$ are listed in Table 24.5.

Analyzing Table 24.5, we first note that, as expected, the code $C_2$ is in systematic form, with the four parity bits appearing at the left of the codewords $Y$ and the three message bits ($X$) appearing at the right. Next, it is easily checked that the bit-wise sum $\oplus$ of any two codewords in $C_2$ belongs to $C_2$; since $C_2$ contains the identity element 0000000, $(C_2, \oplus)$ is a *group*. Then, we observe that the *minimum Hamming distance* (the minimum difference in bit positions between any two codewords) is $d_{\min} = 3$. This indicates that

| Code C1 = (7,4) | x | Code word | Dual code C2 = (7,3) | | | | | | | | Coset | Equals |
|---|---|---|---|---|---|---|---|---|---|---|---|---|
| | | | **0000 000** | **0111 001** | **1011 010** | **1100 011** | **1101 100** | **1010 101** | **0110 110** | **0001 111** | | |
| | 0 | **0000 000** | 0000 000 | 0111 001 | 1011 010 | 1100 011 | 1101 100 | 1010 101 | 0110 110 | 0001 111 | **0 + C2** | |
| | 1 | **0001 111** | 0001 111 | 0110 110 | 1010 101 | 1101 100 | 1100 011 | 1011 010 | 0111 001 | 0000 000 | 1 + C2 | 2 + C2 |
| | 2 | **0010 011** | 0010 011 | 0101 010 | 1001 001 | 1110 000 | 1111 111 | 1000 110 | 0100 101 | 0011 100 | **2 + C2** | |
| | 3 | **0011 100** | 0011 100 | 0100 101 | 1000 110 | 1111 111 | 1110 000 | 1001 001 | 0101 010 | 0010 011 | 3 + C2 | 2 + C2 |
| | 4 | **0100 101** | 0100 101 | 0011 100 | 1111 111 | 1000 110 | 1001 001 | 1110 000 | 0010 011 | 0101 010 | 4 + C2 | 2 + C2 |
| | 5 | **0101 010** | 0101 010 | 0010 011 | 1110 000 | 1001 001 | 1000 110 | 1111 111 | 0011 100 | 0100 101 | 5 + C2 | 2 + C2 |
| | 6 | **0110 110** | 0110 110 | 0001 111 | 1101 100 | 1010 101 | 1011 010 | 1100 011 | 0000 000 | 0111 001 | 6 + C2 | 0 + C2 |
| | 7 | **0111 001** | 0111 001 | 0000 000 | 1100 011 | 1011 010 | 1010 101 | 1101 100 | 0001 111 | 0110 110 | 7 + C2 | 0 + C2 |
| | 8 | **1000 110** | 1000 110 | 1111 111 | 0011 100 | 0100 101 | 0101 010 | 0010 011 | 1110 000 | 1001 001 | 8 + C2 | 2 + C2 |
| | 9 | **1001 001** | 1001 001 | 1110 000 | 0010 011 | 0101 010 | 0100 101 | 0011 100 | 1111 111 | 1000 110 | 9 + C2 | 2 + C2 |
| | 10 | **1010 101** | 1010 101 | 1101 100 | 0001 111 | 0110 110 | 0111 001 | 0000 000 | 1100 011 | 1011 010 | 10 + C2 | 0 + C2 |
| | 11 | **1011 010** | 1011 010 | 1100 011 | 0000 000 | 0111 001 | 0110 110 | 0001 111 | 1101 100 | 1010 101 | 11 + C2 | 0 + C2 |
| | 12 | **1100 011** | 1100 011 | 1011 010 | 0111 001 | 0000 000 | 0001 111 | 0110 110 | 1010 101 | 1101 100 | 12 + C2 | 0 + C2 |
| | 13 | **1101 100** | 1101 100 | 1010 101 | 0110 110 | 0001 111 | 0000 000 | 0111 001 | 1011 010 | 1100 011 | 13 + C2 | 0 + C2 |
| | 14 | **1110 000** | 1110 000 | 1001 001 | 0101 010 | 0010 011 | 0011 100 | 0100 101 | 1000 110 | 1111 1111 | 14 + C2 | 2 + C2 |
| | 15 | **1111 111** | 1111 111 | 1000 110 | 0100 101 | 0011 100 | 0010 011 | 0101 010 | 1001 001 | 1110 000 | 15 + C2 | 2 + C2 |

**Figure 24.6** Cosets of $C_2 = (7, 3)$ in $C_1 = (7, 4)$. The codewords of $C_2$ are listed in the top row, while the codewords of $C_1$ (called $x = 0, 1, 2, \ldots, 15$) are listed in the left column. The elements of $C_1$ that are in $C_2$ are highlighted with a dark background. The two rightmost columns show the coset names and identify their equivalences, the only two different cosets being captured in $0 + C_2$ and $2 + C_2$.

the code $C_2$ has the same correction capability as $C_1$, namely $t = (d_{\min} - 1)/2 = 1$ bit (see Section 11.2). Then comparing the codes listed in Table 24.4 ($C_1 = (7, 4)$) and Table 24.5 ($C_2 = C_1^{\perp} = (7, 3)$), we observe that *the second is included in the first*, namely the property $C_2 \subset C_1$, i.e., $C_2$ is a *subgroup* of $C_1$. According to the previous section, such a condition (along with the condition that $C_2$ must have the same error-correction capability $t$ as $C_1$) makes the dual code $C_2$ fully eligible to construct a CSS$(C_1, C_2)$ quantum code of dimension $2^{k_1 - k_2} = 2^{4-3} = 2$.

The construction of CSS$(C_1, C_2)$ is made by applying the definition in Eq. (24.36), which, given $x \in C_1$, requires one to determine all the $|C_2| = 8$ binary codewords $x + y$ forming the cosets $x + C_2$, and to sum up the $|x + y\rangle$ quantum states thus defined. Figure 24.6 provides the full list of cosets of $C_2$ in $C_1$. The top row lists the codewords of $C_2$, while the left column lists the codewords of $C_1$ (called $x = 0, 1, 2, \ldots, 15$) and whose elements that are in $C_2$ have been highlighted with a dark background. From the table information, it is readily observed that all the common elements $x \in C_1, C_2$ belong to the same coset $0 + C_2$, while all elements $x' \in C_1, x' \notin C_2$ belong to the same coset $2 + C_2$ (which could as well be called $15 + C_2$, for instance). Therefore, taking any codeword pair $x, x'$ with such properties is sufficient to generate and define the two codewords of CSS$(C_1, C_2)$, which we shall call $|\widehat{0}\rangle, |\widehat{1}\rangle$. For instance, we can use $x = 0000000$ ($x \in C_1, C_2$) and $x' = 1111111 (x' \in C_1, x' \notin C_2)$. We, thus, obtain:

$$
\begin{aligned}
|\widehat{0}\rangle &\equiv |0000000 + C_2\rangle \\
&= \frac{1}{\sqrt{8}} \begin{pmatrix} |0000000\rangle + |0111001\rangle + |1011010\rangle + |1100011\rangle \\ |1101100\rangle + |1010101\rangle + |0110110\rangle + |0001111\rangle \end{pmatrix},
\end{aligned}
\tag{24.53}
$$

**Table 24.6** Defining a rule for the codeword bits of $C_2$ (Table 24.5); see text for description.

| 1 | 2 | 3 | 4 | 5 | 6 | 7 |
|---|---|---|---|---|---|---|
| $x_1+x_2$ | $x_1+x_3$ | $x_2+x_3$ | $x_1+x_2+x_3$ | $x_1$ | $x_2$ | $x_3$ |
| 0 | 0 | 0 | 0 | 0 | 0 | 0 |
| 0 | 1 | 1 | 1 | 0 | 0 | 1 |
| 1 | 0 | 1 | 1 | 0 | 1 | 0 |
| 1 | 1 | 0 | 0 | 0 | 1 | 1 |
| 1 | 1 | 0 | 1 | 1 | 0 | 0 |
| 1 | 0 | 1 | 0 | 1 | 0 | 1 |
| 0 | 1 | 1 | 0 | 1 | 1 | 0 |
| 0 | 0 | 0 | 1 | 1 | 1 | 1 |

$$\begin{aligned}|\widehat{1}\rangle &\equiv |1111111 + C_2\rangle \\ &= \frac{1}{\sqrt{8}}\begin{pmatrix}|1111111\rangle + |1000110\rangle + |0100101\rangle + |0011100\rangle \\ |0010011\rangle + |0101010\rangle + |1001001\rangle + |1110000\rangle\end{pmatrix}.\end{aligned} \tag{24.54}$$

To recall, to construct the Steane code $\{|\widehat{0}\rangle, |\widehat{1}\rangle\}$ we have used a generator matrix $\hat{G} = \tilde{H}^{\mathrm{T}}$ in *systematic form*, as shown in Eq. (24.52). As previously mentioned, there exist several other possibilities for defining the parity-check matrix $\tilde{H}^{\mathrm{T}}$ of the (7, 4) Hamming code. It is left as an easy exercise to determine the Steane code $\{|\widehat{0}\rangle, |\widehat{1}\rangle\}$ obtained with a matrix $\hat{G}' = \tilde{H}'\mathrm{T}$ different from that used in this chapter. It does not matter that the resulting codewords $|\widehat{0}\rangle, |\widehat{1}\rangle$ differ from those defined in Eqs. (24.53) or (24.54), because in any $(n, k)$ block code the choice of generator and parity-check matrices is only a matter of convention.

How can we realize a quantum circuit capable of encoding the above-defined $|\widehat{0}\rangle, |\widehat{1}\rangle$ Steane codewords, which represent finely crafted superpositions of eight 7-qubits? Short of a rule to predict the values of the individual qubits within each of these 7-qubit superpositions, the task of designing such a circuit seems rather complex! But here comes a bit of magic in the way we can define a rule for the two codewords. As a starting point, we have seen that each 7-qubit $|Y\rangle$ in the definition $|\widehat{0}\rangle$ refers to the block codeword $Y$, which is an element of the coset $0 + C_2$. To find the rule governing the bit values of $Y$, we first list the codewords in seven columns, as shown in Table 24.6. Because the code is in systematic form, the first four columns 1234 represent parity bits, while the last three columns 567 represent the message bits. If we call $x_1, x_2, x_3$ these three bits, we observe from the table that the parity bits in columns 1234 are defined by the sums $x_1 + x_2$, $x_1 + x_3$, $x_2 + x_3$, and $x_1 + x_2 + x_3$, respectively.

The Steane codeword $|\widehat{0}\rangle$ is, thus, defined by the three message bits $x_1, x_2, x_3$, as:

$$\begin{aligned}|\widehat{0}\rangle &= \frac{1}{\sqrt{8}}\sum_{y\in C_2}|x+y\rangle \\ &\equiv \frac{1}{\sqrt{8}}\sum_{x_1,x_2,x_3}|x_1+x_2\rangle\,|x_1+x_3\rangle\,|x_2+x_3\rangle\,|x_1+x_2+x_3\rangle\,|x_1\rangle\,|x_2\rangle\,|x_3\rangle.\end{aligned} \tag{24.55}$$

As shown in Eq. (24.54), the other Steane codeword $|\hat{1}\rangle$ is generated by the bitwise addition of $z = 1111111$ to each of the elements in $C_2/0 + C_2$, which yields the alternate definition:

$$
\begin{aligned}
|\hat{1}\rangle &= \frac{1}{\sqrt{8}} \sum_{y \in C_2} |x + y + z\rangle \\
&\equiv \frac{1}{\sqrt{8}} \sum_{x_1, x_2, x_3 = 0,1} (|x_1 + x_2 + 1\rangle |x_1 + x_3 + 1\rangle |x_2 + x_3 + 1\rangle \\
&\quad \otimes |x_1 + x_2 + x_3 + 1\rangle |x_1 + 1\rangle |x_2 + 1\rangle |x_3 + 1\rangle).
\end{aligned}
\tag{24.56}
$$

Letting the bit variable $a = 0, 1$ we finally obtain a common definition for the Steane codewords:

$$
\begin{aligned}
|\hat{a}\rangle &\equiv \frac{1}{\sqrt{8}} \sum_{x_1, x_2, x_3 = 0,1} (|x_1 + x_2 + a\rangle |x_1 + x_3 + a\rangle |x_2 + x_3 + a\rangle \\
&\quad \otimes |x_1 + x_2 + x_3 + a\rangle |x_1 + a\rangle |x_2 + a\rangle |x_3 + a\rangle).
\end{aligned}
\tag{24.57}
$$

We may simplify the above definition even further by setting the bit variables $y_1 = x_1 + a$, $y_2 = x_2 + a$ and $y_3 = x_3 + a$, which gives

$$
\begin{aligned}
|\hat{a}\rangle &\equiv \frac{1}{\sqrt{8}} \sum_{y_1, y_2, y_3 = 0,1} (|y_1 + y_2 + a\rangle |y_1 + y_3 + a\rangle |y_2 + y_3 + a\rangle \\
&\quad \otimes |y_1 + y_2 + y_3\rangle |y_1\rangle |y_2\rangle |y_3\rangle).
\end{aligned}
\tag{24.58}
$$

(where the modulo-2 addition property $u + u = 0$ is applied for any bit $u = x_i, y, a = 0, 1$). As we shall see next, the general definition of $|\hat{a}\rangle$ in Eq. (24.57) represents the "magic formula" needed to build a complete *Steane encoder* circuit!

Indeed, consider the first qubit in $|\hat{a}\rangle$, namely $|y_1 + y_2 + a\rangle$. The most elementary subcircuit that can be implemented to generate $|y_1 + y_2 + a\rangle$ with $y_1, y_2 = 0, 1$ is shown in Figure 24.7. It is seen from the figure that the two Hadamard gates produce two $|y_i\rangle = |+\rangle$ states that can act as control qubits on the target qubit $|a\rangle$ through CNOT gates (recalling that $|y\rangle\,\mathrm{CNOT}|x\rangle = |y \oplus x\rangle \equiv |x + y\rangle$). We observe that $|a\rangle$, which represents the originator's message qubit to be encoded, is not limited to the basis states $|0\rangle$ or $|1\rangle$; most generally, it can be of the form $|a\rangle = \alpha|0\rangle + \beta|1\rangle$, where, as usual, $\alpha, \beta$ are complex amplitudes satisfying $|\alpha|^2 + |\beta|^2 = 1$.

Based on the above, it is not at all difficult to conceive the full Steane-code encoding circuit shown in Fig. 24.8, which represents only one possible implementation out of many other variants, let alone the flexibility associated with the many other choices for generator matrix $\hat{G}$ in the $(7, 4)$ code. Note that the same circuit, as traversed from right to left, can be used for *decoding*, i.e., retrieving the message qubit $|a\rangle = \alpha|0\rangle + \beta|1\rangle$ from the 7-qubit $|\hat{a}\rangle$, as obtained from the recipient after implementing error correction.

This concludes this chapter on quantum error correction. An eerie feeling may rise from realising that the above-described CSS codes only represent another conceptual subspace within a grander space of the so-called *stabilizer codes*. Such codes, which are, by and large, not limited to error correction but rather are at the root of advanced

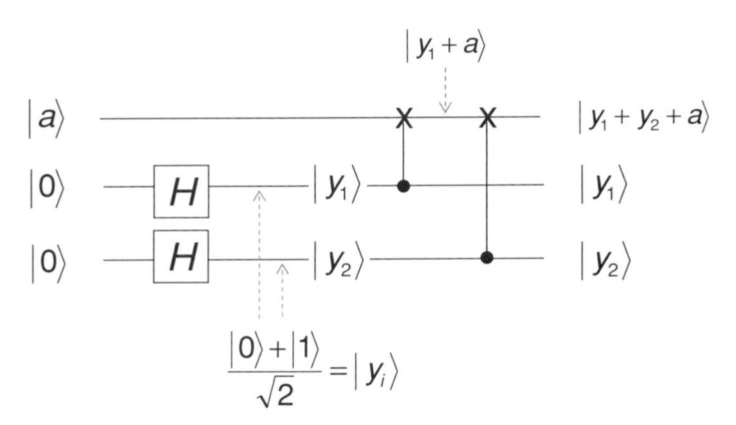


**Figure 24.7** Elementary circuit to generate the quantum state $|y_1 + y_2 + a\rangle$ given bit values $\alpha, y_1, y_2 = 0, 1$, as based on $H$ (Hadamard) and CNOT gates. The states obtained at intermediate stages are shown and pointed to by dotted arrows.

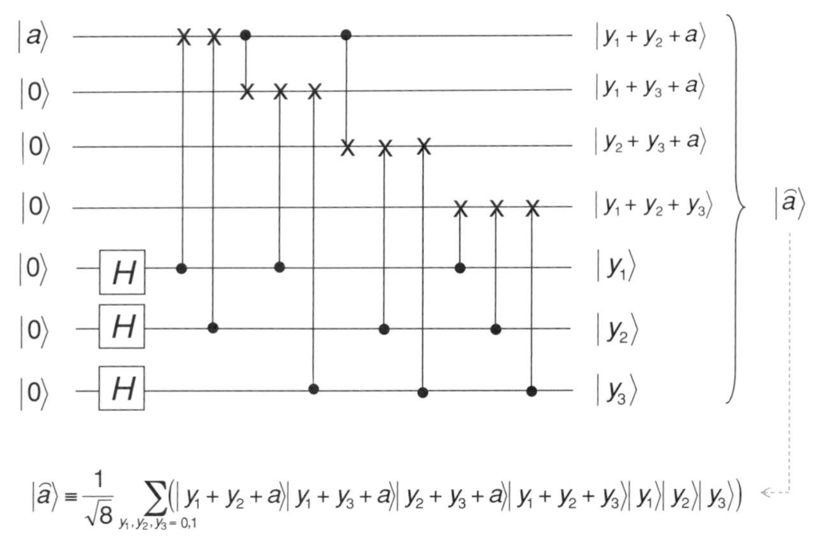


$$|\hat{a}\rangle \equiv \frac{1}{\sqrt{8}} \sum_{y_1, y_2, y_3 = 0,1} \left(|y_1 + y_2 + a\rangle |y_1 + y_3 + a\rangle |y_2 + y_3 + a\rangle |y_1 + y_2 + y_3\rangle |y_1\rangle |y_2\rangle |y_3\rangle\right)$$

**Figure 24.8** Possible circuit implementation for encoding the $|\hat{a}\rangle = |\hat{0}\rangle, |\hat{1}\rangle$ Steane code example, as defined in Eq. (24.58).

quantum computing, represent yet another field by itself. Within the framework of these chapters, hopefully the reader has grasped the essentials to exercise his or her curiosity towards this direction. If the reader has gone this far in these chapters, there are no more conceptual difficulties involved, except for the never-ending work of specialization in the field. The last chapter, concerning *quantum cryptography* should complete our basic training towards the impossible or ill-defined grander picture, whether or not we elect to become a specialist!

## 24.5 Exercises

**24.1** (B): Establish the following three operator definitions

$$\begin{aligned}
Z_{12} &= Z \otimes Z \otimes I = (|00\rangle\langle 00| + |11\rangle\langle 11|) \otimes I - (|01\rangle\langle 01| + |10\rangle\langle 10|) \otimes I, \\
Z_{23} &= I \otimes Z \otimes Z = I \otimes (|00\rangle\langle 00| + |11\rangle\langle 11|) - I \otimes (|01\rangle\langle 01| + |10\rangle\langle 10|), \\
Z_{13} &= Z \otimes I \otimes Z = |0\rangle\langle 0| \otimes I \otimes |0\rangle\langle 0| + |1\rangle\langle 1| \otimes I \otimes |1\rangle\langle 1| \\
&\qquad - (|0\rangle\langle 0| \otimes I \otimes |1\rangle\langle 1| + |1\rangle\langle 1| \otimes I \otimes |0\rangle\langle 0|),
\end{aligned}$$

where $Z$ is the Pauli matrix,

$$Z = \begin{pmatrix} 1 & 0 \\ 0 & -1 \end{pmatrix}.$$

**24.2** (M): Given the 9-qubit quantum error-correction (Shor) code

$$\begin{aligned}
|\hat{\hat{0}}\rangle &= \left.\frac{|000\rangle + |111\rangle}{\sqrt{2}}\right|_A \left.\frac{|000\rangle + |111\rangle}{\sqrt{2}}\right|_B \left.\frac{|000\rangle + |111\rangle}{\sqrt{2}}\right|_C \\
|\hat{\hat{1}}\rangle &= \left.\frac{|000\rangle - |111\rangle}{\sqrt{2}}\right|_A \left.\frac{|000\rangle - |111\rangle}{\sqrt{2}}\right|_B \left.\frac{|000\rangle - |111\rangle}{\sqrt{2}}\right|_C,
\end{aligned}$$

show that a single qubit phase-flip error in any of the three blocks $A, B, C$ can be detected through the successive measurements

$$\begin{aligned}
X_{123456} &= X_1 X_2 X_3 X_4 X_5 X_6, \\
X_{456789} &= X_4 X_5 X_6 X_7 X_8 X_9,
\end{aligned}$$

where $X$ is the Pauli matrix,

$$X = \begin{pmatrix} 0 & 1 \\ 1 & 0 \end{pmatrix}.$$

**24.3** (M): Assuming a 3-qubit entangled block with the second qubit being flipped,

$$|\hat{\hat{q}}\rangle = \frac{|010\rangle + |101\rangle}{\sqrt{2}},$$

show that the syndrome measurement $Z_{13}$ of the last two qubits yields the expectation value

$$\langle Z_{13} \rangle = +1.$$

**24.4** (T): Establish that the set $C$ of 16 elements shown in Table 24.7 (underscore "–" being for reading clarity), together with bitwise addition $\oplus$, form a group $(C, \oplus)$. Then determine the different subgroups of $(C, \oplus)$ and possible cosets of $C'$ in $C$.

**Table 24.7** Elements for Exercise 24.4.

| Number | Element | Number | Element |
|---|---|---|---|
| 0 | 0000_000 | 8 | 1000_110 |
| 1 | 0001_111 | 9 | 1001_001 |
| 2 | 0010_011 | 10 | 1010_101 |
| 3 | 0011_100 | 11 | 1011_010 |
| 4 | 0100_101 | 12 | 1100_011 |
| 5 | 0101_010 | 13 | 1101_100 |
| 6 | 0110_110 | 14 | 1110_000 |
| 7 | 0111_001 | 15 | 1111_111 |

**24.5** (T): Given the $2 \times 2$ matrix,

$$U = \begin{pmatrix} u_{11} & u_{12} \\ u_{21} & u_{22} \end{pmatrix},$$

with coefficients having the bit values $u_{ij} = 0, 1$ and a 2-qubit state $|x\rangle$, where $x$ is a two-bit codeword, determine a quantum circuit that achieves the transformation

$$|x\rangle|0\rangle \rightarrow |x\rangle\,|Ux\rangle\,,$$

uniquely based on CNOT gates. Show that a more compact circuit can be elaborated with CCNOT gates.

**24.6** (T): Given a linear block code $C$ with codewords $x = x_1x_2 \ldots x_n$, show that for any $y$ such that $y \in C^\perp$ (dual of $C$), we have

$$(a) \sum_{x \in C} (-1)^{x \cdot y} = |C|,$$

and otherwise ($y \notin C^\perp$)

$$(b) \sum_{x \in C} (-1)^{x \cdot y} = 0,$$

where $x \cdot y = x_1y_1 + x_2y_2 + \cdots + x_ny_n$ is the binary scalar product.

**24.7** (M): Given the parity-check matrix of the $C = (7, 4)$ Hamming code,

$$\tilde{H} = \begin{pmatrix} 0 & 0 & 0 & 1 & 1 & 1 & 1 \\ 0 & 1 & 1 & 0 & 0 & 1 & 1 \\ 1 & 0 & 1 & 0 & 1 & 0 & 1 \end{pmatrix},$$

determine the codewords of the corresponding $C^\perp = (7, 3)$ dual code, as generated by the matrix $\hat{G} = \tilde{H}^{\mathrm{T}}$.

Then determine the two codewords $|\hat{0}\rangle$, $|\hat{1}\rangle$ of the CSS($C$, $C^\perp$) code.

# 25 Classical and quantum cryptography

This final chapter concerns *cryptography*, the principle of securing information against access or tampering by third parties. *Classical cryptography* refers to the manipulation of classical bits for this purpose, while *quantum cryptography* can be viewed as doing the same with *qubits*. I describe these two approaches in the same chapter, as in my view the field of cryptography should be understood as a whole and appreciated within such a broader framework, as opposed to focusing on the specific applications offered by the quantum approach. I, thus, begin by introducing the notions of *message encryption*, *message decryption*, and *code breaking*, the action of retrieving the message information contents without knowledge of the code's *secret algorithm* or *secret key*. I then consider the basic algorithms to achieve encryption and decryption with binary numbers, which leads to the early IBM concept of the *Lucifer cryptosystem*, which is the ancestor of the first *data encryption standard* (DES). The principle of *double-key encryption*, which alleviates the problem of *key exchange*, is first considered as an elegant solution but it is unsafe against code-breaking. Then the revolutionary principles of *cryptography without key exchange* and *public-key cryptography* (PKC) are considered, the latter also being known as *RSA*. The PKC–RSA cryptosystem is based on the extreme difficulty of *factorizing large numbers*. This is the reason for the description made earlier in Chapter 20 concerning *Shor's factorization algorithm*. Here, the discussion is expanded to qualify the difficulty of the factorization problem by classical computing means, with a detailed analysis of the time and memory requirements, yielding predictions for solving the *various RSA challenges*. I then briefly describe DES with its double- and triple-implementation variants (2DES, 3DES), and the innovative features of the *advanced encryption standard* (AES). Then follows a discussion of the other applications of modern cryptography, including *signature* and *authentication*, which are based on the principle of *one-way encryption* or *hashing*. I then move to *quantum cryptography*, which is now placed in the global context of *communication security*, rather than addressed as a separate field.

*Quantum cryptography* is widely believed to provide "*provably secure*" means to exchange information, based on fundamental quantum-mechanics principles. This is the introductory concept of *secure communication channel*. As we shall see in this chapter, the concept boils down to that of *quantum key distribution* (QKD), solving the age-long issue of *key exchange*, without interception risks, as an absolute, "quantum" competitor to PKC–RSA. I then describe the three hero protocols, referred to as *BB84*, *B92*, and *EPR* protocols for QKD. As it turns out, and contrary to widespread belief, none of these protocols is provably safe unless extra security steps are implemented. Indeed, classical-coding means, "esoterically" referred to as *information reconciliation*, *privacy amplification*, or *secret key distillation* (SKD), are necessary to achieve "ultimate" (albeit exponential) security confidence in the protocol. The closing section discusses the eventual vulnerability of quantum cryptosystems, when replaced in the realistic network environment, with its various forms of attack. The conclusion is that assuming all attacks are possible (as opposed to sheltering under "postulates" that these remain out of the picture), quantum cryptography represents just another, albeit very elegant, technique for key exchange. Despite its exponential degree of safety, QKD must be situated within the grander domain of *global network security*, where there cannot be "absolute" confidence in any cryptosystem.

## 25.1 Message encryption, decryption, and code breaking

This section reviews the elementary basics of cryptography principles, definitions, and jargon.

There exist several instances where some private, confidential, sensitive, or even classified information must be conveyed through a communication channel. Examples abound, from online tax declaration and business reports on company's intranets to military and government communications. The driving concept for protecting such contents is cryptography, which comes from the Greek *kryptos* (secret) and *graphein* (writing). Thus, cryptography is the art of transforming a message into a self-contained secret, while ensuring a practical technique for the destinee to retrieve the information contents.

The action of *encryption* is to transform an original text, or *plaintext*, into a *ciphertext*. Nowadays, "plaintexts" are raw computer data, namely blocks of bits, but the spirit is exactly similar. The word *cipher* refers to the coding algorithm or method of encryption. An equivalent, fancier, definition of cipher is *cryptosystem*. The reverse operation, i.e., resolving a ciphertext into the original plaintext, is called *decryption*. Decryption is based on the principle of a *secret key*. Throughout cryptographic history, the secret key could have been any method of encryption and decryption based on correspondence *tables*, *squares*, or *grids*, kept secret between the corresponding parties. Alternatively, the secret key could have been a codeword (named *keyword*), usually simple to pass on and memorize, but more or less difficult to guess. Nowadays, full and public knowledge of the algorithm used in cryptographic standards poses no threat whatsoever to security. Rather, secrecy is based entirely on the secret key, its type, length, means of exchange, and rapid renewal (the last two points to be addressed later).

A third party who may try to make sense of a ciphertext without the key must resolve the enigma through more or less extensive labor. Such a task is referred to as *cryptoanalysis* (or, for short, *cryptoanalysis*). The goal of cryptanalysis, also called *code breaking*, is to "break" the cipher or ciphertext (familiarly called "code"), meaning recovering the plaintext by practical, inexpensive, hopefully rapid, and effortless methods. Note that code breaking should not be confused with the nobler, academic notion of *code cracking*.[1] In any case, the task of trying either to "break" or "crack" a code is referred to as an *attack*, giving a military flavor. Codes may be attacked either by sophisticated cryptanalysis algorithms or, more directly, by sheer "brute force" (hence the name *brute-force attack* in crypto jargon). This latter approach refers to various techniques of guessing keys at random, exploring what is referred to as the *keyspace*. For an $N$-bit key size, the keyspace corresponds to $2^N$ possibilities. In code breaking, rapidity of execution is essential, because the plaintext is generally not only content-sensitive but also time-sensitive. This is the reason why successive generations of cryptosystems and key sizes have been designed to make it virtually impossible to a third party, even a state with powerful cryptanalysis and computing means, to break them. With a key size of $N = 64$ bits, the keyspace is $2^{64} = 10^{64 \log 2/\log 10} = 1.8 \times 10^{19}$. At a rate of trying out one billion keys per second, a brute-force attack would, thus, require $1.8 \times 10^{10}$ seconds, or some 570 years to explore the full keyspace! Clearly, each extra bit of key size doubles the number of key possibilities. Nowadays, some cryptosystems may use key sizes as large as $N = 512$ bits or $N = 1024$ bits, corresponding to huge keyspaces of $10^{154}$ and $10^{308}$, respectively. However, one should not conclude that cryptosystems become absolutely safe against attacks when the key size is sufficiently large. This subject being beyond the scope of this chapter, suffice it to state here that one should not underestimate the power of cryptanalysis, and also that the generation of truly random keys is not technically simple. If a code breaker can guess the algorithm used to generate the random keys, his or her task is simplified. But the most straightforward approach is to intercept the key itself, at the time the communicating parties are doing what is referred to as the *key exchange*.

Indeed, the weakness of cryptography lies in the fact that, in order to work, the secret key must be communicated one way or another to the recipient. Such an exchange is not without risk of interception, no matter how rapid and periodically changing. Clearly, the worst situation is to be confident in the safety of a cryptosystem, while the channel used to exchange the keys is in fact being wired! As a matter of fact, cryptography faces a double security problem: transmitting the secret key without any measurable risk of interception, and remaining virtually invulnerable to cryptanalysis.

The history of cryptography, or *cryptology*, represents a succession of increasingly sophisticated cryptoalgorithms, which have invariably been broken at some point by sheer cryptanalysis.[2] Most ancient and pre-computer-era algorithms have been based

[1] The concept of *code cracking* is somewhat different. It could be defined as breaking a code *for the first time*. Put simply, code crackers are "champion" code breakers. The goal of code crackers is not so much to retrieve secret messages (for which they usually cannot care less) as to prove to history that they have been able to "crack" a previously reputed "invulnerable" or "unbreakable" cipher. Code cracking may be viewed as a noble academic attempt to explore the limits of cryptosystem security, with a view to improve the algorithms, or just as an engineering challenge to test security and standards.

[2] For fascinating accounts of cryptology, from ancient to modern times, see, for instance: D. Kahn, *The Code Breakers* (New York: Scribner, 1967); S. Singh, *The Code Book* (New York: Anchor Books, 1999); S. Levy, *Crypto*, (New York: Penguin Books, 2001).

on the principle of *alphabetic substitution*.[3] We may simplistically view any alphabetic-substitution code as representing a one-to-one table correspondence between plaintext and ciphertext letter symbols. There exist 26! possible permutations between the 26 letters, namely $26! = 1 \times 2 \times 3 \times \ldots \times 25 \times 26 = 4.0 \times 10^{26}$. Assuming that each code possibility could be checked every second, breaking the code would take some 13 billion billion years, i.e., about *one billion times the age of our Universe*! And with a rate of one billion checks per second, we still have the age of the Universe. Hence brute-force attack is not a code-breaking solution. Older cryptanalysis would, rather, attempt to guess which table, square, or grid algorithm and keys were used, based on the fact that human beings at the time needed to keep the encryption and decryption process simple and practical (as done by paper and pencil!), and the secret keys easy to memorize. No matter how sophisticated, however, with mainframe computer power, such alphabetic-substitution codes may now be broken in milliseconds or seconds. This is because all possible algorithms and table, square, or grid combinations of past history can be systematically explored, and also mainly because the keyword sizes remain relatively small. Overall, cryptanalysis is facilitated by the detection of *frequencies*, the repeated occurrences of certain ciphertext symbols and associations. The frequency properties of the English language were described in Chapter 9. We have seen that the two most likely letter associations are TH and HE, making THE the most likely three-letter word in English. Thus, any frequent two-symbol or three-symbol occurrence in a ciphertext is likely to correspond to these plaintext letter associations. In cryptography jargon, the word fragments, identified are called *cribs*. Based on a few cribs, the rest of the guess can progress very fast, should the context of the message (most likely words, e.g., in wartime) also be known to the code breaker. Frequency analysis and identification of cribs are basically the spirit of cryptanalysis as far as ancient (alphabetic-substitution) cryptosystems are concerned.

Cryptography is over 2000 years old. The ingenuity and skills that were deployed through the ages to conceive and then break ever-complex ciphers are both phenomenal and mind boggling. It is beyond the scope of this chapter to make even a brief survey of cipher development. I shall just mention here the intriguing *Enigma machine*, which was used by the German army during World War II.[4] It represented a major turn in the history of cryptography, not solely because it was the first "cryptographic computer," but because of its highly sophisticated encryption and decryption algorithm. Probably one had never believed so much in cryptosystem invulnerability! The Enigma cipher used only 26 typewriter characters with a three-rotor system, yielding as many as 10 586 916 764 424 000 or *10 million billion* (or 10 peta) polyalphabetic substitution

[3] For a detailed and illustrated review of historical alphabetic-substitution algorithms, see, for instance, my previous work: E. Desurvire, *Wiley Survival Guide in Global Telecommunications, Broadband Access, Optical Components and Networks, and Cryptography* (New York: J. Wiley & Sons, 2004), Ch. 3, pp. 345–477.

[4] There exist many excellent websites dedicated to the Enigma machine, see for instance:
http://en.wikipedia.org/wiki/Enigma_machine,
www.codesandciphers.org.uk/enigma/index.htm,
www.mlb.co.jp/linux/science/genigma/enigma-referat/enigma-referat.html,
www.enigmahistory.org/simulators.html.

possibilities. Yet the code was eventually cracked in 1932, again illustrating the unlimited resource of human ingenuity!

A second historical turn in cryptography history was marked by the advent of *digital computers* and, hence, the possibility not only to *substitute letter symbols* into each other, but to *intermingle their bits*, and intermingle all the bits of all plaintext symbols, notwithstanding the extended keyboard alphabet of $2^8 = 256$ ASCII word symbols. Moreover, *binary operations* can be introduced (such as XOR and polynomial multiplication) to make cryptanalysis even more intractable. These new classes of cryptosystem yield ciphertexts under the form of near-perfect pseudo-random bit sequences, thus, betraying no specific patterns or block frequencies on which previous cryptanalysis fed.

A third historical turn was marked by the invention of *public-key cryptography* or PKC (see further), which is basically cryptography *without key exchange*. As we shall see, the strength or confidence in PKC is based on the difficulty of factorizing large numbers into primes. But, as we already know, the difficulty could be alleviated, should practical quantum computing be developed, with *Shor's factorization algorithm*, which was described in Chapter 20.

The above illustrates that the notion of *cryptosystem invulnerability* is all relative, especially considering the power of supercomputers and the ever-increasing sophistication of cryptanalysis.

## 25.2 Encryption and decryption with binary numbers

This section reviews the basics of what could be called "modern" cryptography. As we shall see, the principles used in modern cryptography are not just the same older ones implemented with powerful computers and bigger keys, like "super-Enigma" machines. Rather, and as mentioned earlier, they represent a third phase of history with the introduction of new algorithms, leading to ciphers that are hopelessly hard to crack with current-generation supercomputers. Although digital cryptography may appear a tricky subject (and it really is!), the elementary basics are surprisingly simple to describe. Here, I will first look at how encryption and decryption work with *binary numbers*, from simple transposition to more complex algorithms, which opened the way to modern standards.

Alphanumeric characters, including spaces, punctuation, numbers, and other signs, which we use as a written language, are seen by computers in the form of standard binary-number codes. As mentioned earlier, the two main alphanumerical codes used in computer communications are ASCII (*American standard code for information interchange*),[5] and its extension EBCDIC (*extended binary coded decimal interchange code*). These two standard codes use seven-bit and eight-bit (one byte) words, respectively. The

[5] Standard ASCII tables providing the corresponding alphanumeric/binary conversion can be found, for instance, in:
http://en.wikipedia.org/wiki/ASCII,
www.webopedia.com/quick_ref/asciicode.asp,
www.neurophys.wisc.edu/comp/docs/ascii/.

**Table 25.1** Ciphering by switching bit positions by pairs.

| Plaintext | I | m | i | s | s | y | o | u |
|---|---|---|---|---|---|---|---|---|
| **PT-ASCII** | 1001001 | 1101101 | 1101001 | 1110011 | 1110011 | 1111001 | 1101111 | 1110101 |
| **CT-ASCII** | 0110001 | 1110011 | 1110001 | 1101101 | 1101101 | 1111001 | 1110111 | 1101011 |
| **Ciphertext** | 1 | s | q | m | m | y | w | k |

**Table 25.2** Ciphering by taking characters by pairs and swapping their center bits.

| Plaintext | I | m | i | s | s | y | o | u |
|---|---|---|---|---|---|---|---|---|
| **PT-ASCII** | 1001001 | 1101101 | 1101001 | 1110011 | 1110011 | 1111001 | 1101111 | 1110101 |
| **CT-ASCII** | 1001001 | 1101101 | 1100001 | 1111011 | 1111011 | 1110001 | 1100111 | 1111101 |
| **Ciphertext** | I | m | a | { | { | q | g | } |

ASCII and EBCDIC codes have $2^7 = 128$ and $2^8 = 256$ possible alphanumerical character and other keyboard-command possibilities, respectively. For cryptography purposes, this represents a huge extension of the previous 26-letter alphabet from earlier times.

How to construct ciphers from binary-numbers? The answer is that there is an unlimited number of ways to proceed, even in the simplest cases. Instead of 26 symbol characters, we now have only two symbols, namely 0 and 1. Since each group of seven bits (ASCII) represents one plaintext character, there is a wide variety of possible schemes for transposition, substitution, or permutations. The number of ways in which any single seven-bit word can be modified into any other seven-bit word is $2^7 = 128$, as opposed to only 26 with the ordinary alphabet. But the interesting feature of the binary system is that now we can code the bits not only within a single seven-bit block, but over the entire message sequence, which scrambles the blocks between themselves. For instance, the (four bit per character) sequence

$$a_3a_2a_1a_0b_3b_2b_1b_0c_3c_2c_1c_0d_3d_2d_1d_0\ldots$$

can be transformed by switching the bit positions by pairs, which gives

$$a_2a_3a_0a_1b_2b_3b_0b_1c_2c_3c_0c_1d_2d_3d_0d_1\ldots$$

Alternatively, we can take characters by pairs and swap their center bits, i.e., for five bit per character sequences:

$$a_4a_3b_2a_1a_0b_4b_3a_2b_1b_0c_4c_3d_2c_1c_0d_4d_3c_2d_1d_0\ldots$$

Clearly, the decryption algorithms consist of performing the reverse substitutions from the above ciphertexts. Let us see now what these two approaches produce as cipher texts (CT). Consider the example in Tables 25.1 and 25.2 (ASCII, seven-bit characters), noting that we can now use capital characters in the alphabetical plaintext (PT).

In the first case, we observe that the last bit ($a_0$) of each character is unchanged (since the number of bits is odd), and that the character y is invariant. In the second case, the first two characters, I and m, are invariant. But if we were to swap more bits up to a complete permutation algorithm, we would achieve most perfect scrambling. For an eight-character or 56-bit codeword, the number of possible permutations is $56! = 7.1 \times 10^{74}$, although with a sufficiently complex substitution algorithm, this approach would seemingly yield undecipherable messages. But the drawback is that its key (the scrambling algorithm) is fixed once and for all. No matter how complex, a fixed-key cryptoalgorithm is, in fact, very vulnerable to attack. The key can be intercepted during its communication (over the Internet, for instance) without the two parties being ever aware of it. The key can be retrieved from the software used to encrypt or decrypt, should one of the terminal computers be intruded by a third party. The alternative is to use a different key (referred to as a *one-time-pad cipher*) for each new message. This means that the two parties have agreed beforehand on a list of different keys (here, substitution algorithms) to be used only once. This approach is not at all practical for any routine use, such as in an army central command. If the list of keys is to be communicated over the Internet (or any other wide-range transmission medium), or listed in a book (as in the case of the Enigma machine), it is also extremely vulnerable. The solution to this issue, which is not trivial, is to be addressed in the next two subsections. In the following, we will prepare ourselves further by considering how *secret keys* can be used in the binary system.

In the binary system, a secret key is made of a single codeword of any desirable length, which must remain known only to the two parties. In contrast, the encryption and decryption algorithm can be known to anyone. If the secret key is to be memorized, so as to leave no paper or computer-file trace of it, it must be plain English (or any language for that matter); for instance, a "codeword," a "passphrase" or even the first paragraph or page of a book text commonly used and secretly agreed on. In the last case, the key may simply be the page number of the book, which is easy to memorize. By itself, this requirement of easy memorization or that the key be in plain English represents a built-in weakness, because it exposes the cryptosystem to frequency analysis. Nowadays, this is no longer an issue, since random binary keys are used. The remaining issue is how to safely exchange and renew such keys, which is discussed later.

Mixing the secret key with the plaintext provides a means for both encryption and decryption. Such mixing is performed by simple *Boolean operations*. In the binary system, the four operations, called Boolean, are NO (logical inversion), AND (multiplication), OR (addition or subtraction), and XOR (exclusive OR). The rules of these operations on one (NO) or two (AND, OR, XOR) bits or "operands" have been introduced in Chapter 15. We note that with binary numbers, addition and subtraction are the same ($1 + 1 = 1 - 1 = 0$, $1 + 0 = 1 - 0 = 0 + 1 = 0 - 1$). We also note that XOR is the same as OR, except for the rule $1 + 1 = 0$, which is similar to $9 + 1 = 0$ in the ordinary digital system. The operators (NO, AND, OR, and XOR) are also noted by mathematicians $\neg$, $\wedge$, $\vee$, and $\oplus$, respectively. Thus NO $a$, $a$ AND $b$, and $a$ XOR $b$ are noted $\neg a$, $a \wedge b$, and $a \oplus b$, respectively. The other basic Boolean operators are $\leq$ (noted $\rightarrow$) and $=$ (noted $\leftrightarrow$). Thus, $a \rightarrow$ b is zero if a > b, and one otherwise. Similarly, $a \leftrightarrow b$ is zero if $a \neq b$, and one if $a = b$. It is easily checked that the binary result of the operation $(a \rightarrow b) \wedge (b \leftrightarrow c)$ is unity if and only if the condition $a \leq b = c$ is fulfilled. In this chapter, we will not be concerned by Boolean logic other than just using XOR.

**Table 25.3** Ciphering with the (case-sensitive) key sAXOPHON.

| Plaintext | I | m | i | s | s | y | o | u |
|---|---|---|---|---|---|---|---|---|
| **Key** | s | A | X | O | P | H | O | N |
| **PT-ASCII** | **1**001001 | 1101101 | 1101001 | 1110011 | 1110011 | 1111001 | 1101111 | 1110101 |
| **K-ASCII** | **1**110011 | 1000001 | 1011000 | 1001111 | 1010000 | 1001000 | 1001111 | 1001110 |
| **CT-ASCII** | **0**111010 | 0101100 | 0110001 | 0111100 | 0100011 | 0110001 | 0100000 | 0111011 |
| **Ciphertext** | ; | ‘ | 1 | < | # | 1 | space | ; |

An interesting property of XOR is that its double application to $a$ with the same operand $b$ restores operand $a$, i.e., $a \oplus b \oplus b = a$. We can, thus, use this property for the purpose of encryption with a key. Indeed, if we perform XOR between each bit of the plaintext and the key, we obtain a ciphertext. If we repeat the operation, this time between the ciphertext (CT) and the key (K), we obtain the plaintext (PT). Table 25.3 shows an example, with sAXOPHON as a (case-sensitive) key.

We see from this example that the same ciphertext character may correspond to different plaintext characters, namely the ;s corresponds to I and u and the 1s corresponds to i and y, respectively. We may, therefore, rapidly conclude that this XOR operation with a key as long as the plaintext (or sufficiently long and repeated) and as many as 128 possible characters would produce quite a strong cipher. But this is not the case. To provide here a simple explanation, consider indeed the first codeword of the ciphertext, 0111010. Because of the XOR operation, it is certain that

(1) For bit numbers 7, 3, and 1: the operands must be of the same parity, namely coming from the operations 0 XOR 0 or 1 XOR 1;
(2) For bit numbers 6, 5, 4, and 2: the operands must be of opposite parities, namely, coming from the operations 0 XOR 1 or 1 XOR 0.

Since there are two operand choices for each bit, the number of possibilities is $2^7 = 128$. But we can reduce this number by correlating the two different conditions for each bit. To establish such a correlation, we can tabulate all correspondence possibilities between each seven-bit ciphertext block and key block in the form of a $128 \times 128 = 16\,384$ element array, where the cases that violate the two conditions are left empty. A straightforward frequency analysis of both plaintext and key, searching for the most common letters, digrams, trigrams, and other most commonly used English words will allow a small computer to finish the job in seconds! This description suggests that in the binary system, encryption with a secret key on a one-to-one character basis is unsafe, unless complex random keys are used.

As we saw earlier, the binary system offers a tremendous advantage, which is the possibility of mingling the characters together in the coding process, on a bit-by-bit or block-by-block basis (a *block* being a string of bits of any prescribed length). This is the principle used by the first attempt at a standard cipher, developed in the 1960s by IBM and called *Lucifer* (as a pun on *cipher*). Figure 25.1 shows the Lucifer coding procedure. In the first step, the plaintext is converted into a string of binary digits (not necessarily ASCII). Then the string is split into 64-bit blocks (step 2) which are

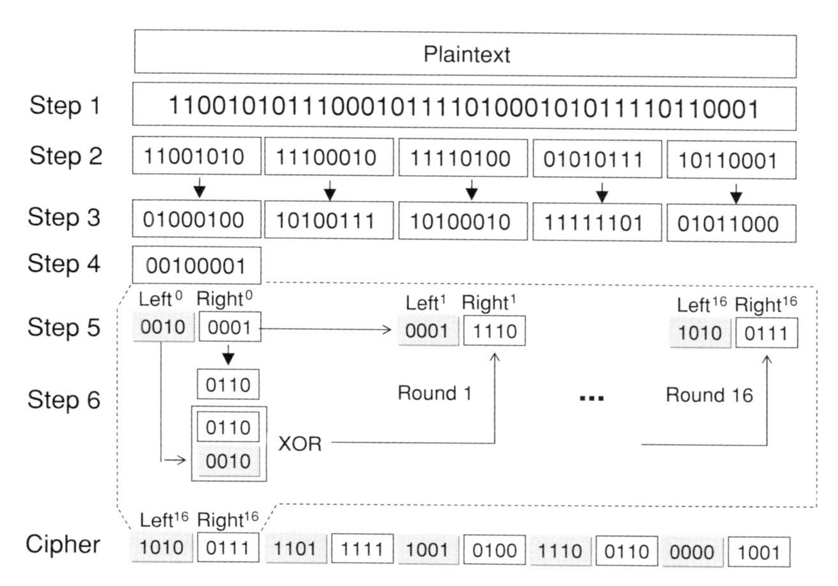


**Figure 25.1** Principle of the Lucifer encryption algorithm, as described in the text.

individually encrypted (step 3). The resulting blocks are shuffled (step 4), then split into two 32-bit elements called *left*$^0$ and *right*$^0$ (step 5). The next step is called a *round*, and this round will be repeated sixteen times. Call these two elements L0 and R0, respectively. The round first passes R0 through a *mangler*, which performs a complex bit substitution according to a *mangling function*. The result is XORed with L0 to give R1. The original R0 becomes L1. The next round does the same with the pair (L1, R1), and so on until one gets (L16, R16). The cipher is then formed with the concatenated string of (L16, R16) elements.

We see that this new approach for encryption is quite complex and definitely harder to break than any previous scheme. Even with public knowledge of the encryption algorithm, the code breaker must know which mangling function (the key) has been used for the 16 rounds, which shuffling algorithm is used in step 4, and which encryption algorithm is used in step 3. The combination of these three unknowns presents a formidable obstacle to code-breaking attempts. But when these different elements are known, decryption is "straightforward" for any computer. The cipher is decomposed into 32-bit block pairs representing as many L16, R16 groups. The element L16 becomes R15. The element R16 is then XORed with L15, then the result is passed through the inverse mangling function (the key), which yields R15. And so on until L0 and R0 are obtained. The L0 and R0 elements are then concatenated into a 64-bit block.

The bits are then reordered, according to the inverse of the shuffling algorithm. The resulting block is decrypted. Then all decrypted blocks are concatenated together and converted into plaintext. We can infer that the whole decrypting operation can be performed in a few milliseconds of computer time, depending on the complexity of the step-3 cipher and the mangling function. As I have stated it, the mangling function represents the key to the encryption and decryption process. Such a key can be defined by a *number*, which represents the different integer parameters to be used for generating the mangling function. Thus, the number of possible keys is virtually infinite. To prevent any possibility that this code became *absolutely* unbreakable, it was agreed that the number of keys should be $2^{15} \approx 72 \times 10^{15}$, which corresponds to a 56-bit key. Twenty-five years later, the principle of the Lucifer system with a 56-bit key was officially endorsed in the USA to become the *data encryption standard* (DES), which is still widely in use today, albeit superseded by its own variants and other standards, such as AES (see further).

## 25.3 Double-key encryption

In spite of the tremendous levels of protection offered by algorithms like DES and Lucifer, a major weakness remains, which is the obligation for the originator to communicate the *key* to the destinee. Like the Germans with Enigma, both parties could keep a secret copy of the same book, where the key to be used is indicated for each day of a month or year. A new set of books must be created and exchanged on this monthly or yearly basis. But a malicious third party may intercept the book and rapidly copy the contents, while making sure that no one notices anything. As a result, all secret communications based on this book's keys could be read "in the open" by this third party, the worst being that the two ends are unaware of it.

In cryptography, the tradition has it to call the originator *Alice* and the destinee *Bob*. The malicious interceptor, who eavesdrops, is *Eve*. From this point, I will use these three nicknames. For Alice, the encryption task is like putting a message into a locked steel box and sending it to Bob through the mail. If Bob does not have a copy of Alice's key to open the lock, then he can't retrieve Alice's message. Alice must then provide Bob a copy of her key, for instance by meeting regularly in some agreed time and place. But this way of proceeding amounts to the same thing as passing on and hiding key books, notwithstanding the time and effort required.

This intractable problem of *key exchange* caused cryptographers to search for new directions in the way message "boxes" could be locked and unlocked. They found that one simple solution could be to use *two locks* instead of one! To visualize how this can work, assume that both Alice and Bob have different keys of their own. Alice locks her message box and sends it to Bob (step 1). Bob cannot open it, since he does not have Alice's key. What he does is to put a second lock on the box, lock it with his own key, and send the box back to Alice (step 2). Then Alice opens the first lock with her key. She then sends back the box to Bob (step 3). Now Bob can open the box, since he has the key of the remaining lock (step 4). Such a process is referred to as *double-key encryption*. The remarkable result is that the message is safely communicated without Alice and Bob exchanging any key whatsoever. Consider now a practical illustration. Table 25.4 shows the different steps of double-key encryption, using the principle of adding (XOR) a same-length key (since the purpose is just to prove the point through successive XOR additions, I have chosen random keys for Alice and Bob).

We see from the above example that at the end of this double-key encryption process, Bob is able to retrieve the plaintext unaltered (step 4). We note again that in the different

**Table 25.4** Double-key encryption.

| Plaintext | I | m | i | s | s | y | o | u |
|---|---|---|---|---|---|---|---|---|
| **PT** | 1001001 | 1101101 | 1101001 | 1110011 | 1110011 | 1111001 | 1101111 | 1110101 |
| **Alice's key** | 0001001 | 1010100 | 0101011 | 0111101 | 0111010 | 0001001 | 1001001 | 1100100 |
| **Step 1** | 1000000 | 0111001 | 1000010 | 1001110 | 1001001 | 1110000 | 0100110 | 0010001 |
| **Bob's key** | 1011001 | 1000111 | 0111010 | 1111101 | 0101110 | 0011001 | 1001000 | 0001011 |
| **Step 2** | 0011001 | 1111110 | 1111000 | 0110011 | 1100111 | 1101001 | 1101110 | 0011010 |
| **Alice's key** | 0001001 | 1010100 | 0101011 | 0111101 | 0111010 | 0001001 | 1001001 | 1100100 |
| **Step 3** | 0010000 | 0101010 | 1010011 | 0001110 | 1011101 | 1100000 | 0100111 | 1111110 |
| **Bob's key** | 1011001 | 1000111 | 0111010 | 1111101 | 0101110 | 0011001 | 1001000 | 0001011 |
| **Step 4** | 1001001 | 1101101 | 1101001 | 1110011 | 1110011 | 1111001 | 1101111 | 1110101 |
| **Plaintext** | I | m | i | s | s | y | o | u |

**Table 25.5** Intercepting the message in Table 25.4.

| | | | | | | | | |
|---|---|---|---|---|---|---|---|---|
| **Step 1** | 1000000 | 0111001 | 1000010 | 1001110 | 1001001 | 1110000 | 0100110 | 0010001 |
| **Step 2** | 0011001 | 1111110 | 1111000 | 0110011 | 1100111 | 1101001 | 1101110 | 0011010 |
| **XOR** | 1011001 | 1000111 | 0111010 | 1111101 | 0101110 | 0011001 | 1001000 | 0001011 |
| **Step 3** | 0010000 | 0101010 | 1010011 | 0001110 | 1011101 | 1100000 | 0100111 | 1111110 |
| **XOR** | **1001001** | **1101101** | **1101001** | **1110011** | **1110011** | **1111001** | **1101111** | **1110101** |
| **Plaintext** | I | m | i | s | s | y | o | u |

steps, Alice and Bob only used their own keys. The explanation of this successful retrieval is the following. Call $M$ the plaintext message block and $A$, $B$ the keys from Alice and Bob. The cipher in step 1 is, thus, $M$ XOR $A$, or $M \oplus A$. The cipher in step 2 is $M \oplus A \oplus B$. That in step 3 is $M \oplus A \oplus B \oplus A = M \oplus B$ (using the above-described property of XOR). Finally, the result in step 4 is $M \oplus B \oplus B = M$, which is the plaintext.

While the previous double-key encryption surely works without any key exchange, it is absolutely unsafe! Indeed, if Eve is ever able to intercept all three messages (ciphers in step 1, step 2, and step 3 in the above), the only thing she has to do to retrieve the plaintext is to XOR these together, as shown in Table 25.5.

There is nothing surprising in this result if we consider from the previous definitions that what Eve does is the operation

$$(M \oplus A) \oplus (M \oplus A \oplus B) \oplus (M \oplus B) = M \oplus M \oplus M = M.$$

Since Eve can so effortlessly retrieve the plaintext, without even any cryptanalysis, we conclude that this "no key exchange" approach based on the addition of independent keywords is, by and large, the worst possible and riskiest way to proceed.

A safer solution for Alice and Bob could consist in using *substitution* algorithms instead. But the problem is that substitution algorithms have no reason to be commutative. If Alice first encrypts the message with her own substitution-key algorithm, then Bob does the same with his own, and then Alice and Bob again, the end result would be nonsense. A fancier approach had to be devised, which led to the breakthroughs of *cryptography without key exchange*, and *public-key cryptography* (PKC), which are described in the next two subsections.

## 25.4 Cryptography without key exchange

The seemingly intractable problem of cryptography *without* key exchange was eventually solved in the mid 1970s. The ground-breaking idea was to use the principle of *one-way functions* in modular arithmetic. Here is the first proposed algorithm (now referred to as *Diffie–Hellman* or *Diffie–Hellman–Merkle*, after the inventors).

Step 0 Alice and Bob call each other (or send an email) to agree on a choice of two numbers $(m, n)$, regardless of the possibility that Eve could have wired the line; they can even make this choice openly public.

Step 1 Alice chooses a *large* integer $A$ and keeps it secret. Then she computes

$$a = m^A \bmod n$$

and sends the result (e.g., by phone or email) to Bob.

Step 2 Bob chooses a *large* integer $B$ and keeps it secret. Then he computes

$$b = m^B \bmod n$$

and sends the result (e.g., by phone or email) to Alice;

Step 3 At her end, Alice computes

$$k = b^A \bmod n.$$

Step 4 At his end, Bob computes

$$k' = a^B \bmod n.$$

As it turns out, the results of these last two computations are equal, i.e., $k = k' = m^{AB}$ (modulo $n$). This means that both Alice and Bob have the same number, which they can use as a common key. Eve is not able to get this information, even if she has intercepted the information on the $(a, b)$ values that Alice and Bob communicated to each other. Although Eve also knows $m, n$, she has no clue as to the exponent $AB$ of the key. And since Alice and Bob chose large secret numbers, $AB$ is also very large! For instance, if $A$ and $B$ are of the order of 1000, the exponent of the key is of the order of 1 000 000, representing a 20-bit key in the binary system. They might as well choose numbers in the order of 10 000, which generate 100 000 000 different key possibilities (26-bit key). Note that from Alice's or Bob's ends, the whole procedure actually involves only two steps (1 and 3 and 2 and 4, respectively), which I have decomposed here for clarity.

It is completely remarkable and counterintuitive that two persons can mutually agree on a secret key by openly exchanging information via a public communication channel (even an eavesdropped line!) without this secret ever being known. The Diffie–Hellman–Merkle algorithm of "key generation without key exchange" can even be generalized to parties of three or more. Consider the following sequence with Alice, Bob, and Cindy, now grouped into simultaneous steps:

Step 1 Alice chooses a large integer number $A$ and keeps it secret; she computes $a = m^A \bmod n$ and sends the result to Bob;
Bob chooses a large integer number $B$ and keeps it secret; he computes $b = m^B \bmod n$ and sends the result to Cindy;
Cindy chooses a large integer number $C$ and keeps it secret; she computes $c = m^C \bmod n$ and sends the result to Alice.

Step 2 Alice computes $c' = c^A \bmod n$ and sends the result to Bob;
Bob computes $a' = a^B \bmod n$ and sends the result to Cindy;
Cindy computes $b' = b^C \bmod n$ and sends the result to Alice.

Step 3 Alice computes $k = (b')^A \bmod n$;
Bob computes $k' = (c')^A \bmod n$;
Cindy computes $k = (a')^A \bmod n$.

As immediately checked, the three results $k, k', k''$, computed separately by Alice, Bob, and Cindy, are all identical to $m^{ABC}$, which now represents their secret key. Note that the initial secret numbers $A, B, C$ don't need to be as large as in the previous case. Numbers of the order of 1000 generate as many as 1 billion keys. But there is no reason not to use greater numbers to give a trillion possibilities.

Here is another variant of the Diffie–Hellman–Merkle algorithm (referred to as *Hughes*), where Alice wants the secret key to be some specific number, $k$.

Step 1 Alice chooses a large integer number $A$ and keeps it secret; she computes

$$k = m^A \bmod n.$$

Step 2 Bob chooses a large integer number $B$ and keeps it secret; he computes

$$b = m^B \bmod n$$

and sends the result to Alice.

Step 3 Alice computes

$$a = b^A \bmod n$$

and sends the result to Bob.

Step 3 Bob computes

$$c = 1/B$$

and

$$k' = a^c \bmod n.$$

The result of Bob's operation is $k' = a^{1/B} = a^{1/B} = (b^A)^{1/B} = (m^{AB})^{1/B} = m^A = k$ (modulo $n$). Thus, Alice has successfully communicated her preferred key choice to Bob. The advantage of the Hughes algorithm is that Alice can initiate the communication of encrypted messages herself, not only to Bob but to any group of persons. She can send them the key at a later time.

These approaches concern *secret-key cryptography*, which, except for the absence of key exchange, resort to the same classical approach when it comes to encryption and decryption. Furthermore, the two parties share the same (secret) key, therefore, they can both encrypt and decrypt with that key. In the communication channel, their relationship is symmetric, hence, the name *symmetric-key* cryptography. An interesting alternative would be to have different keys for encryption and decryption: Bob can send Alice encrypted messages using a shared key, but only Alice is able to decrypt them, using her own secret key. Bob's and Alice's communication being, thus, asymmetric, they are using a new form of *asymmetric-key* cryptography. In the next section, we shall see that modular arithmetic opened the path to asymmetric cryptography, better known as *public-key cryptography* (PKC).

## 25.5 Public-key cryptography and RSA

The approach of *public-key cryptography* (PKC) led to the RSA standard, named after its inventors R. Rivest, A. Shamir, and L. Adleman. A detailed description of PKC and RSA was presented in Chapter 20 concerning *Shor's (quantum) factorization algorithm*, along with modular-algebra basics shown in Appendix T. The reason for describing PKC and RSA in that earlier chapter is to illustrate that the cryptoalgorithm would be readily "cracked" overnight should, one day, Shor's factorization be successfully implemented in *quantum computers*. Fortunately for PKC and RSA, such a horizon still lies in the far distant future, but in the timescale of cryptology, how "distant" is that? Here, I assume that the reader is already familiar with PKC and RSA, or will refer back to Chapter 20 and, therefore, I shall only further develop the discussion about factorizing large numbers.

The PKC and RSA cryptosystem is, indeed, based on the fact that it is extremely difficult to factorize large numbers. To recall, given a number $n$, factorization consists of finding the unique set of prime numbers $(p_1, p_2, \ldots, p_k)$ whose product $p_1 \times p_2 \times \ldots \times p_k$ is equal to $n$. Note that these prime factors are not generally all different. For instance, the number 1000 is factorized into $2 \times 2 \times 2 \times 5 \times 5 \times 5 = 2^3 \times 5^3$. We can then use just two sufficiently large and different prime numbers $(p, q)$ to generate a bigger number $n = pq$, knowing that someone who knows $n$ would have a hard time figuring out the two primes $(p, q)$, and this is basically the essence of the PKC and RSA safety assumption. For instance, consider $n = 62\,615\,533$. To find its factorization, we need to divide it by prime numbers, trying them out one after another. In this specific case, the answer is $p = 7919$ and $q = 7907$, which are the highest two primes in the first-thousand prime list.[6] This choice made factorization easy with any pocket calculator.

[6] See lists of the first 1000 primes in, for instance:
http://primes.utm.edu/lists/small/1000.txt
www.math.utah.edu/~pa/math/primelist.html.

But what about $n = 15\,773\,077$? The answer ($p = 2383$, $q = 6619$) is less immediate, since it takes 200 division tests to find $q$ in this first-thousand-prime list, but this is still within reach of a home computer having this list as a database. Assume next that we select $(p, q)$ from a huge list of known primes, for instance up to $10^9$, yielding numbers of the order of $10^{18}$. Even at a rate of $10^9$ division tests per second (a computing power that only a few states are able to afford), this would take about $10^9$ seconds or *31.7 years* to check!

With large PKC and RSA key sizes (e.g., $n \approx 2^{64} \approx 10^{19}$, $n \approx 2^{128} \approx 10^{38}$), the problem of factorization is no longer within easy reach, and rather requires both extended computing facilities and time effort. To promote the advancement and security of cryptography, the company *RSA Security* offers challenges to the public, which take the form of huge modulus numbers to factorize (namely, given $n$, find the two primes $p$ and $q$ such that $n = pq$).[7] The current challenges concern different numbers with lengths from 704 to 2048 bits. Here is the RSA-2048 number, which has 617 decimal digits:[8]

2519590847565789349402718324004839857142928212620403202777713783604366202070
7595556264018525880784406918290641249515082189298559149176184502808489120072
84499268739280728777673597141834727026189637501497182469116507761337985909570
00973304597488084284017974291006424586918171951187461215151726546322822168699
8754918242243363725908514186546204357679842338718477444792073993423658482382
42811981638150106748104516603773060562016196762561338441436038339044149526344
32190114657544454178424020924616515723350778707749817125772467962926386356373
28991215483143816789988504044536402352738195137863656439121201039712282
2120720357

Two of the earlier challenges, named RSA-140 (140 decimal digits, 465 bits) and RSA-155 (155 decimal digits, 512 bits), were successfully met by international teams of researchers from both academy and industry in February and August 1999, respectively. Note that RSA-155 was renamed RSA-512 so that the number defines bit size rather than decimal-digit size.

Factorizing RSA-140 required 125 workstations (175 MHz) and 60 personal computers (300 MHz) to run for about one month, a combined 8.9 years of CPU time. Including the preparation work, the total elapsed time was actually 2.2 months.

Factorizing RSA-512 (previously RSA-155) required 160 workstations (175–400 MHz) and 120 personal computers (300–450 MHz) to run for 3.7 months, representing a combined 35.7 years of CPU time. The total elapsed time was about 7.4 months.

The two latest challenges, which were successfully met in 2003 and 2005, concerned RSA-576 and RSA-640, respectively. The latter took a combined 30 CPU years, representing 5 months of elapsed time at a 2.2 GHz computing rate per machine! As for the RSA-2048, it may remain unchallenged not only in this current decade (2008–2020), but possibly over this century, as the estimates below illustrate.

[7] See: www.rsa.com/rsalabs/node.asp?id=2093.

[8] Exact reproduction in the printed form of this book is not guaranteed.

The general method of "attack" for the factorization problem (and also finding logarithms $A$ from $a = m^A \bmod n$) is known as GNFS (*general number field sieve*), or NFS. The word "sieve" comes from *Erathosthenes* (240 BC). This is the same person who estimated the circumference of the Earth (tens of centuries before the Earth was proven to be a sphere), based on the observation that, at the same time of year, the sun casts shadows with different angles for different latitudes. His nonintuitive "sieving algorithm" for sorting out prime numbers can be described as follows: "List all integers less than $n$ and remove multiples of all primes less than or equal to the square root of $n$; the numbers that are left are the primes below $n$." The NFS algorithm is implemented in two stages.[9] The first stage, called "sieving," requires moderate computation power (hence, the use of multiple personal computers and workstations in parallel), and leads to a set of equations referred to as the "matrix." The second stage requires a supercomputer with massive internal CPU-memory use (e.g., 0.8 Mbytes for RSA-140 and 3.2 Gbytes for RSA-155–RSA-512). Finding the solution of the matrix might take as much time as the sieving stage, or just a fraction thereof. Once found, this solution leads to instant factorization or logarithm definition. The two parameters of interest for attacking a number $n$ (as defined by its bit length, also referred to as "field size") are the requirements in time ($L$) and memory space ($S$). It can be shown that these requirements are defined by

$$L(n) = \exp\left[C \times (\log n)^{\frac{1}{3}} (\log \log n)^{\frac{2}{3}}\right], \tag{25.1}$$

$$S(n) = \sqrt{L(n)}, \tag{25.2}$$

where $C$ is a constant. Knowing these requirements for a smaller number $n'$, we get the relative estimates

$$l_{n/n'} = \frac{L(n)}{L(n')} = \exp\left\{\text{const} \times \left[\begin{array}{l}(\log n)^{\frac{1}{3}} (\log \log n)^{\frac{2}{3}} \\ - (\log n')^{\frac{1}{3}} (\log \log n')^{\frac{2}{3}}\end{array}\right]\right\}, \tag{25.3}$$

$$s_{s/n'} = \frac{S(n)}{S(n')} = \sqrt{\frac{L(n)}{L(n')}}. \tag{25.4}$$

Table 25.6 provides the numbers $(l_{n/n'}, s_{n/n'})$ corresponding to the different values of $(n, n')$. The first row corresponds to the data provided by RSA,[10] i.e., where estimates are relative to $n' = 512$. Analyzing these data, the constant $C$ involved in the definition in Eq. (25.1) is observed to follow a heuristic rule $C = 64.11 \times (n/n')^{0.2542}$. We can use this rule to compute other estimates with $n' = 576$, 640, 704, 768, 1024, and 2048. The new dataset we, thus, obtain heuristically provides some estimates for the time and memory size increase factors required to reach larger-number sizes from the best previous achievement. We, thus, see from the table that the time increase from RSA-512 to RSA-576 represents about an 11-fold factor, corresponding to a raw computing time of $11 \times 7.4$ months $= 81.4$ months, or 6 years and 9 months! But if about 1000

[9] See *RSA Bulletin* 13 (2000), www.rsa.com/rsalabs/node.asp?id=2088.
[10] See *RSA Bulletin* 13 (2000), www.rsa.com/rsalabs/node.asp?id=2088.

**Table 25.6** Relative time and memory space increases for factoring number $n$ or solving a logarithm problem of same field size (top row) in reference to the corresponding data for smaller number $n'$ (left columns). The numbers are expressed in bit size.

| | Time increase | | | | | | Memory-space increase | | | | | |
|---|---|---|---|---|---|---|---|---|---|---|---|---|
| | 576 | 640 | 704 | 768 | 1024 | 2048 | 576 | 640 | 704 | 768 | 1024 | 2048 |
| **512** | 10.9 | 101 | 835 | 6000 | $7\times10^6$ | $9\times10^{15}$ | 3.3 | 10 | 29 | 77 | 2650 | $9\times10^7$ |
| **576** | | 1.1 | 8.2 | 390 | $3\times10^5$ | $2\times10^{14}$ | | 2.8 | 7.7 | 20 | 550 | $1\times10^7$ |
| **640** | | | 6.5 | 39 | $3\times10^{10}$ | $3\times10^{12}$ | | | 2.5 | 6 | $2\times10^5$ | $2\times10^6$ |
| **704** | | | | 5.5 | 2400 | $1\times10^{11}$ | | | | 2.5 | 50 | $4\times10^5$ |
| **768** | | | | | 335 | $9\times10^9$ | | | | | 20 | $9\times10^4$ |
| **1024** | | | | | | $3\times10^6$ | | | | | | 1750 |

workstations and personal computers were used along with (say) five supercomputers, this delay could be reduced to a single year. The corresponding increase factor for the CPU memory space is 3.3, corresponding to $3.3 \times 3.2$ Gbytes $= 10.5$ Gbytes. Consider now that RSA-576 was solved in 2003. We see from our estimates that to reach the ballpark of RSA-768, the time and memory-space requirements are increased by about 400-fold and 20-fold, respectively. With the conventional approach, this would require some 400 000 workstations and 100 supercomputers. As far as RSA-1024 and RSA-2048 are concerned, it is hard to make any sense of the projections in both time and memory space!

Another way to analyze time projections for the factorization or logarithm problem is to consider the progress actually realized over the last 30 years. Remarkably, the progress in key size and time is very closely linear, like another *Moore's law*. Note, however, that Moore's law is linear but in a logarithmic performance scale, unlike in the present case. From the aforementioned reference, the law can be expressed using the phenomenological formula

$$\text{size}_{\text{dec}} = 4.23 \times (Y - 1970) + 23, \tag{25.5}$$

where $Y$ is the current year and the modulus size is expressed in the number of decimal digits. This linear law has been verified over the past 30 years. The projected years for solving 1024-bit keys (309 decimal digits) and 2048-bit keys (617 decimal digits) are 2037 and 2110, respectively. Other investigators have suggested a model, according to which the size should be dictated by a cubic law

$$\text{size}_{\text{dec}} = \left(\frac{Y - 1928}{13.25}\right)^3, \tag{25.6}$$

which gives the years 2017 or 2018 for reaching 1024-bit keys, and 2041 for reaching 2048-bit keys. We see that both laws predict that 1024-bit keys won't be solved before 20 years at the very least. Keys as large as 2048 bits might take 40 years (2040) or over a century (2110). It is important to note that such estimates are only based on *publicly available* data. Throughout history, it has always been assumed that government agencies have always been ahead of any progress in this field, and the market implications of cryptography do not exclude the possibility that private, yet unpublicized, efforts, have already reached substantially higher performance. One final remark: there is no mathematical proof that breaking RSA absolutely requires factorization of the modulus $n$, it is just what is currently conjectured, until new breakthroughs may happen. But even this remark represents a conjecture by itself.

The above facts, data, and estimates help one to grasp (at the very least) what the problem of factorization of large numbers (or finding logarithms) mean in terms of resource, time, and effort. The fact that teams were able to complete the factorization of RSA-140 and RSA-155/512 does not mean that the RSA cryptosystem is now broken or "cracked." Instead, these successful experiments provided valuable information on the state of the art in factorization, from algorithms and methods to hardware and time requirements. It is still safe for Alice and Bob to use 512-bit RSA keys (even more so with 1024-bit keys), knowing the costs in personnel, capital investment, and time that Eve would have to support. Another important consideration is the *lifetime* of the data to be protected. This lifetime can range from the scale of a single day (e.g., certified signature, restricted-access broadcast, stock-exchange instructions) to several years (e.g., business and financial contracts, long-term strategy).

With very large scale integration (VSLI), integrated components (IC) technology has produced chips capable of performing RSA encryption and decryption with various modulus sizes (namely 32, 120, 256, 272, 298, 512, 593, 1024). Compared with DES (see the next section), RSA is 1000 times slower from the hardware perspective. From the software perspective, RSA is only 100 times slower. Encryption is always faster than decryption. Using a workstation, the encryption and decryption tasks require 30 ms and 160 ms, respectively, for 512 bits and 80 ms and 930 ms for 1024 bits.[11] The process can be speeded up by an appropriate choice of *public key*. Appropriate values are those which have binary words of value $2^N + 1$, which in binary have the form $(100\ldots01)$ with $N + 1$ bits and only two 1 bits. This choice reduces the operation of exponentiation to only $N + 1$ successive multiplications. Referring back to the PKC parameters introduced in Section 20.7, the standard recommendations are $e = 3(11)$, $e = 17(10001)$, and $e = 65537(100001)$. Such a selection avoids $e = 5(101)$ and $e = 9(1001)$ because of the fact that the public key must be relatively prime with $\phi$. And as it nicely turns out, the numbers 3, 17, and 64 537 are all primes, therefore, more likely to be relatively prime with $\phi$. To complicate Eve's task, the recommendation is also that the private key be sufficiently large and that the choice of modulus be broad ("sufficiently" and "broad" being here left to professional appreciation). It is not recommended to use the same private key for encrypted signature and authentication. It can be shown that malicious Eve can retrieve Alice's private key by sending her a specially prepared document and asking her candidly to sign or certify it with her private key! Several variants and extensions of PKC and asymmetric-key cryptography can be also found in the footnote reference.

[11] See B. Schneir, *Applied Cryptography*, 3rd edn. (New York: John Wiley and Sons, 1996).

## 25.6 Data encryption standard (DES) and advanced encryption standard (AES)

The previous sections have introduced the reader to some aspects of *cryptography standardization*, which concern the encryption and decryption algorithms as well as the formats to be used for keys and ciphers. In this section, I shall briefly describe two leading standard cryptosystems, referred to as DES (*data encryption standard*) and AES (*advanced encryption standard*).

As we have seen, the story began in the 1960s under IBM's Lucifer initiative, which set the grounds for the adoption, in the 1970s, of DES. The initial DES version had a 56-bit key, a reduction from the proposed 128-bit format, which was meant to facilitate cracking the code should government authorities need access for national-security reasons. Rumors have circulated about the existence in DES of embedded *trapdoor functions*, which could enable direct decryption by state-class parties, but such rumors did not prevent its eventual standardization and widespread adoption. At the time, it was considered that cracking DES by *brute-force* attack, i.e., trying all possible keys at random over the entire $2^{56} - 1$ keyspace, would require about 1000 years of CPU time. The rapid progress in computer speed proved this belief utterly wrong. In 1997, the company *RSA Security* issued a public challenge to crack DES. The challenge was successfully met that same year after only 96 days of computation, using a network of no less than 14 000 computers. In the two following years, other teams succeeded in cracking DES in 41 days, then in 56 hours, then in 22 hours and 15 minutes! The machine used for the 56-hour record, nicknamed *Deep Crack*, used 27 parallel motherboards, each one with 64 processor chips (1728 total) capable altogether of performing 90 billion key tests per second.

The simple way to improve DES strength and security was to achieve *double* and *triple* encryption with two and three independent secret keys, respectively. These *double-DES* (2DES) and *triple-DES* (3DES) approaches, which bring the effective key size to $2 \times 56 = 112$ bits and $3 \times 56 = 168$ bits, respectively, were endorsed in 1999 by NIST (*National Institute of Standards and Technologies*), the new name of the original standardization body that launched DES.[12]

In *2DES*, the DES algorithm is implemented just twice in a row with two independent secret keys, which brings the effective key length to $56 + 56 = 112$. Note that doubling the key length squares the *complexity* (the size of the keyspace), namely, $(2^{56})^2 = 2^{112} = 10^{33.7} \approx 5 \times 10^{33}$. Thus, *Deep Crack*, which is able to check out $10^9$ keys per second would now require $5 \times 10^{24}$ seconds of CPU time, corresponding to $1 \times 10^{17}$ years or about 15 centuries! Even after an improbable increase of CPU power by one billionfold, this brute-force attack would require $1 \times 10^8$ years or four centuries! But Diffie and Hellman once proved that the effective number of keys is not given by the above figures; the 2DES keyspace is merely doubled, namely $2^{57}$, which does not increase safety so much with respect to standard DES.

[12] Note that DES was also approved by ANSI (American Standards Institute) under the name of *data encryption algorithm* (DEA) and by ISO (International Standards Organization) under the name of DEA-1.

In *3DES*, three independent secret keys are used; one can simply iterate DES with three keys. A different approach, referred to by ANSI as TDEA (*triple data encryption algorithm*) is to use a combination of encryption and decryption with the three keys: if $K_1, K_2, K_3$ are the three keys and $E_K$ and $D_K$ the encryption and decryption functions with a given key $K$, the cipher $C$ and message $M$ are defined by $C = E_{K3}\{D_{K2}[E_{K1}(M)]\}$ and $M = D_{K1}\{E_{K2}[D_{K3}(C)]\}$, respectively. With 3DES, the complexity is really a whopping $2^{112}$.

In spite of the improvements introduced by double and triple encryption, DES further evolved into the *advanced encryption standard* (AES), with the prospect of lasting use in future decades. It is beyond the scope of this chapter to enter into the algorithmic features of DES and AES. For a rapid but detailed introduction of the DEA and AES, see, for instance, my earlier publication.[13] However, it is worthwhile highlighting here some of the innovating features of AES. Basically, AES is a symmetric encryption algorithm using blocks of 128 bit (twice the size of DES), and key sizes of 128, 192 or 256 bits (twice or four times the 64-bit DES key size). The ground-breaking innovation in AES is the introduction of *byte-to-byte multiplication*, which comes as a supplemental operation resource to the Boolean XOR addition of previous DES. For the bytes to keep a constant size under multiplication, the operation requires the bytes to be expressed in a *polynomial representation*, and the multiplication to be performed *modulo some irreducible polynomial*. Although we will not have any use of polynomial multiplication in this chapter, its principle is described in Appendix Y, for the sake of education and also for the curious or demanding. The AES algorithm is completed by various stages of mingling together the plaintext and key "sub-bytes," shifting the result by rows and columns, and effecting complex permutations thereof through an "S-box" look-up table. Cryptanalysts have shown that with a 256-bit key, AES has a complexity of $2^{100}$ (and not $2^{256}$), which comes close, so to speak, to that of 3DES, which is about three orders of magnitude greater ($2^{112}$).

As a concluding statement, it should be stressed herewith that *cryptography is hardly limited to encryption, decryption and key exchange protocols*. Indeed, to communicate messages safely through any of the above-described cryptosystems is not all that matters in real cryptospace. There exists a broad catalog of other severe issues and more immediate threats, which academics focusing on quantum alternatives generally tend to overlook, while these represent far more serious exposures to attacks, even at basic network layers. For instance, how can Bob be confident that Alice *is* the real Alice, and the reverse? How are they confident that their messages, as received, are the ones they intended to share? This is the realm of *digital signatures* and message *authentication*. As the name indicates, a digital signature is a way of certifying, within reasonable confidence, that the message originator Alice is really the person claimed, or that her signature is technically *unforgeable*. Authentication is different. It ensures that nothing in the message was altered, even to the level of a single punctuation mark. This protection is important in all matters pertaining to official records, such as titles, patents, or

[13] E. Desurvire, *Wiley Survival Guide in Global Telecommunications, Broadband Access, Optical Components and Networks, and Cryptography* (New York: John Wiley and Sons, 2004), Ch. 3, pp. 345–477.

tax declarations (contents and time stamping), contracts, authorizations, and confidential orders, for instance. The key to these processes is the use of *hash functions*, which can be viewed as *one-way encryption*. For short, and to convey a flavor, a 10 000-page plaintext book can be "hashed" into a 512-bit signature (the "hash"). The same plaintext with a single comma being altered would yield a fully different hash, immediately betraying any forgery. The hashing algorithm also makes it extremely difficult for Eve to forge a plaintext while yielding the same hash (referred to as *hash attack*). For a basic introduction to PKC-based signature, authentication, and hash algorithms, see my earlier book.[14] As also outlined in the reference, there exist many futuristic applications of cryptography, which are likely to mark the next generations, starting with present times: *certified email, anonymous digital cash, simultaneous secret exchange, and contract signature* and *global electronic polling or voting*, to quote a few. Such not-so-distant applications of cryptography are only a hint of the vastness and complexity of the subject, especially when it comes to the border of individual rights and privacy protection issues. It would be naive to trust that rights and privacy would be better protected if under the full control of computing systems. The two examples of electronic voting and digital cash point to future controversies and new forms of potential abuse. A so-called perfect electronic voting or digital cash system could be hacked, causing the nullity of elections or cash transactions, and putting democracy and private rights at an unprecedented level of risk. And this observation remains true and accurate despite any means of "provably secure" transmission of data!

*Quantum cryptography*, to be described in the next section, should, therefore, be addressed within the full scope of communications security and its applications, a field that relies essentially on *classical* cryptoalgorithms. While quantum cryptography represents a true revolution in the field, *it only solves some of the many issues faced by communications security*. This is the correct perception I seek to convey here by having made this chapter a comprehensive description of cryptography from classical to quantum, rather than exclusively focusing on the second, which would have missed the global application context.

## 25.7 Quantum cryptography

In this section, and the following ones, we shall venture into the intriguing and tricky subject of *quantum cryptography* (QC). After having reviewed the vast array of possibilities offered by classical cryptosystems (DES, AES, PKC) and being convinced that a would-be "Eve" would face extreme difficulties in code-breaking attempts, we may first wonder about or challenge the usefulness of QC approaches. It is possible to answer such a question straight away. First, from any current knowledge, it must be stated that *QC is not a means of encrypting and decrypting information; rather, it is a means of distributing secret keys over a public channel*; moreover, *such an approach, referred to*

[14] E. Desurvire, *Wiley Survival Guide in Global Telecommunications, Broadband Access, Optical Components and Networks, and Cryptography* (New York: John Wiley and Sons, 2004), Ch. 3, pp. 345–477.

*as "quantum key distribution" (QKD), is provably, absolutely secure according to the laws of quantum mechanics.*

The above important statements will serve as a conceptual background for the following description of QC and QKD and avoid raising other expectations. Here, we shall not venture into the debate of assessing whether or not absolute security (vs. that offered by PKC, for instance) is at all required in any secret-key exchange, or its impact on global communications security. The purpose of this chapter is only to convey the principle of how QKD works and not to try measuring the pros and cons of the approach from industrial and realistic network-environment standpoints. This being said, *provable absolute security* in the process of key exchange may only be seen as a desirable "extra" feature, especially if proven practical to implement and compatible with higher-layer network standards. The main difficulty is that academics involved in PKC may fail to recognize several other forms of network-security threats, particularly at the level of Alice and Bob, who just manipulate classical bits at a bottom network-layer level. The main argumentation in favor of PKC is the impossibility of Eve's capturing the secret key without Alice's and Bob's awareness, and this is true as long as Eve's attack only concerns the quantum channel over which the key is exchanged. It is assumed that Alice's and Bob's computers and local networks cannot be approached by Eve and be physically tapped, including by wireless means. This is the main assumption to keep in mind while discovering the extraordinary elegance of QKD and accepting the "absoluteness" of its security.

In the forthcoming sections, we shall consider several QKD approaches, referred to as QKD *protocols*. Since QKD is about key exchange, the quantum channel must be a communication channel involving physical distance. Thus, its implementation requires one to use *light* and its quantum constituents, *photons*. It is important, therefore, to describe beforehand the properties of electromagnetic waves and the associated quantum measurements, which are both at the root of QKD. In the description, I shall also develop the necessary parallels with QIT and qubit and operator formalism.

## 25.8 Electromagnetic waves, polarization states, photons, and quantum measurements

In this section, we review the basic physical properties of light, and how these properties can be conceptually, and quite nicely, related to that of quantum states.

The elementary particles of light, or equivalently, light *energy quanta*, are called *photons*. The particle-like characteristics of light only appear at relatively low light powers, where the number of quanta approaches the order of unity.[15] At macroscopic scales, with large numbers of quanta, such photon granularity vanishes, and light appears to the physical observer in the form of an *electromagnetic* (EM) *wave*. The EM wave is

[15] More specifically, the average number of photons $\langle n \rangle$ emitted per second (units of photon/s) by a light source of power $P$ (units of W) at electromagnetic frequency $\nu$ (units of Hz or $s^{-1}$) is defined as $\langle n \rangle = P/(h\nu)$, where $h = 6.62 \times 10^{-34}$ J/s is Planck's constant.

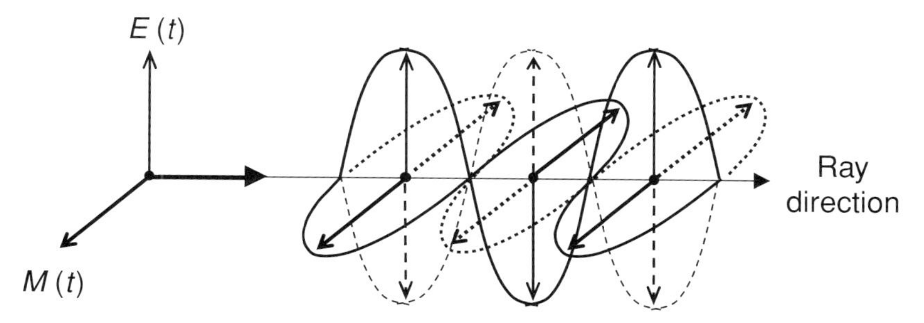


**Figure 25.2** Electromagnetic wave associated with a light ray, showing electric (E) and magnetic (M) components oscillating in orthogonal planes

made up of two oscillating field components, one electrical (E) and one magnetic (M), as illustrated in Fig. 25.2. The two EM field components oscillate in planes orthogonal to the direction of the EM-wave propagation, which is that of the associated *light ray*. The direction pointed to by either the E-field or the M-field oscillation is referred to as *light polarization*. When the E-field direction is orthogonal to the ray direction, one refers to the light beam as being *transverse-electric*, or *TE-polarized*. In the other case, where the M-field is orthogonal to the ray, the light is called *transverse-magnetic*, or *TM-polarized*. When the polarization is either TE or TM, light is said to be *linearly polarized*. For short, we can say that linearly polarized light is either *vertically* (TE) or *horizontally* (TM) *polarized*. As a first hint of quantum mechanics, one refers to the TE and TM possibilities as *polarization states*. Consistently, the TE and TM polarization states are said to be *orthogonal*. Occasionally, I shall designate the TE and TM states by the symbols $\updownarrow$ and $\leftrightarrow$, respectively.

Linear polarization, in fact, represents a special case where the E-field component oscillates with a constant direction in space, meaning that its oscillation plane is fixed. In the most general case, we may conceive of the E-field as resulting from the superposition of two orthogonally polarized components $E_1$, $E_2$, each one being characterized by its own phase, $\varphi_1, \varphi_2$. If the two components are in phase ($\varphi_1 = \varphi_2$), then the resulting E-field is linearly polarized. If not ($\varphi_1 \neq \varphi_2$), the resulting E-field direction rotates over time. To see this more clearly, define an orthonormal basis $\vec{u}_1, \vec{u}_2$ associated with the two E-field components. It is then possible to express the resulting E-field in this basis as a vector with two complex coordinates $E_k \mathrm{e}^{\mathrm{i}\varphi_k}$, $k = 1, 2$ according to:

$$E = \begin{pmatrix} E_1 \mathrm{e}^{\mathrm{i}\phi_1} \\ E_2 \mathrm{e}^{\mathrm{i}\phi_2} \end{pmatrix} \mathrm{e}^{\mathrm{i}\omega t}, \tag{25.7}$$

where $\omega = 2\pi\nu$ is the oscillation frequency in radian/s. In the case $\varphi_1 = \varphi_2 \equiv \varphi$, we have from the above definition

$$E = \mathrm{e}^{\mathrm{i}\phi} \begin{pmatrix} E_1 \\ E_2 \end{pmatrix} \mathrm{e}^{\mathrm{i}\omega t}, \tag{25.8}$$

which corresponds to a time-invariant direction, (or oscillation plane) of the E-field. With the appropriate basis rotation (the choice being conventional), we may also write

$$E = \frac{e^{i\phi}}{|E|}\begin{pmatrix}1\\0\end{pmatrix}e^{i\omega t}, \tag{25.9}$$

which corresponds to a TE-polarized state. Such a representation is immediately reminiscent of a quantum state, namely here the qubit $|0\rangle$ from the basis $\{|0\rangle, |1\rangle\}$. We may, thus, identify the TE polarization state with the quantum state $|0\rangle$. Consistently, we identify the TM polarization with the quantum state or qubit $|1\rangle$, but with an E-field relative phase to be determined later on.

Consider next the case $\varphi_1 \neq \varphi_2$ or $\delta\varphi = \varphi_1 - \varphi_2 \neq 0$. Let us rewrite Eq. (25.7) as

$$E = e^{i\frac{\varphi_1+\varphi_2}{2}}\begin{pmatrix}E_1 e^{i\frac{\delta\varphi}{2}}\\E_2 e^{-i\frac{\delta\varphi}{2}}\end{pmatrix}e^{i\omega t}, \tag{25.10}$$

where we have overlooked the common phase factor $e^{i(\varphi_1+\varphi_2)/2}$. Taking the real part of the E-field, we obtain:

$$E = \begin{bmatrix}E_1\cos\left(\omega t + \dfrac{\delta\varphi}{2}\right)\\[2ex] E_2\cos\left(\omega t - \dfrac{\delta\varphi}{2}\right)\end{bmatrix}. \tag{25.11}$$

The result in Eq. (25.11) shows that the two E-field components alternatively reach maximal amplitudes at times $t = 2n\pi - \delta\varphi/(2\omega)$ and $t' = 2n\pi + \delta\varphi/(2\omega)$, respectively, with $n$ being an integer. The E-field direction, thus, periodically rotates about the light ray axis at the frequency $\omega = 2\pi\nu$. Consider the special case $\delta\varphi = \pi/2$, and for simplicity assume $E_1 = E_2 = 1$. From Eq. (25.11), and after basic trigonometry we obtain (within an arbitrary phase $\varphi$):

$$E \equiv \begin{bmatrix}\cos(\omega t + \varphi)\\ \sin(\omega t + \varphi)\end{bmatrix}. \tag{25.12}$$

The result in Eq. (25.12) shows that the E-field periodically rotates about the ray axis at the frequency $\omega = 2\pi\nu$, the vector end describing a corkscrew-like trajectory. It is easily established that for an observer looking at an incoming light ray, the E-field appears to rotate in the *clockwise* (cw) direction. In the case $\delta\varphi = -\pi/2$, we obtain a similar conclusion, except that the E-field rotates in the counterclockwise (ccw) direction. In both cases, light is said, therefore, to be *circularly polarized*. Conventionally, one refers to the ccw case as representing *right-circular polarization* and the cw case as representing *left-circular polarization*.[16] As in the linear-polarization case, we may view the two circular polarizations as representing two *orthogonal states* (orthogonality to be shown

[16] This convention is easy to remember when looking at one's right or left hands, the thumb pointing up and the fingers bent. With the right hand, the fingers indicate a ccw direction, while with the left hand, they indicate a cw direction.

hereafter). Occasionally, I shall designate the right- and left-circular polarization states by the symbols $\cup$ and $\cap$, respectively.

Back to complex notation, we may represent the circularly polarized E-fields according to the decomposition:

$$\begin{aligned} E &= \frac{1}{|E|}\begin{pmatrix} e^{i\frac{\delta\varphi}{2}} \\ e^{-i\frac{\delta\varphi}{2}} \end{pmatrix} e^{i\omega t} \\ &= \frac{e^{i\frac{\delta\varphi}{2}}}{|E|}\begin{pmatrix} 1 \\ 0 \end{pmatrix} e^{i\omega t} + \frac{e^{-i\frac{\delta\varphi}{2}}}{|E|}\begin{pmatrix} 0 \\ 1 \end{pmatrix} e^{i\omega t} \\ &= e^{i\varphi}\left[\frac{1}{|E|}\begin{pmatrix} 1 \\ 0 \end{pmatrix} e^{i\omega t} + \frac{e^{-i\delta\varphi}}{|E|}\begin{pmatrix} 0 \\ 1 \end{pmatrix} e^{i\omega t}\right], \end{aligned} \tag{25.13}$$

where $\varphi = \delta\varphi/2$ is an arbitrary phase factor. For circularly polarized E-fields we have

$$\begin{aligned} E^{\pm} &= e^{i\varphi}\left[\frac{1}{|E|}\begin{pmatrix} 1 \\ 0 \end{pmatrix} e^{i\omega t} + \frac{e^{\pm i\frac{\pi}{2}}}{|E|}\begin{pmatrix} 0 \\ 1 \end{pmatrix} e^{i\omega t}\right] \\ &\equiv e^{i\varphi}\left[\frac{1}{|E|}\begin{pmatrix} 1 \\ 0 \end{pmatrix} e^{i\omega t} \pm i\frac{1}{|E|}\begin{pmatrix} 0 \\ 1 \end{pmatrix} e^{i\omega t}\right]. \end{aligned} \tag{25.14}$$

The decomposition in Eq. (25.13) is immediately reminiscent of a quantum state superposition! Indeed, if we define the second term in the decomposition as $\pm|1\rangle$, the superposition is equal to $|0\rangle + |1\rangle \equiv \sqrt{2}\,|+\rangle$ in the upper case (+, or cw, or left-circular polarization, or $\cap$), and to $|0\rangle - |1\rangle \equiv \sqrt{2}\,|-\rangle$ in the lower case (−, or ccw, or right-circular polarization, or $\cup$). Within a normalization factor of $1/\sqrt{2}$, we see that the two circularly polarized states $\cap$, $\cup$ are equivalent to the quantum states or qubits $|+\rangle, |-\rangle$. We can now check the states orthogonality by effecting the (Hilbert-space) scalar product:

$$\begin{aligned} E^{+}\cdot\left(E^{-}\right)^{*} &= \frac{e^{i\varphi}}{|E|}\left[\begin{pmatrix} 1 \\ 0 \end{pmatrix} + i\begin{pmatrix} 0 \\ 1 \end{pmatrix}\right] e^{i\omega t}\cdot\left\{\frac{e^{i\varphi}}{|E|}\left[\begin{pmatrix} 1 \\ 0 \end{pmatrix} - i\begin{pmatrix} 0 \\ 1 \end{pmatrix}\right] e^{i\omega t}\right\}^{*} \\ &= \frac{1}{|E|^{2}}\begin{pmatrix} 1 \\ i \end{pmatrix}\cdot\begin{pmatrix} 1 \\ i \end{pmatrix} \\ &= \frac{1}{|E|^{2}}\left(1 + i^{2}\right) \\ &\equiv 0, \end{aligned} \tag{25.15}$$

which is the expected result, just as $|+\rangle, |-\rangle$ are orthogonal states. It is left as an easy exercise to show that the superposition of two circularly polarized E-fields having opposite directions (i.e., $E^{+} \pm E^{-}$) yield linearly polarized E-fields. This result is expected, since within a $1/\sqrt{2}$ normalization factor, we know for a fact that $|+\rangle + |-\rangle = |0\rangle$ and $|+\rangle - |-\rangle = |1\rangle$. Thus, *circular polarization states can be viewed as the superposition of two linear polarization states (oscillating in quadrature), and the reverse.*

The above demonstration, albeit somewhat tedious, was very well worth it: it made it possible to reach a major conclusion: *there exist four special states of polarization for the classical E-field, which form two orthogonal bases, namely* $\{\updownarrow, \leftrightarrow\}$ *and* $\{\cap, \cup\}$, *respectively. Furthermore, there exists a one-to-one correspondence between these two bases and the quantum bases* $\{|0\rangle, |1\rangle\}$ *and* $\{|+\rangle, |-\rangle\}$, *respectively*. We must notice that in order to reach this conclusion we did not need to make any assumption about the quantum nature of light. Such an assumption will come into the picture later, when considering single-photon measurements. It will turn out, however, that the EM field associated with the photon is essentially described by the same $\{\updownarrow, \leftrightarrow\}$, $\{\cap, \cup\}$ bases and, therefore, the above conclusion remains fully valid at quantum scales.

Next, I shall introduce an optical component referred to as a *quarter-wave plate* (QWP). As the name suggests, it is a piece of flat material, and it is transparent to light. The material, however, is a special type of crystal exhibiting the property of *birefringence*. A material is said to be birefringent if the speed of light varies according to the polarization orientation of the incident light ray (assumed linearly polarized). Thus, the speed of light is faster in some polarization direction (referred to as the *fast axis*) and slower in the orthogonal direction (referred to as the *slow axis*). If the incident light ray is parallel to either the fast or slow axes, the light polarization remains unchanged. However, if the fast (or slow) axis forms a 45° angle with the incident polarization, the E-field component that projects onto the fast axis propagates faster than the E-field component that projects onto the slow axis, thus, introducing a phase delay $\Delta\varphi$ between the two components and, hence, making the state of polarization of the ray evolve as it traverses the plate. The additional feature of the QWP is that its thickness is precisely calculated in order for this net phase delay to be $\Delta\varphi = \pi/2$ (corresponding to a quarter wavelength, hence, the name). Intuitively, we may already infer that *the QWP transforms the input polarizations from linear to circular and the converse*, and this inference is absolutely correct! Let us prove such a property now. Assuming a linearly polarized incident E-field, as defined in Eq. (25.8), and the QWP axes oriented at 45° (cw) from the vertical axis, the incident E-field is projected along the QWP fast and slow axes according to the definition:

$$E_{\text{in}} = e^{i\phi}\frac{1}{\sqrt{2}}\begin{pmatrix} E_1 + E_2 \\ E_1 - E_2 \end{pmatrix} e^{i\omega t}. \tag{25.16}$$

After traversing the QWP, the output E-field has become

$$\begin{aligned} E_{\text{out}} &= e^{i\phi}\frac{1}{\sqrt{2}}\begin{bmatrix} E_1 + E_2 \\ e^{i\Delta\varphi}(E_1 - E_2) \end{bmatrix} e^{i\omega t} \\ &\equiv e^{i\phi}\frac{1}{\sqrt{2}}\begin{bmatrix} E_1 + E_2 \\ i(E_1 - E_2) \end{bmatrix} e^{i\omega t}. \end{aligned} \tag{25.17}$$

Substituting $E_1 = 1$, $E_2 = 0$ (incident ray in the $\updownarrow$ orientation, or aligned with the fast axis, or, conventionally, $|0\rangle$), we obtain:

$$\begin{aligned} E_{\text{out}} &= e^{i\phi}\frac{1}{\sqrt{2}}\begin{pmatrix} 1 \\ i \end{pmatrix} e^{i\omega t} \\ &= e^{i\phi}\frac{1}{\sqrt{2}}\left[\begin{pmatrix} 1 \\ 0 \end{pmatrix} + i\begin{pmatrix} 0 \\ 1 \end{pmatrix}\right] e^{i\omega t}, \end{aligned} \tag{25.18}$$

which is immediately identified as the left-circular polarization state, or $\cap$, or $|+\rangle$. Clearly, the case $E_1 = 0$, $E_2 = 1$, corresponding to an incident ray in the $\leftrightarrow$ orientation (or aligned with the slow axis, or, conventionally, $|1\rangle$), yields the right-circular polarization state, or ∪, or $|-\rangle$. Thus, *the QWP converts linear polarization states into circular polarization states, according to the transformations* $|0\rangle \to |+\rangle$ *and* $|1\rangle \to |-\rangle$.

Let us show next the converse operation. Assume an incident ray that is circularly polarized, according to the base ∩ or ∪ (equivalently, $|+\rangle$ or $|-\rangle$). It suffices it to substitute $E_1 = 1$ and $E_2 = \pm\mathrm{i}$ in Eqs. (25.16) and (25.17) to obtain for the output E-field:

$$E_{\text{out}} = \mathrm{e}^{\mathrm{i}\phi}\frac{1}{\sqrt{2}}\begin{bmatrix}1\pm\mathrm{i}\\ \mathrm{i}(1\mp\mathrm{i})\end{bmatrix}\mathrm{e}^{\mathrm{i}\omega t} = \mathrm{e}^{\mathrm{i}\phi}\frac{1}{\sqrt{2}}\begin{bmatrix}1\pm\mathrm{i}\\ \mathrm{i}\pm 1\end{bmatrix}\mathrm{e}^{\mathrm{i}\omega t}$$
$$= \begin{cases} \mathrm{e}^{\mathrm{i}\phi}\dfrac{1}{\sqrt{2}}\begin{pmatrix}1+\mathrm{i}\\ 1+\mathrm{i}\end{pmatrix}\mathrm{e}^{\mathrm{i}\omega t} \equiv \mathrm{e}^{\mathrm{i}\phi+\frac{\pi}{4}}\begin{pmatrix}1\\ 1\end{pmatrix}\mathrm{e}^{\mathrm{i}\omega t} \\ \mathrm{e}^{\mathrm{i}\phi}\dfrac{1}{\sqrt{2}}\begin{bmatrix}1-\mathrm{i}\\ -(1-\mathrm{i})\end{bmatrix}\mathrm{e}^{\mathrm{i}\omega t} \equiv \mathrm{e}^{\mathrm{i}\phi-\frac{\pi}{4}}\begin{pmatrix}1\\ -1\end{pmatrix}\mathrm{e}^{\mathrm{i}\omega t}. \end{cases} \tag{25.19}$$

As expressed in the vertical or horizontal basis (i.e., after rotating the reference axes by 45° ccw), the output E-field is, finally,

$$E_{\text{out}} \equiv \begin{cases} \mathrm{e}^{\mathrm{i}\phi+\frac{\pi}{4}}\begin{pmatrix}1\\ 0\end{pmatrix}\mathrm{e}^{\mathrm{i}\omega t} \\ \mathrm{e}^{\mathrm{i}\phi-\frac{\pi}{4}}\begin{pmatrix}0\\ 1\end{pmatrix}\mathrm{e}^{\mathrm{i}\omega t}, \end{cases} \tag{25.20}$$

which (within arbitrary phase factors) corresponds to the linearly polarized states ↕ (top) and ↔ (bottom), respectively. Thus, *the QWP converts circular polarization states into linear polarization states, according to the transformations* $|+\rangle \to |0\rangle$ *and* $|-\rangle \to |1\rangle$, as previously announced.

We may summarize all of the above conclusions by recalling from Chapter 16 that the *Hadamard matrix*, *H*, precisely achieves the four transformations

$$\begin{cases} H\,|0\rangle = |+\rangle \\ H\,|1\rangle = |-\rangle \\ H\,|+\rangle = |0\rangle \\ H\,|-\rangle = |1\rangle. \end{cases} \tag{25.21}$$

This observation establishes that a QWP device transforms the polarized EM states (↕, ↔, ∩, ∪) in a way strictly equivalent to that of the Hadamard matrix, *H*, on the quantum states ($|0\rangle$, $|1\rangle$ $|+\rangle$, $|-\rangle$).

From the above description, we have seen that it is possible to manipulate the polarization states of the EM field. To progress further, we must, at this stage, take into account the quantum nature of light, namely the particle-like behavior of the photon. Such behavior can be readily understood by considering the two experiments illustrated in Figs. 25.3 and 25.4. In the first experiment (Fig. 25.3), we divide an incident EM light beam with E-field $E$ into two beams of equal E-field amplitudes; this is achieved through a 50:50 beamsplitter. Since the incident EM power is given by $P = |E|^2$, the E-field amplitudes of the two output beams are $E/\sqrt{2}$. Two detectors placed on the

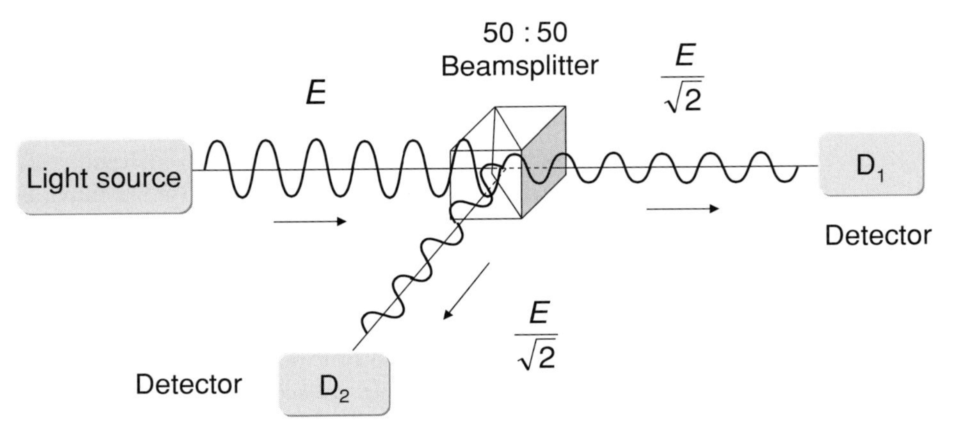


**Figure 25.3** Dividing an EM light beam (electrical field $E$, power $P = |E|^2$) through a 50:50 beamsplitter, resulting in two beams (electrical fields $E/\sqrt{2}$) of detected powers both equal to $P' = P/2$.

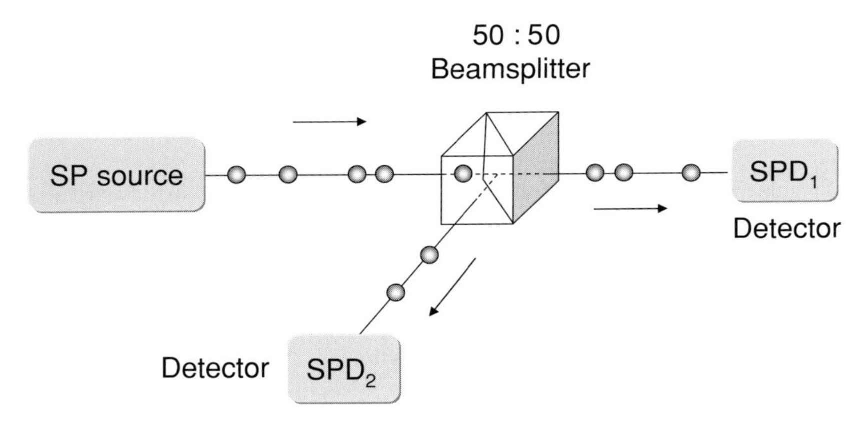


**Figure 25.4** Same experimental apparatus as Fig. 25.3 with single photons (SP) being emitted and detected.

output beam paths measure equal EM powers of $P' = P/2$, as expected. The picture becomes very different if we reduce the EM power to the point of reaching the level of photon granularity, i.e., the number of photons emitted per second by the light source is of the order of unity. The experimental apparatus shown in Fig. 25.4 is basically identical to the previous one, except that now the light source produces single photons[17] (SP) and that the two detectors are capable of detecting such single photons. Since photons are particle-like energy quanta, they cannot be split. On reaching the 50:50 beamsplitter, therefore, they "choose" at random to follow one of the two possible beam paths.[18]

[17] The sequence of single photons emitted by a light source can be visualized as individual drops falling in a sink from a leaking tap, or individual cars passing on a highway lane: one at time, but at random times.

[18] The choice for photons to take one path or the other is dictated by the random fluctuations of the "vacuum field," which enters through the fourth or unused port of the 50:50 beamsplitter.

The two choices have exact and equal probabilities of $P = 0.5$. As seen from the figure, the key difference from the previous experiment is that the photon stream is randomly partitioned. Each single-photon detector (SPD) receives one photon at a time, which generates a "ping" count. A ping from detector $SPD_1$ means that the photon has chosen the straight-through path, while a ping from detector $SPD_2$ means that it has chosen the reflection path, as shown. If the straight-through and the reflection paths have strictly equal lengths, there is no possibility of the detectors $SPD_1$, $SPD_2$ generating two pings or counts simultaneously. The two-count histogram is the same as in a coin-flipping experiment (see, for instance, Fig. 1.3). Over a sufficient period of time, the numbers of counts from $SPD_1$ and $SPD_2$ become about equal. This is equivalent to saying that with a sufficiently large number of photons, the light beam has been divided into two beams of equal power, as in the previous experiment with classical light. The second experiment, which is routinely done in the laboratory, represents one of the many physical proofs of the quantum nature of light and its photon granularity.

From this point on, we shall assume that we are dealing with single photons, and that the associated E-fields are in any of the four polarization states $\{\updownarrow, \leftrightarrow\}$, $\{\cap, \cup\}$ or equivalently $\{|0\rangle, |1\rangle\}$, $\{|+\rangle, |-\rangle\}$, respectively. For short, we may say that the photon "is" in any of these quantum states. Next, I introduce two other components: the *half-wave* plate (HWP) and the *polarization beamsplitter* (PBS).

As its name indicates, the HWP is similar to the previously described QWP, except that the net phase delay experienced by two orthogonal E-field components is now $\pi$ or one half of a wavelength. As in the QWP case, the fast axis of the plate must be oriented at 45° of the incident E-field polarization direction, assumed linear. Since the phase shift corresponds to a factor $e^{i\pi} = -1$, the sign of one of the two polarization components is reversed, and the result is a 90° rotation of the incident linear polarization. Thus, the HWP swaps the basis states $\{\updownarrow, \leftrightarrow\}$ into each other, which is equivalent to the transformations $|0\rangle \to |1\rangle$ and $|1\rangle \to |0\rangle$, corresponding to *the action of the Pauli matrix $X$ on the states* $|0\rangle, |1\rangle$ (see Chapter 16). If the incident E-field polarization is circular, the HWP axis orientation is unchanged, but the direction of rotation is reversed. Thus, the effect of the HWP is to swap the basis states $\{\cap, \cup\}$ into each other, which is equivalent to the transformations $|+\rangle \to |-\rangle$ and $|-\rangle \to |+\rangle$, corresponding to *the action of the Pauli matrix $Z$ on the states* $|+\rangle, |-\rangle$ (see Chapter 16). Thus, placing a HWP next to a linearly polarized SP source and orienting it at either 0° or 45° makes it possible to generate single photons into either the $|0\rangle$ or the $|1\rangle$ linear-polarization state. The PBS is a special assembly of birefringent crystal prisms whose effect, as the name indicates, is to separate an incident light beam into two orthogonally polarized components. As shown in Fig. 25.5, a detector $SPD_1$ placed in the straight-through path only detects vertically polarized, or $|0\rangle$ photons, while a detector $SPD_2$ placed in the "reflection" path only detects horizontally polarized, or $|1\rangle$ photons. The PBS–$SPD_1$–$SPD_2$ set-up, thus, constitutes *a quantum measurement apparatus to determine the state of linearly polarized photons*. If a count is obtained from $SPD_1$ or from $SPD_2$, we may attribute the values +1 or −1, respectively, to these two possible measurements. Recall from Chapter 16 that $|0\rangle$, $|1\rangle$, and $\pm 1$ are the eigenvectors and eigenvalues of the *Pauli*

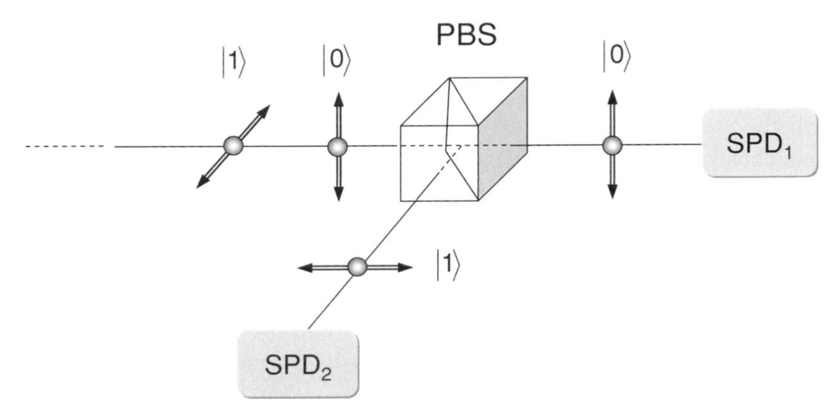


**Figure 25.5** Splitting a light beam of single, linearly polarized photons through a polarization beamsplitter (PBS). The apparatus corresponds to a measurement of photon states in the $Z$ basis.

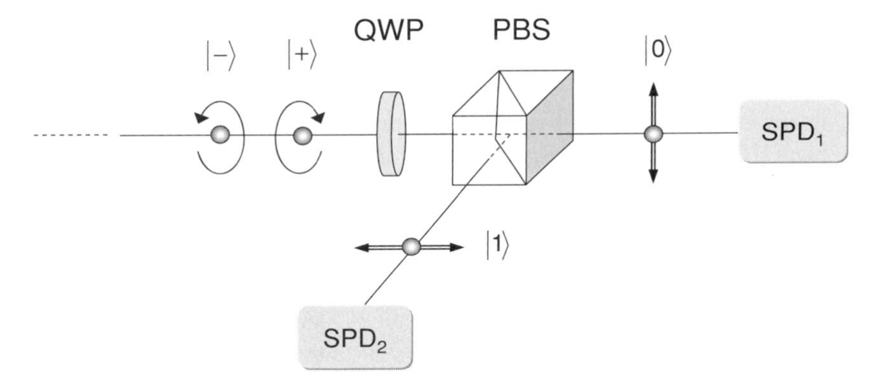


**Figure 25.6** Splitting a light beam of single, circularly polarized photons through a polarization beamsplitter (PBS) preceded by a quarter-wave plate (QWP). The apparatus corresponds to a measurement of photon states in the $X$ basis.

*matrix* $Z$, according to:

$$\begin{cases} Z|0\rangle = |0\rangle \leftrightarrow \begin{pmatrix} 1 & 0 \\ 0 & -1 \end{pmatrix}\begin{pmatrix} 1 \\ 0 \end{pmatrix} = \begin{pmatrix} 1 \\ 0 \end{pmatrix} \\ Z|1\rangle = -|1\rangle \leftrightarrow \begin{pmatrix} 1 & 0 \\ 0 & -1 \end{pmatrix}\begin{pmatrix} 0 \\ 1 \end{pmatrix} = -\begin{pmatrix} 0 \\ 1 \end{pmatrix}. \end{cases} \tag{25.22}$$

Thus, *measuring linearly polarized photons is equivalent to "observing"* $Z$, or equivalently, *to operating in the measurement basis* $\{|0\rangle, |1\rangle\}$, or, for short, *in the* $Z$ *basis*.

A different set-up, which involves *circularly polarized* photons, is shown in Fig. 25.6. The generation of left- or right-circularly polarized photons is achieved by placing a QWP (quarter-wave plate) next to a linearly polarized SP source, with its fast axis at 45° cw or ccw, respectively. A way to tell whether a photon is left- or right-circularly polarized is to convert its polarization into a linear polarization, followed by the

**Table 25.7** Probabilities $p(\pm 1)$ of measuring eigenvalues $\pm 1$ (counts in detectors $SPD_1$ or $SPD_2$) according to input photon state ($\updownarrow$, $\leftrightarrow$, $\cap$, $\cup$ or $|0\rangle$, $|1\rangle$, $|+\rangle$, $|-\rangle$) and measurement apparatus ($X$, $Z$).

| **Input photon state** | $\updownarrow$ or $\|0\rangle$ | | $\leftrightarrow$ or $\|1\rangle$ | | $\cap$ or $\|+\rangle$ | | $\cup$ or $\|-\rangle$ | |
|---|---|---|---|---|---|---|---|---|
| **Measurement basis** | $X$ | $Z$ | $X$ | $Z$ | $X$ | $Z$ | $X$ | $Z$ |
| $p(+1)$ | 0.5 | **1** | 0.5 | **0** | **1** | 0.5 | **0** | 0.5 |
| $p(-1)$ | 0.5 | **0** | 0.5 | **1** | **0** | 0.5 | **1** | 0.5 |

PBS–$SPD_1$–$SPD_2$ apparatus, as illustrated in the figure. The QWP–PBS–$SPD_1$–$SPD_2$ set-up, thus, constitutes *a quantum measurement apparatus to determine the state of circularly polarized photons*. If a count is obtained from $SPD_1$ or from $SPD_2$, we may attribute the values $+1$ or $-1$, respectively, to these two possible measurements. Recall from Chapter 16 that $|+\rangle$, $|-\rangle$, and $\pm 1$ are the eigenvectors and eigenvalues of the *Pauli matrix* $X$, according to:

$$\begin{cases} X|+\rangle = |+\rangle \leftrightarrow \begin{pmatrix} 0 & 1 \\ 1 & 0 \end{pmatrix} \frac{1}{\sqrt{2}} \begin{pmatrix} 1 \\ 1 \end{pmatrix} = \frac{1}{\sqrt{2}} \begin{pmatrix} 1 \\ 1 \end{pmatrix} \\ X|-\rangle = -|-\rangle \leftrightarrow \begin{pmatrix} 0 & 1 \\ 1 & 0 \end{pmatrix} \frac{1}{\sqrt{2}} \begin{pmatrix} 1 \\ -1 \end{pmatrix} = -\frac{1}{\sqrt{2}} \begin{pmatrix} 1 \\ -1 \end{pmatrix}. \end{cases} \quad (25.23)$$

Thus, *measuring circularly polarized photons is equivalent to "observing"* $X$, or equivalently, *to operate in the measurement basis* $\{|+\rangle, |-\rangle\}$, or, for short, *in the* $X$ *basis*.

Assume, next, that we have no knowledge of the input photon states. For the measurement, we have the possibility of using either the $X$ or the $Z$ bases. What happens if we choose the wrong one (namely $X$ for linearly polarized photons, and $Z$ for circularly polarized photons)? There are two ways to answer such a question. The first one is a *classical* answer. It has been established for a fact that circular polarization is the superposition of two orthogonal linear polarizations (oscillating in quadrature), and the reverse. Thus, a circularly polarized E-field incident on a PBS is decomposed into its two linear-polarization components. Likewise, a linearly polarized E-field incident on a QWP–PBS apparatus is first transformed by the QWP into a circularly polarized E-field, and then decomposed by the PBS into its two linear components. But this explanation is valid only for classical E-fields! Indeed, consider now the case of polarized single photons. A circularly polarized photon incident onto a PBS cannot be physically split into two linearly polarized photons. Since the photon is associated with two linear-polarization states (oscillating in quadrature) it has to "choose" one or the other, and such a choice comes with a 50% probability, as in the photon-beam partitioning experiment described earlier. The same conclusion applies to the case of linearly polarized photons incident on a QWP–PBS apparatus. The key conclusion is that *measuring photons with the wrong basis*, i.e., circularly polarized photons with basis $Z$ or linearly polarized photons with basis $X$, *produces equiprobable, random outcomes*. The different measurement possibilities are summarized in Table 25.7.

We now have all the necessary tools to describe the principle of quantum cryptography, or, more precisely, the QKD protocols. Before proceeding to a formal description of such protocols, it is useful to consider a communication link between Alice and Bob, and how they may be able to communicate "secret" information by means of $X$- and $Z$-basis photon-state preparations and measurements; this is described in the next section.

## 25.9 A secure photon communication channel

In this section, we assume that Alice and Bob can send light signals to each other by means of free-space-optics or an optical fiber link, which we refer to as a *photon communication channel*. Both have single-photon sources that are capable of producing photons in any polarization states ($\updownarrow, \leftrightarrow, \cap, \cup$ or $|0\rangle, |1\rangle, |+\rangle, |-\rangle$), and independent photon-measurement apparatus in the $X$ and $Z$ bases. What is possible for Alice is the same as for Bob, so here we only need to consider the case of Alice transmitting photons to Bob.

As a starting point, Alice generates a *random* 16-bit sequence from her end, say, for instance:

$$1010\ 0011\ 1110\ 0100.$$

Then Alice prepares a sequence of 16 photons, which she randomly polarizes according to the following selection rule (for instance):

For the 1 bit, use either $\updownarrow$ or $\cap$ polarizations at random;

For the 0 bit, use either $\leftrightarrow$ or $\cup$ polarizations at random.

Consistently with the above bit sequence and random-selection rule, Alice's preparation yields (for instance) the following polarized-photon sequence:

$$\updownarrow\leftrightarrow\ \cap\cup\ \leftrightarrow\leftrightarrow\ \cap\ \updownarrow\updownarrow\ \cup\cap\ \leftrightarrow\ \cup\ \cup\ \cap\ \updownarrow\ .$$

Alice then inputs the above photon sequence through the link. At the other end, Bob receives the photon sequence and makes 16 corresponding measurements, according to *his own random choice* of $X$ and $Z$ basis. Table 25.8 shows Bob's basis choices and the outcome of each of Bob's measurements. *Conventionally, and unless otherwise specified, a* $+1$ *eigenvalue measurement will be now called a 1 bit of information, and a* $-1$ *measurement will be called a 0 bit of information, respectively*. The table shows that Bob's random choices of $X$ and $Z$ measurement bases are either "right" or "wrong." The right basis choices correspond to measuring $\updownarrow$, $\leftrightarrow$ with $Z$, and $\cap$, $\cup$ with $X$. Comparing the top and bottom lines of the table, we notice that the right choices give Bob information bits that are *always* correct with respect to Alice' initial sequence. In the other case of wrong basis choice, Bob's information bits are inherently random and, therefore, not correlated with those of Alice's sequence.

**Table 25.8** Alice's 16-bit and corresponding photon-state sequence, Bob's choice of $X$ and $Z$ measurement basis, and the outcome of the measurements in terms of eigenvalues and corresponding 1 or 0 information bits. The outcomes shown in bold correspond to Bob's accidentally correct choices of measurement basis.

| **Alice's bit sequence** | 1010 | 0011 | 1010 | 0011 |
|---|---|---|---|---|
| **Alice's photon state sequence** | ↕↔ ∪∩ | ↔↔ ∩ ↕ | ↕ ∪∩ ↔ | ∪ ∪ ∩ ↕ |
| **Bob's measurement basis choice** | $ZXZX$ | $XZZX$ | $XZXX$ | $XXZZ$ |
| **Bob's measurement outcome (eigenvalue)** | **+1** +1 −1 **−1** | +1 **−1** +1+ 1 | −1 −1**+1**+1 | **−1** **−1** −1 **+1** |
| **Bob's measurement outcome (bit)** | **1** 1 0 **0** | 1 **0** 1 1 | 0 0 **1** 1 | **0** **0** 0 **1** |

Bob's next move is to communicate to Alice, using a public channel like telephone or email, not the bits he has measured, but *his measurement-basis sequence*, namely from Table 25.8:

$$ZXZX\ XZZX\ XZXX\ XXZZ.$$

Such information will tell Alice Bob's right choices in the measurement, since she is the only one to know how she has coded the photon polarizations. She then uses the same public channel to give Bob her answer, which comes in the form:

YesNoNoYes   NoYesNoNo   NoNoYesNo   YesYesNoYes.

This information tells Bob where his measurements were right and, therefore, which information bits are valid and which other ones are to be discarded.[19] Thus, Bob and Alice keep as valid the following bit sequence (− meaning discarded bit):

$$1 - -0/ - 0 - -/ - -1 - /00 - 1 \equiv 1001001,$$

which amounts to nine valid bits altogether. Since the probability of Bob choosing the right measurement basis is 50%, it is clear that for longer sequences Bob would obtain about 50% valid bits. Thus, if Alice sends a 1000 bit sequence to Bob, the whole operation will result in obtaining a valid bit sequence of a length of nearly 500 bits. Such a sequence can then be used by both Alice and Bob as a shared *secret key* to encrypt and decrypt messages over a public channel, using DES or AES cryptoalgorithms, for instance.[20]

[19] Note that the same result can be obtained the other way around: over a public channel, Alice tells Bob which bases she used to code each of the polarized photons, namely $ZZXX$ $ZZXZ$ $ZXXZ$ $XXXZ$, and then Bob tells Alice which of his measurements were right. What matters is that both Alice and Bob know which bits are to be discarded in the sequence.

[20] Here, I am not addressing the issue of photon loss due to effects such as atmospheric or fiber *absorption*. Put simply, all photons must be generated within a certain time slot. The absence of photons in a given time slot does not mean a measurement failure, but rather an absence of measurement (or a global, zero $SPD_1$–$SPD_2$ count). It is a basic engineering issue to predict the probability that, given the link budget and the photon generation and measurement apparatus used by Alice and Bob, a polarized photon is successfully measured from either end. Alice and Bob, thus, have an accurate knowledge of the photon-transmission capability of their link, namely, how many photons can be successfully transmitted on average. Hence, any observed deviation from this communication quality immediately betrays any wiretapping action on the link.

One may wonder how "safe" the above-described key-distribution approach is. How can Alice and Bob be certain that no malicious Eve is able to intercept the key? The answer is, in fact, quite simple: *there is no possibility whatsoever for Eve to eavesdrop the key-distribution channel*. The proof of this statement is also quite simple: *photons cannot be split*. The photon communications channel is, therefore, reputed "absolutely invulnerable," according to this fundamental law of quantum mechanics. There are many subtleties associated with such a concept of "invulnerability," which I reserve for the reader to the end, see Section 25.13.

## 25.10 The BB84 protocol for QKD

In this section, I describe the *BB84 protocol*, named after its inventors C. H. Bennet and G. Brassard, and the year of publication, 1984. The description made in the previous section is essentially that of the BB84 protocol. Here, it just needs to be formalized further and generalized to any key sizes. Note that the role of Alice and Bob can be interchanged, each one being capable of generating a QKD session, which is also a potential means for authentication and other security tests.

Alice uses a random-bit generator to produce two sequences $A = (a_1, a_2, \ldots)$ and $B = (b_1, b_2, \ldots)$, each of $N = (4 + \Delta)n$ bits length, with $\Delta$ being a small integer number playing the role of a reserve, and $2n$ being the desired key length (the explanation for the factor of two comes later). Then she produces an $N$-*qubit* block sequence $|\psi\rangle$, which can be expressed in the form:

$$\begin{aligned} |\psi\rangle &= \left|\psi_{a_1 b_1}\right\rangle \otimes \left|\psi_{a_2 b_2}\right\rangle \otimes \cdots \otimes \left|\psi_{a_N b_N}\right\rangle \\ &\equiv \bigotimes_{k=1}^{N} \left|\psi_{a_k b_k}\right\rangle . \end{aligned} \tag{25.24}$$

For each cbit pair $a_k, b_k$ the corresponding single qubit $\left|\psi_{a_k b_k}\right\rangle$ is generated according to the following (possible) convention:

$$\begin{cases} |\psi_{00}\rangle = |0\rangle \\ |\psi_{01}\rangle = |+\rangle \\ |\psi_{10}\rangle = |1\rangle \\ |\psi_{11}\rangle = |-\rangle . \end{cases} \tag{25.25}$$

We observe that the random value of the bit $b_k$ determines the choice of either state, $|0\rangle, |+\rangle$ (or $\updownarrow, \cap$ with polarized photons) to represent $a_k = 0$, and the same with $|1\rangle, |-\rangle$ (or $\leftrightarrow, \cup$ with polarized photons) to represent $a_k = 1$. We may also state that $b_k = 0$ encodes the $a_k$ bit in the quantum basis $Z$, and $b_k = 1$ encodes the $a_k$ bit in the quantum basis $X$. This generalizes the description made in the previous section.

The $N$-qubit block $|\psi\rangle$ is then input to the quantum communication channel by Alice, and Bob receives the state $\rho = \varepsilon\left(|\psi\rangle\langle\psi|\right)$. Such a formulation recalls that the channel is characterized by a *quantum operation* $\varepsilon$, which includes any effects such as quantum noise and Eve's wiretapping.

At his end, Bob uses a random-bit generator to produce a sequence $B' = (b'_1, b'_2, \ldots)$ of $N = (4 + \Delta)n$ bits length. This is the random sequence he uses to determine the choice of measurement basis, $X$ or $Z$, for each of the received qubit. For instance, Bob can choose conventionally that if $b'_k = 0$, the measurement basis is $Z$, and if $b'_k = 1$ the measurement basis is $X$. In any case, there is a 50% chance of this measurement choice being "right" or "wrong," according to the description in the previous section. The outcome of Bob's measurement is, thus, a random sequence $A' = (a'_1, a'_2, \ldots)$ of length $N = (4 + \Delta)n$.

Using a public channel, either Bob announces to Alice his sequence of state-measurement bases (*ZXZX* ...), or Alice announces to Bob her sequence of state-preparation bases (*ZZXX* ...). The other person then tells which bits are to be discarded, according to the fact that either measurement or preparation was made in the "wrong" basis. The wrong bits being discarded, the two bit sequences $A$ (from Alice) and $A'$ (from Bob) are, thus, reconciled. With a sufficiently large bit sequence length $N = (4 + \Delta)n$, there is a high probability that there are at least $2n$ valid bits in $A, A'$. In the contrary event, the protocol session must be aborted and restarted with new random sequences $A, B, B'$.

We may conclude too quickly that the above protocol wholly suffices to guarantee the strict equality of Alice's and Bob's bit strings, or $A = A'$. Indeed, there is yet a possibility that the quantum channel used by Alice and Bob is corrupted by the effects of random *quantum noise* and also Eve's tampering actions (e.g., suppressing, injecting, or modifying *some* of the qubits). In this last event, Eve is capable of obtaining partial information about the secret key and, therefore, a higher level of security is required from Alice and Bob. This is where the techniques of *secret key distillation* (SKD) through *information reconciliation* and *privacy amplification* come into the picture. It is beyond the scope of this chapter to venture into the mathematical details of SKD algorithms. Here, it will suffice to convey a rough idea of the underlying principle.

At this point, Alice and Bob have validated two $2n$ bit sequences, $A, A'$, but have no reason to believe that these do perfectly match. Further, they suspect that some of the bits might have been corrupted by quantum noise and eavesdropping. Their next step is to effect "information reconciliation." Alice selects a random subset $X$ of $n$ bits from her sequence $A$, sends $X$ to Bob over a *public channel* (assumedly error-free) and also informs Bob which bits she has selected. Bob receives a sequence $Y$, which he is then able to compare with the corresponding $n$-bit selection from his $A'$ sequence. Bob then deduces the error syndrome $E$ (as defined by $Y = X + E$), deducts the number of errors $e$ and tells Alice the value over the public channel. If $e$ is over a certain commonly agreed threshold $t$, the QKD protocol is immediately aborted. In the favorable case ($e \leq t$), Alice and Bob use a secret classical code, $C_1$, having an error-correction capability of $t$, and unknown to Eve. This makes it possible for each to determine a common codeword $W = X' = Y'$ corresponding to the correction of $X$ and $Y$ under this code. Yet it can be shown that Eve is able to possess partial information about $W$. Then the next step is for Alice and Bob to perform "privacy amplification," which consists of extracting from $W$ a smaller secret key $S$ of size $m < n$, based on their remaining $n$ bits. How they achieve this and why, in this process, Eve's information about the reduced key $S$ is lost to an arbitrarily high level of confidence, is beyond the scope of this chapter to describe. The lesson to retain from the obligation for Alice and Bob to perform SKD is that *QKD alone is not as "absolutely secure" a protocol as is commonly believed.*

## 25.11 The B92 protocol

A simpler version of the BB84 protocol, referred to as *B92* (after its inventor, C. H. Bennet, and the year, 1992), uses only two reference qubits. Alice uses a random-bit generator to produce a single sequence $A = (a_1, a_2, \ldots)$ of $N = (4 + \Delta)n$ bits length. Then she produces an $N$-*qubit* block sequence $|\psi\rangle$, which can be expressed in the form:

$$\begin{aligned} |\psi\rangle &= |\psi_{a_1}\rangle \otimes |\psi_{a_2}\rangle \otimes \cdots \otimes |\psi_{a_N}\rangle \\ &\equiv \bigotimes_{k=1}^{N} |\psi_{a_k}\rangle. \end{aligned} \tag{25.26}$$

For each cbit $a_k$ the corresponding single qubit $|\psi_{a_k}\rangle$ is generated according to the following convention:

$$\begin{cases} |\psi_0\rangle = |0\rangle \\ |\psi_1\rangle = |+\rangle = \dfrac{|0\rangle + |1\rangle}{\sqrt{2}}. \end{cases} \tag{25.27}$$

On receiving the block $|\psi\rangle$, Bob uses a random-bit generator to produce a sequence $A' = (a'_1, a'_2, \ldots)$ of $N = (4 + \Delta)n$ bits length. Conventionally, Bob chooses that if $a'_k = 0$, the qubit measurement basis is $Z$, and if $a'_k = 1$ it is $X$. The outcome of Bob's eigenvalue measurement is a random-bit sequence $B = (b_1, b_2, \ldots)$ of length $N = (4 + \Delta)n$. Here, I shall use the convention according to which the eigenvalue $+1$ corresponds to $b_k = 0$ and $-1$ corresponds to $b_k = 1$.[21] Bob's different measurement outcomes (eigenvalues and cbits $b_k$ as functions of cbits $a_k, a'_k$) are summarized in Table 25.9. We first observe from the table that if the "right" bases for Bob to measure $|\psi_0\rangle$ and $|\psi_1\rangle$ are $Z$ and $X$, respectively, the outcome is deterministic, since in either cases Bob certainly obtains $b_k = 0$, corresponding to $a_k = a'_k$. Therefore, contrary to the BB84 protocol, the information about which are the "right" bases is surely *not* to be publicly exchanged! Consider now the use of the "wrong" bases, which correspond to the cases $a'_k = 1 - a_k$. We observe from the table that Bob's measurements randomly yield the cbits $b_k = 0$ or $b_k = 1$ with equal chances. Thus, it is a matter of Bob communicating his cbit sequence $B = (b_1, b_2, \ldots b_N)$ to Alice over a public channel, while keeping $A' = (a'_1, a'_2, \ldots a'_N)$ secret. Only the bit pairs $a_k, a'_k$ yielding $b_k = 1$ are retained by Alice and Bob. Clearly, such public knowledge does not tell Eve anything about the values of $a_k, a''_k = 1 - a_k$, which have equal chances of being 0 or 1. This operation yields two common subsets of $A$, $A''$ with at least $2n$ valid bits (Bob needs to complement his own bits, according to $a''_k = 1 - a'_k = a_k$, so that his subset matches Alice's). The

[21] The opposite convention, introduced earlier, is applicable to the B92 protocol if Alice chooses instead $|\psi_0\rangle = |1\rangle$.

**Table 25.9** Outcomes of Bob's measurements (eigenvalues and cbits) according to Alice's qubits and Bob's measurement basis $Z, X$.

| | | Alice | |
|---|---|---|---|
| | | $a_k = 0$<br>$\vert\psi_0\rangle = \vert 0\rangle$ | $a_k = 1$<br>$\vert\psi_1\rangle = \vert +\rangle$ |
| **Bob** | $a'_k = 0$<br>$Z$ | $Z\vert 0\rangle = \vert 0\rangle$<br>Eigenvalue $+1$, $b_k = 0$ | $Z\vert +\rangle = \vert 0\rangle$ or $\vert 1\rangle$, 50% chances<br>$b_k = 0, 1$ |
| | $a'_k = 1$<br>$X$ | $X\vert 0\rangle = \vert 0\rangle$ or $\vert 1\rangle$, 50% chances<br>$b_k = 0, 1$ | $X\vert +\rangle = \vert +\rangle$<br>Eigenvalue $+1$, $b_k = 0$ |

rest of the B92 protocol follows the same path as that of BB84, i.e., using information reconciliation and privacy amplification or SKD.

## 25.12 The EPR protocol

The *EPR protocol*, named after entangled EPR pairs or Bell states (see Chapter 16) was invented by A. K. Eckert in 1991. The idea is that Alice and Bob share, over a quantum channel, an ensemble of $N = (4 + \Delta)n$ such EPR pairs, based, for instance, on one of the four EPR–Bell states:[22]

$$|\beta_{00}\rangle = \frac{|00\rangle + |11\rangle}{\sqrt{2}}. \tag{25.28}$$

Prior to any measurement, the quantum state of Alice's and Bob's system is, thus, defined by the $N$-EPR tensor:

$$\begin{aligned} |\psi\rangle &= |\beta_{00}\rangle_1 \otimes |\beta_{00}\rangle_2 \otimes \cdots \otimes |\beta_{00}\rangle_N \\ &\equiv \bigotimes_{k=1}^{N} |\beta_{00}\rangle_k \\ &= \frac{1}{\sqrt{2^N}} (|00\rangle + |11\rangle)^{\otimes N}. \end{aligned} \tag{25.29}$$

As with previous protocols, Alice and Bob use random-bit generators to produce bit sequences $A = (a_1, a_2, \ldots)$ and $A' = (a'_1, a'_2, \ldots a'_N)$, of length $N = (4 + \Delta)n$ bits. The bit values $a_k, a'_k$ determine their choice of measurement basis, for instance $Z$ for $a_k = 0$ (Alice's side) or $a'_k = 0$ (Bob's side), and $X$ in the other case, as applicable to the first (Alice) and second (Bob) qubits, respectively, of any of the EPR pairs from the above $N$-EPR tensor. The cbit outcomes of Alice or Bob measurements generate random bit sequences $B = (b_1, b_2, \ldots, b_N)$ and $B = (b'_1, b'_2, \ldots, b'_N)$, respectively.

[22] EPR–Bell states can be physically generated through pairs of *entangled photons*. See, for instance: http://physicsworld.com/cws/article/print/11360/1/smallphotons, http://physicsworld.com/cws/article/news/24358, www.quantum.at/research/quantum-teleportation-communication-entanglement/entangled-photons-over-144-km.html.

How does an eigenvalue measurement with $X$ or $Z$ apply in the case of a single EPR pair, such as $|\beta_{00}\rangle$? The answer is simple, if one considers that the entangled state can also be written in the form[23]

$$|\beta_{00}\rangle = \frac{|++\rangle + |--\rangle}{\sqrt{2}}. \tag{25.30}$$

If Alice uses $X$ for her first qubit measurement of $|\beta_{00}\rangle$, she obtains equally likely eigenvalues $\pm 1$ with corresponding post-measurement or collapsed states $|\psi\rangle_B = |+\rangle$ or $|-\rangle$, and cbits $b_k = 0$ or $b_k = 1$, respectively (using the same convention as in B92); or

If Alice uses $Z$ for her first qubit measurement of $|\beta_{00}\rangle$, she obtains equally likely eigenvalues $\pm 1$ with corresponding post-measurement or collapsed states $|\psi\rangle_B = |0\rangle$ or $|1\rangle$, and cbits $b_k = 0$ or $b_k = 1$, respectively.

We note here that Alice's measurements of the first qubits of the EPR pair $|\beta_{00}\rangle$ result in the *instant* collapse of the state into one of the qubits, $|\psi\rangle_B = |+\rangle$, $|-\rangle$, $|0\rangle$, or $|1\rangle$, depending on the measurement type and its eigenvalue outcome. Such a collapse is instant indeed, as a result of the intriguing property of entangled states called *nonlocality*. This term is attached to any widely separated systems, which cannot be treated as behaving independently, regardless their physical separation. It is thought-provoking that the action of Alice (on her qubit) instantly affects the system (hence, Bob's qubit) without consideration of wave propagation subjected to the speed of light, $c$. As in *quantum teleportation* (Chapter 18), there is no violation of Einstein's theory of special relativity, according to which information cannot travel faster than $c$. The nonlocality of entangled states, which permits their instant collapse, does

[23] To recall, the states $|\pm 1\rangle$ are given by the qubit superpositions

$$|+\rangle = \frac{1}{\sqrt{2}}(|0\rangle + |1\rangle),\ |-\rangle = \frac{1}{\sqrt{2}}(|0\rangle - |1\rangle).$$

Thus,

$$|0\rangle = \frac{\sqrt{2}}{2}(|+\rangle + |-\rangle) = \frac{1}{\sqrt{2}}(|+\rangle + |-\rangle),\ |1\rangle = \frac{\sqrt{2}}{2}(|+\rangle - |-\rangle) = \frac{1}{\sqrt{2}}(|+\rangle - |-\rangle)$$

and, hence,

$$\begin{aligned}
|\beta_{00}\rangle &= \frac{1}{\sqrt{2}}(|00\rangle + |11\rangle) \\
&= \frac{1}{\sqrt{2}}(|0\rangle\,|0\rangle + |1\rangle\,|1\rangle) \\
&= \frac{1}{\sqrt{2}}\left[\left(\frac{|+\rangle + |-\rangle}{\sqrt{2}}\right)\otimes\left(\frac{|+\rangle + |-\rangle}{\sqrt{2}}\right) + \left(\frac{|+\rangle - |-\rangle}{\sqrt{2}}\right)\otimes\left(\frac{|+\rangle - |-\rangle}{\sqrt{2}}\right)\right] \\
&= \frac{1}{\sqrt{2}}\left[\left(\frac{|++\rangle + |+-\rangle + |-+\rangle + |--\rangle}{2}\right) + \left(\frac{|++\rangle - |+-\rangle - |-+\rangle + |--\rangle}{2}\right)\right] \\
&\equiv \frac{|++\rangle + |--\rangle}{\sqrt{2}}.
\end{aligned}$$

not violate such a principle, because actually no information is exchanged in the process.[24]

Next comes the clever thing about the EPR protocol. Indeed, if Bob then performs a measurement on his remaining qubit (the post-measurement state) $|\psi\rangle_B$, incidentally at random *with the same basis choices* as Alice's, there is absolute certainty that his eigenvalues or cbits ($b'_k$) exactly match those of Alice's ($b_k$), just as with the BB84 protocol. Over a public channel, Alice and Bob only need to compare the bases they used (namely the cbits $a_k$, $a'_k$), and retain as valid only their *secret* cbits $b_k$, $b'_k$ corresponding to the matching cases $a_k = a'_k$.

The same conclusion applies should Bob perform his measurement *before* Alice, or even at the same time! Indeed, the outcome of their measurements is independent of their sequence order, as is an easy exercise to verify. In this perspective, the EPR protocol is not a matter of mere key "exchange," since, in fact, the key remains fully undetermined until both measurements from Alice and Bob have been duly performed, results compared, and valid cbits finally identified.

In view of making an inventory of common secret bits, by sharing measurement bases over a public channel, the EPR protocol appears to be very similar to BB84. The difference, however, is that Alice and Bob must share beforehand a wealth of *entangled states*. This allows them to perform independent measurements, regardless of time sequence and their roles as originator or recipient in the quantum channel. With BB84, one or the other must agree on their roles (originator Alice or recipient Bob) and about the protocol sequence, and only single qubits are "exchanged." Although it is more complex to implement, because of the need for entangled EPR–Bell states as opposed to single qubits, EPR can be viewed as the conceptual "crown jewel" of QKD. Like the other protocols, however, information reconciliation and privacy amplification by SKD is required, to eliminate the effects of quantum noise and eavesdropping.

Another major issue, which concerns all QKD protocols, is *distance*. The different quantum states that are exchanged (BB84, B92) or shared (EPR) by Alice and Bob must, throughout the process, remain immune to various forms of physical perturbation, so as to keep their *coherence*, a term to mean their initial qubit description. Those perturbations introduce *decoherence*, whose nature is to ruin the protocol implementation and its

[24] Here is a classical analogy (or *Gedanken experiment*) to illustrate the concept of nonlocality, which may provoke animated discussions. Assume that Alice prepares two different envelopes, each including either a picture of a cat or that of a dog, and seals them. Alice asks a third party to pick up one of the two envelopes. This envelope is to be sent to Bob, an astronaut on a mission somewhere in the Solar System (meaning several light-hours away). We may conceive of the system represented by the two envelopes altogether forming the entangled superposition $|\beta\rangle = \left(|\text{cat}\rangle_A |\text{dog}\rangle_B + |\text{dog}\rangle_A |\text{cat}\rangle_B\right)/\sqrt{2}$, where only the qubit $|\cdot\rangle_X$ ($X = A, B$) is accessible and measurable by Alice ($A$) or Bob ($B$). The outcome of either measurement is equally likely to be cat or dog. Alice may wait until Bob receives his envelope to open hers, or alternatively she may decide to do it beforehand, hours ahead or just a few minutes before. Opening her envelope, Alice finds the dog! Then she gets the *instant* knowledge that Bob will find, or has already found, the cat. And this despite the need for the information to propagate for several hours through space. It may well be that Alice and Bob perform their openings at the same universal time. They both instantly know who has the cat and who has the dog, without exchanging information. This is an illustration of the principle of *nonlocality* of the combined system. But it took Bob's envelope to travel at (somewhere under) the speed of light to reach him, therefore, putting a physical limit to the information flow from Alice to Bob, which is consistent with relativity.

magic. This is where *quantum error correction* (Chapter 24) may come to the rescue, but as an additional burden to an approach otherwise presumed to be fairly straightforward in implementation. It is important to recall here that error-correction in the quantum domain requires fully operational quantum-gate circuits with qubit dimensions of seven to nine in the quantum channel (as applicable to BB84, B92). Other techniques have been proposed to *purify*, *distil*, or *swap* EPR pairs, making it possible to expand the channel reach in the presence of quantum noise. Such considerations illustrate that it is theoretically possible to implement both local and nonlocal QKD algorithms at global scales, but at the expense of increased complexity in quantum processing and gate circuits. In contrast, classical cryptography can pretend to *globality* – the key requirement of truly seamless network security – while lacking the "provable invulnerability" of the quantum cryptosystem.

## 25.13 Is quantum cryptography "invulnerable?"

In this concluding section, I shall analyze how a malicious Eve may be able to tamper with Alice's and Bob's secure photon communication (or QKD) channel and, hence, threaten the concept of absolute secrecy in key exchange and message communication.

Assume, first, that a malicious Eve puts a "wire tap" somewhere inside Alice and Bob's photon communication channel. To access the information originally destined for Bob, Eve must perform photon measurements. But like Bob, she does not know, for each of the incoming photons, which measurement basis, $X$ or $Z$, could be the correct one. Eve may perform the measurement anyway, but she now has to generate new photons in replacement, so that Bob does not notice anything. But this is where Eve is tricked: she does not know in which basis-state Alice generated the input photon ($X$ or $Z$), so she has to make a guess with a 50% chance of being right. When Bob measures Eves' photon substitute, he also has a 50% chance of choosing the right measurement basis. At the end of the process, when Bob communicates his measurements to Alice over the public channel, it immediately appears that he has been only 25% successful! And this is the signature of Eve's wiretap. Eve may try to outsmart Alice and Bob by not measuring all of Alice's photons, but only a few at random, in the hope that her tampering is not detected, and retrieve at least some part of the secret key. But even this approach cannot succeed. This is because Alice will immediately detect that Bob does not get the right proportion of correct bits. The widely known conclusion is that *it is not possible to eavesdrop a photon communication channel without Alice and Bob becoming immediately aware of it*. To remain undetectable, however, Eve may choose instead to tamper with only a few photons amongst a long sequence. She may even occasionally luck out in intercepting and re-emitting some of the photons in the right bases. This approach is referred to as a *man-in-the-middle attack* (MIMA), see more later. It may be justly argued that MIMA would not give Eve sufficient information to figure out the whole key. However, the key is weakened to some extent, and this fact justifies the need for SKD. Thus, contrary to widespread acceptance (except in the expert community), *QKD alone is not absolutely secure, while QKD + SKD provably is*, as far as the key-exchange protocol is concerned. See further for more discussion on how such a notion as "absolute security" may be otherwise challenged.

If Eve is capable of eavesdropping Alice's and Bob's *public* communication channel, can she derive any useful information about their secret key? At a first level of analysis, the answer is definitely no. All the information that Eve has access to is *ZXZX* ... from Bob's side, and YesNoNoYes... from Eve's side. Such information does not reveal in any way the outcome of Bob's valid measurements ($\pm 1$ eigenvalues, or 1 and 0 bits). *The fact that Eve knows which of Bob's measurements were right does not tell her, in any way, which are the measured and valid bits*. At a second level of analysis, however, the tapped information makes it possible for Eve to analyze the possible algorithms according to which Alice and Bob choose their random basis and measurement sequences. Thus, rather than trying to wiretap the key, Eve may figure out Alice's and Bob's random-sequence generation algorithms and, thus, reconstruct or predict, with some potential success, the key exchange. Should Eve access the technology used by Alice and Bob for random-key generation, she does not need to eavesdrop their photon communication channel anymore, having a powerful tool for *key attacks* on the classical DES and AES ciphers. This is referred to as *random-number generator attac*k (RNGA).

A second possibility for Eve is to *impersonate* Bob, which is the spirit of MIMA. Eve, a central office (CO) or point-of-presence (POP) network employee, may, indeed, figure out how to take over the public communication channel used by Alice and Bob to finalize their key exchange. Basically, the operation results in a key exchange *K between Alice and Eve* (and not Alice and Bob), all without Alice's and Bob's awareness. Eve may then impersonate Alice by proceeding to a key exchange $K'$ with Bob, again without their awareness. Eve must also be able to intercept the encrypted messages, but this is an easier task, considering her CO or POP access prerogatives. Assuming that Eve has been successful in implementing such a complex arrangement altogether, she has become the "man in the middle" (so to speak). Consider, indeed, Eve's following course of action:

Intercept Alice's encrypted message $M$, along with $K$ as the secret key (that Alice thought she was sharing with Bob), call it $E_K^{\text{Alice}}(M)$, and perform decryption, i.e., generate $M = D_K^{\text{Eve}}(E_K^{\text{Alice}}(M))$;

Possibly change Alice's message $M$ into some other message $M'$ and perform re-encryption with Eve's new key $K'$, i.e., generate $E_{K'}^{\text{Eve}}(M')$;

Perform key exchange with Bob, resulting in shared secret key $K'$;

Send Bob the encrypted message $E_{K'}^{\text{Eve}}(M')$.

The above appears as a perfect MIMA of the reputedly "invulnerable" channel. No quantum-mechanical principles have been violated in the process, only the confidence that Alice and Bob are the only, exclusive, communicating parties. To achieve such a success in "breaking" the QKD channel and value chain, however, Eve must completely control both photon and public communication channels, to prevent them to "compare notes" and detect any anomaly. This would assume another level of verification protocol, which can, in turn, be attacked by malicious Eve.

It is noteworthy to mention the *giant-pulse attack* (GPA). Alice's and Bob's transmitter and receiver photon-measurement terminals are obviously protected from any optical reflections in their shared link, a concept referred to as a *return loss*. Such a return loss, which expresses the probability that a single photon will be reflected might range between $10^{-2}$ and $10^{-6}$. A "giant" light pulse, with $10^2 - 10^6$ photons, may provide Eve some information about Alice's and Bob's choices of $X$ and $Z$ bases, as based on the polarization state (↕, ↔, ∩, ∪) of the reflected pulse. Additionally, such a probing is performed at a wavelength different from that used by Alice and Bob, so that Eve's probing action may remain essentially unnoticed. This type of "side-channel" attack illustrates that QKD's absolute security may be challenged by classical means, and that great care must be taken in any physical inplementation of QKD to eliminate the possibility of any side channels.

Another form of attack, which is far more straightforward, is for Eve to sever the photon communication channel physically. This is referred to as the *denial-of-service attack* (DoSA). The contingency plan for Alice and Bob would be to resort to classical cryptosystems and network means, exposing themselves to ordinary forms of classical security attack. Finally, the weakest point in a quantum cryptosystem is not the link, which, as we have seen, cannot be tampered with, without triggering alarms, but the terminals themselves. Since these terminals must be connected to a network of some kind, they are potentially exposed to attacks, for instance "spy" viruses, which can detect the keys that are exchanged between Alice and Bob. Considering these possibilities, is it possible to state that a quantum cryptosystem is *absolutely* secure? The answer is yes, but only within a certain set of assumptions regarding the security of the other elements in which the cryptosystem is embedded. The worst situation for Alice and Bob would be to trust, in absolute confidence, a system that could be wired without their awareness.

An element that remains central to the discussion about cryptosystem security is the *criticality* of the application: what information must be protected, and how critical is the communication success? In situations of conflict, where all the communications means (civilian or military) may be disabled, denied, or destroyed, there must always remain one way or another for communicating critical information. The cryptosystem must be able to borrow multiple, if not redundant, paths, just as with the Internet protocol. It must also be able to reach Alice or Bob anywhere they may happen to be, supposedly not in a predefined place. Notwithstanding its inherent strength, QKD remains a point-to-point, local cryptosystem whose extension at global scales and possibilities for path redundancy seem impractical. Furthermore, a classical communication channel is always required for Alice and Bob to compare measurements and agree on the secret key. The main assumption of QKD is that such a channel is always available, and resilient against any form of attack, and in realistic conflict situations this fact cannot be taken for granted!

Finally, it is important to stress that despite the availability of provably-secure QKD protocols, the core cryptosystems eventually used in any classical message/ciphertext channels (DES, AES, and future upgrades) remain 100% exposed to conventional attacks (cryptanalysis, code-cracking . . .). Thus, channel security ultimately rests upon the classical notion of "code invulnerability", which represents a "reasonable conjecture" within a cryptosystem's lifetime.

This discussion leads to the closing conclusion that despite its awesome conceptual elegance, quantum cryptography (or QKD) only represents a supplemental technique of information protection, to be situated somewhere within the grander domain of global network security, where there exists *no such a thing* as "absolute" confidence in any cryptosystems.

# Appendix A (Chapter 4) Boltzmann's entropy

## Task 1

Show that the number of ways $W$ of arranging $N$ particles into $m$ boxes with populations $N_i$ is given by

$$W = \frac{N!}{N_1!N_2!\ldots N_m!}. \tag{A1}$$

Let us proceed as follows: we first apply the property according to which the number of ways of selecting $n$ objects out of $m$ objects ($m \geq n$) is $C_m^n = \frac{m!}{n!(m-n)!}$, with ! representing the factorial function:

$$1! = 1,\ 2! = 1 \times 2,\ 3! = 1 \times 2 \times 3,\ \ldots,\ n! = 1 \times 2 \times \cdots (n-1) \times n,$$

with, by convention, $0! = 1$. The number of ways of selecting $N_1$ particles out of $m$ particles is, thus,

$$C_m^{N_1} = \frac{m!}{N_1!(m-N_1)!}. \tag{A2}$$

The number of ways of selecting $N_2$ particles out of $m - N_1$ particles is then

$$C_{m-N_1}^{N_2} = \frac{(m-N_1)!}{N_2!(m-N_1-N_2)!}, \tag{A3}$$

and so on until the box of energy $E_{m-1}$, which has a number of ways

$$\begin{aligned} C_{m-N_1-N_2-\ldots N_{m-2}}^{N_{m-1}} &= \frac{(m-N_1-N_2-\cdots-N_{m-2})!}{N_{m-1}!(m-N_1-N_2-\cdots-N_{m-1})!} \\ &= \frac{(m-N_1-N_2-\cdots-N_{m-2})!}{N_{m-1}!N_m!} \end{aligned} \tag{A4}$$

of being filled up. Multiplying all these possibilities together, we get

$$W = \frac{m!}{N_1!(m-N_1)!}\frac{(m-N_1)!}{N_2!(m-N_1-N_2)!}\cdots\frac{(m-N_1-N_2-\cdots-N_{m-2})!}{N_{m-1}!N_m!} \tag{A5}$$

and crossing out terms appearing in both numerator and denominator, two by two, leads to the result in Eq. (A1).

## Task 2

Show that $(1/N)\log W$ takes the following limit when the number of particles $N$ ($N = \sum_{i=1}^{m} N_i$) becomes infinite:

$$\lim_{N\to\infty} \frac{\log W}{N} = H, \tag{A6}$$

where

$$H = \sum_{i=0}^{m} p_i \log p_i, \tag{A7}$$

and $p_i = N_i/N$ is the probability of finding $N_i$ particles in the microstate of energy $E_i$.

To demonstrate the above result, notice first that from Eq. (A1), we have

$$\begin{aligned}\frac{\log W}{N} &= \frac{1}{N}\log\left(\frac{N!}{N_1!N_2!\cdots N_m!}\right)\\ &= \frac{1}{N}\log\left(\frac{N!}{\prod_{i=1}^{m} N_i!}\right).\end{aligned} \tag{A8}$$

Assuming $x = N, N_i$, is large, we use Stirling's approximation formula:

$$x! = x^x \mathrm{e}^{-x}\sqrt{2\pi x}\left(1 + \frac{1}{12x} + \frac{a}{x^2} + \cdots\right), \tag{A9}$$

where the series in parenthesis can be approximated to unity. Thus,

$$\begin{aligned}\frac{\log W}{N} &\approx \frac{1}{N}\log\left(\frac{N^N \mathrm{e}^{-N}\sqrt{2\pi N}}{\prod_{i=1}^{m} N_i^{N_i}\mathrm{e}^{-N_i}\sqrt{2\pi N_i}}\right)\\ &= \frac{1}{N}\log\left(N^N\mathrm{e}^{-N}\sqrt{2\pi N}\right) - \frac{1}{N}\log\left(\prod_{i=1}^{m} N_i^{N_i}\mathrm{e}^{-N_i}\sqrt{2\pi N_i}\right)\\ &= \frac{\log(N^N)}{N} + \frac{\log(\mathrm{e}^{-N})}{N} + \frac{\log(\sqrt{2\pi N})}{N}\\ &\quad - \frac{1}{N}\sum_{i=1}^{m}\left\{\log\left(N_i^{N_i}\right) + \log(\mathrm{e}^{-N_i}) + \log(\sqrt{2\pi N_i})\right\}\end{aligned}$$

$$
\begin{aligned}
&= \log N \left(1 + \frac{1}{2N}\right) - 1 + \frac{\log(2\pi)}{2N} \\
&\quad - \sum_{i=1}^{m} \left\{ \frac{N_i}{N} \log N_i - \frac{N_i}{N} + \frac{\log N_i}{2N} + \frac{\log(2\pi)}{N} \right\} \\
&\approx \log N - 1 - \sum_{i=1}^{m} \frac{N_i}{N} \log N_i + \sum_{i=1}^{m} \frac{N_i}{N} - \sum_{i=1}^{m} \frac{\log N_i}{2N} \\
&= - \sum_{i=1}^{m} \frac{N_i}{N} \log \frac{N_i}{N} - \sum_{i=1}^{m} \frac{\log N_i}{2N} \\
&\approx - \sum_{i=1}^{m} \frac{N_i}{N} \log \frac{N_i}{N} \\
&\equiv - \sum_{i=1}^{m} p_i \log p_i . \qquad \text{(A10)}
\end{aligned}
$$

# Appendix B (Chapter 4) Shannon's entropy[1]

Consider a source with $N$ elements of occurrence probability $p_i$ $(i = 1 \ldots N)$. We look for an information measure $H$, which should meet three conditions:

(1) $H = H(p_1, p_2, \ldots, p_N)$ is a continuous function of the probability set $p_i$;
(2) If all probabilities were equal (namely $p_i = 1/N$), the function $H$ should increase monotonously with $N$;
(3) If any occurrence breaks down into two successive possibilities, the original $H$ should break down into a weighed sum of the corresponding individual values of $H$.

## Step 1

We first define $A(N) = H(\frac{1}{N}, \frac{1}{N}, \ldots, \frac{1}{N})$, which is the value taken by $H$ when probabilities are equal. Consider now $A(2) = H(\frac{1}{2}, \frac{1}{2})$, which is the mean information of two equiprobable events. After condition (3), if the second event represents two equal possibilities, the new information measure is given by the weighted sum

$$\begin{aligned} H\left(\frac{1}{2}, \frac{1}{4}, \frac{1}{4}\right) &= H\left(\frac{1}{2}, \frac{1}{2}\right) + \frac{1}{2} H\left(\frac{1}{2}, \frac{1}{2}\right) \\ &= A(2) + \frac{1}{2} A(2). \end{aligned} \tag{B1}$$

The first term in the left-hand side represents the information contribution of the two events, and the second term represents the extra information contained in the second event (which occurs half of the time). If the first event also represents two equal possibilities, we have

$$\begin{aligned} H\left(\frac{1}{4}, \frac{1}{4}, \frac{1}{4}, \frac{1}{4}\right) &= H\left(\frac{1}{2}, \frac{1}{4}, \frac{1}{4}\right) + \frac{1}{2} H\left(\frac{1}{2}, \frac{1}{2}\right) \\ &= A(2) + \frac{1}{2} A(2) + \frac{1}{2} A(2) \\ &\leftrightarrow A(4) = 2A(2) \\ &\leftrightarrow A(2^2) = 2A(2), \end{aligned} \tag{B2}$$

[1] This is a more detailed description of Shannon's demonstration. See C. E. Shannon, A mathematical theory of communication. *Bell Syst. Tech. J.*, **27** (1948), 79–423, 623–56. This paper can be downloaded from http://cm.bell-labs.com/cm/ms/what/shannonday/paper.html.

where we used the result in Eq. (B1). We can indefinitely repeat this operation of breaking down each of the single events into two other possibilities, and finally obtain the general rule:

$$A(2^m) = mA(2). \tag{B3}$$

Consider next $A(S) = H(\frac{1}{S}, \frac{1}{S}, \ldots, \frac{1}{S})$, representing the information contained in $S$ equiprobable events. It is clear from the previous demonstration that we also have the property:

$$A(S^m) = mA(S), \tag{B4}$$

meaning that the *gain of information* obtained by splitting the $S$ initial events $m$ times is precisely $m$.

Since the result in Eq. (B4) applies for any integer $S$, we also have, for any integer $T \neq S$ and $n \neq m$,

$$A(T^n) = nA(T). \tag{B5}$$

## Step 2

With a given choice of $n$ (sufficiently large), it is *possible* to find $m$ for which we have the double inequality

$$S^m \leq T^n < S^{m+1}. \tag{B6}$$

To convince ourselves of the validity of this result, we take the logarithm of both inequalities to obtain the following condition on the existence of $m$ (given $n, S, T$):

$$u - 1 < m \leq u, \tag{B7}$$

with $u = n \log T / \log S$. If $n$ is large enough, we have $u > 0$, regardless of the values of $S$ and $T$. According to Eq. (B7); the only two possibilities for $m$ are

(i) $u$ is an integer: $m = u$,
(ii) $u$ is not an integer: $m = E(u)$,

where $E(x)$ means the integer part of real $x$ (e.g., $E(2.75) = 2$). The first case is straightforward. The second case is demonstrated by first setting $u = E(u) + x$ and $u - 1 = E(u - 1) + x$ where $x$ is a real number satisfying $0 < x < 1$. Substituting these two definitions in Eq. (B7) yields $0 < x < m - E(u - 1) \leq 1 + x$. Since $m - E(u - 1)$ is nonzero, we have $m > E(u - 1)$. Since $m - E(u - 1)$ is an integer; we also have $m - E(u - 1) \leq 1$. The only integer solution is $m = E(u)$, since $E(u) - E(u - 1) = 1$. The general solution of Eq. (B6) is, thus, $m = E(u) = E(n \log T / \log S)$.

Having shown that the property in Eq. (B6) is always valid for $n$ sufficiently large, we perform the following operations:

$$\begin{aligned} m \log S \leq n \log T &< (m+1) \log S \\ \frac{m}{n} \leq \frac{\log T}{\log S} &< \frac{m}{n} + \frac{1}{n}. \end{aligned} \tag{B7}$$

Since $n$ can be chosen arbitrarily large, the last result means that

$$0 \leq \frac{\log T}{\log S} - \frac{m}{n} < \varepsilon, \tag{B8}$$

with $\varepsilon$ being made arbitrarily small ($\varepsilon = 1/n \to 0$).

We now use the property that the function $A(N)$ is monotonically increasing, as per requirement (2). With Eq. (B9), this property gives

$$A(S^m) \leq A(T^n) < A(S^{m+1}) \tag{B9}$$

and with Eq. (B5):

$$\begin{aligned} m A(S) \leq n A(T) &< (m+1) A(S) \\ \leftrightarrow \frac{m}{n} \leq \frac{A(T)}{A(S)} &< \frac{m}{n} + \frac{1}{n} \\ \leftrightarrow 0 \leq \frac{A(T)}{A(S)} - \frac{m}{n} &< \varepsilon, \end{aligned} \tag{B10}$$

$\varepsilon$ being made arbitrarily small. Combining Eqs. (B8) and (B10), we get

$$\begin{cases} 0 \leq \dfrac{\log T}{\log S} - \dfrac{m}{n} < \varepsilon \\ 0 \leq \dfrac{A(T)}{A(S)} - \dfrac{m}{n} < \varepsilon. \end{cases} \tag{B11}$$

This result shows that, as $n$ becomes large ($\varepsilon \to 0$), the distance between the function $m/n$ and the two quantities $\log T/ \log S$ and $A(T)/A(S)$ vanishes. The only possibility of verifying this property is that we have:

$$\frac{\log T}{\log S} \equiv \frac{A(T)}{A(S)}, \tag{B12}$$

or

$$A(T) = K \log T, \tag{B13}$$

where $K$ is an arbitrary constant, which must be positive to satisfy condition (2). We can also derive this result more formally using the previously established relation $m = E(u) = E(n \log T/ \log S)$, which gives:

$$\begin{cases} \text{(i) } 0 \leq \dfrac{\log T}{\log S} - \dfrac{1}{n} E\left(n \dfrac{\log T}{\log S}\right) < \varepsilon \\ \text{(ii) } 0 \leq \dfrac{A(T)}{A(S)} - \dfrac{1}{n} E\left(n \dfrac{\log T}{\log S}\right) < \varepsilon, \end{cases} \tag{B14}$$

There exist two possible cases:

- $u$ is an integer: then $E(u) = u = n \log T/\log S$, and the first inequality in Eq. (B12) intrinsically holds regardless of the size of $n$. The second inequality converts to

$$\frac{A(T)}{A(S)} - \frac{\log T}{\log S} < \varepsilon, \tag{B15}$$

which, in the limit of large $n$, leads to the conclusion expressed in Eqs. (B12) and (B13);
- $u$ is not integer: then $E(u) = u + x = n \log T/\log S + x$, where $0 < x < 1$. In this case, the first inequality intrinsically holds regardless of the size of $n$, and the second inequality converts to

$$\frac{A(T)}{A(S)} - \frac{\log T}{\log S} < \varepsilon(1 + x) < 2\varepsilon \tag{B16}$$

which leads to the same conclusion.[2]

## Step 3

Assume that the source now contains $N$ equiprobable possibilities. Its information measure is, therefore, $H(1/N, 1/N, \ldots, 1/N) = A(N) = K \log N$. We can arbitrarily group these possibilities into $m$ subgroups having $n_i$ elements each, and whose associated probabilities are $p_i = n_i/N$. According to condition (3), if we define a first partition of two subgroups of length $n_1$ and $n_2 = N - n_1$, with probabilities $p_1 = n_1/N$ and $p_2 = n_2/N$, respectively, we have

$$A(N) = H(p_1, p_2) + p_1 A(n_1) + p_2 A(n_2). \tag{B17}$$

In the right-hand development of Eq. (B17), the first term corresponds to the information provided by this partition and the second two terms corresponds to the information contained in the two subgroups (weighted by their respective occurrence probabilities $p_1, p_2$). Note that if any subgroup has only one element (e.g., $n_1 = 1$), the corresponding information vanishes, since $A(1) = 0$.

We can then continue this partitioning by splitting the second subgroup into two new subgroups with $n_2$ and $n_3$ elements ($n_2 + n_3 = N - n_1$) and respective probabilities $p_2 = n_2/N$, $p_3 = n_3/N$, yielding the information decomposition:

$$A(N) = H(p_1, p_2, p_3) + p_1 A(n_1) + p_2 A(n_2) + p_3 A(n_3). \tag{B18}$$

[2] It is interesting to note that the Appendix in C. E. Shannon, A mathematical theory of communication. *Bell Syst. Tech. J.*, **27** (1948), 79–423, 623–56 proceeds differently: Eqs. (B8) and (B11) are written in the alternative forms $|\log T/\log S - m/n| < \varepsilon$ and $|A(T)/A(S) - m/n| < \varepsilon$, respectively. From there, it is directly concluded that $|A(T)/A(S) - \log T/\log S| < 2\varepsilon$, which is far from obvious (as the reader might easily check). As a matter of fact, this last inequality summarizes at once the results in Eqs. (B15) and (B16) without providing any of the details. This apparent omission could be explained by the author's concern to save room in a very mathematically intensive paper or, as another possibility, to challenge the reader with the proof.

Iteration of the above partitioning into $m$ subgroups under the condition $\sum_{i=1}^{m} n_i = N$ finally yields:

$$A(N) = H(p_1, p_2, \ldots, p_m) + \sum_{i=1}^{m} p_i A(n_i). \tag{B19}$$

A general definition of the function $H(p_1, p_2, \ldots, p_m)$ is, thus, obtained from Eq. (B19):

$$\begin{aligned} H(p_1, p_2, \ldots, p_m) &= A(N) - \sum_{i=1}^{m} p_i A(n_i) \\ &= K \log N - K \sum_{i=1}^{m} p_i \log n_i \\ &= -K\left(\log \frac{1}{N} + \sum_{i=1}^{m} p_i \log n_i\right) \\ &= -K \sum_{i=1}^{m} p_i \log \frac{n_i}{N} \\ &\equiv -K \sum_{i=1}^{m} p_i \log p_i. \end{aligned} \tag{B20}$$

For an $m$-symbol source, the only function that satisfies the requirements (1) to (3) is, therefore:

$$H = -K \sum_{i=1}^{m} p_i \log p_i. \tag{B21}$$

# Appendix C (Chapter 4) Maximum entropy of discrete sources

In this appendix, I show first that for a discrete source of $k$ independent events, $X = \{x_1, x_2, \ldots, x_k\}$, the source entropy is maximized when all events are equiprobable, corresponding to the *uniform distribution*. Second, I derive the discrete distribution for which entropy is maximized when there is a constraint on the mean $N = \langle x \rangle$. With this constraint, I show that in the case where the events take integer values ($x_1 = 0, x_2 = 1, \ldots$ with $k \to \infty$) the entropy is maximized for the *discrete exponential distribution* of mean $N$ (*Bose–Einstein* or *thermal distribution*). In the case of a discrete source of finite size, $X = \{x_1, x_2, \ldots, x_k\}$, where the events take nonnegative real values, I show that the distribution maximizing entropy is the *Maxwell–Boltzmann distribution*. I then analyze the effect of other additional constraints in the determination of maximum entropy and of the corresponding distributions.

## Uniform distribution solution

By definition, the entropy of the source $X = \{x, x_2, \ldots, x_k\}$ is:

$$H(X) = -\sum_{i=1}^{k} p(x_i) \log p(x_i), \tag{C1}$$

or, with simplified notation:

$$H(X) = -\sum_{i} p_i \log p_i. \tag{C2}$$

For convenience, I shall use natural logarithms; this does not affect the generality of the following demonstrations.

Using the method of *Lagrange multipliers*, we first define the function with parameter $\lambda$ as:

$$f = H(X) + \lambda \sum_{i} p_i, \tag{C3}$$

while assuming the constraint:

$$s_0 = 1 - \sum_{i} p_i = 0. \tag{C4}$$

This constraint ensures that the sum of all probabilities $p_i$ is equal to unity. The task is to minimize $f$ with respect to $p_i$, namely, to find the solution of:

$$\frac{\mathrm{d}f}{\mathrm{d}p_j} = \frac{\mathrm{d}f}{\mathrm{d}p_j}\left[H(X) + \lambda \sum_i p_i\right] = 0, \tag{C5}$$

which yields:

$$\begin{aligned}\frac{\mathrm{d}f}{\mathrm{d}p_j} &= \frac{\mathrm{d}}{\mathrm{d}p_j}\left(-\sum_i p_i \log p_i + \lambda \sum_i p_i\right)\\ &= \frac{\mathrm{d}f}{\mathrm{d}p_j}(-p_j \log p_j) + \lambda\\ &= -\log p_j - 1 + \lambda\\ &= 0,\end{aligned} \tag{C6}$$

which yields the solution $p_j = \exp(\lambda - 1)$. Since $\lambda$ is a constant, the distribution is uniform. Substituting this result into the constraint $s_0$ defined in Eq. (C4) gives

$$\begin{aligned}s_0 &= 1 - \sum_i p_i\\ &= 1 - \sum_{i=1}^{k} = \exp(\lambda - 1)1 - k\exp(\lambda - 1)\\ &= 0,\end{aligned} \tag{C7}$$

which yields the solution $\exp(\lambda - 1) = 1/k$ (or $\lambda = 1 - \log k$), which finally gives $p_i = 1/k$. The conclusion is that the PDF that maximizes the entropy of any discrete source of $k$ independent events, $X = \{x_1, x_2, \ldots, x_k\}$, is the uniform distribution defined by $p(x_i) = 1/k$.

## Discrete-exponential (Bose–Einstein) distribution solution

We consider next a second problem, which can be formulated as follows: what is the discrete distribution of given mean $N$ that maximizes entropy? We note that if a mean $N$ is specified, the source events must correspond to a discrete set of real values, i.e., $x_1 = m_1, x_2 = m_2, \ldots, x_k = m_k$ with $m_i \neq m_j$ for $i \neq j$. Whether the values $m_i$ are real or integer, or equally spaced or ordered (i.e., $m_{i+1} > m_i$) is not important in the derivation of the general solution to this maximization problem.

The solution comes again from the Lagrange-multipliers method, this time with the following function to be minimized:

$$f = H(X) + \lambda \sum_i p_i + \mu \sum_i x_i p_i, \tag{C8}$$

and with the additional constraint:

$$s_1 = N - \sum_i x_i p_i = 0, \tag{C9}$$

which ensures that the PDF mean is $\langle x \rangle = N$. The minimum of $f$ is found by solving:

$$\frac{\mathrm{d}f}{\mathrm{d}p_j} = \frac{\mathrm{d}}{\mathrm{d}p_j}\left[H(X) + \lambda \sum_i p_i + \mu \sum_i x_i p_i +\right] = 0, \tag{C10}$$

or

$$\begin{aligned}\frac{\mathrm{d}f}{\mathrm{d}p_j} &= \left[-\sum_i p_i \log p_i + \lambda \sum_i p_i + \mu \sum_i x_i p_i +\right] \\ &= -\log p_j - 1 + \lambda + \mu x_j \\ &= 0,\end{aligned} \tag{C11}$$

which yields

$$\begin{aligned}\frac{\mathrm{d}f}{\mathrm{d}p_j} &= \left[-\sum_i p_i \log p_i + \lambda \sum_i p_i + \mu \sum_i x_i p_i +\right] \\ &= -\log p_j - 1 + \lambda + \mu x_j \\ &= 0,\end{aligned} \tag{C12}$$

and the PDF solution

$$\begin{aligned}p_j &= \exp(\lambda - 1 + \mu m_j) \\ &= \exp(\lambda - 1)[\exp \mu]^{m_j} \\ &\equiv P Q^{m_j},\end{aligned} \tag{C13}$$

with $P = \exp(\lambda - 1)$ and $Q = \exp \mu$. To define $P, Q$, one must then find the two unknown parameters $\lambda, \mu$ by substituting the result in Eq. (C13) into the two constraints in Eqs. (C4) and (C9):

$$s_0 = 1 - \sum_i p_i = 1 - P \sum_i Q^{m_i} = 0, \tag{C14}$$

$$s_1 = N - P \sum_i m_i Q^{m_i} = 0, \tag{C15}$$

which, considering that $Q > 0$, yields

$$P = \frac{1}{\sum_i Q^{m_i}}, \tag{C16}$$

$$\sum_i (m_i - N) Q^{m_i} = 0. \tag{C17}$$

Given an arbitrary set of real or integer values $m_1, m_2, \ldots, m_k$, the solution for $Q$ in must be computed numerically, since Eq. (C17) is a *transcendental* equation of the form

$$a_1 Q^{m_1} + a_2 Q^{m_{12}} + \cdots + a_k Q^{m_{1k}} = 0, \tag{C18}$$

where $a_i = m_i - N$. Here, I will not discuss the conditions for which a real solution $Q > 0$ may exist in the general case. The only conclusion we can reach is that if such a solution $Q$ exists, the PDF defined in Eq. (C13) takes the form:

$$p_i = \frac{Q^{m_j}}{\sum_i Q^{m_j}}. \tag{C19}$$

I show next that this PDF matches the *discrete exponential distribution* (see Chapter 1), provided we assume that:

(a) $m_1, m_2, \ldots, m_k$ are ordered integers with $m_1 = 0, m_2 = 1, \ldots, m_i = i - 1$;
(b) $k$ is infinite;
(c) $\mu < 0$ (for which $Q < 1$, the condition for the denominator in Eq. (C19) to be finite).

With these assumptions, the solution is $Q = N/(N + 1)$ and $P = 1/(N + 1)$, which gives

$$p_i = \frac{1}{N+1}\left(\frac{N}{N+1}\right)^i, \tag{C20}$$

which is known as the *Bose–Einstein* (BE) distribution (see Chapter 1).[1] It is shown in particular that in the limit of large means ($N \gg 1$), we have $H \approx \log N$. Incidentally, this is the same entropy as that of a uniform distribution of $N$ discrete events ($N$ an integer). The BE distribution corresponds to the photon statistics of incoherent light, such as those emitted from thermal sources (e.g., candle, light bulb, Sun, stars). It is also characteristic of spontaneous emission in optical amplifiers.[2]

## Maxwell–Boltzmann distribution solution

I consider next another case of interest for the solution in Eq. (C19), where:

[1] With these assumptions, all summations in the definitions carry from $i = 0$ to infinity. We then use the property of the geometrical series $\sum_{i=0}^{\infty} Q^i = \frac{1}{1-Q}$ to get the result $p_i = (1 - Q)Q^i$ from Eq. (C19). To solve for $Q$, we calculate the mean as defined by $N = \sum i p_i$, which develops into:

$$\begin{aligned} N &= (1-Q)\sum_{i=0}^{\infty} iQ^i \\ &= (1-Q)Q\sum_{i=0}^{\infty} iQ^{i-1} = (1-Q)Q\frac{\mathrm{d}}{\mathrm{d}Q}\left(\sum_{i=0}^{\infty} Q^i\right) \\ &= (1-Q)Q\frac{\mathrm{d}}{\mathrm{d}Q}\left(\frac{1}{1-Q}\right) \\ &= (1-Q)Q\frac{1}{(1-Q)^2} = \frac{Q}{1-Q}, \end{aligned}$$

which yields $Q = \frac{N}{N+1}$, then $P = \frac{1}{N+1}$ and proves Eq. (C20). The entropy of the BE distribution is defined by: $H = -\sum PQ^i \log(PQ^i)$. Elementary calculation and substitution of the definitions of $P$ and $Q$ yields $H = (N+1)\log(N+1) - N\log N$, which can also be written $H = \log N\lfloor(1 + \frac{1}{N})^{N+1}\rfloor$. In the limit of large $N$ (or $1/N = \varepsilon \to 0$), we have $H \approx \log\{N[1 + (N+1)\varepsilon]\} \approx \log N$.

[2] E. Desurvire, *Erbium-Doped Fiber Amplifiers, Principles and Applications* (New York: John Wiley & Sons, 1994), Ch. 3, p. 154. E. Desurvire, How close to maximum entropy is amplified coherent light? *Opt. Fiber Technol.*, **6** (2000), 357.

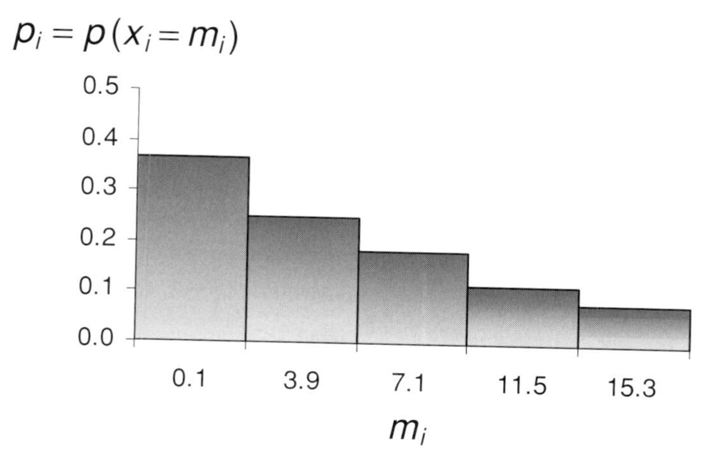


**Figure C1** PDF obtained with $|\mu| = 0.1$, $k = 5$, and $m_1 = 0.1$, $m_2 = 3.9$, $m_3 = 7.1$, $m_4 = 11.5$, $m_5 = 15.3$.

(a) $m_1, m_2, \ldots, m_k$ are ordered positive real numbers with $m_1 = 0$;
(b) $k$ is *finite*;
(c) $\mu < 0$ (again to ensure denominator convergence in Eq. (C19).

In this case, the PDF solution is

$$p_i = \frac{e^{-m_j|\mu|}}{\sum_{i=1}^{k} e^{-m_j|\mu|}}. \tag{C21}$$

Figure C1 shows a plot of this PDF obtained, for example, with $|\mu| = 0.1$, $k = 5$, and $m_1 = 0.1$, $m_2 = 3.9$, $m_3 = 7.1$, $m_4 = 11.5$, $m_5 = 15.3$.

In physics, this distribution characterizes the atomic populations of electrons within a set of $k$ energy levels, when the atoms are in a state of thermal equilibrium. To be specific, let $\mu = -h\nu/k_B T$, with:

$h\nu$ = *photon energy* at frequency $\nu$ ($h$ = Planck's constant),
$k_B T$ = *phonon energy* at absolute temperature $T$ ($k_B$ = Boltzmann's constant),
$m_i$ = energy of atomic level $i$ divided by $h\nu$.

The distribution in Eq. (C21) then takes the form:

$$p_i = \frac{e^{-m_j \frac{h\nu}{k_B T}}}{\sum_{i=1}^{k} e^{-m_j \frac{h\nu}{k_B T}}}, \tag{C22}$$

which is known as the *Maxwell–Boltzmann distribution*. This distribution shows that at thermal equilibrium, the electron population is highest in the lowest energy level, and decreases exponentially as the energy of the level increases. In particular, the population ratio between two atomic levels $i$ and $j$ is given by

$$\frac{p_i}{p_j} = \frac{\mathrm{e}^{-m_j \frac{h\nu}{k_B T}}}{\mathrm{e}^{-m_j \frac{h\nu}{k_B T}}} = \mathrm{e}^{-(m_j - m_j)\frac{h\nu}{k_B T}} \equiv \mathrm{e}^{-\frac{\Delta E_{ij}}{k_B T}}, \tag{C23}$$

where $\Delta E_{ij} = (m_i - m_j)h\nu$ is the energy difference between the two levels. The interesting conclusion of this analysis is that at thermal equilibrium electrons randomly occupy the atomic energy levels according to a law of *maximal entropy*. It can be shown that:[3]

$$\begin{aligned} H_{\max} &= \langle m \rangle \frac{h\nu}{k_B T} \\ &= \frac{\dfrac{h\nu}{k_B T}}{\exp\left(\dfrac{h\nu}{k_B T}\right) - 1}. \end{aligned} \tag{C24}$$

This results expresses in *nats* (1 nat = 1.44 bit) the average information contained in an atomic system at temperature $T$ with energy levels separated by $\Delta E = h\nu$, with $h\nu \gg k_B T$. The same result holds for a two-level atomic system ($k = 2$). The quantity $\langle m \rangle h\nu$ represents the mean thermal energy stored in the atomic system. The ratio $\langle m \rangle h\nu / k_B T$ represents the mean number of thermal phonons required to bring the atom into this mean-energy state.

[3] By definition, the source entropy is $H = -\sum_i p_i \log p_i$. By convention, we take here the natural logarithm so that the unity of entropy is the nat. With solution $p_i = Q^{m_i}/P$ (and $P = \sum_i Q^{m_i}$), we obtain:

$$\begin{aligned} H &= -\sum_i \frac{Q^{m_i}}{P} \log \frac{Q^{m_i}}{P} = -\sum_i \frac{Q^{m_i}}{P} (\log Q^{m_i} - \log P) \\ &= -\sum_i \frac{Q^{m_i}}{P} m_i \log Q + \sum_i \frac{Q^{m_i}}{P} \log P, \end{aligned}$$

which gives $H = -\langle m \rangle \log Q + \log P$. We now develop the second term:

$$\begin{aligned} \log P &= \log\left(\sum_{i=1}^{k} Q^{m_i}\right) \\ &= \log(Q^{m_1} + Q^{m_2} + \cdots + Q^{m_k}) \\ &= \log[Q^{m_1}(1 + Q^{m_2 - m_1} + \cdots + Q^{m_k - m_1})]. \end{aligned}$$

Substituting $Q = \exp(-\frac{h\nu}{k_B T})$ and $\Delta E_{ij} = (m_i - m_j)h\nu$ into the preceding, we obtain:

$$H = \langle m \rangle \frac{h\nu}{k_B T} + \log\left\{\exp\left(-m_1 \frac{h\nu}{k_B T}\right)\left[1 + \exp\left(-\frac{\Delta E_{21}}{k_B T}\right) + \cdots + \exp\left(-\frac{\Delta E_{k1}}{k_B T}\right)\right]\right\}.$$

For simplicity, we can assume that the energy levels are all equidistant, i.e. $\Delta E_{ij} \equiv \Delta E = h\nu$ and $\Delta E_{k1} \equiv (k-1)h\nu$. Using the geometric series formula $1 + q + q^2 + \cdots + q^{k-1} = (1 - q^k/1 - q)$, the entropy is:

$$H = \langle m \rangle \frac{h\nu}{k_B T} + \log\left[\frac{1 - \exp\left(-k\dfrac{h\nu}{k_B T}\right)}{1 - \exp\left(-\dfrac{h\nu}{k_B T}\right)}\right].$$

## Discrete distributions maximizing entropy under additional constraints

In this last section, we consider a general method of deriving discrete PDFs, which maximize entropy while being subject to an arbitrary number of constraints.

Assume, for instance, that the constraints correspond to the different PDF moments $\langle x^0 \rangle = 1$, $\langle x \rangle = N$, $\langle x^2 \rangle, \ldots, \langle x^n \rangle$, which can be expressed according to:

$$\begin{cases} s_0 = 1 - \sum_i p_i = 0 \\ s_1 = N - \sum_i x_i p_i = 0 \\ s_1 = \langle x^2 \rangle - \sum_i x_i^2 p_i = 0 \\ \ldots \\ s_n = \langle x^n \rangle - \sum_i x_i^n p_i = 0. \end{cases} \tag{C25}$$

The functional $f$ to be minimized with the Lagrange multipliers $\lambda_0, \lambda_1, \lambda_2, \ldots, \lambda_n$ is then

$$f = H(X) + \lambda_0 \sum_i p_i + \lambda_1 \sum_i x_i p_i + \lambda_2 \sum_i x_i^2 p_i + \cdots + \lambda_n \sum_i x_i^n p_i. \tag{C26}$$

Taking the derivative of $f$ with respect to $p_j$ yields the development:

$$\begin{aligned} \frac{\mathrm{d}f}{\mathrm{d}p_j} &= \frac{\mathrm{d}}{\mathrm{d}p_j}\left\{ H(X) + \lambda_0 \sum_i p_i + \lambda_1 \sum_i x_i p_i + \lambda_2 \sum_i x_i^2 p_i + \cdots + \lambda_n \sum_i x_i^n p_i \right\} \\ &= -\frac{\mathrm{d}}{\mathrm{d}p_j} \sum_i p_i \log p_i + \lambda_0 \frac{\mathrm{d}}{\mathrm{d}p_j} \sum_i p_i + \lambda_1 \frac{\mathrm{d}}{\mathrm{d}p_j} \sum_i x_i p_i + \lambda_2 \frac{\mathrm{d}}{\mathrm{d}p_j} \sum_i x_i^2 p_i \\ &\quad + \cdots + \lambda_n \frac{\mathrm{d}}{\mathrm{d}p_j} \sum_i x_i^n p_i \\ &= -\log p_j - 1 + \lambda_0 + \lambda_1 x_j + \lambda_2 x_j^2 + \cdots + \lambda_n x_j^n \\ &= 0, \end{aligned} \tag{C27}$$

Considering a two-level atomic system ($k = 2$), we have, in particular;

$$\begin{aligned} H &= \langle m \rangle \frac{h\nu}{k_\mathrm{B} T} + \log \left[ \frac{1 - \exp\left(-2\dfrac{h\nu}{k_\mathrm{B} T}\right)}{1 - \exp\left(-\dfrac{h\nu}{k_\mathrm{B} T}\right)} \right] \\ &= \langle m \rangle \frac{h\nu}{k_\mathrm{B} T} + \log\left[1 + \exp\left(-\frac{h\nu}{k_\mathrm{B} T}\right)\right]. \end{aligned}$$

If we assume as well that $h\nu/(k_\mathrm{B}T)$ is large enough that the exponential can be neglected, we obtain $H \approx \langle m \rangle \frac{h\nu}{k_\mathrm{B} T}$, which is valid for all systems with $k \geq 2$. It is then straightforward to determine $\langle m \rangle$, from the definition $\langle m \rangle = m_1 p_1 + m_2 p_2$. The result is the well known "mean occupation number" of Boltzmann's distribution:

$$\langle m \rangle = \frac{1}{\exp\left(\dfrac{h\nu}{k_\mathrm{B} T}\right) - 1}.$$

which corresponds to the general PDF solution

$$p_j = \exp\left(\lambda_0 - 1 + \lambda_1 x_j + \lambda_2 x_j^2 + \cdots + \lambda_n x_j^n\right), \tag{C28}$$

or

$$p_j = A_0 A_1^{x_j} A_2^{x_j^2} \ldots A_n^{x_j^n} \tag{C29}$$

with

$$\begin{cases} A_0 = \exp(\lambda_0 - 1) \\ A_1 = \exp(\lambda_1) \\ A_1 = \exp(\lambda_2) \\ \ldots \\ A_n = \exp(\lambda_n). \end{cases} \tag{C30}$$

The solutions in Eq. (C29) and (C30) for $\lambda_0, \lambda_1, \lambda_2, \ldots, \lambda_n$ using the $n+1$ constraints in Eq. (C25) can only be found numerically. An example of a resolution method and its PDF solution in the case $n = 2$, the event space $X$ being the set of integer numbers, can be found in.[4] In these references, it is shown that the photon statistics of optically amplified coherent light (i.e., laser light passed through an optical amplifier) is very close to the PDF solution of maximal entropy.

It is straightforward to show that in the general case, the maximum entropy is given by the following analytical formula:[5]

$$\begin{aligned} H_{\max} &= 1 - (\lambda_0 + \lambda_1 \langle x \rangle + \lambda_2 \langle x^2 \rangle + \cdots + \lambda_n \langle x^n \rangle) \\ &= 1 - \sum_{i=0}^{n} \lambda_i \langle x^i \rangle. \end{aligned} \tag{C31}$$

Further discussion and extensions of the continuous PDF case of the entropy-maximization problem can be found in.[6]

[4] E. Desurvire, How close to maximum entropy is amplified coherent light? *Opt. Fiber Technol.*, **6** (2000), 357. E. Desurvire, *Erbium-Doped Fiber Amplifiers, Device and System Developments* (New York: John Wiley & Sons, 2002), Ch. 3, p. 202.

[5] We have

$$\begin{aligned} H &= -\sum_j p_j \log p_j \\ &= -\sum p_j \log\left(A_0 A_1^{x_j} A_2^{x_j^2} \ldots A_n^{x_j^n}\right) \\ &= -\sum p_j \left(\log A_0 + x_j \log A_1 + x_j^2 \log A_2 + \cdots + x_j^n \log A_n\right) \\ &= -\left(\log A_0 \sum p_j + \log A_1 \sum x_j p_j + \log A_2 \sum x_j^2 p_j + \cdots + \log A_n \sum x_j^n p_j\right) \\ &= -(\lambda_0 - 1 + \lambda_1 \langle x \rangle + \lambda_2 \langle x^2 \rangle + \cdots + \lambda_n \langle x^n \rangle) \end{aligned}$$

[6] T. M. Cover and J. A. Thomas, *Elements of Information Theory* (New York: John Wiley & Sons, 1991), Ch. 11, p. 266.

# Appendix D (Chapter 5) Markov chains and the second law of thermodynamics

In this appendix, I shall first introduce the concept of *Markov chains*, then use it with the results of Chapter 5 concerning relative entropy (or Kullback–Leibler distance) to describe the *second law of thermodynamics*.

## Markov chains and their properties

Consider a source $X$ of $N$ random events $x$ with probability $p(x)$. If we look at a succession of these events over time, we then observe a series of individual outcomes, which can be labeled $x_i$ $(i = 1 \ldots n)$, with $x_i \in X^n$. The resulting series, which is, thus, denoted $x_1 \ldots x_n$, forms what is called a *stochastic process*.

Such a process can be characterized by the *joint probability distribution* $p(x_1, x_2, \ldots, x_n)$. In this definition, the first argument $x_1$ represents the outcome observed at time $t = t_1$, the second represents the outcome observed at time $t = t_2$, and so on, until observation time $t = t_n$. Then $p(x_1, x_2, \ldots, x_n)$ is the probability of observing $x_1$, then $x_2$, etc., until $x_n$. If we repeat the observation of the $n$ events, but now starting from any time $t_q (q > 1)$, we shall obtain the series labeled $x_{1+q} \ldots x_{n+q}$, which corresponds to the joint distribution $p(x_{1+q}, x_{2+q}, \ldots, x_{n+q})$. By definition, the stochastic process is said to be *stationary* if for any $q$ we have

$$p(x_{1+q}, x_{2+q}, \ldots, x_{n+q}) = p(x_1, x_2, \ldots, x_n), \tag{D1}$$

meaning that the joint distribution is invariant with time translation. Note that such an invariance does not mean that $x_{1+q} = x_1, x_{2+q} = x_2$, and so on! The property only means that the *joint probability* is time invariant, or does not depend at what time we start the observation and which time intervals we use between two observations.

What is a *Markov process*? Simply defined, it is a chain process where the event outcome at time $t_{n+1}$ is *only* a function of the outcome at time $t_n$, and *not* of any other preceding events. Such a property can be written formally as:

$$p(x_{n+1}|x_n, x_{n-1}, \ldots, x_1) \equiv p(x_{n+1}|x_n). \tag{D2}$$

This means that the event $x_{n+1}$ is *statistically independent*, in the strictest sense, from all preceding events but $x_n$. Using Bayes's formula and the above property, we get:

$$\begin{cases} p(x_1, x_2) = p(x_2|x_1)p(x_1) \\ p(x_1, x_2, x_3) = p(x_3|x_1, x_2)p(x_1, x_2) = p(x_3|x_2)p(x_2|x_1)p(x_1) \\ \text{etc.,} \end{cases} \tag{D3}$$

and consequently

$$p(x_1, x_2, \ldots, x_n) = p(x_n|x_{n-1})p(x_{n-1}|x_{n-2})\ldots p(x_2|x_1)p(x_1). \tag{D4}$$

A Markov chain is said to be *time invariant* if the conditional probabilities $p(x_n|x_{n-1})$ do not depend on the time index $n$, i.e., they are themselves time invariant. For instance, if $a$ and $b$ are two specific outcomes, we have $p(x_n = b|x_{n-1} = a) = p(x_{n-1} = b|x_{n-2} = a)$ $= \cdots = p(x_2 = b|x_1 = a)$. If we recall the property of conditional probabilities:

$$p(y) = \sum_{x\in X} p(y|x)p(x), \tag{D5}$$

then we have for time-invariant Markov chains (as applying to any time $t_{n+1}$):

$$p(x_{n+1}) = \sum_{x_n\in X} p(x_{n+1}|x_n)p(x_n), \tag{D6}$$

or equivalently

$$p(x_{n+1}) = \sum_{x_n\in X} p(x_n)P_{x_n x_{n+1}}, \tag{D7}$$

where we define $P_{x_n x_{n+1}} \equiv p(x_{n+1}|x_n)$ as being the coefficients of a certain *transition matrix* $P$ (note the reverse order of the coefficient subscripts). Such a transition matrix uniquely defines the Markov chain, and defines the evolution of *any* other probability distribution $q$, namely:

$$q(x_{n+1}) = \sum_{x_n\in X} q(x_n)P_{x_n x_{n+1}}. \tag{D8}$$

The expressions in Eqs. (D7) or (D8) correspond to a matrix-vector equation. The matrix $P$ is, thus, applied to transform the $N$-vector of coordinates $p(x = x_n), x \in X$, which we call $\mu$. The result of such a transformation is an $N$-vector of coordinates $p(x = x_{n+1})$, $x \in X$, which we call $\mu'$. The matrix-vector equation (Eq. (D7)) is, thus, summarized in the form:

$$\mu' = \mu P. \tag{D9}$$

To take a practical example, consider the $2 \times 2$ transition matrix that corresponds to a *two-state* Markov chain ($X$ being made of two events):

$$P = \begin{pmatrix} P_{11} & P_{12} \\ P_{21} & P_{22} \end{pmatrix} = \begin{pmatrix} \alpha & 1-\alpha \\ \beta & 1-\beta \end{pmatrix}, \tag{D10}$$

where $\alpha, \beta$ are real constants. This means that this Markov process is time invariant. Replacing this definition in Eq. (D10), we obtain:

$$\begin{aligned}\mu' &\equiv (\mu_1', \mu_2') \\ &= \mu P \\ &= (\mu_1, \mu_2)\begin{pmatrix}\alpha & 1-\alpha \\ \beta & 1-\beta\end{pmatrix} \\ &\equiv [\alpha\mu_1 + \beta\mu_2, (1-\alpha)\mu_1 + (1-\beta)\mu_2].\end{aligned} \tag{D11}$$

Since the input coordinates satisfy $\mu_1 + \mu_2 = 1$ (being probabilities), we observe that the sum of the output coordinates is also unity, $\mu_1' + \mu_2' = \mu_1 + \mu_2 = 1$, which justifies our choice for the time-invariant transition matrix $P$ (it is easily shown that this is actually the only one).

This example will help us to illustrate yet another important concept. We have seen that a Markov process can be time invariant, meaning that the transition matrix has constant or unchanging coefficients. But this time invariance does not mean that the probability distribution does not change over time: we have just seen from our previous example that in the general case $\mu' \neq \mu$, which means that $p(x)$ at time $t_{n+1}$ is generally different from $p(x)$ at time $t_n$. But nothing forbids the distribution from remaining unchanged over time. By definition, we shall say that the distribution $\mu$ is *stationary* if the following property is satisfied:

$$\mu' = \mu P = \mu. \tag{D12}$$

With the previous example, it is easily established that the stationary solution satisfies:

$$\begin{cases} \mu_1 = p(x_1) = \dfrac{\beta}{1-\alpha+\beta} \\ \mu_2 = p(x_2) = \dfrac{1-\alpha}{1-\alpha+\beta}, \end{cases} \tag{D13}$$

with the condition $\alpha - \beta \neq 1$. Such a distribution is of the type $p(x_1) = \beta/M$ and $p(x_2) = 1 - \beta/M$, where $M = 1 - \alpha + \beta$, meaning that it is generally *nonuniform*. The specific case of a *uniform stationary distribution* is given by $\beta = M/2$, which gives $p(x_1) = p(x_2) = 1/2$.

The lesson learnt from the above example is that *time-invariant Markov processes have stationary solutions*. If the process is initiated at time $t_1$ with a stationary solution, then the process is also stationary, meaning that the probability distribution $p(x)$ at time $t_{n+1}$ is the same as at time $t_n$ or $t_{n-1}$ or $t_1$. Note that such a stationary solution is not necessarily unique. Two conditions for uniqueness of the stationary solution,[1] which we will assume here without demonstration, are:

(a) The process is *aperiodic* (i.e., the evolution of $p(x)$ does not show periodic oscillations with equal or increasing amplitudes);
(b) There exists a nonzero probability that the variable $x \in X$ will be reached within a finite number of steps (the process is then said to be *irreducible*).

[1] T. M. Cover and J. A. Thomas, *Elements of Information Theory* (New York: John Wiley & Sons, 1991), Ch. 2.

Under these two conditions, the stationary solution is unique. Moreover, the distribution at time $t_n$ in the limit $n \to \infty$ asymptotically converges towards the stationary solution, *regardless* of the initial distribution at time $t_1$. This property will be demonstrated in the second part of this appendix.

Assuming that the conditions of uniqueness are satisfied in the previous example, the *entropy* $H(X)_{t=t_n}$ converges towards the limit:

$$\begin{aligned} H(X)_{t=t_\infty} &\equiv H_\infty \\ &= -\mu_1 \log \mu_1 - \mu_2 \log \mu_2 \\ &= -\frac{\beta}{M} \log \frac{\beta}{M} - \left(1 - \frac{\beta}{M}\right) \log \left(1 - \frac{\beta}{M}\right). \end{aligned} \tag{D14}$$

It is easily verified that when the stationary solution is uniform ($\beta = M/2$), then $H_\infty = H_{\max} = \log 2 \equiv 1$ bit/symbol, which represents the maximum possible entropy for a two-state distribution (Chapter 4). In the general case where the stationary solution is nonuniform ($\beta \neq M/2$), we have, therefore, $H_\infty < H_{\max}$. This means that the system evolves towards an entropy limit that is lower than the maximum. Here comes the interesting conclusion for this first part of the appendix: assuming that the initial distribution is uniform and the stationary solution nonuniform, the entropy will converge to a value $H_\infty < H_{\max} = H(X)_{t=t_1}$. This result means that the entropy of the system *decreases* over time, in apparent contradiction with the *second law of thermodynamics*. Such a contradiction is lifted by the argument that a real physical system has no reason to be initiated with a uniform distribution, giving maximum entropy for initial conditions. In this case, and if the stationary distribution is uniform, then the entropy will grow over time, which represents a simplified version of the second law, as we shall see in the second part. Note that the stationary distribution does not need to be uniform for the entropy to increase. The condition $H_\infty > H(X)_{t=t_1}$ is sufficient, and it is in the domain of physics, not mathematics, to prove that such a condition is representative of real physical systems.

## Proving the second law of thermodynamics

The second part of this appendix provides an elegant information-theory proof of the second law of thermodynamics.[2] The tool used to establish this proof is the concept of *relative entropy*, also called the *Kullback–Leibler distance*, which was introduced in Chapter 5.

Considering two joint probability distributions $p(x, y), q(x, y)$, the relative entropy is defined as the quantity:

$$\begin{aligned} D[p(x, y) \| q(x, y)] &= \left\langle \log \frac{p(x, y)}{q(x, y)} \right\rangle_{X,Y} \\ &= \sum_{x \in X} \sum_{y \in Y} p(x, y) \log \frac{p(x, y)}{q(x, y)}. \end{aligned} \tag{D15}$$

[2] T. M. Cover and J. A. Thomas, *Elements of Information Theory* (New York: John Wiley & Sons, 1991), Ch. 2.

In particular, it was shown that the relative entropy obeys the chain rule:

$$\begin{aligned} D[p(x,y)\|q(x,y)] &= D[p(x)\|q(x)] + D[p(y|x)\|q(y|x)] \\ &= D[p(y)\|q(y)] + D[p(x|y)\|q(x|y)], \end{aligned} \tag{D16}$$

where $D[p(.|.)\|q(.|.)]$ is a *conditional relative entropy*. Finally, an important property is that the relative entropy is *always positive* (regardless of the arguments being joint or conditional probabilities), except in the specific case $p = q$, where it is zero (thus, $D[p\|q] > 0$ if $p \neq q$ and $D[p\|p] = D[q\|q] = 0$).

We shall apply the above properties to the case of Markov chains. In this analysis, the variables $x_n$ and $x_{n+1}$ are substituted for the variables $x$ and $y$, which define the system events from a single source $X$ that can be observed at two successive instants ($x_n, x_{n+1} \in X$). Let us assume now that the system evolution is characterized by a time-invariant Markov process. Such a process is defined by a unique transition probability matrix $R$, which has the time-independent elements $R_{x_n x_{n+1}} = r(x_{n+1}|x_n)$. Consistently with the property in Eq. (D2), the conditional probabilities are uniquely defined for $p$ and $q$:

$$\begin{cases} p(x_{n+1}|x_n) \equiv r(x_{n+1}|x_n) \\ q(x_{n+1}|x_n) \equiv r(x_{n+1}|x_n), \end{cases} \tag{D17}$$

Next, we apply the chain rule in Eq. (D16):

$$\begin{aligned} D[p(x_{n+1},x_x)\|q(x_{n+1},x_x)] &= D[p(x_x)\|q(x_x)] + D[p(x_{n+1}|x_n)\|q(x_{n+1}|x_n)] \\ &= D[p(x_{x+1})\|q(x_{x+1})] + D[p(x_n|x_{n+1})\|q(x_n|x_{n+1})]. \end{aligned} \tag{D18}$$

Substituting Eq. (D17) in Eq. (D18), we obtain

$$\begin{aligned} &D[p(x_x)\|q(x_x)] + D[r(x_{n+1}|x_n)\|r(x_{n+1}|x_n)] \\ &\quad = D[p(x_{x+1})\|q(x_{x+1})] + D[p(x_n|x_{n+1})\|q(x_n|x_{n+1})], \end{aligned} \tag{D19}$$

or equivalently, since $D[r\|r] = 0$:

$$D[p(x_{x+1})\|q(x_{x+1})] = D[p(x_x)\|q(x_x)] - D[p(x_n|x_{n+1})\|q(x_n|x_{n+1}). \tag{D20}$$

Considering the property $D[p\|q] \geq 0$, Eq. (D20) shows that $D[p(x_{x+1})\|q(x_{x+1})] \leq D[p(x_x)\|q(x_x)]$. This result means that in a time-invariant Markov process, *the relative entropy or distance between any two distributions can only decrease over time.*

In particular, we can choose $q = q_{\text{st}}$ to be a *stationary solution* of the Markov process. If this solution is unique, then its distance for any other distribution $p$ decreases over time. This means that $p$ converges to the asymptotic limit defined by $q_{\text{st}}$ (it can be shown, although it is not straightforward, that $D[p\|q_{\text{st}}] = 0$ or $p \approx q_{\text{st}}$ in this limit).

Assume next that the stationary solution of the Markov process is a uniform distribution, which we shall call $u_{\text{st}}$ (namely, $u_{\text{st}}(x) = 1/N$, $x \in X$). From the definition of distance (Eq. (D15) applied to single-variable distributions), we obtain

$$\begin{aligned} D[p\|u_{\rm st})] &= \sum_{x\in X} p \log \frac{p}{1/N} \\ &= \sum_{x\in X} p \log p + \sum_{x\in X} p \log N \\ &\equiv H_{\max} - H(X), \end{aligned} \tag{D21}$$

with $H_{\max} = \log N$. Since the distance decreases over time while staying positive, the above result means that *the system entropy $H(X)$ increases over time towards the upper limit $H_{\max}$*.

This demonstration could be considered to represent one of several possible proofs of the second law of thermodynamics. We should not conclude that the second law implies that the stationary solution of any physical system must be uniform! What was shown is simply that this condition is sufficient, short of being necessary.

# Appendix E (Chapter 6) From discrete to continuous entropy

In this appendix, we find how the two entropy definitions in the discrete-source and continuous-source cases connect.[1] For the discrete case, we have, by definition:

$$H(X) = -\sum_{x_i \in X} p(x_i)\log p(x_i), \tag{E1}$$

and for the continuous case:

$$H(X') = -\int_X p'(x)\log p'(x)\mathrm{d}x, \tag{E2}$$

where the discrete and continuous distributions, which relate to the sources $X$ (discrete) and $X'$ (continuous) are called $p(x_i)$ and $p'(x)$, respectively.

Considering the continuous function $p(x)$, we can decompose its integration domain into small bins of width $\Delta$, which we label with the index $j$. The variable $x$ belongs to the bin if the condition $j\Delta \leq x < (j+1)\Delta$ is satisfied. The width $\Delta$ is chosen small enough so that in any bin $j$ there exists a value $x_j$, for which

$$\int_{j\Delta}^{(j+1)\Delta} p'(x)\mathrm{d}x \equiv p'(x_j)\Delta \tag{E3}$$

(note that there is no functional relation between the discrete distribution $p(x_i)$ and the continuous distribution $p'(x_j)$ at points $x_j$ of the integration domain). Equation (E3), however, defines a *discrete* distribution, which we shall call $p''(x_j) = p'(x_j)\Delta$. Such a distribution carries over the discrete set $X'' = \{x_j\}$, for which Eq. (E3) is satisfied. According to the additivity property of the integrals, this distribution satisfies:

$$\sum_{x_j \in X''} p''(x_j) = \int_{X'} p'(x)\mathrm{d}x = 1 \tag{E4}$$

(we note that the left-hand side of this equation is called the *Riemann integral* of the continuous function $p'(x)$). We have, thus, obtained a strict equivalence between the continuous distribution $p'(x)$ and a discrete distribution $p''(x_j)$. We, thus, expect that

[1] See also: T. M. Cover and J. A. Thomas, *Elements of Information Theory* (New York: John Wiley & Sons, 1991). D. Feldmann, *A Brief Tutorial On: Information Theory, Excess Entropy and Statistical Complexity* (2002), available online at http://hornacek.coa.edu/dave/Tutorial/index.html.

their respective entropies become very nearly equal if the bin interval $\Delta$ is chosen sufficiently small. The surprise is that it this not all the case, as I show next.

The entropy $H(X'')$ of the discrete distribution $p''$ related to the source $X''$ is defined according to Eq. (E1):

$$H(X'') = -\sum_{x_j \in X''} p''(x_j) \log p(x_j) \tag{E5}$$

Substituting in Eq. (E5) the definition $p''(x_j) = p'(x_j)\Delta$, and using the property in Eq. (E4), we obtain:

$$\begin{aligned} H(X'') &= -\sum_{x_j \in X''} p'(x_j)\Delta \log[p'(x_j)\Delta] \\ &= -\Delta \sum_{x_j \in X''} p'(x_j) \log p'(x_j) - \log \Delta \sum_{x_j \in X''} p'(x_j)\Delta \\ &\equiv -\Delta \sum_{x_j \in X''} p'(x_j) \log p'(x_j) - \log \Delta. \end{aligned} \tag{E6}$$

In Eq. (E6), the first term represents the Riemann integral of the function $-p(x) \log p'(x)$. This Riemann integral converges to the integral $-\int p(x) \log p'(x)\mathrm{d}x = H(X')$ as the bin size is made to vanish ($\Delta \to 0$), which is the continuous entropy. But in such a limit, the second term in $-\log \Delta$ becomes infinite! This divergence reflects the fact that what we have done is an $n$-bit quantization of the continuous function $p'(x)$. In such a quantization, the number of bits is $n = -\log \Delta$ (or $\Delta = 2^{-n}$). Thus, the result in Eq. (E6) shows that the relation between the discrete entropy $H(X'')$ and what we have defined to be the continuous entropy $H(X')$ is actually

$$H(X'') = H(X') + n. \tag{E7}$$

The quantity $H(X') + n$ can be interpreted as the number of bits required (on average) to describe the *continuous* random variable $X'$ with $n$-bit accuracy. To understand what "$n$-bit accuracy" means, consider that $\Delta = 1$, or $n = 0$, corresponds to integers. Then $\Delta = 1/2$, or $n = 1$, corresponds to numbers increasing by steps of 0.5. It takes 1 bit of extra accuracy to specify whether $x$ is rounded to an integer $I$ or to $I + 0.5$. Then $\Delta = 1/4$ or $n = 2$ corresponds to numbers increasing by steps of 0.25. It takes 2 bits of extra accuracy to specify the value of $x$ out of the four rounded cases $I$, $I + 0.25$, $I + 0.5$, or $I + 0.75$. Thus, $n$-bit accuracy corresponds to the number of extra bits required to be accurate within any incremental power of ½. The divergence in $-\log \Delta$, which makes the discrete-case entropy infinite, reflects the fact that it takes an infinite number of bits to discretize a continuous set of variables accurately.

# Appendix F (Chapter 8) Kraft–McMillan inequality

Assume a prefix code made of a set of $N$ codewords $c_1, c_2 \ldots c_N$ of various lengths $l_1, l_2 \ldots l_N$ satisfying $l_1 \leq l_2 \leq \cdots \leq l_N$.

If the code is binary, the *Kraft–McMillan inequality* is

$$\sum_{k=1}^{N} 2^{-l_k} \leq 1, \tag{F1}$$

and for $M$-ary code:

$$\sum_{k=1}^{N} M^{-l_k} \leq 1. \tag{F2}$$

To prove this inequality, I shall use the demonstration of Proakis (2001),[1] one of several other variations to be found in the literature. For simplicity, we consider the case $M = 2$, but the arguments developed are valid for all $M$.

We first construct a binary tree of order $l = l_N$, as shown in Fig. F1 (assuming here, for instance, $l = 4$).

As the figure illustrates, the tree has $2^l$ terminal nodes or "leaves" (here $2^4 = 16$ terminal nodes). The tree also has $2^4 - 1 = 15$ branching nodes. Each node is located on a uniquely defined path, labeled by the 0 and 1 signs at each splitting. For instance, the location of terminal node A is defined by the path labeled 1011. We call the node order the length of its path label (e.g., node A is of order four). The idea is to assign each of the codewords $c_k$ (length $l_k$) to any of the nodes of the tree (branching or terminal), which is of order $l_k$. Such an assignment is continued until there is no codeword left. Assume that there are only five codewords ($N = 5$), with lengths $l_1 = 1, l_2 = 2, l_3 = 3$, and $l_4 = l_5 = 4$. As shown in Fig. F2, the codeword $c_1$ could, thus, be assigned (for instance) to the order-1 node defined by the path 1.

Because the code is a prefix code, $c_1$ cannot be the prefix to any other codewords, meaning that this choice eliminates all the subsequent nodes found in the path beginning with 1 (connected in the figure by dashed lines). We continue by assigning $c_2$ to any available order-2 node, and so on until $c_5$, as illustrated in the figure. Based on this example, it is easy to observe that each assignment of codeword $c_k$ with length $l_k$ eliminates $2^{l-l_k}$ terminal nodes. The total number of terminal nodes eliminated is,

[1] J. G. Proakis, *Digital Communications*, 4th edn. (New York: McGraw Hill, 2001).

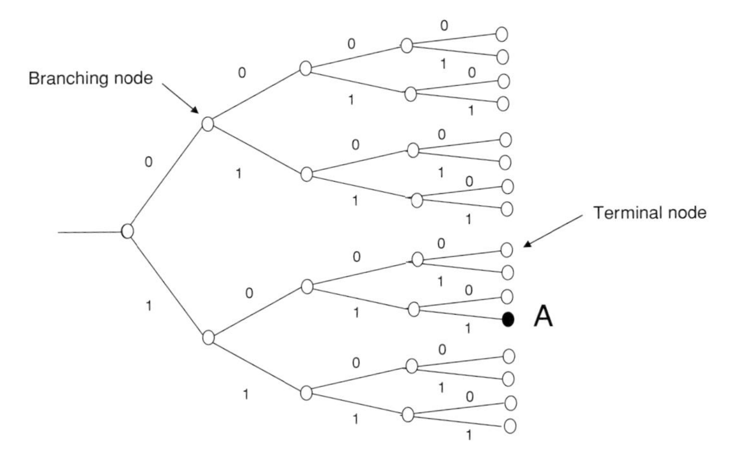


**Figure F1** Binary tree of order $l = l_N$.

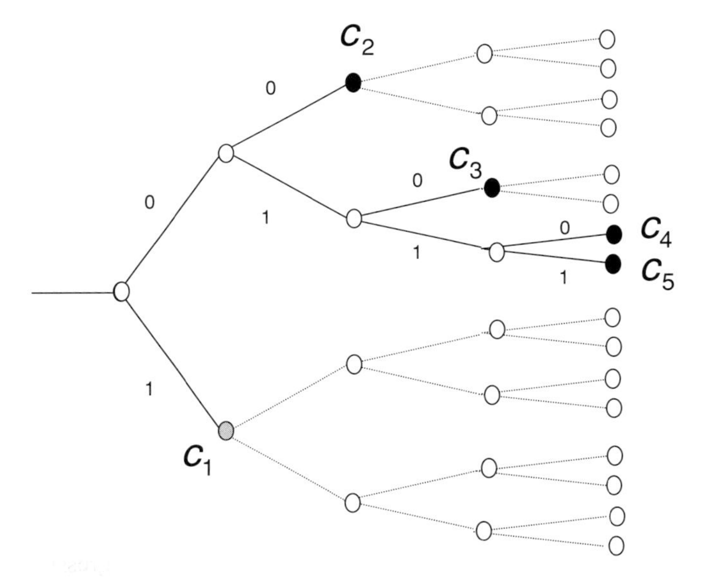


**Figure F2** Binary tree for codewords $c_1$, $c_2$, $c_3$, $c_4$, and $c_5$.

therefore,

$$\sum_{k=1}^{N} 2^{l-l_k} \leq 2^l, \tag{F3}$$

where $2^l$ (to recall) is the total number of terminal nodes. The result of (F3) gives

$$\sum_{k=1}^{N} 2^{-l_k} \leq 1, \tag{F4}$$

which is the Kraft–McMillan inequality.

# Appendix G (Chapter 9) Overview of data compression standards

This appendix provides a brief overview of common data compression standards used for sounds, texts, files, images, and videos. The description is just meant to be introductory and makes no pretense of comprehensively defining the actual standards and their current updated versions. The list of selected standards is also indicative, and does not reflect the full diversity of those available in the market, as freeware, shareware, or under license. It is a tricky endeavor to attempt a description here in a few pages of a subject that would fill entire bookshelves. The hope is that the reader will get a flavor and will be enticed to learn more about this seemingly endless, yet fascinating subject. Why put this whole matter into an appendix, and not a fully fledged chapter? This is because this set of chapters is primarily focused on *information theory*, not on *information standards*. While the first provides a universal and slowly evolving background reference, like science, the second represents practically all the reverse. As we shall see through this appendix, however, information standards are extremely sophisticated and "intellectually smart," despite being just an application field for the former. And there are no telecom engineers or scientists who may ignore or will not benefit from this essential fact and truth!

## Sounds

*Speech* is historically the first type of information that has been subject to coding and compression. The need for speech coding has first come from the development of telephony, with the introduction of digitally sampled voice progressively replacing the old analog telephone service. The benefits of digital voice are essentially twofold: (a) a better sound quality for the users, perceived as free from background noise and interference, and (b) owing to the high fidelity of the transmission, and in particular to *error-correction coding*, the possibility for the telephone operator to compress and multiplex together several voice channels in the same time slot (referenced to as time-division multiplexing, or TDM). But there are even more important benefits for the telephone operator, for managing the voice traffic, in terms of switching, multiplexing, provisioning, and servicing. Another key application is the possibility to mix voice and computer data in the same telephone line, which first appeared under the name of ISDN (integrated services digital networks), as the precursor of our current Internet.

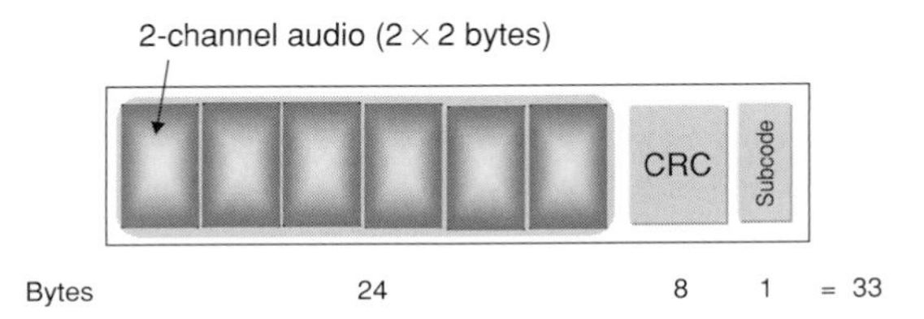


**Figure G1** Schematic representation of standard audio-CD frame.

The standard of digital voice for telephony, which was released by ITU-T in 1972, is referred to as G.711.[1] Analog-to-digital (A–D) voice conversion is based on the technique of *pulse-code modulation* (PCM).[2] It is an uncompressed, lossless code, which converts 8000 samples per second into eight-bit codewords, resulting in a channel rate of $8 \times 8000 = 64$ kbit/s. The two main sampling algorithms used in PCM are the *A-law*,[3] as used in Europe, and the *$\mu$-law*,[4] as used in North America and Japan. My earlier work gives more details about these sampling algorithms and their elaborated variants.[5] This book also describes the digital-voice multiplexing (TDM) standards used by telephone operators, from the early *plesiosynchronous digital hierarchy* (PDH), to the current *synchronous digital hierarchy* (SDH or SONET). While the original bit rate has been set to 64 kbit/s, voice channels can also be encoded into 16–32 kbit/s, with the same perceived sound quality, and, using more sophisticated algorithms, even down to some 2 kbit/s, with reasonably acceptable quality. There exist many algorithms to perform digital-voice compression, for instance removing the silences in phone conversation (a 40% bandwidth saving!), or predicting coding of voice patterns. Nowadays, the majority of developed countries use digital telephony, whether based on wireline or wireless network systems. With today's Internet, digital telephony is almost a mere commodity, which a number of operators offer for free, regardless of connection time and reach.

*Music* represents a second major application area of digital coding. The revolution, called by some the "big bang" in digital audio, came in the late 1970s with the audio *CD* (compact disk), as originated from a joint standardization effort of Sony and Philips.[6] The resulting audio-CD standard has been published in the *Red Book*.[7] As with speech, the A–D conversion is achieved with PCM, here with a two-channel (stereo effect), each made of 16-bit or two-byte codewords, at a sampling rate of 44.1 kHz. The first level of block code is called a *frame*. As illustrated in Fig. G1, the audio-CD frame includes six stereo samplings ($2 \times 2 \times 6 = 24$ bytes), an eight-byte error-correction field (CRC), a "subcode" byte for control and display purposes, e.g., telling to the CD player which

[1] See, for instance: http://en.wikipedia.org/wiki/G.711.

[2] See, for instance: http://en.wikipedia.org/wiki/Pulse-code_modulation.

[3] See, for instance: http://en.wikipedia.org/wiki/A-law_algorithm.

[4] See, for instance: http://en.wikipedia.org/wiki/Mu-law_algorithm.

[5] E. Desurvire, *Wiley Survival Guides in Global Telecommunications, Signaling Principles, Network Protocols, and Wireless Systems* (New York: J. Wiley & Sons, 2004).

[6] See, for instance: http://en.wikipedia.org/wiki/Compact_disk; http://searchstorage.techtarget.com/sDefinition/0,,sid5_gci503642,00.html; www.answers.com/topic/compact-disc-2.

[7] See, for instance: www.mpeg.org/MPEG/DVD/Red_Book/CD.html.

song or track is currently being read. This makes the frame $24 + 8 + 1 = 33$ bytes long altogether, as seen from the figure. The frame is then processed according to the following. Each audio byte is first converted into a 14-bit *EFM codeword* (eight-to-fourteen modulation, also called 8/14 code).[8] The EFM code expansion ensures that each 1 bit to be physically recorded on the CD is surrounded by at least two 0 bits (up to a maximum of ten 0 bits), which is a necessary condition for mechanical tracking, phase-locking, and synchronization purposes.[9] The 14-bit EFM codeword is then interleaved with a three-bit merging word, and finally appended with a unique, 24- or 27-bit synchronization codeword, acting as a frame delimiter. The whole frame conversion results in $33 \times (14 + 3) + 27 = 588$ bits. Like any computer disk, the CD is organized into sectors, each sector containing 98 frames. The aforementioned one-byte subcode, thus, provides eight channels with 98 bit/sector to ensure various functions, like track monitoring and timing or indexing information within tracks. The standard reading speed being set to 75 sectors per second, we obtain the channel rate:

$$588\,\text{bit/frame} \times 98\,\text{frame/sector} \times 75\,\text{sector/second} = 4.32180\,\text{Mbit/s}.$$

After EFM demodulation and decoding, overhead removal, and error correction, the user (or payload) channel rate is reduced to $2 \times 16\,\text{bit} \times 44.1\,\text{kHz} = 1.4112\,\text{Mbit/s}$. The physically recorded program area is $86.05\,\text{cm}^2$, with a track pitch of $p = 1.5$–$1.6\,\mu\text{m}$. Thus, the full length of the track (referred to as the "recordable spiral") is $l = 86.05\,\text{cm}^2/p \times 10^{-4}\,\text{cm} = 5.73$–$5.38\,\text{km}$. With a standard scanning velocity of $v = 1.2\,\text{m/s}$, the corresponding playing time is $t = l/v = 75$–$80$ minutes. At the user rate, this corresponds to $1.470\,(\text{Mbit/s}) \times t(s) = 6615$–$7056$ Mbit or 826–882 Mbyte, which, after some extra sector or error-correction overhead (representing 13%),[10] yielding 720–767 Mbytes, is quite close to the storage capacity offered by current CD-ROM vendors (750–800 Mbytes). The CD-ROM standard will be described later under the heading "Files." Why this 75–80 minute duration for audio CDs? One alleged explanation is that CDs had to be able to play the entire Beethoven's 9th symphony, which took exactly 74 minutes in the slowest recording at the time.[11]

Any program that performs digital encoding and decoding is referred to as a *codec* (short for coder–decoder, compressor–decompressor, or compression–decompression algorithm). Thus, audio codecs represent the family of codes and algorithms that generate digital audio files and restore the sound to the human ear. As we have seen earlier, the audio codec used in CDs is based on PCM with 44.1 kHz sampling rate and $2 \times 16$-bit codewords (the factor of two being for stereo-effect purposes). To recall, PCM is an uncompressed, lossless code, which yields maximum sound quality (referred to as *CD-quality*) for professionals or audiophiles, but necessarily comes with relatively large sizes for the digital-audio files. Microsoft and IBM have adapted this uncompressed-PCM audio codec for use in home computers, under the brand name *WAV* (short for

[8] See, for instance: http://en.wikipedia.org/wiki/Eight-to-Fourteen_Modulation.

[9] See, for instance: www.physics.udel.edu/~watson/scen103/efm.html.

[10] See, for instance: http://en.wikipedia.org/wiki/A-law_algorithm.

[11] http://en.wikipedia.org/wiki/Compact_disk; http://searchstorage.techtarget.com/sDefinition/0,,sid5_gci503642,00.html; www.answers.com/topic/compact-disc-2.

waveform). The data are recognized by the *.wav* filename extension. Any audio file can be compressed using *lossy codecs*, typically up to an 80% compression rate,[12] with the original sound quality being lost forever, yet producing acceptable sound restitution, depending on the music, audience, and utilization. See further on for this topic. In contrast, *lossless audio compression* fully preserves the original CD quality. Two key examples of lossless compression codes that can be used for music are the popular *ZIP* (see further), which can achieve 10–20% compression, and *FLAC* (short for free lossless audio codec), which can achieve over 40% compression.[13] In short, FLAC is based on two algorithms: *linear predictive coding* (LPC)[14] and *run-length encoding (RLE)*.[15] The first makes it possible to decompose acoustic spectra into a reduced list of parameters, which in the reverse implementation can faithfully reproduce the original sounds. Linear predictive coding was originally developed for speech analysis, low bit-rate compression and re-synthesis. Key applications of LPC include cellular telephony (GSM), speech recognition, and electronically synthesized music. The principle of RLE is to replace sequences of repeated codewords (called "runs") with the codewords preceded by their counts. For instance, the sequence XXXXXXXXXXYYYYYY is readily compressed into 10X6Y. This provides additional sound compression, for instance with silent or monotone passages. The LPC parameters are stored by means of *Rice–Golomb codes*, which are described in Chapter 10. Next to FLAC, there actually exists a wealth of lossless audio codecs with compression rates near 40% or better, such as: *WavPack*, *ALAC*, *Monkey's Audio*, *OptimFROG*, *Shorten*, *WMA*, *LA*, *TTA*, *LPAC*, *MPEG4 ALS*, *Real Lossless, Shorten, MUSICompress/WaveZIP*, *AudioZip, WaveArc*, *Pegasus SPS (ELS-Ultra)*, *Sonarc*, *WavPack*, and *RKAU*. Several comparative-merit lists for these different codecs are available.[16]

*Lossy* data compression[17] turns out most useful in any application where the full integrity of the original information does not have to be preserved. The key benefits that outweigh integrity are manifold: reduced file sizes (or fuller use of available storage or disk space); faster file transmission (or up- and downloading on the Internet); faster encoding and decoding processing for real-time or streaming applications. Key applications of lossy compression concern audio, images, and video files. Concerning audio files, an underlying principle of lossy compression is based on *psychoacoustic*

[12] Meaning that the size $S$ of the compressed data is 20% of the size $U$ of the uncompressed, source data. Note that the general convention is to define the compression as the ratio $S/U$, not $1 - S/U$, which scales the opposite way as the ratio.

[13] The compression rate is in fact dependent on the type of music source. It can be 30–40% for pop, rock, techno, and other loud, noisy music, and 40–60% for quieter choral and orchestral pieces, see: www.firstpr.com.au/audiocomp/lossless/#Links.

[14] There exist many Internet sites and tutorials for LPC, see for instance: www.data-compression.com/speech.shtml; www.answers.com/topic/linear-predictive-coding; http://cnx.org/content/m12473/latest/.

[15] See: http://flac.sourceforge.net/features.html; www.answers.com/topic/flac; http://en.wikipedia.org/wiki/FLAC.

[16] http://wiki.hydrogenaudio.org/index.php?title=Lossless_comparison; http://flac.sourceforge.net/comparison.html; www.compression-links.info/Lossless_Audio_Coding; http://members.home.nl/w.speek/comparison.htm.

[17] http://en.wikipedia.org/wiki/Lossy_data_compression.

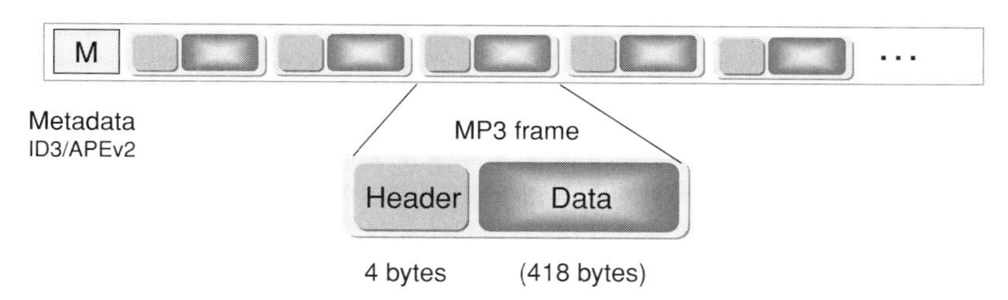


**Figure G2** MP3 frame sequence.

*analysis*,[18] which capitalizes on the limitations of the human ear and the subjective perception and interpretation of sounds. For instance, sounds can obscure or mask each other in both time (forward or backward time masking) and frequency (frequency masking). Also, weak sounds in the frequency spectrum may be masked by louder sounds at other frequencies. In all cases, the information concerning masked sounds can safely be removed without any audible loss of quality. The compression algorithm may also give priority to the sounds situated well within the audible range, which conveys an idea of the power of *psychoacoustic modeling*. Most lossy audio codecs, which include *MP3*, *Ogg Vorbis*, *WMA*, *Musicam*, *LAME*, and *ATRAC*, for instance, are based on such principles. The resulting compressed files typically represent 10–12% of the original audio recordings. Such an unparalleled feature opened new perspectives for broadcasting, downloading, and peer-to-peer sharing music over the Internet (overlooking, here, tricky issues of copyright, illegal copying, and piracy!), and in consumer electronics, such as portable music players. The key differences between these codecs are not only measured in compression and quality performance, or by their operating-system compatibility, but also in the fact that they are either "proprietary," to be implemented under license, or alternatively offered as open-source programs, left to any manufacturer or private user to implement freely or even to modify.

As of today (2006), the patented and standardized *MP3*,[19] which appeared in the mid 1990s, is still by far the most popular. The name is short for *MPEG-1/2 audio layer 3*. The encoded MP3 files (having the filename extension *.mp3*) are made of a sequence of independent of MP3 frames, as illustrated in Fig. G2. A given sequence may be "encapsulated" by heading *metadata file* (whose format is referred to as *ID3* or *APEv2*). This tag contains information, such as the music title, author, artist, and track number. Each frame has an MP3 header field and an MP3 data field. The four-byte header contains a sync-marker word (12 bits), the number of the MPEG and layer versions (3 bits), the bit rate (4 bits, e.g., $1010 = 160\,\text{bit/s}$, for instance), the sampling frequency (2 bits, e.g., $00 = 44.1\,\text{kHz}$), and other size information. There are lots of available bit rates in the two MPEG-1/2 standards, with $B = 128\,\text{kbit/s}$ or $B = 192\,\text{bit/s}$ being most often chosen as de facto values. The bit rate can also be varied from

[18] http://en.wikipedia.org/wiki/Psychoacoustic_model; www.sfu.ca/sonic-studio/handbook/Psychoacoustics.html; www.binaural.com/serendipity/index.php?/archives/62-Tutorial-The-Psychoacoustics-of-Multichannel-Audio.html.

[19] See, for instance: http://en.wikipedia.org/wiki/MP3; www.mp3-tech.org/; www.mpeg.org/MPEG/mp3.html#overview; www.pcmag.com/encyclopedia_term/0,,t=mp3&i=47286,00.asp.

one MP3 frame to the next, which allows one to allocate more bits to the most dynamic music segments (i.e., with more complex spectral movements) and fewer bits in the less dynamic ones. Thus, a rate of 224 kbit/s would be used for a symphonic orchestra, while 48 kbit/s is sufficient for music made of pure frequency tones. These figures are to be compared with the bit rate of uncompressed CD-quality recording, which, as we have seen is 1.411 Mbit/s, which explains the aforementioned $10\times$ to $12\times$ average compression improvement. As for the MP3 data field, the size depends on the number of time-frequency samples or the sampling rate. For the $S = 44.1$ kHz sampling-rate case, there are exactly 1152 one-byte samples in the data field.[20] The byte size is eventually given by the formula $1152 \times B/(8 \times S)$, or 418 bytes with $B = 128$ kbit/s. However, the data are then compressed through *Huffman coding* (see Chapter 9), which is optimized for each individual frame's payload. This does not allow one to predict the actual byte size of the resulting MP3 data field, or the actual compression rate, which is payload dependent.

## Datafiles

Unlike with digital audio, image, and video files, which can suffer lossy compression without loss of perceived quality, most computer data require 100% fidelity in compression processing. Hence, the importance of lossless compression, each with algorithms being ideally suited to the type of source for the most efficient file squeezing. We are now talking about file archiving and packaging, to make the best use of our computer memory space and save time in loading and transmitting files, in particular when it comes to *email*. The inventory of existing licensed or open-source datafile compression standards is quite substantial.[21] Most home-computer users, regardless of operating systems, are familiar with the shareware *ZIP* (file extension *.zip* and many other variants), which has evolved from a complex IP-litigation history.[22] The 1989-originated root algorithm, *PKZIP*, is now used in programs with mutually supported formats called *WinZip*, *BOMArchiveHelper*, *KGB Archiver*, *PicoZip*, *Info-Zip*, *WinRar*, *7-Zip*, *Izarc*, and *ALZip*, to quote only a few.[23] Zooming in further, PKZIP is based on the algorithms *DEFLATE* (for compression) and *INFLATE* (for decompression), which both use a combination of *Lempel–Ziv coding* (LZ77, see Chapter 10) and *Huffman coding* (see Chapter 9). Based on DEFLATE, the file archiver *GZIP* (short for GNU ZIP, with filename extensions *.gz*, *.tgz*, *.tar.gz*, not to be confused with ZIP)[24] was developed in the early 1990s as a freeware, circumventing its patented predecessors. Its format includes a ten-byte header (version, timestamp), some optional extra headers (e.g., to include the original file name), a DEFLATE body of compressed data, and an eight-byte

[20] See for instance: www.compuphase.com/mp3/sta013.htm#MP3FRAMEHDR.

[21] See, for instance: http://en.wikipedia.org/wiki/List_of_archive_formats; for a full list, see: http://en.wikipedia.org/wiki/List_of_file_archivers.

[22] See, for instance: http://en.wikipedia.org/wiki/ZIP_%28file_format%29.

[23] See, for instance: http://en.wikipedia.org/wiki/ZIP_file_format.

[24] See, for instance: http://en.wikipedia.org/wiki/Gzip.

CRC trailer for error correction. The archive freeware *IZArc* supports an impressive list of archive file formats, including applications for CD and DVD images. It also provides 256-bit AES encryption, repairing corrupted archives, and several other advanced features.[25]

Understandably, the achievable compression rate of any file-archiving programs is strongly dependent on the datafile type (e.g., text, slides, tabulated data, HTML web page, executables, etc.) and anything of a different format that the file might also contain (e.g., raw or uncompressed pictures, equation fields, and other types of embedded additions). An approximate comparison is shown in.[26] This study indicates that pure text files can be zipped down to 19–27% of their original sizes (compression rates of 73–81%), the leaders seemingly being 7-ZIP and RAR, with 19–20% rates. For executables, 7-ZIP champions with 27% squeezing (the other competitors confined to 36–40% ratios), and for raw images this performance is reduced to 50–60%. These figures must, however, be weighted against the coding and decoding times, obviously coming with a higher tax for the champions, but not systematically. Thus, each case is a tricky matter of finding the right trade-off between minimal archival size, the time required to squeeze the file, and the time taken to recover the data as uncompressed. Ideally, compression should be performed as a systematic background routine, in such a way as not to slow down other computer tasks.

The external and permanent storage of computer files is popularly based on the *CD-ROM* (*compact-disk, read-only memory*),[27] which is wholly similar to the previously described audio-CD, both in terms of looks and storage space (700–800 Mbytes). A key difference is that the CD-ROM is primarily designed for computer data and, hence, it includes the function of *error-correction*, which is achieved through *Reed–Solomon* (RS) encoding (see Chapter 11). The CD-ROM can be "burnt" according to three preset modes: *audio* (for music tracks or copying audio-CDs), and *mode 1* and *mode 2* for PC data. The CD-ROM has 333 000 sectors of 2352 bytes length (which gives a capacity 783.2 Mbytes). In the audio mode, no error correction is used, thus, the full sector length (2352 bytes) can be used for storing audio files. Both mode 1 and mode 2 use a 16-byte header in each sector, for the purposes of synchronization and identification. Unlike in mode-1, mode-2 does not include error correction, which leaves $2352 - 16 = 2336$ bytes for payload. In mode 1, the 288-byte trailer for error correction leaves a payload of $2336 - 288 = 2048$ bytes. CD-ROMs are characterized by three different possible read and write speeds: A = write-once, B = rewrite and C = read-only. By definition, the $1\times$ speed rating corresponds to 150 kbyte/s. Thus, a $32\times$ write-only speed corresponds to 4.8 Mbyte/s. A 700 Mbyte CD-ROM can, thus, be copied in approximately $700/4.8 = 146$ s or approximately 2 min 30 s. The different speed ratings for the A, B, and C functions are then identified under the generic label $A\times, B\times, C\times$, for instance $12\times, 10\times, 32\times$.[28]

[25] See, for instance: http://en.wikipedia.org/wiki/Izarc; www.izarc.org/.

[26] See, for instance: http://en.wikipedia.org/wiki/Comparison_of_file_archivers.

[27] See, for instance: http://en.wikipedia.org/wiki/CD-ROM.

[28] See, for instance: http://en.wikipedia.org/wiki/Izarc; www.izarc.org/.

**Figure G3** Effect of JPEG compression from 0% to 100% with close-ups shown at right.

## Images

With the development and popularity of digital photography, and the multiplication of Internet websites, the compression of *images* is playing an increasingly important role. Not only does it save memory space, but its also make it possible to speed up the downloading of web pages or email picture attachments, up to the point of instant gratification. The most commonly used standard, *JPEG*, is the creation of the *Joint Photographic Expert Group*, which was launched in the mid 1980s under the ISO standardization body.[29] The JPEG standard, which is defined through various filename extensions *.jpg*, *.jpeg*, *.jpe*, *.jfif*, and *.jif*, is a *lossy* compression codec. This feature is illustrated in Fig. G3. The original file size corresponding to the top-left image (as converted for simplicity into a grayscale one) is 683 kbyte. Compressing the same image through a photo editor by command factors of 50%, 75%, and 100% reduced the file size to 102 kbyte, 65 kbyte, and 15 kbyte, respectively. A look at the close-ups reveals that the effect of lossy compression is not noticeable at a 50% factor (nicely, the file size has, however, been reduced to 15% of the original). At a 75% factor and above, the loss of pixel information becomes apparent, to the point of yielding an image of poor quality, with observed "blocky and blurry" artifacts. Clearly, the amount of compression is a matter of subjectively determining the trade-off between file size and image quality, which depends on the final application, for instance either a screen wallpaper, a web banner, or a file icon.

[29] See, for instance: http://en.wikipedia.org/wiki/JPEG; www.imaging.org/resources/jpegtutorial/index.cfm.

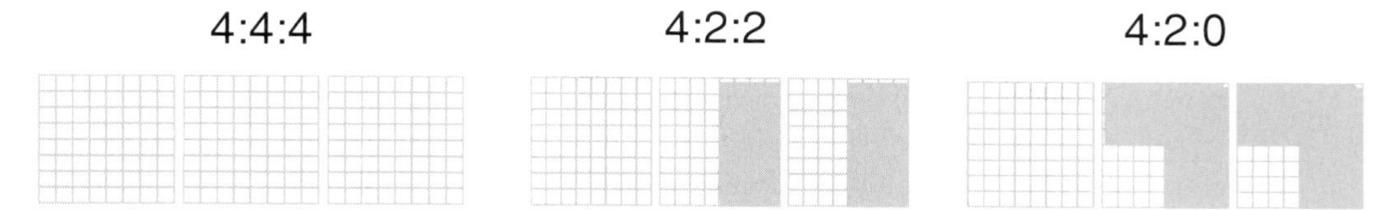


**Figure G4** Compressing the two 8 × 8 chroma blocks of 4:4:4, resulting in $\frac{1}{3}$ size (4:2:2) and $\frac{1}{2}$ size (4:2:0) reductions.

It is beyond the scope of this appendix to run into the complex details of image analysis and the JPEG encoding algorithm. Here, I shall just provide a brief summary of the main concepts and features of JPEG. Digital images are made of two-dimensional (2D) *pixel arrays*, sometimes referred to as *bitmaps*. Each pixel is defined through $3 \times 8 = 24$-bit codewords, with eight bits defining the intensity of each of the red, blue, and green (RBG) components, on a 0–255 scale. This code makes up to $256 \times 256 \times 256 = 16\,777.216$ or about *16.7 million* possible colors! This original pixel is then analyzed and decomposed into a new color space, which considers three components: one for luminance (or brightness, or "luma"), and two for *chrominance* (or "chroma"). Chrominance is another way of labeling colors according to their *hue* (position in a linear color scale) and *saturation* (intensity). The reason for such a conversion is that the luminance and chrominance data offer more possibilities for compression, as we shall see. The JPEG algorithm first transforms the 2D-RBG pixel array into three 2D arrays, one for luminance and two for chrominance. Each of these arrays is then decomposed into blocks of 64 pixels, which equivalently form 8 × 8 pixel arrays (each pixel being eight bits). As the human eye sees more details in luminance, it is possible to throw out some information in the chrominance blocks. This is referred to as *downsampling* or *chroma subsampling*.[30] There exist many different possibilities of achieving downsampling, of which I will only mention two. Let us name 4:4:4 the original three-block set, as shown in Fig. G4. As seen from the figure, halving the horizontal data in the last two 8 × 8 (chroma) blacks results in a set called 4:2:2, which is one-third smaller.[31] Halving the chroma data in both horizontal and vertical directions results in the set 4:2:0, which is one-half smaller. This 4:2:0 compression is the scheme used in most JPEG images, and also in *digital video* (DV) and *high-definition DV* (HDV), in *MPEG* (including MPEG-2 for the *digital video disk* or DVD).

However, the JPEG compression algorithm does not stop there! The 8 × 8 blocks (or their reduced versions) are then submitted to *discrete cosine transform* (DCT), which is analogous to a 2D discrete Fourier transform.[32] The DCT maps each block into its frequency version containing the Fourier coefficients. The coefficients are then subjected to a *quantizer*, which uses variable step or "quantum" sizes (smaller for low frequencies, and the reverse). Each coefficient is scanned through a zigzag pattern, then divided by

[30] See for instance: http://en.wikipedia.org/wiki/YUV_4:2:2.

[31] The figure does not have the purpose of explaining how the halving in both horizontal and vertical directions is actually performed.

[32] See: www.imaging.org/resources/jpegtutorial/jpgdct1.cfm; http://en.wikipedia.org/wiki/Discrete_cosine_transform; http://rnvs.informatik.tu-chemnitz.de/~jan/MPEG/HTML/mpeg_tech.html.

its quantum size and the result rounded off to an integer. It is not unusual that more than half of the resulting coefficients end up being zero. Then it is possible to further compress the blocks through *run-length encoding* (RLE), as described earlier with sound compression. The RLE algorithm outputs the number of previous zeros and the nonzero coefficient amplitude, forming a sequence of *pairs*. Each pair is then encoded through a lossless, variable-length codeword such as generated by *Huffman* coding (see Chapter 9). The more powerful arithmetic coding (see Chapter 10) is also possible, but is covered by patents. On completion, JPEG writes an end-of-block codeword, and after processing the three blocks, an end-of-sequence codeword. Decoding the JPEG image is basically the same succession of operation in the inverse-function order.

The JPEG standard later evolved to *JPEG-2000* (filename extensions *.jp2* and *.j2c*).[33] The key difference from the former JPEG is the use of the *discrete wavelet transform* (DWT) algorithm[34] instead of DCT. The interest of DWT is not only the possibility of achieving greater compression rates (typically 20% higher) while making imperceptible the characteristic "blocky and blurry" artifacts of the JPEG described earlier. As a major optional feature, JPEG2000 also provides *lossless* compression, owing to the reversibility of DWT. The long list of many other benefits can be explored in the references given in the footnotes.

What about other image-coding standards? Let me briefly describe first *GIF* and *PNG*. Short for *graphic interchange format*,[35] GIF (filename extension *.gif*) is a *lossless* image-compression codec, which is also popular in websites. Its inherent simplicity stems from the limited palette of 256 (eight-bit) preset colors extracted from the 24-bit RBG space. While such a palette is too limited for image-photography applications, it is fully adequate for continuous-color pictures, such as line-art, graphics, and logos, and all sorts of animations or advertising, which abound in Internet pages or email attachments, such as e-cards. The key feature of GIF is the use of *Lempel–Ziv–Welch* (LZW) for compression algorithm, see description in Chapter 10, which is far more efficient than *run-length encoding* (RLE). As LZW turned out later to be bound to certain patents (the issue having become obsolete, from 2006), the PNG format, short for *portable network graphics*,[36] was then developed as freeware, eventually to become the third most-preferred image file for Internet use, despite varieties in web-browser support, versions, and related compatibility problems. Like its GIF predecessor, *PNG* (filename extension *.png*) is a *lossless* image-compression codec. Its RBG color palette is, however, increased from 24 bits to 32 bits and 48 bits. One of the interesting features of PNG is object transparency (also available to some more limited extent in GIF), i.e., the possibility of making certain image objects appear translucent over their background, as achieved through a supplementary "alpha" channel. The core PNG compression algorithm is based on the license-free DEFLATE which, as we have seen earlier, combines LZ77 and

[33] See, for instance: http://en.wikipedia.org/wiki/JPEG_2000. See also, the JPEG-2000 official website: www.jpeg.org/jpeg2000/.

[34] See, for instance: http://en.wikipedia.org/wiki/Wavelet; http://en.wikipedia.org/wiki/Discrete_wavelet_transform.

[35] See, for instance: http://en.wikipedia.org/wiki/GIF.

[36] See, for instance: http://en.wikipedia.org/wiki/Portable_Network_Graphics.

Huffman coding. Under its *MNG* extension, PNG also supports graphic animation,[37] although it is more complex and not widely supported on the Internet.

Two other image formats are *TIFF* (tagged image file format)[38] and *BMP* (short for bitmap)[39]. In short, TIFF (filename extension *.tiff*) was originally designed as a common standard for image scanners; with a 32-bit uncompressed format albeit with the option of lossless LZW (Lempel–Ziv–Welch) compression. Its specificity, hence, its name, is the capability of handling multiple images and data in a single file with as many header "tags." The format BMP (filename extensions *.bmp* or *.dib*) offers a variety of coloring depths (1, 2, 4, 8, 16, or 32 bits) with an uncompressed image format. Owing to their redundancy, BMP files lend themselves to efficient lossless compression, for instance through ZIP (see above).

## Video

The most generic standard to encode video data is known as *MPEG* (short for motion pictures expert group),[40] more accurately defined under *MPEG-1*, *MPEG-2*, and *MPEG-4*. It is beyond the scope of this overview to approach such standards in the appropriate level of detail but, as in the previous sections, I shall focus on the key features and on the central role of data compression algorithms.

## MPEG-1

This is the early standard for *digital video* (DV), and is also videoCD/DVD compatible (see later).[41] It achieves audio and video multiplexing, synchronization, and data compression. The video quality is primarily defined according to the pixel resolution of the 2D image frame, namely $352 \times 240$ (NTSC system, used in North America) and $352 \times 288$ (PAL-SECAM system, used in Europe). The corresponding frame rates are $29.97 \approx 30$ frame/s and 25 frame/s, respectively. There are four types of MPEG-1 frame: *I*, for intra-coded frames, which are individually compressed; *P*, for predictive frames, which code the frame image as a difference from that of the previous one in the sequence, *B*, for bidirectionally predictive frames, which codes the image as a difference from both the previous and the next image, and *D*, which uses block coding on an average image from several successive frames. This type of low-quality resolution is only used for fast-forwarding visualization. A group of pictures (GOP)[42] is then typically formed by the 12-frame sequence IBBPBBPBBPBB, which always begins with an I-frame. The above sequencing makes it possible to encode movie pictures at rates

[37] See, for instance: http://en.wikipedia.org/wiki/Multiple-image_Network_Graphics.

[38] See, for instance: http://en.wikipedia.org/wiki/Tagged_Image_File_Format.

[39] See, for instance: http://en.wikipedia.org/wiki/Windows_bitmap.

[40] See: www.mpeg.org; www.chiariglione.org/mpeg/.

[41] See, for instance: http://en.wikipedia.org/wiki/MPEG-1; www.chiariglione.org/mpeg/standards/mpeg-1/mpeg-1.htm.

[42] See, for instance: http://en.wikipedia.org/wiki/Group_of_pictures; http://rnvs.informatik.tu-chemnitz.de/~jan/MPEG/HTML/mpeg_tech.html.

between 1.2 Mbit/s and 1.5 Mbit/s. A main drawback of MPEG-1 is that it is based on the principle of *progressive* (or noninterlaced) *image scanning*. Images are scanned line by line, which provides good vertical resolution, but does not allow for smooth motion restitution in the case of fast-changing scenes. As seen earlier, in the sound compression section, MPEG-1 remains the underlying standard for the most popular MP3 (short for MPEG-1/2 layer 3).

## MPEG-2

This comes as an enhancement of MPEG-1 in both audio and video.[43] To recall from the previous section on sound, the MPEG-2 audio layer 3 is also an underlying standard for MP3, which allows us here to encode more than just one stereo track. As for the video, just like in MPEG-1, the I-frames are processed according to the JPEG image-compression algorithm (see previous description in the image section). The P-frame offers more compression potential than the I-frame, because it only encodes successive image differences. The comparison algorithm considers macroblocks of 16 × 16 pixels. The differences are then described by the translation of similar groups of pixels within these macroblocks, characterized by a "motion vector." If the algorithm fails satisfactorily to code the transition with the appropriate motion vector and additional predictive corrections, then the output frame is an unaltered I-frame. The B-frame encoding works according to the same principle as the P-frame, except that it uses the information of the two surrounding images. A supplemental feature is that (unless otherwise specified) frames are interlaced on display or scanning (unlike in MPEG-1), which allows smooth transitions in the case of fast-moving pictures. Here, we shall overlook the supplemental notions of "transport stream" and "program stream," which are more application focused. In terms of image resolution, frame rate and bit rate, MPEG-2 offers a broad variety of possible configurations. One first refers to an MPEG-2:

- *Profile*, through (a) the type of frames in the GOP [I-P or I-P-B], (b) the JPEG luma or chroma sampling [4:2:0 or 4:2:2] used for the images, and (c) the stream range [1, 1–2, or 1–3]. The simplest profile, called SP, is thus defined by [I, P][4:2:0][1], and the highest one, called HP, by [I, P, B][4:2:2][1–3];
- *Level*, through (a) the number of pixels per line [325 to 1920], (b) the number of lines [288 to 1152], at a rate of 30 frames per second. The lowest level, called LL, is [352][288], and the highest one, called HL, is [1920][1152].
- *Profile@level,* through (a) the 2D motion-picture pixel resolution [176 × 144 to 1920 × 1080 with many intermediate variants], (b) the frame rate [15 to 60], (c) the sampling [4:2:0 or 4:2:2], and (d) the bit rate [0.096 to 300 Mbit/s]. The lowest option, called *SP@LL*, is, thus, [176 × 144][15][4:2:0][0.096], while the highest one (a potential for future applications), called *422P@HL*, is either [1920 × 1080][30][4:2:2][300] or [1280 × 720][60][4:2:2][300].

[43] See, for instance: http://en.wikipedia.org/wiki/MPEG-2.

It is clear that each of the standard profile@level options has been designed to meet specific consumer-electronics applications and requirements, such as: *wireless mobile handsets* (3G cellular telephony), *personal digital assistants* (PDA), *set-top boxes* (STB), *digital versatile disk* (DVD), *digital video broadcast* (DVB), *high-definition video* (HDV), *high-definition television* (HDTV), and many other professional-imaging applications, present and future. A key application of MPEG-2, which is interesting to consider here as an illustrative example, is the DVD.[44] As users know, commercial DVDs now come in several capacity options, including read-and-write versions (known as DVD $\pm$ R, DVD $\pm$ RW, DVD-RAM),[45] namely, 4.7/9.4 Gbytes (single-layer) and 8.5/17.1 Gbytes (double layer). The video channel has a variable bit rate that peaks at 9.8 Mbit/s, yielding a total bit rate (audio + video) varying between 300 kbit/s and 10.08 Mbit/s. A 4.7 Gbyte DVD can store movies with 90 $-120$ min running times.

## MPEG-4

This is a format patented by ISO/IEC, which includes no less than 23 standards or "parts."[46] These allow for audio and video compression enhancement and a wide range of novel multimedia applications, including object-oriented coding, virtual-reality modeling, user graphic interactivity, computer-generated video and audio, and digital rights management and protection. The MPEG-4 audio codec, which is described in *Part 3*, is referred to as *advanced audio coding* (AAC).[47] The key intent of AAC is to provide better sound compression without affecting quality, or better quality with a given stream rate. The claim is that in sound quality a 48 kbit/s AAC stream would outperform a 128 kbit/s MP3 stream, while the consensus concerns more the AAC superiority over any other 32–64 kbit/s audio codec, and its improved compression with respect to MP3. The MPEG-4 video codecs are described in *Part 2* and *Part 10*. The first (Part 2) concerns image compression through DCT in view of the simplest implementation "profiles," see further. The second (Part 10), is also referred to as *H.264* (the ITU-T denomination) or *advanced video coding* (AVC).[48] The key intent of this advanced codec is to reduce the bit rate without affecting the picture's end quality. This is enabled by several feature, of which only a few can be mentioned or make sense within the scope of this overview. For instance, inter-picture prediction is not limited to B-frames (as in MPEG-2) but up to 32 neighboring reference frames. A variable macroblock size ranging from $4 \times 4$ to $16 \times 16$ enables more precise mapping and tracking of moving regions in still backgrounds. Compression is introduced by *context-adaptive binary arithmetic coding* and *variable-length (Huffman) coding* (Chapter 10), referred to as CABAC and CAVLC,

[44] See, for instance: http://en.wikipedia.org/wiki/DVD.
[45] See, for instance: http://en.wikipedia.org/wiki/DVD_Formats.
[46] See, for instance: http://en.wikipedia.org/wiki/MPEG-4; www.chiariglione.org/mpeg/standards/mpeg-4/mpeg-4.htm; www.apple.com/quicktime/technologies/mpeg4/.
[47] See, for instance: http://en.wikipedia.org/wiki/MPEG-4_Part_3; www.mpeg.org.
[48] See, for instance: http://en.wikipedia.org/wiki/H.264.

respectively, which are completed by default with *exponential-Golomb coding*.[49] The key result is that the combined features of H.264/AVC make it possible to obtain the same quality as MPEG-2 at half the bit rate or even less. As with MPEG-2, there exist a range of “profiles” and “levels” adapted for each consumer-electronics or professional application: 3G mobile, video telephony, videoconferencing, broadcast television, Internet video streaming, CD storage, etc. The controversial format *DivX*, which has been dubbed “the MP3 of video,” was designed in early 2000. It capitalizes on the aforementioned MPEG-4 Part 2 video compression features, balancing video quality against file size.[50] The end result is the possibility of squeezing an entire DVD content into a (700–800 Mbyte) data CD, without perceptible loss of quality.

[49] See, for instance: http://en.wikipedia.org/wiki/Exponential-Golomb_coding.

[50] See, for instance: http://en.wikipedia.org/wiki/DivX.

# Appendix H (Chapter 10) Arithmetic coding algorithm

In this appendix, I derive a generic algorithm[1] to compute the real intervals $[u, v)_N$ corresponding to the *arithmetic coding* of any string of $N$ symbols defined by $a_1 a_2 \ldots a_N$, where $a_k$ belongs to the $n$-events source $X = \{x_1, x_2, \ldots, x_n\}$ with associated probability distribution $p_X = \{p(x_1) \ldots p(x_n)\}$. I will use index $k$ to designate the symbol position in the string, and index $i$ to designate the value $x_i$ of this symbol. By convention, the symbol $x_n$ is exclusively used for signaling string termination. For easier reading, I shall use the notation $\langle u, v\rangle$ to designate the intervals instead of the cumbersome $[u, v)$.

To build the algorithm, consider first the simple example of a source $X = \{x_1, x_2, x_3\}$ and strings with lengths up to $N = 2$. According to the arithmetic coding, and as shown in Fig. H1, the initial interval $I = [0, 1) \equiv \langle 0, 1\rangle$ is divided into the three subintervals

$$I_1 = \langle 0, p(x_1)\rangle, \tag{H1}$$

$$I_2 = \langle p(x_1), p(x_1) + p(x_2)\rangle, \tag{H2}$$

$$I_3 = \langle p(x_1) + p(x_2), 1\rangle, \tag{H3}$$

with $I_1 + I_2 + I_3 = I$.

By convention, all complete strings should end with the symbol $x_3$. The three possible strings are, therefore, $x_3$, $x_2x_3$, and $x_1x_3$. Figure H1 shows how each of the intervals $I_1$ and $I_2$ is divided in turn into three subintervals ($I_{11}, I_{12}, I_{13}$ and $I_{21}, I_{22}, I_{23}$). The subintervals of interest for complete strings, $I_3$, $I_{23}$, and $I_{13}$, are highlighted in Fig. H1. For future use, we define the interval width of $I = \langle u, v\rangle$ as $w(I) = v - u$. In particular,

$$w(I) = 1, \tag{H4}$$

$$w(I_1) = p(x_1), \tag{H5}$$

$$w(I_3) = p(x_2). \tag{H6}$$

The single-symbol string, $x_3$, which, according to our convention, is useless or has zero message payload, is, thus, defined by the interval $I_3 = \langle u, v\rangle$. The start and stop (values of this interval are, according to Eq. (H3):

$$\begin{cases} u = p(x_1) + p(x_2) \\ v = p(x_1) + p(x_2) + p(x_3). \end{cases} \tag{H7}$$

[1] D. J. C. MacKay, *A Short Course in Information Theory* (1995), www.inference.phy.cam.ac.uk/mackay/info-theory/course.html; see also D. J. C. MacKay, *Information Theory, Inference and Learning Algorithms* (Cambridge, UK: Cambridge University Press, 2003), p. 113.

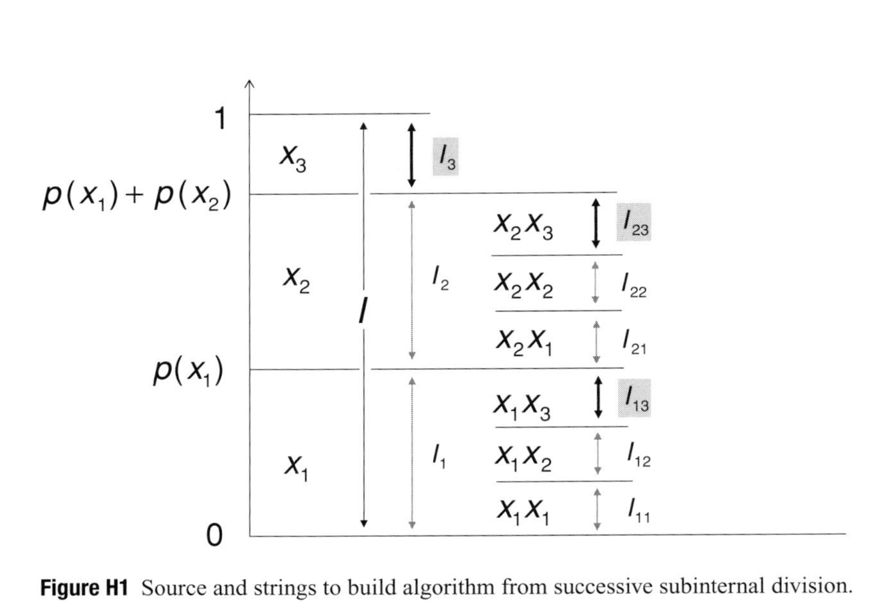


**Figure H1** Source and strings to build algorithm from successive subinternal division.

Although it looks a bit useless and complicated, it is accurate to rewrite the definition in Eq. (H7) in the general form:

$$\begin{cases} u(I_3) = p(x_1|s)p(s) + p(x_2|s)p(s) \equiv u(I) + w(I)\sum_{i=1}^{2} p(x_i|s) \\ v(I_3) = p(x_1|s)p(s) + p(x_2|s)p(s) + p(x_3|s)p(s) \equiv u(I) + w(I)\sum_{i=1}^{3} p(x_i|s), \end{cases} \tag{H8}$$

where the argument $s$ refers to all preceding symbols in the string (since there are none here, $p(s) = 1 = w(I)$ and $p(x_{1,2,3}|s) \equiv p(x_{1,2,3})$), where $u(I) = 0$ is the start value of the interval corresponding to all preceding symbols ($I$).

Consider next the interval $I_{23}$, which is the last subdivision of $I_2$ and corresponds to the string $x_2x_3$, as shown in Fig. H1. Consistently with the subdivision, the start and the stop values of this interval are:

$$\begin{cases} u = p(x_1) + p(x_1|x_2)p(x_2) + p(x_2|x_2)p(x_2) \\ v = p(x_1) + p(x_1|x_2)p(x_2) + p(x_2|x_2)p(x_2) + p(x_3|x_2)p(x_2), \end{cases} \tag{H9}$$

which can be rewritten in the form:

$$\begin{cases} u(I_{23}) \equiv u(I_2) + w(I_2)\sum_{i=1}^{2} p(x_i|x_2) \\ v(I_{23}) = u(I_2) + w(I_2)\sum_{i=1}^{3} p(x_i|x_2). \end{cases} \tag{H10}$$

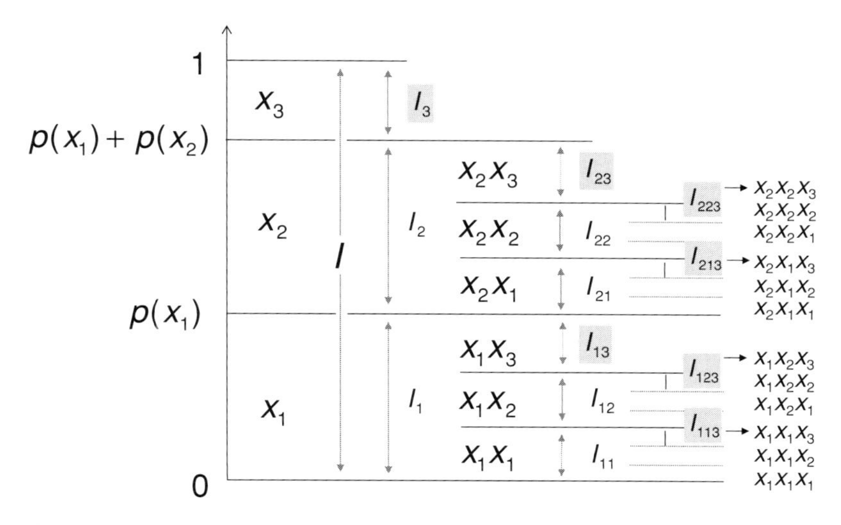


**Figure H2** Symbol strings with lengths up to $N = 3$.

The same results as in Eqs. (H9) and (H10) can be obtained for the interval $I_{13}$, corresponding to the string $x_1x_3$ (Fig. H1), with the argument $I_2$ changed into $I_1$.

Consider next the case of symbol strings with lengths up to $N = 3$, which correspond to all the intervals that have been highlighted in Fig. H2.

The strings with lengths $N = 3$, namely $x_1x_1x_3, x_1x_2x_3, x_2x_1x_3$, and $x_2x_2x_3$, are associated with the subintervals $I_{113}, I_{123}, I_{213}$, and $I_{223}$. We do not need to consider all of them to prove the general relation that defines the start and stop values of each of these subintervals. Take, for instance, $I_{223}$. Looking at Fig. H2, we readily obtain:

$$\begin{cases} u(I_{223}) = p(x_1) + p(x_1|x_2)p(x_2) + p(x_1|x_2,x_2)p(x_2,x_2) + p(x_2|x_2,x_2)p(x_2,x_2) \\ \qquad = p(x_1) + p(x_1,x_2) + p(x_2,x_2)\left[p(x_1|x_2,x_2) + p(x_2|x_2,x_2)\right] \\ \qquad \equiv u(I_{22}) + w(I_{22})\displaystyle\sum_{i=1}^{2} p(x_i|x_2,x_2) \\ v(I_{223}) = p(x_1) + p(x_1|x_2)p(x_2) + p(x_1|x_2,x_2)p(x_2,x_2) + p(x_2|x_2,x_2)p(x_2,x_2) \\ \qquad + p(x_3|x_2,x_2)p(x_2,x_2) \\ \qquad = p(x_1) + p(x_1,x_2) + p(x_2,x_2)[p(x_1|x_2,x_2) + p(x_2|x_2,x_2) + p(x_3|x_2,x_2)] \\ \qquad \equiv u(I_{22}) + w(I_{22})\displaystyle\sum_{i=1}^{3} p(x_i|x_2,x_2). \end{cases} \tag{H11}$$

In this derivation, we used the property $p(x_2, x_2) = w(I_{22}) = v(I_{22}) - u(I_{22})$, according to which the width of the subinterval corresponding to a preceding string sequence $x_ix_j$ is equal to the joint probability $p(x_i, x_j) = p(x_j|x_i)p(x_i)$. We also used the fact that, by definition, $u(I_{22}) = p(x_1) + p(x_1, x_2)$.

The previous analysis makes it possible to rewrite Eq. (H11) concerning any three-symbol string of the type $x_m x_n x_3$ $(m, n = 1 \ldots 2)$ in the general form:

$$\begin{cases} u(I_{mn3}) = u(I_{mn}) + w(I_{mn}) \sum_{i=1}^{2} p(x_i | x_m, x_n) \\ v(I_{mn3}) = u(I_{mn}) + w(I_{mn}) \sum_{i=1}^{3} p(x_i | x_m, x_n). \end{cases} \quad \text{(H12)}$$

In turn, Eq. (H12) suggests a general recurrence formula defining the interval $I = \langle u_N, v_N \rangle$ corresponding to any string $a_1 a_2 \ldots a_k \ldots a_N$ of length $N$ made from a source $X = \{x_1, x_2, \ldots, x_i, \ldots, x_n\}$ of size $n$, with the string being terminated by symbol $x_n$. The start and stop values of the corresponding interval are computed by successive iterations from $k = 2$ to $k = N$ according to the following recurrence relations:

$$\begin{cases} u_k = u_{k-1} + w_{k-1} Q(X, a_1, a_2, \ldots, a_{k-1}) \\ v_k = u_{k-1} + w_{k-1} R(X, a_1, a_2, \ldots, a_{k-1}) \\ w_k = v_k - u_k, \end{cases} \quad \text{(H13)}$$

with

$$\begin{cases} Q(X, a_1, a_2, \ldots, a_{k-1}) = \sum_{i=1}^{n-1} p(x_i | a_1, a_2, \ldots, a_{k-1}) \\ R(X, a_1, a_2, \ldots, a_{k-1}) = \sum_{i=1}^{n} p(x_i | a_1, a_2, \ldots, a_{k-1}) = 1, \end{cases} \quad \text{(H14)}$$

and with the initial conditions

$$\begin{cases} \text{given } a_1 = x_j: \\ u_1 = \begin{cases} 0; \ j = 1 \\ \sum_{i=1}^{j-1} p(x_i); \ j > 1 \end{cases} \\ v_1 = u_1 + p(x_j). \end{cases} \quad \text{(H15)}$$

We note that $R(X, a_1, a_2, \ldots, a_{k-1}) = 1$, by application of the property $\sum_{x \in X} p(x|y) = 1$ for any event $y$ (see Chapter 5), therefore, the sum involved in the definition of $R$ does not need to be computed. From this property, we also have $1 = R(X, a_1, a_2, \ldots, a_{k-1}) = Q(X, a_1, a_2, \ldots, a_{k-1}) + p(x_n | a_1, a_2, \ldots, a_{k-1})$, which eliminates the need to compute the sum involved in the function $Q$ as well. We can then redefine the algorithm in Eq. (H13) according to the much simpler form:

$$\begin{cases} u_k = u_{k-1} + w_{k-1} \left[1 - p(x_n | a_1, a_2, \ldots, a_{k-1})\right] \\ v_k = w_{k-1} \\ w_k = v_k - u_k. \end{cases} \quad \text{(H16)}$$

It is easily verified that Eqs. (H15) and (H16) yield the correct values $u_N$, $v_N$ for the strings of lengths $N = 1 \ldots 3$ previously analyzed. Note that the algorithm defined by these equations is valid only for the case of interest where the last symbol of the message string, $a_N$, is equal to $x_n$. This algorithm represents a simpler variant of that described in the reference[2], which is valid for any symbol values of $a_N$ (namely all subintervals). The generalization of the above algorithm to the general case ($a_N = x_j$) simply consists of modifying Eq. (H14) to:

$$\begin{cases} Q(X, a_1, a_2, \ldots, a_{k-1}) = \sum_{i=1}^{j-1} p(x_i | a_1, a_2, \ldots, a_{k-1}) \\ R(X, a_1, a_2, \ldots, a_{k-1}) = \sum_{i=1}^{j} p(x_i | a_1, a_2, \ldots, a_{k-1}), \end{cases} \tag{H17}$$

with, by convention, $Q$ being set to zero when $j = 1$.

[2] D. J. C. MacKay, *A Short Course in Information Theory* (1995), www.inference.phy.cam.ac.uk/mackay/info-theory/course.html; see also D. J. C. MacKay, *Information Theory, Inference and Learning Algorithms* (Cambridge, UK: Cambridge University Press, 2003), p. 113.

# Appendix I (Chapter 10) Lempel–Ziv distinct parsing

In this appendix I show, based on the demonstration of Cover and Thomas (1991)[1] and a few useful explanations, that for a given string sequence (or message) of length $n$, *the number of distinct-parsing phrases* $c(n)$ is bounded according to:

$$c(n) \leq \frac{n}{(1-\varepsilon_n)\log n}, \tag{I1}$$

where $\varepsilon_n$ vanishes as $n \to \infty$.

## Distinct parsing

This is the action of partitioning a given message into different substrings, or phrases, with variable length $m$. Assume that the maximum phrase length allowed is $K$. The number of distinct phrases having $m$ bits is $2^m$. The length $L_K$ of a sequence made from concatenating all possible phrases of length $\leq K$ is, therefore,

$$L_K = \sum_{m=1}^{K} m2^m. \tag{I2}$$

In a note at the end of this appendix, I show that this length is equal to

$$L_K = (K-1)2^{K+1} + 2. \tag{I3}$$

Consider first the simple case where the message length is $n = L_K$. The *number of phrases* $c(n)$ in this message is maximized when the phrases are as short as possible, meaning that there are $2^m$ phrases of size $m$ for each $m$. This yields an upper bound for $c(n)$, which can be developed according to:

$$\begin{aligned} c(n) &\leq \sum_{m=1}^{K} 2^m \\ &= \frac{1-2^{K+1}}{1-2} - 2^0 \\ &= 2^{K+1} - 2, \end{aligned} \tag{I4}$$

[1] T. M. Cover and J. A. Thomas, *Elements of Information Theory* (New York: John Wiley & Sons, 1991), p. 320–1.

$$
\begin{aligned}
&< 2^{K+1} \\
&\leq \frac{L_K - 2}{k - 1} \\
&< \frac{L_K}{k - 1}
\end{aligned}
$$

where $K > 1$ is assumed.

Consider next the more complex case where the message length satisfies $L_K \leq n < L_{K+1}$. We can write $n = L_K + \Delta_K$, where $\Delta_K$ is bounded by the difference $L_{K+1} - L_K = K2^{K+2} + 2 - [(K - 1)2^K + 2] = (K + 1)2^K$, or $\Delta_K < (K + 1)2^K$. The parsing generates a certain number of phrases of length $\leq K$, with the rest being of length $K + 1$. The maximum number of phrases of length $K + 1$ is $2^{K+1}$, which is strictly greater than $\Delta_K/(K + 1)$. Thus, the upper bound for $c(n)$ is now given by:

$$
\begin{aligned}
c(n) &< \frac{L_K}{k - 1} + \frac{\Delta_K}{K + 1} \\
&= \frac{L_K + \Delta_K}{k - 1} + \frac{\Delta_K}{K + 1} - \frac{\Delta_K}{K - 1} \\
&= \frac{L_K + \Delta_K}{k - 1} + \frac{\Delta_K(K - 1) - \Delta_K(K + 1)}{K^2 - 1} \\
&= \frac{L_K + \Delta_K}{k - 1} - \frac{2\Delta_K}{K^2 - 1} \\
&\leq \frac{L_K + \Delta_K}{k - 1} \\
&\equiv \frac{n}{k - 1}.
\end{aligned}
\tag{I5}
$$

The next step is to bound $K$ (find the maximum value of $K$) with respect to $n$. Assume most generally that $L_K \leq n \leq L_{K+1}$. The first inequality with Eq. (I3) gives

$$
\begin{aligned}
n &\geq L_k \\
&= (K - 1)2^{K+1} + 2 \\
&\geq 2^K.
\end{aligned}
\tag{I6}
$$

The lower bound $2^K$ in Eq. (I6) is found by setting $K = 1$ then $K > 1$. There is equality in the first case, and in the second case we have

$$
(K - 1)2^{K+1} + 2 > 2^{K+1} + 2 > 2^{K+1} > 2^K.
$$

The result in Eq. (I6) implies:

$$
K \leq \log n. \tag{I7}
$$

The second inequality with Eqs. (I3) and (I7) gives

$$\begin{aligned} n &\le L_{k+1} \\ &= K2^{K+2} + 2 \\ &\ge K2^{K+2} + 2.2^{K+2} \\ &= (K+2)2^{K+2} \\ &\le (\log n + 2)2^{K+2}. \end{aligned} \tag{I8}$$

Hence,

$$K + 2 \ge \log \frac{n}{\log n + 2}, \tag{I9}$$

which, with the condition $n > 1$, develops into

$$\begin{aligned} K - 1 &\ge \log \frac{n}{\log n + 2} - 3 \\ &= \log n - \log(\log n + 2) - 3 \\ &= \left(1 - \frac{\log(\log n + 2) + 3}{\log n}\right) \log n. \end{aligned} \tag{I10}$$

If we assume $n \ge 4$, we have $\log n + 2 \le 2 \log n$ and, consequently, $\log(\log n + 2) \le \log(2 \log n) = \log \log n + 1$. Substituting this inequality in Eq. (I10) yields

$$\begin{aligned} K - 1 &\ge \left(1 - \frac{\log(\log n) + 4}{\log n}\right) \log n \\ &\equiv (1 - \varepsilon_n) \log n, \end{aligned} \tag{I11}$$

with

$$\varepsilon_n = \frac{\log(\log n) + 4}{\log n}. \tag{I12}$$

We now require that $K - 1 > 0$ (a condition which has been overlooked in the reference).[2] Since we already assumed $n \ge 4$ ($\log n \ge 2$), this extra requirement is equivalent to $\varepsilon_n < 1$. Letting $n = 2^q$ ($q > 0$ and real) and substituting in Eq. (I12), we obtain the new condition $\log q < q - 4$. As easily checked, a sufficient condition is $q \ge 8$ or $n \ge 128$ (the exact condition being $n \ge 109$).

Under the condition $\varepsilon_n < 1$ (or $K - 1 > 0$), substituting the inequality in Eq. (I11) into that in Eq. (H5) to obtain the final result:

$$\begin{aligned} c(n) &\le \frac{n}{K - 1} \\ &\le \frac{n}{(1 - \varepsilon_n) \log n}. \end{aligned} \tag{I13}$$

[2] T. M. Cover and J. A. Thomas, *Elements of Information Theory* (New York: John Wiley & Sons, 1991), p. 320–1.

## Proof of Eq. (I3)

Setting $q = 2$, we write the sum in Eq. (I2) in the form

$$\begin{aligned} L_k &= \sum_{m=1}^{K} mq^m \\ &= q\sum_{m=1}^{K} mq^{m-1} \\ &= q\frac{d}{\mathrm{d}q}\sum_{m=1}^{K} q^m \\ &= q\frac{d}{\mathrm{d}q}\left(\sum_{m=0}^{K} q^m\right). \end{aligned}$$

Since the sum involved in the last right-hand side is equal to $(1-q^{K+1})/(1-q)$, we get

$$\begin{aligned} L_k &= q\frac{d}{\mathrm{d}q}\left(\frac{1-q^{K+1}}{1-q}\right) \\ &= \frac{-(K+1)q^K(1-q)-(1-q^{K+1})(-1)}{(1-q)^2} \\ &\underset{q=2}{\equiv} q[1-q^{K+1}+(K+1)q^K] \\ &= q[1-q^{K+1}+(K+1)q^K \\ &= q+q^{K+1}(1-q)+Kq^{K+1} \\ &\underset{q=2}{\equiv} 2+(K+1)q^{K+1}. \end{aligned}$$

# Appendix J (Chapter 11) Error-correction capability of linear block codes

In this appendix, I show that linear block codes have the capability of correcting any error patterns of Hamming weight $w(E) = e$, or any number of $e$ errors in the received block code, provided that

$$e \leq \left\{ \frac{d_{\min} - 1}{2} \right\}, \tag{J1}$$

where $d_{\min}$ is the minimum Hamming distance, and with the brackets corresponding to the integer-floor definition, i.e., $\{(d_{\min} - 1)/2\} \equiv \lfloor (d_{\min} - 1)/2 \rfloor$ corresponding to the highest integer contained in the argument.[1]

To show this, assume a block code $C$, with a minimum Hamming distance $d_{\min}$. To recall, $d_{\min}$ represents *the minimum number of bit positions for which two codewords differ in* $C$. Let $X$ be any codeword belonging to $C$, and $Y$ any received block code, which may contain errors. Define the Hamming distance between vectors $X$ and $Y$ as $d(X, Y)$. It is clear that if $d(X, Y) < d_{\min}$, or equivalently,

$$d(X, Y) \leq d_{\min} - 1, \tag{J2}$$

the block code $Y$ does not belong to $C$. In this case, any possible error pattern is detected by the code.

Assume next that $X$ and $Y$ are the transmitted and received vectors, respectively. Since the minimum Hamming distance $d_{\min}$ is either an odd or an even integer, we can write

$$2t + 1 \leq d_{\min} \leq 2t + 2, \tag{J3}$$

where $t \geq 0$ is an integer. I am now going to show that *the code is capable of correcting all error patterns having* $t$ *or fewer errors*. Let $Z$ be another vector in $C$. Using the distance property referred to as *triangle inequality* (see Chapter 5), we have

$$d(X, Z) \leq d(X, Y) + d(Y, Z). \tag{J4}$$

Suppose now that the number of errors in the received vector $Y$ is $w(E) = e$. This means that

$$d(X, Y) = e. \tag{J5}$$

[1] The following demonstration is inspired and adapted from S. Lin and D. J. Costello, *Error Control Coding* (Englewood Cliffs, NJ: Prentice-Hall, 1983).

Since both vectors $X$ and $Z$ belong to the code, their Hamming distance satisfies $d(X, Z) \geq d_{\min}$, or using the left-hand side inequality in Eq. (J3):

$$d(X, Z) \geq d_{\min} \geq 2t + 1. \tag{J6}$$

Combining Eqs. (J4), (J5), and (J6), we then obtain:

$$\begin{aligned} 2t + 1 &\leq d_{\min} \\ &\leq d(X, Z) \\ &\leq e + d(Y, Z) \\ \Rightarrow d(Y, Z) &\geq 2t + 1 - e \\ \Rightarrow d(Y, Z) &> 2t - e. \end{aligned} \tag{J7}$$

To interpret the last inequality, consider the two possible cases: (a) $e \leq t$, and (b) $e > t$.

In case (a), we obtain $2t - e > t$, thus, $d(Y, Z) > t$. Since in this case and from Eq. (J5) we also have $d(Y, X) = e \leq t$, this means that the received vector $Y$ is closer to the transmitted vector $X$ than any other vector $Z$ in the code. The principle of *maximum likelihood decoding* (see Chapter 11) ensures that the code will necessarily correct the received block $Y$ to the transmitted block $X$. Using Eq. (J3), with $e \leq t$, we also have

$$\begin{aligned} 2e + 1 &\leq 2t + 1 \\ &\leq d_{\min} \\ \Rightarrow e &\leq \frac{d_{\min} - 1}{2} \\ \Rightarrow e &\leq \left\lfloor \frac{d_{\min} - 1}{2} \right\rfloor. \end{aligned} \tag{J8}$$

The last inequality in the above result, thus, establishes that the code is capable of correcting all error patterns $E$ having up to $\lfloor (d_{\min} - 1)/2 \rfloor$ errors.

Consider next case (b), where $e > t$. I am going to show that the code is not capable of correcting all the corresponding possible error patterns, because there exists at least one case where the received vector $Y$ is closer to an incorrect vector $Z$ than to the transmitted vector $X$. To show this, let $X$ and $Y$ be the transmitted and received vectors, respectively, and assume that:

(1) $Z$ is another vector in the code $C$ such that $d(X, Z) = d_{\min}$;
(2) $E_1$ and $E_2$ are two different error patterns satisfying the properties:
  (i) $E_1 + E_2 = X + Z$,
  (ii) $E_1$ and $E_2$ do not have nonzero bits in the same positions.

Defining $w(U)$ as the Hamming weight of the block or vector $U$, the consequences of the above assumptions are summarized by:

$$\begin{cases} w(E_1 + E_2) = w(X + Z) \\ w(E_1 + E_2) = w(E_1) + w(E_2) \Leftrightarrow w(E_1) + w(E_2) = w(X + Z) = d(X, Z) = d_{\min} \\ w(X + Z) = d(X, Z) = d_{\min}. \end{cases} \tag{J9}$$

The first and second equalities in Eq. (J9) obviously stem from properties (i) and (ii), respectively. The third equality, $w(X + Z) = d(X, Z)$, comes from the definition of the Hamming distance, which represents the number of bit positions by which two blocks or vectors differ from each other; each of the nonzero bits in the vector $X + Z$ represents such positions, which number as $w(X + Z)$.

Assume next that $E_1$ is the error vector of the transmitted vector $X$. The received vector is, therefore:

$$Y = X + E_1. \tag{J10}$$

The Hamming distances $d(X, Y)$ and $d(X, Z)$ are found using results in Eq. (J10) and property (i), respectively, as follows:

$$d(X, Y) = w(X + Y) = w(X - Y) = w(E_1), \tag{J11}$$

$$d(Y, Z) = w(Y + Z) = w(X + E_1 + Z) = w(X - E_1 + Z) = w(E_2). \tag{J12}$$

We assume next that the error pattern associated with $E_1$ corresponds to a number of errors $w(E_1) = e$ strictly greater than $t$, i.e., $e > t$ or $e \geq t + 1$. Based on this assumption and after Eqs. (J3) and (J9), we have

$$\begin{cases} d_{\min} = w(E_1) + w(E_2) \leq 2t + 2 \\ t + 1 + w(E_2) \leq w(E_1) + w(E_2) \leq 2t + 2 \end{cases} \Rightarrow w(E_2) \leq t + 1. \tag{J13}$$

Substituting $w(E_1) \geq t + 1$ and $w(E_2) \leq t + 1$ in Eqs. (J11) and (J12), we obtain finally:

$$\begin{cases} d(X, Y) \geq t + 1 \\ d(Y, Z) \leq t + 1 \end{cases} \Rightarrow d(Y, Z) \leq d(X, Y). \tag{J14}$$

The above result indicates that there exists an error pattern with $e > t$ errors, which results in a transmitted vector $Y$ closer to an incorrect vector $Z$ than the transmitted vector $X$. As a consequence, based on the principle of maximum-likelihood decoding, the code will output an incorrectly decoded block. This provides the proof that the code is capable of correcting any error pattern with number of errors $e \leq t$, or $e \leq \{(d_{\min} - 1)/2\}$, where the brackets indicate the integer contained in the argument.

# Appendix K (Chapter 13) Capacity of binary communication channels

In this appendix, I provide the solution of the maximization problem for mutual information, which defines the channel capacity:

$$C = \max_{p(x)} H(X;Y), \tag{K1}$$

as applicable to the general case of a *binary communication channel* (symmetric or asymmetric), whose transition matrix is defined according to

$$\begin{aligned} P(Y|X) &= \begin{pmatrix} p(y_1|x_1) & p(y_1|x_2) \\ p(y_2|x_1) & p(y_2|x_2) \end{pmatrix} \\ &= \begin{pmatrix} a & 1-b \\ 1-a & b \end{pmatrix}, \end{aligned} \tag{K2}$$

where $a, b$ are real numbers in the interval $[0, 1]$. To recall, the channel mutual information $H(X;Y)$ is given by

$$H(X;Y) = H(Y) - H(Y|X). \tag{K3}$$

Thus, we must first calculate the output-source entropy $H(Y)$ and the equivocation entropy $H(Y|X)$. For this, we need the output probability distribution $p(y_1)$, $p(y_2)$, which is obtained from the transition-matrix elements in Eq. (K2) as follows:

$$\begin{aligned} p(y_1) &= p(y_1|x_1)p(x_1) + p(y_1|x_2)p(x_2) \\ &= aq + (1-b)(1-q) \\ p(y_2) &= p(y_2|x_1)p(x_1) + p(y_2|x_2)p(x_2) \\ &= (1-a)q + b(1-q) \\ &\equiv 1 - p(y_1). \end{aligned} \tag{K4}$$

In Eq. (K4), we have defined the input probability distribution according to $p(x_1) = q$ and $p(x_2) = 1 - q$. From Eq. (K4), we can calculate the output-source entropy:

$$\begin{aligned} H(Y) &= -p(y_1)\log p(y_1) - p(y_2)\log p(y_2) \\ &= -p(y_1)\log p(y_1) - [1 - p(y_1)]\log[1 - p(y_1)] \\ &= f[p(y_1)] \\ &= f[aq + (1-b)(1-q)], \end{aligned} \tag{K5}$$

where the function $f$ is defined by

$$f(u) = f(1-u) \\ = -u\log u - (1-u)\log(1-u), \tag{K6}$$

noting that, as usual, the logarithms are in base two.

The next step consists of calculating the equivocation $H(Y|X)$. For this, we need the joint distribution $p(x, y)$. Using Bayes's theorem and the conditional probabilities shown in Eq. (K2) we obtain:

$$\begin{cases} p(y_1, x_1) = p(y_1|x_1)p(x_1) = aq \\ p(y_1, x_2) = p(y_1|x_2)p(x_2) = (1-b)(1-q) \\ p(y_2, x_1) = p(y_2|x_1)p(x_1) = (1-a)q \\ p(y_2, x_2) = p(y_2|x_2)p(x_2) = b(1-q). \end{cases} \tag{K7}$$

The results in Eq. (K5) now make it possible to calculate the equivocation $H(Y|X)$:

$$\begin{aligned} H(Y|X) &= -\sum_{i=1}^{2}\sum_{j=1}^{2} p(x_i, y_j)\log p(y_j|x_i) \\ &= -[aq\log a + (1-a)q\log(1-a) \\ &\quad + (1-b)(1-q)\log(1-b) + b(1-q)\log b] \\ &\equiv qf(a) + (1-q)f(b). \end{aligned} \tag{K8}$$

Substituting Eqs. (K7) and (K8) in Eq. (K3), we obtain the mutual information $H(X;Y)$:

$$H(X;Y) = f[aq + (1-b)(1-q)] - qf(a) - (1-q)f(b). \tag{K9}$$

Setting $q = 0$ or $q = 1$ in the result in Eq. (K9) yields, in both cases, $H(X;Y) = 0$. The case $a = 1-b$ (or $a + b = 1$) corresponds to $H(X;Y) = 0$, as can easily be verified from Eq. (K9). This means that, regardless of the input probability distribution, the mutual information is zero. As discussed in the main text, such a channel is *useless*. The condition $a = 1-b$ in the transition matrix (Eq. (K2)) represents the most general definition of useless channels, i.e., including but not limited to the case $a = b = \varepsilon = 1/2$.

In the general case, we have $H(X;Y) \geq 0$, and the function is concave (meaning that a cord between any two points is always below the maximum). Therefore, the maximum of $H(X;Y) \geq 0$ is given by the root of the derivative $\mathrm{d}H(X;Y)/\mathrm{d}q$. Thus, we must solve

$$\begin{aligned} \frac{\mathrm{d}H(X;Y)}{\mathrm{d}q} &= \frac{\mathrm{d}}{\mathrm{d}q}\{f[aq + (1-b)(1-q)] - qf(a) - (1-q)f(b)\} \\ &= 0. \end{aligned} \tag{K10}$$

Using the definition of $f$ in Eq. (K8) and going through elementary calculations yields the following solution for $q$:

$$q = \frac{1}{a+b-1}\left(b - 1 + \frac{1}{1+2^W}\right), \tag{K11}$$

with

$$W = \frac{f(a) - f(b)}{a + b - 1}. \tag{K12}$$

Concerning the continuity of the above solution in the case $a + b = 1$, see the note at the end of this appendix.

The optimal distribution defined in Eqs. (K11) and (K12) can now be substituted into Eq. (K10). After elementary calculation, we obtain:

$$\begin{aligned} C &= \max_q H(X;Y) \\ &= \log(1 + 2^W) - W\frac{2^W}{1 + 2^W} - q[f(b) - f(a)] - f(b) \\ &= \log(1 + 2^W) + \frac{(1-a)f(b) - bf(a)}{a + b - 1}. \end{aligned} \tag{K13}$$

Define

$$U = \frac{(1-a)f(b) - bf(a)}{a + b - 1} \tag{K14}$$

and substitute $U$ in Eq. (K13) to get the channel capacity:

$$\begin{aligned} C &= \log(1 + 2^W) + U \\ &= \log[(1 + 2^W)2^U] \\ &= \log(2^U + 2^{U+W}) \\ &\equiv \log(2^U + 2^V), \end{aligned} \tag{K15}$$

with

$$V = U + W = \frac{(1-b)f(a) - af(b)}{a + b - 1}. \tag{K16}$$

## Note

The optimal probability distribution $q = p(x_1) = 1 - p(x_2)$ defined in Eqs. (K11) and (K12) seemingly has a pole in $a + b = 1$. I show herewith that it is actually defined over the whole plane $a, b \in [0, 1]$, namely, that it is analytically defined in the limiting case $a + b = 1$. I shall establish this by first setting $a + b = 1 - \varepsilon$ in Eq. (K11), then using the Taylor expansion of the function $f(a + \varepsilon)$ to the second order, i.e.,

$$f(a + \varepsilon) = f(a) + \varepsilon \log\frac{1-a}{a} + \varepsilon^2\frac{1}{2a(1-a)},$$

which gives

$$W = \log\frac{1-a}{a} + \varepsilon\frac{1}{2a(1-a)}.$$

By substituting this result in Eq. (K11), and expanding the exponential term, one easily finds $1/(1 + 2^W) \approx a - \varepsilon/2$, which gives the limit $q = 1/2$, corresponding to the uniform distribution. However, such a distribution does not maximize the mutual information $H(Y; X)$, since we have seen that in the limiting case $a + b = 1$ we have $H(Y; X) = 0$.

I show next that the channel capacity is also defined over the plane $a, b \in [0, 1]$, including the limiting case $a + b = 1$. Indeed, using the same Taylor expansion as previously, we easily obtain

$$\begin{aligned} U &= (1 - a) \log \frac{a}{1 - a} - f(a) + \frac{\varepsilon}{2a} \\ &= \log a + \frac{\varepsilon}{2a} \\ &\approx \log a. \end{aligned}$$

Substituting this result and $1/(1 + 2^W) \approx a - \varepsilon/2 \approx a$ into Eq. (K15) yields $C = \log(1 + 2^W) + U \approx -\log a + \log a = 0$, which is the expected channel capacity in the limiting case $a + b = 1$. The function $V$ is also found to take the limit $V \approx \log(1 - a)$, which, from Eq. (K15), also gives $C = \log(2^U + 2^V) \approx 0$.

# Appendix L (Chapter 13) Converse proof of the channel coding theorem

This appendix provides the converse proof of the CCT.[1] The converse proof must show that to achieve transmission with arbitrary level accuracy (or error probability), the condition $R \leq C$ must be fulfilled. The demonstration seeks to establish two properties, which I shall name A and B.

## Property A

To begin with, we must establish the following property, referred to as *Fano's inequality*, according to which:

$$H(X^n|Y^n) \leq 1 + p_e nR, \tag{L1}$$

where $p_e$ is the error probability of the code ($p_e = 1 - \tilde{p}$), i.e., the probability that the code will output a codeword that is different from the input message codeword.

To demonstrate Fano's inequality, we define $W = 1 \ldots 2^{nR}$ as the integer that labels the $2^{nR}$ possible codewords in the input-message codebook. We can view the code as generating an output integer label $W'$, to which the label $W$ of the input message codeword may or may not correspond. We can, thus, write $p_e = p(W' \neq W)$. We then define a binary random variable $E$, which tells whether or not a codeword error occurred: $E = 1$, if $W' \neq W$, and $E = 0$, if $W' = W$. Thus, we have $p(E = 1) = p_e$ and $p(E = 0) = 1 - p_e$. We can then introduce the conditional entropy $H(E, W|Y^n)$, which is the average information we have on $E, W$, given the knowledge of the output codeword source, $Y^n$. Referring back to the *chain rule* in Eqs. (5.22) and (5.23), we can expand $H(E, W|Y^n)$ in two different ways:

$$\begin{aligned} H(E, W|Y^n) &= H(E|Y^n) + H(W|E, Y^n) \\ &= H(W|Y^n) + H(E|W, Y^n). \end{aligned} \tag{L2}$$

Since $E$ is given by the combined knowledge of label $W$ and output source $Y^n$, we have $H(E|W, Y^n) = 0$. We also have $H(E|Y^n) = H(E)$, since the only knowledge of $Y^n$ does not condition the knowledge of $E$. Substituting these results into Eq. (L2), we

[1] Inspired from M. Cover and J. A. Thomas, *Elements of Information Theory* (New York: John Wiley & Sons, 1991), pp. 203–9.

obtain

$$H(W|Y^n) = H(E) + H(W|E, Y^n). \tag{L3}$$

We shall now find an upper bound for $H(W|Y^n)$. First, we can decompose the second term in the right-hand side in Eq. (L3) as follows:

$$\begin{aligned} H(W|E, Y^n) &= p(E = 0)H(W|Y^n, E = 0) + p(E = 1)H(W|Y^n, E = 1) \\ &= (1 - p_e)H(W|Y^n, E = 0) + p_e H(W|Y^n, E = 1). \end{aligned} \tag{L4}$$

We have $H(W|Y^n, E = 0) = 0$, since knowing $Y^n$ and that there is no codeword error is equivalent to knowing the input codeword label $W$. Second, we have $H(W|Y^n, E = 1) \leq \log(2^{nR} - 1) < \log(2^{nR}) = nR$, since knowing $Y^n$ and that there is a codeword error, the number of mistaken codeword possibilities is $2^{nR} - 1$, with uniform probability $q = 1/(2^{nR} - 1)$; hence the entropy can be upper bounded by $H' = -\log(q) < nR$. Finally, we have $H(E) \leq 1$, since $E$ is a binary random variable. Substituting the two upper bounds into Eq. (L3) yields:

$$H(W|Y^n) \leq 1 + p_e nR \tag{L4}$$

and

$$H(X^n|Y^n) \leq 1 + p_e nR, \tag{L5}$$

since the knowledge of the input message source $X^n$ and the codeword label $W$ are equivalent. Equation (L5) is *Fano's inequality*.

## Property B

The second tool required for the converse proof of the CCT is the property according to which *the channel capacity per transmission is not increased by passing through the data several times*. Note that this property is true if one assumes that the channel is *memoryless*, namely that there is no possible correlation between errors concerning successive bits or successive codewords.

As I established earlier, the channel capacity for an $n$-bit codeword is $nC$, where $C$ is the capacity of the binary channel, corresponding to a single bit transmission. The proposed new property can be restated as

$$H(X^n; Y^n) \leq nC. \tag{L6}$$

The corresponding proof of Eq. (L6) is relatively straightforward. Indeed, we have, by definition

$$H(X^n; Y^n) = H(Y^n) - H(Y^n|X^n). \tag{L7}$$

Let us now develop the second term in the right-hand side in Eq. (L7) as follows:

$$\begin{aligned} H(Y^n|X^n) &= H(y_1|X^n) + H(y_2|y_1, X^n) + \cdots + H(y_n|y_1, y_2, \ldots, y_{n-1}, X^n) \\ &= \sum_{i=1}^{n} H(y_i|y_1, y_2, \ldots y_{i-1}, X^n) \\ &= \sum_{i=1}^{n} H(y_i|x_i). \end{aligned} \tag{L8}$$

The first equality stems from substituting the extended output source $Y^n = y_1, Y^{n-1}$, $Y^{n-1} = y_2, Y^{n-2}$, etc., (here $y_k$ means the binary source of bit or rank $k$ in the codeword), into the chain rule (Eq. (5.22)) as follows:

$$\begin{aligned} H(Y^n|X^n) &\equiv H(y_1, Y^{n-1}|X^n) \\ &= H(y_1|X^n) + H(Y^{n-1}|y_1, X^n) \\ &= H(y_1|X^n) + H(y_2 Y^{n-2}|y_1, X^n) \\ &= H(y_1|X^n) + H(y_2|y_1, X^n) + H(Y^{n-2}|y_1, y_2, X^n) \\ &= \cdots \\ &\equiv H(y_1|X^n) + H(y_2|y_1, X^n) + \cdots + H(y_n|y_1, y_2, \ldots, y_{n-1}, X^n). \end{aligned} \tag{L9}$$

The last equality in Eq. (L8) stems from the fact that, assuming a memoryless channel, all received bits $y_1, y_2, \ldots, y_{i-1}$ are uncorrelated with the received bit $y_i$, hence $H(y_i|y_1, y_2, \ldots, y_{i-1}, X^n) = H(y_i|X^n)$. Furthermore, the knowledge of $y_i$ is not conditioned to that of the input message bits $x_1, x_2, \ldots, x_n$ except for the bit $x_i$ of the same rank $i$ in the codeword, hence $H(y_i|y_1, y_2, \ldots, y_{i-1}, X^n) = H(y_i|x_i)$. Substituting the result in Eq. (L8) into Eq. (L7) yields:

$$H(X^n; Y^n) = H(Y^n) - \sum_{i=1}^{n} H(y_i|x_i). \tag{L10}$$

Next, we look for an upper bound in the right-hand side in Eq. (L8). We observe that $H(Y^n) \leq \sum_{i=1}^{n} H(y_i)$, the equality standing if the distribution $p(y_i)$ was uniform, in which case we would have $H(y_i) \equiv H(Y)$ and $H(Y^n) = nH(Y)$. Applying the inequality to Eq. (L10) we finally obtain

$$\begin{aligned} H(X^n; Y^n) &\leq \sum_{i=1}^{n} H(y_i) - \sum_{i=1}^{n} H(y_i|x_i) \\ &= \sum_{i=1}^{n} [H(y_i) - H(y_i|x_i)] \\ &\equiv \sum_{i=1}^{n} H(x_i; y_i). \end{aligned} \tag{L11}$$

This result shows that the mutual information between the two transmitted or received codeword sources is less than or equal to the sum of mutual information between the transmitted or received bits in the binary channel. Since, by definition, $H(x_i; y_i) \leq$

max $H(x_i; y_i) = C$, where $C$ is the binary-channel capacity, we also have

$$H(X^n; Y^n) \leq nC. \tag{L12}$$

This result establishes that the mutual information between extended sources $X^n$, $Y^n$, where bits are passed through the channel $n$ times, is no greater than $n$ times the binary-channel capacity.

The two properties A and B can now be used (finally!) to establish the converse proof of the CCT. As before, we assume that the $2^{nR}$ input message codewords are chosen at random with a uniform distribution, hence $H(X^n) = nR$. By definition of the mutual information, $H(X^n; Y^n)$, and introducing the majoring properties A and B, we obtain

$$\begin{aligned} H(X^n; Y^n) &= H(X^n) - H(X^n|Y^n) \\ &\leftrightarrow \\ nR &= H(X^n) = H(X^n; Y^n) + H(X^n|Y^n) \\ nR &\leq nC + 1 + p_e nR \\ &\leftrightarrow \\ R &\leq p_e R + \frac{1}{n} + C. \end{aligned} \tag{L13}$$

Since the starting assumption is that the error probability of the code, $p_e$, vanishes for $n \to \infty$, the above result asymptotically becomes:

$$R \leq C, \tag{L14}$$

which represents a necessary condition and, hence, proves the converse of the CCT.

# Appendix M (Chapter 16) Bloch sphere representation of the qubit

In this appendix, I show that qubits can be represented by a unique point on the surface of a sphere, referred to as a *Bloch sphere*.

As seen from the main text, any qubit can be represented as the vector linear superposition

$$|q\rangle = \alpha|0\rangle + \beta|1\rangle, \tag{M1}$$

where $|0\rangle$, $|1\rangle$ form an orthonormal basis in the 2D vector space, and $\alpha$, $\beta$ are complex numbers, which represent the qubit coordinates in this space. The length of the qubit vector $|q\rangle$ is, therefore, given by $|\alpha|^2 + |\beta|^2$. Since any two complex numbers $\alpha$, $\beta$ can be defined in the exponential representation as $\alpha = |a|e^{i\theta_1}$ and $\beta = |\beta|e^{i\theta_2}$, one can write, from Eq. (M1):

$$|q\rangle = \alpha|0\rangle + \beta|1\rangle = |a|e^{i\theta_1}|0\rangle + |\beta|e^{i\theta_2}|1\rangle \tag{M2}$$

Assume next that the qubit vector is unitary, i.e., $|\alpha|^2 + |\beta|^2 = 1$. Substituting this property and with the definition $\tan(\theta/2) = |\beta|/|\alpha|$, we obtain from Eq. (M2):

$$\begin{aligned} |q\rangle &= \frac{|\alpha|}{\sqrt{|\alpha|^2 + |\beta|^2}} e^{i\theta_1}|0\rangle + \frac{|\beta|}{\sqrt{|\alpha|^2 + |\beta|^2}} e^{i\theta_2}|1\rangle \\ &= \frac{1}{\sqrt{1 + \tan^2(\theta/2)}} e^{i\theta_1}|0\rangle + \frac{\tan^2(\theta/2)}{\sqrt{1 + \tan^2(\theta/2)}} e^{i\theta_2}|1\rangle \\ &\equiv \cos\frac{\theta}{2} e^{i\theta_1}|0\rangle + \sin\frac{\theta}{2} e^{i\theta_2}|1\rangle, \end{aligned} \tag{M3}$$

with $0 \leq \theta \leq \pi/2$.

Finally, introducing $\gamma = \theta_1$ and $\varphi = \theta_2 - \theta_1$, the qubit takes the form

$$|q\rangle = e^{i\gamma}\left(\cos\frac{\theta}{2}|0\rangle + \sin\frac{\theta}{2} e^{i\varphi}|1\rangle\right). \tag{M4}$$

Overlooking the argument $\gamma$, which only represents an arbitrary (said "unobservable") phase shift, we finally obtain

$$|q\rangle = \cos\frac{\theta}{2}|0\rangle + \sin\frac{\theta}{2} e^{i\varphi}|1\rangle. \tag{M5}$$

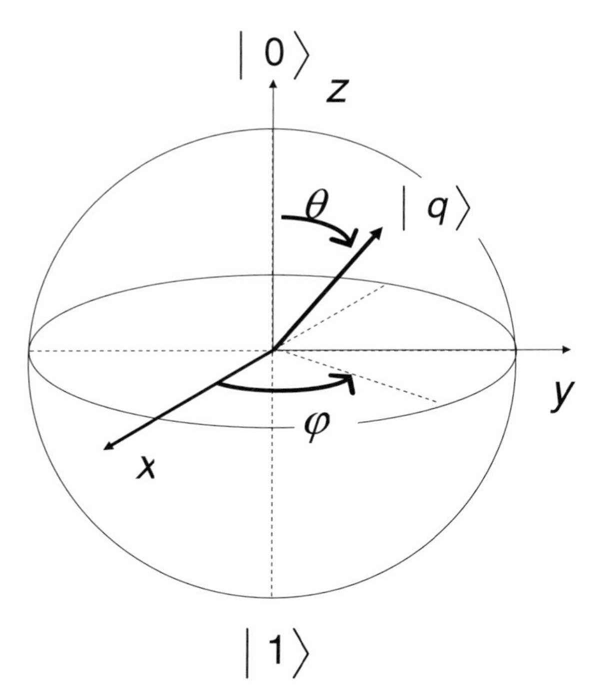


**Figure M1** Qubit represented as point of coordinates $(\theta, \varphi)$ on Bloch sphere.

The qubit can, thus, be represented through two *angular coordinates* $\theta$, $\varphi$, which define the unique position of a point on a sphere of unit radius, which is the *Bloch sphere* illustrated in Fig. M1.

It is seen from Fig. M1 that the pure qubits $|0\rangle$ or $|1\rangle$ correspond to $\theta = 0$ or $\theta = \pi$, respectively, which occupy the north and south poles of the Bloch sphere. The qubit representation of quantum information, thus, corresponds to an *infinity of states* located on the surface of the Bloch sphere. See more on this topic in Appendix N.

# Appendix N (Chapter 16) Pauli matrices, rotations, and unitary operators

In this appendix, I provide further conceptual background concerning *Pauli matrices*. It is shown that Pauli matrices make it possible to define any $2 \times 2$ *unitary* matrix $U$ (namely, satisfying the property $U^+U = I$, where the upper symbol ($^+$) stands for *Hermitian conjugation*, and $I$ is the identity matrix. The different results or theorems obtained in this appendix will be usefully applied to other chapters.

Recall first the definitions of the four Pauli matrices:

$$\begin{aligned} I &\equiv \sigma_0 = \begin{pmatrix} 1 & 0 \\ 0 & 1 \end{pmatrix} \qquad & X \equiv \sigma_x = \sigma_1 = \begin{pmatrix} 0 & 1 \\ 1 & 0 \end{pmatrix} \\ Y &\equiv \sigma_y = \sigma_2 = \begin{pmatrix} 0 & -\mathrm{i} \\ \mathrm{i} & 0 \end{pmatrix} \qquad & Z \equiv \sigma_z = \sigma_3 = \begin{pmatrix} 1 & 0 \\ 0 & -1 \end{pmatrix}. \end{aligned} \tag{N1}$$

Note that the different notations $\sigma_{0,1,2,3}$, $\sigma_{0,x,y,z}$, and $I, X, Y, Z$ are equivalent. We progress in several steps, which recall the properties of unitary operators, their complex-exponential representation, and, finally, the representation of unitary operators through rotations on the Bloch sphere.

## Hermitian conjugation and unitary matrices

Given an operator $A$, whose matrix representation is characterized by the complex coefficients $a_{ij}$, the *Hermitian conjugate* of $A$, called $A^+$, is defined by the coefficients $\bar{a}_{ji}$, corresponding to the complex-conjugate, transposed matrix. By definition, an operator or matrix is called *unitary* if it satisfies the property $A^+A = I$.

## Exponential representation of complex numbers

A complex number $z = a + ib$ has modulus or length $|z| = zz^* = \sqrt{a^2 + b^2}$. It is possible to represent a complex number in the form

$$\begin{aligned} z &= |z| \left( \frac{a}{\sqrt{a^2 + b^2}} + \mathrm{i} \frac{b}{\sqrt{a^2 + b^2}} \right) \\ &\equiv |z| (\cos\theta + \mathrm{i}\sin\theta), \end{aligned} \tag{N2}$$

where $\theta = \tan^{-1}(b/a)$ is the argument of $z$. One may also write:

$$z = |z|(\cos\theta + \mathrm{i}\sin\theta) \equiv |z|\,\mathrm{e}^{\mathrm{i}\theta}, \tag{N3}$$

where $\mathrm{e}^{\mathrm{i}\theta} = \cos\theta + \mathrm{i}\sin\theta$ is a complex number called an *imaginary exponential*. Such notation is justified if one considers the definition of the *real* exponential $\exp(x) = \mathrm{e}^x$:

$$\begin{aligned} \mathrm{e}^x &= \sum_{n=0}^{\infty} \frac{x^n}{n!} \\ &= 1 + x + \frac{x^2}{2!} + \frac{x^3}{3!} + \cdots + \frac{x^n}{n!} + \cdots \end{aligned} \tag{N4}$$

Substituting $x = \mathrm{i}\theta$ into Eq. (N4) yields

$$\begin{aligned} \mathrm{e}^{\mathrm{i}\theta} &= \sum_{n=0}^{\infty} \frac{(\mathrm{i}\theta)^n}{n!} = 1 + \mathrm{i}\theta + \frac{(\mathrm{i}\theta)^2}{2!} + \frac{(\mathrm{i}\theta)^3}{3!} + \cdots + \frac{(\mathrm{i}\theta)^n}{n!} + \cdots \\ &= 1 + \mathrm{i}\theta - \frac{\theta^2}{2!} - \mathrm{i}\frac{\theta^3}{3!} + \cdots \\ &= \sum_{n=0}^{\infty} \frac{(-1)^n \theta^{2n}}{(2n)!} + \mathrm{i}\sum_{n=0}^{\infty} \frac{(-1)^n \theta^{2n+1}}{(2n+1)!} \\ &\equiv \cos\theta + \mathrm{i}\sin\theta, \end{aligned} \tag{N5}$$

where the two series corresponding to the real and imaginary parts of $\mathrm{e}^{\mathrm{i}\theta}$ have been substituted with their exact function definitions $\cos\theta$ and $\sin\theta$, respectively. Noteworthy are the two identities

$$\exp(\mathrm{i}\pi) = -1, \tag{N6}$$

$$\exp(\mathrm{i}\pi/2) = \mathrm{i}. \tag{N7}$$

## Exponential operator

Assume an operator $A$ satisfying the property $A^2 = I$. We then define the *exponential operator* $\exp(\mathrm{i}A\theta)$ according to

$$\exp(\mathrm{i}A\theta) = \cos(\theta)I + \mathrm{i}\sin(\theta)A. \tag{N8}$$

Note in Eq. (N8) that the real part of the exponential involves the operator $I$ and the imaginary part involves the operator $A$. One proves the above development by substituting the argument $x = \mathrm{i}A\theta$ in the series expansion in Eq. (N4) and using the property $A^2 = I$:

$$
\begin{aligned}
e^{iA\theta} &= \sum_{n=0}^{\infty} \frac{(iA\theta)^n}{n!} \\
&= I + iA\theta + \frac{(iA\theta)^2}{2!} + \frac{(iA\theta)^3}{3!} + \cdots + \frac{(iA\theta)^n}{n!} + \cdots \\
&= I + iA\theta - \frac{\theta^2 I}{2!} - i\frac{\theta^3 A}{3!} + \cdots \\
&= \sum_{n=0}^{\infty} \frac{(-1)^n \theta^{2n}}{(2n)!} I + i \sum_{n=0}^{\infty} \frac{(-1)^n \theta^{2n+1}}{(2n+1)!} A \\
&\equiv \cos(\theta) I + i \sin(\theta) A
\end{aligned}
\tag{N9}
$$

(in the above, the two series expansions are recognized to correspond to the functions $\cos\theta$ and $\sin\theta$, respectively).

## Rotation operators

We now have the mathematical tools to introduce a new class of unitary operators generated by the three Pauli matrices $X, Y, Z$, and called *rotation operators*. Such operators are defined according to:

$$
\begin{cases}
R_x(\gamma) = e^{-i\frac{\gamma}{2}X} = e^{-i\frac{\gamma}{2}\sigma_1} \\
R_y(\gamma) = e^{-i\frac{\gamma}{2}Y} = e^{-i\frac{\gamma}{2}\sigma_2} \\
R_z(\gamma) = e^{-i\frac{\gamma}{2}Z} = e^{-i\frac{\gamma}{2}\sigma_3}.
\end{cases}
\tag{N10}
$$

The above exponential-operator definitions are relevant, since Pauli matrices satisfy the condition $X^2 = Y^2 = Z^2 = \sigma_i^2 = 1$. Substituting the definitions in Eqs. (N1) and (N8) into Eq. (N10) yields (as easily checked):

$$
\begin{cases}
R_x(\gamma) = \cos\frac{\gamma}{2} I - i\sin\frac{\gamma}{2} X = \begin{pmatrix} \cos\frac{\gamma}{2} & -i\sin\frac{\gamma}{2} \\ -i\sin\frac{\gamma}{2} & \cos\frac{\gamma}{2} \end{pmatrix} \\[2ex]
R_y(\gamma) = \cos\frac{\gamma}{2} I - i\sin\frac{\gamma}{2} Y = \begin{pmatrix} \cos\frac{\gamma}{2} & -\sin\frac{\gamma}{2} \\ \sin\frac{\gamma}{2} & \cos\frac{\gamma}{2} \end{pmatrix} \\[2ex]
R_z(\gamma) = \cos\frac{\gamma}{2} I - i\sin\frac{\gamma}{2} Z = \begin{pmatrix} \exp\left(-i\frac{\gamma}{2}\right) & 0 \\ 0 & \exp\left(i\frac{\gamma}{2}\right) \end{pmatrix}.
\end{cases}
\tag{N11}
$$

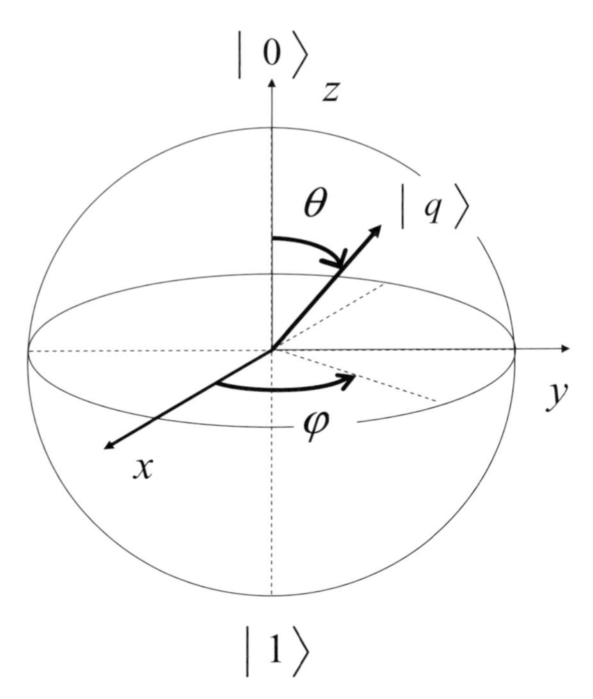


**Figure N1** Qubit represented as point of coordinates $(\theta, \varphi)$ on Bloch sphere.

I shall now interpret the effect of the three above rotation operators on qubits. As we have seen from the main text, the general definition of a qubit $|q\rangle$ is

$$|q\rangle = \alpha|0\rangle + \beta|1\rangle = \cos\frac{\theta}{2}|0\rangle + \sin\frac{\theta}{2}e^{i\varphi}|1\rangle. \tag{N12}$$

The above definition shows that any qubit can be represented by a single point of angular coordinates $(\theta, \varphi)$ on the surface of the *Bloch sphere* (with $\theta = 0 \to \pi$ and $\varphi = 0 \to 2\pi$), which is show in Fig. N1 for convenience.

It is left as an exercise to show that the three operators $R_k(\gamma)$ $(k = x, y, z)$ rotate the qubit by an angle $\gamma$ about the axis $k$, in the counterclockwise direction.

Consider next a more general form of the rotation operator. First, we define the *Pauli vector* in the Cartesian reference frame $(\vec{x}, \vec{y}, \vec{z}))$ according to

$$\begin{aligned}\vec{\sigma} &= \sigma_1\vec{x} + \sigma_2\vec{y} + \sigma_3\vec{z}\\ &= (\sigma_1, \sigma_2, \sigma_3).\end{aligned} \tag{N13}$$

Given a unitary vector $\vec{n} = (n_x, n_y, n_z)$, the scalar product $\vec{n}\cdot\vec{\sigma}$ corresponds to the matrix operator defined by $U = \vec{n}\cdot\vec{\sigma} = n_x\sigma_1 + n_y\sigma_2 + n_z\sigma_3$. It is easily established that $U$ is unitary or $U^2 = I$. Indeed, we have

$$\begin{aligned}U^2 &= UU\\ &= (n_x\sigma_1 + n_y\sigma_2 + n_z\sigma_3)(n_x\sigma_1 + n_y\sigma_2 + n_z\sigma_3)\\ &= \left(n_x^2 + n_y^2 + n_z^2\right)I + n_x n_y(\sigma_1\sigma_2 + \sigma_2\sigma_1) + n_y n_z(\sigma_2\sigma_3 + \sigma_3\sigma_2)n_z n_x(\sigma_3\sigma_1 + \sigma_1\sigma_3)\\ &= I.\end{aligned} \tag{N14}$$

In the above result, we have applied the property in Eq. (16.19) from the text, namely $\sigma_i\sigma_j = -\sigma_j\sigma_i$ for $i \neq j$ and our assumption that $\vec{n}$ is unitary ($n_x^2 + n_y^2 + n_z^2 = 1$). Since $U^2 = 1$, the exponential-operator definition in Eq. (N9) is valid for the operator $U$, and we have, for any real $\theta$:

$$\begin{aligned} \exp(\mathrm{i}U\theta) &= I\cos\theta + \mathrm{i}U\sin\theta \\ &\leftrightarrow \\ \exp[\mathrm{i}(\vec{n}\cdot\vec{\sigma})\theta] &= I\cos\theta + \mathrm{i}(\vec{n}\cdot\vec{\sigma})\sin\theta. \end{aligned} \tag{N15}$$

We recognize in this result a more general expression from which the elementary rotation operators $R_k(\gamma)$ ($k = x, y, z$) defined in Eq. (N10) can be derived, i.e., by setting $\vec{n} = \vec{x}$, $\vec{n} = \vec{y}$, or $\vec{n} = \vec{z}$ with $\theta = -\gamma/2$. The general expression corresponds to a qubit rotation of (counterclockwise) angle $\gamma$ about the axis defined by the unit vector $\vec{n} = (n_x, n_y, n_z)$.

Since the result of any $2 \times 2$ unitary transformation $A$ is to move a qubit on the surface of the Bloch sphere, there exists a unique rotation associated with such a move (like taking a direct flight from city to city on the Earth). This unique rotation is defined through the operator $\exp[\mathrm{i}(\vec{n}\cdot\vec{\sigma})\theta] = A$ (within an unobservable phase factor). We can make this operator more explicit, by developing the definition in Eq. (N15), according to

$$\begin{aligned} \exp[\mathrm{i}(\vec{n}\cdot\vec{\sigma})\theta] &= I\cos\theta + \mathrm{i}(\vec{n}\cdot\vec{\sigma})\sin\theta \\ &= \sigma_0\cos\theta + \mathrm{i}\sin\theta(n_x\sigma_1 + n_y\sigma_2 + n_z\sigma_3) \end{aligned} \tag{N16}$$

or

$$\begin{aligned} A &\equiv \mathrm{e}^{\mathrm{i}\delta}\exp[\mathrm{i}(\vec{n}\cdot\vec{\sigma})\theta] \\ &\equiv \sum_{i=0}^{3}\mu_i\sigma_i. \end{aligned} \tag{N17}$$

where $\mu_i$ are complex numbers defined by

$$\mu_0 = \cos\theta,\ \mu_1 = \mathrm{i}n_x\sin\theta,\ \mu_2 = \mathrm{i}n_y\sin\theta,\ \mu_3 = \mathrm{i}n_z\sin\theta, \tag{N18}$$

which, in particular, satisfy $\sum_i |\mu_i|^2 = 1$. To summarize this result, *any* $2 \times 2$ *unitary transformation, or rotation on the Bloch sphere, can be expressed as a linear complex expansion of Pauli matrices*. It is left as an elementary exercise to determine the parameters $\vec{n}, \theta$ associated with any unitary transformation $A$.

## Euler's theorem

This theorem states that every $2 \times 2$ *unitary* matrix $U$ can be expressed from the two rotation operators $R_y$, $R_z$ and a set of four real numbers $\alpha, \beta, \gamma, \delta$, according to the product:

$$U = \mathrm{e}^{\mathrm{i}\delta}R_z(\alpha)R_y(\beta)R_z(\gamma). \tag{N19}$$

To show this, we first substitute the corresponding definitions of $R_y$, $R_z$ from Eq. (N11) to obtain:

$$U = e^{i\delta}\begin{pmatrix} e^{-i\frac{\alpha+\gamma}{2}}\cos\frac{\beta}{2} & -e^{-i\frac{\alpha-\gamma}{2}}\sin\frac{\beta}{2} \\ e^{i\frac{\alpha-\gamma}{2}}\sin\frac{\beta}{2} & e^{i\frac{\alpha+\gamma}{2}}\cos\frac{\beta}{2} \end{pmatrix}. \quad \text{(N20)}$$

It is left as an exercise to prove that the unitary condition $U^{+}U = 1$ is both necessary and sufficient for the above matrix coefficients to represent any unitary matrix $U$.

Euler's theorem, as defined in Eq. (N19), is referred to as *Z-Y decomposition of rotations*. In fact *Z-X decomposition*, according to $U = e^{i\delta}R_z(\alpha)R_x(\beta)R_z(\gamma)$, is also possible, as well as any decomposition involving two different Pauli operators. Most generally, given any two nonparallel unitary vectors $\vec{n}$, $\vec{m}$

$$U = e^{i\delta}R_{\vec{n}}(\alpha)R_{\vec{m}}(\beta)R_{\vec{n}}(\gamma), \quad \text{(N21)}$$

where $R_{\vec{p}}(\theta) = \exp[i(\vec{p}\cdot\vec{\sigma})\theta]$ is the rotation operator of angle $\theta$ about the axis defined by the unitary vector $\vec{p}$.

## Decomposition of 2 × 2 unitary matrices

A general property stemming from Euler's theorem is that any 2 × 2 unitary matrix can be decomposed according to

$$U = e^{i\delta}AXBXC, \quad \text{(N22)}$$

where $A$, $B$, $C$ are 2 × 2 unitary matrices satisfying $ABC = I$ and $X = \sigma_1$.

To prove this property, we must find at least one set of three matrices $A$, $B$, $C$ that satisfies $ABC = I$ and correspond to the unique decomposition of $U$ in Eq. (N22). Assume heuristically, for instance,

$$\begin{cases} A = R_z(2\psi)R_y(-2\chi) \\ B = R_y(2\chi)R_z(2\omega) \\ C = R_z(2\phi), \end{cases} \quad \text{(N23)}$$

where $\phi$, $\chi$, $\psi$ are rotation angles to be determined (the factor of two will lighten the calculations). It is clear that Eq. (N23) satisfies the condition $ABC = I$, or

$$\begin{aligned} ABC &= R_z(2\psi)R_y(-2\chi)R_y(2\chi)R_z(2\omega)R_z(2\phi) \\ &\equiv R_z(2\psi)R_z(2\omega)R_z(2\phi) \\ &= I, \end{aligned} \quad \text{(N24)}$$

if we impose the angle condition

$$\psi + \omega + \phi = 0. \quad \text{(N25)}$$

We shall now develop the product $AXBXC$ to obtain

$$
\begin{aligned}
AXBXC &= AXBXR_z(2\phi) \\
&= AXB\begin{pmatrix} 0 & 1 \\ 1 & 0 \end{pmatrix}\begin{pmatrix} \mathrm{e}^{-\mathrm{i}\phi} & 0 \\ 0 & \mathrm{e}^{\mathrm{i}\phi} \end{pmatrix} \\
&= AXB\begin{pmatrix} 0 & \mathrm{e}^{\mathrm{i}\phi} \\ \mathrm{e}^{-\mathrm{i}\phi} & 0 \end{pmatrix} \\
&= AXR_y(2\chi)R_z(2\omega)\begin{pmatrix} 0 & \mathrm{e}^{\mathrm{i}\phi} \\ \mathrm{e}^{-\mathrm{i}\phi} & 0 \end{pmatrix} \\
&= AXR_y(2\chi)\begin{pmatrix} \mathrm{e}^{-\mathrm{i}\omega} & 0 \\ 0 & \mathrm{e}^{\mathrm{i}\omega} \end{pmatrix}\begin{pmatrix} 0 & \mathrm{e}^{\mathrm{i}\phi} \\ \mathrm{e}^{-\mathrm{i}\phi} & 0 \end{pmatrix} \\
&= AXR_y(2\chi)\begin{pmatrix} 0 & \mathrm{e}^{-\mathrm{i}(\omega-\varphi)} \\ \mathrm{e}^{\mathrm{i}(\omega-\varphi)} & 0 \end{pmatrix} \\
&= AX\begin{pmatrix} \cos\chi & -\sin\chi \\ \sin\chi & \cos\chi \end{pmatrix}\begin{pmatrix} 0 & \mathrm{e}^{-\mathrm{i}(\omega-\varphi)} \\ \mathrm{e}^{\mathrm{i}(\omega-\varphi)} & 0 \end{pmatrix} \\
&= AX\begin{pmatrix} -\sin\chi\,\mathrm{e}^{\mathrm{i}(\omega-\varphi)} & \cos\chi\,\mathrm{e}^{-\mathrm{i}(\omega-\varphi)} \\ \cos\chi\,\mathrm{e}^{\mathrm{i}(\omega-\varphi)} & \sin\chi\,\mathrm{e}^{-\mathrm{i}(\omega-\varphi)} \end{pmatrix} \\
&= A\begin{pmatrix} 0 & 1 \\ 1 & 0 \end{pmatrix}\begin{pmatrix} -\sin\chi\,\mathrm{e}^{\mathrm{i}(\omega-\varphi)} & \cos\chi\,\mathrm{e}^{-\mathrm{i}(\omega-\varphi)} \\ \cos\chi\,\mathrm{e}^{\mathrm{i}(\omega-\varphi)} & \sin\chi\,\mathrm{e}^{-\mathrm{i}(\omega-\varphi)} \end{pmatrix} \qquad \text{(N26)} \\
&= A\begin{pmatrix} \cos\chi\,\mathrm{e}^{\mathrm{i}(\omega-\varphi)} & \sin\chi\,\mathrm{e}^{-\mathrm{i}(\omega-\varphi)} \\ -\sin\chi\,\mathrm{e}^{\mathrm{i}(\omega-\varphi)} & \cos\chi\,\mathrm{e}^{-\mathrm{i}(\omega-\varphi)} \end{pmatrix} \\
&= R_z(2\psi)R_y(-2\chi)\begin{pmatrix} \cos\chi\,\mathrm{e}^{\mathrm{i}(\omega-\varphi)} & \sin\chi\,\mathrm{e}^{-\mathrm{i}(\omega-\varphi)} \\ -\sin\chi\,\mathrm{e}^{\mathrm{i}(\omega-\varphi)} & \cos\chi\,\mathrm{e}^{-\mathrm{i}(\omega-\varphi)} \end{pmatrix} \\
&= R_z(2\psi)\begin{pmatrix} \cos\chi & \sin\chi \\ -\sin\chi & \cos\chi \end{pmatrix}\begin{pmatrix} \cos\chi\,\mathrm{e}^{\mathrm{i}(\omega-\varphi)} & \sin\chi\,\mathrm{e}^{-\mathrm{i}(\omega-\varphi)} \\ -\sin\chi\,\mathrm{e}^{\mathrm{i}(\omega-\varphi)} & \cos\chi\,\mathrm{e}^{-\mathrm{i}(\omega-\varphi)} \end{pmatrix} \\
&= R_z(2\psi)\begin{pmatrix} [\cos^2\chi - \cos^2\chi]\mathrm{e}^{\mathrm{i}(\omega-\varphi)} & 2\cos\chi\,\sin\chi\,\mathrm{e}^{-\mathrm{i}(\omega-\varphi)} \\ -2\cos\chi\,\sin\chi\,\mathrm{e}^{\mathrm{i}(\omega-\varphi)} & [\cos^2\chi - \cos^2\chi]\mathrm{e}^{-\mathrm{i}(\omega-\varphi)} \end{pmatrix} \\
&= R_z(2\psi)\begin{pmatrix} \cos 2\chi\,\mathrm{e}^{\mathrm{i}(\omega-\varphi)} & \sin 2\chi\,\mathrm{e}^{-\mathrm{i}(\omega-\varphi)} \\ -\sin 2\chi\,\mathrm{e}^{\mathrm{i}(\omega-\varphi)} & \cos 2\chi\,\mathrm{e}^{-\mathrm{i}(\omega-\varphi)} \end{pmatrix} \\
&= \begin{pmatrix} \mathrm{e}^{-\mathrm{i}\psi} & 0 \\ 0 & \mathrm{e}^{\mathrm{i}\psi} \end{pmatrix}\begin{pmatrix} \cos 2\chi\,\mathrm{e}^{\mathrm{i}(\omega-\varphi)} & \sin 2\chi\,\mathrm{e}^{-\mathrm{i}(\omega-\varphi)} \\ -\sin 2\chi\,\mathrm{e}^{\mathrm{i}(\omega-\varphi)} & \cos 2\chi\,\mathrm{e}^{-\mathrm{i}(\omega-\varphi)} \end{pmatrix} \\
&= \begin{pmatrix} \cos 2\chi\,\mathrm{e}^{\mathrm{i}(\omega-\varphi-\psi)} & \sin 2\chi\,\mathrm{e}^{-\mathrm{i}(\omega-\varphi+\psi)} \\ -\sin 2\chi\,\mathrm{e}^{\mathrm{i}(\omega-\varphi+\psi)} & \cos 2\chi\,\mathrm{e}^{-\mathrm{i}(\omega-\varphi-\psi)} \end{pmatrix} \\
&\equiv \begin{pmatrix} \cos 2\chi\,\mathrm{e}^{-2\mathrm{i}(\varphi+\psi)} & \sin 2\chi\,\mathrm{e}^{2\mathrm{i}\varphi} \\ -\sin 2\chi\,\mathrm{e}^{-2\mathrm{i}\varphi} & \cos 2\chi\,\mathrm{e}^{2\mathrm{i}(\varphi+\psi)} \end{pmatrix}.
\end{aligned}
$$

Identifying this result with the definition in Eq. (N20), we obtain

$$\begin{pmatrix} e^{-i\frac{\alpha+\gamma}{2}}\cos\frac{\beta}{2} & -e^{-i\frac{\alpha-\gamma}{2}}\sin\frac{\beta}{2} \\ e^{i\frac{\alpha-\gamma}{2}}\sin\frac{\beta}{2} & e^{i\frac{\alpha+\gamma}{2}}\cos\frac{\beta}{2} \end{pmatrix} = \begin{pmatrix} \cos 2\chi\, e^{-2i(\phi+\psi)} & \sin 2\chi\, e^{2i\phi} \\ -\sin 2\chi\, e^{-2i\phi} & \cos 2\chi\, e^{2i(\phi+\psi)} \end{pmatrix} \tag{N27}$$

and, thus,

$$\begin{aligned} &2\chi = -\frac{\beta}{2}, \quad 2\phi + 2\psi = \frac{\alpha+\gamma}{2} = -2\omega, \quad -2\phi = \frac{\alpha-\gamma}{2} \\ &\leftrightarrow \\ &2\chi = -\frac{\beta}{2}, \quad 2\psi = \alpha, \quad 2\omega = -\frac{\alpha+\gamma}{2}, \quad 2\phi = \frac{\gamma-\alpha}{2}. \end{aligned} \tag{N28}$$

The three unitary operators are then fully defined according to

$$\begin{cases} A = R_z(\alpha)R_y\left(\frac{\beta}{2}\right) \\ B = R_y\left(-\frac{\beta}{2}\right)R_z\left(-\frac{\alpha+\gamma}{2}\right) \\ C = R_z\left(\frac{\gamma-\alpha}{2}\right). \end{cases} \tag{N29}$$

## Commutation properties of rotation operators

Two rotation operators $R_i(\theta)$, $R_j(\varphi)$ do not commute except when $i = j$. It is left as an easy exercise to verify that

$$[R_i(2\theta), R_j(2\varphi)] = -2i\varepsilon_{ijk}\sigma_k \sin\theta \sin\varphi. \tag{N30}$$

Consider, for instance, the triple rotation $A = R_i(-\theta)R_j(\varphi)R_i(\theta)$, with $i \neq j$. We would intuitively think that the last rotation $R_i(-\theta)$ cancels the effect of the first rotation $R_i(\theta)$ and, therefore, that $A = R_j(\varphi)$ but, as we shall see, this is not the case. Indeed:

$$\begin{aligned} A &= R_i(-\theta)R_j(\varphi)R_i(\theta) \\ &= R_i(-\theta)\{R_i(\theta)R_j(\varphi) + [R_j(\varphi), R_i(\theta)]\} \\ &= R_i(-\theta)R_i(\theta)R_j(\varphi) + R_i(-\theta)[R_j(\varphi), R_i(\theta)] \\ &= R_j(\varphi) - 2i\varepsilon_{ijk}R_i(-\theta)\sigma_k, \end{aligned} \tag{N31}$$

which proves that, except for $i = j$ ($\varepsilon_{ijk} = 0$), in the general case $A \neq R_j(\varphi)$.

It is also straightforward to show that the commutation of the rotation operators with the Pauli matrices satisfy

$$[R_i(2\theta), \sigma_j] = 2i\varepsilon_{ijk}\sin\theta\sigma_k. \tag{N32}$$

This result expresses the fact that the two operators $R_i(2\theta)$, $\sigma_j$ do not commute, except in the specific cases $i = j$ ($\varepsilon_{iik} = 0$) and $2\theta = 0, 2\pi$ ($R_i(0) = R_i(2\pi) = I$).

# Appendix O (Chapter 17) Heisenberg uncertainty principle

In this appendix, I provide a demonstration of the Heisenberg uncertainty principle.[1] According to this principle, the uncertainties $\Delta A$, $\Delta B$ associated with two observables $A$, $B$, measured in the state $|\psi\rangle$, must satisfy the inequality:

$$\Delta A \Delta B \geq \frac{1}{2}|\langle\psi|[A, B]|\psi\rangle|. \tag{O1}$$

To prove Eq. (O1), define

$$|\varphi\rangle = (A + \mathrm{i}\lambda B)\,|\psi\rangle, \tag{O2}$$

where $\lambda$ is a real parameter. We then obtain

$$\begin{aligned}
\langle\varphi|\varphi\rangle &= \langle\psi|(A + \mathrm{i}\lambda B)^{+}(A + \mathrm{i}\lambda B)|\psi\rangle \\
&= \langle\psi|(A^{+} - \mathrm{i}\lambda B^{+})(A + \mathrm{i}\lambda B)|\psi\rangle \\
&= \langle\psi|A^{+}A|\psi\rangle + \mathrm{i}\lambda\langle\psi|(A^{+}B - B^{+}A)|\psi\rangle + \langle\psi|B^{+}B|\psi\rangle \\
&\equiv \langle\psi|A^{2}|\psi\rangle + \mathrm{i}\lambda\langle\psi|[A, B]|\psi\rangle + \lambda^{2}\langle\psi|B^{2}|\psi\rangle \\
&\geq 0,
\end{aligned} \tag{O3}$$

where we used the commutator definition $[A, B] = AB - BA$ and the fact that the observables are Hermitian. The property $\langle\varphi|\varphi\rangle \geq 0$ (with $\langle\varphi|\varphi\rangle$ real) is inherent to the definition of inner product. We also have the property $\langle\psi|X^2|\psi\rangle \geq 0$ for any observable $X = A, B$ and state $|\psi\rangle$, see note. Hence, $\mathrm{i}\langle\psi|\,[A, B]\,|\psi\rangle$ must be real, with $\mathrm{i}\langle\psi|\,[A, B]\,|\psi\rangle = \pm|\langle\psi|\,[A, B]\,|\psi\rangle|$, and Eq. (O3) is of the polynomial form $P(\lambda) = a\lambda^2 \pm b\lambda \geq c \geq 0$ with real coefficients $a, b, c$ satisfying $a, b, c \geq 0$. Since $P(\lambda) \geq 0$, it can only have zero or one root, which corresponds to a discriminant $\delta = b^2 - 4ac \leq 0$. We, thus, have

$$\begin{aligned}
\delta &= |\langle\psi|\,[A, B]\,|\psi\rangle|^2 - 4\langle\psi|B^2|\psi\rangle\langle\psi|A^2|\psi\rangle \\
&\leq 0,
\end{aligned} \tag{O4}$$

[1] See, for instance: C. Cohen-Tannoudji, B. Diu, and F. Laloe, *Quantum Mechanics* (Paris: Hermann, 1977), Vol. 1, pp. 28–7.

which can be put in the inequality form:

$$\begin{aligned}\langle\psi|A^2|\psi\rangle\langle\psi|B^2|\psi\rangle &\equiv \langle A^2\rangle\langle B^2\rangle \\ &\geq \frac{1}{4}|\langle\psi|[A, B]|\psi\rangle|^2.\end{aligned} \quad (O5)$$

Now define the new observables $\tilde{A}$, $\tilde{B}$ from the previous observables $A$, $B$ according to

$$\begin{cases} A = \tilde{A} - \langle\psi|\tilde{A}|\psi\rangle \equiv \tilde{A} - \langle\tilde{A}\rangle \\ B = \tilde{B} - \langle\psi|\tilde{B}|\psi\rangle \equiv \tilde{B} - \langle\tilde{B}\rangle. \end{cases} \quad (O6)$$

It is easily verified that $[A, B] = [\tilde{A}, \tilde{B}]$. Substituting the definitions in Eq. (O6) into Eq. (O5), we obtain

$$\langle A^2\rangle\langle B^2\rangle = \langle(\tilde{A} - \langle\tilde{A}\rangle)^2\rangle\langle(\tilde{B} - \langle\tilde{B}\rangle)^2\rangle, \quad (O7)$$

or

$$\Delta\tilde{A}^2\Delta\tilde{B}^2 \geq \frac{1}{4}|\langle\psi|[\tilde{A}, \tilde{B}]|\psi\rangle|^2, \quad (O8)$$

or

$$\Delta\tilde{A}\Delta\tilde{B} \geq \frac{1}{2}|\langle\psi|[\tilde{A}, \tilde{B}]|\psi\rangle|. \quad (O9)$$

It is also easily verified that $\Delta X^2 = \Delta\tilde{X}^2$ for any operator $X = A, B$ defined by Eq. (O6). Hence, we finally obtain the *Heisenberg uncertainty relation*:

$$\Delta A\Delta B \geq \frac{1}{2}|\langle\psi|[A, B]|\psi\rangle|. \quad (O10)$$

## Note

The property $\langle\psi|X^2|\psi\rangle \geq 0$ for any observable $X$ and state $|\psi\rangle$ is shown as follows. Because $X$ is an observable, there exists an orthonormal or eigenstate base $\{|\lambda_i\rangle\}$, whose state elements satisfy $X|\lambda_i\rangle = \lambda_i|\lambda_i\rangle$, where $\lambda_i$ are the corresponding real eigenvalues. The state $|\psi\rangle$ can, thus, be uniquely decomposed into this base with coordinates $\mu_i$, according to

$$|\psi\rangle = \sum_i x_i|\lambda_i\rangle.$$

Hence, we obtain:

$$\begin{aligned}\langle\psi|X^2|\psi\rangle &= \left(\sum_i \bar{\mu}_i\langle\lambda_i|\right) XX \left(\sum_j \mu_j|\lambda_j\rangle\right) \\ &= \sum_i\sum_j \bar{\mu}_i\mu_j\lambda_i\lambda_j\delta_{ij} \\ &= \sum_i \lambda_i^2|\mu_i|^2 \\ &\geq 0.\end{aligned}$$

# Appendix P (Chapter 18) Two-qubit teleportation

In this appendix, I formally demonstrate the possibility of *simultaneously teleporting two qubits* from Alice's location to Bob's. The proposed six-qubit quantum circuit, shown in Fig. P1, is an original, symmetrical variant of that described in Gottesman and Chuang (1999).[1] Such an example also represents a test case for the analysis of quantum circuits and Bell measurements, as a full illustration of the concepts and formalism described in Chapter 18, hence, the detailed calculations presented here.

In the circuit shown in Fig. P1, the boxes $B$, $B'$ stand for Bell-state measurements and $X^n$, $X^{n'}$, $Z^n$, $Z^{n'}$ are Pauli gates controlled by classical bits $n, m, n'm'$. The inputs $|q\rangle_1$, $|q'\rangle_6$ are two qubits from Alice, who accesses the quantum wires 1, 2, 5, and 6. Bob only has access to the quantum wires 3 and 4, where he retrieves the teleported qubits under the tensor state $|\psi\rangle = \text{CNOT}|q\rangle_3|q'\rangle_4 = C_{43}|q\rangle_3 \otimes |q'\rangle_4$, as illustrated in Fig. P1 (applying a second CNOT, allowing Bob to retrieve Alice's individual qubits $|q\rangle_3$, $|q'\rangle_4$). For the teleportation, Alice and Bob share a 4-qubit entangled state $|\chi\rangle$, defined by

$$|\chi\rangle = \frac{1}{4}(|0000\rangle + |0111\rangle + |1100\rangle + |1011\rangle)_{2345} \tag{P1}$$

(the circuit used to generate $|\chi\rangle$, which entangles two Bell states $|\beta_{00}\rangle$ is shown in Fig. 18.11).

We proceed now to the formal demonstration of the 2-qubit teleportation effect that is achieved through the above-described circuit. To analyze the qubit evolution in pre- and post-measurement stages, we need first to detail the Bell-state measurement circuits $B$, $B'$. Such circuits are shown in Fig. 18.5, and reproduced in Fig. P2 according to the circuit notations.

In the figure, the term $C_{ij}$ designates CNOT gates where $i$ is the control qubit, $j$ is the target qubit, and $H_k$ are Hadamard gates placed on the quantum wire $k$. The two Bell-measurement apparatuses $B$, $B'$ output the post-measurement classical bits $n, m$ and $n'm'$, respectively, which correspond to pure states $|nmn'm'\rangle_{1256}$, to be defined later. Our task now is to calculate the 6-qubit state situated just past the two Hadamard gates in $B$, $B'$ and ahead of Alice's two measurements. This will allow us to know what post-measurement states $|nmn'm'\rangle \otimes |*\rangle_{23}$ are to be expected, and hence, the action of the $X^n$, $X^{n'}$, $Z^n$, $Z^{n'}$ gates based on Bob's knowledge of the four classical bits $n, m, n'm'$.

[1] D. Gottesman and I. L. Chuang, Quantum teleportation is a universal computational primitive. *Nature*, **402** (1999), 390–3, http://arxiv.org/PS_cache/quant-ph/pdf/9908/9908010v1.pdf.

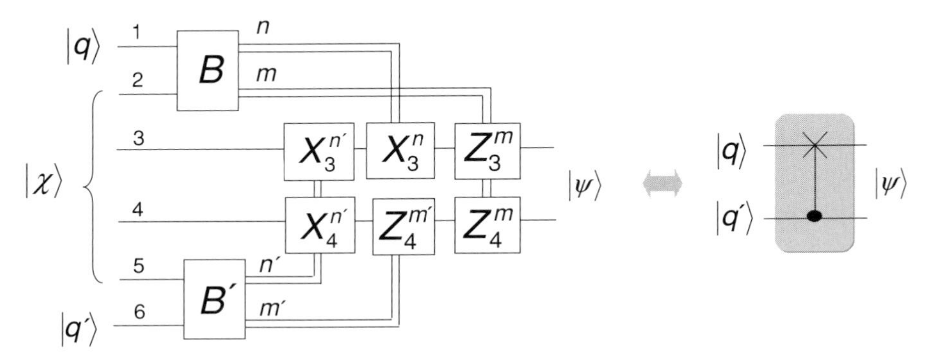


**Figure P1** Six-qubit quantum circuit for teleporting two qubits.

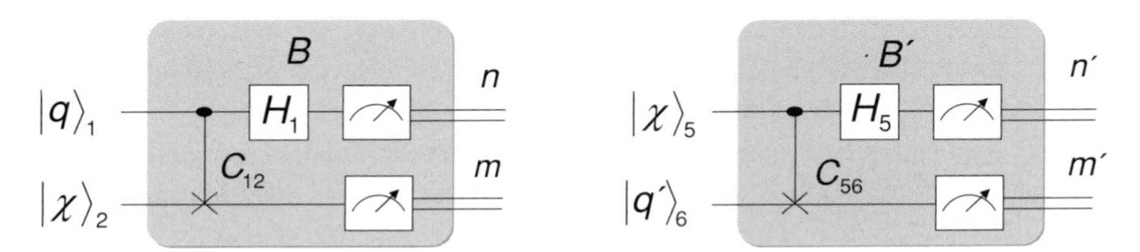


**Figure P2** Principle of Bell measurement or measurement in Bell state.

We then define Alice's two qubits as:

$$\begin{cases} |q\rangle_1 \equiv \alpha|0\rangle + \beta|1\rangle \\ |q'\rangle_6 \equiv \alpha'|0\rangle + \beta'|1\rangle, \end{cases} \tag{P2}$$

and based on the definition of $|\chi\rangle$ in Eq. (P1), we develop the 6-qubit tensor $|\Phi\rangle$ that is input to the circuit, as follows (overlooking normalization factors):

$$\begin{aligned}
|\Phi\rangle &= |q\rangle_1 \otimes |\chi\rangle_{2345} \otimes |q'\rangle_6 \\
&= (\alpha|0\rangle_1 + \beta|1\rangle_1) \otimes (|0000\rangle + |0111\rangle + |1100\rangle + |1011\rangle)_{2345} \otimes |q'\rangle_6 \\
&= \begin{pmatrix} \alpha|00000\rangle + \alpha|00111\rangle + \alpha|01100\rangle + \alpha|01011\rangle + \\ \beta|10000\rangle + \beta|10111\rangle + \beta|11100\rangle + \beta|11011\rangle \end{pmatrix}_{12345} \otimes |q'\rangle_6 \\
&= \begin{pmatrix} \alpha|00000\rangle + \alpha|00111\rangle + \alpha|01100\rangle + \alpha|01011\rangle + \\ \beta|10000\rangle + \beta|10111\rangle + \beta|11100\rangle + \beta|11011\rangle \end{pmatrix}_{12345} \otimes (\alpha'|0\rangle_6 + \beta'|1\rangle_6) \\
&= \begin{Bmatrix} \alpha\alpha'(|000000\rangle + |001110\rangle + |011000\rangle + |010110\rangle) + \\ \alpha\beta'(|000001\rangle + |001111\rangle + |011001\rangle + |010111\rangle) + \\ \alpha'\beta(|100000\rangle + |101110\rangle + |111000\rangle + |110110\rangle) + \\ \beta\beta'(|100001\rangle + \beta|101111\rangle + |111001\rangle + |110111\rangle) \end{Bmatrix}_{123456}.
\end{aligned} \tag{P3}$$

We must then apply the action of the two CNOT gates $C_{12}$, $C_{56}$ in $B$, $B'$, noted $C_{12}$, $C_{56}$ respectively, with the first and second indices representing the control and target qubits.

For any binary combinations $a, b, c, d, e, f = \{0, 1\}$ in a 6-qubit state $|abcdef\rangle$, it is clear that the gates have the corresponding action:

$$\begin{cases} C_{12}|0abcde\rangle = |0abcde\rangle \\ C_{12}|1abcde\rangle = |1\bar{a}bcde\rangle \\ C_{56}|abcd0e\rangle = |abcd0e\rangle \\ C_{56}|abcd1e\rangle = |abcd1\bar{e}\rangle. \end{cases} \tag{P4}$$

Effecting the above rules in Eq. (P3), we obtain $|\Phi'\rangle = C_{56}C_{12}|\Phi\rangle$ as:

$$|\Phi'\rangle = \begin{Bmatrix} \alpha\alpha'(|000000\rangle + |001111\rangle + |011000\rangle + |010111\rangle) + \\ \alpha\beta'(|000001\rangle + |001110\rangle + |011001\rangle + |010110\rangle) + \\ \alpha'\beta(|110000\rangle + |111111\rangle + |101000\rangle + |100111\rangle) + \\ \beta\beta'(|110001\rangle + |111110\rangle + |101001\rangle + |100110\rangle) \end{Bmatrix}_{123456}. \tag{P5}$$

Next, we must calculate the action of the subsequent Hadamard gates $H_1$, $H_5$ acting on qubits 1 and 5, respectively, recalling first that

$$\begin{cases} H|0\rangle = |+\rangle = \dfrac{|0\rangle + |1\rangle}{\sqrt{2}} \\ H|1\rangle = |-\rangle = \dfrac{|0\rangle - |1\rangle}{\sqrt{2}}, \end{cases} \tag{P6}$$

and, hence, for any binary combination $a, b, c, d, e = \{0, 1\}$:

$$\begin{cases} H_1|0abcde\rangle = |+abcde\rangle = \dfrac{|0abcde\rangle + |1abcde\rangle}{\sqrt{2}} \\ H_1|1abcde\rangle = |-abcde\rangle = \dfrac{|0abcde\rangle - |1abcde\rangle}{\sqrt{2}} \\ H_5|abcd0e\rangle = |abcd+e\rangle = \dfrac{|abcd0e\rangle + |abcd1e\rangle}{\sqrt{2}} \\ H_5|abcd1e\rangle = |abcd-e\rangle = \dfrac{|abcd0e\rangle - |1abcd1e\rangle}{\sqrt{2}}. \end{cases} \tag{P7}$$

Applying the two Hadamard gates $H_1$, $H_5$ onto $|\Phi'\rangle$ in Eq. (P5), and overlooking the normalization factor, we obtain $|\Phi\rangle = H_5H_1|\Phi'\rangle$, as:

$$|\Phi\rangle = \begin{Bmatrix} \alpha\alpha'(|+000+0\rangle + |+011-1\rangle + |+110+0\rangle + |+101-1\rangle) + \\ \alpha\beta'(|+000+1\rangle + |+011-0\rangle + |+110+1\rangle + |+101-0\rangle) + \\ \alpha'\beta(|-100+0\rangle + |-111-1\rangle + |-010+0\rangle + |-001-1\rangle) + \\ \beta\beta'(|-100+1\rangle + |-111-0\rangle + |-010+1\rangle + |-001-0\rangle) \end{Bmatrix}_{123456}, \tag{P8}$$

where the states $|\pm\rangle$ are defined in Eq. (P6). Substituting these definitions in Eq. (P8), but in two steps to avoid mistakes, the result decomposes itself into 64 6-qubits:

$$|\Phi''\rangle = \left\{ \begin{array}{l} \alpha\,\alpha' \begin{pmatrix} |0000+0\rangle + |1000+0\rangle + \\ |0011-1\rangle + |1011-1\rangle + \\ |0110+0\rangle + |1110+0\rangle + \\ |0101-1\rangle + |1101-1\rangle \end{pmatrix} + a\beta' \begin{pmatrix} |0000+1\rangle + |1000+1\rangle + \\ |0011-0\rangle + |1011-0\rangle + \\ |0110+1\rangle + |1110+1\rangle + \\ |0101-0\rangle + |1101-0\rangle \end{pmatrix} \\ +\alpha'\beta \begin{pmatrix} |0100+0\rangle - |1100+0\rangle + \\ |0111-1\rangle - |1111-1\rangle + \\ |0010+0\rangle - |1010+0\rangle + \\ |0001-1\rangle - |1001-1\rangle \end{pmatrix} + \beta\,\beta' \begin{pmatrix} |0100+1\rangle - |1100+1\rangle + \\ |0111-0\rangle - |1111-0\rangle + \\ |0010+1\rangle - |1010+1\rangle + \\ |0001-0\rangle - |1001-0\rangle \end{pmatrix} \end{array} \right\}_{123456}$$

$$= \left\{ \begin{array}{l} \alpha\,\alpha' \begin{pmatrix} |000000\rangle + |000010\rangle + |100000\rangle + |100010\rangle + \\ |001101\rangle - |001111\rangle + |101101\rangle - |101111\rangle + \\ |011000\rangle + |011010\rangle + |111000\rangle + |111010\rangle + \\ |010101\rangle - |010111\rangle + |110101\rangle - |110111\rangle \end{pmatrix} \\ +\alpha\,\beta' \begin{pmatrix} |000001\rangle + |000011\rangle + |100001\rangle + |100011\rangle + \\ |001100\rangle - |001110\rangle + |101100\rangle - |101110\rangle + \\ |011001\rangle + |011011\rangle + |111001\rangle + |111011\rangle + \\ |010100\rangle - |010110\rangle + |110100\rangle - |110110\rangle \end{pmatrix} \end{array} \right\}_{123456}$$

$$+ \left\{ \begin{array}{l} \alpha'\,\beta \begin{pmatrix} |010000\rangle + |010010\rangle - |110000\rangle - |110010\rangle + \\ |011101\rangle - |011111\rangle - |111101\rangle + |111111\rangle + \\ |001000\rangle + |001010\rangle - |101000\rangle - |101010\rangle + \\ |000101\rangle - |000011\rangle - |100101\rangle + |100111\rangle \end{pmatrix} \\ +\beta\,\beta' \begin{pmatrix} |010001\rangle + |010011\rangle - |110001\rangle - |110011\rangle + \\ |011100\rangle - |011110\rangle - |111100\rangle + |111110\rangle + \\ |001001\rangle + |001011\rangle - |101001\rangle - |101011\rangle + \\ |000100\rangle - |000110\rangle - |100100\rangle + |100110\rangle \end{pmatrix} \end{array} \right\}_{123456}. \tag{P9}$$

Note that the above development is far more easily obtained using a math-equation editor than through handwriting calculation. Yet, the operation requires extreme care in the different character substitutions. I shall now regroup the 64 terms in Eq. (P9) according to the 16 possible base elements $|mnm'n'\rangle_{1256}$ for the combined Bell-measurement bases, which, for clarity, I order in binary progression of the indices $mnm'n' = 0000, 0001, 0010, 0011\ldots$, as:

$$|mn\rangle_{12} \otimes |m'n'\rangle_{56} \equiv |mnm'n'\rangle_{1256}$$

$$= \left\{ \begin{array}{l} |0000\rangle_{1256}, |0001\rangle_{1256}, |0010\rangle_{1256}, |0011\rangle_{1256}, \\ |0100\rangle_{1256}, |0101\rangle_{1256}, |0110\rangle_{1256}, |0111\rangle_{1256}, \\ |1000\rangle_{1256}, |1001\rangle_{1256}, |1010\rangle_{1256}, |1011\rangle_{1256}, \\ |1100\rangle_{1256}, |1101\rangle_{1256}, |1110\rangle_{1256}, |1111\rangle_{1256}, \end{array} \right\}. \tag{P10}$$

Note that I choose here to label the base states according to $|mnm'n'\rangle_{1256}$, which is arbitrary but will make sense further on. To lighten the notations, in the following I shall write

$$|mnabm'n'\rangle_{123456} \equiv |mnm'n'\rangle|ab\rangle_{34}. \tag{P11}$$

Factoring the terms, first according to $|ab\rangle_{34}$, and then according to $|mnm'n'\rangle$ we obtain:

$$
|\Phi\rangle = \left\{\begin{matrix}
\alpha\,\alpha'\begin{pmatrix}
(|0000\rangle + |0010\rangle + |1000\rangle + |1010\rangle)|00\rangle_{34} + \\
(|0001\rangle + |0011\rangle + |1001\rangle + |1011\rangle)|11\rangle_{34} + \\
(|0100\rangle + |0110\rangle + |1100\rangle + |1110\rangle)|10\rangle_{34} + \\
(|0101\rangle + |0111\rangle + |1101\rangle + |1111\rangle)|01\rangle_{34}
\end{pmatrix} \\
+\,\alpha\beta'\begin{pmatrix}
(|0001\rangle + |0011\rangle + |1001\rangle + |1011\rangle)|00\rangle_{34} + \\
(|0000\rangle + |0010\rangle + |1000\rangle + |1010\rangle)|11\rangle_{34} + \\
(|0101\rangle + |0111\rangle + |1101\rangle + |1111\rangle)|01\rangle_{34} + \\
(|0100\rangle + |0110\rangle + |1100\rangle + |1110\rangle)|01\rangle_{34}
\end{pmatrix} \\
\alpha'\beta\begin{pmatrix}
(|1000\rangle + |0110\rangle + |1100\rangle + |1110\rangle)|00\rangle_{34} + \\
(|0101\rangle + |0111\rangle + |1101\rangle + |1111\rangle)|11\rangle_{34} + \\
(|0000\rangle + |0010\rangle + |1000\rangle + |1010\rangle)|10\rangle_{34} + \\
(|0001\rangle + |0011\rangle + |1001\rangle + |1011\rangle)|01\rangle_{34}
\end{pmatrix} \\
+\,\beta\,\beta'\begin{pmatrix}
(|0101\rangle + |0111\rangle + |1101\rangle + |1111\rangle)|00\rangle_{34} + \\
(|0100\rangle + |0110\rangle + |1100\rangle + |1110\rangle)|11\rangle_{34} + \\
(|0001\rangle + |0011\rangle + |1001\rangle + |1011\rangle)|10\rangle_{34} + \\
(|0000\rangle + |0010\rangle + |1000\rangle + |1010\rangle)|01\rangle_{34}
\end{pmatrix}
\end{matrix}\right]
$$

$$
= \left\{\left[\begin{pmatrix}
|0000\rangle(\alpha\,\alpha'|00\rangle + \alpha\,\beta'|11\rangle + \alpha'\beta|10\rangle + \beta\,\beta'|01\rangle_{34}) + \\
|0001\rangle(\alpha\,\beta'|00\rangle + \alpha\,\alpha'|11\rangle + \beta\,\beta'|10\rangle + \alpha'\,\beta|01\rangle_{34}) + \\
|0010\rangle(\alpha\,\alpha'|00\rangle - \alpha\,\beta'|11\rangle + \alpha'\beta|10\rangle + \beta\,\beta'|01\rangle_{34}) + \\
|0011\rangle(\alpha\,\beta'|00\rangle - \alpha\,\alpha'|11\rangle + \beta\,\beta'|10\rangle + \alpha'\,\beta|01\rangle_{34}) + \\
|0100\rangle(\alpha'\,\beta|00\rangle + \beta\,\beta'|11\rangle + \alpha\,\alpha'|10\rangle + \alpha\,\beta'|01\rangle_{34}) + \\
|0101\rangle(\beta\,\beta'|00\rangle + \alpha'\,\beta|11\rangle + \alpha\,\beta'|10\rangle + \alpha\,\alpha'|01\rangle_{34}) + \\
|0110\rangle(\alpha'\,\beta|00\rangle - \beta\,\beta'|11\rangle + \alpha\,\alpha'|10\rangle + \alpha\,\beta'|01\rangle_{34}) + \\
|0111\rangle(\beta\,\beta'|00\rangle - \alpha'\,\beta|11\rangle + \alpha\,\beta'|10\rangle + \alpha\,\alpha'|01\rangle_{34})
\end{pmatrix}\right]\right\}
$$

$$
+ \left\{\left[\begin{pmatrix}
|0000\rangle(\alpha\,\alpha'|00\rangle + \alpha\,\beta'|11\rangle - \alpha'\beta|10\rangle - \beta\,\beta'|01\rangle_{34}) + \\
|0001\rangle(\alpha\,\beta'|00\rangle + \alpha\,\alpha'|11\rangle - \beta\,\beta'|10\rangle - \alpha'\,\beta|01\rangle_{34}) + \\
|0010\rangle(\alpha\,\alpha'|00\rangle - \alpha\,\beta'|11\rangle - \alpha'\beta|10\rangle + \beta\,\beta'|01\rangle_{34}) + \\
|0011\rangle(\alpha\,\beta'|00\rangle - \alpha\,\alpha'|11\rangle - \beta\,\beta'|10\rangle + \alpha'\,\beta|01\rangle_{34}) + \\
|1100\rangle(-\alpha'\,\beta|00\rangle - \beta\,\beta'|11\rangle + \alpha\,\alpha'|10\rangle + \alpha\,\beta'|01\rangle_{34}) + \\
|1101\rangle(-\beta\,\beta'|00\rangle - \alpha'\,\beta|11\rangle + \alpha\,\beta'|10\rangle + \alpha\,\alpha'|01\rangle_{34}) + \\
|1110\rangle(-\alpha'\,\beta|00\rangle + \beta\,\beta'|11\rangle + \alpha\,\alpha'|10\rangle - \alpha\,\beta'|01\rangle_{34}) + \\
|1111\rangle(-\beta\,\beta'|00\rangle + \alpha'\,\beta|11\rangle + \alpha\,\beta'|10\rangle - \alpha\,\alpha'|01\rangle_{34})
\end{pmatrix}\right]\right\} \quad \text{(P12)}
$$

(for clarity, I have divided the 16 lines into two groups, $m = 0$ and $m = 1$, appearing within the two columns in brackets [], respectively).

The result in Eq. (P12) reveals the details of the 16 post-measurement states, which have the form

$$
|mnm'n'\rangle_{1256}|\theta\rangle_{34} = |mnm'n'\rangle_{1256}(w|00\rangle + x|11\rangle + y|10\rangle + z|01\rangle)_{34}. \quad \text{(P13)}
$$

Thus, each measurement from Alice that yields $|mnm'n'\rangle_{1256}$ results in a corresponding system collapse on Bob's two wires $|\theta\rangle_{34}$. Last but not least, we must show that the 16 collapsed states $|\theta\rangle_{34}$ indeed correspond to the proposed circuit shown in Fig. P1, and involve the gates $X^n$, $X^{n'}$, $Z^n$, $Z^{n'}$ corresponding to the cbits $m, n, m', n'$. Formally, the circuit corresponds to a tensor operator $U_{mnm'n'}$ effecting the transformation

$$U_{mnm'n'}|\theta\rangle_{34} = |\psi\rangle \tag{P14}$$

and defined as:

$$U_{mnm'n'} = Z_4^m Z_3^m Z_4^{m'} X_3^n X_4^{n'} X_3^{n'}. \tag{P15}$$

We must show that after passing through these gates, the resulting qubit is $|\psi\rangle = C_{43}|q\rangle_3 \otimes |q'\rangle_4$. Substituting $|q\rangle_3$ and $|q'\rangle_4$ from Eq. (P2) into the definition of $|\psi\rangle$, we obtain:

$$\begin{aligned}|\psi\rangle &= C_{43}|q\rangle_3 \otimes |q'\rangle_4\\ &= C_{43}(\alpha|0\rangle + \beta|1\rangle)_3 \otimes |\alpha'|0\rangle + \beta'|1\rangle\rangle_4\\ &= C_{43}(\alpha\alpha'|00\rangle + \alpha\beta'|01\rangle + \alpha'\beta|10\rangle + \beta\beta'|11\rangle)_{34}\\ &\equiv (\alpha\alpha'|00\rangle + \alpha\beta'|11\rangle + \alpha'\beta|10\rangle + \beta\beta'|01\rangle)_{34}.\end{aligned} \tag{P16}$$

It is then easily verified that $U_{mnm'n'}|\theta\rangle_{34} = |\psi\rangle$ for any of the 16 post-measurement states:

$$\begin{aligned}U_{mnm'n'}|\theta\rangle_{34} &= Z_4^0 Z_3^0 Z_4^0 X_3^0 X_4^0 X_3^0(\alpha\alpha'|00\rangle + \alpha\beta'|11\rangle + \alpha'\beta|10\rangle + \beta\beta'|01\rangle)_{34}\\ &= Z_4^0 Z_3^0 Z_4^0 X_3^0 X_4^1 X_3^1(\alpha\beta'|00\rangle + \alpha\alpha'|11\rangle + \beta\beta'|10\rangle + \alpha'\beta|01\rangle)_{34}\\ &= Z_4^0 Z_3^0 Z_4^1 X_3^0 X_4^0 X_3^0(\alpha\alpha'|00\rangle - \alpha\beta'|11\rangle + \alpha'\beta|10\rangle - \beta\beta'|01\rangle)_{34}\\ &= Z_4^0 Z_3^0 Z_4^1 X_3^0 X_4^1 X_3^1(\alpha\beta'|00\rangle - \alpha\alpha'|11\rangle + \beta\beta'|10\rangle - \alpha'\beta|01\rangle)_{34}\\ &= Z_4^0 Z_3^0 Z_4^0 X_3^1 X_4^0 X_3^0(\alpha'\beta|00\rangle + \beta\beta'|11\rangle + \alpha\alpha'|10\rangle + \alpha\beta'|01\rangle)_{34}\\ &= Z_4^0 Z_3^0 Z_4^0 X_3^1 X_4^1 X_3^1(\beta\beta'|00\rangle + \alpha'\beta|11\rangle + \alpha\beta'|10\rangle + \alpha\alpha'|01\rangle)_{34}\\ &= Z_4^0 Z_3^0 Z_4^1 X_3^1 X_4^0 X_3^0(\alpha'\beta|00\rangle - \beta\beta'|11\rangle + \alpha\alpha'|10\rangle - \alpha\beta'|01\rangle)_{34}\\ &= Z_4^0 Z_3^0 Z_4^1 X_3^1 X_4^1 X_3^1(\beta\beta'|00\rangle - \alpha'\beta|11\rangle + \alpha\beta'|10\rangle - \alpha\alpha'|01\rangle)_{34}\\ &= Z_4^1 Z_3^1 Z_4^0 X_3^0 X_4^0 X_3^0(\alpha\alpha'|00\rangle + \alpha\beta'|11\rangle - \alpha'\beta|10\rangle - \beta\beta'|01\rangle)_{34}\\ &= Z_4^1 Z_3^1 Z_4^0 X_3^0 X_4^1 X_3^1(\alpha\beta'|00\rangle + \alpha\alpha'|11\rangle - \beta\beta'|10\rangle - \alpha'\beta|01\rangle)_{34}\\ &= Z_4^1 Z_3^1 Z_4^1 X_3^0 X_4^0 X_3^0(\alpha\alpha'|00\rangle - \alpha\beta'|11\rangle - \alpha'\beta|10\rangle + \beta\beta'|01\rangle)_{34}\\ &= Z_4^1 Z_3^1 Z_4^1 X_3^0 X_4^1 X_3^1(\alpha\beta'|00\rangle - \alpha\alpha'|11\rangle - \beta\beta'|10\rangle + \alpha'\beta|01\rangle)_{34}\\ &= Z_4^1 Z_3^1 Z_4^0 X_3^1 X_4^0 X_3^0(-\alpha'\beta|00\rangle - \beta\beta'|11\rangle + \alpha\alpha'|10\rangle + \alpha\beta'|01\rangle)_{34}\\ &= Z_4^1 Z_3^1 Z_4^0 X_3^1 X_4^1 X_3^1(-\beta\beta'|00\rangle - \alpha'\beta|11\rangle + \alpha\beta'|10\rangle + \alpha\alpha'|01\rangle)_{34}\\ &= Z_4^1 Z_3^1 Z_4^1 X_3^1 X_4^0 X_3^0(-\alpha'\beta|00\rangle + \beta\beta'|11\rangle + \alpha\alpha'|10\rangle - \alpha\beta'|01\rangle)_{34}\\ &= Z_4^1 Z_3^1 Z_4^1 X_3^1 X_4^1 X_3^1(-\beta\beta'|00\rangle + \alpha'\beta|11\rangle + \alpha\beta'|10\rangle - \alpha\alpha'|01\rangle)_{34} = |\psi\rangle.\end{aligned} \tag{P17}$$

The first equality in Eq. (P17), corresponding to $|mnm'n'\rangle = |0000\rangle$ is immediately established by comparison with Eq. (P16). All the other equalities can be verified with the understanding that they hold within an occasional (unobservable) phase factor $e^{i\pi} = -1$. As an illustration of the verification, consider, for instance, the last equality in Eq. (P16), and apply the Pauli operators as defined:

$$\begin{aligned}
U_{mnm'n'}|\theta\rangle_{34} &= Z_4^1 Z_3^1 Z_4^1 X_3^1 X_4^1 X_3^1(-\beta\beta'|00\rangle + \alpha'\beta|11\rangle + \alpha\beta'|10\rangle - \alpha\alpha'|01\rangle)_{34} \\
&= Z_4^1 Z_3^1 Z_4^1 X_3^1(-\beta\beta'|11\rangle + \alpha'\beta|00\rangle + \alpha\beta'|01\rangle - \alpha\alpha'|10\rangle)_{34} \\
&= Z_4^1 Z_3^1 Z_4^1(-\beta\beta'|01\rangle + \alpha'\beta|10\rangle + \alpha\beta'|11\rangle - \alpha\alpha'|00\rangle)_{34} \\
&= Z_4^1 Z_3^1(\beta\beta'|01\rangle + \alpha'\beta|10\rangle - \alpha\beta'|11\rangle - \alpha\alpha'|00\rangle)_{34} \\
&= Z_4^1(\beta\beta'|01\rangle - \alpha'\beta|10\rangle + \alpha\beta'|11\rangle - \alpha\alpha'|00\rangle)_{34} \\
&= (-\beta\beta'|01\rangle - \alpha'\beta|10\rangle - \alpha\beta'|11\rangle - \alpha\alpha'|00\rangle)_{34} \\
&= -(\alpha\alpha'|00\rangle + \alpha\beta'|11\rangle + \alpha'\beta|10\rangle + \beta\beta'|01\rangle)_{34} \\
&\equiv e^{i\pi}|\psi\rangle.
\end{aligned} \tag{P18}$$

The operation of the quantum circuit, together with Alice transmitting her four measured cbits $m, n, m', n'$ to Bob, thus, result in the successful teleportation of the tensor state $|\psi\rangle$, as defined in Eq. (P16). As mentioned at the beginning of this appendix, Bob can retrieve Alice's qubits $|q\rangle, |q'\rangle$ by effecting a final CNOT operation on $|\psi\rangle$, as described by the operator $C_{34}$. Indeed,

$$\begin{aligned}
|\psi_{out}\rangle &= C_{43}|\psi'\rangle \\
&= C_{43}(\alpha\alpha'|00\rangle + \alpha\beta'|11\rangle + \alpha'\beta|10\rangle + \beta\beta'|01\rangle)_{34} \\
&= (\alpha\alpha'|00\rangle + \alpha\beta'|01\rangle + \alpha'\beta|10\rangle + \beta\beta'|11\rangle)_{34} \\
&= (\alpha|0\rangle + \beta|1\rangle)_3 \otimes |\alpha'|0\rangle + \beta'|1\rangle\rangle_4 \\
&\equiv |q\rangle_3 \otimes |q'\rangle_4,
\end{aligned} \tag{P19}$$

which shows that $|\psi'\rangle$ is the unique tensor product of the qubits $|q\rangle, |q'\rangle$.

# Appendix Q (Chapter 19) Quantum Fourier transform circuit

In this appendix, I develop a simplified tensor-product expression for the components resulting from the *quantum Fourier transform* (QFT) of $N$-qubits. This expression leads to the conception of a corresponding QFT circuit based on simple $2 \times 2$ quantum gates.

The starting point is the fundamental QFT definition (Eq. (19.16)), which transforms the orthonormal basis $|n\rangle$ according to

$$\text{QFT}\,|n\rangle = \frac{1}{\sqrt{N}} \sum_{k=0}^{N-1} e^{ik\frac{2n\pi}{N}}\,|k\rangle. \tag{Q1}$$

The goal is to reduce the right-hand side in this expression into a tractable tensor product, from which a simple QFT gate circuit can be constructed. I focus on the imaginary-exponential terms in Eq. (Q1), namely $\exp(2i\pi kn/N)$, where $k, n = 0 \ldots N-1$. From this point on, we assume that $N$ is exactly a power of two, i.e., $N = 2^K$, where $K \geq 1$ is an integer. We now expand the integer $k$ according to the possible binary representation $k \equiv k_1 2^{K-1} + k_2 2^{K-2} + \cdots + k_K 2^0$, where $k_p = 0, 1$ and $k = 1, 2 \ldots K$, and express the ratio $k/N$ according to

$$\begin{aligned} \frac{k}{N} &= \frac{k}{2^K} \\ &= k_1 2^{-1} + k_2 2^{-2} + \cdots + k_K 2^{-K} \\ &= \sum_{p=1}^{K} \frac{k_p}{2^p}. \end{aligned} \tag{Q2}$$

Substituting this representation into the exponential in Eq. (Q1), and introducing next the tensor notation $|k\rangle = |k_1 k_2 \ldots k_K\rangle$ we obtain

$$\text{QFT}|n\rangle = \frac{1}{2^{K/2}} \sum_{k=0}^{2^K-1} \exp\left[2i\pi n \sum_{p=1}^{K} \frac{k_p}{2^p}\right] |k\rangle, \tag{Q3}$$

$$\begin{aligned}\mathrm{QFT}|n\rangle &= \frac{1}{2^{K/2}} \sum_{k_1,k_2,\ldots,k_K=0,1} \exp\left[2i\pi n \sum_{p=1}^{K} \frac{k_p}{2^p}\right] |k_1 k_2 \ldots k_K\rangle \\ &= \frac{1}{2^{K/2}} \sum_{k_1=0,1} \sum_{k_2=0,1} \cdots \sum_{k_K=0,1} \exp\left[2i\pi n \sum_{p=1}^{K} \frac{k_p}{2^p}\right] |k_1 k_2 \ldots k_K\rangle \\ &\equiv \frac{1}{2^{K/2}} \sum_{k_1=0,1} e^{2i\pi n \frac{k_1}{2^1}} |k_1\rangle \otimes \sum_{k_2=0,1} e^{2i\pi n \frac{k_2}{2^2}} |k_2\rangle \otimes \cdots \otimes \sum_{k_K=0,1} e^{2i\pi n \frac{k_K}{2^K}} |k_K\rangle.\end{aligned} \tag{Q4}$$

This result shows that QFT $|n\rangle$ can be decomposed indeed into a $K$-tensor product. In this decomposition, each term, indexed by $p$, is actually made of the two contributions:

$$\sum_{k_p=0,1} e^{2i\pi n \frac{k_p}{2^p}} |k_p\rangle = |0\rangle + e^{\frac{2i\pi n}{2^p}} |1\rangle. \tag{Q5}$$

Hence, we can develop Eq. (Q5) according to:

$$\mathrm{QFT}|n\rangle = \frac{1}{2^{K/2}} \left(|0\rangle_1 + e^{\frac{2i\pi n}{2^1}} |1\rangle_1\right) \otimes \left(|0\rangle_2 + e^{\frac{2i\pi n}{2^2}} |1\rangle_2\right) \otimes \cdots \otimes \left(|0\rangle_K + e^{\frac{2i\pi n}{2^K}} |1\rangle_K\right). \tag{Q6}$$

The next step is to simplify further each of the exponential terms in Eq. (Q6). To do this, we substitute the binary expansion for integers $n$, namely, $n = n_1 2^{K-1} + n_2 2^{K-2} + \cdots + n_{K-1} 2^1 + n_K 2^0$, where $n_q = 0, 1$, in each of the exponents, which gives

$$\begin{cases} 2i\pi \dfrac{n}{2^1} = 2i\pi \left[\left(n_1 2^{K-2} + n_2 2^{K-3} + \cdots + n_{K-1}\right) + \dfrac{n_K}{2^1}\right] \\ 2i\pi \dfrac{n}{2^2} = 2i\pi \left[\left(n_1 2^{K-3} + n_2 2^{K-4} + \cdots + n_{K-2}\right) + \dfrac{n_{K-1}}{2^1} + n\dfrac{k_K}{2^2}\right] \\ \vdots \\ 2i\pi \dfrac{n}{2^{K-1}} = 2i\pi \left[(n_1) + \dfrac{n_2}{2^1} + \cdots + \dfrac{n_{K-1}}{2^{K-2}} + \dfrac{n_K}{2^{K-1}}\right] \\ 2i\pi \dfrac{n}{2^K} = 2i\pi \left[\dfrac{n_1}{2^1} + \dfrac{n_2}{2^2} + \cdots + \dfrac{n_{K-1}}{2^{K-1}} + \dfrac{n_K}{2^K}\right]. \end{cases} \tag{Q7}$$

In the above, I have put in parentheses all the integer terms, to separate them from the other terms that are defined as fractions of powers of two. Clearly, the integer terms do not contribute to the imaginary exponential. The consequence of this can be generalized under the formula

$$\begin{cases} \exp\left(2i\pi \dfrac{n}{2^m}\right) = \exp\left(2i\pi\, \Omega_m\right) \\ \Omega_m = \displaystyle\sum_{l=1}^{m} \dfrac{n_{K-m+l}}{2^l}. \end{cases} \tag{Q8}$$

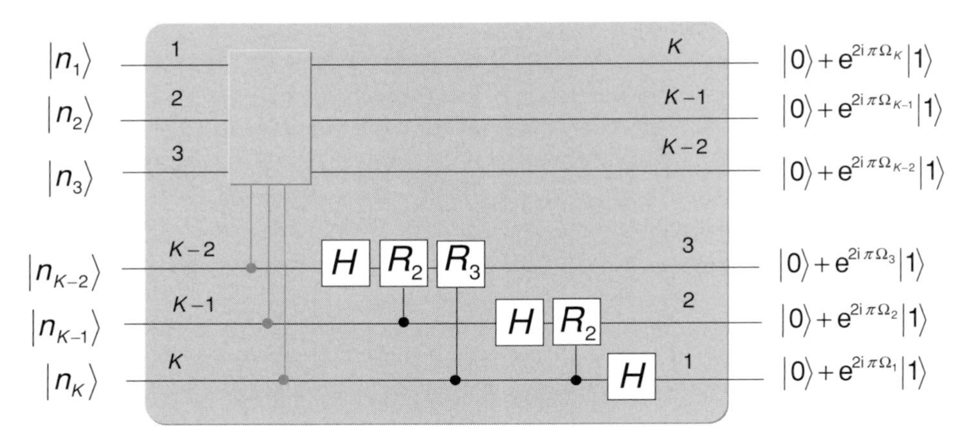


**Figure Q1** Gate circuit corresponding to first three output qubits.

From Eqs. (Q6)–(Q8), we finally obtain a nice closed-formed expression for the QFT tensor product:

$$\mathrm{QFT}|n\rangle = \frac{1}{2^{K/2}}\left(|0\rangle_1 + e^{2i\pi\Omega_1}|1\rangle_1\right) \otimes \left(|0\rangle_2 + e^{2i\pi\Omega_2}|1\rangle_2\right) \otimes \cdots \otimes \left(|0\rangle_K + e^{2i\pi\Omega_K}|1\rangle_K\right). \tag{Q9}$$

With $|n\rangle = |n_1 n_2 \ldots n_K\rangle$, we have the following relation for each of the gate *outputs*

$$\mathrm{QFT}|n\rangle_p = \frac{1}{2^{K/2}}\left(|0\rangle_p + e^{2i\pi\Omega_p}|1\rangle_p\right). \tag{Q9}$$

Considering the first output qubits we have

$$\begin{aligned}\mathrm{QFT}|n\rangle_1 &= \frac{1}{2^{K/2}}(|0\rangle_1 + e^{2i\Omega_1}|1\rangle_1)\\ &= \frac{1}{2^{K/2}}\left(|0\rangle_1 + e^{i\pi n_K}|1\rangle_1\right),\end{aligned} \tag{Q10}$$

which shows that (within the normalization factor $2^{K/2}$) the QFT circuit wire corresponds either to the identity gate $I$ (with input $n_K = 0$) or the Hadamard gate $H$ (with input $n_K = 1$). We can then proceed to construct the QFT circuit from bottom to top, numbering the output qubits from 1 to $K$, accordingly. As for the second output qubit, we have $2i\pi\Omega_2 = i\pi(n_{K-1} + n_K/2)$ and then:

$$\mathrm{QFT}|n\rangle_2 = \frac{1}{2^{K/2}}\left(|0\rangle_2 + e^{2i\Omega_2}|1\rangle_2\right) = \frac{1}{2^{K/2}}\left(|0\rangle_2 + e^{i\pi n_{K-1}}e^{i\frac{\pi}{2}n_K}|1\rangle_2\right). \tag{Q11}$$

We can immediately identify the above transformation as built from a Hadamard gate on the output wire 2 with input $n_{K-1}$, and a *controlled-phase* gate (call it $R_2$), with phase $\pi n_K/2$ and $n_K$ as the control qubit. The next output qubit is

$$\begin{aligned}\mathrm{QFT}|n\rangle_3 &= \frac{1}{2^{K/2}}\left(|0\rangle_3 + e^{2i\Omega_3}|1\rangle_3\right)\\ &= \frac{1}{2^{K/2}}\left(|0\rangle_3 + e^{i\pi n_{K-2}}e^{i\frac{\pi}{2}n_{K-1}}e^{i\frac{\pi}{4}n_K}|1\rangle_3\right).\end{aligned} \tag{Q12}$$

We can immediately identify the above transformation as built from a Hadamard gate on the output wire 3 with input $n_{K-2}$, the controlled-phase gate $R_2$, with $n_{K-1}$ as the control qubit, followed by a second gate (call it $R_3$), with phase $\pi n_K/4$ and $n_K$ as the control qubit.

The gate circuit corresponding to the first three output qubits is illustrated in Fig. Q1. It is clear from the above description and the definition of $\Omega_m$ in Eq. (Q8) that the whole QFT circuit represents a generalization of that shown in Fig. Q1, with a sequence of controlled-phase gates defined by the matrix

$$R_n = \begin{pmatrix} 1 & 0 \\ 0 & e^{2i\frac{\pi}{2^n}} \end{pmatrix}. \tag{Q13}$$

On the top quantum wire, the gate sequence from left to right is thus: $HR_2R_3 \ldots R_K$ (see Fig. 19.4 in main text for a complete view of the QFT circuit).

# Appendix R (Chapter 20) Properties of continued fraction expansion

This appendix provides the proof of three key properties of the *continued fraction expansion*.

First, define the real number $x_n$ through a suite of positive real numbers $[a_0, a_1, \ldots, a_n]$ according to the continued fraction

$$x_n = a_0 + \cfrac{1}{a_1 + \cfrac{1}{a_2 + \cfrac{1}{a_3 + \cdots \cfrac{1}{a_n}}}}, \tag{R1}$$

with $n \leq N$. Each real number in the set $\{x_0, x_1, \ldots, x_{N-1}, x_N\}$ is called a *convergent* of $x_N$, while $x_n$ is called the $n$th *convergent* of $x_N$.

## Property 1

The finite suite $[a_0, a_1, \ldots, a_n]$ of positive real numbers corresponds to the ratio $x_n = p_n/q_n$, as defined through:

$$\begin{cases} p_n = a_n p_{n-1} + p_{n-2} \\ q_n = a_n q_{n-1} + q_{n-2}, \end{cases} \tag{R2}$$

with $n \geq 2$, $p_0 = a_0$, $q_0 = 1$, $p_1 = 1 + a_0 a_1$, and $q_1 = a_1$ for $n = 0, 1$.

The proof of these relations comes by induction. Indeed, for $n = 0, 1$, we have:

$$\begin{cases} p_0 = a_0 \\ q_0 = 1 \end{cases} \rightarrow \frac{p_0}{q_0} = \frac{a_0}{1} \equiv a_0, \tag{R3}$$

$$\begin{cases} p_1 = 1 + a_0 a_1 \\ q_1 = a_1 \end{cases} \rightarrow \frac{p_1}{q_1} = \frac{1 + a_0 a_1}{a_1} \equiv a_0 + \frac{1}{a_1}. \tag{R4}$$

We then observe that:

$$\begin{cases} x_2 = a_0 + \cfrac{1}{a_1 + \cfrac{1}{a_2}} \\ x_3 = a_0 + \cfrac{1}{a_1 + \cfrac{1}{a_2 + \cfrac{1}{a_3}}}, \end{cases} \tag{R5}$$

which illustrates that $[a_0, a_1, a_2]$ and $[a_0, a_1, a_2 + 1/a_3]$ correspond to $x_2$ and $x_3$, which are the 2nd and the 3rd convergents of $x_N$, respectively. Assume next that the relations in Eq. (R2) apply to any order $n \geq 1$. We must then show that the relations also apply to the order $n + 1$. Then given the $n$th convergent $x_n$, i.e., $[a_0, a_1, \ldots, a_{n-1}, a_n]$, the $(n + 1)$th convergent $x_{n+1}$ is of the form $[a_0, a_1, \ldots, a_{n-1}, a_n + 1/a_{n+1}]$. Applying Eq. (R2), to the latter, we obtain

$$\tilde{x}_{n+1} = \frac{\left(a_n + \dfrac{1}{a_{n+1}}\right) p_{n-1} + p_{n-2}}{\left(a_n + \dfrac{1}{a_{n+1}}\right) q_{n-1} + q_{n-2}}. \tag{R6}$$

Multiplying the numerator and denominator in Eq. (R6) by $a_{n+1}$ and reordering terms yields

$$\begin{aligned}\tilde{x}_{n+1} &= \frac{a_{n+1}\left(a_n p_{n-1} + p_{n-2}\right) + p_{n-1}}{a_{n+1}\left(a_n q_{n-1} + q_{n-2}\right) + q_{n-1}} \\ &= \frac{a_{n+1} p_n + p_{n-1}}{a_{n+1} q_n + q_{n-1}} \\ &\equiv \frac{p_{n+1}}{q_{n+1}} \\ &= x_{n+1}.\end{aligned} \tag{R7}$$

The result in Eq. (R7) shows that $\tilde{x}_{n+1} = x_{n+1}$, corresponding to the suite $[a_0, a_1, \ldots, a_{n+1}]$, which proves that the relations in Eq. (R2) apply at order $n + 1$.

If the suite $[a_0, a_1, \ldots, a_N]$ is made of positive integers, then $p_n, q_n$ are also positive integers and $x_n = p_n/q_n$ is a *rational* number for $0 \leq n \leq N$.

## Property 2

The real numbers $p_n, q_n$ defined in Eq. (R2) are *co-prime*, and satisfy the relation $(n \geq 1)$:

$$q_n p_{n-1} - p_n q_{n-1} = (-1)^n. \tag{R8}$$

It is left as an easy exercise first to prove the above by induction. Next, we must show that $p_n, q_n$ are co-prime, meaning that their greatest common divisor (GCD) is one.

Assume that $k \geq 1$ is some common divisor of $p_n, q_n$, i.e., $p_n = k\tilde{p}_n$ and $q_n = k\tilde{q}_n$, with $p_n/q_n = \tilde{p}_n/\tilde{q}_n$. Likewise, assume that $k' \geq 1$ is some common divisor of $p_{n-1}, q_{n-1}$, i.e., $p_{n-1} = k'\tilde{p}_{n-1}$ and $q_{n-1} = k'\tilde{q}_{n-1}$, with $p_{n-1}/q_{n-1} = \tilde{p}_{n-1}/\tilde{q}_{n-1}$. Substituting these definitions into Eq. (R8) yields:

$$kk'\left(\tilde{q}_n \tilde{p}_{n-1} - \tilde{p}_n \tilde{q}_{n-1}\right) = (-1)^n. \tag{R9}$$

Clearly, $\tilde{p}_n/\tilde{q}_n$ has the same continued fraction expansion as $p_n/q_n$ at all orders $n \geq 1$, meaning that from Eq. (R8) we also have $\tilde{q}_n\tilde{p}_{n-1} - \tilde{p}_n\tilde{q}_{n-1} = (-1)^n$ and, hence, from Eq. (R9):

$$kk' = 1. \tag{R10}$$

Consider next the case $n = 1$. Since $q_0 = 1$, we have $k' = 1$ (hence $p_0, q_0$ are co-prime) and from Eq. (R10), we have $k = 1$ (hence $p_1, q_1$ are co-prime). The same conclusion applies to the case $n = 2$ ($p_2, q_2$ are co-prime, since $p_1, q_1$ are co-prime), and so on, which proves Property 2, according to which $p_n, q_n$ are co-prime at all orders $n \geq 1$.

## Property 3

Given a *rational* number $x$, if two integers $p, q$ are such that

$$\left|\frac{p}{q} - x\right| \leq \frac{1}{2q^2}, \tag{R11}$$

then $p/q$ is a convergent of $x$.

We can prove this property as follows. First, we know from the above description that there exists a finite fraction expansion of the rational number $x$, as defined by the convergents $x_n = p_n/q_n$ ($n = 0, 1, \ldots, N$). We shall assume, without loss of generality, that $n$ is even. We then define the error $0 \leq \delta \leq 1$ corresponding to the convergent $x_n$ according to

$$x - \frac{p_n}{q_n} = \frac{\delta}{2q_n^2}. \tag{R12}$$

We introduce

$$\lambda = 2\frac{q_n p_{n-1} - p_n q_{n-1}}{\delta} - \frac{q_{n-1}}{q_n}, \tag{R13}$$

for which (according to Eq. (R13)) we have

$$x = \frac{\lambda p_n + p_{n-1}}{\lambda q_n + q_{n-1}}. \tag{R14}$$

The result in Eq. (R14) shows that $\lambda = a_{n+1}$ in the continued fraction expansion $[a_0, a_1, \ldots, a_n, \lambda]$ that *exactly* defines $x$. Applying Property 2 in Eq. (R13), with $n$ being even, we obtain

$$\begin{aligned} \lambda &= \frac{2}{\delta} - \frac{q_{n-1}}{q_n} \\ &> 2 - \frac{q_{n-1}}{q_n}. \end{aligned} \tag{R15}$$

Since by definition $q_{n-1}/q_n$ is a rational number and also $q_{n-1}/q_n < 1$, the result in Eq. (R15) implies that $\lambda > 1$, and also that $\lambda$ *is a rational number*. Consequently, $\lambda$ can be exactly expressed through a finite, continued fraction expansion $[b_0, b_1, \ldots, b_M]$, and, thus, the fraction expansion $[a_0, a_1, \ldots, a_n, b_0, b_1, \ldots, b_M]$ *exactly* defines $x$. The key conclusion is that *the condition in* Eq. (R9) *is necessary and sufficient for any rational number $p/q$ to be convergent on $x$.*

I shall now illustrate Property 3 through a numerical example, which links to the problem to be solved in the main text. As described in the text, the phase measurement circuit yields the value $x = \tilde{\varphi}$, which is a fair approximation of the exact phase $\varphi = p/q$, where $p, q$ are integers. The problem is that given $x$, we must find the exact integer $q$.

Assume, for instance, that the exact phase $\varphi = p/q$ is

$$\begin{aligned}\varphi &= \frac{711}{413}\\ &\equiv 1.72154963680387,\end{aligned}$$

of which we obtain the following measurement value, and which we know (by choice of the register size) to be accurate to $10^{-6}$:[1]

$$\tilde{\varphi} = 1.721549(\ldots).$$

Our task is now to obtain the exact value $q = 413$ by implementing the continuous fraction expansion algorithm for the measurement $\tilde{\varphi}$. For reference purposes, let us expand $\varphi$ first. It is straightforward to obtain:

$$\varphi = \frac{711}{413} = 1 + \cfrac{1}{1 + \cfrac{1}{2 + \cfrac{1}{1 + \cfrac{1}{1 + \cfrac{1}{2 + \cfrac{1}{4 + \cfrac{1}{5}}}}}}},$$

which shows that the suite $[1, 1, 2, 1, 1, 2, 4, 5]$ is the expansion of $\varphi$. We shall now expand $\tilde{\varphi}$ step by step, being careful to effect no truncation in the successive results. For the first three steps, we obtain

$$\begin{aligned}\tilde{\varphi} &= 1.721549\\ &= 1 + 0.721549\\ &= 1 + \cfrac{1}{\cfrac{1}{0.721549}}\\ &= 1 + \frac{1}{1.385907263401380}\end{aligned}$$

[1] More precisely, the corresponding accuracy is $\varepsilon = |\varphi - \tilde{\varphi}| = 3.69901547059293 \times 10^{-7}$, representing in the binary system (to be actually used for the algorithm implementation), an accuracy of up to 10 bits, since $2^{-11} = 1.2 \times 10^{-7} < \varepsilon < 2^{-10} = 5.4 \times 10^{-7}$.

**Table R1** "Split and divide" expansion up to 12 steps.

| $n$ | $a_n$ | $u_n = 1/(u_{n-1} - a_{n-1})$ | $p_n$ | $q_n$ | $p_n/q_n$ | $\Delta$ | $1/2q_n^2$ | $1/2q_n^2 - \Delta$ | Valid |
|---|---|---|---|---|---|---|---|---|---|
| 0 | 1 | 1.721549000000000 | 1 | 1 | 1.00000000000000 | $7.215 \times 10^{-1}$ | $5.00 \times 10^{-1}$ | $-2.215 \times 10^{-1}$ | No |
| 1 | 1 | 1.385907263401380 | 2 | 1 | 2.00000000000000 | $2.785 \times 10^{-1}$ | $5.00 \times 10^{-1}$ | $2.215 \times 10^{-1}$ | Yes |
| 2 | 2 | 2.591296134687970 | 5 | 3 | 1.66666666666667 | $5.488 \times 10^{-2}$ | $5.56 \times 10^{-2}$ | $6.726 \times 10^{-4}$ | Yes |
| 3 | 1 | 1.691199961128960 | 7 | 4 | 1.75000000000000 | $2.845 \times 10^{-2}$ | $3.13 \times 10^{-2}$ | $2.800 \times 10^{-3}$ | Yes |
| 4 | 1 | 1.446759340620720 | 12 | 7 | 1.71428571428571 | $7.264 \times 10^{-3}$ | $1.02 \times 10^{-2}$ | $2.940 \times 10^{-3}$ | Yes |
| 5 | 2 | 2.238341561276870 | 31 | 18 | 1.72222222222222 | $6.726 \times 10^{-4}$ | $1.54 \times 10^{-3}$ | $8.706 \times 10^{-4}$ | Yes |
| 6 | 4 | 4.195659349727770 | 136 | 79 | 1.72151898734177 | $3.065 \times 10^{-5}$ | $8.01 \times 10^{-5}$ | $4.947 \times 10^{-5}$ | Yes |
| 7 | 5 | 5.110923660900250 | 711 | 413 | 1.72154963680387 | $0.000 \times 10^{0}$ | $2.93 \times 10^{-6}$ | $2.931 \times 10^{-6}$ | Yes |
| 8 | 9 | 9.015209125664090 | 6535 | 3796 | 1.72154899894626 | $6.379 \times 10^{-7}$ | $3.47 \times 10^{-8}$ | $-6.032 \times 10^{-7}$ | No |
| 9 | 65 | 65.749999183803000 | 425486 | 247153 | 1.72154900001214 | $6.368 \times 10^{-7}$ | $8.19 \times 10^{-12}$ | $-6.368 \times 10^{-7}$ | No |
| 10 | 1 | 1.333334784351880 | 432021 | 250949 | 1.72154899999602 | $6.368 \times 10^{-7}$ | $7.94 \times 10^{-12}$ | $-6.368 \times 10^{-7}$ | No |
| 11 | 2 | 2.999986940889930 | 1289528 | 749051 | 1.72154900000133 | $6.368 \times 10^{-7}$ | $8.91 \times 10^{-13}$ | $-6.368 \times 10^{-7}$ | No |

$$
\begin{aligned}
&= 1 + \frac{1}{1 + 0.385907263401380} \\
&= 1 + \cfrac{1}{1 + \cfrac{1}{\cfrac{1}{0.385907263401380}}} \\
&= 1 + \cfrac{1}{1 + \cfrac{1}{2.591296134687970}} \\
&= 1 + \cfrac{1}{1 + \cfrac{1}{2 + 0.591296134687970}} \\
&= \ldots
\end{aligned}
$$

The results of this "split and divide" expansion up to 12 steps are summarized in Table R1. For each step $n$, the table lists the expansion coefficient $a_n = \lfloor u_n \rfloor$, the ratio $u_n = 1/(u_{n-1} - a_{n-1})$, starting from $u_0 = x$, the integer numbers $p_n, q_n$ as defined by Eq. (R2), the ratio $p_n/q_n$, the error $\Delta = |p_n/q_n - \varphi|$, the upper bound $1/2q_n^2$ in Eq. (R11), and the difference $1/2q_n^2 - \Delta$. If this difference is negative, this means that the condition in Eq. (R11) is not valid, and hence that $p_n/q_n$ is not convergent on $\varphi$. We observe from Table R1 that for $n \geq 1$ the condition is valid up to the order $n = 7$, which yields six convergents. The last convergent, $p_7/q_7 = 711/413$, is the exact definition of $\varphi$. The algorithm has, thus, made it possible to determine $\varphi = p/q$ unambiguously given the approximation $\tilde{\varphi}$ and its known accuracy.

# Appendix S (Chapter 20) Computation of inverse Fourier transform in the factorization of $N = 21$ through Shor's algorithm

In this appendix, I detail the computation of the inverse Fourier transform involved in the factoring of $N = 15$ through Shor's algorithm. To recall the parameters used in Section 20.5, we have $K = 11$ for the first register size and, thus, $M = 2^{11} = 2048$. After a measurement in the second register ($z = 8$), the state $|u_2\rangle$ of the first register, which is input to the inverse Fourier transform circuit, is

$$|u_2\rangle = \frac{1}{\sqrt{4M}}(|1\rangle + |5\rangle + |9\rangle + |13\rangle + \cdots). \tag{S1}$$

By definition (see Chapter 19) the inverse Fourier transform $\text{FT}^+$ acts on the $M$-qubit state $|n\rangle$ according to

$$\text{FT}^+|n\rangle = \frac{1}{\sqrt{4M}} \sum_{k=0}^{M-1} \mathrm{e}^{-k\frac{2i\pi n}{M}} |k\rangle. \tag{S2}$$

Thus, from the above definition of $|u_2\rangle$, we obtain and develop the transformation as:

$$\begin{aligned}
\text{FT}^+|u_2\rangle &= \tfrac{1}{\sqrt{4M}}\text{FT}^+(|1\rangle + |5\rangle + |9\rangle + |13\rangle + \cdots) \\
&= \frac{1}{2M}\sum_{k=0}^{M-1}(\mathrm{e}^{-k\frac{2i\pi 1}{M}}|k\rangle + \mathrm{e}^{-k\frac{2i\pi 5}{M}}|k\rangle + \mathrm{e}^{-k\frac{2i\pi 9}{M}}|k\rangle + \mathrm{e}^{-k\frac{2i\pi 13}{M}}|k\rangle + \cdots) \\
&= \frac{1}{2M}\sum_{k=0}^{M-1}(\mathrm{e}^{-k\frac{2i\pi 1}{M}} + \mathrm{e}^{-k\frac{2i\pi 5}{M}} + \mathrm{e}^{-k\frac{2i\pi 9}{M}} + \mathrm{e}^{-k\frac{2i\pi 13}{M}} + \cdots)|k\rangle \\
&= \frac{1}{2M}\sum_{k=0}^{M-1}\mathrm{e}^{-k\frac{2i\pi}{M}}\left[(\mathrm{e}^{-k\frac{8i\pi}{M}})^0 + (\mathrm{e}^{-k\frac{8i\pi}{M}})^1 + (\mathrm{e}^{-k\frac{8i\pi}{M}})^2 + (\mathrm{e}^{-k\frac{8i\pi}{M}})^3 + \cdots\right]|k\rangle \\
&= \frac{1}{2M}\sum_{k=0}^{M-1}\mathrm{e}^{-k\frac{2i\pi}{M}}\sum_{m=0}^{M-1}(\mathrm{e}^{-k\frac{8i\pi}{M}})^m|k\rangle \\
&\equiv \frac{1}{2M}\sum_{k=0}^{M-1}\mathrm{e}^{-k\frac{2i\pi}{M}}\frac{1-\mathrm{e}^{-8i\pi k}}{1-\mathrm{e}^{-k\frac{8i\pi}{M}}}|k\rangle \\
&= \frac{1}{2M}\sum_{k=0}^{M-1}\mathrm{e}^{-k\frac{2i\pi}{M}}\frac{1-\mathrm{e}^{-8i\pi k}}{\mathrm{e}^{-k\frac{4i\pi}{M}}(\mathrm{e}^{k\frac{4i\pi}{M}} - \mathrm{e}^{-k\frac{4i\pi}{M}})}|k\rangle
\end{aligned}$$

$$\begin{aligned}
&\equiv \frac{1}{4iM}\sum_{k=0}^{M-1} e^{k\frac{2i\pi}{M}} \frac{1-e^{-8i\pi k}}{\sin\left(\dfrac{4\pi k}{M}\right)}|k\rangle \\
&= \frac{1}{4iM}\sum_{k=0}^{M-1} e^{k\frac{2i\pi}{M}} \frac{e^{-4i\pi k}\left(e^{4i\pi k}-e^{-4i\pi k}\right)}{\sin\left(\dfrac{4\pi k}{M}\right)}|k\rangle \\
&\equiv \frac{1}{2M}\sum_{k=0}^{M-1} e^{-k\frac{2i\pi}{M}(2M-1)} \frac{\sin(4\pi k)}{\sin\left(k\dfrac{4\pi}{M}\right)}|k\rangle.
\end{aligned} \tag{S3}$$

We can rewrite the result in Eq. (S3) in the form

$$\mathrm{FT}^{+}|u_2\rangle = \sum_{k=0}^{M-1} \alpha_k |k\rangle, \tag{S4}$$

where the amplitude coefficient $\alpha_k$ is defined by

$$\alpha_k = \frac{1}{2M} e^{-k\frac{2i\pi}{M}(2M-1)} \frac{\sin(4\pi k)}{\sin\left(k\dfrac{4\pi}{M}\right)}. \tag{S5}$$

The corresponding probability distribution $p(k) = |\alpha_k|^2$ corresponding to the output measurement $k$ is, therefore:

$$p(k) = \frac{1}{4M^2} \frac{\sin^2(4\pi k)}{\sin^2\left(k\dfrac{4\pi}{M}\right)}. \tag{S6}$$

We observe that for any integer $k = 0, 1, \ldots, M-1$, the numerator in Eq. (S6) is zero. So is the denominator whenever $4\pi k/M$ is an integer multiple of $\pi$, or

$$\begin{aligned}
&\frac{4\pi k}{M} = n \times \pi \\
&\qquad \leftrightarrow \\
&k = n \times \frac{M}{4} = n \times 2^7 = n \times 512,
\end{aligned} \tag{S7}$$

where $n$ is an integer. In this last case, the numerator: denominator ratio is analytically undetermined, but setting $k = nM/4 + \varepsilon$ and taking the limit:

$$\lim_{\varepsilon\to 0} \frac{\sin^2(4\pi k)}{\sin^2\left(k\dfrac{4\pi}{M}\right)} = \lim_{\varepsilon\to 0} \frac{\sin^2\left[4\pi\left(\dfrac{nM}{4}+\varepsilon\right)\right]}{\sin^2\left[\dfrac{4\pi}{M}\left(\dfrac{nM}{4}+\varepsilon\right)\right]}$$

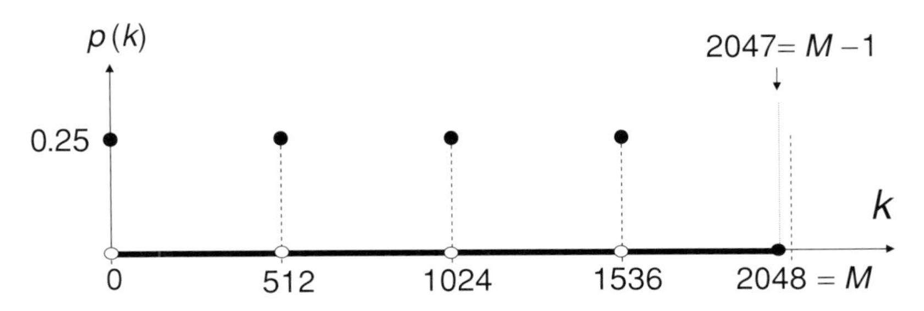


**Figure S1** Plot of $p(k)$.

$$
\begin{aligned}
&= \lim_{\varepsilon \to 0} \frac{\sin^2(nM\pi + 4\pi\varepsilon)}{\sin^2\left(n\pi + \dfrac{4\pi}{M}\varepsilon\right)} \\
&= \lim_{\varepsilon \to 0} \frac{\sin^2(4\pi\varepsilon)}{\sin^2\left(\dfrac{4\pi}{M}\varepsilon\right)} \\
&= \frac{(4\pi\varepsilon)^2}{\left(\dfrac{4\pi}{M}\varepsilon\right)^2} \\
&\equiv M^2,
\end{aligned}
\tag{S8}
$$

which, from Eq. (S6), yields $p_{\max} = 1/4$. In the range $k = 0, 1, 2, \ldots, M-1$, the maxima of $p(k)$ are thus located at $k = 0$ $(n = 0)$, $k = 512$ $(n = 1)$, $k = 1024$ $(n = 2)$ and $k = 1536$ $(n = 3)$. The plot of $p(k)$ is shown in Fig. S1.

# Appendix T (Chapter 20) Modular arithmetic and Euler's theorem

In this appendix, we shall review some basic principles and properties of *modular arithmetic*. As Section 20.7 shows, modular arithmetic is at the root of *public key cryptography* (PKC).

Let $m$ and $x$ be two integers, with $m$ being nonzero. Define $[x/m]$ as the integer part of the division of $x$ by $m$. If we consider, for instance, $m = 3$ and the ratios $2/3 = 0.66$, $4/3 = 1.33$, $6/3 = 2$ and $11/3 = 3.66$, we, thus, obtain $[2/3] = 0$, $[4/3] = 1$, $[6/3] = 2$, and $[11/3] = 3$. It is then possible to express any integer number $x$ in the form:

$$x = m[x/m] + r, \tag{T1}$$

where $r$ stands for *residue*. With the previous examples, we have:

$$\begin{cases} 2 = 3[2/3] + 2 \\ 4 = 3[4/3] + 1 \\ 6 = 3[6/3] + 0 \\ 11 = 3[11/3] + 2. \end{cases} \tag{T2}$$

We observe that the residues are positive integers that range from $r = 0$ to $r = m - 1$. We can define such residues in the form

$$r = x - m[x/m]. \tag{T3}$$

By convention, one designates the residues according to any of the following equivalent notations:

$$\begin{cases} r = |x|_m \\ r = x \bmod m \\ x \equiv r[m] \\ x = r(\bmod m), \end{cases} \tag{T4}$$

which reads as, "$r$ equals $x$ modulo $m$," or "$x$ equals $r$ modulo $m$," it being understood which one is the residue. Here, we shall mainly use the second definition, with the convention that "mod $m$" carries over the entire right-hand side expression or even the full line (this to avoid a parenthesis inflation, unless parentheses are introduced otherwise). The following are basic examples:

$$\begin{cases} (3+4) \bmod 6 = 7 \bmod 6 = 1 \\ (8+7) \bmod 15 = 15 \bmod 15 = 0 \\ (7-4) \bmod 4 = 3 \bmod 4 = 3 \\ (2 \times 3) \bmod 4 = 6 \bmod 4 = 2 \\ (7 \times 9) \bmod 11 = 63 \bmod 11 = 8. \end{cases} \tag{T5}$$

In modular arithmetic, addition and multiplication are *commutative*, *associative*, and *distributive* operations, meaning the following properties:

$$(a \pm b) \bmod m = (a \bmod m) \pm (b \bmod m) \bmod m, \tag{T6}$$

$$(a \times b) \bmod m = (a \bmod m) \times (b \bmod m) \bmod m, \tag{T7}$$

$$[a \times (b + c)] \bmod m = (a \times b) \bmod m + (a \times c) \bmod m) \bmod m. \tag{T8}$$

Here are two other obvious or less trivial properties ($k$ = integer):

$$(-a) \bmod m = (m - a) \bmod m = [(m-1) \times a] \bmod m, \tag{T9}$$

$$(k \times m) \bmod m = 0, \tag{T10}$$

$$(k \times a) \bmod (k \times m) = k \times (a \bmod m). \tag{T11}$$

While addition and multiplication are straightforward in modular arithmetic, *division* is not. For instance, what is the residue $r$ as defined through $r = (11/5) \bmod 3$? In modular arithmetic, $r$ should be the number that satisfies $5 \times r = 11 \bmod_3 = 2$. The only solution to this equation is $r = 1$, since $5 \times 1 \bmod 3 = 5 \bmod 3 = 2$. We, thus, see that $|11/5|_3 = 1$, which is not at all intuitive. One can also define the number $(11/5) \bmod 3$ as the product $(11 \times 1/5) \bmod 3$ or, equivalently, $(11 \bmod 3) \times (1/5 \bmod 3) = 2 \times (1/5 \bmod 3) \bmod 3$. According to the same procedure, we find that $|1/5|_3 = 2$, since $5 \times 2 = 10 \bmod_3 = 1$. We, thus, obtain $2 \times (1/5 \bmod 3) \bmod 3 = 2 \times 2 \bmod 3 = 4 \bmod 3 = 1 = (11/5) \bmod 3$, which is consistent with the previous definition of division. Since inverses can be defined in modular arithmetic, there is no need to use the division operation. However, a restriction applies to the relation between the number to be inverted and the modulus: they should be co-prime, meaning that they should have one as their greatest common divider (see further).

The conversion between $x$ and its residue $r$ (modulo $m$) is called *modular reduction*. Such a reduction operation corresponds to a *one-way function*. This term refers to the fact that the correspondence that is established is not reversible, since there is an infinity of numbers with the same residue. Powers and exponentials of the type $y = x^p$ and $y = a^x$ ($a, p$ = integers) are also one-way functions, since the result $y$ is associated with an infinity of possible values $x$. Here is an important property concerning the commutation of exponents, which is the same as in ordinary arithmetic, and which we will have the opportunity to exploit:

$$(a^p \bmod m)^q \bmod m = (a^q \bmod m)^p \bmod m = a^{pq} \bmod m. \tag{T12}$$

Having introduced modular arithmetic basics, I can describe two fundamental theorems that have been derived by *Fermat* and *Euler*.

Two numbers are said to be *relatively prime*, or *co-prime* to each other, if they do not divide each other, or if their greatest common divider (GCD) is unity. For instance, (3,4), (2,9), and (5,12) are co-prime. Clearly, two prime numbers are co-prime. Primes and co-primes play a central role in modular arithmetic, as the *Fermat* and *Euler* theorems and applications illustrate.

*Fermat's theorem* states that if $m$ is prime and $(a, m)$ are co-prime, then

$$a^{m-1} \bmod m = 1, \tag{T13}$$

which can be put into the equivalent form: $a^m \bmod m = a$. Thus, we have $a \bmod 2 = 1$ or $a^2 \bmod 3 = 1$ for any $a$ that is prime with number two or number three, and so on with any prime modulus, for instance $11^{23-1} \bmod 23 = 1$. We see that as a first benefit, Fermat's theorem makes it possible to considerably simplify calculations in modular arithmetic.

*Euler's theorem* (named after the Swiss mathematician) states that if $p$ and $q$ are two primes and $N = pq$ is their product, then we have for any integer $a$ which is prime with $n$:

$$a^{(p-1)(q-1)} \bmod N = 1. \tag{T14}$$

The factor $(p-1)(q-1)$ with $pq = N$ is also called $\phi(N)$. A consequence of Euler's theorem is that the inverse $1/a$ is readily defined as

$$\frac{1}{a} = a^{-1} = a^{\phi(n)-1} \bmod N. \tag{T15}$$

For instance, what is the inverse of 5 modulo 21? We have $21 = 3 \times 7$, so $\phi(21) = (3-1) \times (7-1) = 12$ and $1/5 = 5^{12-1} \bmod 21 = 48\,828\,125 \bmod 21 = 17$. Thus, $|1/5|_{21} = 17$, which is readily verified as $(17 \times 5) \bmod 21 = 85 \bmod 21 = 1$. We see that the calculation of an inverse through Euler's theorem is quite straightforward, compared with the method previously described. But the requirement is that the modulus $N$ is defined by the product of two primes, which represents a special or fortuitous case.

It is important to note that if $(a, N)$ are *not* co-prime to each other, the inverse $a^{-1}$ does not exist. If the two are co-primes but Euler's theorem cannot apply, the inverse $a^{-1}$ can be found using the *extended Euclidian algorithm*. Here, I shall explain this algorithm through two representative examples: for instance, find the inverse of 7 in modulus 39, and the inverse of 5 in modulus 16 272, i.e., the values of $7^{-1} \bmod 39$ and $5^{-1} \bmod 16\,272$. Taking the first example, the algorithm proceeds as shown in Table T1, with the steps as numbered in the first column.

In steps 1 and 2, we fill out the table as shown. We first divide 39 by 7, to find that the integer part of the result is $[39/7] = 5$. In step 3, we multiply line 2 by 5, and in step 4 we subtract the result of line 3 from that of line 1. This completes the first round. Consider next the ratio $7/4$, whose integer part is $[7/4] = 1$. In step 5, we multiply the results of line 4 by 1, and in step 6, we subtract the result of line 5 from that of line 2. This completes the second round. Consider next the ratio $4/3$, whose integer part is

**Table T1** Extended Euclidian algorithm to find $7^{-1}$ mod 39.

| | | |
|---|---|---|
| 1 | 39 | 39 |
| 2 | 7 | 1 |
| 3 | $5 \times 7 = 35$ | $5 \times 1 = 5$ |
| 4 | $39 - 35 = 4$ | $39 - 5 = 34$ |
| 5 | $1 \times 4 = 4$ | $1 \times 34 = 34$ |
| 6 | $7 - 4 = 3$ | $1 - 34 = -33$ |
| 7 | $1 \times 3 = 3$ | $1 \times (-33) = -33$ |
| 8 | $4 - 3 = 1$ | $34 - (-33) = 67 = 28$ |

**Table T2** Extended Euclidian algorithm to find $5^{-1}$ mod 16 272.

| | | |
|---|---|---|
| 1 | 16 272 | 16 272 |
| 2 | 5 | 1 |
| 3 | $3\,254 \times 5 = 16\,270$ | $3\,254 \times 1 = 3\,254$ |
| 4 | $16\,272 - 16\,270 = 2$ | $16\,272 - 3254 = 13\,018$ |
| 5 | $2 \times 2 = 4$ | $2 \times 13\,018 = 26\,036$ |
| 6 | $5 - 4 = 1$ | $1 - 26\,036 = -26\,035$ |

$[4/3] = 1$. In step 7, we multiply the results of line 6 by 1, and in step 8, we subtract the result of line 7 from that of line 4. If we have obtained 1 in the first column (line 8), we have finished. When positive, the number to the right is the inverse we have been looking for. Here, we find that the inverse of 7 is 28. Proof: $7 \times 28 = 196 = 1$ (modulo 39).

Consider next the example in Table T2, which concerns the search of $5^{-1}$ mod 16 272.

In steps 1 and 2, we fill out the table as shown. We first divide 8064 by 7, to find that the integer part of the result is $[16\,272/5] = 3254$. In step 3, we multiply line 2 by 1612, and in step 4 we subtract the result of line 3 from that of line 1. This completes the first round. Consider next the ratio $5/2$, whose integer part is $[5/2] = 2$. In step 5, we multiply the results of line 4 by 2, and in step 6, we subtract the result of line 5 from that of line 2. This completes the search, since we have obtained 1 in the first column (line 6). However, the number in the right column, $b = -26\,035$ is negative. In this case, the inverse is the complement $km + b$, namely $5^{-1} = 2 \times 16\,272 - 26\,035 = 6509$. The proof is that $5 \times 6509 = 32\,545 = (2 \times 16\,272) + 1 = 1$ (modulo 16 272). This numerical value will be used in the main text concerning my numerical illustration of the PKC algorithm.

# Appendix U (Chapter 21) Klein's inequality

This appendix provides a basic demonstration of *Klein's inequality*, according to which the relative entropy $S(\rho\|\sigma)$ between two quantum systems characterized by the density operators $\rho$ and $\sigma$ is *nonnegative*.

Assume the following eigenstate-basis decompositions for the density operators:

$$\begin{cases} \rho = \sum_i p_i |i\rangle\langle i| \\ \sigma = \sum_k p_k |k\rangle\langle k|. \end{cases} \tag{U1}$$

Substituting the first of the above definitions into the relative entropy, we obtain

$$\begin{aligned} S(\rho\|\sigma) &= \mathrm{tr}(\rho\log\rho) - \mathrm{tr}(\rho\log\sigma) \\ &= \sum_i \langle i|\rho\log\rho|i\rangle - \sum_i \langle i|\rho\log\sigma|i\rangle \\ &= \sum_i p_i \log p_i - \sum_i \langle i|\rho\log\sigma|i\rangle. \end{aligned} \tag{U2}$$

The last term in the right-hand side can then be developed using the eigenstate property $\langle i|\rho = \langle i|p_i$, which gives $\langle i|\rho\log\sigma|i\rangle = p_i\langle i|\log\sigma|i\rangle$. Furthermore, since $\sigma$ is diagonal we have, from the second definition in Eq. (U1), $\log\sigma = \sum_k \log q_k|k\rangle\langle k|$ and, therefore,

$$\begin{aligned} \langle i|\log\sigma|i\rangle &= \langle i|\left(\sum_k \log q_k|k\rangle\langle k|\right)|i\rangle \\ &= \sum_k \langle i|k\rangle\langle k|i\rangle\log q_k \\ &= \sum_k |\langle i|k\rangle|^2\log q_k. \end{aligned} \tag{U3}$$

Substituting these results into Eq. (U2), we obtain

$$\begin{aligned} S(\rho\|\sigma) &= \sum_i p_i\log p_i - \sum_i p_i \sum_k |\langle i|k\rangle|^2\log q_k \\ &= \sum_i p_i\left(\log p_i - \sum_k |\langle i|k\rangle|^2\log q_k\right). \end{aligned} \tag{U4}$$

Then we use the property according to which $\langle \log y \rangle \leq \log\langle y \rangle$ for any function $y$ or, in the discrete case, $\sum x_k \log y_k \leq \log \sum x_k y_k$, where $\{x_k\}$ is a probability distribution. Such a property stems from the concavity of the logarithm function (see Exercise 5.10). Letting $x_k(i) = |\langle i|k\rangle|^2$, we note that $x_k(i) \geq 0$ and $\sum_k x_k(i) = 1$, which makes $x_k(i)$ a probability distribution. Thus, we have

$$\begin{aligned} \sum_k |\langle i|k\rangle|^2 \log q_k &\leq \log\left(\sum_k |\langle i|k\rangle|^2 q_k\right) \\ &\equiv \log r_i. \end{aligned} \tag{U5}$$

Combining, then, Eqs. (U4) and (U5) we finally obtain

$$\begin{aligned} S(\rho\|\sigma) &\geq \sum_i p_i(\log p_i - \log r_i) \\ &= \sum_i p_i \log \frac{p_i}{r_i} \\ &\geq 0. \end{aligned} \tag{U6}$$

In the term $\sum p_i \log(p_i/r_i)$, we recognize the classical definition of *relative entropy* or the *Kullback–Leibler distance* $D(p\|r)$ between two discrete probability distributions $\{p_i\}$, $\{r_i\}$. As we have seen in Chapter 5, $D(p\|r)$ is nonnegative, which justifies the last inequality introduced in the right-hand side in Eq. (U6). The above, thus, proves Klein's inequality or the nonnegativity of $S(\rho\|\sigma)$.

# Appendix V (Chapter 21) Schmidt decomposition of joint pure states

This appendix describes the so-called *Schmidt decomposition*. Such a decomposition applies to the case of a *composite* system made of two subsystems $A$, $B$, which is in a *pure joint state* $|\psi\rangle$, namely, $|\psi\rangle = |\psi_A\rangle \otimes |\psi_B\rangle \equiv |\psi_A\rangle|\psi_B\rangle \equiv |\psi_A\psi_B\rangle$. According to Schmidt decomposition, the joint state can be expressed in the form

$$|\psi\rangle = \sum_i x_i |i_A\rangle|i_B\rangle, \tag{V1}$$

where $x_i$ $(i = 1 \ldots n)$ are nonnegative numbers, called *Schmidt coefficients*, satisfying the property $\sum_i x_i^2 = 1$, and where $\{|i_A\rangle\}$, $\{|i_B\rangle\}$ are some orthonormal basis for the subsystems $A$, $B$.

The proof that the decomposition in Eq. (V1) exists and is unique proceeds as follows. Let $\{|j\rangle\}$ and $\{|k\rangle\}$ be two orthonormal bases for the subsystems $A$ and $B$, respectively. The joint state can most generally be defined through the expansion

$$|\psi\rangle = \sum_{jk} \omega_{jk} |j\rangle|k\rangle, \tag{V2}$$

where $\omega_{jk}$ are the complex coefficients of some matrix $\Omega$. According to the principle of *singular value decomposition*,[1] it is possible to express $\Omega$ in the form $\Omega = UDV$, where $U$, $V$ are *unitary matrices*, and $D$ is a *diagonal matrix* having *nonnegative elements*. The diagonal elements $x_i = d_{ii}$ of $D$ are called the *singular values* of $\Omega$. The matrix elements of $\Omega$, thus, decompose as follows

$$\omega_{jk} = \sum_i u_{ji} d_{ii} v_{ik} = \sum_i x_i u_{ji} v_{ik}, \tag{V3}$$

which we substitute into Eq. (V2) to obtain:

$$\begin{aligned} |\psi\rangle &= \sum_{ijk} x_i u_{ji} v_{ik} |j\rangle|k\rangle \\ &= \sum_i x_i \left( \sum_j u_{ji}|j\rangle \sum_k v_{ik}|k\rangle \right). \end{aligned} \tag{V4}$$

[1] See, for instance: http://mathworld.wolfram.com/SingularValueDecomposition.html; http://en.wikipedia.org/wiki/Singular_value_decomposition.

We then introduce the definitions

$$\begin{cases} |i_A\rangle = \sum_j u_{ji}|j\rangle \\ |i_B\rangle = \sum_k v_{ik}|k\rangle. \end{cases} \tag{V5}$$

The states $|i_A\rangle$, $|i_B\rangle$ form orthonormal sets, because of the unitarity of the matrices $U$, $V$. Considering $|i_A\rangle$, for instance, we have

$$\begin{aligned} \langle l_A|i_A\rangle &= \left(\sum_k u_{kl}^*\langle k|\right)\left(\sum_j u_{ji}|j\rangle\right) \\ &= \sum_{jk} u_{kl}^* u_{ji}\langle k|j\rangle \\ &= \sum_{jk} u_{kl}^* u_{ji}\delta_{kj} \\ &= \sum_j u_{jl}^* u_{ji} \\ &= (U^+U)_{li} \\ &= I_{li} \\ &= \delta_{li}, \end{aligned} \tag{V6}$$

which proves the orthonormality of the set $\{|i_A\rangle\}$, and likewise for $\{|i_B\rangle\}$. Substituting Eq. (V5) into Eq. (V4) yields Eq. (V1), i.e., the Schmidt decomposition of $|\psi\rangle$. Because $|\psi\rangle$ is a pure state, we also have $\langle\psi|\psi\rangle = \sum_i x_i^2 = 1$, which completes the proof.

Here are two examples of Schmidt decomposition according to Eq. (V1):

$$\begin{aligned} |\psi\rangle &= \frac{1}{\sqrt{2}}(|10\rangle + |01\rangle) \\ &\equiv 0 \times |0_A\rangle|0_B\rangle + \frac{1}{\sqrt{2}}|0_A\rangle|1_B\rangle + \frac{1}{\sqrt{2}}|1_A\rangle|0_B\rangle + 0 \times |1_A\rangle|1_B\rangle \end{aligned} \tag{V7}$$

and

$$\begin{aligned} |\psi\rangle &= \frac{1}{\sqrt{2}}(|00\rangle - |11\rangle) \\ &\equiv \frac{1}{\sqrt{2}}|0_A\rangle|\tilde{0}_B\rangle + 0 \times |0_A\rangle|\tilde{1}_B\rangle + 0 \times |1_A\rangle|\tilde{0}_B\rangle + \frac{1}{\sqrt{2}}|1_A\rangle|\tilde{1}_B\rangle, \end{aligned} \tag{V8}$$

with the other choice of orthonormal basis $\{|\tilde{0}_B\rangle, |\tilde{1}_B\rangle\} = \{|0_B\rangle, -|1_B\rangle\}$ for the subsystem $B$.

# Appendix W (Chapter 21) State purification

This appendix describes the operation of *state purification*. The operation consists of associating a pure joint state $|\psi_{AR}\rangle$ with a system $A$ that is in a mixed state $|\psi_A\rangle$. The benefit of the operation is the possibility of deriving the density operator $\rho_A$ directly from the *partial trace* of the pure-state density operator $\rho_{AR} = |\psi_{AR}\rangle\langle\psi_{AR}|$. The term *state purification*, thus, means expanding the system $A$ in mixed state $|\psi_A\rangle$ into a composite system $AR$, whose joint state $|\psi_{AR}\rangle$ is a pure state. The state $|\psi_{AR}\rangle$ is referred to as the purification of $|\psi_A\rangle$, or the purification of $\rho_A$.

The principle of state purification is relatively simple to explain. First, assume for the mixed state $|\psi_A\rangle$ the following decomposition over the orthonormal basis $\{|i_A\rangle\}$:

$$|\psi_A\rangle = \sum_i \sqrt{p_i}|i_A\rangle. \tag{W1}$$

Second, assume a reference system $R$ of same dimension of $A$ and same basis $\{|i_R\rangle\}$, meaning $\langle i_R|j_A\rangle = \delta_{ij}$. Then define the joint state $|\psi_{AR}\rangle$:

$$|\psi_{AR}\rangle = \sum_i \sqrt{p_i}|i_A\rangle|i_R\rangle. \tag{W2}$$

The above-defined joint state $|\psi_{AR}\rangle$ is a *pure state*, because it cannot be defined by any other possible state $|\phi_{AR}\rangle$ of $AR$. We can show this by using the most general definition for $|\phi_{AR}\rangle$:

$$|\phi_{AR}\rangle = \sum_i \sum_{j\neq i} \alpha_{ij}|i_A\rangle|j_R\rangle. \tag{W3}$$

Then we have

$$\begin{aligned}
\langle\psi_{AR}|\phi_{AR}\rangle &= \left(\sum_k \sqrt{p_k}\langle k_A|\langle k_R|\right)\left(\sum_i \sum_{j\neq i} \alpha_{ij}|i_A\rangle|j_R\rangle\right) \\
&= \sum_k \sum_i \sum_{j\neq i} \alpha_{ij}\langle k_A|i_A\rangle\langle k_R|j_R\rangle \\
&= \sum_k \sum_i \sum_{j\neq i} \alpha_{ij}\delta_{ki}\delta_{kj} \\
&= \sum_i \sum_{j\neq i} \alpha_{ij}\delta_{ij} \\
&\equiv 0,
\end{aligned} \tag{W4}$$

which establishes that $|\psi_{AR}\rangle$, $|\phi_{AR}\rangle$ are, indeed, orthogonal.

Since $|\psi_{AR}\rangle$ is a pure state, the density operator of the composite system $AR$ is

$$\begin{aligned}\rho_{AR} &= |\psi_{AR}\rangle\langle\psi_{AR}| \\ &\equiv \sum_{ij}\sqrt{p_i p_j}|i_A\rangle\langle j_A| \otimes |i_R\rangle\langle j_R|.\end{aligned} \tag{W5}$$

As the next step, we execute the partial tracing over $R$ according to the generic definition in Eq. (21.25), which here reads ($B = R$):

$$\begin{aligned}\mathrm{tr}_R U &= \mathrm{tr}_R(|i_A\rangle\langle j_A| \otimes |k_R\rangle\langle l_R|) \\ &\equiv |i_A\rangle\langle j_A| \times \mathrm{tr}(|k_R\rangle\langle l_R|),\end{aligned} \tag{W6}$$

hence,

$$\begin{aligned}\mathrm{tr}_R(\rho_{AR}) &= \mathrm{tr}_R\left(\sum_{ij}\sqrt{p_i p_j}|i_A\rangle\langle j_A| \otimes |i_R\rangle\langle j_R|\right) \\ &= \sum_{ij}\sqrt{p_i p_j}\,\mathrm{tr}(|i_A\rangle\langle j_A| \otimes |i_R\rangle\langle j_R|) \\ &= \sum_{ij}\sqrt{p_i p_j}|i_A\rangle\langle j_A|\times\mathrm{tr}(|i_R\rangle\langle j_R|) \\ &= \sum_{ij}\sqrt{p_i p_j}|i_A\rangle\langle j_A| \times \delta_{ij} \\ &= \sum_{i} p_i|i_A\rangle\langle i_A| \\ &\equiv \rho_A.\end{aligned} \tag{W7}$$

This result shows that the partial trace over $R$ of the composite system $AR$ yields the density operator $\rho_A$, which corresponds to the definition of *state purification*.

# Appendix X (Chapter 21) Holevo bound

In this appendix, I formally establish that the mutual information $H(X;Y)$ in a quantum communication channel is bounded by a maximum, $\chi$, referred to as *Holevo bound*, and defined by

$$\begin{aligned} H(X;Y) &\leq S(\rho) - \sum_x p_x S(\rho_x) \\ &= \chi \end{aligned} \tag{X1}$$

where $\rho = \sum_x p_x \rho_x = \langle \rho \rangle$. In the above definition, $\rho_x$ represents the density matrix of any quantum state used for the transmission, as selected from an "alphabet" $X$ according to some probability distribution $\{p_x\}$. The retrieval of information is made by POVM measurements, which define a source of random outcomes $Y$ of size $n$.

The formal proof of the Holevo bound requires one to assume three quantum systems, $P$, $Q$, $R$. The first system, $P$, corresponds to the quantum source used by the originator to "prepare" the states $\rho_x$, namely with density operator:

$$\rho_P = \sum_x p_x |x\rangle\langle x|, \tag{X2}$$

with $\{|x\rangle\}$ being the orthonormal basis where $\rho$ is diagonal. The composite system $PQ$, as described by the tensor product

$$\rho_{PQ} = \sum_x p_x |x\rangle\langle x| \otimes \rho_x, \tag{X3}$$

represents the joint state of the system after the preparation of the states $\rho_x$, based on the definition of partial tracing, i.e.,

$$\begin{aligned} \mathrm{tr}_P(\rho_{PQ}) &= \mathrm{tr}_P\left(\sum_x p_x |x\rangle\langle x| \otimes \rho_x\right) \\ &= \sum_x \mathrm{tr}_P(p_x |x\rangle\langle x| \otimes \rho_x) \\ &= \sum_x \mathrm{tr}(p_x |x\rangle\langle x|) \times \rho_x \\ &= \sum_x p_x \rho_x \\ &\equiv \rho, \end{aligned} \tag{X4}$$

$$
\begin{aligned}
\mathrm{tr}_Q(\rho_{PQ}) &= \mathrm{tr}_Q\left(\sum_x p_x|x\rangle\langle x|\otimes\rho_x\right)\\
&= \sum_x \mathrm{tr}_Q\left(p_x|x\rangle\langle x|\otimes\rho_x\right)\\
&= \sum_x p_x|x\rangle\langle x|\mathrm{tr}\left(\rho_x\right)\\
&= \sum_x p_x|x\rangle\langle x|\\
&\equiv \rho_P.
\end{aligned} \tag{X5}
$$

The third system, $R$, is a fictitious auxiliary system representing, together with $Q$, the measuring apparatus of the recipient. We assume that before measurement, $R$ is in the pure state $|0\rangle$, corresponding to $\rho_R = |0\rangle\langle 0|$. Thus, the initial state of the whole system (or quantum communication channel) is described by the density operator

$$
\rho_{PQR} = \sum_x p_x|x\rangle\langle x|\otimes\rho_x\otimes|0\rangle\langle 0|. \tag{X6}
$$

Next, we need to define the effect of the recipient's POVM measurement. As seen in Chapter 17, the POVM measurement is defined as a set of Hermitian operators, $\{E_y\}$, which satisfy the completeness relation (Eq. (17.62)), and which correspond to a set of *measurement operators*, $\{M_y = \sqrt{E_y}\}$, satisfying the property $M_y M_y^+ = E_y$. Applying any single measurement labeled $y$ to a system in state $\rho$ is a *quantum operation* that leaves the system in the *post-measurement state* $\rho' = M_y\rho M_y^+$. Here, we shall use a somewhat different quantum operation. It consists of effecting the POVM measurement in system $Q$ and storing the result in system $R$. The corresponding quantum operation leaves the composite system $QR$ in the post-measurement state

$$
\rho'_{QR} = \sum_y M_y\rho_x M_y^+\otimes|y\rangle\langle y|. \tag{X7}
$$

This definition is justified by the partial tracing over system $Q$ to yield, for system $R$:

$$
\begin{aligned}
\rho'_R &= \mathrm{tr}_Q(\rho'_{QR})\\
&= \mathrm{tr}_Q\left(\sum_y M_y\rho_x M_y^+\otimes|y\rangle\langle y|\right)\\
&= \sum_y \mathrm{tr}_Q\left(M_y\rho_x M_y^+\otimes|y\rangle\langle y|\right)\\
&= \sum_y \mathrm{tr}\left(M_y\rho_x M_y^+\right)\times|y\rangle\langle y|\\
&= \sum_y \mathrm{tr}\left(M_y^+ M_y\rho_x\right)\times|y\rangle\langle y|\\
&= \sum_y \mathrm{tr}\left(E_y\rho_x\right)\times|y\rangle\langle y|\\
&\equiv \sum_y \langle E_y\rangle_x|y\rangle\langle y|,
\end{aligned} \tag{X8}
$$

where $\langle E_y \rangle_x = \mathrm{tr}(E_y \rho_x)$ represents the expectation value of the measurement $E_y$, *given the input state* $\rho_x$. Note that $\mathrm{tr}(\rho'_R) \neq 1$. To get the actual final state of $R$ we need to effect the normalization

$$\tilde{\rho}'_R = \frac{\mathrm{tr}_Q(\rho'_{QR})}{\mathrm{tr}[\mathrm{tr}_Q(\rho'_{QR})]} \equiv \frac{\sum_y \langle E_y \rangle_x |y\rangle\langle y|}{\sum_{y'} \langle E_{y'} \rangle_x}, \tag{X9}$$

or, in the matrix form in base $\{|y\rangle\}$:

$$\tilde{\rho}'_R = \frac{1}{\sum_y \langle E_y \rangle_x} \begin{pmatrix} \langle E_1 \rangle_x & 0 & 0 & 0 & 0 \\ 0 & \langle E_2 \rangle_x & 0 & \cdots & 0 \\ 0 & 0 & \ddots & \ddots & 0 \\ \vdots & \vdots & \ddots & \ddots & \vdots \\ 0 & 0 & \cdots & 0 & \langle E_n \rangle_x \end{pmatrix}. \tag{X10}$$

The coefficients in this density-matrix definition, thus, provide the conditional probability of obtaining $\langle E_y \rangle_x$ as the outcome of the measurement, *given the input state* $\rho_x$. This justifies the definition in Eq. (X7) of the proposed quantum operation (not normalized, for simplicity).

In summary, our measurement operation leaves the composite system $PQR$ in the state

$$\rho'_{PQR} = \sum_{xy} p_x |x\rangle\langle x| \otimes M_y \rho_x M_y^+ \otimes |y\rangle\langle y|. \tag{X11}$$

We may now trace out the system over $Q$ to obtain $\rho'_{PR}$ and, hence, to be able to derive later on the mutual information $S(\rho_P; \rho_R)$ existing after the measurement. This tracing gives

$$\begin{aligned} \rho'_{PR} &= \mathrm{tr}_Q \left( \sum_{xy} p_x |x\rangle\langle x| \otimes M_y \rho_x M_y^+ \otimes |y\rangle\langle y| \right) \\ &= \sum_{xy} p_x |x\rangle\langle x| \times \mathrm{tr}(M_y \rho_x M_y^+) \otimes |y\rangle\langle y| \\ &= \sum_{xy} p_x |x\rangle\langle x| \times \mathrm{tr}(E_y \rho_x) \otimes |y\rangle\langle y|. \end{aligned} \tag{X12}$$

In the above, we may substitute $\mathrm{tr}(E_y \rho_x) = p(y|x)$ as a *conditional probability*, because $\{E_y\}$ form a POVM operator set satisfying the completeness relation. Using *Bayes's theorem* (see Chapter 1), we have$p(x, y) = p(y|x)p_x$ for the joint probability, which, from Eq. (X12), yields:

$$\begin{aligned}\rho'_{PR} &= \sum_{xy} p_x p(y|x)|x\rangle\langle x|_P \otimes |y\rangle\langle y|_R \\ &= \sum_{xy} p(x, y)|x\rangle\langle x|_P \otimes |y\rangle\langle y|_R.\end{aligned} \tag{X13}$$

Because of the above tensor-operator form, we might as well express the post-measurement joint state in the tensor-state form:

$$|\psi'\rangle_{PR} = \sum_{xy} p(x, y)|xy\rangle. \tag{X14}$$

It is clear, then, that the *joint VN entropy* $S'(\rho_P, \rho_R)$ and *mutual information* $S'(\rho_P; \rho_R)$ of the post-measurement system are simply given by

$$\begin{cases} S'(\rho_P, \rho_R) \equiv H(X, Y) \\ S'(\rho_P; \rho_R) \equiv H(X; Y), \end{cases} \tag{X15}$$

just as in a *classical communication channel* with an originator source $X$ and a recipient source $Y$, correlated together with joint probability $p(x, y)$.

The next and final step is to relate the post-measurement mutual information $S'(\rho_P; \rho_R)$ to that characterizing the system before the measurement, which we call $S(\rho_P; \rho_Q)$. To this effect, we shall first assume the following inequality (the proof being provided at the end of this appendix):

$$S'(\rho_P; \rho_R) \leq S(\rho_P; \rho_Q). \tag{X16}$$

This property states that in quantum channels, *mutual information may not necessarily be conserved, neither be increased by measurement*. We obtain $S(\rho_P; \rho_Q)$ as:

$$S(\rho_P; \rho_Q) = S(\rho_P) + S(\rho_Q) - S(\rho_P, \rho_Q), \tag{X17}$$

with (from Eqs. (X2) and (X3)):

$$\begin{cases} S(\rho_P) = S\left(\sum_x p_x|x\rangle\langle x|\right) \equiv H(X) \\ S(\rho_Q) = S\left[\mathrm{tr}_P\left(\sum_x p_x|x\rangle\langle x| \otimes \rho_x\right)\right] = S\left[\sum_x p_x|x\rangle\langle x| \times \mathrm{tr}\,(\rho_x)\right] \\ \qquad = S\left(\sum_x p_x|x\rangle\langle x|\right) \equiv S(\rho) \\ S(\rho_P, \rho_Q) = S\left(\sum_x p_x|x\rangle\langle x| \otimes \rho_x\right) \equiv H(X) + \sum_x p_x S(\rho_x). \end{cases} \tag{X18}$$

[1]

[1] To obtain this result, define $R_x = |x\rangle\langle x| \otimes \rho_x$. We have, then,

$$S\left(\sum_x p_x|x\rangle\langle x| \otimes \rho_x\right) = S\left(\sum_x p_x R_x\right).$$

From Eqs. (X17) and (X18), we obtain

$$S(\rho_P; \rho_Q) = S(\rho) - \sum_x p_x S(\rho_x). \tag{X19}$$

Combining the results in Eqs. (X15), (X16), and (X19), we finally obtain

$$H(X; Y) \leq S(\rho) - \sum_x p_x S(\rho_x) = \chi, \tag{X20}$$

which establishes the proof of the *Holevo bound* stated in Eq. (X1).

To reach such a conclusion, the inequality in Eq. (X16), namely, $S'(\rho_P; \rho_R) \leq S(\rho_P; \rho_Q)$, was made as a key assumption. The following discussion, destined for the demanding, provides the key elements proving such an assumption. This proof is, however, not complete, because it eventually rests on an additional property, referred to as *strong subadditivity*. Proving strong subadditivity requires no less than two theorems, of which the proofs are quite mathematically involved! Therefore, and for the purpose of these chapters, such a property shall be taken here as a given postulate.

## Proof of the inequality in Eq. (X16), with discussion

In the following, I simplify the notation by setting $S(\rho_A) \equiv S(A)$, $S(\rho_A; \rho_B) \equiv S(A; B)$, and the like for $A, B = P, Q, R$. By definition, $S(A; B, C)$ refers to the mutual information between system $A$ and composite system $BC$. Then I shall observe and discuss the three following properties:

(i) $S(P; Q) = S(P; Q, R)$, since prior to measurement, $R$ is uncorrelated to $PQ$ and, thus, does not contribute to $PQ$ mutual information;
(ii) $S(P; Q, R) \geq S'(P; Q, R)$, since the measurement between $P$ and $R$ has no reason to increase the mutual information between $P$ and $QR$;
(iii) $S'(P; Q, R) \geq S'(P; R)$, since discarding the reference system $Q$ between $P$ and $R$ has no reason to increase the mutual information between $P$ and $R$.

The first property (i) is obvious. The second property (ii) can be established as follows. Let $S(A; B)$ and $S'(A; B)$ be the mutual information before and after the measurement between any systems $A$ and $B$ (here, $A$ plays the role of system $P$ and $B$ plays the role of the composite system $QR$). We introduce another system $C$ in

By application of the property in Eq. (21.47), we have

$$S\left(\sum_x p_x R_x\right) \equiv \sum_x p_x S(R_x) + H(X).$$

And because $R_x$ is a tensor operator, we have

$$S(R_x) = S(|x\rangle\langle x| \otimes \rho_x) = S(|x\rangle\langle x|) + S(\rho_x) = 0 + S(\rho_x) \equiv S(\rho_x),$$

which, finally, yields

$$S\left(\sum_x p_x |x\rangle\langle x| \otimes \rho_x\right) = H(X) + \sum_x p_x S(\rho_x).$$

a pure state (e.g., $|0\rangle$ or $|1\rangle$). We then have $S(A; B) = S(A; B, C)$, because of property (i). A measurement on $B$ with the information being transferred to $C$ must not change mutual information between $A$ and $BC$, therefore, $S(A; B, C) = S'(A; B, C)$. But on discarding $C$, we must have $S'(A; B) \leq S'(A; B, C)$, according to property (iii), which states that discarding a system cannot increase mutual information. We must now focus on this property (iii), rewriting the inequality as $S(A; B) \leq S(A; B, C)$ for any systems $A, B, C$. According to the definition of mutual information, we have $S(A; B) = S(A) + S(B) - S(A, B)$ and $S(A; B, C) = S(A) + S(B, C) - S(A, B, C)$, which yields by substitution $S(A, B, C) + S(B) \leq S(A, B) + S(B, C)$. This last property is referred to as *strong subadditivity inequality*. Its demonstration requires no less than two additional theorems (including the so-called *Lieb's theorem*), both being rather mathematically involved! Therefore, here we shall take the "strong subadditivity" property as a given postulate, which closes the loop. Combining properties (i)–(iii) we finally obtain $S(P; Q) \geq S'(P; R)$, which is the inequality in Eq. (X16).

# Appendix Y (Chapter 25) Polynomial byte representation and modular multiplication

The novelty introduced by the AES cryptosystem is the use of modular *multiplication* by byte blocks, which is described in this appendix. For multiplied bytes to remain within a one-byte size, this multiplication must be defined modulo a certain "prime" number. This requires one to introduce another representation for binary numbers, which is the *polynomial (byte) representation*.

A *polynomial* is an element that can be defined according to the expression $m(x) = c_n x^n + c_{n-1} x^{n-1} + \cdots + c_1 x + c_0$, where $c_n, c_{n-1} \ldots c_1, c_0$ are called the polynomial *coefficients* and $n$ is the polynomial's *degree*. Since a byte is the eight-bit word $c_7 c_6 c_5 c_4 c_3 c_2 c_1 c_0$ ($c_k = 0$ or 1), its equivalent polynomial representation is $m(x) = c_7 x^7 + c_6 x^6 \cdots + c_1 x + c_0$. Let's look now at the XOR operation between two polynomials. Consider the two numbers $\{1010\ 0010\} = \{\mathrm{A3}\}$ and $\{0101\ 1111\} = \{5\ \mathrm{F}\}$. In binary, hexadecimal, and polynomial representations, we get

$$\{1010\ 0010\} \otimes \{0101\ 1111\} = \{1111\ 1101\}, \tag{Y1}$$

$$\{\mathrm{A3}\} \otimes \{\mathrm{5F}\} = \{\mathrm{FD}\}, \tag{Y2}$$

$$\begin{aligned} &\{x^7 + x^5 + x\} \oplus \{x^6 + x^4 + x^3 + x^2 + x + 1\} \\ &= \{x^7 + x^6 + x^5 + x^4 + x^3 + x^2 + 1\}. \end{aligned} \tag{Y3}$$

The operation of *polynomial multiplication*, modulo a polynomial $m(x)$, which we label with the symbol $\bullet$, is no more difficult to grasp. The idea is that the reduction modulo $m(x)$ should give a polynomial whose degree (here) is strictly less than eight, so that the result of the multiplication can always be represented by a one-byte word. The modulus $m(x) = x^8 + x^4 + x^3 + x + 1$, which is used in the AES cryptosystem, meets such a requirement. This specific polynomial is also *irreducible*, meaning that it has no dividers other than 1 and itself. It is the conceptual equivalent of a *prime number* in the polynomial field. The polynomial dividers of $m(x)$ are any numbers $u(x)$, $v(x)$ that verify $u(x) \times v(x) = m(x)$, where $\times$ is the usual algebraic multiplication. Let us look now at how multiplications $\times$ and $\bullet$ work through a simple example. Take $u(x) = x^3 + x^2 + 1$ and $v(x) = x^2 + x$. Following usual algebra, we first get the product

$$\begin{aligned} u(x) \times v(x) &= (x^3 + x^2 + 1)(x^2 + x) \\ &= (x^5 + x^4 + x^2) \oplus (x^4 + x^3 + x) \\ &= x^5 + x^3 + x^2 + x. \end{aligned} \tag{Y4}$$

To obtain this result, we use $\oplus$ to perform the addition between the polynomial coefficients of the same order, since this is the operation used in the corresponding binary representation. The second task is to reduce the above result modulo a given polynomial, say, $m(x) = x^4 + 1$. By definition, we have for any polynomial $p(x)$,

$$p(x) = q(x) \times m(x) + r(x), \tag{Y5}$$

where $q(x)$ is the *quotient* and $r(x)$ is the *remainder*, also noted $p(x) \bmod m(x)$. Note that the degree of the remainder must be lower than that of $m(x)$, otherwise it could divide it. Taking the previous example, we must find the polynomials $q(x)$ and $r(x)$ that satisfy

$$x^5 + x^3 + x^2 + x = p(x) \times (x^4 + 1) + q(x). \tag{Y6}$$

From the above relation, we must conclude that $p(x) = ax + b$ and $q(x) = cx^3 + dx^2 + ex + f$, where $a, b, c, d, e, f$ are the unknowns. Substituting these two definitions into the right-hand side yields $a = c = d = 1$, $b = e = f = 0$, which gives $p(x) = x$ and $q(x) = x^3 + x^2$. Thus,

$$x^3 + x^2 = x^5 + x^3 + x^2 + x \bmod x^4 + 1, \tag{Y7}$$

which illustrates how modulus reduction is performed.

With the modulus-reduction tool in hand, we can now consider an example that uses the irreducible polynomial $m(x) = x^8 + x^4 + x^3 + x + 1$. Take, for instance, the two numbers {4A} and {18}. What is $p(x) = \{4A\} \bullet \{19\}$? The conversion into polynomial representation gives

$$\begin{aligned} p(x) &= \{01001010\} \bullet \{00011001\} \\ &= (x^6 + x^3 + x) \times (x^4 + x^3 + 1) \bmod m(x) \\ &= x^{10} + x^9 + x^7 + x^5 + x^4 + x^3 + x \bmod m(x) \\ &= x^7 + x^6 + x^5 \bmod m(x) \\ &= \{11100000\} \\ &= \{E0\}, \end{aligned} \tag{Y8}$$

which is the result of the operation. Since we have defined a multiplication law for polynomials (modulo $m(x)$), we must also introduce the corresponding concept of *inverse polynomials*. The inverse polynomial of a nonzero polynomial, $p(x)$, is noted $p^{-1}(x)$ and is defined by the equivalent identities

$$\begin{cases} p(x) \bullet p^{-1}(x) = 1 \\ p^{-1}(x) = p(x) \bmod m(x). \end{cases} \tag{Y9}$$

We also know that the inverse exists because $m(x)$ is irreducible, i.e., $p(x)$ and $m(x)$ do not have a common divider. As with modular arithmetic, the search for the inverse can be performed by the *extended Euclidian algorithm* (EEA).

Let me explain the EEA through two representative examples: for instance, find (1) the inverse of 7 in modulus 39, i.e., the value of $7^{-1} \bmod 39$ and (2) the inverse of 5 in modulus 16 272, i.e., the value of $5^{-1} \bmod 16\,272$. Considering example (1),

**Table Y1** Extended Euclidian algorithm to find $7^{-1}$ mod 39.

| | | |
|---|---|---|
| 1 | 39 | 39 |
| 2 | 7 | 1 |
| 3 | $5 \times 7 = 35$ | $5 \times 1 = 5$ |
| 4 | $39 - 35 = 4$ | $39 - 5 = 34$ |
| 5 | $1 \times 4 = 4$ | $1 \times 34 = 34$ |
| 6 | $7 - 4 = 3$ | $1 - 34 = -33$ |
| 7 | $1 \times 3 = 3$ | $1 \times (-33) = -33$ |
| 8 | $4 - 3 = 1$ | $34 - (-33) = 67 = 28$ |

**Table Y2** Extended Euclidian algorithm to find $5^{-1}$ mod 16 272.

| | | |
|---|---|---|
| 1 | 16 272 | 16 272 |
| 2 | 5 | 1 |
| 3 | $3254 \times 5 = 16\,270$ | $3254 \times 1 = 3254$ |
| 4 | $16\,272 - 16\,270 = 2$ | $16\,272 - 3254 = 13\,018$ |
| 5 | $2 \times 2 = 4$ | $2 \times 13\,018 = 26\,036$ |
| 6 | $5 - 4 = 1$ | $1 - 26\,036 = -26\,035$ |

the algorithm proceeds as shown in Table Y1, with the steps as numbered in the first column.

In steps 1 and 2, we fill out the table as shown. We first divide 39 by 7, to find that the integer part of the result is $[39/7] = 5$. In step 3, we multiply line 2 by 5, and in step 4 we subtract the result of line 3 from that of line 1. This completes the first round. Consider next the ratio $7/4$, whose integer part is $[7/4] = 1$. In step 5, we multiply the results of line 4 by 1, and in step 6, we subtract the result of line 5 from that of line 2. This completes the second round. Consider next the ratio $7/3$, whose integer part is $[4/3] = 1$. In step 7, we multiply the results of line 6 by 1, and in step 8, we subtract the result of line 7 from that of line 4. If we have obtained 1 in the first column (line 8), we have finished. When positive, the number to the right is the inverse that we have been looking for. Here, we find that the inverse of 7 is 28. Proof: $7 \times 28 = 196 \equiv 1$ (modulo 39).

Consider the second example, which concerns the search of $5^{-1}$ mod 16 272. We obtain Table Y2.

In steps 1 and 2, we fill out the table as shown. We first divide 16 272 by 5, to find that the integer part of the result is $[16\,272/5] = 3254$. In step 3, we multiply line 2 by 3254, and in step 4 we subtract the result of line 3 from that of line 1. This completes the first round. Consider next the ratio $5/2$, whose integer part is $[5/2] = 2$. In step 5, we multiply the results of line 4 by 2, and in step 6, we subtract the result of line 5 from that of line 2. This completes the search, since we have obtained 1 in the first column (line 6). However, the number in the right column, $b = -26\,035$ is negative. In this case, the inverse is the complement $km + b$ ($k > 1$), namely $5^{-1} = 16\,272 - 26\,035 = 6509$. Proof: $5 \times 6509 = 32\,545 = (2 \times 16\,272) + 1 \equiv 1$ (modulo 16 272).

I shall now implement the EEA with polynomials modulo $m(x)$. For instance, we want the inverse of $p(x) = x^5 + x^3 + 1$ modulo $m(x) = x^8 + x^4 + x^3 + x + 1$. Table Y3 shows the different calculation steps:

**Table Y3** The steps in calculating $p^{-1}(x)$.

| | | |
|---|---|---|
| 1 | $x^8 + x^4 + x^3 + x + 1$ | $x^8 + x^4 + x^3 + x + 1$ |
| 2 | $x^5 + x^3 + 1$ | $1$ |
| 3 | $(x^3 + x)(x^5 + x^3 + 1) = x^8 + x^4 + x^3 + x$ | $(x^3 + x) \times 1 = x^3 + x$ |
| 4 | $(x^8 + x^4 + x^3 + x + 1) \oplus (x^8 + x^4 + x^3 + x)$ $= 1$ | $(x^8 + x^4 + x^3 + x + 1) \oplus (x^3 + x)$ $= x^8 + x^4 + 1$ |

Since we obtained 1 in the last line of left column, the result in the right column, $x^8 + x^4 + 1$, is the inverse $p^{-1}(x)$. We can readily check that

$$\begin{aligned} p(x) \bullet p^{-1}(x) &= (x^5 + x^3 + 1) \times (x^8 + x^4 + 1) \bmod m(x) \\ &= x^{13} + x^{11} + x^9 + x^8 + x^7 + x^5 + x^4 + x^3 + x \bmod m(x) \\ &\equiv 1 \bmod m(x). \end{aligned}$$

# Index